Akil, 1985

CRC Handbook of HPLC for the Separation of Amino Acids, Peptides, and Proteins

Volume II

Editor

William S. Hancock, Ph.D.

Reader in Chemistry
Department of Chemistry,
Biochemistry, and Biophysics
Massey University
Palmerston North
New Zealand

CRC Press, Inc.
Boca Raton, Florida

Library of Congress Cataloging in Publication Data
Main entry under title:

Handbook of HPLC for the separation of amino acids,
peptides, and proteins.

Bibliography: p.
Includes indexes.
1. Amino acids--Separation--Handbooks, manuals, etc.
2. Peptides--Separation--Handbooks, manuals, etc.
3. Proteins--Separation--Handbooks, manuals, etc.
4. High performance liquid chromatography--Handbooks,
manuals, etc. I. Hancock, William S. II. Title:
Handbook of H.P.L.C. for the separation of amino acids,
peptides, and proteins. [DNLM: 1. Amino acids--Analysis.
2. Chromatography, High pressure liquid. 3. Peptides--
Analysis. 4. Proteins--Analysis. QU60 H236]
QD431.7.H36 1984 547.7'5 83-14908
ISBN 0-8493-3510-8 (v. 1)
ISBN 0-8493-3511-6 (v. 2)

Direct all inquiries to CRC Press, Inc., 2000 Corporate Blvd., N.W., Boca Raton, Florida, 33431.

International Standard Book Number 0-8493-3510-8 (v. 1)
International Standard Book Number 0-8493-3511-6 (v. 2)

Library of Congress Card Number 83-14908
Printed in the United States

PREFACE

The growth in application of high performance liquid chromatography (HPLC) to the life sciences can be judged from the massive increase in papers applied to the separation of macromolecules that were published recently in the *Journal of Chromatography, Analytical Biochemistry,* or the *Journal of Liquid Chromatography*. In addition more specialist journals such as *Brain Research, Journal of Endocrinology,* and *Biochemistry* are now publishing reports that make extensive use of HPLC separations. The study of neuropeptides provides a powerful example of the potential of the technique when it is extensively applied to a new area of research. Unfortunately many other areas of research in the life sciences have been slow to follow this lead. A major difficulty is often that polypeptide separations require careful optimization, which can be a daunting prospect to the researcher uninitiated in the subtilities of liquid chromatography. Another difficulty is a lack of knowledge of the biologist in the properties of siliconaceous supports and in the instrumentation required for a high efficiency separation. This Handbook was planned with these difficulties in mind, with a range of chapters that will both introduce the technique and then lead on to the detailed optimization of the chromatographic parameters required for a given separation.

At the same time even the experienced chromatographer is faced with a difficulty, that is with the rapid advances in instrumentation and separation conditions. For example, recent advances in the understanding and preparation of reversed phases means that dramatically improved separations can be achieved by the correct choice of the parent silica used to manufacture the reversed phase. It is a goal of the Handbook to present initial examples of promising new applications in HPLC in the hope that it will stimulate further studies. An example of such an application can be found in ligand-exchange chromatography, which as the reader will find in the later chapters, is a technique of great potential.

At Massey University our interest in HPLC has always been problem-orientated, so that currently we are studying the role of apolipoproteins in heart disease, and using HPLC as an analytical and preparative technique. I hope that the Handbook will reflect this practical orientation and that it will be of value to other researchers in the biological fields, who like us, find their studies are inextricably linked to the joys of chromatography.

I would like to acknowledge my gratitude for the assistance given by my research colleague and friend, Dr. David Harding in this task. Also the continued support by Professors Dick Batt and Geoff Malcolm of my research activities has made possible my interest in high performance liquid chromatography and its application to protein chemistry. I am grateful for the strong support from the members of the Advisory Board and for the many hours of careful preparation that individual researchers put into the preparation of their reports. The finished Handbook bears strong testimony to the skills of the secretaries at Massey University and the publishing staff at CRC Press. Most of all I am indebted to my wife Elizabeth for her encouragement and patient tolerance of my labors. To her this Handbook is affectionately dedicated.

Bill Hancock
Palmerston North
New Zealand

THE EDITOR

Dr. William Hancock graduated in Organic Chemistry and Biochemistry from the Adelaide University in South Australia and then continued with a Ph.D. in Natural Product Chemistry at the same University. This work was supervised by Drs. Massy-Westropp and Mander and involved synthetic organic chemistry. Even then chromatography on silica acid and alumina played a vital role in characterization and purification of the reaction products.

After graduation in 1970 he worked as a post-doctoral fellow in the laboratories of Professors Vagelos and Marshall at the Washington University School of Medicine in St. Louis. The research involved the total chemical synthesis of Acyl Carrier Protein by the Merrifield solid phase method. Again the synthetic prducts required extensive purification and this time gel filtration, ion-exchange chromatography, and affinity chromatography techniques were used.

In 1972 he was appointed Lecturer in Chemistry in the joint department of Chemistry, Biochemistry, and Biophysics at Massey University with the goal of establishing a peptide synthesis group. At that time, New Zealand Universities were undergoing a period of rapid expansion and it was decided to recruit researchers in different areas of protein chemistry in this department. As it was found in St. Louis, the purification of synthetic peptides was an important challenge as traditional chromatographic methods based on polysaccharide matrices were inefficient. On July 23, 1973, Joel Morrisett a colleague from Houston, Texas noted in a letter that his department had just purchased a liquid chromatograph for purification of their peptides. He then made prophetic statement "We feel LC is going to provide the ultimate criteria of purity". The editor was impressed by this information and with a grant of $9,800 from a New Zealand Scientific Research Committee (funded by a very popular local lottery known as the "Golden Kiwi") purchased an HPLC in late 1974. The equipment then traveled by slow boat from Milford, USA to the antipodies and was installed in mid-1975. Research in the laboratory soon demonstrated that peptides were extremely difficult to chromatograph on reversed-phase columns, often with long and irreproducible retention times. In the following year Dr. Hancock chanced on a stray comment by Reg Adams (then of the Perkin Elmer Corporation) that phosphoric acid could suppress the active sites on the silica and thus allow the more efficient chromatography of peptides. Although the reason for use of phosphoric acid did not allow for the complex ionic structures of peptides, it was nonetheless a key suggestion which allowed the laboratory to rapidly chromatograph a variety of peptide and protein samples by reversed-phase HPLC. A gratifying feature of this development was that the stray comment was made at the Lord Mayor's banquet held in honor of an IUPAC congress held in Dunedin, New Zealand (August, 1976). Thus social functions can also have a useful scientific function!

The next 6 years allowed a rapid development of the technique with some 50 papers on the subject. During this time Dr. Hancock became interested in the study of the role of the protein components of lipoproteins in heart disease. Therefore current research is directed at the synthesis and study of the interaction of model lipid binding peptides with reversed-phase columns. In addition he has co-authored with Dr. Sparrow of Houston, Texas a review on "The Separation of Proteins by Reversed Phase HPLC" in *High Performance Liquid Chromatography, Advances and Perspectives,* Volume 3, (C. Horvath, Editor). With the same co-author he has written *A Laboratory Manual on the Separation of Biological Materials by HPLC,* which is currently in press. In fact both of these publications arose from a most profitable sabbatical spent in 1980 at the Baylor College of Medicine in the department where the prophetic statement was made about the potential of reversed-phase HPLC. It is a tribute to the durability of modern liquid chromatographs that the original system mentioned in the 1973 letter was still functional and was used extensively by Dr. Hancock.

Other career details about Dr. Hancock include promotion to Senior Lecturer in 1977 and Reader in Chemistry in 1982. Also he is a member of the New Zealand Institutes of Chemistry and Biochemistry, Endocrinology and Immunology Societies, and a Fellow of the American Heart Association.

In addition to an interest in lipoproteins, he has studied the separation by LC of small peptides such as enkephalins, releasing factors and angiotensin and of protein hormones such as insulin and growth hormone. Another recent area of research has been the development of preparative separations (multigram amount) of peptides and proteins with volatile mobile phases such as perfluoroalkanoic acids or ammonium bicarbonate. A related interest has been the development of highly specific matrices for affinity chromatography based on the use of 1,1′-carbonyldiimidazole rather than cyanogen bromide as the activating reagent.

ADVISORY BOARD

CONTRIBUTORS

Monika Abrahamsson, M.Pharm.
Kabi Vitrum AB R & D
Analytical Chemistry Department
Stockholm, Sweden

Nicholas M. Alexander, Ph.D.
Professor of Pathology
University of California School of Medicine
San Diego, California

M. Andre, Ph.D.
Quality Control
Biochemie GmbH
Kundl, Austria

Hitoshi Aoshima, Ph.D.
Associate Professor
Department of Chemistry
Yamaguchi University
Yamaguchi, Japan

B. G. Archer, Ph.D.
Beckman Instruments, Inc.
Berkeley, California

Jeffrey M. Becker, Ph.D.
Professor of Microbiology
University of Tennessee
Knoxville, Tennessee

Jack P. Bell, Ph.D.
Varian Associates
Walnut Creek Instrument Division
Walnut Creek, California

J. Claude Bennett, M.D.
Professor and Chairman
Department of Medicine
University of Alabama
Birmingham, Alabama

C. A. Bishop
Alphatech Systems, Ltd.
Auckland, New Zealand

Ajit S. Bhown, Ph.D.
Assistant Professor (Research)
Department of Medicine
Division of Clinical Immunology
and Rheumatology
University of Alabama
Birmingham, Alabama

Dennis D. Blevins, Ph.D.
Department of Chemistry
University of Arizona
Tucson, Arizona

J. P. H. Burbach, Ph.D.
Rudolf Magnus Institute for Pharmacology
University of Utrecht
Utrecht, The Netherlands

Michael F. Burke, Ph.D.
Associate Professor
Department of Chemistry
University of Arizona
Tucson, Arizona

L. E. Burnworth
Beckman Instruments, Inc.
Berkeley, California

M. T. Campbell, Ph.D.
Senior Scientific Officer
Department of Histology and Embryology
Sydney, Australia

P. R. Carnegie, Ph.D.
Professor of Animal Sciences
La Trobe University
Bundoora, Australia

Marcel H. Caude, Ph.D.
Laboratoire de Chimie Analytique
Centre National de la Recherche Scientifique
Paris, France

David Chung, B.S.
Laboratory of Molecular Endocrinology
University of California
San Francisco, California

E. E. Codd, M.D.
Rudolf Magnus Institute for Pharmacology
University of Utrecht
Utrecht, The Netherlands

Henri Colin, Ph.D.
Ecole Polytechnique
Laboratoire de Chimie Analytique Physique
Palaiseau, France

Nelson H. C. Cooke, Ph.D.
Manager
Chemical Research and Development
Altex Scientific, Inc.
Berkeley, California

P. H. Corran, D. Phil.
National Institute for
Biological Standards and
Control
London, England

David H. Coy, Ph.D.
Research Professor
Department of Medicine
Tulane University School of Medicine
New Orleans, Louisiana

Jacques Crommen, Ph.D.
Assistant Professor
Institute of Pharmacy
University of Liege
Liege, Belgium

Vadim A. Davankov, D.Sc.
Professor
Nesmeyanov Institute of Organo-Element
Compounds
Academy of Sciences
Moscow, USSR

Stanley Norris Deming, Ph.D.
Associate Professor of Chemistry
University of Houston
Houston, Texas

Pier Giorgio Desideri, Ph.D.
Institute of Analytical Chemistry
University of Florence
Florence, Italy

D. M. Desiderio, Ph.D.
Professor of Neurology (Chemistry)
Director, Charles B. Stout Neuroscience
Mass Spectrometry Laboratory
University of Tennessee
Center for Health Sciences
Memphis, Tennessee

Miral Dizdaroglu, Ph.D.
Research Chemist
Center for Radiation Research
National Bureau of Standards
Washington, D.C.

Celina Edelstein, B.A.
Assistant Professor
Department of Medicine
University of Chicago
Chicago, Illinois

D. Fourmy
Attaché de Recherche
INSERM
Toulouse, France

Bengt Fransson
Pharmacist
Research Assistant
Institute of Biochemistry
University of Uppsala
Uppsala, Sweden

Sadaki Fujimoto, Ph.D.
Assistant Professor
Department of Biochemistry
Kyoto College of Pharmacy
Kyoto, Japan

David D. Gay, Ph.D.
Medical Research Associate
Psychopharmacology Medical Research
The Upjohn Company
Kalamazoo, Michigan

M. Judith Gemski
Research Chemist
Walter Reed Army Institute of Research
Washington, D.C.

Emanuel Gil-Av, Ph.D.
Professor
Department of Organic Chemistry
The Weizmann Institute of Science
Rehovot, Israel

Jay A. Glasel, Ph.D.
Professor
Department of Biochemistry
University of Connecticut Health Center
Farmington, Connecticut

Tyge Greibrokk, Ph.D.
Associate Professor
Department of Chemistry
University of Oslo
Oslo, Norway

Kerstin Gröningsson, Ph.D.
Department of Analytical Chemistry
Astra Läkemedel AB
Research and Development Laboratories
Södertälje, Sweden

K. Gstrein, Ph.D.
Quality Control
Biochemie GmbH
Kundl, Austria

Anne Guyon-Gruaz
charge de Recherche
U113 INSERM
Paris, France

Istvan Halasz, Ph.D.
Professor of Applied Physical Chemistry
University of Saarland
Saarbrucken, West Germany

William S. Hancock, Ph.D.
Reader
Department of Chemistry, Biochemistry, and Biophysics
Massey University
Palmerston North, New Zealand

Ichiro Hara, Ph.D.
Emeritus Professor
Laboratory of Chemistry
Tokyo Medical and Dental University
Chiba, Japan

David R. K. Harding, Ph.D.
Senior Research Officer
Department of Chemistry, Biochemistry, and Biophysics
Massey University
Palmerston North, New Zealand

Daniela Heimler
Institute of Analytical Chemistry
University of Florence
Florence, Italy

Dennis W. Hill, Ph.D.
Associate Professor
Microchemistry Laboratory
University of Connecticut
Storrs, Connecticut

John C. Hodgin, M.S.
Applications Chemist
Micromeritics Instrument Corporation
Norcross, Georgia

Patrick Y. Howard, Ph.D.
Marketing Manager
Micromeritics Instrument Corporation
Norcross, Georgia

Victor J. Hruby, Ph.D.
Professor
Department of Chemistry
University of Arizona
Tucson, Arizona

Rex S. Humphrey, Ph.D.
Research Officer
New Zealand Dairy Research Institute
Palmerston North, New Zealand

W. Jeffrey Hurst, M.S.
Group Leader, Analytical Research
Analytical Research and Laboratory Services
Hershey Foods Technical Center
Hershey, Pennsylvania

Taiji Imoto, Ph.D.
Professor
Faculty of Pharmaceutical Sciences
Kyushu University
Fukuoka, Japan

Ken Inouye, Ph.D.
Head, Peptide Chemistry Research Group
Shiongi Research Laboratories
Osaka, Japan

Susumu Ishimitsu
Instructor
Department of Biochemistry
Kyoto College of Pharmacy
Kyoto, Japan

P. S. L. Janssen
Group Leader
Scientific Development Group
Organon International BV
Oss, The Netherlands

Alain P. Jardy
Assistant Professor
Laboratoire de Chimie Analytique
ESPCI
Paris, France

Yoshio Kato, Ph.D.
Senior Chemist
Central Research Laboratory
Toyo Soda Manufacturing Company, Inc.
Yamaguchi, Japan

Yukio Kimura, Ph.D.
Professor
Faculty of Pharmaceutical Sciences
Mukogawa Women's University
Hyogo, Japan

Claude B. Klee, M.D.
Chief, Macromolecular Interactions Section
Laboratory of Biochemistry
National Institutes of Health
Bethesda, Maryland

Ryusei Konaka, Ph.D.
Head, Analytical Chemistry Research Group
Shionogi Research Laboratories
Osaka, Japan

Ante M. Krstulovic, Ph.D.
Laboratoire de Chimie Analtique Physique
Ecole Polytechnique
Palaiseau, France

Robert A. Lahti, Ph.D.
Senior Research Associate
CNS Diseases Research
The Upjohn Company
Kalamazoo, Michigan

Michal Lebl, Ph.D.
Scientist
Peptide Chemistry Group
Institute of Organic Chemistry and Biochemistry
Czechoslovak Academy of Sciences
Prague, Czechoslovakia

Robert Lehrer, Ph.D.
Supervisor, Training
Instrument Systems Division
Du Pont Company
Wilmington, Delaware

Juhani Leppäluoto, M.D.
Associate Professor
Department of Physiology
University of Oulu
Oulu, Finland

Luciano Lepri, Ph.D.
Associate Professor of Chemical Qualitative Analysis
Institute of Analytical Chemistry
Firenze, Italy

Randolph V. Lewis, Ph.D.
Assistant Professor of Biochemistry
University of Wyoming
Laramie, Wyoming

Choh Hao Li, Ph.D.
Laboratory of Molecular Endocrinology
University of California
San Francisco, California

Jen-Kun Lin, Ph.D.
Professor
Institute of Biochemistry
College of Medicine
National Taiwan University
Taipei, Taiwan
Republic of China

S. Linde, Ph.D.
Scientist
Hagedorn Research Laboratory
Gentofte, Denmark

J. G. Loeber, Ph.D.
Scientist
Laboratory for Endocrinology
National Institute of Public Health
Bilthoven, The Netherlands

Elsa Lundanes, Ph.D.
Research Associate
Department of Chemistry
University of Oslo
Oslo, Norway

Hisao Mabuchi, M.D.
Medical Doctor
Department of Internal Medicine and Nephrology
Nishijin Hospital
Kyoto, Japan

Allan S. Manalan, M.D.
Laboratory of Biochemistry
National Cancer Institute
National Institutes of Health
Bethesda, Maryland

C. McMartin, Ph.D.
Ciba-Geigy Pharmaceuticals Division
West Sussex, England

F. Nachtmann, Ph.D.
Quality Control
Biochemie GmbH
Kundl, Austria

Fred Naider, Ph.D.
Professor of Chemistry
College of Staten Island
Staten Island, New York

Hisamitsu Nakahashi, M.D.
Medical Doctor
Department of Internal Medicine and Nephrology
Nishijin Hospital
Kyoto, Japan

Toshio Nambara, Ph.D.
Professor
Pharmaceutical Institute
Tohoku University
Sendai, Japan

Akira Ohara, Ph.D.
Professor
Department of Biochemistry
Kyoto College of Pharmacy
Kyoto, Japan

Eiji Okada
Project Manager, Liquid Chromatography
Tokyo Research and Application Laboratory
Shimadzu Corporation
Tokyo, Japan

Mitsuyo Okazaki, Ph.D.
Assistant Professor
Laboratory of Chemistry
Tokyo Medical and Dental University
Chiba, Japan

C. Olieman, Ph.D.
Analytical Chemist
Institute for Dairy Research
Ede, The Netherlands

Stephen Oroszlan
Director
Laboratory of Molecular Virology and Carcinogenesis; and
Head, Immunochemistry Section
National Cancer Institute
Frederick Cancer Research Facility
LBI-Basic Research Program
Frederick, Maryland

Wayne R. Peterson
Department of Food Science and Human Nutrition
University of Florida
Gainesville, Florida

Petro E. Petrides, M.D.
Department of Neurobiology
Stanford University School of Medicine
Stanford, California

John K. Pollak, Ph.D.
Reader in Histology and Embryology
University of Sydney
Sydney, Australia

L. Pradayrol, Ph.D.
Charge de Recherche
INSERM
Toulouse, France

R. L. Prestidge, Ph.D.
Research Fellow
Department of Immunobiology
Medical School
Auckland, New Zealand

Ulf Ragnarsson, Ph.D.
Institute of Biochemistry
University of Uppsala
Uppsala, Sweden

D. Raulais, Ph.D.
Maitre de Recherche
LA 163 C.N.R.S.
Paris, France

A. Ribet, Ph.D.
Charge de Recherche
INSERM
Toulouse, France

L. G. Richards
Beckman Instruments, Inc.
Berkeley, California

Pierre Rivaille, Ph.D.
Maitre de Recherche
LA 163 C.N.R.S.
Paris, France

Robert H. Rosset, Ph.D.
Professor of Analytical Chemistry
ESPCI; and
Director
ESPCI Analytical Chemistry Laboratory
Paris, France

S. Sakakibara, Ph.D.
Director
Peptide Institute
Protein Research Foundation

Stephen I. Sallay, Ph.D.
Professor of Chemistry, Director
Research Institute for Cancer Detection
Purdue University
Fort Wayne, Indiana

Tatsuru Sasagawa
Department of Biochemistry
University of Washington
Seattle, Washington

Angelo M. Scanu, M.D.
Professor
Departments of Medicine and Biochemistry
University of Chicago
Chicago, Illinois

David H. Schlesinger, Ph.D.
Professor of Medicine and Cell Biology
New York University Medical Center
New York, New York

John A. Schmit
Manager, Marketing Technical
Instrument Systems Division
Du Pont Company
Wilmington, Delaware

Walter A. Schroeder, Ph.D.
Senior Research Associate in Chemistry
Division of Chemistry and Chemical Engineering
California Institute of Technology
Pasadena, California

Bernard Sebille, Ph.D.
Professor of Chemistry
University of Paris XII
Créteil, France

G. G. Skellern, Ph.D.
Lecturer
Drug Metabolism Research Unit
Department of Pharmacy
University of Strathclyde
Glasgow, Scotland

K. A. Smolenski
Ph.D. Student
Bristol University
Bristol, England

Alvin N. Starratt, Ph.D.
Research Scientist
Research Centre, Agriculture Canada
London, Ontario, Canada

Alvin Steinfeld, Ph.D.
Adjunct Associate Professor
Senior Research Associate
Department of Chemistry
College of Staten Island
Staten Island, New York

Alvin S. Stern, Ph.D.
Senior Scientist
Department of Molecular Genetics
Hoffmann-La Roche, Inc.
Nutley, New Jersey

Mary E. Stevens
Research Technician
Research Centre, Agriculture Canada
London, Ontario, Canada

M. Patricia Strickler, Ph.D.
Waters Associates
Washington, D.C. Office
Rockville, Maryland

James D. Stuart, Ph.D.
Associate Professor
Department of Chemistry
University of Connecticut
Storrs, Connecticut

Keith Sugden, M.Sc., C.CHEM.
Pharmaceutical Division
Reckitt and Colman, Ltd.
Kingston-Upon-Hull
N. Humberside, England

Marjorie E. Svoboda, Ph.D.
Research Associate
Department of Pediatrics
University of North Carolina
Chapel Hill, North Carolina

David C. Teller
Department of Biochemistry
University of Washington
Seattle, Washington

Nicole Thuaud, Ph.D.
Research Associate
University of Paris
Créteil, France

E. Tomlinson, Ph.D.
Professor of Pharmaceutical Chemistry
University of Amsterdam
Amsterdam, The Netherlands

J. A. D. M. Tonnaer, Ph.D.
Research Associate
Scientific Development Group
Organon International BV
Oss, The Netherlands

K. Unger, Ph.D.
Professor of Chemistry
Johannes Gutenberg University
Mainz, West Germany

J. W. van Nispen, Ph.D.
Research Chemist
Scientific Development Group
Organon International BV
Oss, The Netherlands

Judson J. Van Wyk, M.D.
Division Chief
Pediatric Endocrinology
Kenan Professor of Pediatrics
University of North Carolina
Chapel Hill, North Carolina

J. Verhoef, Ph.D.
Rudolf Magnus Institute for Pharmacology
University of Utrecht
Utrecht, The Netherlands

Adolf von Wurttemberg, B.S.
Applications Chemist
Micromeritics Instrument Corporation
Norcross, Georgia

D. Voskamp, Ph.D.
Organic Chemist
Laboratory of Organic Chemistry
Delft, The Netherlands

Olli Vuolteenaho, M.D.
Research Associate
Department of Physiology
University of Oulu
Oulu, Finland

Joseph J. Warthesen, Ph.D.
Associate Professor of Food Chemistry
Department of Food Science and Nutrition
University of Minnesota
St. Paul, Minnesota

Kunio Watanabe
Staff, Peptide Chemistry Research Group
Shionogi Research Laboratories
Osaka, Japan

C. Timothy Wehr, Ph.D.
Manager, HPLC Applications Laboratory
Varian Associates
Walnut Creek Instrument Division
Walnut Creek, California

Shulamith Weinstein, Ph.D.
Senior Scientist
Department of Organic Chemistry
The Weizmann Institute of Science
Rehovot, Israel

Benny S. Welinder, Ph.D.
Scientist
Hagedorn Research Laboratory
Gentofte, Denmark

J. M. Wilkinson, Ph.D.
Research Fellow
Department of Biochemistry
Birmingham University
Birmingham, England

A. Witter, Ph.D.
Rudolf Magnus Institute for Pharmacology
University of Utrecht
Utrecht, The Netherlands

Julianne C. Wood-Rethwill
Group Leader
New Food Product Division
Armour Dial Company
Scottsdale, Arizona

Hidenori Yamada, Ph.D.
Assistant Professor
Faculty of Pharmaceutical Sciences
Kyushu University
Fukuoka, Japan

Joan M. Zanelli, Ph.D.
National Institute for Biological Standards and Control
London, England

Örjan Zetterqvist, Ph.D.
Lecturer of Medical and Physiological Chemistry
University of Uppsala
Uppsala, Sweden

TABLE OF CONTENTS

Volume I

Detection Methods

Separation of Free Amino Acids

Analysis of Biologically Active Peptides

Use of HPLC in Protein Sequencing

Protein Separations

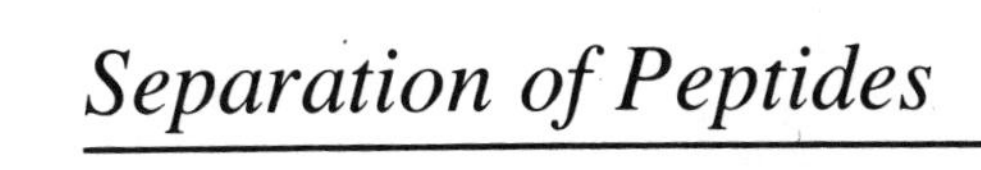

Separation of Peptides

REVIEW OF SEPARATION CONDITIONS FOR PEPTIDES

William S. Hancock and David R. K. Harding

INTRODUCTION

Different peptides exhibit a wide range of polarities and consequently a single set of mobile-phase conditions would not be expected to achieve chromatographic analysis of all molecules on reversed-phase (RP) HPLC. In addition the early applications of HPLC to the analysis of peptides were not entirely successful with the use of reversed-phase columns. Poor resolution was frequently observed to be associated with peak broadening and long retention times.[1] These features are not altogether unexpected when one considers the complex ionic equilibria that these amphoteric compounds can undergo (see the chapters by Tomlinson and Deming).

The addition of almost any salt to the mobile phase results in an improvement in the chromatographic profile obtained for a peptide on a reversed-phase column (see Figure 1).[2] This improvement can be attributed to the blocking of interactions of the sample with the silanol groups present in the reversed-phase packing material. An example of the effect of an added electrolyte can be seen in the report by Yamada and Imoto on the use of dilute hydrochloric acid as a mobile phase.

Various salts when added to the mobile phase can, however, give rise to significant differences in peak shapes and retention times for a given sample. Figure 2 shows the elution profile obtained for the peptide, Leu-Trp-Met-Arg in the absence of an ion-pairing reagent (part A), in the presence of 5 m*M* sodium hexanesulfonate, pH 7.1 (part B) and pH 2.1 (part C), and 0.1% phosphoric acid, pH 2.5 (part D). The chapters by Tomlinson and Deming show that a variety of effects occur when different ions are added to the mobile phase and that the separation mechanism can be complex. This review of separation conditions will, however, concentrate on the practical aspects of peptide separations.

THE USE OF MOBILE PHASES THAT CONTAIN PHOSPHORIC ACID OR PHOSPHATES

We found that the addition of phosphoric acid to the eluant allows the facile analysis of a range of peptides.[3] In addition the use of phosphoric acid allows the UV detection of eluted peptides at wavelengths in the range of 200 to 220 nm, and hence monitoring of the separation at high sensitivity. Figure 3 shows the elution profile of three peptides with a mobile phase that contains 0.1% phosphoric acid. Also, several reports in the handbook describe the separation of peptides with mobile phases that contain phosphate salts, for examples see the review chapter by Bishop as well as chapters by Inouye et al. (synthetic peptide analogs), Lebl (carba analogs), Tonnaer (angiotensins), Nachtmann and Gstrein (oxytocin intermediates), Guyon-Gruaz et al. (LHRH), and Groningsson and Abrahamsson (somatostatin).

The phosphoric acid-containing mobile phase has proved to be particularly useful for the analysis of synthetic peptides produced by solid-phase or solution peptide synthesis.[4] Figure 4 shows the resolution of three separate peptides related to angiotensin II, Val-Tyr-Ile-His-Pro-Phe, Val-Ile-His-Pro-Phe, and Tyr-Ile-His-Pro-Phe. The samples represent crude mixtures obtained from solid-phase peptide synthesis; Val-Ile-His-Pro-Phe represents a failure peptide caused by an incomplete coupling of tyrosine during the synthesis. The retention times obtained indicate that very good resolution of similar peptides is possible. This technique can thus be used to detect minor deletion products from peptide synthesis. Also HPLC

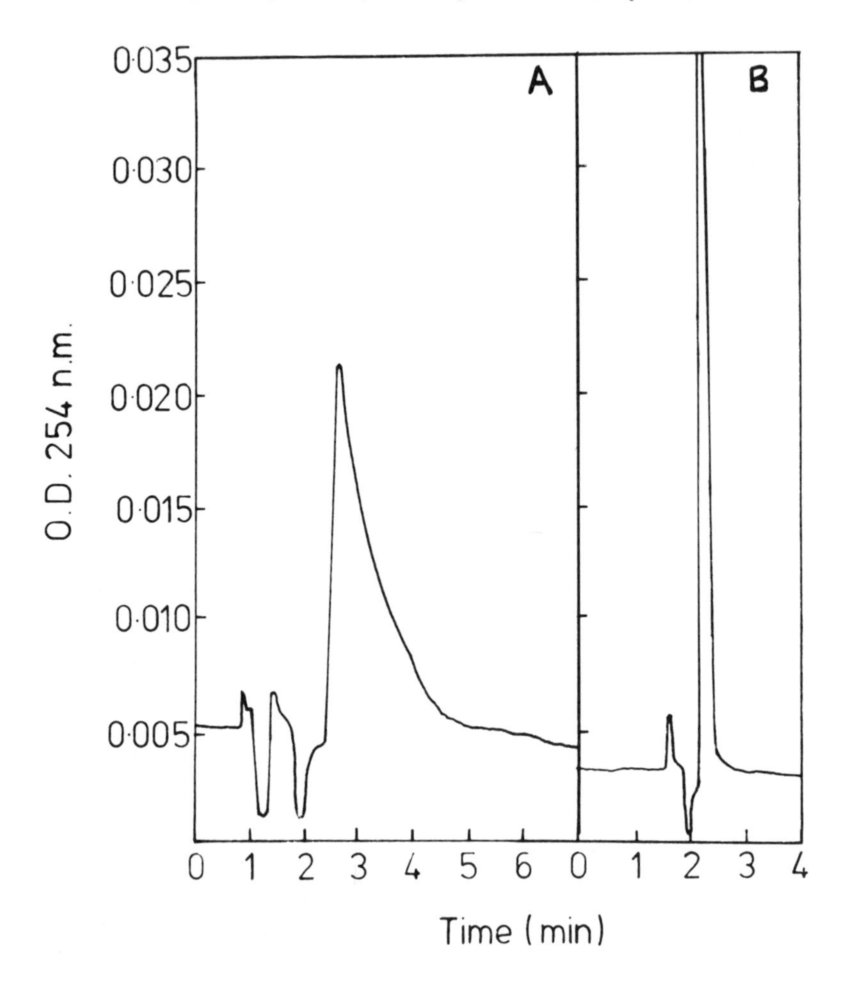

FIGURE 1. The elution profile of Gly-Leu-Tyr on a μBondapak®-alkylphenyl column in the absence (A) and presence (B) of 5 m*M* sodium hexanesulfonate, pH 7.1. The mobile phase was methanol-water (1:1). A flow rate of 1.5 mℓ/min was used. (From Hancock, W. S., Bishop, C. A., Meyer, L. J., Harding, D. R. K., and Hearn, M. T. W., *J. Chromatogr.*, 161, 291, 1978. With permission.)

analysis can be used to monitor the purification of a synthetic peptide using classical techniques. Figure 5 gives the elution profiles obtained for the crude and purified tetrapeptide, Cys-Ala-Gly-Tyr, which indicated that gel filtration separations had produced a highly purified product. In addition the chapters by Ragnarsson and Fransson, Inouye et al., Lebl, and Nachtmann and Gstrein give numerous examples of this application of RP-HPLC.

THE USE OF MOBILE PHASES THAT CONTAIN NONPOLAR, ANIONIC ION-PAIRING REAGENTS

The combination of a small polar anion such as $H_2PO_4^-$ with the cationic groups (RNH_3^+) of a peptide has allowed the rapid analysis of a variety of samples (see Figure 3 as an example). The hydrophilic ion pair affected an increase in polarity of the sample with a consequent decrease in retention on a reversed-phase column (see the chapter by Bishop). While this system was particularly useful for hydrophobic peptides, it was found that many peptide samples were not adequately retained on a C_{18} column in the presence of a phosphate-

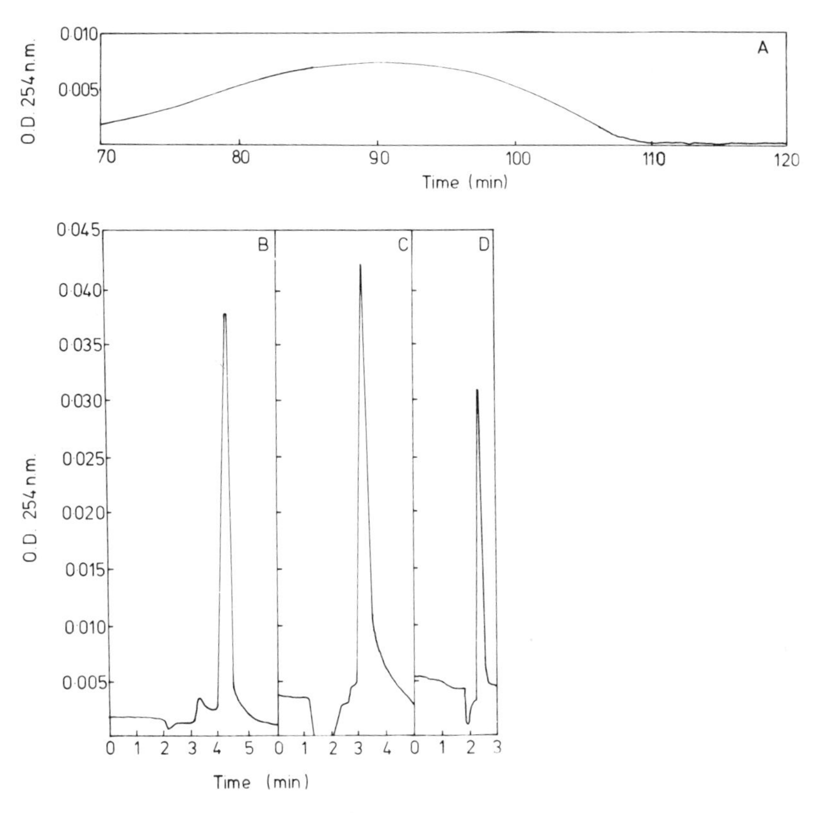

FIGURE 2. The elution profile of Leu-Trp-Met-Arg on a μBondapak®-alkylphenyl column with methanol-water (1:1) as eluent. A, no ion-pairing reagent was added; B, sodium hexanesulfonate, pH 7.1; C, hexanesulfonic acid; pH 2.1; D, phosphoric acid, pH 2.5. (From Hancock, W. S., Bishop, C. A., Meyer, L. J., Harding, D. R. K., and Hearn, M. T. W., *J. Chromatogr.*, 161, 291, 1978. With permission.)

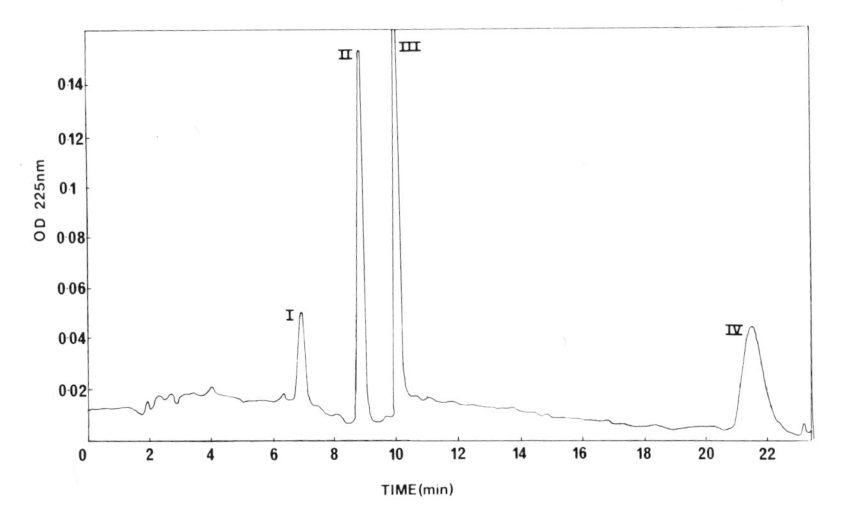

FIGURE 3. Results of an analysis of a mixture of peptides and benzoic acid on a μBondapak®-Fatty acid analysis column with 25% acetonitrile and 75% water and 0.1% H_3PO_4 as the mobile phase. Peak 1 corresponds to Met-Arg-Phe-Ala, peak II to Leu-Trp-Met-Arg, peak III to benzoic acid, and peak IV to Leu-Trp-Met-Arg-Phe; 1 μg of each component, made up in the mobile phase, was loaded in a volume of 25 μℓ; the flow rate was 1.5 mℓ/min, the pressure 144 atm, and the temperature 22°C. (From Hancock, W. S., Bishop, C. A., Prestidge, R. L., Harding, D. R. K., and Hearn, M. T. W., *Science*, 200, 1168, 1978. With permission.)

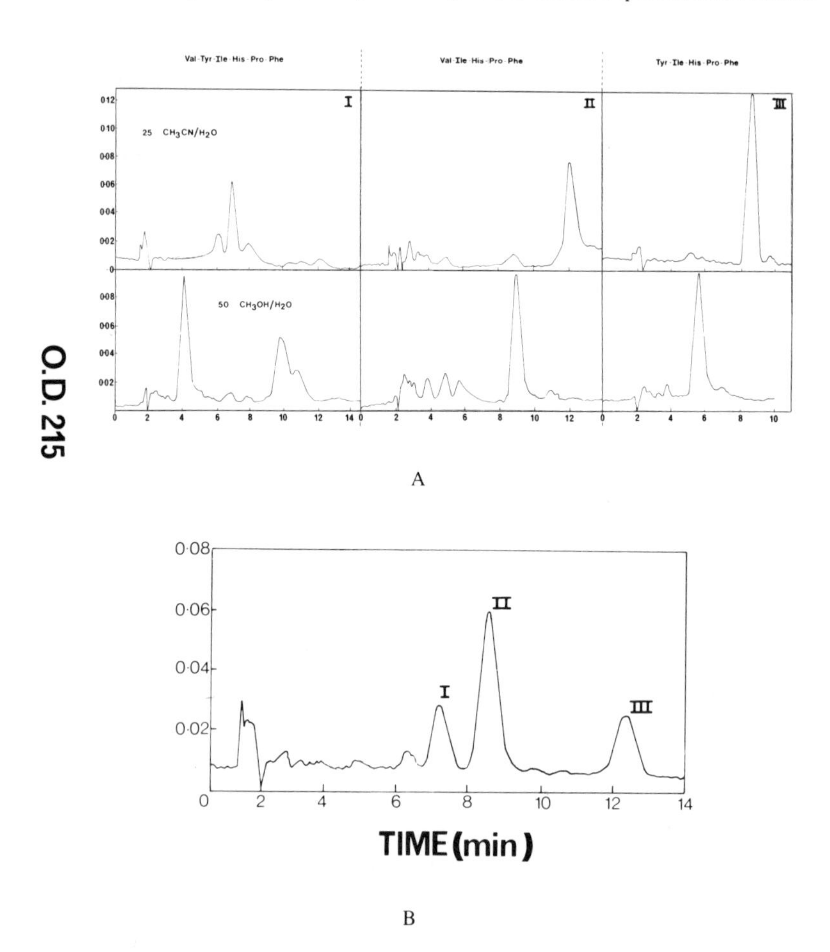

FIGURE 4. (A) Elution profiles of three synthetic angiotensin fragments, Val-Tyr-Ile-His-Pro-Phe (L), Val-Ile-His-Pro-Phe(II), and Tyr-Ile-His-Pro-Phe(III), on a μBondapak®-Fatty acid analysis column. The mobile phase consisted of either 25% acetonitrile and 75% water with 0.1% H_3PO_4, or 50% methanol and 50% water with 0.1% H_3PO_4. This figure illustrates the marked alteration in retention times resulting from the use of different solvent compositions. (B) The elution profile of a mixture of the same three angiotensin fragments on a μBondapak®-Fatty acid analysis column. The mobile phase was 25% acetonitrile and 75% water, with 0.1% H_3PO_4. (From Hancock, W. S., Bishop, C. A., Prestidge, R. L., Harding, D. R. K., and Hearn, M. T. W., *Science*, 200, 1168, 1978. With permission.)

containing mobile phase (see the chapters on thyrotropin releasing factor [Vuolteenaho and Leppaluoto], angiotensins [Tonnaer], and basic hydrophilic peptides [Ragnarsson and Fransson] for examples). The chromatographic properties of several peptides on a μBondapak®-alkylphenyl column in the presence of hydrophobic anionic reagents were investigated. These studies indicated that hydrophobic ion-pairing reagents can be used to significantly increase the retention of polar peptides. Table 1 shows the result of an analysis of seven different peptides by RP-HPLC with a hydrophilic (H_3PO_4), or a moderately hydrophobic ($CH_3(CH_2)_5SO_3Na$) and an extremely hydrophobic ($CH_3(CH_2)_{11}SO_4Na$) ion-pairing reagent. Clearly the more hydrophobic the ion-pairing reagent the greater the observed retention. Perhaps the most striking result from these studies is the dramatic changes in the selectivity of the chromatographic system that one can achieve by the use of different ion-pairing reagents (Figure 6) or pH (Table 2). In Figure 6 it can be seen that the elution order of

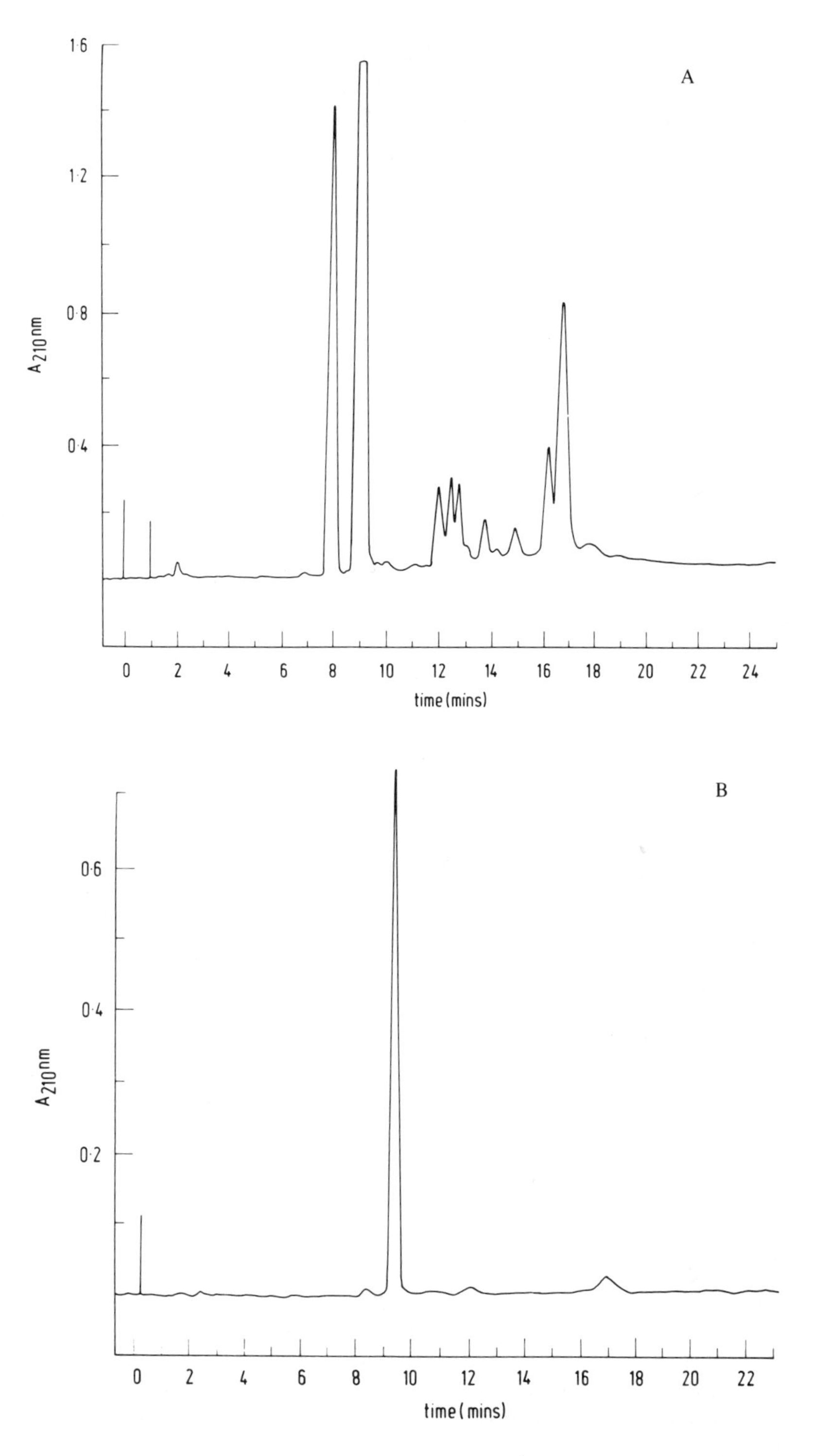

FIGURE 5. (A) Elution profile of crude solid-phase synthetic Cys-Ala-Gly-Tyr product. Column: μBondapak®-C_{18}, flow rate 2 mℓ/min. A 10-min linear gradient of water, 0.1% phosphoric acid to 40% methanol-water-0.1% phosphoric acid was used. The gradient was started 1 min after injection of 4 μg of sample. (B) Elution profile obtained under similar conditions after repeated Sephadex® G15 and Bio-Gel® P_2 gel permeation chromatography of the desired tetrapeptide. (From Hearn, M. T. W. and Hancock, W. S., *Trends Biochem. Sci.*, N58, 1979. With permission.)

Table 1
THE EFFECT OF HYDROPHOBIC ION PAIRING ON THE RETENTION TIME OF PEPTIDES ON A μBONDAPAK®-ALKYLPHENYL COLUMN[a]

Eluent: methanol-water (1:1); the values given are the retention times in minutes

	Ion-pairing reagent			
Peptide[a]	None	H_3PO_4 (pH 2.4, 5 m*M*)	Hexanesulfonate sodium salt (pH 6.5, 5 m*M*)	SDS (pH 7.15, 5 m*M*)
L-W-M-R	112	2.3	4.0	16.2
L-W-M-R-F	>120	5.1	10.2	40.5
G-F	2.3	2.3	2.3	2.5
G-G-Y	64.5	1.9	3.4	6.2
M-R-F	32.5	2.4	3.6	>58
G-L-Y	2.5	2.4	2.4	2.7
R-F-A	>48	2.05	3.0	33.2

[a] The code for amino acid is used by M. O. Dayhoff in "Atlas of Protein Sequence and Structure," National Biomedical Research Foundation, Silver Spring, MD, A, alanine; D, aspartic acid; F, phenylalanine; G, glycine; K, lysine; L, leucine; M, methionine; R, arginine; S, serine; W, tryptophan; Y, tyrosine.

From Hancock, W. S., Bishop, C. A., Meyer, L. J., Harding, D. R. K., and Hearn, M. T. W., *J. Chromatogr.*, 161, 291, 1978. With permission.

three peptides, Gly-Gly-Tyr,Gly-Leu-Tyr, and Arg-Phe-Ala can be reversed by the substitution of phosphoric acid with SDS in the mobile phase. The examination of the elution profile of a peptide in the presence of several different ion-pairing reagents should provide excellent evidence of purity, just as thin-layer chromatography (TLC) in different solvent systems is presently used as a criterion of purity.

In a similar manner the separation of other polar peptides has been achieved with the addition of a suitable alkyl sulfonate to the mobile phase, for example, see the chapters by Ragnarsson and Fransson, Vuolteenaho and Leppaluoto, and Tonnaer.

In initial studies aimed at examining the effect of phosphoric acid solutions on the elution of peptides from reversed-phase columns, the use of acetic acid and trifluoroacetic (TFA) acid as mobile-phase additives was examined. It was observed that acetic acid gave broad peaks with significantly greater retention times[3] and was generally not suitable for the analysis of peptides. TFA, however, gave excellent results that were comparable to those obtained with phosphoric acid solutions. The major disadvantages of this mobile phase when compared to phosphoric acid were the greater acidity of the fluorinated acid which resulted in decreased column lifetime and a somewhat greater absorbance at 210 nm. A major advantage, however, was the excellent volatility of TFA solutions which conveniently allowed preparative separations (see the chapter by Olieman and Voskamp). However polar peptides are often not sufficiently retained for efficient separations on nonpolar columns when complexed with trifluoracetate, even in the absence of organic modifier in the mobile phase. However, as is shown in Table 3, perfluoroalkanoic acids can be used in a similar manner to the alkyl sulfonates for the separation of polar peptides.[7] In addition these reagents have been shown to give excellent peak shapes.[7] The chapter by Olieman and Voskamp shows the potential of perfluorinated carboxylic acids in peptide separations. Also the chapter by Starrett and Stevens uses heptafluorobutyric acid in the purification of proctolin.

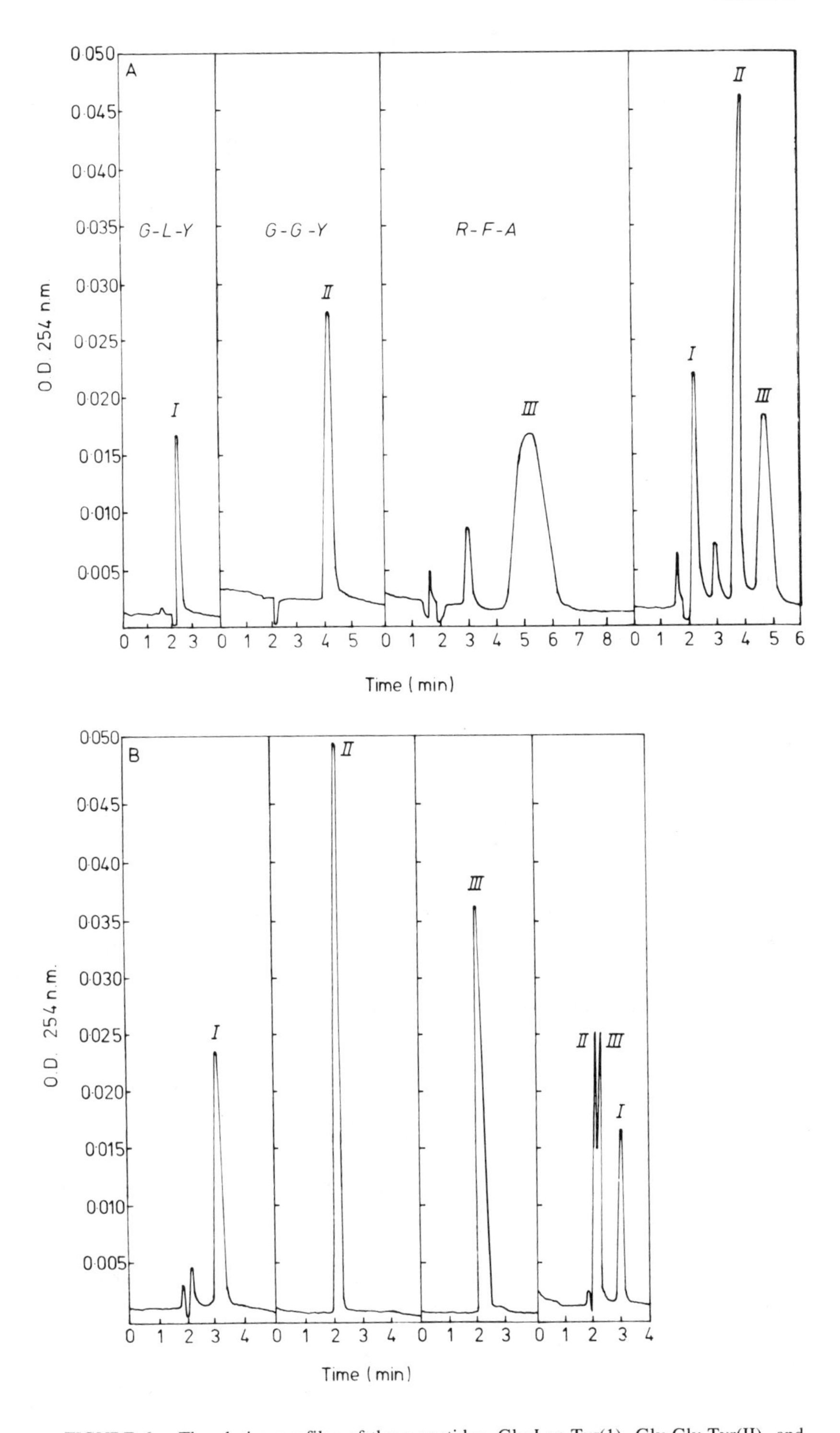

FIGURE 6. The elution profiles of three peptides, Gly-Leu-Tyr(1), Gly-Gly-Tyr(II), and Arg-Phe-Ala(III) on a μBondapak®-alkylphenyl column with methanol-water (1:1) as eluent. (A) 1-m*M* SDS was added to the eluent; (B) 0.1% H_3PO_4. (From Hancock, W. S., Bishop, C. A., Meyer, L. J., Harding, D. R. K., and Hearn, M. T. W., *J. Chromatogr.*, 161, 291, 1978. With permission.)

Table 2
THE EFFECT OF pH ON THE RETENTION TIME OF PEPTIDES ON A μBONDAPAK®-ALKYLPHENYL COLUMN IN THE PRESENCE OF AN ION-PAIRING REAGENT

Eluent: methanol-water (1:1); the values given are the retention times in minutes

	Ion-pairing reagent			
	$CH_3(CH_2)_5SO_3Na$		$CH_3(CH_2)_{11}SO_4Na$	
Peptide	**pH 2.1**	**pH 6.5[a]**	**pH 2.9**	**7.15[a]**
L-W-M-R	3.2	3(−)	>25	16.2(−)
L-W-M-R-F	5.6	10.2(+)	—	40.5
G-F	2.5	2.3(−)	8.9	2.5(−)
G-G-Y	2.1	2.4(+)	5.5	6.2(+)
M-R-F	3.1	3.6(+)	—	>58
G-L-Y	2.5	2.3(−)	9.3	2.7(−)
R-F-A	2.5	3(+)	—	33.2

[a] The + or − sign indicates whether an increased or decreased retention time is caused by an increase in pH.

From Hancock, W. S., Bishop, C. A., Meyer, L. J., Harding, D. R. K., and Hearn, M. T. W., *J. Chromatogr.*, 161, 291, 1978. With permission.

THE USE OF MOBILE PHASES THAT CONTAIN CATIONIC ION-PAIRING REAGENTS

In a similar manner to the results obtained with anionic reagents, Tables 4 and 5 show that addition of a cationic reagent to the mobile phase can give a useful modification of the chromatographic behavior of a peptide. The results indicate that depending on the cationic reagent involved, the retention time of the peptide can either be significantly increased (e.g., tetraethylammonium ion) or decreased (e.g., dodecylammonium ion) on a reversed-phase column.

As is described in the chapter by Deming there are several possible interactions between any ionic modifier added to the mobile phase and sample molecules and/or the stationary phase. Amines are particularly effective in deactivating interactions between silanol groups and sample molecules (see the chapter by Unger). This deactivation is probably an important factor in the success of triethylammonium phosphate as an ionic modifier for the separation of peptide and protein samples.[9] Also alkylamines with a long alkyl chain probably interact with the reversed phase to form a dynamic coating (see the chapters by Deming and Tomlinson). These solvent-generated stationary-phase modifications can permit ionic interactions between the peptide and the stationary phase. This effect appears to be the dominant mechanism with anionic detergents like sodium dodecylsulfate and probably also for cationic detergents like dodecylammonium acetate. At pH 4 a peptide would be predominantly protonated and thus show minimal interaction with a cationic stationary phase. This will be reflected in short retention times as can be seen, for example with dodecylammonium acetates in Tables 4 and 6 and the decrease in retention time observed for alkylamines of increasing C-chain length (see Table 5). Table 6 shows that the retention time for a given peptide can be dramatically affected by the addition of a nonpolar anionic or cationic reagent to the mobile phase.

Table 3
EFFECT OF PERFLUOROALKANOIC ACIDS ON THE RETENTION TIME OF VARIOUS PEPTIDES

Peptide[a]	Retention time		
	CF_3COOH	C_2F_5COOH	C_3F_7COOH
	Mobile Phase—Water		
G-L-A	2.35	2.4	2.5
G-G-D	2.15	2.3	2.4
G-G-R	2.15	2.95	4.0
G-G-E	2.15	2.5	2.65
G-G-K	2.15	2.6	3.0
L-G-G-G	2.6	3.9	4.8
	Mobile Phase — 15% Acetonitrile — Water		
R-F-A	2.4	2.8	3.2
G-L-Y	2.8	3.0	3.35
	Mobile Phase — 22.5% Acetonitrile — Water		
L-W-M-R	2.65	3.2	3.4
L-W-M-R-F	5.6	7.8	9.1
M-R-F-A	2.4	2.8	2.9

Note: The analyses were carried out on a μBondapak®-C_{18} column with a 5-m*M* solution of the perfluoroalkanoic acid in the mobile phase.

[a] The code for amino acids is G, Gly; L, Leu; A, Ala; D, Asp; R, Arg; E, Glu; K, Lys; F, Phe; Y, Tyr; W, Trp; M, Met.

From Harding, D. R. K., Bishop, C. A., Tarttelin, M. F., and Hancock, W. S., *Int. J. Peptide Protein Res.*, 18, 214, 1981. With permission.

The increase and then decrease in retention time for the tetraalkylammonium ion series (Table 4) could be explained by considering the possible balance between the ion-pairing and ion-exchange effects of the reagents added to the mobile phase. It is possible that the tetramethyl- and tetraethylammonium ions act predominantly by ion-pairing effects while the tetrabutylammonium ion undergoes little ion pairing with the solute due to the steric bulk of the cation but acts by forming a surface coating on the stationary phase. Several of the peptides showed two peaks with the intermediate case of the tetrapropylammonium ion (see Table 5), an observation which could be explained on the basis of these competing effects. The data in Tables 4 and 5 also suggest that an alkyl chain of a significant number of C-atoms and surface bulk is required for an effective coating of the stationary phase. In the chapter by Mabuchi a three-step purification system based on ion-pairing reactions and dynamic modification of the stationary phase is described.

PREPARATIVE SEPARATIONS OF PEPTIDES

Figure 7 shows the separation of a synthetic pentadecapeptide on a Radial Pak-C_{18} column with a mobile phase which contains triethylammonium phosphate.[10] This example shows that useful amounts of peptide can be separated on an analytical column and in fact up to 50 mg of crude synthetic peptides have been separated on this column. A major disadvantage

Table 4
COMPARISON OF THE EFFECT OF TETRAALKYLAMMONIUM SALTS ON THE RETENTION TIME OF PEPTIDES[a,b]

	Retention time (min)				
	Ammonium salt ($\overset{+}{R_4N}\overset{-}{OAc}$)				
Peptide[3]	R = H	R = CH_3	R = CH_2CH_3	R = $(CH_2)_2CH_3$	R = $(CH_2)_3CH_3$
L-W-M-R	2.8	3.0	3.05	3.5	2.7
L-W-M-R-F	5.7	6.8	6.5	8.7	4.0
G-F	2.4	2.4	2.6	2.7	2.05
G-G-Y	2.0	2.0	2.2	1.84, 2.05[c]	1.7
M-R-F	2.6	2.7	2.8	3.0	1.7
G-L-Y	2.4	2.5	2.8	2.7	2.1
R-F-A	2.3	2.3	2.6	2.1, 2.7	1.8

Note: The other eluent parameters are described under Methods.

[a] The separations were achieved on a μBondapak®-alkylphenyl column with 1:1, methanol:water as the mobile phase. Each cationic reagent was added as a 2-m*M* solution, with acetate as the anion and the pH adjusted to 4.0. The flowrate used was 1.5 mℓ/min.

[b] A, alanine; D, aspartic acid; F, phenylalanine; G, glycine; K, lysine; L, leucine; M, methionine; R, arginine; W, tryptophan; Y, tyrosine.

[c] Two peaks were observed.

From Hancock, W. S., Bishop, C. A., Battersby, J. E., Harding, D. R. K., and Hearn, M. T. W., *J. Chromatogr.*, 168, 377, 1979. With permission.

of the mobile phase used is that triethylammonium phosphate is involatile which necessitates a separate desalting step to obtain the peptide in a salt-free form.

The preparative separation of peptides by reversed-phase chromatography has become a practical method since the introduction of volatile mobile phases. TFA is currently the most popular volatile, mobile-phase additive (see the chapter by Olieman and Voskamp). Another factor that facilitates preparative separations is the availability of a preparative HPLC such as the Waters Prep-500 System which allows the proportionate scale-up of an analytical separation. An example of this scale-up is the preparative separation of a 5-g sample of a tetrapeptide Leu-$(Gly)_3$ on a 5.7 cm × 30 cm Prep Pak-500-C_{18} cartridge (75 μm).[11] The mobile phase contained 0.05% TFA dissolved in water:methanol (95:5) and a flowrate of 100 mℓ/min was used (back pressure 100 psi). Figure 8a shows the analytical profile of the crude product from a solution synthesis of the tetrapeptide and Figure 8b shows the profile obtained for an authentic sample. Figure 9 shows the elution profile obtained for a 5-g sample on the preparative instrument. Figure 10 gives the elution profile for the corresponding analytical separations of fractions from the preparative run. The results shown in Figure 10 indicate that the back end of the main peak eluting at approximately 8 min in the preparative run contained pure material. In addition further pure material was obtained from fraction 8 which was collected from the back end of the material recycled from the front end of the main peak. Subsequent experiments showed that the recovery of purified material was approximately 95%.[11] A later publication[12] described the purification of the following synthetic peptides in up to 5-g loadings per preparative run — Gly-Gly-OEt, GlyGlyGlu, GlyGlyLys, PyrHisGly, $(Pro)_3$, and the pentapeptides Leu- and Met-enkephalin.

It was observed in a preparative separation of the polar peptide Gly-Gly-Glu that a mobile

Table 5
COMPARISON OF THE EFFECT OF DIFFERENT ALKYLAMINE SALTS ON THE RETENTION TIME OF PEPTIDES[a]

	Retention time of amine salt ($R\overset{+}{N}H_3$)					
Peptide	R = H	R = CH_3	R = CH_2CH_2OH	R = $(CH_2)_3CH_3$	R = $(CH_2)_5CH_3$	R = $(CH_2)_{11}CH_3$
L-W-M-R	2.7	3.05	3.1	2.9	2.4	1.6
L-W-M-R-F	5.7	7.0	6.1	6.8	4.4	2.1
G-F	2.4	2.5	2.4	2.3	2.25	2.0
G-G-Y	2.0	2.2	2.3	2.0	1.8	1.4
M-R-F	2.6	2.75	2.7	2.5	2.2	1.45
G-L-Y	2.4	2.6	2.6	2.45	2.25	2.0
R-F-A	2.3	2.3	2.45	2.1	2.0	1.3

[a] The same separation conditions were used as described in Table 4.

From Hancock, W. S., Bishop, C. A., Battersby, J. E., Harding, D. R. K., and Hearn, M. T. W., *J. Chromatogr.*, 168, 377, 1979. With permission.

Table 6
COMPARISON OF THE EFFECT OF DODECYLAMINE ACETATE AND SDS ON THE RETENTION TIME OF PEPTIDES

	Retention time (min)	
Peptide	$CH_3(CH_2)_{11}SO_3^-Na^+$ (A)[b]	$CH_3(CH_2)_{11}NH_3^{+-}$ OAc(B)[a]
L-W-M-R	16.2	1.6
L-W-M-R-F	40.5	2.1
G-F	2.5	2.0
G-G-Y	6.2	1.4
M-R-F	>58	1.45
G-L-Y	2.7	2.0
R-F-A	33.2	1.3

[a] The eluent parameters are described in Table 4.
[b] A 2-m*M* solution of this reagent was used in methanol-water, (1:1) (pH 6.5).

From Hancock, W. S., Bishop, C. A., Battersby, J. E., Harding, D. R. K., and Hearn, M. T. W., *J. Chromatogr.*, 168, 377, 1979. With permission.

phase which contained 0.05% TFA gave minimal retention of the peptide on a C_{18} column.[12] For this reason we decided to examine perfluoroalkanoic acids which were significantly more lipophilic than TFA for the preparative separation of polar peptides. Perfluorobutyric acid was found to be particularly useful in the analysis and purification of polar peptides. The addition of a 5-m*M* solution of the reagent to the mobile phase resulted in a significant increase in retention of the peptide relative to that obtained when TFA was added to the mobile phase. In addition, the ammonium salt of the acid could be removed from the purified peptide by extraction with ether, a solvent in which most peptides are insoluble. The usefulness of this mobile phase was demonstrated by the successful purification of the peptide Pyr-His-Gly (Figure 11). In this figure, part B shows the separation achieved when the synthetic product was chromatographed with a mobile phase which contained 0.05% perfluoroacetic acid. In this elution profile there is poor separation between the tripeptide and early eluting polar impurities. Part A shows the same separation except that perfluorobutyric acid is used as the ion-pairing reagent. In this case Pyr-His-Gly is well separated from early and late eluting impurities. This result can be attributed to the increased retention of the solute on the C_{18} column when complexed with the more lipophilic ion-pairing reagent. The analysis of purified material that was isolated from the center of the major peak is shown in Figure 2A.

Ammonium bicarbonate has been a popular solvent for the separation of peptides by conventional chromatographic techniques due to its excellent volatility and the high solubility of many peptides in the buffer. The high apparent pH of this mobile phase (7.7 to 8) precludes its use with siliconaceous supports packed in inflexible columns, due to the generation of column voids caused by dissolution of the silica. The radial compression that is used with the flexible-walled columns fitted in the Radial Compression Module from Waters Associates circumvents this problem as any voids that may be generated are removed during column compression.[13] Provided the column is washed with water and then isopropanol each evening, we have found that an extended lifetime of at least 6 months can be achieved with Radial Pak-C_{18} or -CN columns. As a further precaution a guard column filled with Porasil® B-C_{18} packing material is used. The guard column does not degrade the separation, but does allow for a significant increase in column lifetime due both to removal

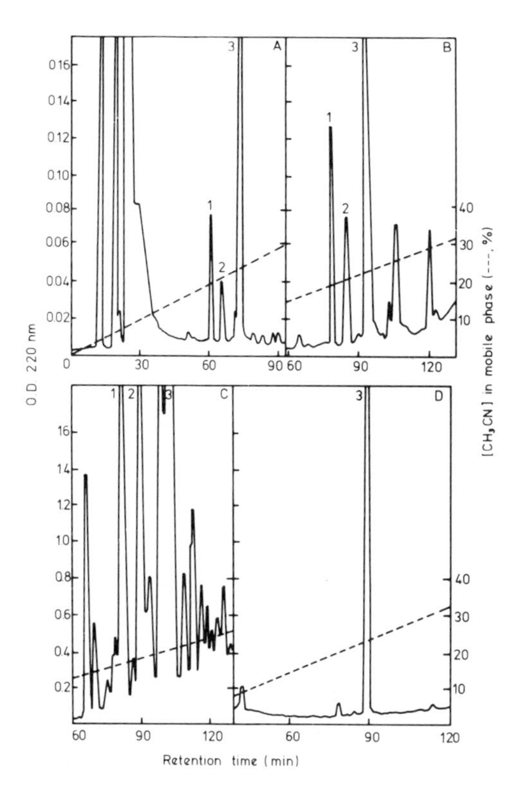

FIGURE 7. The purification of a synthetic analog of the 1-15 segment of human apolipoprotein C-I, in which Phe-14 is replaced with *p*-iodophenylalanine in the chemical synthesis. The mobile phase was purified 1% TEAP with a linear gradient of acetonitrile (see the dashed lines). The flow rate was 1.5 mℓ/min. The loadings in A to C were 0.5, 2 and 8 mg of crude peptide dissolved in 1% TEAP, 3 *M* guanidine hydrochloride at a concentration of 10 mg/mℓ. Peak 3 corresponded to the desired peptide, and this fraction from each run was pooled. The analysis of an aliquot of this pool is shown in D. (From Hancock, W. S. and Sparrow, J. T., *J. Chromatogr.*, 206, 71, 1981. With permission.)

of contaminants from the sample and mobile phase and to dissolution of silica in the guard column (thus partially presaturating the mobile phase with silica).

Figure 12A shows the purification of 350 μg of the synthetic peptide Leu-Glu-Ser-Phe-Leu-Lys-Ser-Trp-Leu-Ser-Ala-Leu-Glu-Gln-Ala-Leu-Lys-Ala. Prior to the HPLC separation, this peptide has been partially purified by gel filtration and ion-exchange chromatography. The sample was loaded in 5 cm^3 of 6 *M* urea as three approximately equal volumes through the U6K sample injector. The large peak was collected (see bar in Figure 12A) and rechromatographed as shown in Figure 12B. Before rechromatography the trapped peak (2 mℓ) was diluted to 6 mℓ with 0.1 *M* NH_4HCO_3 and loaded as three approximately equal volumes using identical conditions as those in Figure 12A. The purified peptides were characterized by amino acid analysis — a typical result was Ser 3.0(3), Glu 3.1(3), Ala 2.9(3), Leu 5.0(5), Phe 1.0(1), Lys 2.0(2), and Trp 1.0(1). The recovery of the purified peptide, as measured by amino acid analysis, was 87%.

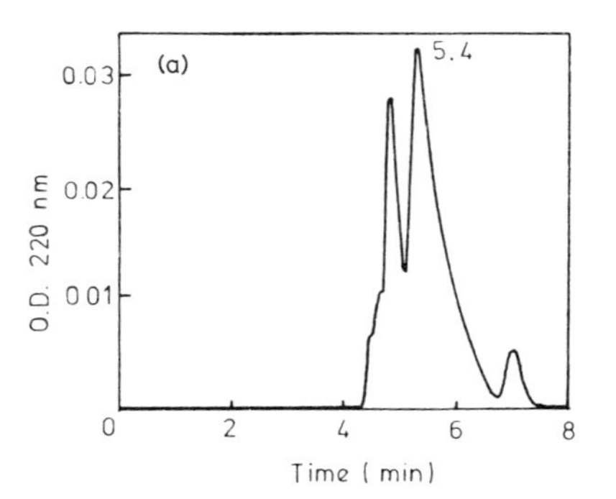

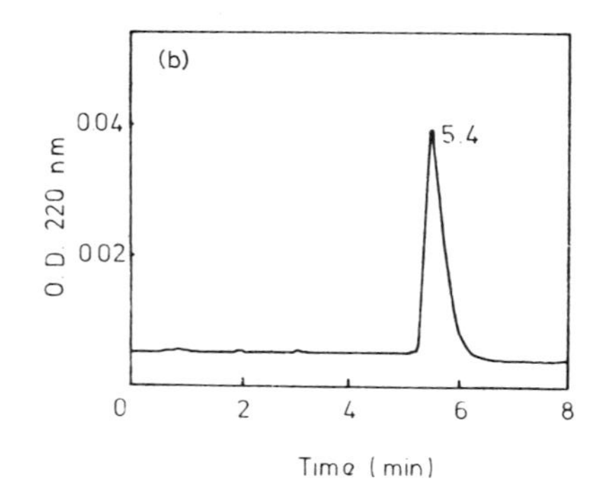

FIGURE 8. The elution profile of the crude synthetic L-Leu$(Gly)_3$ (a) and the commercially obtained standard L-Leu$(Gly)_3$ (b). Chromatographic conditions: column, µBondapak®-C_{18}; mobile phase, water-0.05% TFA, pH 2.3; flow rate, 1.5 mℓ/min. (From Bishop, C. A., Harding, D. R. K., Meyer, L. J., and Hancock, W. S.)

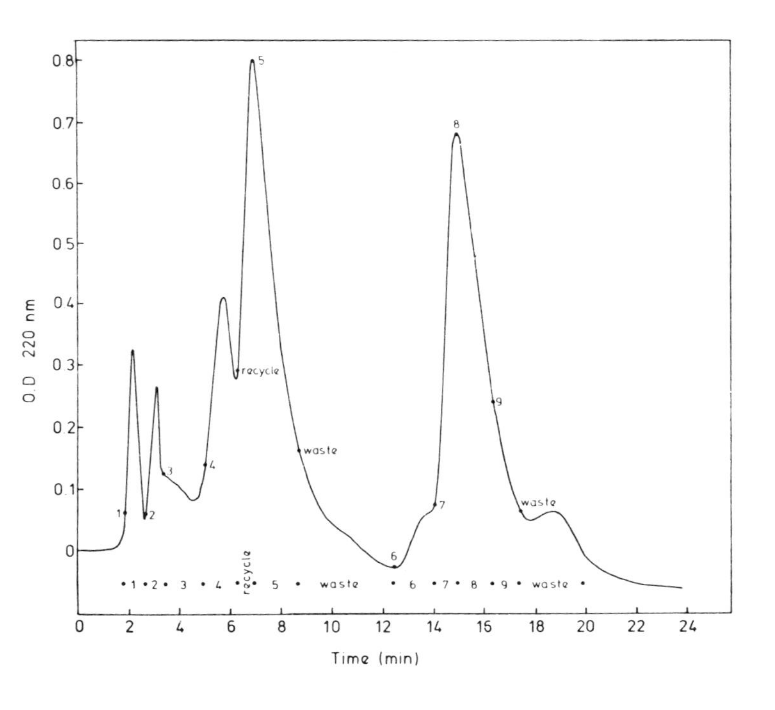

FIGURE 9. The elution profile of the preparative purification of 1 g crude L-Leu$(Gly)_3$. Chromatographic conditions: column, Prep Pak-500/C_{18} cartridge; mobile phase, water-methanol-TFA (95:5:0.05), pH 2.3; flow rate 100 mℓ/min. (From Bishop, C. A., Harding, D. R. K., Meyer, L. J., and Hancock, W. S.)

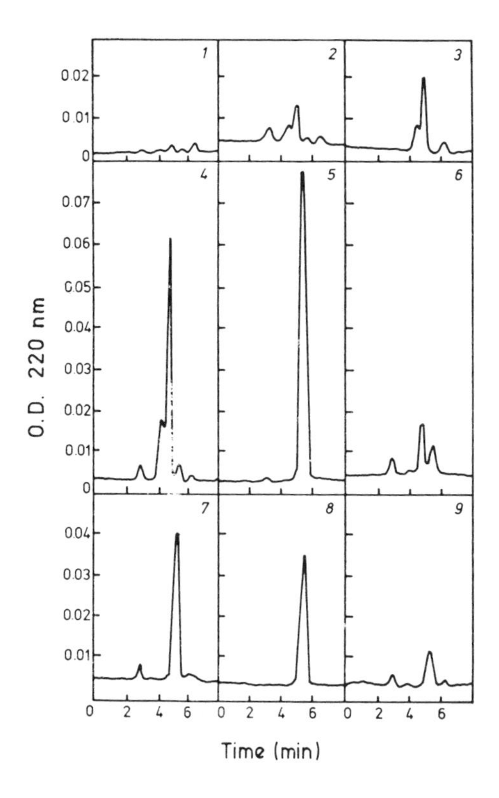

FIGURE 10. The analytical HPLC profiles of the collected fractions (1-9) from the preparative separation of the crude L-Leu$(Gly)_3$. Chromatographic conditions as in Figure 8. (From Bishop, C. A., Harding, D. R. K., Meyer, L. J., and Hancock, W. S.)

USE OF REVERSED-PHASE TLC FOR THE MONITORING OF PREPARATIVE SEPARATIONS

With the advent of preparative reversed-phase separations, it was necessary to have a procedure to assay the purity of the separated fractions. Since an additional liquid chromatograph sometimes is not available to monitor the separation, we have developed a simple TLC method.[14] Figure 13 shows use of the reversed-phase TLC system to follow the semi-preparative purification of a synthetic octadecapeptide, Leu-Glu-Ser-Phe-Leu-Lys-Ser-Trp(CHO)-Leu-Ser-Ala-Leu-Glu-Gln-Ala-Leu-Lys-Ala. The TLC results provided a rapid check of the HPLC separation and also verified that the optical density peaks actually consisted of peptide material (ninhydrin reactive). This peptide contained a large number of hydrophobic residues; therefore, it was strongly retained on a reversed-phase TLC plate. However, satisfactory results were obtained when tetrahydrofuran was added to the mobile phase, and the narrow spots shown in Figure 1B were obtained. Despite differences in the mobile phases, there is a clear correlation between the HPLC and TLC systems; for example, the early eluting material in pool A of the HPLC fractions gave the highest R_F in the reversed-phase TLC system. Thus a reversed-phase TLC system based on Whatman $KC_{18}F$ plates and a mobile phase which contained 3% sodium chloride and 0.2% SDS can be used for the monitoring of fractions from a HPLC separation.

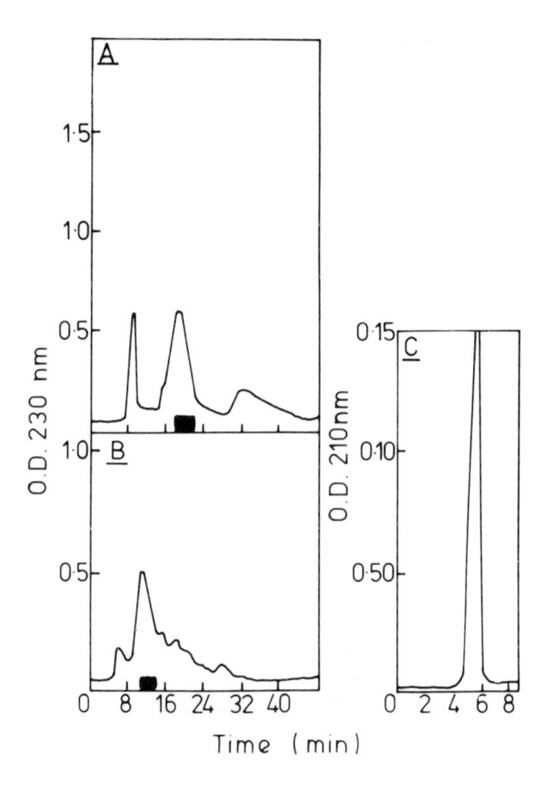

FIGURE 11. Preparative separation of Pyr-His-Gly. A 1-g sample of the crude product was dissolved in the mobile phase and injected onto the C_{18}-column. A flow rate of 100 mℓ/min was used for the preparative separation shown in parts A and B. In part B the mobile phase was 0.05% perfluoroacetic acid, parts A and C perfluorobutyric acid (5 m*M*). The analytical separation shown in part C was carried out on a μBondapak®-C_{18} column, with 5 m*M* perfluorobutyric and as the mobile phase. The materials that were pooled and isolated in the preparative runs were shown by the solid bar. (From Harding, D. R. K., Bishop, C. A., Tarttelin, M. F., and Hancock, W. S., *Int. J. Peptide Protein Res.*, 18, 214, 1981. With permission.)

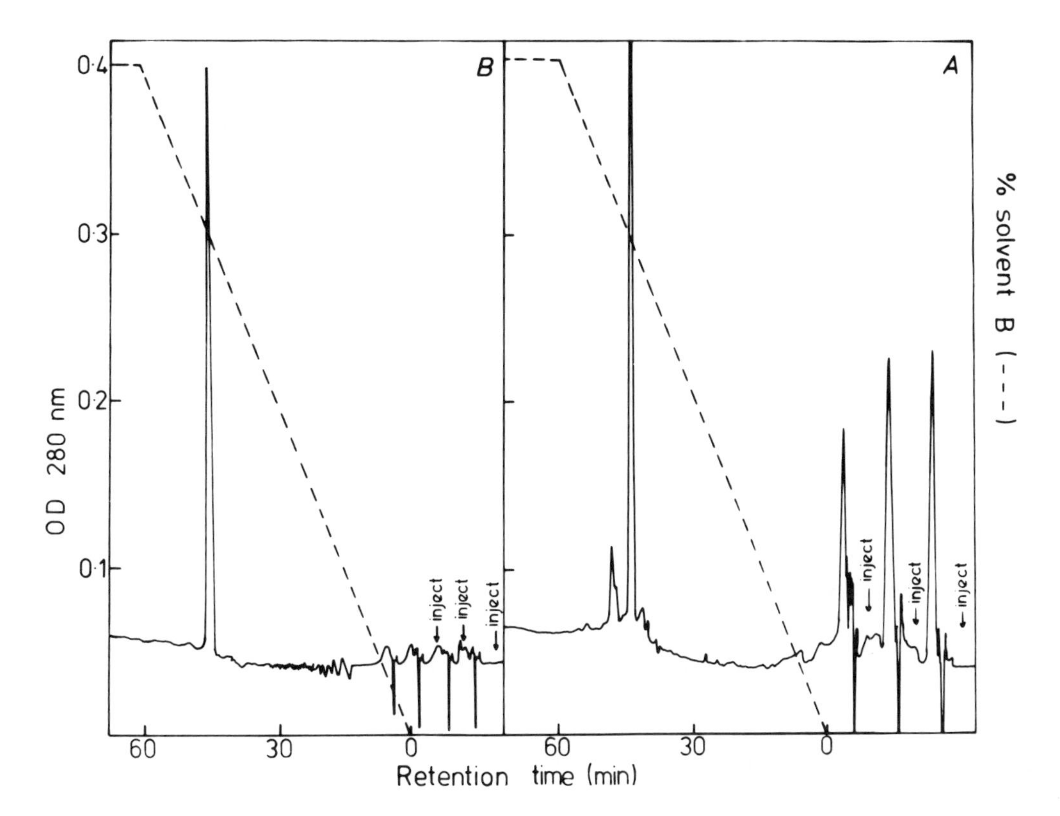

FIGURE 12. The purification of 350 μg of the peptide Leu-Glu-Ser-Phe-Leu-Lys-Ser-Trp-Leu-Ser-Ala-Leu-Gly-Gln-Ala-Leu-Lys-Ala was accomplished using a Radial Pak-CN column with a linear gradient from 0.1 *M* NH_4HCO_3 to i-PrOH:CH_3CN: 0.1 *M* NH_4HCO_3 (3:3:4) at a flow rate of 1.0 mℓ/min. The sample was loaded in three injections from a solution in 5 mℓ of 6 *M* urea. Part A shows the elution profile for the crude peptide mixture. The urea defined by the solid bar was pooled (2 mℓ), diluted to 6 mℓ with 0.1 *M* NH_4HCO_3, and rechromatographed using the same conditions as in Part A. (From Knighton, D. R., Harding, D. R. K., Napier, J. R., and Hancock, W. S., *J. Chromatogr.*, 249, 193, 1982. With permission.)

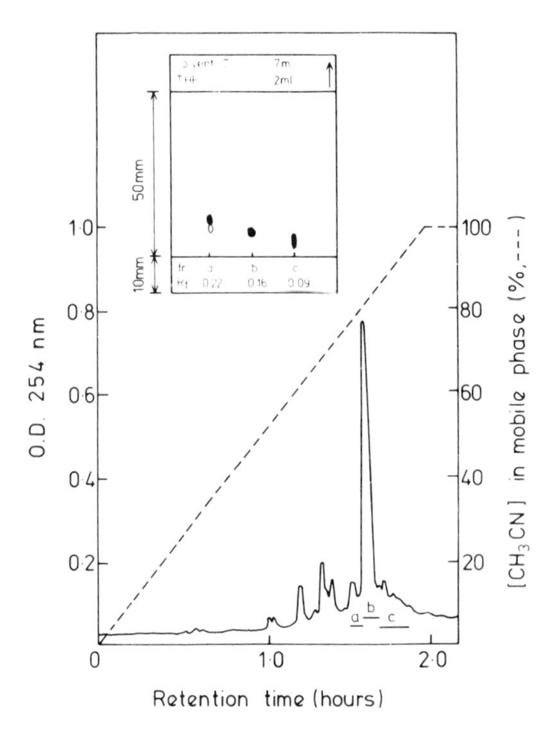

FIGURE 13. The semipreparative purification of a synthetic octadecapeptide, Leu-Glu-Ser-Phe-Leu-Lys-Ser-Trp(CHO)-Leu-Ser-Ala-Leu-Glu-Gln-Ala-Leu-Lys-Ala. The separation was achieved on a Radial PAK-C_{18} column. The initial solvent (Solvent A) consisted of aqueous triethylammonium phosphate (1.5 m*M*, pH 3.2) with a gradient to solvent B, propan-2-ol-acetonitrile-aqueous triethylammonium phosphate (7.5 m*M*), (40:40:20, v/v/v). The flowrate was 1.0 mℓ/min and the gradient (A to B) shown by the dotted line was used. A 5-mg sample of the peptide dissolved in buffer A (0.2 mℓ) was used. The inset shows the corresponding reversed-phase TLC separation of the fractions. The TLC was carried out on a Whatman KC_{18} F reversed-phase TLC plate with 1.5% NaCl and 0.1% SDS dissolved in water: acetonitrile:methanol:tetrahydrofuran, 50:10:10:10. (From Poll, D. J., Knighton, D. R., Harding, D. R. K., and Hancock, W. S., *J. Chromatogr.*, 236, 244,

REFERENCES

1. **Hancock, W. S., Bishop, C. A., and Hearn, M. T. W.,** HPLC in the analysis of underivatised peptides using a sensitive and rapid procedure, *FEBS Lett.*, 72, 139, 1976.
2. **Hancock, W. S., Bishop, C. A., Meyer, L. J., Harding, D. R. K., and Hearn, M. T. W.,** Rapid analysis of peptides by HPLC with hydrophobic ion pairing of amino groups, *J. Chromatogr.*, 161, 291, 1978.
3. **Hancock, W. S., Bishop, C. A., Prestidge, R. L., Harding, D. R. K., and Hearn, M. T. W.,** The use of phosphoric acid in the analysis of underivatised peptides by reversed phase HPLC, *J. Chromatogr.*, 153, 391, 1978.

4. **Hearn, M. T. W., Bishop, C. A., Hancock, W. S., Harding, D. R. K., and Reynolds, G. D.,** Application of reversed phase HPLC in solid phase peptide synthesis, *J. Liquid Chromatogr.,* 2, 1, 1979.
5. **Hancock, W. S., Bishop, C. A., Prestidge, R. L., Harding, D. R. K., and Hearn, M. T. W.,** Reversed phase HPLC of peptides and proteins with ion-pairing reagents, *Science,* 200, 1168, 1978.
6. **Hearn, M. T. W. and Hancock, W. S.,** Ion pair partition reversed phase HPLC, *Trends Biochem. Sci.,* N58, 1979.
7. **Harding, D. R. K., Bishop, C. A., Tarttelin, M. F., and Hancock, W. S.,** Use of perfluoroalkanoic acids as volatile ion-pairing reagents in preparative HPLC, *Int. J. Peptide Protein Res.,* 18, 214, 1981.
8. **Hancock, W. S., Bishop, C. A., Battersby, J. E., Harding, D. R. K., and Hearn, M. T. W.,** The use of cationic reagents for the analysis of peptides by HPLC, *J. Chromatogr.,* 168, 377, 1979.
9. **Rivier, J. E.,** Use of trialkylammonium phosphate (TEAP) in reversed phase HPLC for high resolution and high recoveries of peptides and proteins, *J. Liquid Chromatogr.,* 1, 343, 1978.
10. **Hancock, W. S. and Sparrow, J. T.,** Use of Mixed-mode HPLC for the separation of peptide and protein mixtures, *J. Chromatogr.,* 206, 71, 1981.
11. **Bishop, C. A., Harding, D. R. K., Meyer, L. J., and Hancock, W. S.,** The preparative separation of synthetic peptides on reversed-phase silica packed in radially-compressed flexible walled columns.
12. **Bishop, C. A., Meyer, L. J., Harding, D. R. K., Hancock, W. S., and Hearn, M. T. W.,** *J. Liquid Chromatogr.,* 4, 661, 1981.
13. **Knighton, D. R., Harding, D. R. K., Napier, J. R., and Hancock, W. S.,** The facile, semi-preparative separation of synthetic peptides using ammonium bicarbonate buffers, *J. Chromatogr.,* 249, 193, 1982.
14. **Poll, D. J., Knighton, D. R., Harding, D. R. K., and Hancock, W. S.,** Use of ion-paired, reversed-phase thin-layer chromatography for the analysis of peptides. A simple procedure for the monitoring of preparative reversed phase HPLC, *J. Chromatogr.,* 236, 244, 1982.

SEPARATION OF PEPTIDES BY HIGH-PERFORMANCE ION-EXCHANGE CHROMATOGRAPHY

Miral Dizdaroglu

INTRODUCTION

Ion-exchange chromatography, first introduced by Moore and Stein[1] for the amino acid analysis of proteins and peptides, has played an important role in the past in the separation and purification of peptides. The method of Moore and Stein was later applied to the separation of peptides from the partial hydrolyzate of a protein.[2] Since then, this technique has been used extensively for the separation of peptide fragments obtained by chemical or enzymatic cleavage of proteins as one of the fundamental steps in the determination of their sequences.[3-12] Because of the value of peptide mapping for sequence analysis of proteins, improvements or variations of this technique have been sought. Jones[13] and others[14-16] have shown that the use of automatic column chromatographic equipment can provide highly reproducible separations of peptides, thus allowing peptide mapping to be carried out. Automated ion-exchange chromatography of peptides has also been performed on microbore columns.[17,18]

Various ion-exchange resins have been successfully used for the separation of peptides. Two types of ion exchangers have been generally applied for this purpose: cation exchangers for neutral and basic peptides, and anion exchangers for acidic and neutral peptides. Most ion exchangers have been synthesized from divinylbenzene cross-linked polystyrene.[19] In order to achieve ion-exchange properties, functional groups have then been attached to the polymeric matrix. These microporous resins have low degrees of cross-linking (usually 2 to 8%) to allow permeation of peptides into the matrix to enhance the resolution.[19] Cellulose ion exchangers have also been employed widely for the separation of proteins and peptides.[20,21]

An important improvement in ion-exchange chromatography of peptides was the introduction of volatile buffers, permitting the eluted, salt-free peptides to be directly used for amino acid and sequence analysis.[17,18,22-25] Among these, pyridine-acetic acid buffers have been widely used in combination with ninhydrin-based monitoring systems[16] for the detection of eluted peptides. Usually, ninhydrin has been the reagent of choice for the detection of amino acids and peptides. However, according to the properties of the peptides and of the eluents, several other types of detection systems, such as UV absorption detectors, differential refractometers, radioactivity detectors, and fluorescence methods, have also been used for this purpose.[17,26] A large number of papers have appeared describing the application of ion-exchange column chromatography to the separation problems of peptide chemistry. These papers discuss the use of various ion-exchange resins, eluents, and detection systems mentioned above.[27-60]

During the past decade, high-performance (or high-pressure) liquid chromatography (HPLC) has emerged as an excellent analytical tool for application in diverse areas. The reversed-phase mode of HPLC (RP-HPLC) has become the most popular and broadly utilized technique for many separation problems. The rapid development of this technique has also contributed greatly to the improvement of peptide separations by liquid chromatography (see other chapters of this book).

ION-EXCHANGE HPLC OF PEPTIDES

Up to the present time, ion-exchange HPLC (IE-HPLC) has been used rarely for the separation of peptides.[61] However, new developments in this area have made this technique a powerful and valuable tool for this purpose. Applications of IE-HPLC to peptide separations have used either silica-based or polystyrene-divinylbenzene-based microparticulate porous ion-exchange stationary phases (for an excellent review of modern ion-exchange packings and their comparison, see the book of Snyder and Kirkland[62]). Various mobile phases, both volatile and nonvolatile, have been employed in isocratic and gradient elution modes.

Peptide Separations by IE-HPLC on Polymeric Ion-Exchange Resins

Several authors have reported the separation of peptides from various sources by IE-HPLC using polymeric ion-exchange resins. Van der Rest et al.[63] have described "fingerprints" obtained by IE-HPLC of peptides released from various human type collagens by clostridiopeptidase A (Figure 1). An amino acid analyzer operating at high pressures, a polymeric cation-exchange resin, and sodium citrate buffer system were used for this purpose. Some specific peaks indicated in Figure 1 have also been identified for each type of collagen.

In another instance, an automatic high-pressure peptide analyzer has been applied to the preparative separation of thyrotropin-releasing hormone analogs.[64] Figure 2 shows an example of the separations obtained on this system, which used a polymeric cation-exchange resin and volatile pyridine-acetic acid buffers. Nika and Hultin[65] have also described the use of an automatic peptide analyzer for rapid microscale peptide separations. Separation of the enzymatic digests of ribonuclease A and acidic brain protein S-100 on a polymeric cation exchanger has been demonstrated using volatile pyridine-acetic acid buffers. An improvement of this system was later reported by the same authors.[66] An example that shows the separation of peptic-tryptic peptides of ribonuclease A is given in Figure 3. A sensitivity of 0.1 nmol and a recovery of 90% or more of the eluted peptides were reported for these separations.

Resolution of γ-glutamyl peptides has been achieved by James[67] on an amino acid analyzer using a polymeric anion exchanger and sodium acetate buffer. An amino acid analyzer has also been used to separate methionyl dipeptides.[68] All the papers listed above have used ninhydrin reaction for detection of eluted peptides.

Takahashi et al.[69] have described an HPLC method for analytical peptide mapping as well as for preparative peptide separations using a macroreticular anion-exchange resin of the styrene-divinylbenzene type and a gradient from water to methane-sulfonic acid containing acetonitrile and isopropanol. This elution system allowed direct monitoring of eluting peptides by UV absorption. Excellent resolution of the tryptic digests of some proteins such as calmodulin, bovine brain S-100b and 14-3-2 proteins, and human amyloid protein have been achieved.[69] Two examples of these separations are given in Figures 4 and 5. The assignment of peaks to separated peptides in the total sequence of those proteins has also been performed. Reasonable recoveries ranging from 34 to 99% have been obtained. The same method has also been applied to comparative structural analysis of calmodulin and des(Ala-Lys)-calmodulin isolated from porcine brain.[70] More recently, a cation exchanger of the same type having a high cross-linkage (35%) has been used by Isobe et al.[71] to separate the tryptic digests of various proteins using the same elution system as in Figures 4 and 5. A typical chromatogram obtained with this system is given in Figure 6. Individual peptides have also been assigned to peaks by amino acid analysis. The solvent system used in these investigations[69-71] was not completely volatile. For purposes such as sequence analysis, desalting of the peptides eluted from ion-exchange columns such as these has been carried out by RP-HPLC.[71]

Peptide Separations by IE-HPLC on Silica-Based Ion Exchangers

Most modern HPLC column packings in use today are based on a silica matrix.[62] Silica-based, IE-bonded, porous stationary phases have found broad application in various areas of IE-HPLC. Strong cation exchangers and strong anion exchangers carrying the label SCX

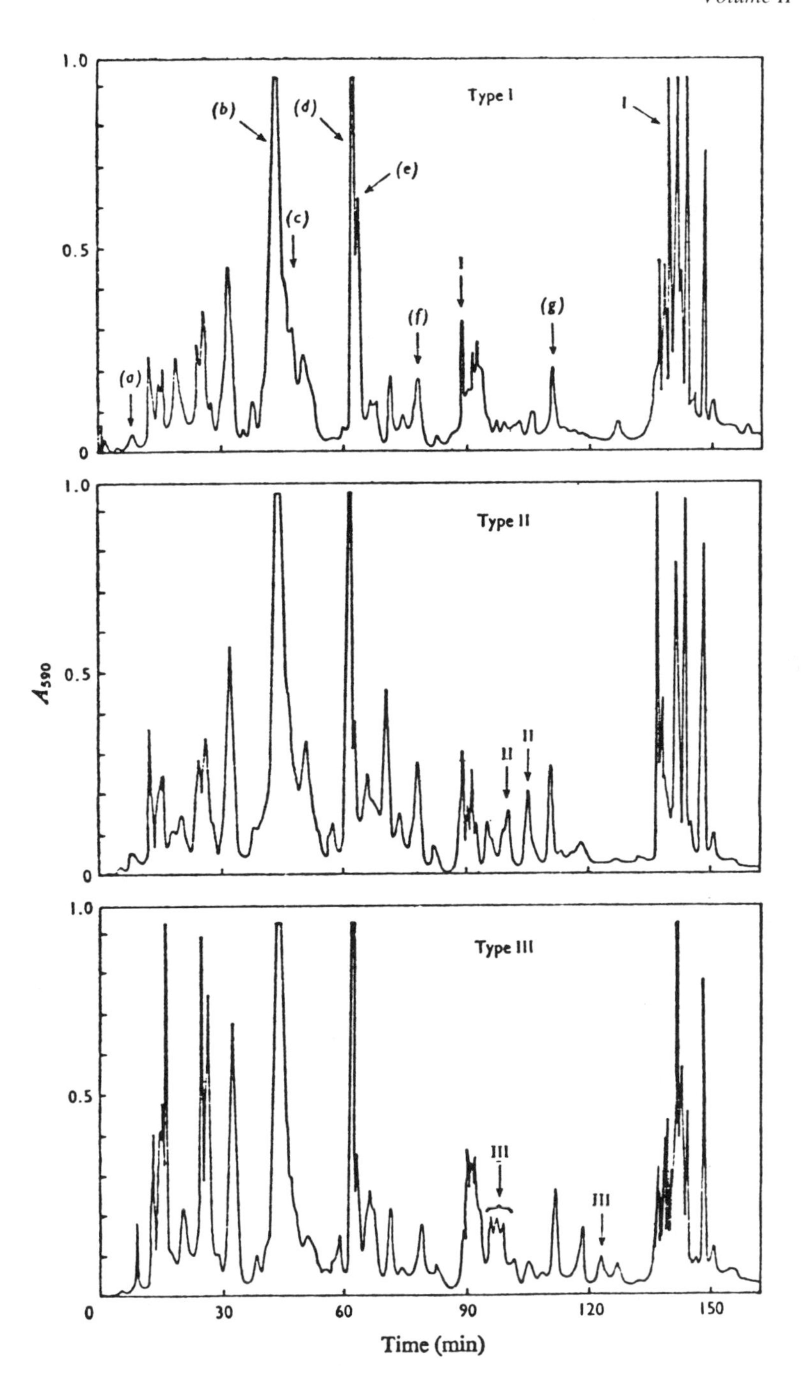

FIGURE 1. HPLC of the various collagens after collagenase digestion. The peptides released by clostridiopeptidase. A digestion of 200 μg of type I, type II, and type III collagens was dissolved in 40 μℓ of sample buffer and chromatographed on the Durrum D-500 amino acid analyser. Specific peaks for type I, II, and III collagens are indicated by the numbers I, II, and III, respectively. Peaks indicated by letters include: urea, 9 min (a); Gly-Pro-Hyp, 43 min (b); Gly-Ala-Ala, 46 min (c); Gly-Pro-Ala, 61 min 15 sec (d); Gly-Pro-Pro, 62 min 20 sec (e); artifact, 78 min (f); NH_3, 111 min (g). (From Van der Rest, M., Cole, W. G., and Glorieux, F. H., *Biochem. J.*, 161, 527, 1977. With permission.)

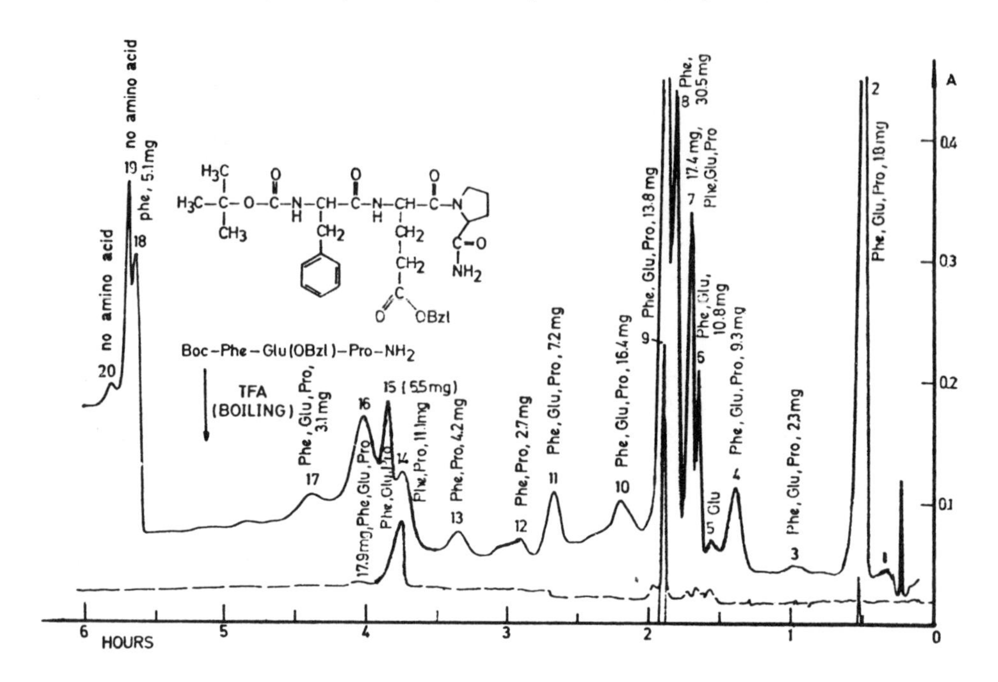

FIGURE 2. Chromatogram of the reaction mixture of Boc-Phe-Glu(OBzl)-Pro-NH_2 after treatment with boiling TFA (2.5 hr). Column: 550 × 9 mm, DC-1A resin. Pyridine-acetic acid buffers: (A) 0.2 *M* pyridine, pH 3.4 (60 min); (B) 0.5 *M* pyridine, pH 4.25 (120 min); (C) 1.0 *M* pyridine, pH 5.0 (120 min); (D) 4.0 *M* pyridine, pH 5.6 (180 min); Flow rate, 1.5 mℓ/min; temperature, 45°C, pressure, 80 to 95 bar; ratio of effluent splitting, 1:22 = 4.4% loss of sample for detection: recorder range, 0 to 0.5 A. Solid line: amount injected, 166 mg; separated material collected, 155 mg; detection after partial hydrolysis. Broken line: amount injected, 109 mg; detection without partial hydrolysis. (From Voelter, W., Bauer, H., Fuchs, S., and Pietrzik, E., *J. Chromatogr.*, 153, 433, 1978. With permission.)

and SAX, respectively, have usually been employed.[62] These types of column packing have also found significant use in IE-HPLC of peptides.[61]

Determination of L-aspartyl-L-phenylalanine methyl ester in various food products and formulations has been carried out by IE-HPLC on a silica-based strong cation exchanger.[71] Radhakrishnan et al.[73] have also applied a SCX for the separation of peptides using volatile pyridine-acetic acid buffers and an automated fluorescamine-column monitoring system. As an example of this, Figure 7 shows the separation of some synthetic peptides. This method has also been used for purification of some biologically active peptides using slight modifications of the elution system.[74] Furthermore, separation of Met-enkephalin from Leu-enkephalin, β-endorphin, and enkephalin metabolites has been reported by Bohan and Meek[75] using a SCX and sodium phosphate buffers. Nakamura et al.[76] have separated histidine-containing dipeptides on a SCX with lithium citrate buffers and fluorescence detection.

Recently, a difunctional weak anion-exchange bonded stationary phase prepared on porous silica has been introduced for simultaneous analysis of nucleotides, nucleosides, and nucleobases[77] and has also been used for the separation and sequencing of deoxypentanucleotide sequence isomers.[78] This stationary phase was then applied, first by Dizdaroglu and Simic,[79] to dipeptide separations. Underivatized dipeptides were successfully separated by this new IE-HPLC method using mixtures of acetonitrile and triethylammonium acetate (TEAA) buffer as the eluent. This elution system has minimal absorbance in the low UV, allowing gradient elution with sensitive detection of peptides in the 210 to 225 nm range. The TEAA buffer is also volatile, which facilitates easy isolation of peptides for further use. Peptides usually have no or little retention on the weak anion exchanger when only TEAA

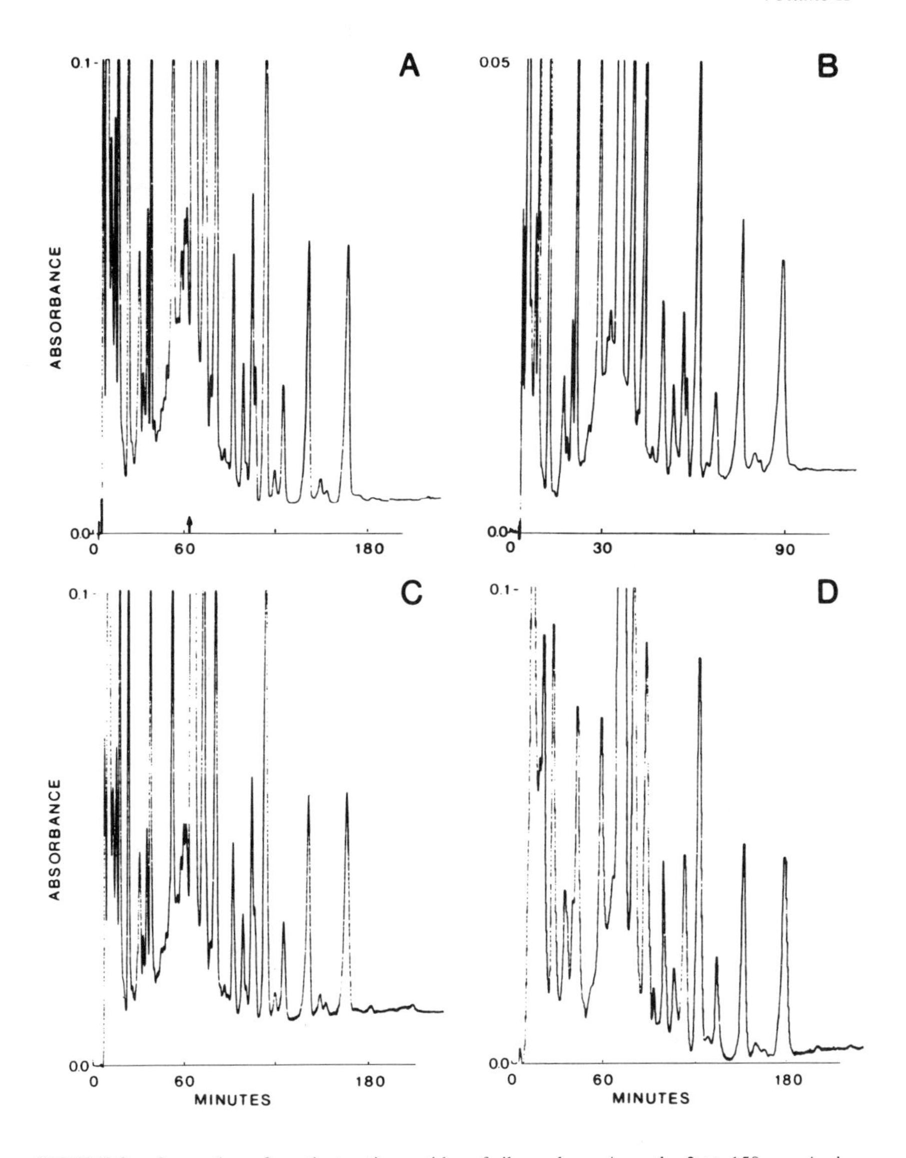

FIGURE 3. Separation of peptic-tryptic peptides of ribonuclease A on the 2 × 150 mm Aminex A-8 column, using a single gradient of 0.1 *M* pyridinium acetate (pH 3.5) and 2.0 *M* pyridine acetate (pH 5.0) for elution. Volume of gradient mixer was 50 mℓ. (A, B) analytical separations of 6.5 nmol of hydrolysate at flow rates of 12 and 24 mℓ/hr, respectively. In B the buffer to ninhydrin ratio was 2:1 (heating time 3 min); (C, D) same conditions as in A, but preparative separations using switching cycles of 4 + 4 sec and 2 + 18 sec, respectively (i.e., peptide recoveries of 40 and 90%). Position of ammonia indicated by arrow. (From Nika, H. and Hultin, T., *Methods Enzymol.*, 91, 359, 1983. With permission.)

buffer is used as the eluent, but addition of an organic solvent such as acetonitrile to the buffer, increases peptide retention and permits separation of multicomponent peptide mixtures. Separation at elevated temperature generally increases peptide retention and affects selectivity as well, and thus the optimal separation temperature will vary for different peptide mixtures.

Separation of Dipeptides

Figure 8 shows the separation of a mixture of some dipeptides by this method. The

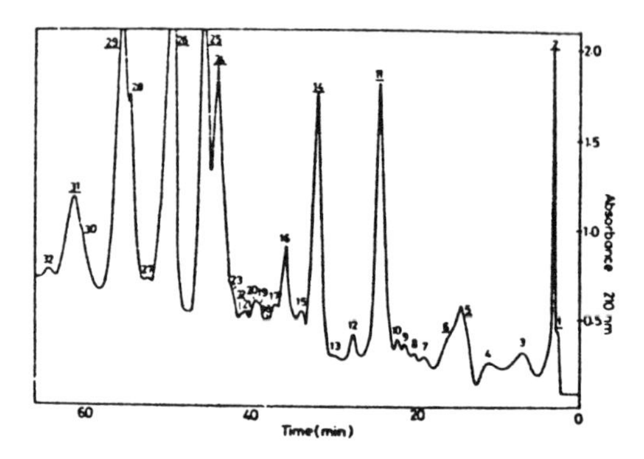

FIGURE 4. Separation of tryptic digest of bovine brain calmodulin. Sample: 50 nmol; column: Diaion CDR-10, 50 × 0.4 cm i.d.; temperature: 22° (~12 min) to 70°C (34 min ~); eluent: water to 0.25 *M* $CH_3SO_3NH_4$ (pH 2.8)-50% CH_3CN-25% $(CH_3)_2CHOH$ by convex curve gradient; flow rate: 0.9 mℓ/min; detector: 2.0 AUFS; recorder: 5 mV. (From Takahashi, N., Isobe, T., Kasai, H., Seta, K., and Okuyama, T., *Anal. Biochem.*, 115, 181, 1981. With permission.)

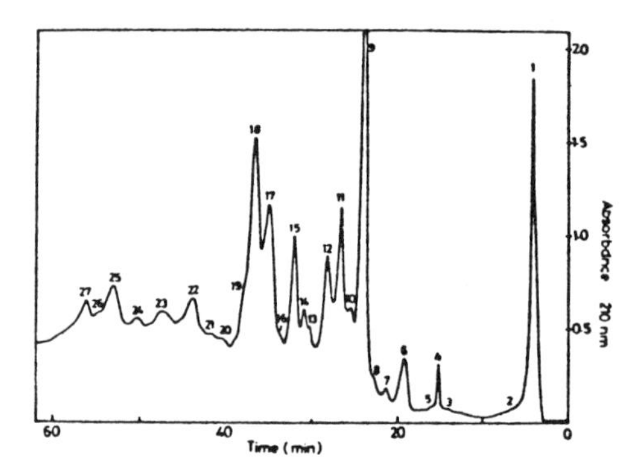

FIGURE 5. Separation of tryptic digest of bovine brain S-100b protein. Sample: 50 nmol. Running conditions are as in Figure 4. (From Takahashi, N., Isobe, T., Kasai, H., Seta, K., and Okuyama, T., *Anal. Biochem.*, 115, 181, 1981. With permission.)

conditions described here do not elute acidic dipeptides containing Asp and Glu residues. These compounds, however, could be eluted and separated by reducing the pH of the eluent (Figure 9). An excellent resolution of sequence isomeric dipeptides has also been achieved, as is shown in Figure 10. Similar chromatographic conditions have also been used for the separation of diastereomeric dipeptides (Figure 11).

All DL,DL-dipeptides examined have been completely resolved into two peaks except DL-Ala-DL-Phe. Because all four possible diastereomers have not been available for these dipeptides, peak assignments have been based on the elution behavior of the four Ala-Ala diastereomers. In this case, the D,L- and L,D-isomers (peak 7 in Figure 11) could be separated from the D,D- and L,L-isomers (peak 11). Based on this result, the conclusion has been drawn that the two peaks of each DL,DL-dipeptide mixture in Figure 11 (except DL-Ala-DL-Phe, peak 2, which apparently does not contain the D,D- and L,L-configurations since L-Ala-L-

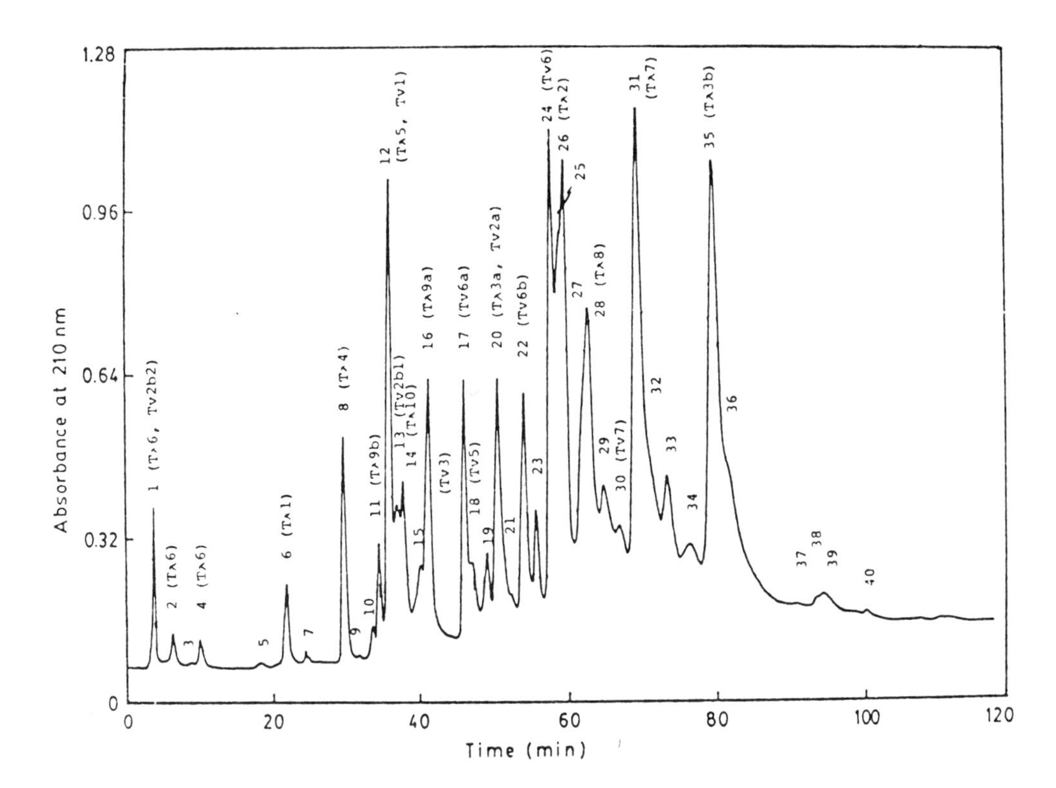

FIGURE 6. Separation of tryptic digest of the S-aminoethyl NIG-58 protein. Sample, 50 nmol; column, Hitachi-Gel 3013C, 25 × 0.4 cm i.d.; temperature, 70°C; eluant, water to 0.4 *M* $NH_4CH_3SO_3$ (pH 6.2) + 50% CH_3CN + 25% $(CH_3)_2CHOH$ by linear gradient; flowrate, 0.5 mℓ/min; detector, 1.28 AUFS; recorder, 1 mV. Numbers in the figure are the peak numbers, and the assignments of the peaks are shown in parentheses. Tλ and Tv refer to the tryptic peptides derived from the constant region and the variable region of the protein, respectively. (From Isobe, T., Takayaso, T., Takai, N., and Okuyama, T., *Anal. Biochem.*, 122, 417, 1982. With permission.)

Phe represented by peak 6 shows a longer retention time) correspond to D,L- and L,D-isomers (shorter retention time) and to L,L- and D,D-isomers (longer retention time), respectively.

Separation of Peptides

The weak anion-exchange (AE) HPLC method introduced by Dizdaroglu and Simic[79] has been applied more recently to the separation of larger peptides with a slight change in the pH of the TEAA buffer.[80] Figure 12 shows the separation of a multicomponent mixture of peptides with a gradient starting from acetonitrile-buffer (75:25) and terminating with 100% buffer. Some peptides that contain a number of acidic amino acids with no compensating basic residues, however, had unacceptably long retention times when the elution system in Figure 12 was used. These peptides have been successfully chromatographed using an isocratic flow of dilute formic acid (pH 2.6) as the eluent instead, as is shown in Figure 13. Dilute formic acid solutions are also compatible with the column packing used and the pumps, and are also volatile, allowing recovery of peptides for further use. As with the TEAA buffer system, peptide recoveries have been determined to be 80% or greater.

Separation of Enzymatic Digests of Peptides

This method has also been applied to separation of peptides resulting from tryptic digestion of some proteins.[80] As an example, Figure 14 shows the separation of the tryptic digest of horse heart cytochrome *c*. The number of peaks detected in this chromatogram correspond

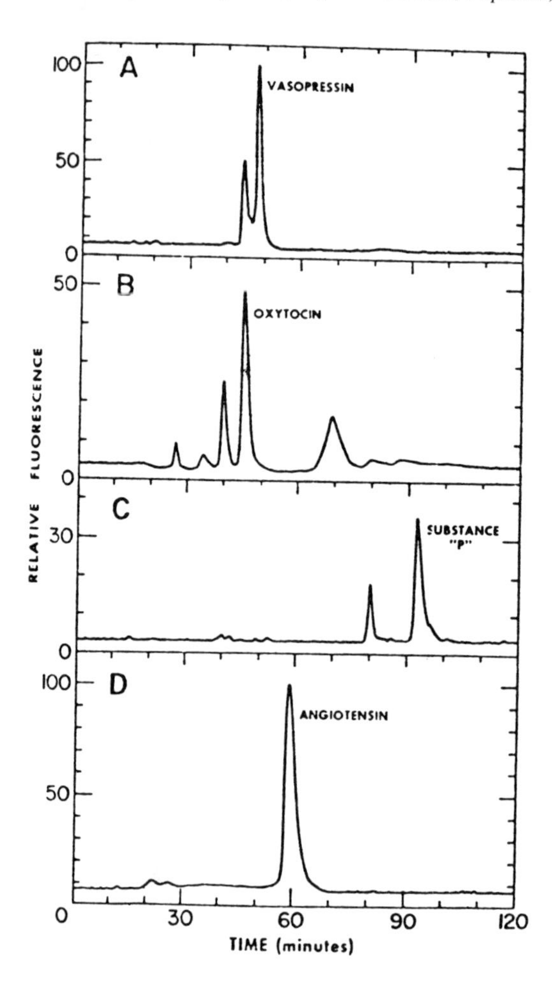

FIGURE 7. Chromatography of synthetic peptides using a 50-min linear gradient from 5 × 10^{-3} *M* pyridine, pH 3.0 to 5 × 10^{-2} *M* pyridine, pH 4.0 followed by a 60-min linear gradient to 5 × 10^{-1} *M* pyridine, pH 5.0. Approximately 10 nmol of each sample were applied to the column and 8% (800 pmol) was utilized for detection. (From Radhakrishnan, A. N., Stein, S., Licht, A., Gruber, K. A., and Udenfriend, S., *J. Chromatogr.*, 132, 552, 1977. With permission.)

closely to the number of the fragments expected from tryptic digestion of cytochrome *c*.[81] Assignment of the peaks to particular peptides, however, has not been carried out. Because some digestion fragments of this protein contain a number of acidic amino acids, the digest has also been chromatographed using formic acid as the eluent, as described above for the separation of acidic peptides, and in addition to the fragments with no or little retention, four other peaks were observed.[80]

Separation of Closely Related Peptides

The weak AE-HPLC method has also found a significant use in separation of closely related peptides such as bradykinins,[80] angiotensins,[82] and neurotensins (NT).[83] Figure 15 shows the resolution of three bradykinins with gradient elution. A good resolution of a hexapeptide from fragments that would be obtained by digestion of the hexapeptide with proteolytic enzymes or chemical cleavage has also been obtained (Figure 16).[80]

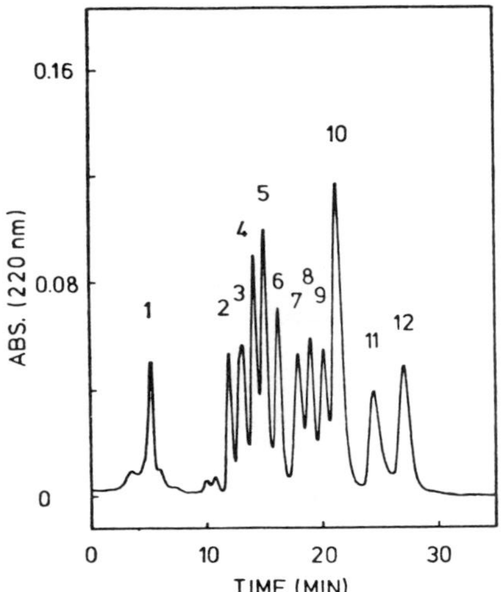

FIGURE 8. Separation of some selected dipeptides. Column, MicroPak AX-10 (10 μm), 30 × 0.4 cm. Temperature, 40°C. Eluent, mixture of 32% 0.01 *M* TEAA (pH 4.3) and 68% acetonitrile. Flow rate, 1 mℓ/min. Peaks: 1, L-Arg-L-Phe; 2, L-Leu-L-Leu; 3, Gly-L-Ile and L-Leu-L-Trp; 4, L-Ala-L-Ile; 5, L-Trp-Gly; 6, L-Trp-L-Phe; 7, L-Val-L-Val and L-Ala-L-His; 8, L-Trp-L-Ala; 9, L-Ala-L-Thr and L-Met-L-Met; 10, Gly-Gly and L-Phe-L-Phe; 11, L-Ser-L-Phe; 12, L-Tyr-L-Tyr. (From Dizdaroglu, M. and Simic, M. G., *J. Chromatogr.*, 195, 119, 1980. With permission.)

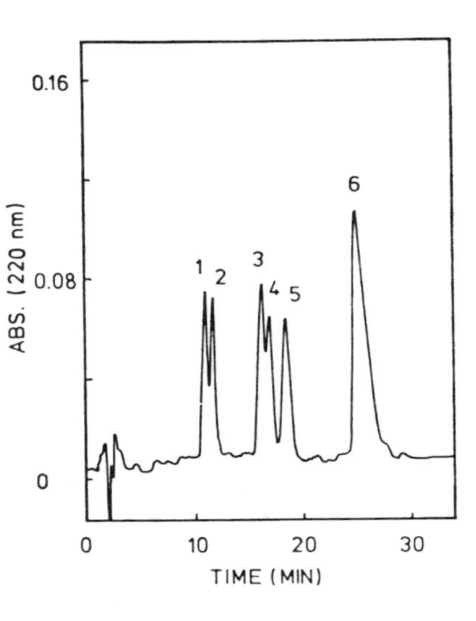

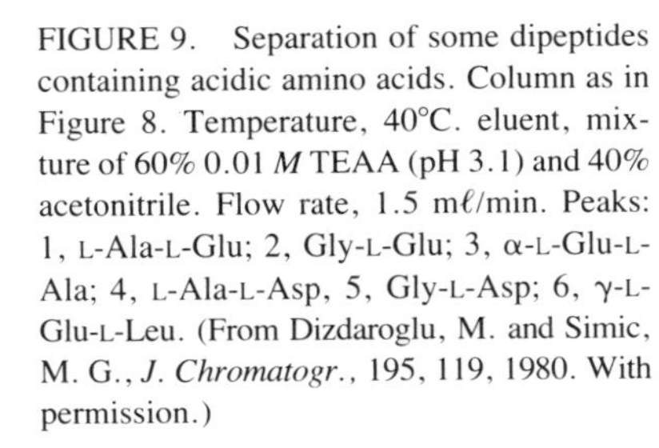

FIGURE 9. Separation of some dipeptides containing acidic amino acids. Column as in Figure 8. Temperature, 40°C. eluent, mixture of 60% 0.01 *M* TEAA (pH 3.1) and 40% acetonitrile. Flow rate, 1.5 mℓ/min. Peaks: 1, L-Ala-L-Glu; 2, Gly-L-Glu; 3, α-L-Glu-L-Ala; 4, L-Ala-L-Asp, 5, Gly-L-Asp; 6, γ-L-Glu-L-Leu. (From Dizdaroglu, M. and Simic, M. G., *J. Chromatogr.*, 195, 119, 1980. With permission.)

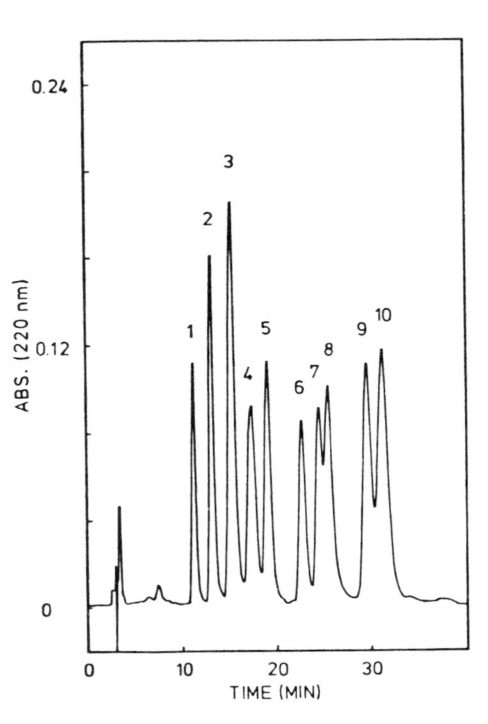

FIGURE 10. Separation of sequence isomeric dipeptides. Column details as in Figure 8. Peaks: 1, DL-Leu-DL-Ala; 2, Gly-L-Phe; 3, L-Ala-L-Leu, Gly-L-Met, and L-Ala-L-Phe; 4, Gly-L-Tyr and DL-Leu-DL-Ala; 5, L-Ala-L-Tyr; 6, L-Phe-Gly; 7, L-Tyr-Gly; 8, L-Met-Gly; 9, L-Phe-L-Ala; 10, L-Tyr-L-Ala. (From Dizdaroglu, M. and Simic, M. G., *J. Chromatogr.*, 195, 119, 1980. With permission.)

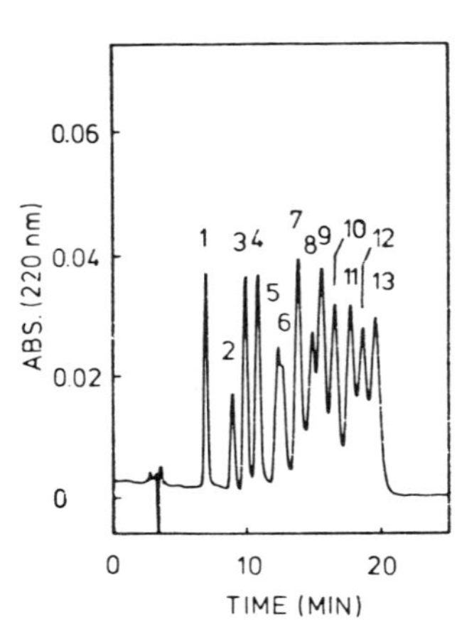

FIGURE 11. Separation of diastereomeric dipeptides. Column as in Figure 8. Temperature, 45°C. Eluent, mixture of 35% 0.01 *M* TEAA (pH 4.3) and 65% acetonitrile. Flow rate, 1 mℓ/min. Peaks: 1, DL-Leu-DL-Phe; 2, DL-Ala-DL-Phe; 3, DL-Leu-DL-Ala; 4, DL-Ala-DL-Val; 5, DL-Leu-DL-Phe; 6, L-Ala-L-Phe; 7, DL-Ala-DL-Ala and DL-Ala-DL-Val; 8, DL-Leu-DL-Ala; 9, DL-Ala-DL-Ser; 10, DL-Ala-DL-Asn; 11, DL-Ala-DL-Ala; 12, DL-Ala-DL-Ser; 13, DL-Ala-DL-Asn. (From Dizdaroglu, M. and Simic, M. G., *J. Chromatogr.*, 195, 119, 1980. With permission.)

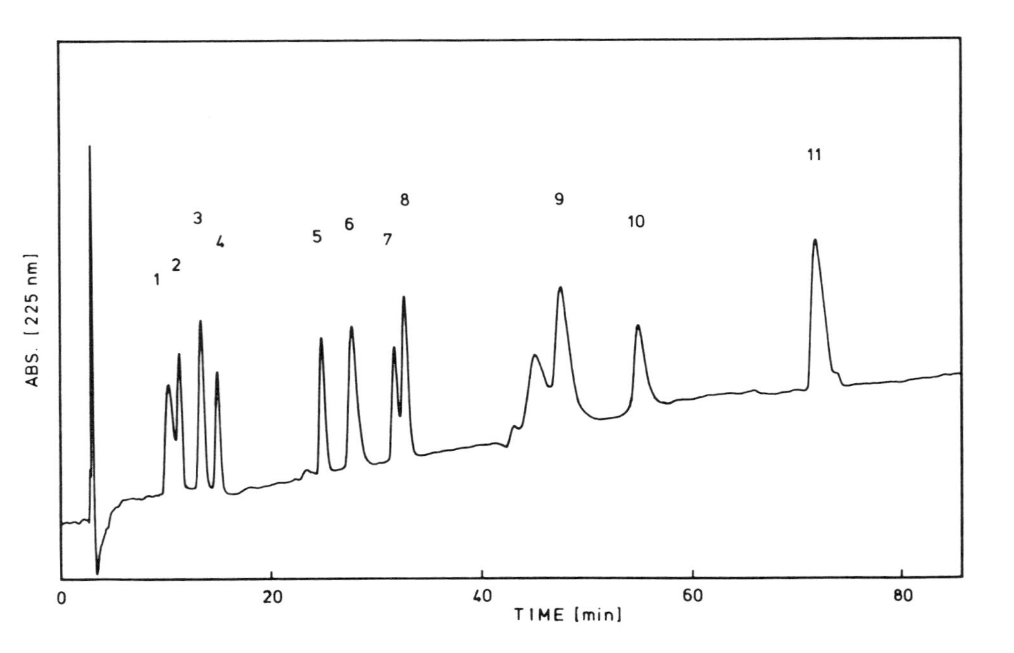

FIGURE 12. Separation of various peptides. Column, MicroPak AX-10 (10 μm), 30 × 0.4 cm. Temperature, 30°C. Eluent: A, acetonitrile, B, 0.01 *M* TEAA (pH 6.0), gradient program: linear starting from 25% B with a rate of 1% B per minute. Flow rate, 1 mℓ/min. Amount of injection, 0.5 to 5 μg per peptide. Peaks: 1, somatostatin; 2, proctolin; 3, NT; 4, Met-enkephalin; 5, bradykinin potentiator c; 6, Lys-Glu-Thr-Tyr-Ser-Lys; 7, α-endorphin; 8, EAE-peptide; 9, glucagon; 10, ribonuclease s-peptide; 11, IgE-peptide. For sequences see Reference 80. (From Dizdaroglu, M., Krutzsch, H. C., and Simic, M. G., *J. Chromatogr.*, 237, 417, 1982. With permission.)

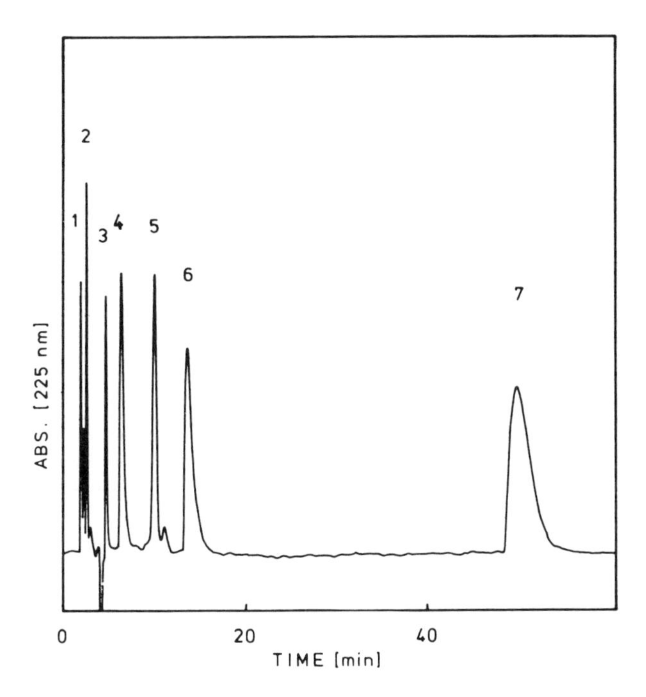

FIGURE 13. Separation of some acidic peptides. Column as in Figure 12. Temperature, 60°C. Eluent, 0.04 *M* formic acid (pH 2.6). Flow rate, 1 mℓ/min. Amount of injection as in Figure 12. Peaks: 1, ribonuclease s-peptide; 2, IgE-peptide; 3, glutathione (oxidized form); 4, Phe-Leu-Glu-Glu-Ile; 5, delta sleep-inducing peptide; 6, γ-Glu-Leu; 7, γ-Glu-Glu. For sequences see Reference 80. (From Dizdaroglu, M., Krutzsch, H. C., and Simic, M. G., *J. Chromatogr.*, 237, 417, 1982. With permission.)

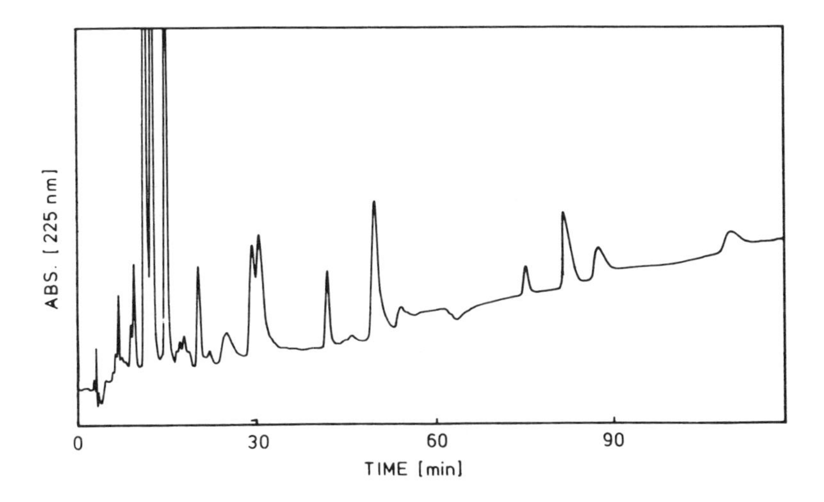

FIGURE 14. Separation of a tryptic digest of horse heart cytochrome *c*. Column details as in Figure 12 except gradient program: linear starting from 25% B with a rate of 0.6% B per minute to 50% then 4.5% B per minute to 100% B. Amount of injection, approximately 10 nmol cytochrome *c*. (From Dizdaroglu, M., Krutzsch, H. C., and Simic, M. G., *J. Chromatogr.*, 237, 417, 1982. With permission.)

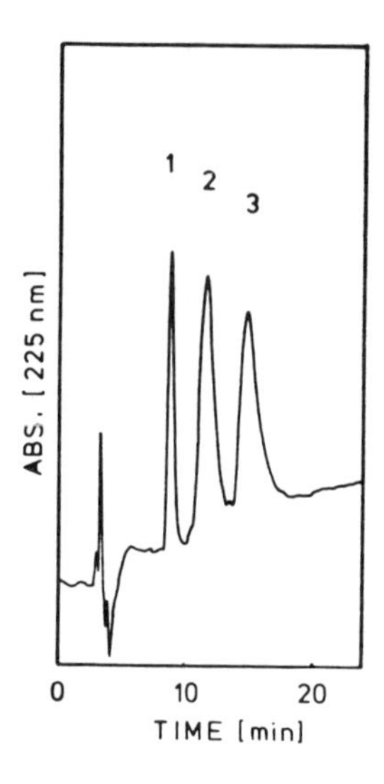

FIGURE 15. Separation of several bradykinins. Column details as in Figure 12 except gradient program: linear starting from 25% B with a rate of 1.7% B per minute. Amount of injection approximately 2 μg per peptide. Peaks: 1, bradykinin; 2, Met, Leu-bradykinin; 3, Lys-bradykinin. For sequences see Reference 80. (From Dizdaroglu, M., Krutzsch, H. C., and Simic, M. G., *J. Chromatogr.*, 237, 417, 1982. With permission.)

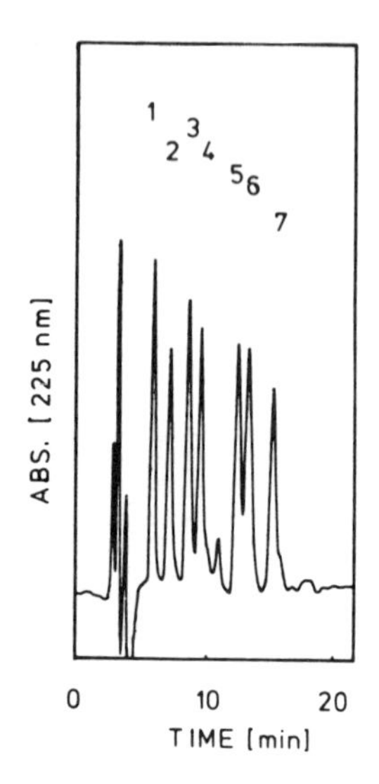

FIGURE 16. Separation of Leu-Trp-Met-Arg-Phe-Ala and its simulated digestion fragments. Column and eluent as in Figure 12. Isocratic elution with 21% B. Flow rate, 1 mℓ/min. Temperature, 50°C. Amount of injection, approximately 1 μg per peptide. Peaks: 1, Leu-Trp-Met-Arg-Phe-Ala; 2, Leu-Trp-Met-Arg-Phe; 3, Leu-Trp-Met-Arg; 4, Met-Arg-Phe-Ala; 5, Met-Arg-Phe; 6, Arg-Phe-Ala; 7, Leu-Trp-Met. (From Dizdaroglu, M., Krutzsch, H. C., and Simic, M. G., *J. Chromatogr.*, 237, 417, 1982. With permission.)

Separation of Angiotensins

Angiotensins (As) are peptide hormones with important biological activities and differ from one another in most instances by only one amino acid residue.[84] The separation of these peptides has usually been carried out by RP-HPLC.[85-87]

The separation of 12 analogs of A using conditions similar to those described above for peptide separations (Figure 12) is shown in Figure 17. A resolution of 11 peaks has been observed. Two A IIs represented by peak 6 could not be resolved from each other at this temperature. Excellent recoveries of 90 to 98% for all As tested have been obtained.[82] In general, recoveries of peptides separated by this method have been determined to be 80% or greater.[80]

Separation of Diastereomers and Analogs of Neurotensin (NT)

NT (pGlu-Leu-Tyr-Glu-Asn-Lys-Pro-Arg-Arg-Pro-Tyr-Ile-Leu) is a peptide hormone with a large spectrum of biological activities.[88] Recently, a large number of fragments, diastereomers, and analogs of NT have been synthesized by St-Pierre et al.[89] and tested for biological activity. Structure-activity studies have shown that substitution of one amino acid by various D-amino acids and various other residues leads to important variations of biological activities of NT.[89] For this reason, separation of these peptides and their purification for studies of their biological activities are very important.

The weak AE-HPLC method has been successfully used for this purpose.[83] A remarkable resolution of some diastereomers of NT from NT itself and from each other has been achieved

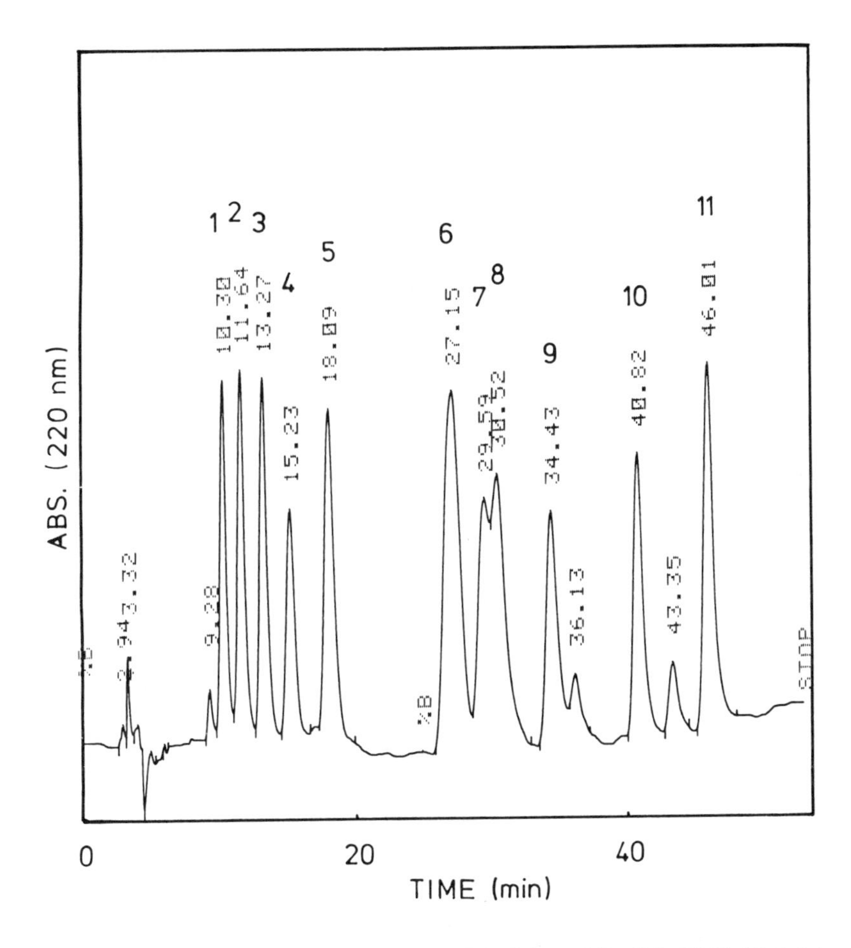

FIGURE 17. Separation of angiotensins by HPLC. Column: MicroPak AX-10, 30 × 0.4 cm; eluent: A, acetonitrile; B, 0.01 *M* TEAA (pH 6.0); gradient elution starting from 24% B with a rate of 0.1% B per minute for 25 min then 0.5% B per minute; column temperature: 26°C; flow rate: 1 mℓ/min; amount of injection per peptide; approximately 1 μg (1 nmol based on A II); AUFS: 0.1 at 220 nm. Peaks: 1, A III; 2, (Val⁴)-A III; 3, A III inhibitor; 4, (Asn¹-Val⁵)-A II; 5, (Sar¹-Ile⁸)-A II; 6, (Sar¹-Ala⁸)-A II and (Sar¹-Gly⁸)-A II; 7, (Sar¹-Thr⁸)-A II; 8, (Sar¹-Val⁵-Ala⁸)-A II; 9, A II; 10, A I; 11, (Val⁵)-A II. For sequences of A I, II, and III see Reference 80. (From Dizdaroglu, M., Krutzsch, H. C., and Simic, M. G., *Anal. Biochem.*, 123, 190, 1982. With permission.)

(Figure 18). Furthermore, five analogs of NT tested have been completely separated from NT (Figure 19). As in the case of As, recoveries of all NTs have been found to be 90% or greater.[83]

Effect of Column Temperature on Retention and Resolution of Peptides

In most instances, the selectivity of the weak anion-exchange stationary phase for peptides is changed by a change in column temperature as mentioned before. Accordingly, this parameter can be varied to obtain an optimal separation of a given mixture of peptides. This, for example, has been excellently demonstrated for the resolution of As[82] and NTs.[83]

In the case of As, an increase in column temperature from 26 to 50°C has been shown to differently affect retention times of individual angiotensins. For instance, retention times of five A IIs containing sarcosine (Figure 17) were decreased by an increase in temperature from 26 to 50°C, whereas retention times of all the remaining angiotensins in this particular mixture became longer (Figures 20 and 21). Consequently, resolution between these As was greatly affected by a change in temperature.

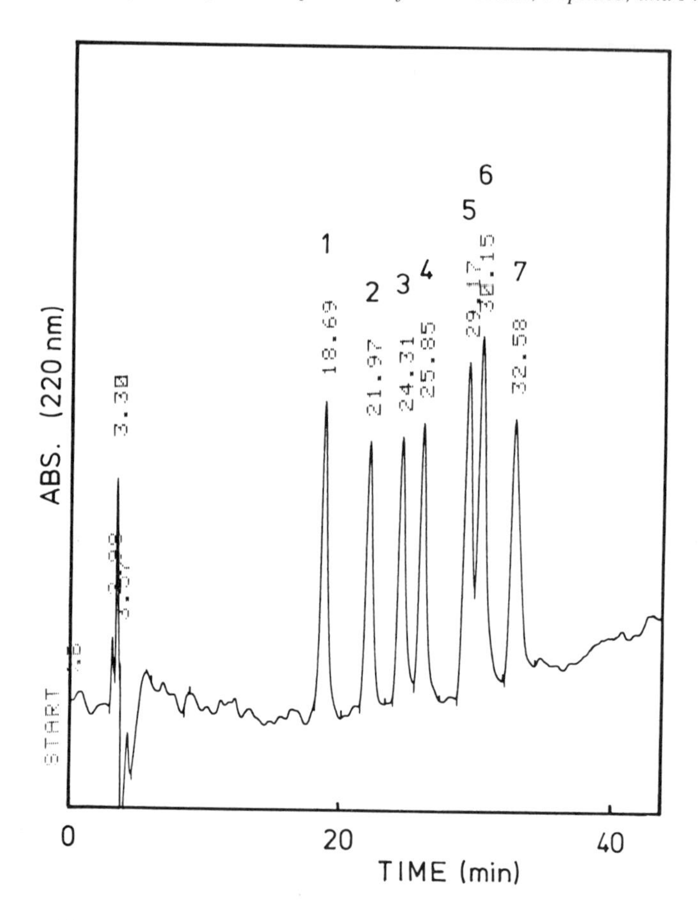

FIGURE 18. Separation of some diastereomers of neurotensin (NT). Column, MicroPak AX-10 (10 μm), 30 × 0.4 cm. Temperature, 50°C. Eluent: A, acetonitrile; B, 0.01 *M* TEAA (pH 6.0), gradient program: linear starting from 23% B with a rate of 0.3% B per minute. Flow rate 1 mℓ/min. Amount of injection per peptide, approximately 1 nmol. AUFS, 0.1 at 220 nm. Peaks: 1, (D-Phe^{11})-NT; 2, (D-Tyr^{11})-NT; 3, (D-Pro^{10})-NT; 4, (Phe^{11})-NT; 5, (D-Arg^{9})-NT; 6, NT; 7, (D-Glu^{4})-NT. (From Dizdaroglu, M., Simic, M. G., Rioux, F., and St-Pierre, S., *J. Chromatogr.*, 245, 158, 1982. With permission.)

More specifically, A IIIs were best separated from each other at 26°C (peaks 1, 2, and 3 in Figure 17). (Asn^{1}-Val^{5})-A II and (Sar^{1}-Ile^{8})-A II (peaks 4 and 5 in Figure 17, respectively) were completely separated at 26°C, while they were only slightly resolved at 40°C (Figure 20) and coeluted at 50°C (Figure 21). However, (Sar^{1}-Ala^{8}-A II and (Sar^{1}-Gly^{8})-A II were not separated at 26°C (peak 6 in Figure 17), but they could be resolved at 40°C (peaks 6 and 7, respectively, in Figure 20), and almost completely separated at 50°C (peaks 5 and 6, respectively, in Figure 21). On the other hand, a slight resolution of (Sar^{1}-Thr^{8})-A II and (Sar^{1}-Val^{5}-Ala^{8})-A II (peaks 7 and 8, respectively, in Figure 17) was obtained at 26°C, whereas these peptides eluted together at 40°C (peak 8 in Figure 20) and 50°C (peak 6 in Figure 21). Also remarkable is the increase in resolution between the group of sarcosine-containing A IIs and A II with increasing temperature (compare peaks 8 and 9 in Figures 17 and 20, and peaks 6 and 7 in Figure 21).

A significant effect of column temperature on the retention of a variety of NTs has also been observed,[83] as Table 1 clearly shows. An increase in temperature from 30 to 50°C

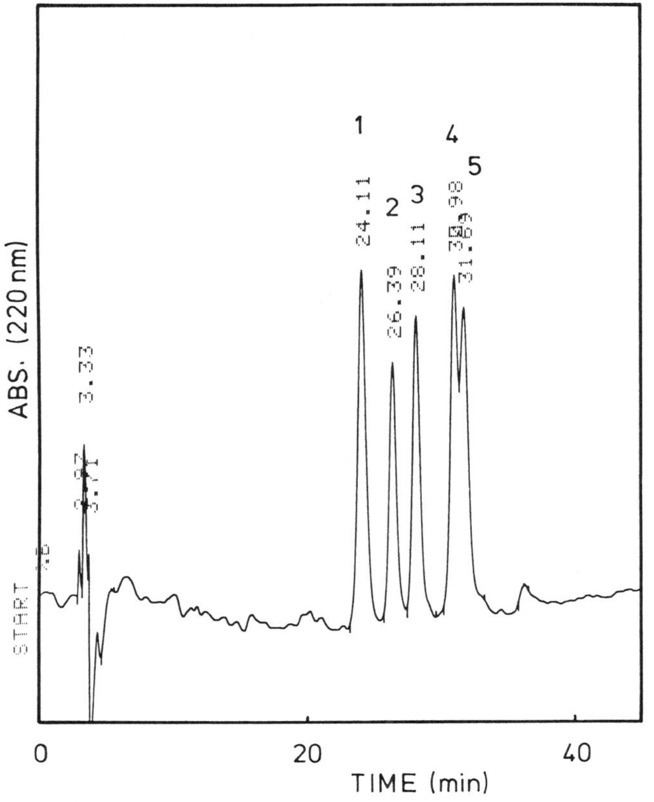

FIGURE 19. Separation of some analogs of NT. Column details as in Figure 18 except temperature: 40°C. Peaks: 1, (Phe[11])-NT and (Trp[11])-NT; 2, (Leu[11])-NT; 3, NT; 4, (Lys[8])-NT; 5, (Lys[9])-NT; (From Dizdaroglu, M., Simic, M. G., Rioux, F., and St-Pierre, S., *J. Chromatogr.*, 245, 158, 1982. With permission.)

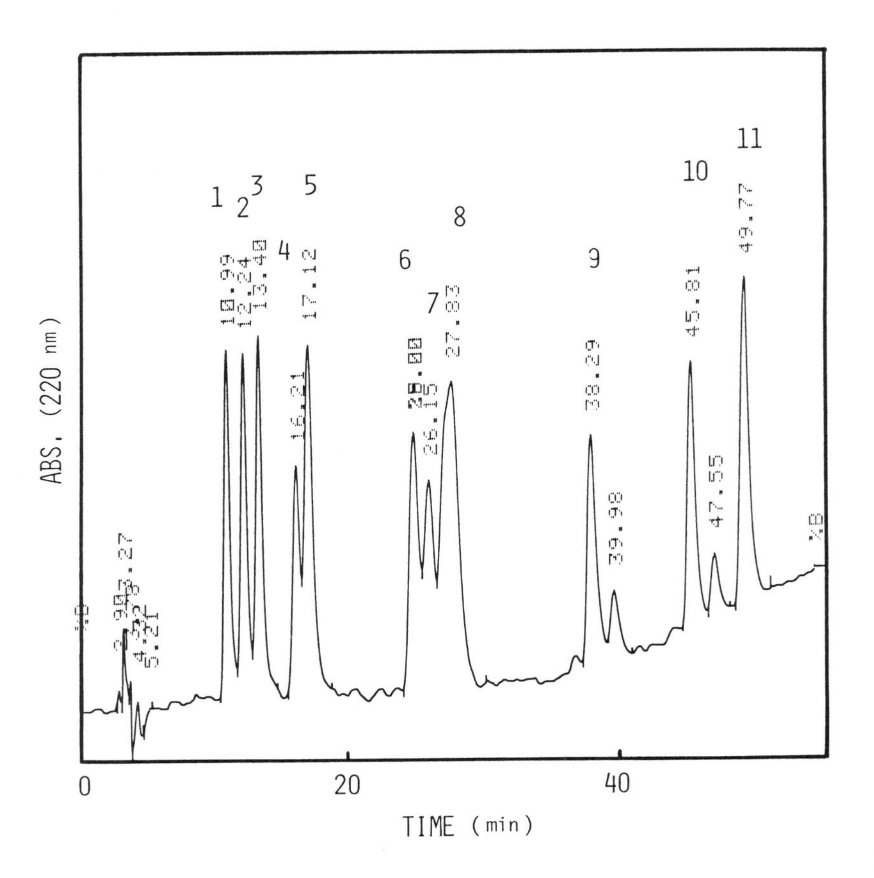

FIGURE 20. Separation of angiotensins by HPLC. Column details as in Figure 17 except temperature: 40°C. Peaks: 1-5 and 9-11 as in Figure 17; 6, (Sar[1]-Ala[8])-A II; 7, (Sar[1]-Gly[8])-A II; 8, (Sar[1]-Thr[8])-A II and (Sar[1]-Val[5]-Ala[8])-A II.

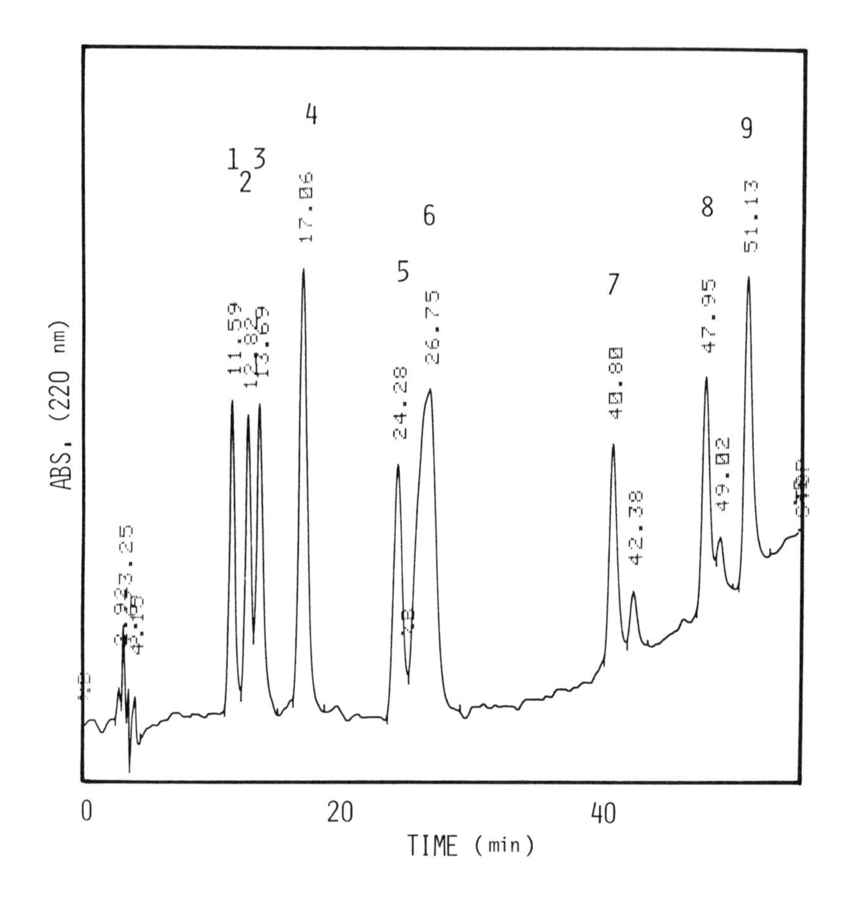

FIGURE 21. Separation of angiotensins by HPLC. Column details as in Figure 17 except temperature: 50°C. Peaks: 1-3 as in Figure 17; 4, (Asn^1-Val^5)-A II and (Sar^1-Ile^8)-A II; 5, (Sar^1-Ala^8)-A II; 6, (Sar^1-Gly^8)-A II, (Sar^1-Thr^8)-A II and (Sar^1-Val^5-Ala^8)-A II; peaks 7-9 as peaks 9-11 in Figure 17, respectively.

increased the retention times of these peptides. However, the retention time of each peptide was differently affected so that the two particular mixtures of NTs were optimally separated at 50 and 40°C (Figures 18 and 19, respectively).

Effect of pH on Retention of Peptides

As would be expected, the pH of TEAA buffer was found to have a significant effect on retention of peptides separated by the method described above. Dipeptides were best separated at pH 4.3,[79] whereas a pH value of 6.0 has been more suitable for separation of larger peptides.[80,82,83] An even stronger pH dependence of retention of acidic peptides has been observed.[79,80] Thus, the pH of the buffer, as well as the column temperature, is an important separation parameter that can be varied to achieve an optimum resolution of peptides.

Further Applications of the Weak AE-HPLC Method

Lemke et al.[90] have applied this method to the further purification of peptides obtained from RP-HPLC of a subtilisin digest of the membrane protein bacteriorhodopsin. Conditions given in Reference 79 were used in this separation. The tryptic digest of a radiolabeled human histocompatibility antigen has also been separated by this method[91] using similar conditions as in Figures 12 and 14. In addition, this method has been extensively used for the separation and isolation of the radiation-induced products of peptides[92,93] and for study

Table 1
DEPENDENCE OF RETENTION TIMES OF NEUROTENSIN ON TEMPERATURE[a]

NT	Retention time (min) (°C) 30	40	50
Figure 18			
(D-Phe[11])-NT	16.1	17.5	18.7
(D-Tyr[11])-NT	19.2	20.7	22.0
(D-Pro[10])-NT	21.8	23.1	24.3
(Phe[11])-NT	23.3	24.3	25.9
(D-Arg[9])-NT	26.5	27.4	29.2
NT	27.6	28.2	30.2
(D-Glu[4])-NT	29.9	31.2	32.6
Figure 19			
(Phe[11])-NT	23.4	24.1	25.9
(Trp[11])-NT	23.4	24.1	25.9
(Leu[11])-NT	25.8	26.4	27.6
NT	27.6	28.1	30.0
(Lys[8])-NT	30.3	31.0	32.8
(Lys[9])-NT	30.9	31.7	33.4

[a] Other column details as in Figure 18.

From Dizdaroglu, M., Simic, M. G., Rioux, F., and St.-Pierre, S., *J. Chromatogr.*, 245, 158, 1982. With permission.

of the digestibility of these products with proteolytic enzymes.[94] Recently, Dizdaroglu and Krutzsch[95] have described a comparison of RP- and weak AE-HPLC methods for peptide separations using the tryptic digest of rat small myelin basic protein isolated from rat brain. In this work, several peptide fragments that could not be resolved by RP-HPLC were separated by weak AE-HPLC.

CONCLUSIONS

This review clearly shows that recent developments have made IE-HPLC a powerful tool for peptide separations. The methodology introduced by Dizdaroglu and Simic[79] appears to be an excellent approach for many separation problems in peptide chemistry. Volatility of buffer used, sensitive and nondestructive detection of eluted peptides at wavelengths in the 210 to 225 nm range, high recoveries and long column life (up to 1 year with several daily injections) are important features of this method. In addition, it has the capability of separating diastereomeric and other closely related peptides, as discussed above in detail. Moreover, the pH of the buffer used and column temperature are important separation parameters that can be changed to improve resolution. As a recent study[95] also suggests, this method could be an important partner of RP-HPLC methods for peptide separations.

REFERENCES

1. **Moore, S. and Stein, W. H.,** Chromatography of amino acids on sulfonated polystyrene resins, *J. Biol. Chem.*, 192, 663, 1951.
2. **Schroeder, W. A., Honnen, L., and Green, F. C.,** Chromatographic separation and identification of some peptides in partial hydrolyzates of gelatin, *Proc. Natl. Acad. Sci. U.S.A.*, 39, 29, 1953.
3. **Moore, S. and Stein, W. H.,** Procedures for the chromatographic determination of amino acids on four percent cross-linked sulfonated polystyrene resins, *J. Biol. Chem.*, 211, 893, 1954.
4. **Hirs, C. H. W., Moore, S., and Stein, W. H.,** Peptides obtained by tryptic hydrolysis of performic acid-oxidized ribonuclease, *J. Biol. Chem.*, 219, 623, 1956.
5. **Bailey, J. L., Moore, S., and Stein, W. H.,** Peptides obtained by peptic hydrolysis of performic acid-oxidized ribonuclease, *J. Biol. Chem.*, 221, 143, 1956.
6. **Ando, T., Ishii, S., and Yamasaki, M.,** Peptides obtained by tryptic digestion of clupeine, *Biochem. Biophys. Acta*, 34, 600, 1959.
7. **Spackman, D. H., Stein, W. H., Moore, S., and Zamoyska, A. M.,** Disulfide bonds of ribonuclease, *J. Biol. Chem.*, 235, 648, 1960.
8. **Edmundson, A. B. and Hirs, C. H. W.,** Structure of sperm whale myoglobin. II. Tryptic hydrolysis of the denaturated protein, *J. Mol. Biol.*, 5, 683, 1962.
9. **Margoliash, E. and Smith, E. L.,** Isolation and amino acid composition of chymotryptic peptides from horse heart cytochrome c, *J. Biol. Chem.*, 237, 2151, 1962.
10. **Light, A. and Smith, E. L.,** Chymotryptic digest of papain. IV. Peptides from the oxidized, carboxy-methylated and denaturated protein, *J. Biol. Chem.*, 237, 2537, 1962.
11. **Edmundson, A. B.,** Separation of peptides on Amberlite IRC-50, *Methods Enzymol.*, 11, 369, 1967.
12. **Hill, R. L., Buettner-Janusch, J., and Buettner-Janusch, V.,** Evolution of hemoglobin in primates, *Proc. Natl. Acad. Sci. U.S.A.*, 50, 885, 1963.
13. **Jones, R. T.,** Structural studies of aminoethylated hemoglobins by automatic peptide chromatography, *Cold Spring Harbor Symp. Quant. Biol.*, 29, 297, 1964.
14. **Nelson, C. A., Noelken, M. E., Buckley, C. E., Tanford, C., and Hill, R. L.,** Comparison of the tryptic peptides from rabbit γ-globulin and two specific rabbit antibodies, *Biochemistry*, 4, 1418, 1965.
15. **Benson, J. V., Jones, R. T., Cormock, J., and Patterson, J. A.,** Accelerated automatic chromatographic analysis of peptides on a spherical resin, *Anal. Biochem.*, 16, 91, 1966.
16. **Hill, R. L. and Delaney, R.,** Peptide mapping with automatic analyzers: use of analyzers and other automatic equipment to monitor peptide separations by column methods, *Methods Enzymol.*, 11, 339, 1967.
17. **Machleidt, W., Otto, J., and Wachter, E.,** Chromatography on microbore columns, *Methods Enzymol.*, 47, 210, 1977.
18. **Herman, A. C. and Vanaman, T. C.,** Automated micro procedures for peptide separations, *Methods Enzymol.*, 47, 220, 1977.
19. **Benson, J. R.,** Improved ion-exchange resins, *Methods Enzymol.*, 47, 19, 1977.
20. **Roy, D. and Konigsberg, W.,** Chromatography of proteins and peptides on diethylaminoethyl cellulose, *Methods Enzymol.*, 25, 221, 1972.
21. **Chin, C. C. Q. and Wold, F.,** Separation of peptides on phosphocellulose and other cellulose ion-exchangers, *Methods Enzymol.*, 47, 204, 1977.
22. **Strid, L.,** Separation of some o-phosphorylated amino acids and peptides on anion-exchange resin, *Acta Chem. Scand.*, 13, 1787, 1959.
23. **Rudloff, V. and Braunitzer, G.,** Concerning hemoglobin. VI. A method for preparative production of naturally occurring peptides. The isolation of the tryptic cleavage products from human hemoglobin A on Dowex 1-X-2 using a ninhydrin-negative volatile buffer, *Z. Physiol. Chem.*, 323, 129, 1961.
24. **Schroeder, W. A.,** Separation of peptides by chromatography on columns of Dowex 50 with volatile developers, *Methods Enzymol.*, 25, 203, 1972.
25. **Schroeder, W. A.,** Separation of peptides by chromatography on columns of Dowex 1 with volatile developers, *Methods Enzymol.*, 25, 214, 1972.
26. **Lai, C. Y.,** Detection of peptides by fluorescence methods, *Methods Enzymol.*, 47, 236, 1977.
27. **Blackburn, S. and Tetley, P.,** New method for the examination of mixtures of diastereoisomeric peptides, *Biochem. Biophys. Acta*, 20, 423, 1956.
28. **Tommel, D. K. J., Vliegenthart, J. F. G., Penders, T. J., and Arens, J. F.,** A method for the separation of peptides and α-amino acids, *Biochem. J.*, 107, 335, 1968.
29. **Noda, K., Okai, H., Kato, T., and Izumaiya, N.,** Studies on separation of amino acids and related compounds. III. Separation of diastereomers of leucyl dipeptides by ion-exchange chromatography, *Bull. Chem. Soc. Jpn.*, 41, 401, 1968.
30. **Manning, J. M. and Moore, S.,** Determination of D- and L-amino acids by ion-exchange chromatography as L,D and L,L dipeptides, *J. Biol. Chem.*, 243, 5591, 1968.

31. **Lindley, H. and Haylett, T.,** Use of ion-exchange cellulose columns with the Technicon auto-analyzer technique for the fractionation of peptides, *J. Chromatogr.*, 32, 192, 1968.
32. **Wall, R. A.,** Separation of peptides on macroreticular ion-exchangers, *Anal. Biochem.*, 35, 203, 1970.
33. **Callahan, P. X., Shepard, J. A., Reilly, T. J., McDonald, J. K., and Ellis, S.,** Separation and identification of dipeptides by paper and column chromatography, *Anal. Biochem.*, 38, 330, 1970.
34. **Hagenmaier, H. and Frank, H.,** A model system for studying peptide synthesis on polymeric supports by ion-exchange chromatography, *J. Chromatogr. Sci.*, 10, 663, 1972.
35. **Heathcote, J. G., Washington, R. J., Keogh, G. J., and Glanville, R. W.,** An improved technique for the analysis of amino acids and related compounds on thin layer of cellulose. VI. The characterization of small peptides by thin-layer and ion-exchange chromatography, *J. Chromatogr.*, 65, 397, 1972.
36. **Haworth, C.,** A study of the chromatographic properties of dipeptides by automatic ion-exchange chromatography, *J. Chromatogr.*, 67, 315, 1972.
37. **Petrova, I. S. and Dolidze, D. A.,** Anion-exchange cellulose separation of free amino acids and peptides from enzymic hydrolyzates of casein and soya bean albumin, *Prikl. Biokhim. Mikrobiol.*, 8, 610, 1972.
38. **Ozawa, Y., Suziki, K., Osama, T., and Koya, M.,** Mapping of peptides in soy sauce by an ion-exchange chromatography, *Agric. Biol. Chem.*, 36, 1371, 1972.
39. **Frank, H. and Hagenmaier, H.,** Use of ion-exchange chromatography in peptide synthesis, *Beckman Rep.*, 1, 12, 1974.
40. **Tsyryapkin, V. A., Shirokov, V. A., and Belikov, V. M.,** Effect of configuration of the amino acid residues on the absorption of peptides of alanine and glycine by a cation-exchange resin, *Izv. Akad. Nauk S.S.S.R., Ser. Khim.*, 11, 2628, 1974.
41. **Frank, H. and Hagenmaier, H.,** Molecular sieve effect in ion-exchange chromatographic separations of oligopeptides, *J. Chromatogr.*, 106, 461, 1975.
42. **Bohlen, P., Stein, S., Stone, J., and Udenfriend, S.,** Automatic monitoring of primary amines in preparative column effluents with fluorescamine, *Anal. Biochem.*, 67, 438, 1975.
43. **Blouquit, Y., Cohen-Solal, M., Braconnier, F., and Rosa, J.,** Automatic assembly for peptide chromatography, *Biochimie*, 57, 113, 1975.
44. **Nys, P. S., Petyushenko, R. M., and Savitskaya, E. M.,** Selection of conditions for the separation of a mixture of peptides using ion-exchangers, *Zh. Fiz. Khim.*, 4, 2330, 1975.
45. **Benson, J. R.,** Fluorescent peptide mapping with microgram quantities of protein, *Anal. Biochem.*, 71, 459, 1976.
46. **Creaser, E. H. and Hughes, G. J.,** Peptide separations using fluorescence detection, *J. Chromatogr.*, 144, 69, 1977.
47. **Kent, S. B. H., Mitchel, A. R., Barany, G., and Merrifield, R. B.,** Test for racemization in model peptide synthesis by direct chromatographic separation of diastereomers of the tetrapeptide leucylalanylglycylvaline, *Anal. Chem.*, 50, 155, 1978.
48. **Neumann, G. and Wallenborg, B.,** Adapted gradient method in ion-exchange and affinity chromatography, *GIT Fachz. Lab.*, 22, 101, 1978.
49. **Sorep, P.,** Purification by DEAE-Sephadex chromatography of three hexapeptides synthesized by the solid-phase technique, *J. Chromatogr.*, 160, 221, 1978.
50. **Kawashiro, K., Morimoto, S., Yoshido, H., and Sugiura, K.,** The synthesis and chromatography of peptide nitriles, in Origin Life, Proc. 2nd ISSOL Meet., 1978, 297.
51. **Skarlat, I. V.,** Ion-exchange chromatography of tryptic hydrolyzates of proteins formed in a cell-free protein synthesizing system, *Fiz.-Khi. Metody Mol. Biol.*, p.119, 1978.
52. **Johnson, P.,** Effective peptide fractionation using an amino acid analyzer ion-exchange resin, *J. Chromatogr. Sci.*, 17, 406, 1979.
53. **Powers, D. A., Fishbein, J. C., Place, A. R., and Sofer, W.,** Solid-phase sequencing in spinning cup sequenators. II. Micromethods for the purification and sequence analysis of proteins and peptides, in Methods Pept. Protein Sequence Anal., Proc. 3rd Int. Conf., 1979, 89, 1980.
54. **Aromatorio, D. K., Parker, J., and Brown, W. E.,** High-resolution analytical and preparative peptide mapping by combination of ion-exchange and thin-layer chromatographies, *Anal. Biochem.*, 103, 350, 1980.
55. **Sampson, B. and Barlow, G. B.,** Separation of peptides and amino acids by ion-exchange chromatography of their copper complexes, *J. Chromatogr.*, 183, 9, 1980.
56. **Salnikow, J.,** Automated fluorogenic detection of peptide effluents in preparative ion-exchange chromatography with volatile buffers, in Methods Pept. Protein Sequence Anal., Proc. 3rd Int. Conf., 1979, 407, 1980.
57. **Bradshaw, R. A., Bates, O. J., and Benson, J. R.,** Peptide separations on substituted polystyrene resins. Effect of cross-linkage, *J. Chromatogr.*, 187, 27, 1980.
58. **Wall, R. A.,** Hydrophobic chromatography with dynamically coated stationary phases. II. Dynamic cation-exchange separations of tyrosinyl peptides, *J. Chromatogr.*, 194, 353, 1980.

59. **Malmstrom, B. M., Nyman, P. O., and Strid, L.,** Improved separation of basic peptides in anion-exchange chromatography, *J. Chromatogr.,* 215, 109, 1981.
60. **Sugihara, J., Imamura, T., Yanase, T., Yamada, H., and Imoto, T.,** Separation of peptides by cellulose-phosphate chromatography for identification of a hemoglobin variant, *J. Chromatogr.,* 229, 193, 1982.
61. **Smith, J. A. and McWilliams, R. A.,** High performance liquid chromatography of peptides, *Am. Lab.,* 12, 25, 1980.
62. **Snyder, L. R. and Kirkland, J. J.,** *Introduction to Modern Liquid Chromatography,* 2nd ed., John Wiley & Sons, New York, 1979, 173 and 419.
63. **Van der Rest, M., Cole, W. G., and Glorieux, F. H.,** Human collagen "fingerprints" produced by clostridiopeptidase A digestion and high-pressure liquid chromatography, *Biochem. J.,* 161, 527, 1977.
64. **Voelter, W., Bauer, H., Fuchs, S., and Pietrzik, E.,** Preparative high-performance liquid chromatography of thyrotropin-releasing hormone analogues, *J. Chromatogr.,* 153, 433, 1978.
65. **Nika, H. and Hultin, T.,** An analyzer and monitor for rapid microscale peptide separations, *Anal. Biochem.,* 98, 178, 1979.
66. **Nika, H. and Hultin, T.,** Analyzer for microscale peptide separations, *Methods Enzymol.,* 91, 359, 1983.
67. **James, L. B.,** Resolution of γ-glutamyl peptides, *J. Chromatogr.,* 172, 481, 1979.
68. **Bachner, L., Boissel, J. P., and Wajcman, H.,** New sensitive technique for the quantitative analysis of initiation peptides, *J. Chromatogr.,* 193, 491, 1980.
69. **Takahashi, N., Isobe, T., Kasai, H., Seta, K., and Okuyama, T.,** An analytical and preparative method for peptide separation by high-performance liquid chromatography on a macroreticular anion-exchange resin, *Anal. Biochem.,* 115, 181, 1981.
70. **Isobe, T., Isioko, N., and Okuyama, T.,** Isolation and characterization of des(Ala-Lys)calmodulin in porcine brain, *Biochem. Biophys. Res. Commun.,* 102, 279, 1981.
71. **Isobe, T., Takayasu, T., and Takai, N., and Okuyama, T.,** High-performance liquid chromatography of peptides on a macroreticular cation-exchange resin: application to peptide mapping of Bence-Jones proteins, *Anal. Biochem.,* 122, 417, 1982.
72. **Fox, L., Anthony, G. D., and Lau, E. P. K.,** High-performance liquid chromographic determination of L-aspartyl-L-phenylalanine methyl ester in various food products and formulations, *J. Assoc. Off. Anal. Chem.,* 59, 1048, 1976.
73. **Radhakrishnan, A. N., Stein, S., Licht, A., Gruber, K. A., and Udenfriend, S.,** High-efficiency cation-exchange chromatography of polypeptides and polyamines in the nanomole range, *J. Chromatogr.,* 132, 552, 1977.
74. **Mabuchi, H. and Nakahashi, H.,** Systematic separation of medium-sized biologically active peptides by high-performance liquid chromatography, *J. Chromatogr.,* 213, 275, 1981.
75. **Bohan, T. P. and Meek, J. L.,** Met-enkephalin: rapid separation from brain extracts using high-pressure liquid chromatography, and quantitation by binding assay, *Neurochem. Res.,* 3, 367, 1978.
76. **Nakamura, H., Zimmerman, C. L., and Pisano, J. J.,** Analysis of histidine-containing dipeptides, polyamines, and related amino acids by high-performance liquid chromatography: application to Guinea pig brain, *Anal. Biochem.,* 93, 423, 1979.
77. **Edelson, E. H., Lawless, J. G., Wehr, C. T., and Abbott, S. R.,** Ion-exchange separation of nucleic acid constituents by high-performance liquid chromatography, *J. Chromatogr.,* 174, 409, 1979.
78. **Dizdaroglu, M., Simic, M. G., and Schott, H.,** Separation and sequencing of the sequence isomers of pyrimidine deoxypentanucleoside tetraphosphates by high-performance liquid chromatography, *J. Chromatogr.,* 188, 273, 1980.
79. **Dizdaroglu, M. and Simic, M. G.,** Separation of underivatized dipeptides by high-performance liquid chromatography on a weak anion-exchange bonded phase, *J. Chromatogr.,* 195, 119, 1980.
80. **Dizdaroglu, M., Krutzsch, H. C., and Simic, M. G.,** Separation of peptides by high-performance liquid chromatography on a weak anion-exchange bonded phase, *J. Chromatogr.,* 237, 417, 1982.
81. **Margoliash, E., Smith, E. L., Kreil, G., and Tuppy, H.,** Amino-acid sequence of horse heart cytochrome c. The complete amino-acid sequence, *Nature (London),* 192, 1125, 1961.
82. **Dizdaroglu, M., Krutzsch, H. C., and Simic, M. G.,** Separation of angiotensins by high-performance liquid chromatography on a weak anion-exchange bonded phase, *Anal. Biochem.,* 123, 190, 1982.
83. **Dizdaroglu, M., Simic, M. G., Rioux, F., and St-Pierre, S.,** Separation of diastereomers and analogues of neurotensin by anion-exchange high-performance liquid chromatography, *J. Chromatogr.,* 245, 158, 1982.
84. **Schwyzer, R.,** Synthetische Analoge des Hypertensins. I. Einleitung, *Helv. Chim. Acta,* 44, 667, 1961.
85. **Molnar, I. and Horvath, C.,** Separation of amino acids and peptides on non-polar stationary phases by high-performance liquid chromatography, *J. Chromatogr.,* 142, 623, 1977.
86. **Margolis, S. A. and Schaffer, R.,** Development of a Standard Reference Material for Angiotensin I, NBS1P 79-1947, National Bureau of Standards, Washington, D.C., 1979.
87. **Guy, M. N., Roberson, G. M., and Barnes, L. D.,** Analysis of angiotensins I, II, III, and iodinated derivatives by high-performance liquid chromatography, *Anal. Biochem.,* 112, 272, 1981.

88. **Carraway, R. and Leeman, S. E.,** The isolation of a new hypotensive peptide, neurotensin, from bovine hypothalami, *J. Biol. Chem.*, 248, 6854, 1973.
89. **St-Pierre, S., Lalonde, J. M., Gendreau, M., Quirion, R., Regoli, D., and Rioux, F.,** Synthesis of peptides by the solid-phase method. VI. Neurotensin, fragments, and analogues, *J. Med. Chem.*, 24, 370, 1981.
90. **Lemke, H. D. Bergmeyer, J., and Oesterhelt, D.,** Determination of modified positions in the polypeptide chain of bacteriohodopsin, *Methods Enzymol.*, 88, 89, 1982.
91. **Van Schravendijk, M. R.,** personal communication, 1982.
92. **Dizdaroglu, M. and Simic, M. G.,** Isolation and characterization of radiation-induced aliphatic peptide dimers, *Int. J. Radiat. Biol.*, 1983, in press.
93. **Gajewski, E., Dizdaroglu, M., Krutzsch, H. C., and Simic, M. G.,** OH Radical-induced racemization and dimerization of methionine peptides, *Int. J. Radiat. Biol.*, 1983, in press.
94. **Dizdaroglu, M., Gajewski, E., and Simic, M. G.,** Enzymatic digestibility of peptides exposed to ionizing radiation, *Int. J. Radiat. Biol.*, 1983, in press.
95. **Dizdaroglu, M. and Krutzsch, H. C.,** A comparison of reversed-phase and weak anion-exchange high-performance liquid chromatographic methods for peptide separations, *J. Chromatogr.*, 264, 223, 1983.

IDENTIFICATION OF KANGAROO AND HORSE MEAT IN PROCESSED MEATS BY IE-HPLC

Patrick R. Carnegie

INTRODUCTION

The substitution of horse and kangaroo meat for beef is a cause for concern in a number of countries. While there are now several routine serological or electrophoretic methods for monitoring the composition of fresh meats, these methods are unsatisfactory with cooked meat products. Recently we developed an HPLC method[1] for the analysis of the histidine dipeptides anserine, balenine, and carnosine, which are present in widely differing proportions in skeletal muscle from different species.

The method utilizes a silica-based ion-exchange column and is an improvement on the system used by Nakamura et al.[2] They used a lithium citrate gradient at an elevated temperature to separate the histidine dipeptides on a Whatman Partisil® -10SCX column. As citrate promotes solubilization of silica the column rapidly deteriorated in performance. This chapter summarizes the application of the improved method to the analysis of meat and meat products.

SEPARATION OF HISTIDINE DIPEPTIDES BY IE-HPLC

Performance of SCX Column

Extraction of meat and meat products with sulfosalicylic acid has been described elsewhere.[1,3]

The extract (5 μℓ) was applied to a Whatman Partisil® -10SCX column maintained at 40°C and the histidine dipeptides were eluted with 0.2 *M* lithium formate buffer, pH 2.9, at 0.7 mℓ/min. Detection was achieved by mixing the eluate with *o*-phthaldialdehyde (OPA) as described by Nakamura et al.[2] Because the fluorescence with OPA was sensitive to changes in temperature the reaction coil was maintained at the optimum temperature of 30°C. The output from the fluorescence detector was automatically integrated by a Hewlett-Packard integrator.

Typical separations of the histidine dipeptides from fresh meat from pig, sheep, horse, and kangaroo are shown in Figure 1. Although the histidine dipeptides are similar in their chemical properties a clear separation was achieved. Carnosine (β-alanylhistidine, Car.) eluted prior to anserine (β-alanyl-1-methylhistidine, Ans.) while from a sulfonated polystyrene column the order was reversed.[3] Balenine (β-alanyl-3-methylhistidine, Bal.) was clearly resolved from anserine on the SCX column. It is suggested that the presence and position of the methyl group contributed to the separation. In other experiments 3-methylhistidine was separated more efficiently from histidine on the SCX column than on sulfonated polystyrene.[5]

Another major advantage of the SCX column compared with a sulfonated polystyrene column was that the free amino acids, including arginine, all eluted prior to the dipeptides, thus no regeneration and equilibration steps were required. Samples were routinely applied at 15-min intervals. Over 500 extracts were applied before the separation between the amino acids and carnosine deteriorated. The manufacturer's recommended procedure for cleaning the column was followed but it did not restore the performance. However by lowering the buffer concentration to 0.1 *M* the separation was restored with the minor disadvantage of a slightly increased time being required for elution. A total of 1000 samples have now been applied to the column.

The cost of buffers and detection reagents for the IE-HPLC method are much cheaper than those required for the separation of the dipeptides on an amino acid analyzer.[3] The

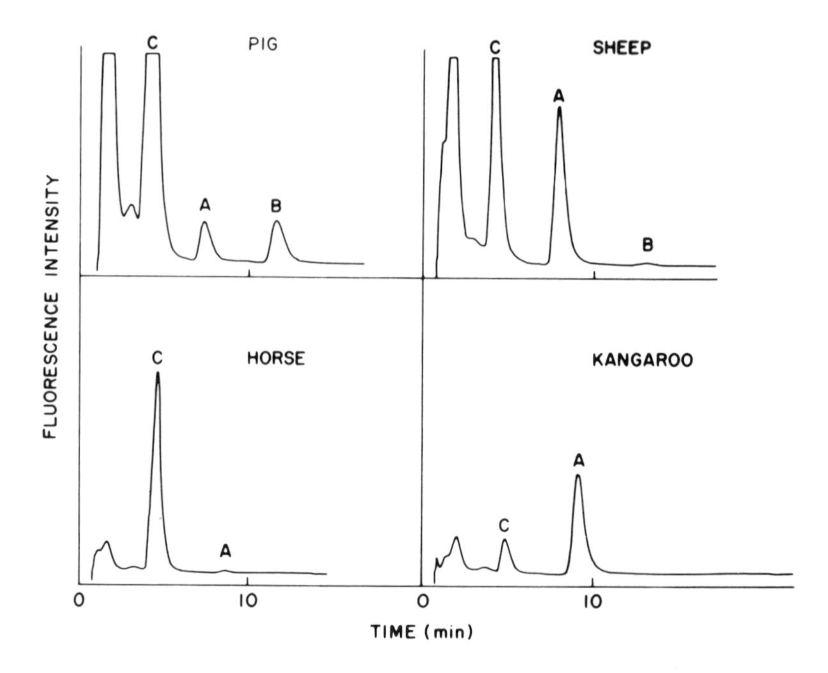

FIGURE 1. Chromatograms of extracts of muscle on Whatman Partisil® -10SCX. C, carnosine; A, anserine; B, balenine. Muscle, 30 g was extracted with 150 mℓ of 8% sulfosalicylic acid, 5 μℓ was applied. The kangaroo extract was diluted fivefold and the horse extract tenfold. (From Carnegie, P. R., Ilic, M. Z., Etheridge, M. O., and Collins, M. G., *J. Chromatogr.*, 261, 153, 1983.)

total operating costs could possibly be further reduced by using a Waters SCX cartridge in a Waters Z module.

Analysis of Fresh Meat

Table 1 summarizes the results from analyses of skeletal muscle from several species. While there is considerable variation between animals within a species, the differences in anserine content and the ratio of the dipeptides is so great that identification of the source of meat can be achieved with considerable confidence.[1]

Skeletal muscle from horse and kangaroo have ratios of histidine dipeptides which are quite characteristic. Carnosine accounts for 99% of the total dipeptides in horse meat and anserine represents 95% of the total in kangaroo meat. Thus, it is possible to detect meat from these species when it is added to beef in amounts greater than 10 to 15%. Despite the large differences in the ratios of dipeptide in different species it is interesting that the total content of these dipeptides does not vary greatly between species.

Analysis of Processed Meats

As the histidine dipeptides are not damaged by cooking it is possible to use the method to identify the meat used in processed meat products. Because of the variation between animals it is not possible to determine the proportions within a mixture to closer than ± 12%. Where pigmeat is present the balenine content is a useful monitor of the amount of pig skeletal muscle present in a product. However, as the balenine content tends to increase with age[3] the presence of meat from culled sows can lead to a higher balenine/anserine ratio.

In Table 2 the results of analyses of typical meat products are given. Over 60 brands of Australian processed meats were analyzed and none were found to contain large amounts of kangaroo or horse meat.[4] Meat pies A, B, and C were estimated to be prepared from

Table 1
HISTIDINE DIPEPTIDES IN FRESH MEAT

Species	μmol/g fresh meat				Ratio		
	Total	Ans.	Car.	Bal.	Ans.	Car.	Bal.
Bovine	16.3	3.2	13.0	0.10	1	4.1	0.03
Ovine	10.3	5.1	5.1	0.1	1	1.0	0.02
Horse	18.8	0.2	18.6	0.0	1	93.0	0.0
Kangaroo	20.7	19.6	1.1	0.0	1	0.06	0.0
Pig	14.2	0.69	12.7	0.78	1	18.4	1.1

Table 2
HISTIDINE DIPEPTIDES IN PROCESSED MEAT PRODUCTS

Product	μmol/g dry matter in filling		Ratio		
	Total	Ans.	Ans.	Car.	Bal.
Meat pie					
A	10.2	1.7	1	5.0	0.02
B	8.3	2.6	1	2.2	0.03
C	12.4	5.0	1	1.5	0.04
Sausage					
D	15.7	1.0	1	12.0	2.1
E	6.3	0.59	1	8.8	0.9
F	3.1	0.76	1	2.6	0.5

beef, 0.5 beef to 0.5 mutton, and 0.2 beef to 0.8 mutton, respectively. All the sausages were labeled "pork" but sausages E and F were found to be typical of mixtures of beef and pork, and mutton and pork, respectively.

The total histidine dipeptide content provides an approximate comparison of the skeletal muscle in a product. One manufacturer stated that beef represented 48% of the filling in their meat pies. The total histidine dipeptides content of their pies indicated a lean meat content of 39%. Many products were found to contain very small amounts of skeletal muscle.[4,5] However for accurate analysis of the skeletal muscle content a determination of 3-methylhistidine in hydrolysates of the product would be more reliable than an analysis of the total histidine dipeptide content.

CONCLUSION

It is suggested that SCX columns could be useful in the separation of other types of small positively charged peptides such as homocarnosine in brain tissue and small peptides obtained during sequencing of proteins.

ACKNOWLEDGMENT

The Australian Meat Research Committee and the Australian Pig Industry Research Committee are thanked for financial support. M. O. Etheridge, M. G. Collins, and M. Z. Ilic are thanked for assistance and discussion.

REFERENCES

1. **Carnegie, P. R., Ilic, M. Z., Etheridge, M. O., and Collins, M. G.,** An improved high performance liquid chromatography method for analysis of histidine dipeptides anserine, carnosine and balenine in fresh meat, *J. Chromatogr.*, 261, 153, 1983.
2. **Nakamura, H., Zimmerman, C. L., and Pisano, J. J.,** Analysis of histidine-containing peptides, polyamines, and related amino acids by high-performance liquid chromatography: application to guinea pig brain, *Anal. Biochem.*, 93, 423, 1979.
3. **Carnegie, P. R., Hee, K. P., and Bell, A. W.,** Ophidine (β-alanyl-L-3-methylhistidine, 'Balenine') and other histidine dipeptides in pig muscles and tinned hams, *J. Sci. Food Agric.*, 33, 795, 1982.
4. **Carnegie, P. R., Collins, M. G., and Ilic, M. Z.,** Use of histidine dipeptides to estimate the proportion of pig meat in processed meats, *Meat Sci.*, in press.
5. **Carnegie, P. R. and Ilic, M. Z.,** unpublished data.

RETENTION OF DIPEPTIDES IN REVERSED-PHASE HPLC

Elsa Lundanes and Tyge Greibrokk

INTRODUCTION

High resolution separation of dipeptides from protein sequencing has been achieved with gas chromatography (after derivatization),[1] as well as with ion-exchange chromatography.[2] The main interest of recent years, however, has been focused on HPLC reversed-phase (RP) systems, not only in connection with protein sequencing, but also for using dipeptides as model substances for gaining knowledge of the separation of peptides in general.

COLUMN PACKINGS

The separation of dipeptides have largely been performed on C_{18}- and C_8-reversed-phase materials, but alkylphenyl,[3] phenyl, and cyanoalkyl columns have also been utilized.[4] On the phenyl columns, dipeptides containing the amino acids phenylalanine or proline, had higher retention compared to C_{18} columns, but most dipeptides were less retained with lower resolution.[4] As a general rule, however, the best resolution of dipeptides has been obtained on C_{18} materials.

MOBILE PHASES

The retention of dipeptides can be regulated by the mobile phase, by altering the concentration of the organic modifier or by changing the pH or by the addition of salts or ion-pairing agents. The capacity factor, k′, decreases with increasing concentration of the organic modifier, with the strongest effect on dipeptides containing the most hydrophobic amino acids.[4-6] Gradient elution is required in complex mixtures of dipeptides.[7]

The retention can be manipulated by the nature of additives which can engage in ion-pair formation or dynamic liquid-liquid ion-exchange interactions.[1,8] The importance of using a buffer in the mobile phase has been emphasized for the purpose of maintaining a constant pH as well as to reduce nonspecific binding to residual silanol groups.[4,5]

With an increasing pH the charged amino group is gradually deprotonated and a free carboxylic terminus becomes ionized resulting in maximal ionization around the isoelectric point of the peptide, leading to minimum retention. With strong acids in the mobile phase, exceptional narrow peaks have been obtained, such as with 0.5 *M* perchloric acid (pH 0.2) at 70°C.[7] At low pH the peak splitting of *cis-trans* isomers of proline peptides disappeared.[9] Another strong acid which results in narrow peaks of dipeptides is trifluoroacetic acid (TFA), as seen in Figure 1.

At higher pH, in the presence of acetate buffers, the retention order of dipeptides with basic amino acids deviates strongly from the order observed at low pH with strong acids (Table 1). With 0.01 *M* ammonium acetate, presumptive interactions with residual silanol groups contributed strongly to the retention of such peptides, leading to significant differences between columns from different manufacturers and of different age (Table 1).[4]

At low pH, hydrophobic anionic reagents result in increased retention, whereas hydrophobic cationic reagents cause decreased retention of dipeptides.[3,8] With 10-μg injections at low pH, a concentration dependent retention was observed with solvents containing less than 25 m*M* of alkylsulfonates.[10] The addition of ion-pairing agents such as pentanoic acid and octanoic acid has been found useful to separate basic dipeptides from acidic dipeptides.[1] It has also been shown that the selectivity can be significantly altered by addition of metal chelates to the mobile phase.[11]

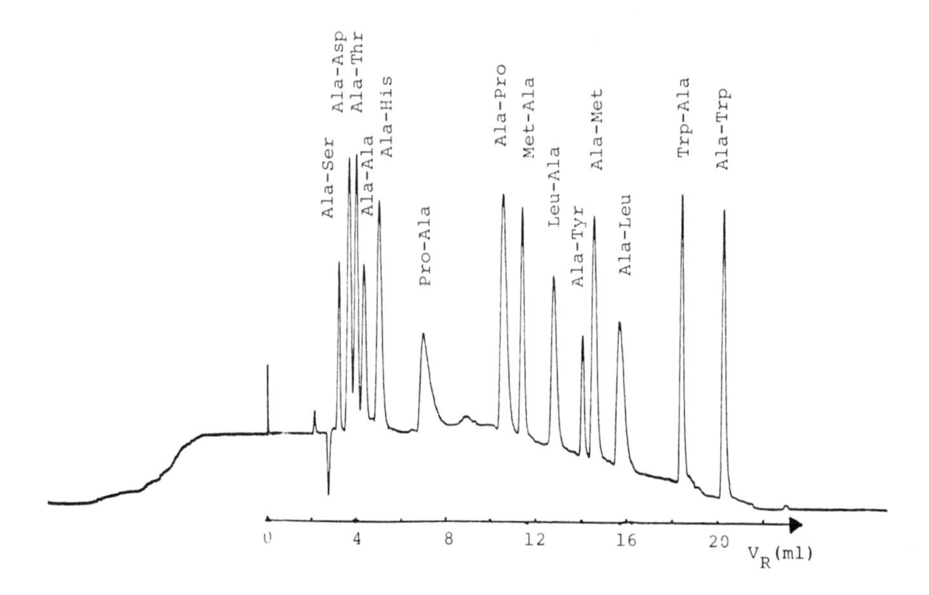

FIGURE 1. Separation of L,L-dipeptides with isocratic elution with 0.08% TFA for 2 min, then a linear 15-min gradient with 0.08% TFA in 0 to 30% acetonitrile, on a Brownlee RP-18 column (4.6 mm × 20 cm), flow rate 1 mℓ/min, UV detection at 210 nm.

END GROUP EFFECTS

Large selectivity differences generally arise as a function of the hydrophobicity of the side chains, with secondary pH-dependent effects involving the α-amino- or the α-carboxylic groups. Attempts have been made to evaluate quantitatively the impact of the different side chains on the retention of the peptides. Using lipophilicity values, retention coefficients or group retention contributions,[12-17] the correlation between predicted and observed retention times for different peptides have varied between satisfactory and inadequate. Generally, the contribution from hydrophobic groups has been shown to have an essentially additive effect on peptide retention,[17] but the participation of competing retention processes involving polar groups can cause deviations from the simple picture of solvophobic effects. Unless the optimal chromatographic conditions are chosen, i.e., with strong acids or suitable counterions at low pH, the competing retention processes may even become dominating. An example of such processes is seen in Table 1, where the contribution from an amide group (vs. a free carboxylic acid) and from the amino acids lysine and histidine became more important than the contribution from most hydrophobic side chains, especially on one of the two columns examined. The large impact of the amide group is partially a result of the absence of an ionizable function, but also a result of polar interactions, as seen from the difference between the two C_{18} columns.

The presence of basic groups, such as the lysinyl residue, at low pH, results in an additional cationic center due to protonation of the side chain amino group. This has the effect of decreasing the hydrophobicity of the lysinyl peptide relative to, say a glycine analog, resulting in a shorter retention time. The converse applies to dipeptides with acidic amino acid residues, e.g., aspartic acid, where protonation causes ionic suppression of the carboxylic groups leading to enhanced retention. The retention order is, however, also dependent on whether the amino acid residue is located at the C-terminal or the N-terminal end. In a peptide containing one hydrophobic subunit mainly responsible for the total retention, a C-terminal location produces higher retention than a N-terminal location, at low

Table 1
RELATIVE CONTRIBUTION TO THE RETENTION OF L,L DIPEPTIDES OF C-TERMINAL AND N-TERMINAL AMINO ACIDS ON TWO DIFFERENT C_{18} COLUMNS WITH 0.01 *M* NH_4OAc (pH 6.5), COMPARED TO THE ORDER OF RELATIVE LIPOPHILICITY ACCORDING TO REKKER,[11] AND THE RELATIVE ORDER OF RETENTION COEFFICIENTS ACCORDING TO MEEK[12]

Relative lipophilicity according to Rekker	Relative order of retention coefficients according to Meek		Relative contribution to retention			
			C-terminal AA		N-terminal AA	
	pH 2.1	pH 7.4	Hypersil	Spherisorb	Hypersil	Spherisorb
Trp	Trp	Trp	Trp	Trp,Phe	Trp	Trp
Phe	Phe	Phe	Phe		Phe	Lys
Ile,Leu	Ile	Ile	Met	Lys	Tyr	His
Tyr	Leu	Leu	Leu	His	Leu	Phe
Val	Tyr	amide	Tyr	Gly-NH_2	Met	Leu
Met	Pro	Tyr,Pro	Ile	Met	Val	Tyr
Pro	Met	Met	Val	Leu	His	Met
Ala	amide	Val,Thr	Gly-NH_2	Ile	Lys	Pro
Lys	Val	Ser	Pro	Tyr	Pro	Ala
Gly	Thr	Ala	His	Pro	Ser	Gly
Asp	His	Lys	Lys	Val	Ala	Ser
Glu	Ala	Gly	Ala	Ala	Gly	Glu
His	Gly	His	Gly	Gly	Glu	
Thr	Asp	Asp	Thr	Thr		
Ser	Lys	Glu	Ser	Ser		
	Ser		Glu,Asp	Glu,Asp		
	Glu					

pH, since the site of the charge is closer to the hydrophobic moiety at the N-terminal position.[6] Under basic conditions the retention order is reversed.[6] In the area of isoelectric pH a N-terminal position has been observed to give slightly higher retention of most, but not all, dipeptides.[4,6] This may partially be explained by the different charge distribution in the peptides at constant pH, caused by interchanging the two subunits.

So far, the value of using group retention coefficients is limited to chromatographic conditions where nonsolvophobic processes are of minor significance. Such conditions are not necessarily compatible with preparative purification of peptides, but usually would cause no problems in analytical procedures. The quantitative impact of the different end groups must also be regarded in view of the considerable effects of the stereochemistry of dipeptides,[4-6] a subject which is treated in another article.

REFERENCES

1. **Lin, S.-N., Smith, L. A., and Capriolo, R. M.,** Analysis of dipeptide mixtures by the combination of ion-pair reversed phase HPLC and gas chromatographic-mass spectrometric techniques, *J. Chromatogr.*, 197, 31, 1980.
2. **Dizdaroglu, M. and Simic, M. G.,** Separation of underivatized dipeptides by HPLC on a weak anion-exchange bonded phase, *J. Chromatogr.*, 195, 119, 1980.
3. **Hancock, W. S., Bishop, C. A., Battersby, J. E., Harding, D. R. K., and Hearn, M. T. W.,** HPLC of peptides and proteins. XI. The use of cationic reagents for the analysis of peptides by HPLC, *J. Chromatogr.*, 168, 377, 1979.
4. **Lundanes, E. and Greibrokk, T.,** Reversed-phase chromatography of peptides, *J. Chromatogr.*, 149, 241, 1978.
5. **Rivier, J. and Burgus, R.,** Application of reversed-phase HPLC to peptides, *Chromatogr. Sci.*, 10, 147, 1979.
6. **Kroeff, E. P. and Pietrzyk, D. T.,** HPLC study of the retention and separation of short chain peptide diastereomers on a C_8 bonded phase, *Anal. Chem.*, 50, 1353, 1978.
7. **Molnár, I. and Horváth, C.,** Separation of amino acids and peptides on non-polar stationary phases by HPLC, *J. Chromatogr.*, 142, 623, 1977.
8. **Hearn, M. T. W., Grego, B., and Hancock, W. S.,** HPLC of amino acids, peptides and proteins, *J. Chromatogr.*, 183, 429, 1979.
9. **Melander, W. R., Jacobson, J., and Horváth, C.,** Effect of molecular structure and conformational change of proline-containing dipeptides in reversed phase chromatography, *J. Chromatogr.*, 234, 269, 1982.
10. **Hearn, M. T. W., Su, S. J., and Grego, B.,** Pairing ion effects in the reversed phase HPLC of peptides in the presence of alkylsulphonates, *J. Liq. Chromatogr.*, 4, 1547, 1981.
11. **Cooke, N. H. C., Viavattene, R. L., Eksteen, R., Wong, W. S., Davies, G., and Karger, B. L.,** Use of metal ions for selective separations in HPLC, *J. Chromatogr.*, 149, 391, 1978.
12. **Rekker, R. F.,** in *The Hydrophobic Fragmental Constant,* Elsevier, New York, 1977, 301.
13. **Meek, J. L.,** Prediction of peptide retention times in HPLC on the basis of amino acid composition, *Proc. Natl. Acad. Sci. U.S.A.*, 77, 1632, 1980.
14. **O'Hare, M. J. and Nice, E. C.,** Hydrophobic HPLC of hormonal polypeptides and proteins on alkylsilane-bonded silica, *J. Chromatogr.*, 171, 209, 1979.
15. **Pliska, V. and Fauchere, J. L.,** in Peptides, Structure and Biological Function, Gross, E. and Meienhofer, J., Eds., Pierce Chemical Company, Rockford, IL., 1979, 249.
16. **Hearn, M. T. W. and Grego, B.,** HPLC of amino acids, peptides and proteins. XXVII. Solvophobic considerations for the separation of unprotected peptides on chemically bonded hydrocarbonaceous stationary phases, *J. Chromatogr.*, 203, 349, 1981.
17. **Su, S.-J., Grego, B., Niven, B., and Hearn, M. T. W.,** Analysis of group retention contributions for peptides separated by reversed phase HPLC, *J. Liq. Chromatogr.*, 4, 1745, 1981.

PREDICTION OF PEPTIDE RETENTION TIMES IN REVERSED-PHASE HPLC

Tatsuru Sasagawa and David C. Teller

INTRODUCTION

Prediction of retention time in reversed-phase high-performance liquid chromatography (RP-HPLC) has been the concern of several recent studies. In 1977, Molanár and Horváth[1] showed that elution order of peptides is predictable under certain experimental conditions. Since then, several peptide retention time prediction methods have been reported[2-7] using C_{18}-columns with linear gradient elution modes. It was assumed that the contribution of each residue to retention be additive and retention time be linearly related to the sum of the contribution of each residue.

$$T_{Ri} = A \sum_{j} D_j n_{ij} + B \tag{1}$$

where T_{Ri} is the retention time of peptide i, D_j (Table 3) is the retention constant of amino acid residue j, n_{ij} is number of residue j in peptide i, and A and B are constants. O'Hare and Nice[2] showed that retention order of small peptides with less than 15 amino acid residues generally correlated with the sum of the Rekker's hydrophobic fragmental constants[8] of the individual amino acids. They also showed that low pH mobile phase was necessary for reproducible results of analysis. Meek[3,4] reported a similar relation for small peptides (20 residues or less) and extended the method to numerical analysis of the retention constants of amino acids.

By increasing both the number of data points and size of peptides, however, it became obvious that the retention time-sum of retention constant relationships is not linear (Figure 1 and 2). Based on this fact, another prediction method was introduced,[9] which assumes a logarithmic relationship between retention time and sum of amino acid retention constants (Figure 3).

$$T_{Ri} = A \ln \left(1 + \sum_{j} D_j n_{ij}\right) + C \tag{2}$$

Based on this model, retention constant D_j (Table 3) was calculated from observed retention times of 100 peptides (Table 1).

This model enabled accurate prediction of retention time for a wide range of peptide size (Table 1 and 5) at several elution gradients of acetonitrile (Figure 5).

Meanwhile, it[10] has become obvious that the free silanol groups of the matrix interact with basic groups of the peptides. Polystyrene resin,[11] which does not exhibit such a secondary interaction, was introduced for the separation of large peptides with mobile phases at several pHs. We have shown that peptide retention time data from a polystyrene column fit this nonlinear model as well (Tables 2 and 4, Figure 4).

METHODS

Retention times were measured on either μBondapak® C_{18}-column (Waters) or PRP-1 column (Hamilton) using a Varian Model 5000 liquid chromatograph. Two mobile-phase systems were used: (1) 0.1% TFA (pH 2) and the mobile-phase modifier was acetonitrile containing 0.07% TFA, and (2) 5 m*M* ammonium bicarbonate (pH 8) and the mobile-phase modifier was acetonitrile. The concentration of the mobile-phase modifier was increased linearly from 0 to 60% over either 30 or 60 min. The flow rate was 2 mℓ/min. The elution was monitored by absorption at either 210 or 216 nm, and retention times were measured from the time at injection to that at the center of the eluting peak.

Regression analyses were performed using FORTRAN-IV programs on both a VAX 11/780 computer and a PDP-12 computer with a floating point processor. The nonlinear least squares method was performed using the Gauss-Newton method.[12] The matrix to be inverted was scaled to unit diagonal prior to inversion, and all operations used double precision arithmetic. The coincidence of results on the two computers indicates that numeric accuracy was not a problem.

When weighted least squares was performed, the weights were $1/N_i^2$, where N_i is the number of amino acids in the peptide, for the μBondapak® C_{18}-column data.

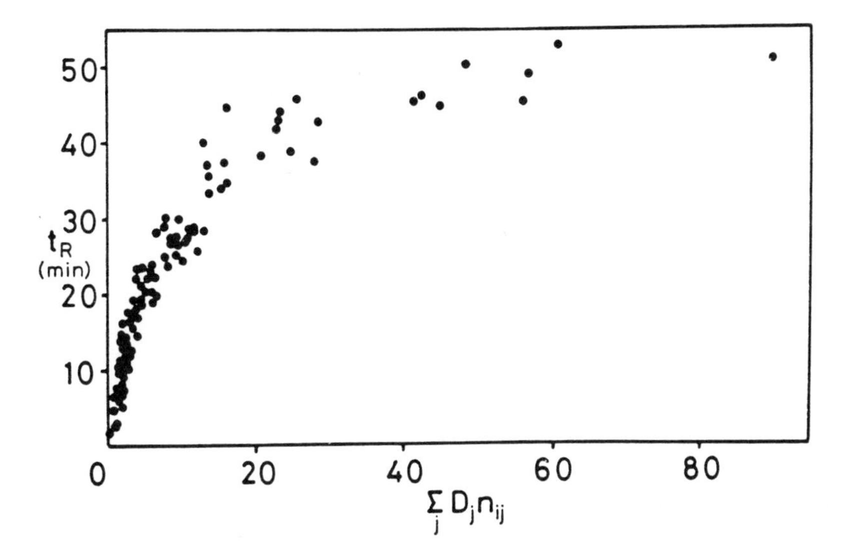

FIGURE 1. Dependence of retention time (t_R) on amino acid composition. The observed retention times (Table 1) were plotted against $\Sigma D_j n_{ij}$, where D_j is modified Rekker's constants (Table 3), and n_{ij} is number of amino acid residue j in peptide i.

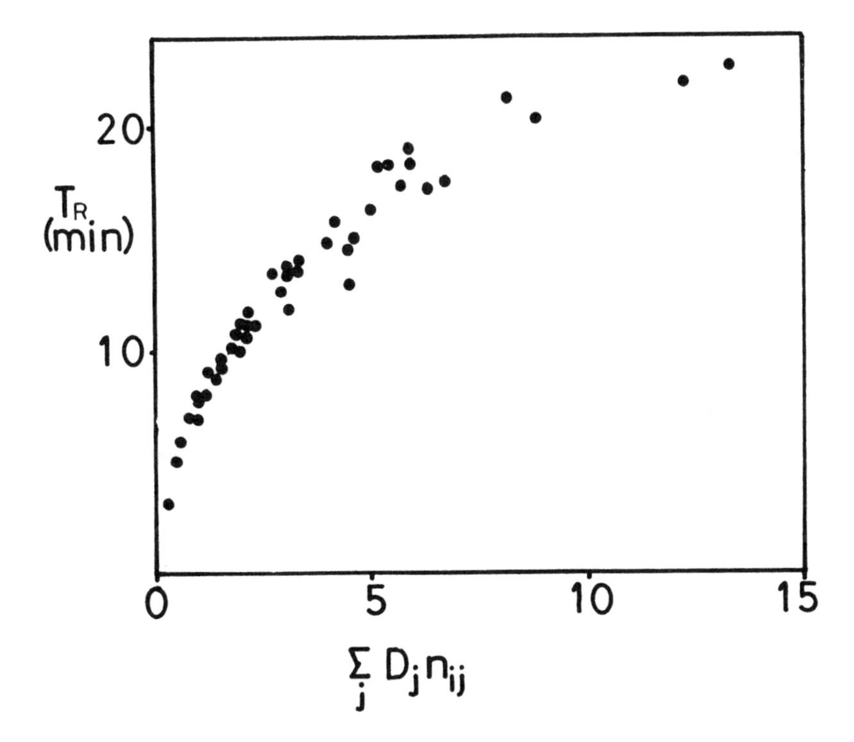

FIGURE 2. Dependence of retention time on amino acid composition. The observed retention times (Table 2, at pH 2) were plotted against $\Sigma D_j n_{ij}$, where D_j is reported in Table 4.

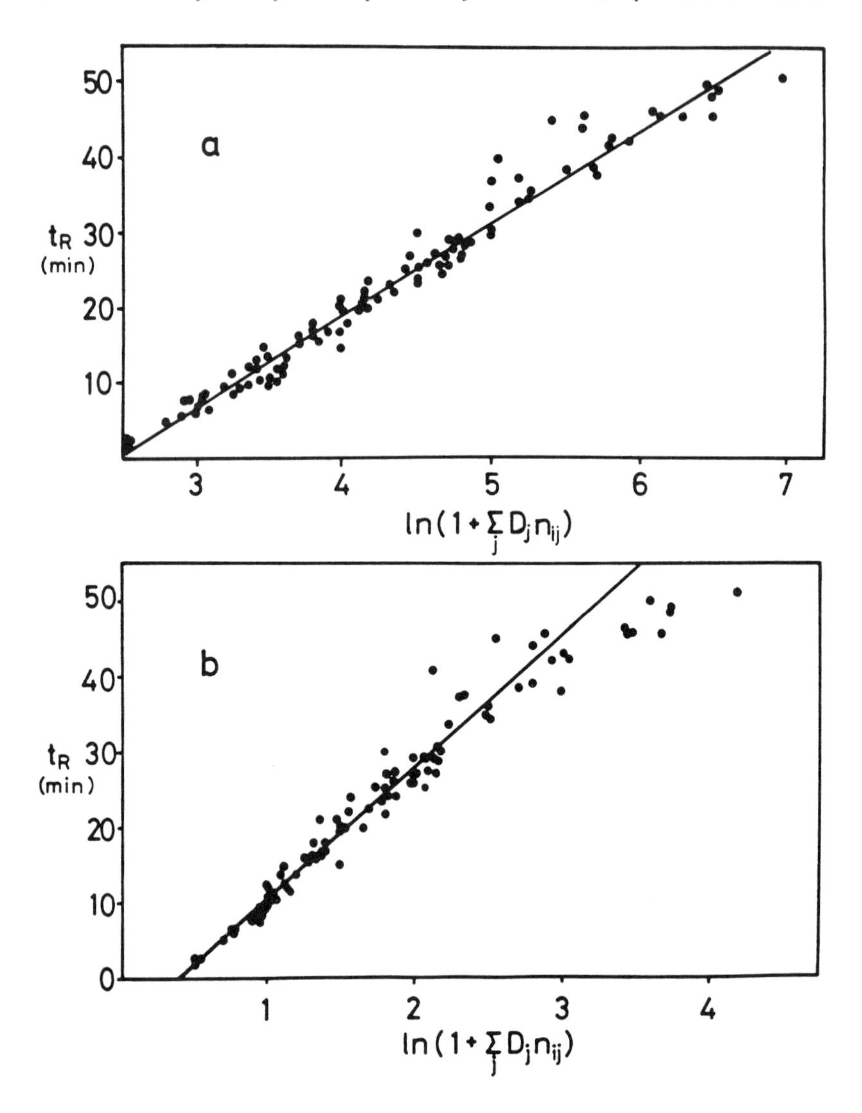

FIGURE 3. Relationship between retention time (t_R) and ln ($1 + \Sigma D_j n_{ij}$). The observed retention times (Table 1) were plotted against ln ($1 + \Sigma D_j n_{ij}$), where D_j is the calculated retention constants either by unweighted or weighted curve fitting. (a) Unweighted fit retention constants were used for D_j. The intercept and the slope of the straight line was −30.3 and 12.4, respectively. The correlation coefficient was 0.984. The mean percent deviation of retention time was 9.9%. (b) Weighted fit retention constants were used for D_j. The intercept and the slope of the straight line was −7.04 and 13.6, respectively. The weighted correlation coefficient was 0.981. The mean percent deviation of retention time was 9.2%.

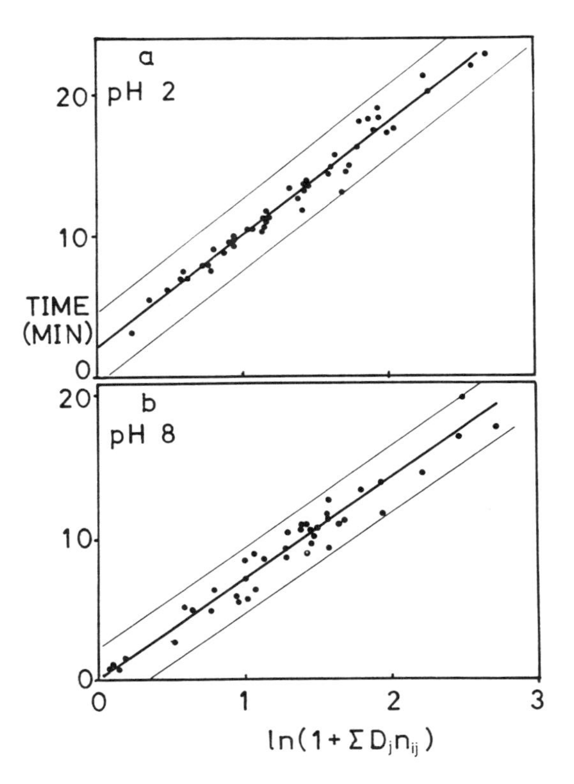

FIGURE 4. Relationship between retention time and ln $(1 + \Sigma D_j n_{ij})$. The observed retention times (Table 2) were plotted against ln $(1 + \Sigma D_j n_{ij})$, where D_j is reported in Table 4. (a) Retention times were observed at pH 2. The intercept and the slope of the straight line was 2.15 and 7.9, respectively. The correlation coefficient was 0.985. The mean percent deviation of retention time was 4.9%. The area indicated by two lines shows 70% confidence interval of prediction. (b) Retention times were observed at pH 8. The intercept and the slope of the straight line was 0.12 and 7.00, respectively. The correlation coefficient was 0.963, the mean percent deviation of retention time was 11.6%.

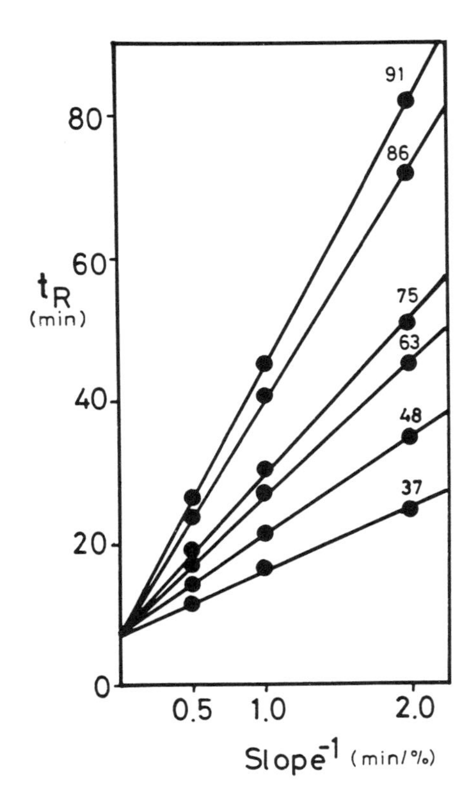

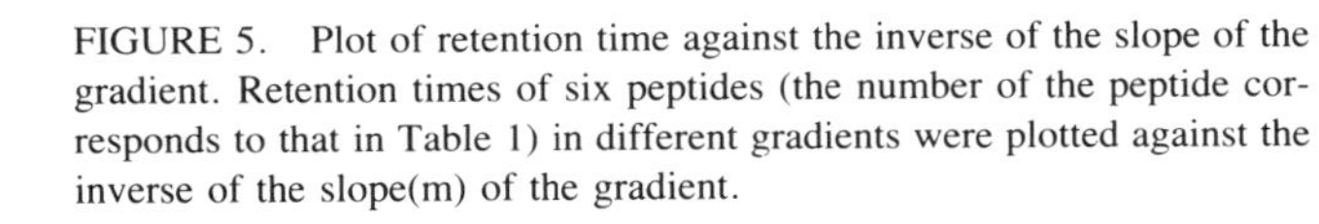

FIGURE 5. Plot of retention time against the inverse of the slope of the gradient. Retention times of six peptides (the number of the peptide corresponds to that in Table 1) in different gradients were plotted against the inverse of the slope(m) of the gradient.

Table 1
COMPARISON OF PREDICTED AND OBSERVED RETENTION TIMES

No.	Peptide	Retention time (min) Observed	Predicted		No.	Peptide	Retention time (min) Observed	Predicted	
1	GGG	1.8	1.3	(1.9)	40	Ac-ADQL	3818.2	16.6	(17.8)
2	PG	2.5	1.3	(2.3)	41	LFK	18.2	19.4	(16.2)
3	ARKM*	2.8	1.0	(2.0)	42	IAEFK	19.5	20.7	(19.6)
4	TEEQ	5.0	4.7	(5.2)	43	ADIDGDGQVNYEE	19.8	21.1	(22.4)
5	GL-NH_2	5.3	5.5	(5.2)	44	VFDKDGNGYI	20.2	19.0	(19.7)
6	Ac-AAA	6.1	6.5	(6.8)	45	ISAAELR	20.3	21.0	(20.1)
7	MTAK	6.5	8.0	(6.4)	46	FESNFNTQATNR	20.3	19.0	(19.4)
8	NLC*	6.6	7.0	(6.9)	47	ELGTVMR	21.2	21.0	(19.2)
9	MARKM*	7.5	10.4	(9.3)	48	GHHEAELK	21.3	18.9	(17.2)
10	MAR	7.8	5.3	(7.2)	49	WWC*NDGR	21.4	22.1	(24.9)
11	YK	8.0	6.0	(7.5)	50	LQDMINE	22.0	21.0	(20.7)
12	TPGSR	8.1	7.6	(7.3)	51	FVQMMTAK	22.5	23.6	(22.8)
13	KYE	8.2	7.7	(9.4)	52	WWC*	23.5	23.4	(24.5)
14	GY	8.5	9.9	(8.8)	53	QIAEFK	23.8	21.3	(20.9)
15	TEAEMK	9.2	10.7	(10.3)	54	Ac-ADQLTEEQIAE	24.0	25.7	(26.0)
16	EY	9.6	9.3	(9.1)	55	RSLGQNPTEAELQDM*	24.0	25.2	(25.1)
17	HLK	9.8	12.9	(10.6)	56	MIREADIDGDGQVNYEE	24.8	27.6	(29.7)
18	FK	9.9	11.4	(10.8)	57	FLTMMAR	25.1	25.3	(24.9)
19	IRE	10.3	12.2	(11.7)	58	VDADGNGTIDFPE	25.3	24.5	(23.8)
20	PL	10.3	13.8	(11.2)	59	HVMTNLGEK*LTDEEVDEM*	25.7	28.1	(27.7)
21	IAE	10.9	13.1	(10.9)	60	LGTVMRSLGQNPTEAE	25.8	27.5	(27.4)
22	GF	11.5	14.1	(11.9)	61	SALLSSDITASVNC*	26.0	26.4	(25.9)
23	KMKDTDSEEE	11.5	9.9	(13.7)	62	NTDGSTDYGILQINSR	26.9	27.6	(27.9)
24	AFR	12.0	13.5	(12.9)	63	VEADVAGHGQDILIR	26.9	29.0	(28.7)
25	DIAAK	12.0	11.7	(13.1)	64	FLTMMARKMKDTDSEEE	27.0	29.1	(30.7)
26	QIAE	12.0	14.2	(11.0)	65	LRHVMTNLGEK*LTDE	27.0	24.7	(24.8)
27	NIPC*	12.4	11.5	(10.6)	66	VTVPLVSDAEC**R	27.3	27.0	(26.1)
28	ASEDLK	13.0	11.9	(12.7)	67	VFDKDGNGYISAAELR	27.5	28.5	(30.1)
29	EAFR	13.5	14.5	(14.3)	68	AFRVFDKDGNGYISAAE	28.6	29.3	(31.2)

30	FDR	13.8	12.6	(12.8)	69	LRHVMTNLGEK*LTDEEVDE	28.6	29.6	(30.8)
31	VFDKDGNGY	14.8	19.0	(19.7)	70	VFDKDGNGYISAAEL	29.0	28.5	(29.5)
32	FKE	15.0	12.5	(12.5)	71	GYSLGNWVC**	29.1	29.1	(29.6)
33	KVFGR	15.6	15.5	(15.6)	72	IREADIDGDGQVNYEEFVQM*	29.2	28.8	(30.5)
34	SLGQNPTEAE	15.8	17.4	(16.9)	73	Ac-ADQLTEEQIAEFK	29.2	27.7	(28.4)
35	GW	16.3	15.5	(15.3)	74	EAFSLFDKDGDGTITTK	30.0	31.4	(31.5)
36	MIRE	16.5	16.9	(16.3)	75	ALELFR	30.2	25.4	(24.6)
37	SHPETLEK	16.7	19.2	(17.5)	76	AFSLFDKDGDGTITTKE	30.4	31.4	(31.5)
38	HGLDNYR	17.0	18.0	(17.6)	77	HVMTNLGEK*LTDEEVDEMIR	33.5	31.0	(32.5)
39	WY-NH_2	17.1	16.6	(17.4)	78	NKALELFRKDIAAKYKELGYQG	34.2	34.0	(37.3)

79	PGYPGVYTEVSYHVDWIK	34.8	34.5	(36.7)
80	DDYGADEIFDSMIC**AGVPEGGK	35.9	34.8	(37.1)
81	EADIDGDGQVNYEEFVQMMTAK	37.2	31.2	(33.5)
82	INEVDADGNGTIDFPEFLTM*	37.5	33.9	(34.1)
83	KDTDSEEEIREAFRVFDKDGNGYISAAELRHVMTNLGEK*LTDEEVDEM*	37.8	40.6	(45.8)
84	IILHENFDYDLLDNDISLLK	38.5	38.1	(40.8)
85	ASSTNLKDILADLIPKEQARIKTFRQQHGNTVVGQITVDM*	39.0	40.3	(43.7)
86	HGVTVLTALGAILK	40.5	32.1	(30.6)
87	Ac-ADQLTEEQIAEFKEAFSLFDKDGDGTITTKELGTVMR	42.1	41.3	(44.6)
88	SQLSAAITALNSESNFARAYAEGIHRTKYWELIYEDC**M*	42.3	42.9	(47.5)
89	Ac-ADQLTEEQIAEFKEAFSLFDKDGDGTITTKELGTVM*	42.8	41.4	(45.4)
90	SLGQNPTEAELQDMINEVDADGNGTIDFPEFLTM	44.0	39.0	(42.3)
91	YLEFISEAIIHVLHSR	45.0	36.5	(38.2)
92	MARKMKDTDSEEEIREAFRVFDKDGNGYISAAELRHVMTNLGEK*LTDEEVDEMIREADIDGDGQVNYEE FVQMMTAK	45.5	46.0	(53.8)
93	NGLAGPLHGLANQEVLVWLTQLQKEVGKDVSDEKLRDYIWNTLNSGRVVPGYGHAVLRKTDPRYTC**QRE FALKHLPHDPM*	45.5	50.3	(53.8)

Table 1 (continued)
COMPARISON OF PREDICTED AND OBSERVED RETENTION TIMES

		Retention time (min)		
		Observed	Predicted	
94	VLSEGEWQLVLHVWAKVEADVAGHGQDILIRLFKSHPETLEKFDRFKHLKTEAEM*	45.6	47.2	(54.1)
95	SLGQNPTEAELQDMINEVDADGNGTIDFPEFLTMMAR	45.8	39.8	(43.6)
96	KMKDTDSEEEIREAFRVFDKDGNGYISAAELRHVMTNLGEK*LTDEEVDEMIREADIDGDGQVN YEEFVQMMTAK	46.2	45.5	(53.1)
97	VDADGNGTIDFPEFLTMMARKMKDTDSEEEIREAFRVFDKDGNGYISAAELRHVMTNLGEK*LTDEE VDEMIREADIDGDGQVNYEEFVQMMTAK	48.3	50.1	(58.8)
98	FKEAFSLFDKDGDGTITTKELGTVMRSLGQNPTEAELQDMINEVDADGNGTIDFPEFLTMMARKMK DTDSEEEIREAFRVFDKDGNGYISAAE	48.9	50.7	(59.0)
99	KASEDLKKHGVTVLTALGAILKKKGHHEAELKPLAQSHATKHKIPIKYLEFISEAIIHVLHSRHPGNFGADAQGAM*	50.0	49.6	(56.2)
100	Ac-ADQLTEEQIAEFKEAFSLFDKDGDGTITTKELGTVMRSLGQNPTEAELQDMINEVDADGNGTIDFPEFLTM MARKMKDTDSEEEIREAFRVFDKDGNGYISAAELRHVMTNLGEK*LTDEEVDEMIREADIDGDGQV NYEEFVQMMTAK	51.0	56.0	(66.7)

Note: The retention times were predicted from Equation 2, where A, C were reported in Figure 2. D_j was reported in Table 3 (column 5,6). The predicted retention times in parentheses were calculated from weighted fit parameters. M*, Homoserine or its lactone; C*, aminoethylcysteine; C**, carboxymethylcysteine; K*, trimethyllysine. Retention times were measured on μBondapak® C_{18}-column, mobile phase was 0.1% TFA, the mobile-phase modifier was acetonitrile containing 0.07% TFA. The slope of the gradient was 1%/min.

Table 2
COMPARISON OF PREDICTED AND OBSERVED RETENTION TIMES ON A POLYSTYRENE COLUMN

No.	Sequence	Retention time			
		pH 2		pH 8	
1	YK	3.1	(4.0)	2.3	(3.9)
2	TPGSR	5.6	(5.0)	5.1	(4.2)
3	HLK	6.2	(5.9)	5.8	(7.3)
4	DIAAK	7.0	(6.8)	1.6	(1.5)
5	FK	7.0	(7.1)	6.5	(7.6)
6	ASEDLKK	7.5	(6.9)	1.3	(0.8)
7	TEAEMK	7.9	(8.3)	0.9	(1.1)
8	FDR	7.9	(7.9)	5.0	(5.5)
9	GF	8.7	(9.1)	8.6	(8.2)
10	DTDSEEEIR	9.1	(8.6)	—	(—)
11	SHPETLEK	9.3	(9.6)	5.0	(4.7)
12	EAFR	9.5	(9.3)	5.5	(6.9)
13	HKIPIK	9.7	(9.6)	10.7	(9.8)
14	HGLDNYR	10.0	(9.3)	8.5	(7.1)
15	LFK	10.2	(11.1)	11.0	(9.9)
16	GHHEAELKPLAQSHATK	10.5	(11.3)	9.7	(10.4)
17	ELGYQG	10.5	(10.5)	6.0	(6.6)
18	GW	10.5	(10.4)	10.5	(9.4)
19	FESNFNTQATNR	11.2	(11.4)	10.1	(9.9)
20	HPGNFGADAQGAMNK	11.2	(11.0)	9.2	(9.0)
21	WY-NH_2	11.2	(11.2)	13.8	(13.8)
22	CELAAMKR	11.8	(11.3)	6.5	(5.8)
23	ELGTVMR	11.8	(13.4)	8.8	(10.2)
24	DGNGYISAAELR	12.7	(13.1)	8.6	(9.1)
25	VEADVAGHGQDILIR	13.0	(15.6)	10.2	(11.3)
26	FF	13.5	(12.6)	11.7	(11.1)
27	NTDGSTDYGILQINSR	13.5	(13.6)	10.7	(10.6)
28	WWCNDGR	13.6	(13.6)	10.7	(10.5)
29	YGGFM	13.6	(13.5)	11.2	(11.9)
30	GTDVQAWIR	14.0	(13.7)	11.3	(11.3)
31	ALELFR	14.5	(14.8)	12.9	(11.1)
32	VFDKDGNGYISAAELR	14.6	(15.6)	11.0	(11.8)
33	Ac-ADQLTEEQIAEFK	14.9	(14.9)	7.2	(7.2)
34	EAFSLFDKDGDGTITTK	15.0	(15.9)	10.3	(10.6)
35	GYSLGNWVCAAK	15.8	(15.1)	13.3	(12.8)
36	HVMTNLGEK*LTDEEVDEMIR	16.3	(16.3)	9.2	(9.2)
37	NLCNIPESALLSSDITASVNC	17.2	(17.9)	11.0	(11.8)
38	NKALELFRKDIAAKYKELGYQG	17.4	(17.2)	17.7	(13.8)
39	IVSDGDGMNAWVAWR	17.6	(18.3)	14.5	(15.7)
40	EADIDGDGQVNYEEFVQM	18.2	(16.6)	9.0	(7.5)
41	HGVTVLTALGAILKK	18.3	(17.4)	—	(—)
42	EADIDGDGQVNYEEFVQMMTAK	18.3	(17.6)	11.0	(10.0)
43	HGVTVLTALGAILKKK	19.0	(18.0)	—	(—)
44	VLSEGEWQLVLHVWAK	20.2	(20.2)	17.2	(17.5)
45	YLEFISEAIIHVLHSR	21.3	(19.7)	19.7	(17.6)
46	SLGQNPTEAELQDMINEVDADGNGTIDFPEFLTM	22.0	(22.6)	11.6	(13.7)
47	VLSEGEWQLVLHVWAKVEADVAGHGQDILIR	22.7	(23.2)	17.8	(19.5)

Note: The retention times were predicted (numbers in parentheses) from Equation 2.A, C were reported in Figure 4; D_j was reported in Table 4. C, carboxymethylcysteine; K*, trimethyllysine. Retention times were measured on a Hamilton PRP-1 column. Mobile phase was either 0.1% TFA or 5 m*M* ammonium bicarbonate; mobile-phase modifier was acetonitrile. The slope of gradient was 2%/min. The flow rate was 2 mℓ/min.

Table 3
RETENTION CONSTANTS (D_j) OF AMINO ACIDS

Amino acid	Rekkers'	Rekkers' (modified)	Meeks'	Present study (D_j)[a] Nonweighted	Weighted
Tryptophan	2.31	2.31	18.1(7)	35.8(12)	2.34(12)
Phenylalanine	2.24	2.24	13.9(18)	31.4(86)	1.71(86)
Isoleucine	1.99	1.99	11.8(4)	27.4(95)	1.38(95)
Leucine	1.99	1.99	10.0(13)	26.4(129)	1.34(129)
Tyrosine	1.70	1.70	8.2(16)	21.0(43)	1.23(43)
Methionine	1.08	1.08	7.1(11)	14.5(64)	0.85(64)
Proline	1.01	1.01	8.0(13)	7.9(33)	0.48(33)
Valine	1.46	1.46	3.3(6)	7.4(89)	0.38(89)
Threonine	−0.26	0.10	1.5(9)	7.4(111)	0.12(111)
Histidine	−0.23	−0.10	0.8(6)	8.8(38)	0.34(38)
Alanine	0.53	0.53	−0.1(8)	2.4(139)	0.13(139)
Glutamine	−1.09	0.20	−2.5(5)	3.2(59)	0.36(59)
Glutamic acid	−0.07	0.20	−7.5(4)	2.7(198)	0.27(198)
Glycine	0.00	0.10	−0.5(20)	4.0(134)	0.22(134)
Serine	−0.56	0.10	−3.7(11)	1.1(62)	0.18(62)
Arginine	—	−0.10	−4.5(10)	0.0(73)	0.26(73)
Aspartic acid	−0.02	0.10	−2.8(7)	−0.1(165)	0.10(165)
Asparagine	−1.05	0.10	−1.6(6)	−11.3(71)	−0.45(71)
Lysine	0.52	−0.10	−3.2(9)	−3.1(98)	0.05(98)
Carboxymethylcysteine	—	0.10	—	32.5(5)	1.57(5)
Homoserine	—	0.10	—	12.3(13)	0.23(13)
Aminoethylcysteine	—	−0.10	—	4.3(5)	0.31(5)
Trimethyllysine	—	−0.10	—	−38.1(9)	−1.38(9)
Acetyl-	—	0.00	3.9(1)	12.4(6)	0.81(6)
Amide-	—	0.00	5.0(8)	−13.2(2)	−0.56(2)

Note: The numbers in parentheses represent the number of amino acids used for each calculation.

[a] Data was obtained on a μBondapak® C_{18}-column (Table 1).

Table 4
RETENTION CONSTANT (D_j) OF AMINO ACIDS

Amino acid	pH 2+	pH 8+
Trp	1.79	2.48
Phe	1.37	1.89
Leu	1.23	1.09
Ile	1.06	1.28
Tyr	0.77	0.67
Met	0.97	1.30
Val	0.73	0.85
Pro	0.55	−0.32
Thr	0.21	−0.07
Ala	0.15	−0.03
Gly	0.04	0.27
Ser	−0.16	0.60
Gln	−0.39	−0.22
Asn	−0.17	−0.25
Glu	0.17	−0.55
Asp	−0.08	−1.04
CMC	0.09	−0.86
His	−0.12	0.66
Lys	−0.50	0.04
Arg	−0.21	0.32
TML*	−2.42	−1.69
Acetyl	0.69	0.71

Note: Data was obtained on Hamilton PRP-1 column (Table 2). *TML, trimethyllysine (parameter determined from single peptide). +, Unweighted parameters for the PRP-1 column.

Table 5
COMPARISON OF DATA FROM LITERATURE WITH PREDICTED RETENTION TIMES

Petides	Number of residues	Retention time (min)	
		Predicted	Observed
Arginine vasotosin	9	12.8	12.0
Lysine vasopressin	9	13.1	13.0
Arginine vasopressin	9	13.9	14.0
ACTH 5-10	6	22.5	17.0
Diphenylalanine	2	20.2	18.0
ACTH 1-18	18	25.8	18.5
Met-enkephalin	5	21.7	19.0
Oxytocin	9	18.6	19.5
ACTH 4-10	7	23.8	20.5
ACTH 1-24	24	28.8	21.5
α-Endorphin	16	27.0	22.0
Leu-enkephalin	5	22.9	22.0
Insulin A (bovine)	21	22.2	22.0
Angiotensin II	8	24.3	23.0
Neurotensin	13	25.1	24.5
α-Melanotropin	13	26.4	26.0
Bombesin	14	25.0	26.0
RNAase	124	36.7	27.5
Triphenylalanine	3	23.5	28.0
Gastrin I	17	29.1	28.5
Substance P	11	26.0	29.0
Substance P 4-11	8	25.7	30.0
ACTH 1-39 (human)	39	32.8	30.5
ACTH 18-39	22	27.8	30.5
ACTH 34-39	6	24.2	31.0
Somatostatin	14	24.6	32.0
Insulin (bovine)	51	34.7	32.0
ACTH 1-39 (porcine)	39	33.5	33.0
Insulin B (bovine)	30	32.6	33.5
β-Endorphin (ovine)	31	31.3	34.0
β-Lipotropin (human)	91	38.4	34.5
Calcitonin (human)	32	31.7	34.5
Cytochrome *c*	104	40.1	35.0
Glucagon	29	31.7	36.0
Tetraphenylalanine	4	25.9	36.5
Calcitonin (salmon)	32	29.4	37.0
Lysozyme	129	40.4	37.5
Myoglobin	153	45.7	45.0
Melittin	25	32.3	46.0

Note: Data were from O'Hare and Nice.[2] Retention times were predicted using Equation 2 and the nonweighted retention constants for amino acids of Table 3. Due to the different solvent system and the presence of pyroglutamic acid and cystine, it was necessary to fit four new constants. The slope and intercept of the straight line was 8.30 and −14.73, respectively. The retention constants for pyroglutamic acid and cystine were found to be −12.2 and −27.3, respectively. The correlation coefficient was 0.81 and the mean percent deviation of retention time was 12.0%.

REFERENCES

1. **Molnár, I. and Horváth, C.,** Separation of amino acids and peptides on non-polar stationary phases by high-performance liquid chromatography, *J. Chromatogr.*, 142, 623, 1977.
2. **O'Hare, M. J. and Nice, E. C.,** Hydrophobic high performance liquid chromatography of hormonal polypeptides and proteins on alkylsilane bonded silica, *J. Chromatogr.*, 171, 209, 1979.
3. **Meek, J. L.,** Prediction of peptide retention times in high-pressure liquid chromatography on the basis of amino acid composition, *Proc. Natl. Acad. Sci. U.S.A.*, 77, 1632, 1980.
4. **Meek, J. L. and Rossetti, Z. L.,** Factors affecting retention and resolution of peptides in high-performance liquid chromatography, *J. Chromatogr.*, 211, 15, 1981.
5. **Su, S. J., Grego, B., Niven, B., and Hearn, M. T. W.,** Analysis of group retention contributions for peptides separated by reversed phase high performance liquid chromatography, *J. Liq. Chromatogr.*, 4, 1745, 1981.
6. **Browne, C. A., Bennett, H. P. J., and Solomon, S.,** The isolation of peptides by high-performance liquid chromatography using predicted elution positions, *Anal. Biochem.*, 124, 201, 1982.
7. **Molnár, I. and Schoeneshoefer, X.,** Structure-retention-relationships and selectivity of peptides and proteins on nonpolar stationary phases, in *High Performance Liquid Chromatography in Protein and Peptide Chemistry*, Lottspeich, F., Henschen, A., and Hupe, K. P. Eds., 1981, 97.
8. **Rekker, R. F.,** *The Hydrophobic Fragmental Constant*, Elsevier, Amsterdam, 1977, 301.
9. **Sasagawa, T., Okuyama, T., and Teller, D. C.,** Prediction of peptide retention times in reversed-phase high-performance liquid chromatography during linear gradient elution, *J. Chromatogr.*, 240, 329, 1982.
10. **Bij, K., Horváth, C., Melander, W., and Nahum, A.,** Surface silanols in silica-bonded hydrocarbonaceous stationary phases, *J. Chromatogr.*, 203, 65, 1981.
11. **Sasagawa, T., Ericsson, L. H., Teller, D. C., Titani, K., and Walsh, K. A.,** in 2nd Int. Symp. HPLC of Proteins, Peptides, and Polynucleotides, 1982, 12.
12. **Bevington, R. P.,** *Data Reduction and Error Analysis for the Physical Science*, McGraw-Hill, New York, 1969.

THREE-STEP HPLC SEPARATION OF PEPTIDES

Hisao Mabuchi and Hisamitsu Nakahashi

INTRODUCTION

High-performance liquid chromatography (HPLC) is an emerging new technology that is of value in the analysis and separation of peptides.[1-3] Until now, tedious and time-consuming traditional methods such as Sephadex® gel filtration followed by ion-exchange chromatography have been used for this purpose. To our knowledge, however, the use of HPLC in the field of the separation of peptides has generally been restricted to the final step of the separation. The progress of HPLC methodology has enabled the systematic separation of peptides such as high-performance gel chromatography (HPGC) followed by reversed-phase HPLC (RP-HPLC).

In this section, we describe a three-step HPLC technique, mainly developed in our laboratory,[4] for the separation of peptides.

STEP 1: HPGC

Gel chromatography (filtration) is an important first step in the separation of peptides. Until now, columns containing cross-linked dextran or polyacrylamide gel have been used for this purpose. However, they are unsuitable for use under conditions employed for HPGC because of the fragility of the gel.

Recently, several column packings for use in HPGC have become commercially available. One of these column packings, TSK-GEL SW (Toyo Soda) type support has been widely used for the separation and molecular weight estimation of proteins with or without denaturing agents such as sodium dodecyl sulfate (SDS) or guanidine hydrochloride.[5-15] Kato et al.[8] have reported that the separation range of TSK-GEL SW type columns was not extended below molecular weights 10,000 even by use of a 2000SW column which has a smaller pore size.

For the separation of peptides having molecular weights below 5000, the existence of a denaturing agent such as SDS is essential. Moreover, the concentration of SDS is an important factor. Our results[4,16] revealed that the addition of 0.3 to 0.5% SDS to phosphate buffer is effective in alleviating the adsorption of peptides on the TSK-GEL 2000SW column. A semilogarithmic plot of molecular weights vs. K_d (distribution coefficient) for several samples is shown in Figure 1. It is possible to estimate the relative molecular weights of proteins and peptides in the range of 1000 to 45,000 despite the deviation of several samples. Strictly speaking, however, other phenomena such as partition are more likely to be predominant in the separation of peptides having molecular weights below 5000.

STEP 2: RP-HPLC

Ion-pair RP-HPLC is an especially useful technique for the separation of peptides.[1] Several compounds have been employed for ion-pairing reagents.[1,17-28] Most recently, volatile and UV transparent solvent systems such as acetonitrile-heptafluorobutyric acid (HFBA) have been developed.[22-24, 26-28]

The peptide samples, dissolved in solutions containing SDS, show different behavior from the samples dissolved in distilled water. When no ion-pairing reagents are used, many samples are strongly retained on the reversed-phase columns. Therefore, direct application of fractions obtained from HPGC to RP-HPLC is difficult. To resolve this problem, we chose SDS as

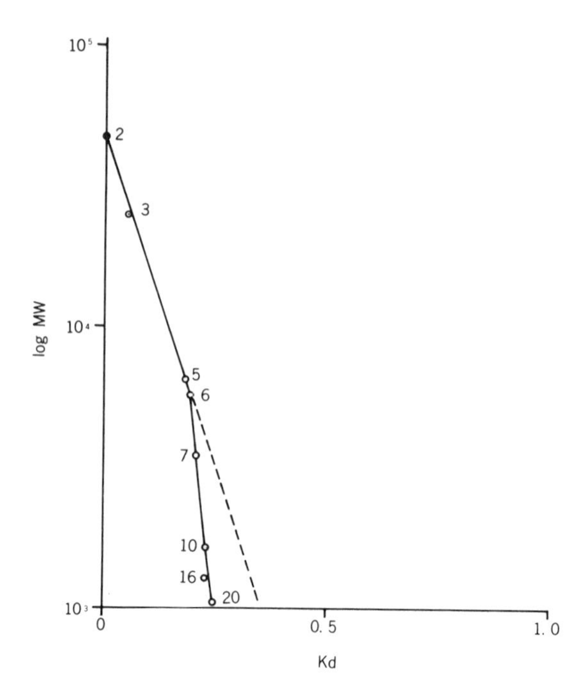

FIGURE 1. Semilogarithmic plot of molecular weight (MW) vs. K_d for several samples chromatographed on a TSK-GEL 2000SW (60 × 0.75 cm) column. Mobile phase, 0.05 *M* sodium phosphate buffer, pH 7.2, containing 0.3% SDS; flow rate, 0.3 mℓ/min; temperature, ambient; detection, UV, 210 nm. Samples are ovalbumin(2), α-chymotrypsinogen A(3), aprotinin (5), insulin(6), glucagon(7), neurotensin(10), angiotensin I(16), vasopressin(20). (From Mabuchi, H. and Nakahashi, H., *J. Chromatogr.*, 213, 275, 1981. With permission.)

an ion-pairing reagent. The usefulness of SDS as an ion-pairing reagent has been described by Hancock et al.[19]

Moreover, phosphate buffer in the fractions obtained from HPGC affect the elution profile of peptides. Addition of phosphoric acid to the mobile phase alleviates this effect (Figure 2). The volatile solvent systems are applicable to RP-HPLC when the removal of SDS and phosphate buffer has been achieved (see Step 3).

STEP 3: REMOVAL OF SDS

SDS is one of the most potent denaturants and solubilizing agents. It has been widely used in the field of protein and peptide chemistry. However, difficulty of removal restricts its application in the separation of peptides. SDS has been used for the molecular weight estimation of peptides using several techniques such as SDS gel chromatography or SDS polyacrylamide electrophoresis and it may be removed by prolonged dialysis. This method is not applicable to peptide samples. Until now, several traditional chromatographic methods have been used for the removal of SDS.[29-32] The chromatographic behavior of SDS shows some different characteristics from that of peptides. These characteristics permit the removal of SDS from peptides by using HPLC techniques.

High-Performance Cation-Exchange Chromatography

The strong cation-exchange (SCX) column, Partisil® SCX (Whatman), has been introduced

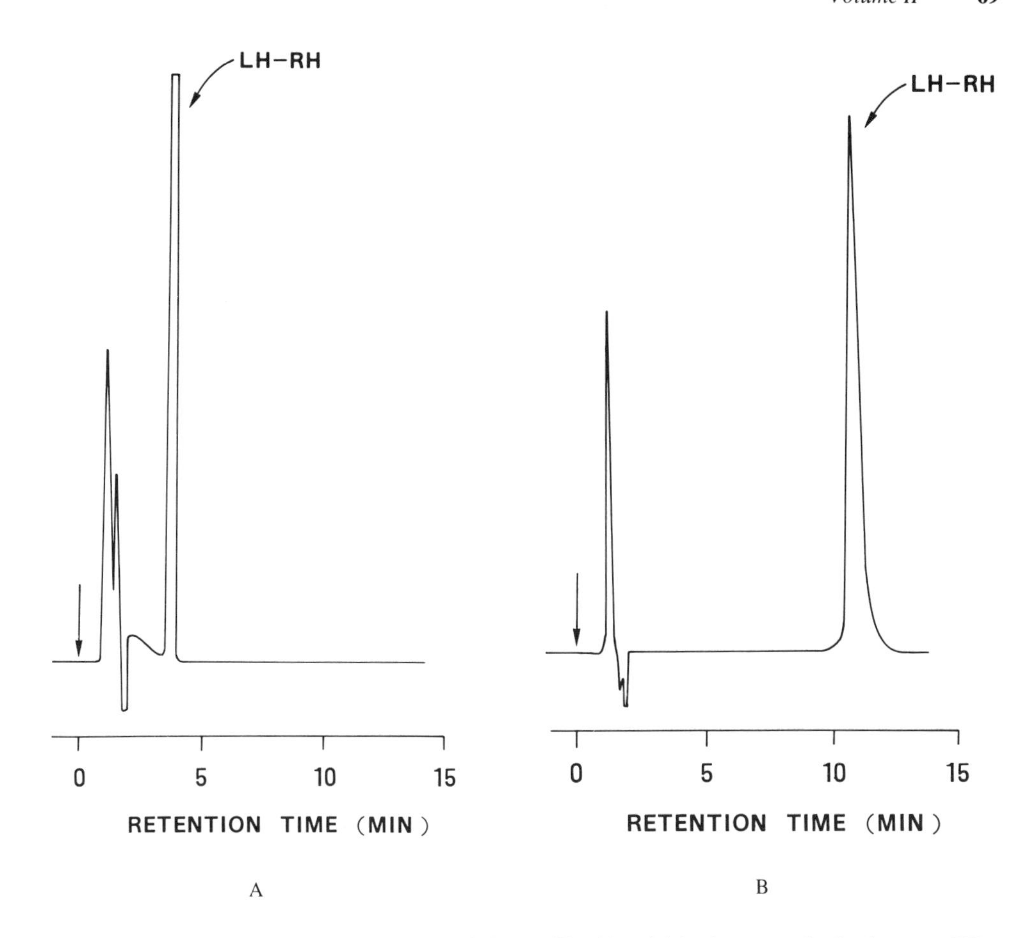

FIGURE 2. Effect of phosphate buffer on the elution profile of leuteinizing hormone releasing hormone (LH-RH). Column, Cosmosil $5C_{18}$ (5-μm, 0.46 × 15 cm); mobile phase, acetonitrile-water (50:50) containing 15 m*M* SDS (A), acetonitrile-water (50:50) containing 15 m*M* SDS and 10 m*M* phosphoric acid (B); flow rate, 1.5 mℓ/min; temperature, ambient; detection, UV, 210 nm, 0.32 AUFS; sample, 5 μg LH-RH dissolved in 500 μℓ, 0.05 *M* sodium phosphate buffer, pH 7.2, containing 0.3% SDS.

for the separation of peptides by Radhakrishan et al.[33] We found that SDS passes unretained through a Partisil® SCX column when eluted with distilled water.[4] This characteristic is applicable to the removal of SDS from peptides using a simple two-step gradient elution with volatile solvents (Figure 3). However, large peptides such as insulin are eluted as two peaks, that is, a SDS-peptide complex peak and a native peptide peak. The dissociation of the SDS-peptide complex may be incomplete in distilled water. For these peptides, instead of distilled water, a low concentration of pyridine-acetate buffer is effective. The solvent system used is readily removed from the eluted samples by lyophilization and facilitates further direct investigation of the separated samples. In addition, both phosphoric acid and acetonitrile can be removed by the above procedure.

Use of ODS Columns or Cartridges

Octadecasilyl-silica (ODS) is a useful solid phase in RP-HPLC. It can be used for efficient desalting and extraction procedures of peptides.[34-39]

SDS is more strongly retained by ODS than most peptides. This characteristic can be applied to the removal of SDS and desalting.[41] All peptides tested having molecular weights below insulin are eluted with 20% *n*-propanol and 1% trifluoroacetic acid (TFA) in water

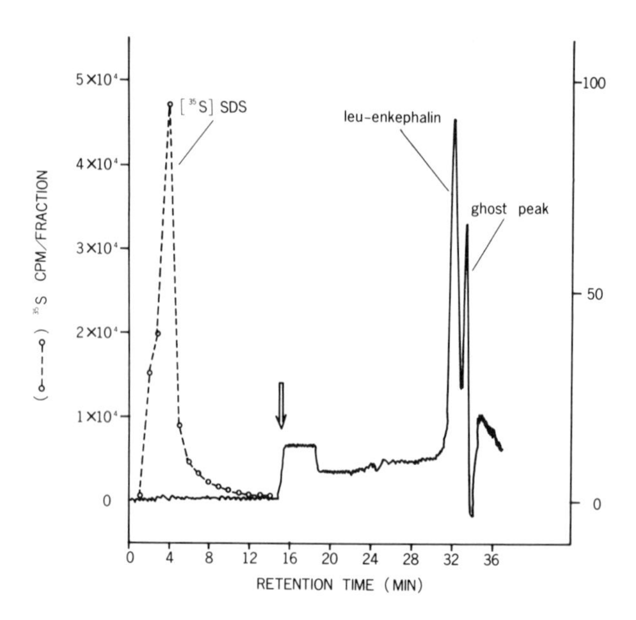

FIGURE 3. Elution profile of Leu-enkephalin dissolved in a solution containing SDS and [^{35}S]SDS chromatographed on a Partisil® SCX (0.46 × 25 cm) column with stepwise gradient elution. Mobile phase, (1) distilled water at a flow rate of 1.5 mℓ/min, 15 min; (2) 3.0 *M* pyridine-0.5 *M* acetic acid at a flow rate of 0.5 mℓ/min, 30 min; temperature, ambient; detection, post-column fluorescence derivatization with fluorescamine. The arrow indicates the beginning of the elution with the second solution. Radioactivity was determined by collecting eluate fractions at 1-min intervals. (From Mabuchi, H. and Nakahashi, H., *J. Chromatogr.*, 213, 275, 1981. With permission.)

but SDS is not eluted. Higher concentrations of organic solvents are necessary to elute the SDS (Figure 4). Large volumes of sample can be loaded on the column by pumping the sample through the column instead of injecting the sample.[36]

For a small volume of sample, a commercially available ODS cartridge (Sep-Pak® C_{18}, Waters) is useful, but it is not reusable. The cartridge is washed with 5 mℓ of methanol followed by 10 mℓ of 1% TFA in water before use. The sample solution is passed through the cartridge then washed with 10 mℓ of 1% TFA in water. Next, the peptides are eluted with 3 mℓ of 20% *n*-propanol and 1% TFA in water. SDS is retained on the cartridge.

SUMMARY

The application of the HPLC techniques mentioned above for the analysis and separation of peptides are summarized in Figure 5. All the steps are versatile. The recovery of peptides is in excess of 90% in all the steps. By using these HPLC techniques we have analyzed small peptides in both normal and uremic serum.[40]

In conclusion, this three-step HPLC technique may have a wide applicability in biological samples.

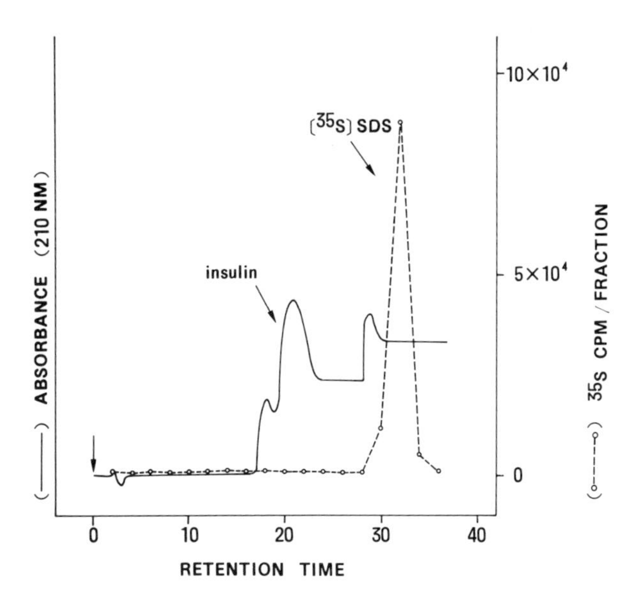

FIGURE 4. Elution profiles of insulin and [^{35}S]SDS chromatographed on a Lichroprep RP-18 (0.8 × 10 cm) column with stepwise gradient elution. Mobile phase, (1) 1% TFA in water, 12 min, (2) 20% *n*-propanol and 1% TFA in water, 12 min, (3) 60% *n*-propanol and 1% TFA in water, 12 min; flow rate, 2.0 mℓ/min; temperature, ambient; detection, UV 210 nm, 1.28 AUFS; sample, monocomponent insulin (2 units) dissolved in 200 μℓ of 1% SDS and [^{35}S]SDS. Radioactivity was determined by collecting eluate fractions at 2-min intervals.

(A) HPGC → RP-HPLC (SDS as an ion-pairing reagent) → REMOVAL OF SDS → PURIFIED PEPTIDE

(B) HPGC → REMOVAL OF SDS → RP-HPLC (volatile solvent system) → PURIFIED PEPTIDE

(C) RP-HPLC (volatile solvent system) → HPGC (for mol. wt. estimation)
↘ FURTHER PURIFICATION

FIGURE 5. The application of a three-step HPLC technique.

REFERENCES

1. **Hearn, M. T. W. and Hancock, W. S.,** Ion pair partition reversed phase HPLC, *Trends Biochem. Sci.*, 4, N58, 1979.
2. **Rubinstein, M.,** Preparative high-performance liquid chromatography of proteins, *Anal. Biochem.*, 98, 1, 1979.
3. **Regnier, F. E. and Gooding, K. M.,** High-performance liquid chromatography of proteins, *Anal. Biochem.*, 103, 1, 1980.

4. **Mabuchi, H. and Nakahashi, H.,** Systematic separation of medium-sized biologically active peptides by high-performance liquid chromatography, *J. Chromatogr.*, 213, 275, 1981.
5. **Fukano, K., Komiya, K., Sasaki, H., and Hashimoto, T.,** Evaluation of new supports for high-pressure aqueous gel permeation chromatography: TSK-GEL SW type columns, *J. Chromatogr.*, 166, 47, 1978.
6. **Rokushika, S., Ohkawa, T., and Hatano, H.,** High-speed aqueous gel permeation chromatography of proteins, *J. Chromatogr.*, 176, 456, 1979.
7. **Ui, N.,** Rapid estimation of the molecular weights of protein polypeptide chains using high-pressure liquid chromatography in 6 M guanidine hydrochloride, *Anal. Biochem.*, 97, 65, 1979.
8. **Kato, Y., Komiya, K., Sasaki, H., and Hashimoto, T.,** Separation range and separation efficiency in high-speed gel filtration on TSK-GEL SW type columns, *J. Chromatogr.*, 190, 297, 1980.
9. **Kato, Y., Komiya, K., Sawada, Y., Sasaki, H., and Hashimoto, T.,** Purification of enzymes by high-speed gel filtration on TSK-GEL SW columns, *J. Chromatogr.*, 190, 305, 1980.
10. **Kato, Y., Komiya, K., Sasaki, H., and Hashimoto, T.,** High-speed gel filtration of proteins in sodium dodecyl sulfate aqueous solution on TSK-GEL SW type, *J. Chromatogr.*, 193, 29, 1980.
11. **Kato, Y., Komiya, K., Sasaki, H., and Hashimoto, T.,** High-speed gel filtration of proteins in 6 *M* guanidine hydrochloride on TSK-GEL SW columns, *J. Chromatogr.*, 193, 458, 1980.
12. **Calam, D. H. and Davidson, J.,** Analysis of glycoprotein hormones and other medically important proteins by high-performance gel filtration chromatography, *J. Chromatogr.*, 218, 581, 1981.
13. **Takagi, T.,** High-performance liquid chromatography of protein polypeptides on porous silica gel columns (TSK-GEL SW) in the presence of sodium dodecyl sulfate: comparison with SDS-polyacrylamide gel electrophoresis, *J. Chromatogr.*, 219, 123, 1981.
14. **Imamura, T., Konishi, K., Yodoyama, M., and Konishi, K.,** High-speed gel filtration of polypeptides in some denaturants, *J. Liq. Chromatogr.*, 4, 613, 1981.
15. **Himmel, M. E. and Squire, P. G.,** High pressure gel permeation chromatography of native proteins on TSK-SW columns, *Int. J. Peptide Protein Res.*, 17, 365, 1981.
16. **Mabuchi, J. and Nadahashi, H.,** Analysis of middle molecular peptide in normal and uremic body fluids by high-performance gel chromatography, *J. Chromatogr.*, 224, 322, 1981.
17. **Rubinstein, M., Stein, S., Gerber, D., and Udenfriend, S.,** Isolation and characterization of the opioid peptides from rat pituitary: β-lipotropin, *Proc. Natl. Acad. Sci. U.S.A.*, 74, 3052, 1977.
18. **Hancock, W. S., Bishop, C. A., Prestidge, R. L., Harding, D. R. K., and Hearn, M. T. W.,** Reversed-phase high-pressure liquid chromatography of peptides and proteins with ion-pairing reagents, *Science*, 200, 1168, 1978.
19. **Hancock, W. S., Bishop, C. A., Meyer, L. J., and Harding, D. R. K.,** High-pressure liquid chromatography of peptides and proteins. VI. Rapid analysis of peptides by high-pressure liquid chromatography with hydrophobic ion-pairing of amino groups, *J. Chromatogr.*, 161, 291, 1978.
20. **Hancock, W. S., Bishop, C. A., Battersby, J. E., and Harding, D. R. K.,** High-pressure liquid chromatography of peptides and proteins. XI. The use of cationic reagents for the analysis of peptides by high-pressure liquid chromatography, *J. Chromatogr.*, 168, 377, 1979.
21. **O'Hare, M. J. and Nice, E. C.,** Hydrophobic high-performance liquid chromatography of hormonal polypeptides and proteins on alkylsilane-bonded silica, *J. Chromatogr.*, 209, 1979.
22. **Starratt, A. N. and Stevens, M. E.,** Ion-pair high-performance liquid chromatography of the insect neuropeptide proctolin and some analogs, *J. Chromatogr.*, 194, 421, 1980.
23. **Bennett, J. P. J., Browne, C. A., and Solomon, S.,** The use of perfluorinated carboxylic acids in the reversed-phase HPLC of peptides, *J. Liq. Chromatogr.*, 3, 1353, 1980.
24. **Mahoney, W. C. and Hermodson, M. A.,** Separation of large denatured peptides by reverse phase high performance liquid chromatography, *J. Biol. Chem.*, 255, 11199, 1980.
25. **Meek, J. L.,** Prediction of peptide retention times in high-pressure liquid chromatography on the basis of amino acid composition, *Proc. Natl. Acad. Sci. U.S.A.*, 77, 1632, 1980.
26. **Harding, D. R. K., Bishop, C. A., Tarttelin, M. F., and Hancock, W. S.,** Use of perfluoroalkanoic acids as volatile ion pairing reagents in preparative HPLC, *Int. J. Peptide Protein Res.*, 18, 214, 1981.
27. **Morris, J. R., Etienne, A. T., Dell, A., and Alburquerque, R.,** A rapid and specific method for the high resolution purification and characterization of neuropeptides, *J. Neurochem.*, 34, 574, 1980.
28. **Wilson, K. J., Honegger, A., and Hughes, G. J.,** Comparison of buffers and detection systems for high-pressure liquid chromatography of peptide mixtures, *Biochem. J.*, 199, 43, 1981.
29. **Lenard, J.,** Rapid and effective removal of sodium dodecyl sulphate from proteins, *Biochem. Biophys. Res. Commun.*, 45, 662, 1971.
30. **Griffith, I. P.,** The use of Sephadex LH-20 to separate dodecyl sulphate and buffer salts from denatured proteins, *J. Chromatogr.*, 109, 399, 1975.
31. **Kapp, O. H. and Vinogradov, S. N.,** Removal of sodium dodecyl sulphate from proteins, *Anal. Biochim.*, 91, 230, 1978.
32. **Amons, R. and Schrier, P. I.,** Removal of sodium dodecyl sulphate from proteins and peptides by gel filtration, *Anal. Biochem.*, 116, 439, 1981.

33. **Radhakrishnan, A. N., Stein, S., Licht, A., Gruber, K. A., and Udenfriend, S.,** High-efficiency cation exchange chromatography of polypeptides and polyamines in the nanomole level, *J. Chromatogr.*, 132, 552, 1977.
34. **Bennett, J. P. J., Hudson, A. M., McMartin, C., and Purdon, G. E.,** Use of octadecasilyl-silica for the extraction and purification of peptides in biological samples. Application to the identification of circulation metabolites of corticotropin-(1-24)-tetracosapeptide and somatostatin in vivo, *Biochem. J.*, 168, 9, 1977.
35. **Bennett, J. P. J., Hudson, A. M., Kelly, L., McMartin, C., and Purdon, G. E.,** A rapid method, using octadecasilyl-silica, for the extraction of certain peptides from tissues, *Biochem. J.*, 175, 1139, 1978.
36. **Böhlen, P., Castillo, F., Ling, N., and Guillemin, R.,** Purification of peptides: an efficient procedure for the separation of peptides from amino acids and salt, *Int. J. Peptide Protein Res.*, 16, 306, 1980.
37. **LaRochelle, F. T., North, J. W. G., and Stein, P.,** A new extraction of arginine vasopressin from blood: the use of octadecasilyl-silica, *Pflügers Arch.*, 387, 79, 1980.
38. **Koehn, J. A. and Canfield, R. E.,** Purification of human fibrino-peptides by high-performance liquid chromatography, *Anal. Biochem.*, 116, 349, 1981.
39. **Gay, D. D. and Lahti, R. A.,** Rapid separation of enkephalins and endorphins on Sep-Pak reverse phase cartridges, *Int. J. Peptide Protein Res.*, 18, 107, 1981.
40. **Mabuchi, J. and Nakahashi, H.,** Analysis of small peptides in uremic serum by high performance liquid chromatography, *J. Chromatogr.*, 228, 292, 1982.
41. **Mabuchi, H. and Nakahashi, H.,** unpublished data.

SEPARATION OF BASIC HYDROPHILIC PEPTIDES BY REVERSED-PHASE HPLC

Ulf Ragnarsson, Bengt Fransson, and Örjan Zetterqvist

INTRODUCTION

The message of the present report is that basic amino acids present in peptides need not necessarily give rise to complications on analysis by reversed-phase HPLC (RP-HPLC). Numerous references could be given to illustrate this, but special sections of this volume on substance P, neurotensin (NT), somatostatin, LH-RH, and other biologically important peptides can serve the same purpose. In all peptides mentioned the content of arginine and/or lysine is matched by the presence of a number of hydrophobic amino acid residues.

The fundamental mechanism governing the retention of peptides on alkylsilane-bonded reversed-phase supports is a hydrophobic interaction. In the peptides mentioned above the influence of the basic residues on peptide retention is counter-balanced by that of the hydrophobic ones. Rekker estimated the relative lipophilicity constants[1] for all amino acids present in proteins (except arginine), and it was found that the retention order of smaller peptides on a reversed-phase column generally correlated with the sum of the constants for strongly hydrophobic residues.[2] Meek derived empirical retention coefficients for amino acid residues and pioneered prediction of retention times for peptides.[3] In this context a set of modified Rekker constants was proposed, which differed most from the original values for the most hydrophilic amino acid residues.[4]

Since basic amino acids are also hydrophilic,[1,4] very poor retention to reversed-phase columns is to be expected for peptides which, in addition to basic residues, contain mainly hydrophilic amino acids.[3,4] However, in such cases, the presence of basic groups can be exploited for ion-pair chromatography, a technique pioneered for drug analysis by Schill and co-workers at Uppsala,[5-7] reviewed by Tomlinson in this volume and, in more detail, elsewhere.[8] Below we will first briefly review applications of this technique to the analysis of basic peptides in general and then proceed to discuss our own results, in which major emphasis will be laid on peptides with essentially only hydrophilic residues, of which at least two are basic.

SEPARATION OF BASIC PEPTIDES BY ION-PAIR CHROMATOGRAPHY — SHORT SELECTIVE LITERATURE REVIEW

Angiotensin II can be classified as a basic, hydrophobic peptide. This compound and some of its analogs were among the first basic peptides to be separated by RP-HPLC under conditions of ion-pair formation.[9] In this case a *hydrophilic* anion, dihydrogen phosphate, was used. In contrast, Hancock et al.[10] were the first to demonstrate that *hydrophobic* ion pairing of unprotected peptide amino groups is a useful technique for the analysis and isolation of peptides by HPLC. They demonstrated that the association of a peptide with a hydrophobic anion in the eluent leads to a less polar complex which has an increased retention time on a reversed-phase column. Thus, depending on the composition and sequence of the peptide, selectivity can be achieved by a proper choice of negatively charged counter ion.[11]

Somatostatin, another basic, fairly hydrophobic peptide, containing 4 aromatic amino acid residues out of 14, has been chromatographed both with hydrophilic[12] and a variety of hydrophobic counter ions.[13,14] Particularly high resolution and recovery were claimed in the first system. In Abrahamsson's and Gröningsson's paper,[13] the influence of different parameters, such as the nature of the support, temperature, organic modifier, pH, buffer concentration, and ion-pairing agent, was carefully investigated, Bennett et al.[14] applied trifluoroacetic acid (TFA), PFPA, HFBA, and UFCA as ion-pairing agents. Another basic peptide, LH-RH, was studied together with somatostatin[12,14] and in connection with adrenocorticotropic hormone (ACTH).[15]

Substance P, which contains one arginine, one lysine, no acidic amino acid, one leucine, and two phenylalanine residues out of 11 amino acids, is a basic, reasonably hydrophobic peptide. Optimal conditions for its separation were developed recently using a C_{18}-column.[16] This peptide requires a short-chain ion-pairing reagent and a high percentage of acetonitrile.

Vasopressin and vasopressin analogs have been analyzed as ion pairs by different authors,[16-18] using both hydrophobic and hydrophilic counter ions. Different C_{18} columns were used. After a careful study of 8 synthetic analogs at pH 2.5 to 6.5, Lindeberg[17] came to the conclusion that methanol and ethanol were more satisfactory than acetonitrile as organic modifier. Schöneshöfer et al.[18] also studied NT among other basic peptides.

ACTH and its N-terminal fragments contain several arginine and lysine residues and have been studied quite extensively.[14,15,18,19] Among ion-pair forming agents applied are TFA,[14,18] other perfluorinated carboxylic acids,[14] phosphate,[19] and butanesulfonate.[15] In all experiments C_{18}-columns were utilized and sharp peaks with moderate retention times were obtained. Nevertheless, for peptides of the size of ACTH, a less hydrophobic support would be worth trying. Glucagon was chromatographed with hydrophilic ion-pairing agents together with ACTH.[18,19]

As indicated above, this literature review is intended to be rather selective in nature. Nevertheless it should be evident that ion-pair techniques have been applied already quite extensively for the analysis of basic peptides by RP-HPLC. This holds true especially of such peptides of medical and pharmaceutical interest.[20] Not only small peptides but also proteins[21] can now be analyzed and studied similarly. Proteins, however, will be treated separately in this volume.

SEPARATION OF BASIC, HYDROPHILIC PEPTIDES BY ION-PAIR CHROMATOGRAPHY — OWN WORK

To study the substrate specificity of cyclic AMP-dependent protein kinase, based on the amino acid sequences of the phosphorylatable sites of rat liver pyruvate kinase[22] and the β-subunit of rabbit muscle phosphorylase kinase,[23] we have synthesized a large number of basic serine peptides,[24,25] most of which contain two arginine residues. As the natural sequences only contain one hydrophobic amino acid in the essential regions, RP-HPLC at first did not appear very promising in our attempts to establish the purity of our synthetic products. However, after preliminary experiments using hexane sulfonate for ion-pair formation had indicated that the retention of our peptides on a C_{18} column could indeed be varied within wide limits and that simultaneously the selectivity was quite good, we decided to explore reversed-phase ion-pair HPLC for the purpose mentioned. This section will open with a thorough discussion about the experiences and results gained so far from our work with peptides derived from rat liver pyruvate kinase. Part of this work has already been published.[26,27]

In connection with the protein kinase experiments outlined above, we became confronted with the problem of establishing the purity of phosphoserine peptides and we decided to apply the same technique also to these peptides. Actually in all cases investigated so far, separation of the phosphoserine peptide from the corresponding nonphosphorylated precursor could be easily accomplished.[27,28] This work will also be reviewed below.

Other applications of reversed-phase ion-pair HPLC to basic, hydrophilic peptides, now in progress in our laboratory, will also be reviewed briefly below. Thus, our first experiments aiming at the separation of diastereoisomeric peptides will be mentioned. We have also applied this technique to a series of vasopressin analogs.

Table 1
PEPTIDE STRUCTURES

M 57	Leu-Arg-Arg-Ala-Ser-Val-Ala
M 66	Arg-Arg-Ala-Ser-Val-Ala
M 67	Arg-Ala-Ser-Val-Ala
M 69	Arg-Arg-Ala-Ser-Val
M 70	Arg-Arg-Ala-Ser
M 72	Leu-Arg-Ala-Ser-Val
M 73	Arg-Leu-Ala-Ser-Val
M 87	Val-Leu-Arg-Arg-Ala-Ser-Val-Ala
M 97	Gly-Val-Leu-Arg-Arg-Ala-Ser-Val-Ala
M 142	His-Arg-Ala-Ser-Val
M 144	Arg-Arg-Ala-Ser-Arg
M 146	Arg-Arg-Ala-D-Ser-Val
M 157	Gva-Arg-Ala-Ser-Val
M 158	Leu-Arg-Arg-Ala-Ser-Val-Arg
M 201	Arg-Arg-D-Ala-Ser-Val
M 308	Arg-Arg-Ala

Note: Gva, δ-Guanidinovaleric acid.

Materials and Methods

Chemicals

For all reversed-phase separations, referred to below, isocratic mobile phases were used, consisting of binary aqueous solvents with acetonitrile, 1-propanol, 2-propanol, or, preferentially, ethanol as organic modifier. Our buffers were prepared from orthophosphoric acid and sodium dihydrogen orthophosphate to an ionic strength of 0.1 *M*. Pentane or hexane sulfonate was added as counter ion to the mobile phases, which were prepared from double glass distilled deionized water, and filtered (0.45 μm, Millipore) prior to use. 1-Pentane and 1-hexane sulfonic acids were obtained as their sodium salts from Eastman-Kodak Co. (Rochester, NY). All solvents and buffer substances were of HPLC, analytical or reagent grade.

Peptides

The structure of the basic serine peptides used as model substances in a previous section, are summarized in Table 1. They were all prepared in this laboratory by the solid-phase method of Merrifield[29,30] as briefly described.[24] Most of them were purified by ion-exchange chromatography on carboxymethyl cellulose and their correct composition verified by amino acid analysis after acid hydrolysis.[24] Their high chromatographical purity has been confirmed in earlier papers[26,27] as well as in the present work. The phosphoserine peptides, studied in a previous section, were obtained by phosphorylation of the corresponding nonphosphorylated, synthetic peptides with ATP and the catalytic subunit of cyclic AMP-dependent protein kinase in the presence of 5 m*M* $MgCl_2$, and isolation by ion-exchange chromatography on carboxymethyl cellulose, as described before.[28,31] Their structures are summarized in Table 2. The vasopressins were a gift from Ferring Läkemedel (Malmö, Sweden). Their structures finally are given in Table 3.

Instrumentation

The liquid chromatographic system consisted of a Model 6000A solvent delivery device, a U6K injector, and a variable wavelength detector (Waters Associates, Inc., Milford, MA). The detection wavelength was in all cases 210 nm, which made possible the detection of peptide quantities down to approximately 20 pmol. The peptides were dissolved in phosphate buffer or mobile phase. Each peptide was sequentially introduced into the injector loop.

Table 2
PHOSPHOPEPTIDE STRUCTURES

P-M 57	Leu-Arg-Arg-Ala-Ser(P)-Val-Ala
P-M 66	Arg-Arg-Ala-Ser(P)-Val-Ala
P-M 87	Val-Leu-Arg-Arg-Ala-Ser(P)-Val-Ala
P-M 97	Gly-Val-Leu-Arg-Arg-Ala-Ser(P)-Val-Ala

Table 3
STRUCTURES OF VASOPRESSINS STUDIED

Code	Amino acid sequence and substituents	
AVP	Cys-Tyr-Phe-Gln-Asn-Cys-Pro-	L-Arg-Gly-NH_2
DAVP		D-Arg
dAVP	Mpa	L-Arg
dDAVP	Mpa	D-Arg
LVP		L-Lys
dLVP	Mpa	L-Lys

Note: Mpa, β-mercaptopropionic acid.

Columns

The separation columns, 150 × 4.6 mm or 250 × 4.6 mm, respectively, were packed by the balanced density slurry technique[32] with Spherisorb, C_{18}, 5 and 10 μm, or C_6, 5 μm, as chromatographic supports for RP-HPLC. The Spherisorb supports were obtained from Phase Separation Ltd. (Queensferry, Clwyd, Great Britain). Separation columns were always preceded by a short guard column, dry packed with μBondapak®, C_{18}/Corasil, 37 to 50 μm (Waters), as support.

The separations were run with flow rates in the range of 0.5 to 1.5 mℓ/min at room temperature (approximately 22°C). In most cases, however, it was 1.0 mℓ/min.

Applications

Peptides Related to the Phosphorylatable Site of Rat Liver Pyruvate Kinase

Most of our results from separation of basic peptides derived from pyruvate kinase have already been reported in detail.[26] Thus, it was demonstrated using three penta- to heptapeptides with one to three arginine residues (M 67, M 66, and M 158) that the retention increases with the number of arginines. This experiment, like all previous ones, was performed using a C_{18} column with 10-μm particles and with hexane sulfonate as ion-pairing reagent. It was also shown that mixtures of three pentapeptides with one arginine (M 67, M 72, and M 73), four hexa- to nonapeptides with two arginines (M 66, M 57, M 87, and M 97) and one pentapeptide and one heptapeptide with three arginines (M 144 and M 158) could be resolved without difficulties. More recently we were able to separate the whole series of peptides with two arginines from the tripeptide M 308 up to the nonapeptide M 97.[27] This last separation was performed at pH 3.1.

At present, phosphate buffer with a pH around 3 and pentane sulfonate are preferred for routine analysis. Figures 1 and 2 illustrate the separation of the above series of peptides containing two arginine residues each (M 308 to M 97) on a 5-μm C_{18}-column. As can be seen the two peaks, corresponding to M 66 and M 69, in this case appear in reversed order compared with earlier findings.[27]

Most of our work so far on separation of basic peptides has been performed on arginine peptides.[26,27] Examples containing lysine peptides are found in a following section. An

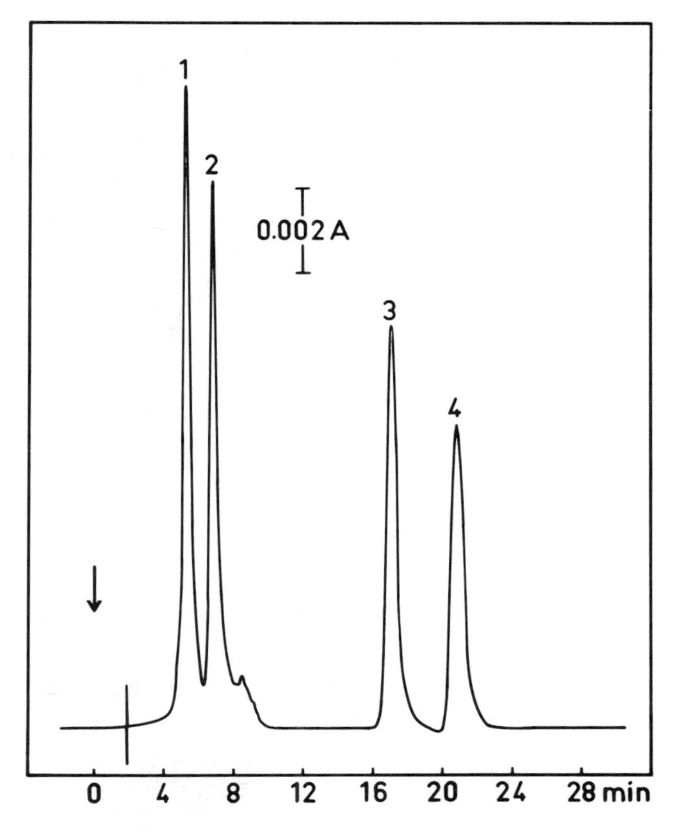

FIGURE 1. Separation of four related basic, hydrophilic peptides containing two arginine residues. Mobile phase: phosphate buffer (pH, 2.9) — ethanol (86:14), flow rate 1.0 mℓ/min. Counter ion: pentane sulfonate (0.015 *M*). Support: Spherisorb C_{18} (5-μm, 150-mm column). Samples: Peak 1 refers to M 70, 2 to M 308, 3 to M 66, and 4 to M 69.

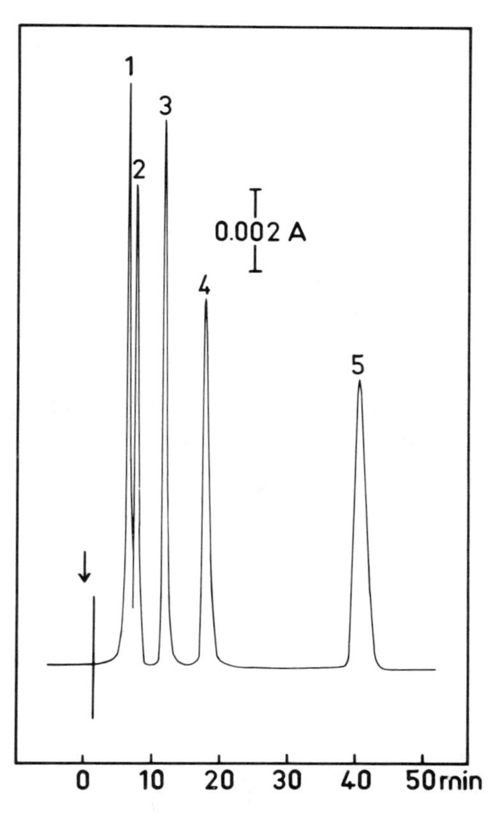

FIGURE 2. Separation of five related basic, hydrophilic peptides containing two arginine residues. Mobile phase: phosphate buffer (pH, 2.9) — ethanol (79:21). Flow rate, counter ion, and support as in Figure 1. Samples: Peak 1 refers to M 66, 2 to M 69, 3 to M 57, 4 to M 87, and 5 to M 97.

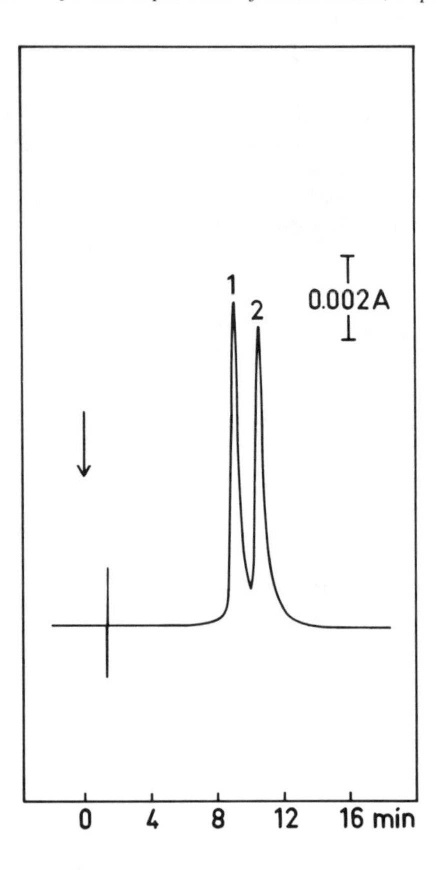

FIGURE 3. Separation of two related basic, hydrophilic pentapeptides one of which contains a histidine residue. Mobile phase, flow rate, counter ion, and support as in Figure 1. Samples: Peak 1 refers to M 142 and 2 to M 69.

example of a histidine peptide is shown in Figure 3. This peptide (M 142) behaves essentially like the corresponding arginine peptide (M 69), but the two peptides can again be separated.

Basic Phosphoserine Peptides

Basic phosphoserine peptides, obtained from synthetic precursors by phosphorylation with ATP and cyclic AMP-dependent protein kinase, have also been studied by this technique. Figures 4 and 5 illustrate the separation of a phosphoserine peptide (P-M 57) from the corresponding nonphosphorylated peptide (M 57) on two different reversed-phase columns under similar conditions.[27,28] Isocratic systems with ethanol as organic modifier were used and in both cases the counter ion was hexane sulfonate. A pH around 3 appears to be important for high selectivity.[28] Under such circumstances also mixtures of four phosphoserine peptides composed of six to nine residues (Table 2) were separated,[28] as demonstrated in Figure 6.

Separations of phosphoserine peptides from the corresponding nonphosphorylated peptides do not seem to be very difficult, although, admittedly, our experience so far is rather limited.[28] The difference in electric charge in such cases at the pH used no doubt has a pronounced effect on ion pairing with hexane sulfonate. The serine phosphate group can, at least in principle, also participate in intramolecular ion pairing.

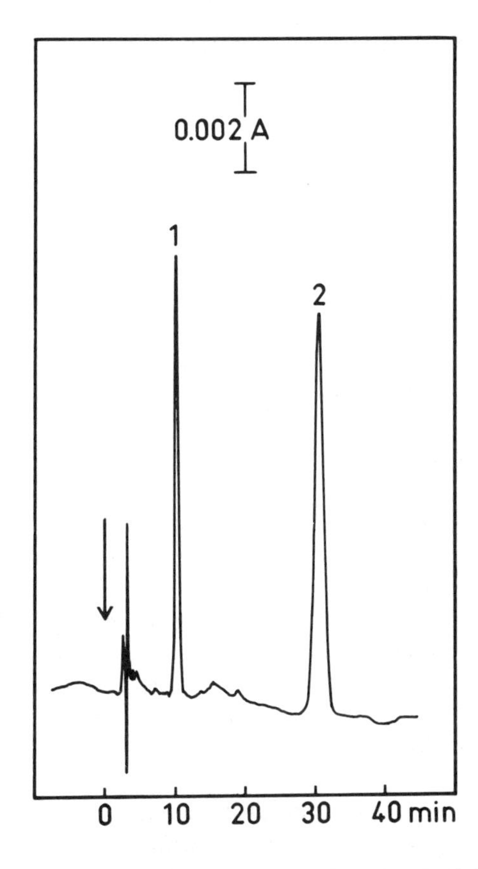

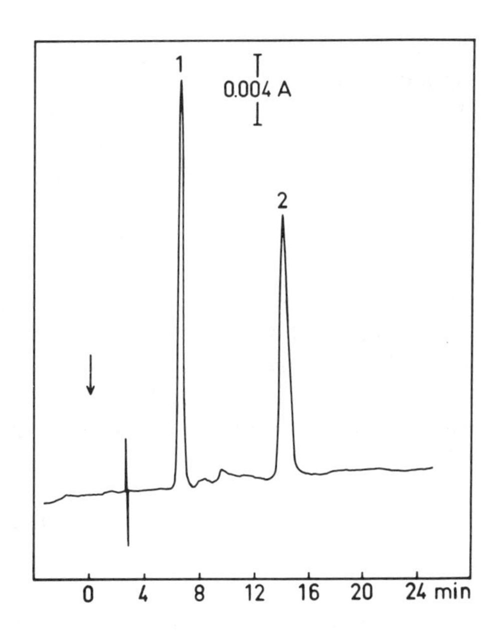

FIGURE 4 and 5. Separation of a basic heptapeptide containing serine and the corresponding phosphoserine peptide. (4) Mobile phase: phosphate buffer (pH, 3.2) — ethanol (75.5:24.5), flow rate 1.0 mℓ/min. Counter ion: hexane sulfonate (0.015 *M*). Support: Spherisorb C_{18} (10-μm, 250-mm column). Samples: Peak 1 refers to P-M 57 and peak 2 to M 57. (5) Mobile phase: phosphate buffer (pH, 3.0) — ethanol (72:28). Flow rate, counter ion, and samples as in Figure 4. Support: Spherisorb C_6 (5-μm, 250-mm column).

This application is expected to play a role in the future in studies of peptide and protein phosphorylation and dephosphorylation. It is estimated that about 100 pmol of basic phosphoserine peptide can at present be detected using a UV detector. The sensitivity is of course lower in this case than when [^{32}P]-labeled phosphate of high specific radioactivity is applied.

Vasopressin Analogs

As further examples of separations of basic peptides by RP-HPLC in conjunction with hydrophobic ion pairing, the results of some experiments with vasopressin analogs are shown below. Although one tyrosine and one phenylalanine residue are present in vasopressin, this peptide can nevertheless be considered fairly hydrophilic.

Figure 7 shows a chromatogram of a mixture of arginine vasopressin (AVP) and lysine vasopressin (LVP), thus demonstrating the selectivity of our system on a peptide pair containing two different basic amino acids. LVP eluted in front of AVP under the conditions used.

To slow down enzymatic degradation of various hormones in vivo, synthetic analogs without terminal amino groups are sometimes prepared. Among the several hundreds of vasopressin analogs prepared, several important substances are the so-called deamino compounds. Figure 8 illustrates the separation of AVP from the corresponding deamino analog (dAVP). As observed earlier for M 157 and M 69 (Figure 6),[26] due to the difference in

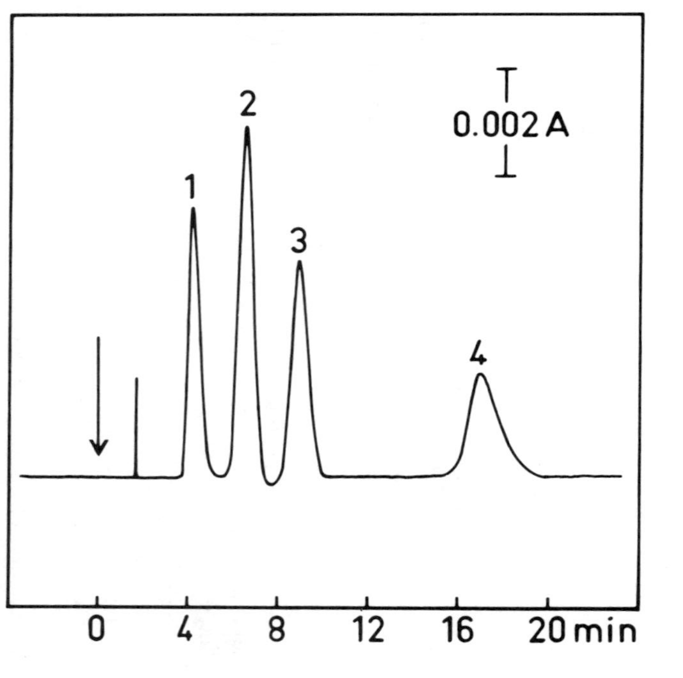

FIGURE 6. Separation of four phosphopeptides composed of six to nine amino acids. Mobile phase: phosphate buffer (pH, 3.2) — ethanol (78:22), flow rate 1.5 mℓ/min. Counter ion: hexane sulfonate (0.015 *M*). Support: Spherisorb C_{18} (10-μm, 250-mm column). Samples: Peak 1 refers to P-M 66, 2 to P-M 57, 3 to P-M 87, and peak 4 to P-M 97.

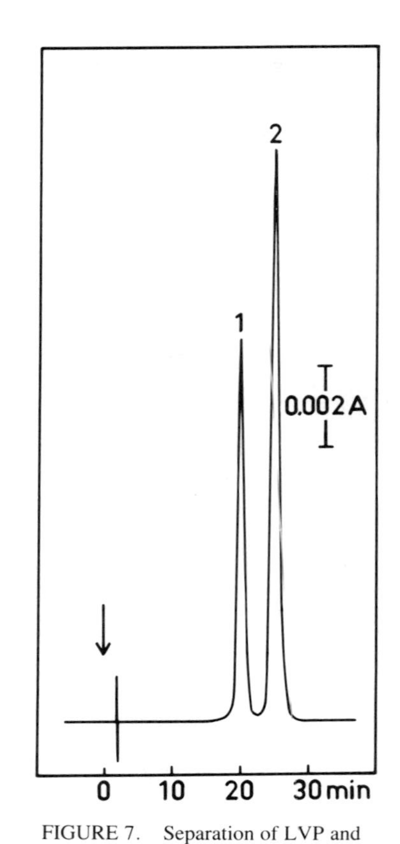

FIGURE 7. Separation of LVP and AVP. Mobile phase: phosphate buffer (pH, 2.9) — ethanol (79:21), flow rate 1.0 mℓ/min. Counter ion: pentane sulfonate (0.015 *M*). Support: Spherisorb C_{18} (5-μm, 150-mm column). Samples: Peak 1 refers to LVP and 2 to AVP.

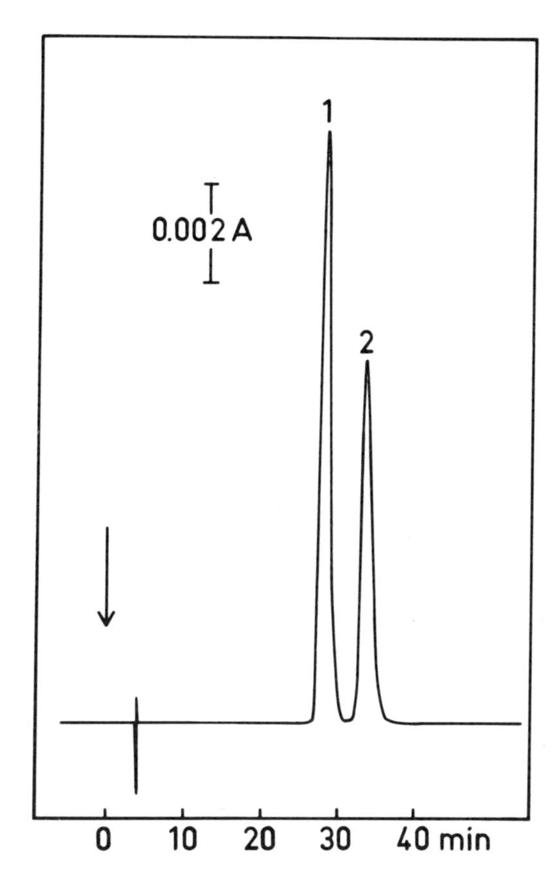

FIGURE 8. Separation of AVP from its corresponding deamino analog dAVP. Conditions as in Figure 7. Samples: Peak 1 refers to AVP and 2 to dAVP.

charge at the pH of the phosphate buffer, separation is to be expected in an ion-pair system like the one used. This is further demonstrated in Figure 9 by the separation of two lysine peptides, LVP and its corresponding analog (dLVP), from each other. In this case, too, the separation is quite satisfactory.

Two further separations from the vasopressin field are shown in the next section.

Diastereoisomeric Peptides

The scope of ion-pair HPLC in the analysis of basic, hydrophilic peptides has been illustrated above. Particularly for peptides differing in electric charge this technique gives excellent results. As the last application of ion-pair HPLC in this context, a few examples from separations of diastereoisomeric peptides will be given.

Figure 10 shows that M 69 and the diastereoisomeric analog with D-serine (M 146) can be separated in less than 10 min under our standard conditions. Partial resolution has so far been accomplished for the corresponding mixture of M 69 and M 201 with alanine of opposite chirality (Figure 11).

Similarly a few pairs of vasopressin diastereoisomers have been investigated. As demonstrated in Figure 12, a mixture of AVP and DAVP can be resolved. The dAVP/dDAVP pair behaves in the same way (Figure 13).

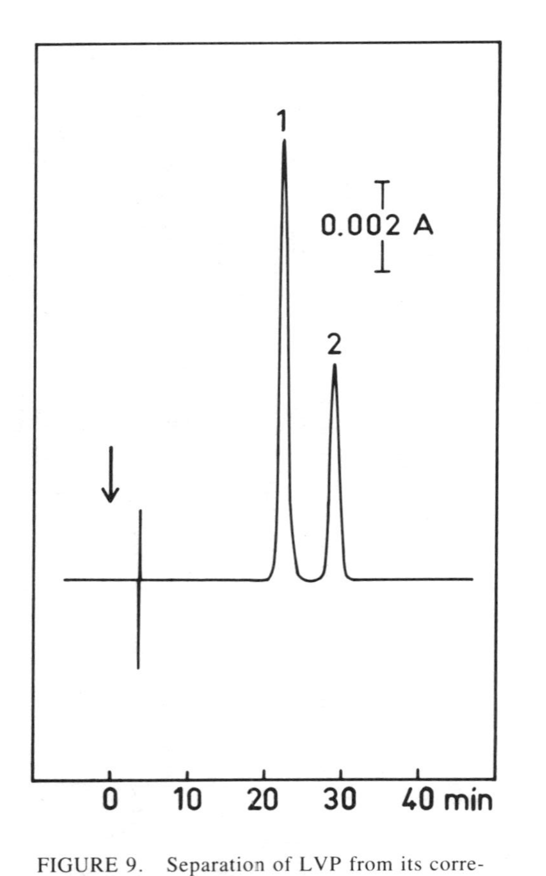

FIGURE 9. Separation of LVP from its corresponding deamino analog dLVP. Conditions as in Figure 7. Samples: Peak 1 refers to LVP and 2 to dLVP.

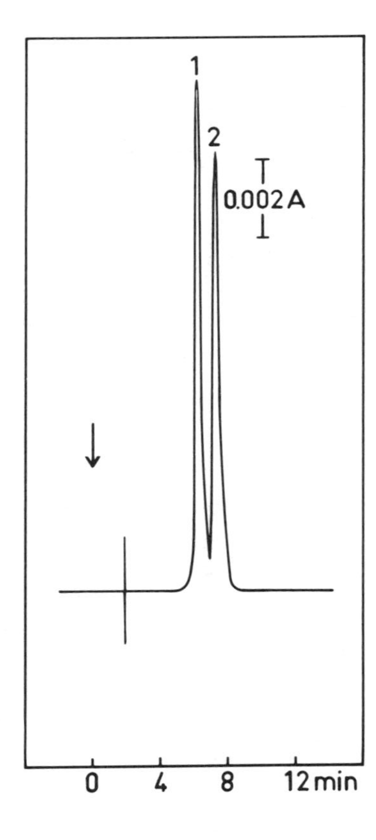

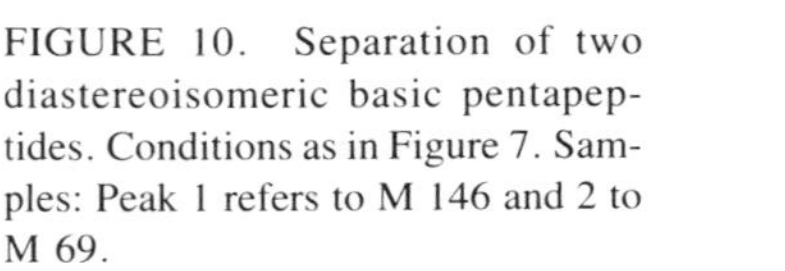

FIGURE 10. Separation of two diastereoisomeric basic pentapeptides. Conditions as in Figure 7. Samples: Peak 1 refers to M 146 and 2 to M 69.

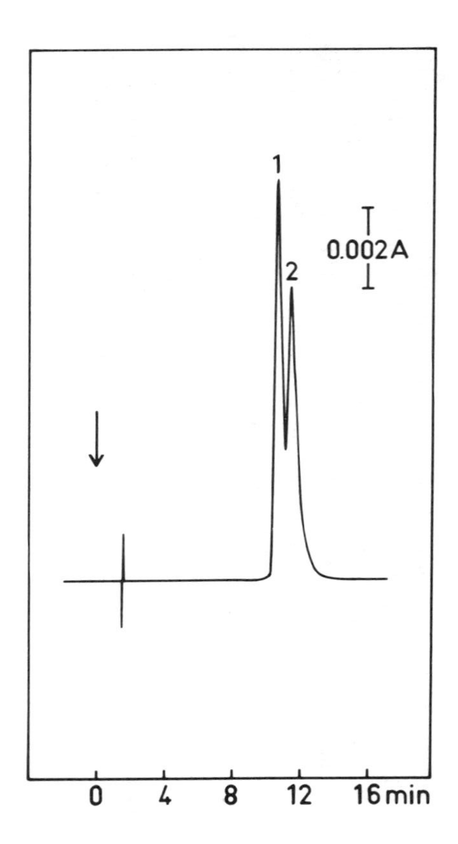

FIGURE 11. Partial separation of two diastereoisomeric basic pentapeptides. Mobile phase: phosphate buffer (pH, 2.9) — ethanol (86:14). Other conditions as in Figure 7. Samples: Peak 1 refers to M 69 and 2 to M 201.

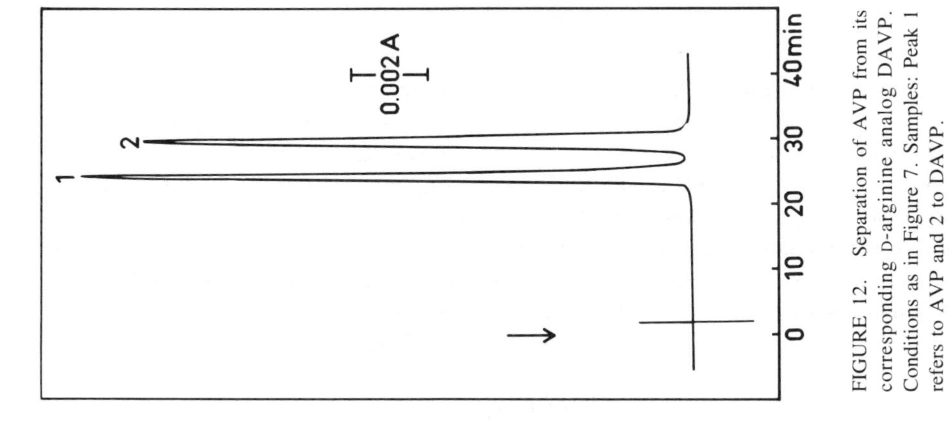

FIGURE 12. Separation of AVP from its corresponding D-arginine analog DAVP. Conditions as in Figure 7. Samples: Peak 1 refers to AVP and 2 to DAVP.

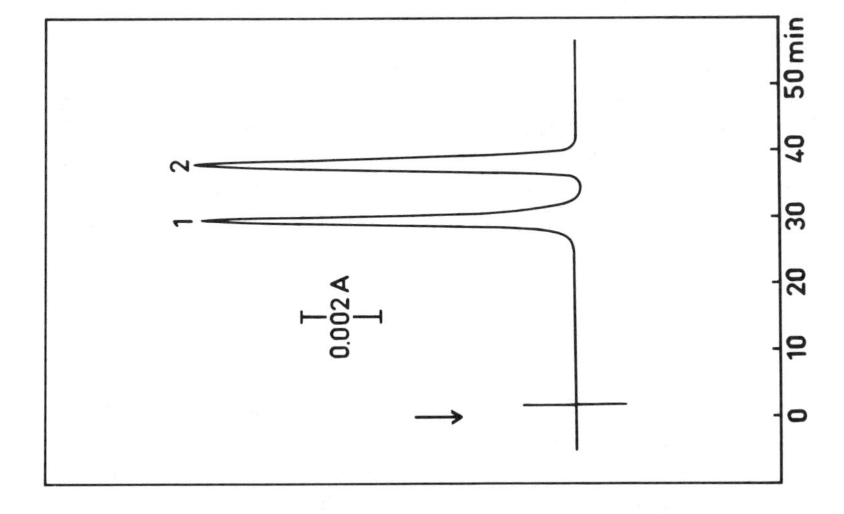

FIGURE 13. Separation of dAVP and dDAVP. Conditions as in Figure 7. Samples: Peak 1 refers to dAVP and 2 to dDAVP.

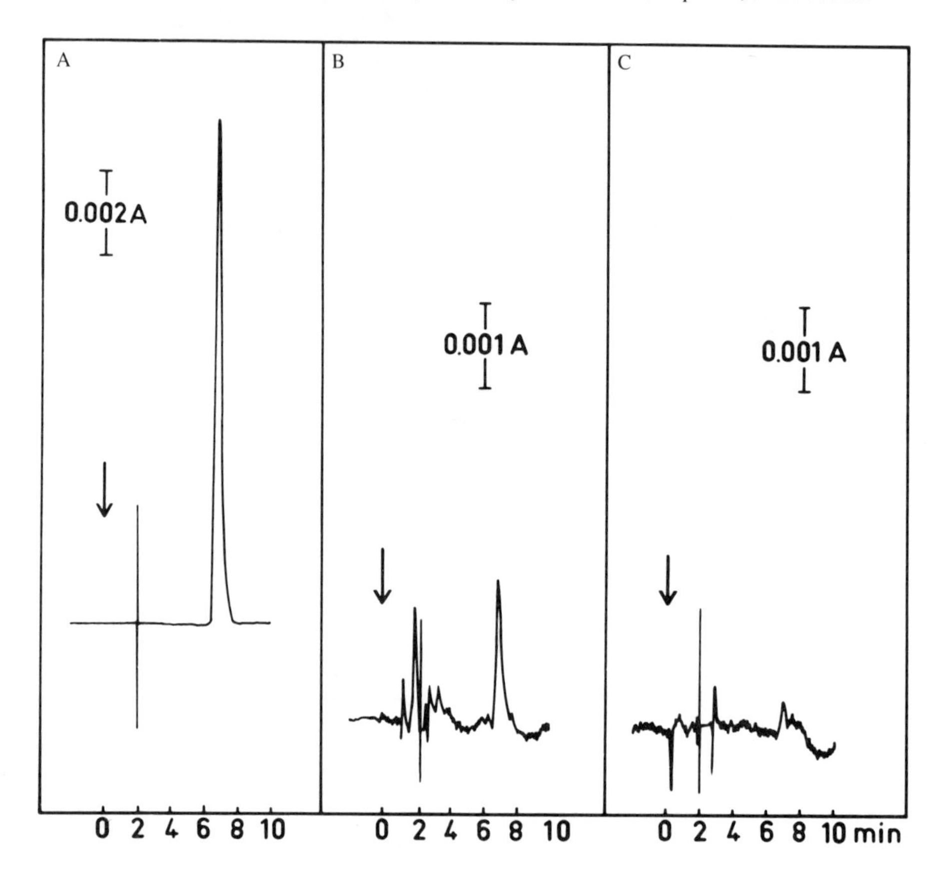

FIGURE 14. Sensitivity of detection. Determination of a pentapeptide containing two arginine residues in the low concentration range. Conditions as in Figure 7. Sample: M 69. Approximate amounts injected: (A) 0.76 nmol, (B) 0.076 nmol, and (C) 19 pmol.

FINAL COMMENTS

As demonstrated above, ion-pair HPLC on reversed-phase columns can be very useful for the separation of basic, hydrophilic peptides. Although some of our peptides are nearly of maximum hydrophilicity (cf. References 1 and 4), no problems whatsoever have been encountered with respect to low retention on the reversed-phase supports. Nevertheless, if such a case would occur for a basic peptide in the future, a sulfonic acid with a larger carbon chain than the ones used so far (pentane and hexane) could probably be applied.

Due to the high UV absorption of peptides at 210 nm, the sensitivity is quite high. For routine analysis we normally apply 1 to 5 nmol of each peptide,[26,28] but considerably smaller amounts of peptide can be detected. Figure 14 demonstrates that, under favorable conditions, peptide quantities down to 10 to 20 pmol can be detected. Amounts in the range of 0.1 to 1 nmol appear to give peaks of a size required for quantitative determinations.

ACKNOWLEDGMENTS

Our own work reported above was made possible by Research Grants 78-7038 from the National Swedish Board for Technical Development, K-KU 3020-108, 109, and 110 from the Swedish Natural Science Research Council, and 13X-4485 from the Swedish Medical Research Council.

REFERENCES

1. **Rekker, R. F.,** *The Hydrophobic Fragmental Constant,* Elsevier, Amsterdam, 1977, 301.
2. **O'Hare, M. J. and Nice, E. C.,** Hydrophobic high-performance liquid chromatography of hormonal polypeptides and proteins on alkylsilane-bonded silica, *J. Chromatogr.,* 171, 209, 1979.
3. **Meek, J. L.,** Prediction of peptide retention times in high-pressure liquid chromatography on the basis of amino acid composition, *Proc. Natl. Acad. Sci. U.S.A.,* 77, 1632, 1980.
4. **Sasagawa, T., Okuyama, T., and Teller, D. C.,** Prediction of peptide retention times in reversed-phase high-performance liquid chromatography during linear gradient elution, *J. Chromatogr.,* 240, 329, 1982.
5. **Wahlund, K.-G. and Gröningsson, K.,** Reversed-phase column chromatography of organic ammonium compounds as ion-pairs, *Acta Pharm. Suecica,* 7, 615, 1970.
6. **Schill, G.,** Uppsala University 500 years, 4, Faculty of Pharmacy at Uppsala University, *Acta Univ. Ups.,* Uppsala 1976, p. 58.
7. **Crommen, J., Fransson, B., and Schill, G.,** Ion-pair chromatography in the low concentration range by use of highly absorbing counter ions, *J. Chromatogr.,* 142, 283, 1977.
8. **Tomlinson, E., Jefferies, T. M., and Riley, C. M.,** Ion-pair high-performance liquid chromatography, *J. Chromatogr.,* 159, 315, 1978.
9. **Molnár, I. and Horváth, C.,** Separation of amino acids and peptides on non-polar stationary phases by high-performance liquid chromatography, *J. Chromatogr.,* 142, 623, 1977.
10. **Hancock, W. S., Bishop, C. A., Meyer, L. J., Harding, D. R. K., and Hearn, M. T. W.,** High-pressure liquid chromatography of peptides and proteins. VI. Rapid analysis of peptides by high-pressure liquid chromatography with hydrophobic ion-pairing of amino groups, *J. Chromatogr.,* 161, 291, 1978.
11. **Hearn, M. T. W. and Hancock, W. S.,** Ion pair partition reversed phase HPLC, *Trends Biochem. Sci.,* 4, N58, 1979.
12. **Rivier, J. E.,** Use of trialkyl ammonium phosphate (TAAP) buffers in reverse phase HPLC for high resolution and high recovery of peptides and proteins, *J. Liq. Chromatogr.,* 1, 343, 1978.
13. **Abrahamsson, M. and Gröningsson, K.,** High-performance liquid chromatography of the tetradecapeptide somatostatin, *J. Liq. Chromatogr.,* 3, 495, 1980.
14. **Bennett, H. P. J., Browne, C. A., and Solomon, S.,** The use of perfluorinated carboxylic acids in the reversed-phase HPLC of peptides, *J. Liq. Chromatogr.,* 3, 1353, 1980.
15. **Terabe, S., Konaka, R., and Inouye, K.,** Separation of some polypeptide hormones by high-performance liquid chromatography, *J. Chromatogr.,* 172, 163, 1979.
16. **Spindel, E., Pettibone, D., Fisher, L., Fernstrom, J., and Wurtman, R.,** Characterization of neuropeptides by reversed-phase, ion-pair liquid chromatography with post-column detection by radioimmunoassay. Application to thyrotropin-releasing hormone, substance P and vasopressin, *J. Chromatogr.,* 222, 381, 1981.
17. **Lindeberg, G.,** Separation of vasopressin analogues by reversed-phase high-performance liquid chromatography, *J. Chromatogr.,* 193, 427, 1980.
18. **Schöneshöfer, M. and Fenner, A.,** Hydrophilic ion-pair reversed-phase chromatography of biogenic peptides prior to immunoassay, *J. Chromatogr.,* 224, 472, 1981.
19. **Hancock, W. S., Bishop, C. A., Prestidge, R. L., Harding, D. R. K., and Hearn, M. T. W.,** Reversed-phase, high-pressure liquid chromatography of peptides and proteins with ion-pairing reagents, *Science,* 200, 1168, 1978.
20. **Krummen, K.,** HPLC in the analysis and separation of pharmaceutically important peptides, *J. Liq. Chromatogr.,* 3, 1243, 1980.
21. **Regnier, F. E. and Gooding, K. M.,** High-performance liquid chromatography of proteins, *Anal. Biochem.,* 103, 1, 1980.
22. **Edlund, B., Andersson, J., Titanji, V., Dahlqvist, U., Ekman, P., Zetterqvist, Ö., and Engström, L.,** Amino acid sequence at the phosphorylated site of rat liver pyruvate kinase, *Biochem. Biophys. Res. Commun.,* 67, 1516, 1975.
23. **Yeaman, S. J., Cohen, P., Watson, D. C., and Dixon, G. H.,** The substrate specificity of adenosine 3′:5′-cyclic monophosphate-dependent protein kinase of rabbit skeletal muscle, *Biochem. J.,* 162, 411, 1977.
24. **Zetterqvist, Ö., Ragnarsson, U., Humble, E., Berglund, L., and Engström, L.,** The minimum substrate of cyclic AMP-stimulated protein kinase, as studied by synthetic peptides representing the phosphorylatable site of pyruvate kinase (type L) of rat liver, *Biochem. Biophys. Res. Commun.,* 70, 696, 1976.
25. **Zetterqvist, Ö. and Ragnarsson, U.,** The structural requirements of substrates of cyclic AMP-dependent protein kinase, *FEBS Lett.,* 139, 287, 1982.
26. **Fransson, B., Ragnarsson, U., and Zetterqvist, Ö.,** Separation of basic, hydrophilic peptides by reversed-phase ion-pair chromatography. I. Analytical applications with particular reference to a class of serine peptide substrates of cyclic AMP-stimulated protein kinase, *J. Chromatogr.,* 240, 165, 1982.

27. **Fransson, B. and Ragnarsson, U.,** Separation of basic, hydrophilic peptides by reversed-phase ion-pair chromatography. III. A progress report, *Peptides,* Bláha, K. and Maloň, P., Eds., Walter de Gruyter & Co., Berlin, 1983, 415.
28. **Fransson, B., Ragnarsson, U., and Zetterqvist, Ö.,** Separation of basic, hydrophilic peptides by reversed-phase ion-pair chromatography. II. Analytical applications with particular reference to phosphoserine peptides, *Anal. Biochem.,* 126, 174, 1982.
29. **Merrifield, R. B.,** Solid phase peptide synthesis. I. The synthesis of a tetrapeptide, *J. Am. Chem. Soc.,* 85, 2149, 1963.
30. **Barany, G. and Merrifield, R. B.,** Solid-phase peptide synthesis, in *The Peptides,* Vol. 2, Gross, E. and Meienhofer, J., Eds., Academic Press, New York, 1980, chap. 1.
31. **Titanji, V. P. K., Zetterqvist, Ö., and Ragnarsson, U.,** Activity of rat-liver phosphoprotein phosphatase on phosphopeptides formed in the cyclic AMP-dependent protein kinase reaction, *FEBS Lett.,* 78, 86, 1977.
32. **Majors, R. E.,** High performance liquid chromatography on small particle silica gel, *Anal. Chem.,* 44, 1722, 1972.

SEPARATION OF OLIGOPEPTIDES BY NORMAL PHASE CHROMATOGRAPHY

Fred Naider, Alvin S. Steinfeld, and Jeffrey M. Becker

INTRODUCTION

The synthesis of peptides using solution-phase procedures often requires the isolation and purification of peptides which contain protecting groups on their amine and carboxyl termini. Indeed the purification of peptide fragments is among the most difficult and tedious steps in the synthesis of biologically active peptides. In addition to their importance in peptide synthesis, protected peptides have been widely exploited as model compounds for protease studies and to explore the effect of amino acid composition and chain length on peptide conformation. For this latter purpose a great variety of homologous series of oligopeptides have been prepared and evaluated.[1] The majority of oligopeptides studied are composed of hydrophobic amino acids and are virtually insoluble in aqueous media. Obviously a prerequisite for conformational analysis is preparation of a chemically and optically pure peptide. When investigating homologous series of peptides, techniques such as thin-layer chromatography (TLC) and elemental analysis become insensitive to increases in chain length and positional isomerism. The purpose of this report is to evaluate HPLC investigations of protected oligopeptides on normal-phase columns. Discussion will be limited to homologous oligopeptides and is restricted predominantly to peptides composed of methionine.

TECHNICAL PROBLEMS

In attempting to apply HPLC to homologous oligopeptides composed of hydrophobic residues the investigator is confronted with the problem of choosing a suitable combination of column and mobile phase. Due to the high insolubility of protected peptides in water, it is difficult to investigate such materials using reversed-phase columns. Thus most studies on fully protected peptides have been conducted on normal-phase microparticulate silica columns. For such studies peptides can be injected in a range of solvents varying from chloroform or methylene chloride on the nonpolar end to methanol and trifluoroethanol on the polar side. In the case of methionine-containing oligopeptides sufficient solubility exists up to the heptapeptide in methylene chloride ($\sim$ 1 mg/mℓ) so that 5 to 25 μg of peptide can be injected in less than 100 $\mu\ell$ of solvent. Similar results have been obtained for homo-oligopeptides of γ-methyl-L-glutamate.[2] However for certain co-oligopeptides of methionine and γ-methyl-L-glutamate with glycine, trifluoroethanol (TFE) was required to obtain reasonable concentrations of hexa-, hepta-, and octapeptides. Injection of the peptide in a strong solvent such as TFE can significantly influence its mobility on silica gel columns. When employing such a strong solvent it is essential to compare different peptides using identical injection volumes.

An additional difficulty encountered with protected peptides composed of most of the 20 naturally occurring amino acid residues is choice of a method for detection of the peptide as it elutes off the column. For peptides of the formula Boc-$(Met)_n$-OMe or co-oligomers in which Met is replaced by Gly,Ala,Val,Leu,Lys,Glu, or Pro detection in the UV region is limited to wavelengths at which the peptide bond absorbs. This eliminates fixed wavelength UV spectrometers which detect at 254 or 280 nm. Excellent results have been obtained using detection at 220 nm where as little as 1 μg of peptide can be observed with a high signal to noise ratio. The use of 220 nm radiation for detection restricts mobile-phase solvents to those with cutoffs below 210 nm thus preventing the application of a host of eluents which

Table 1
MOBILITY OF DI- and TRIPEPTIDES ON μPORASIL®

Peptide	Retention time (min)[a]
Boc-Gly-Met-OMe	8.4
Boc-Ala-Met-OMe	3.8
Boc-Pro-Met-OMe	6.2
Boc-Met-Met-OMe	2.6
Boc-Val-Met-OMe	2.2
Boc-Met-Gly-OMe	6.2
Boc-Met-Ala-OMe	3.2
Boc-Met-Pro-OMe	4.8
Boc-Gly-Met_2-OMe	13.6
Boc-Ala-Met_2-OMe	6.9
Boc-Met_2-Gly-OMe	13.1
Boc-Met_2-Ala-OMe	6.3

[a] Flow rate 2.0 mℓ/min, mobile phase isopropanol/cyclohexane (4:96).

are often employed for the separation of protected peptides. In our laboratory mobile phases containing cyclohexane/isopropanol, cyclohexane/isopropanol/methanol, or cyclohexane/isopropanol/methanol/acetic acid have been employed with excellent success. In the latter system the acetic acid concentration must be below 0.2% or unacceptably high levels of background absorbance are encountered.

In principle, detection is also possible using differential refractometry or IR absorption. No reports appear in the literature on the latter technique and it has not been attempted in our laboratory. Hara et al.[3] reported the use of differential refractometry in a study of the design of binary solvent systems for HPLC of protected peptides. The use of this method of detection allowed them to explore a spectrum of mobile phases including those based on benzene, dichloromethane, chloroform, ethyl acetate, and acetone. Due to their absorption characteristics none of these solvents could be used with UV detection at 220 nm. However, it should be noted that Hara et al. injected 250 μg of peptide per run. Investigation of methionyl-containing peptides revealed that in cyclohexane/isopropanol (96:4) differential refractometry was extremely insensitive and could not detect 5 to 50 μg of peptide. Under identical conditions 220-nm detection gave almost a full-scale deflection for 2 to 20 μg injections of di- and tripeptides at an AUFS setting of 0.1. Since it is likely that 250-μg injections of peptides longer than a pentapeptide would be difficult to achieve due to solubility limitations, it may be a formidable task to apply differential refractometry for detection of a homologous series of peptides.

For the studies summarized herein the μPorasil® column slowly lost activity when mobile phases containing methanol or acetic acid were utilized. The mobilities were also very sensitive to slight changes in the concentration of these polar eluents. In comparing k′ values some differences will be noted from table to table. These reflect runs on different days or different preparations of eluent. Great care was taken to make sure that the data in a given table were obtained on the same day and are internally consistent. The important conclusions from this report involve relative effects and the absolute k′ values should be considered with the above reservation.

SEPARATION OF PROTECTED METHIONINE PEPTIDES ON μPORASIL® SILICA

As indicated in Tables 1 to 3 and Figures 1 to 3, HPLC on normal-phase silica is well

Table 2
MOBILITY OF METHIONINE-CONTAINING PEPTIDES

Peptide	Mobile phase	k′[a]	Mobile phase	k′[a]
Boc-X-Met-Met-OMe	(cyclohexane/isopropanol)		(cyclohexane/isopropanol/ methanol)	
X				
Leu	94:6	0.62	93:6:1	0.54
Val	94:6	0.74	93:6:1	0.62
L-Met	94:6	1.02	93:6:1	0.62
D-Met	94:6	1.10	93:6:1	0.79
Glu(OBzl)	94:6	1.23	93:6:1	0.85
Lys(Z)	94:6	3.15	93:6:1	1.82
Pro	94:6	2.38	93:6:1	1.46
Ala	94:6	1.82	93:6:1	1.18
Gly	94:6	3.51[b]	93:6:1	1.79
Boc-Met-X-Met-OMe				
X				
Leu	94:6	0.59	93:6:1	0.51
Val	94:6	0.74	93:6:1	0.62
Met	94:6	1.02	93:6:1	0.62
Lys(Z)	94:6	3.85	93:6:1	2.10
Pro	94:6	3.62	93:6:1	2.15
Ala	94:6	2.28	93:6:1	1.30
Gly	94:6	7.44	93:6:1	3.18
Boc-Met-X-Met_2OMe				
X				
Leu	94:6	1.02	93:6:1	0.74
Val	94:6	1.49	93:6:1	1.00
L-Met	94:6	2.13	93:6:1	1.18
D-Met	94:6	1.76	93:6:1	1.12
Glu(OBzl)	94:6	1.94	93:6:1	1.23
Lys(Z)	94:6	6.97	93:6:1	2.90
Pro	94:6	5.46	93:6:1	2.60
Ala	94:6	4.97	93:6:1	2.38
Gly	94:6	9.76	93:6:1	3.64

[a] $k' = V_1 - V_0/V_0$ where V_1 = retention volume of component of interest and V_0 = dead volume; Flow rate = 2.0 mℓ/min.

[b] k′ for this compound differs in Tables 2 and 3. The discrepancy is due to runs on different days with columns of different activity. Each compound within a given table was run on same day with the same column.

suited for the separation of hydrophobic peptides composed predominantly of methionine. Separation of a series of peptides composed of methionine and one other amino acid, or of di-, tri-, and tetrapeptides in which the position of the substituent is varied have been achieved.

Effect of Hydrophobicity on Mobility

Studies have been carried out on a variety of di-, tri-, and tetrapeptides of the general formula Boc-X-Met-OMe, Boc-Met-X-OMe, Boc-X-Met-Met-OMe, Boc-Met-X-Met-OMe, and Boc-Met-X-Met-Met-OMe where X = Gly,Ala,Pro,Val,Leu,Met,D-Met, Lys(Z), and Glu(OBzl) (Tables 1 and 2). Using partition coefficients obtained from octanol/water or the free energy of transfer of amino acids from water to dioxane or ethanol, the expected order of hydrophobicity for X is Leu>Val>Met>Pro>Ala>Gly.[4] Although specific data is not available for Lys(Z) and Glu(OBzl) they are known to behave as highly hydrophobic residues

Table 3
MOBILITY OF HOMOLOGOUS SERIES OF METHIONINE PEPTIDES ON μPORASIL®

Peptide	Mobile phase (cyclohexane/isopropanol)	k′[a]	Mobile phase (cyclohexane/isopropanol/methanol)	k′[a]
BocGlyMetOMe	94:6	2.25	92:6:2	1.44
BocGlyMet_2OMe	94:6	3.05[b]	92:6:2	1.71
BocGlyMet_3OMe	94:6	7.23	92:6:2	2.84
BocGlyMet_4OMe	—	—	92:6:2	4.18
BocGlyMet_5OMe	—	—	92:6:2	6.76
BocGlyMet_6OMe	—	—	92:6:2	15.38[c]
BocValMetOMe	92.5:7.5	0.18	—	—
BocValMet_2OMe	92.5:7.5	0.38	—	—
BocValMet_3OMe	92.5:7.5	0.98	—	—
BocValMet_4OMe	92.5:7.5	2.03	—	—
BocValMet_5OMe	92.5:7.5	4.63	—	—
BocMet_2OMe	—	—	92:6:2	0.49
BocMet_3OMe	—	—	92:6:2	0.72
BocMet_4OMe	—	—	92:6:2	1.08
BocMet_5OMe	—	—	92:6:2	1.68
BocMet_6OMe	—	—	92:6:2	2.87
BocMet_7OMe	—	—	92:6:2	8.11[c]
BocGlyMet_5OMe	—	—	92:6:2	6.74
BocMetGlyMet_4OMe	—	—	92:6:2	7.71
BocMet_2GlyMet_3OMe	—	—	92:6:2	5.42
BocMet_3GlyMet_2OMe	—	—	92:6:2	4.43
BocMet_4GlyMetOMe	—	—	92:6:2	7.21
BocMet_5GlyOMe	—	—	92:6:2	6.17
BocAlaMet_5OMe	—	—	92:6:2	4.70
BocMetAlaMet_4OMe	—	—	92:6:2	5.66
BocMet_2AlaMet_3OMe	—	—	92:6:2	5.23
BocMet_3AlaMet_2OMe	—	—	92:6:2	4.37
BocMet_4AlaMetOMe	—	—	92:6:2	4.52
BocMet_5AlaOMe	—	—	92:6:2	3.48

[a] See Table 2, Footnote a.
[b] See Table 2, Footnote b.
[c] Distorted peak shape.

exhibiting very low affinity for water. In general the mobility of these peptides on a μPorasil® column using cyclohexane/isopropanol as the eluent can be correlated with the hydrophobicity of residue X. The more hydrophobic residues resulted in the least retardation of the peptide by the column. Interestingly, proline markedly deviates from the expected order and behaves as a more hydrophilic residue showing a mobility intermediate between those for analogous peptides comprised of Ala and Gly. This behavior is consistent with the high water solubility observed for proline-containing peptides and is undoubtedly due to the inclusion of the proline side chain in a five-membered ring. A surprising deviation from the hydrophobicity trend was observed for peptides containing Lys(Z). Rather than adding to the overall hydrophobicity of the peptide this residue led to a large increase in retention. In fact, of the X residues examined, only Gly resulted in greater retardation on the column. These results indicate that the NH of the urethane protecting group interacts strongly with the silica gel. This is clear from a comparison of peptides containing Lys(Z) (I) and Glu(OBzl) (II) residues.

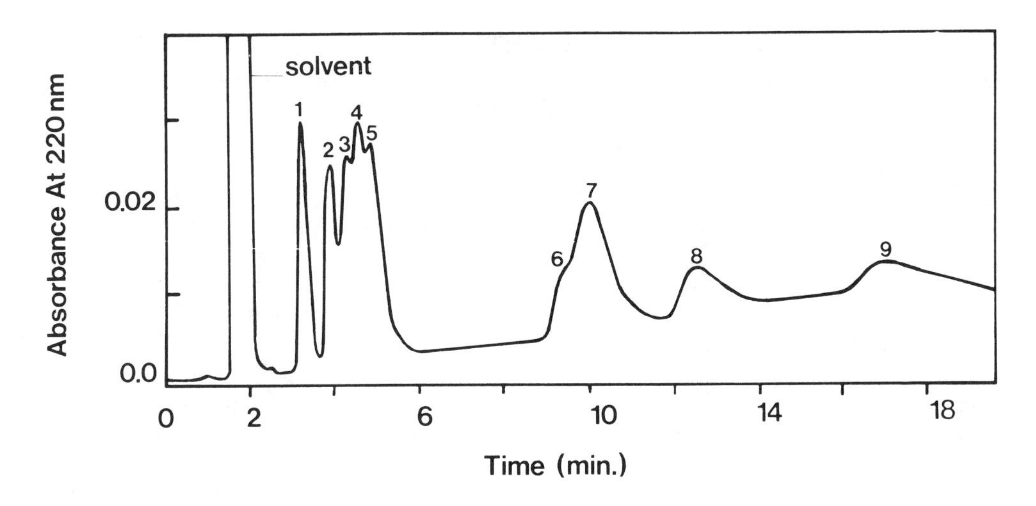

FIGURE 1. Isocratic separation of Boc-Met-X-Met-Met-OMe tetrapeptides on a 30 cm × 3.9 mm i.d. μPorasil® column using cyclohexane/isopropanol (94:6), a flow rate of 2.0 mℓ/min and ambient temperatures. X = 1 to 9; 1, Leu; 2, Val; 3, D-Met; 4, Glu(OBzl); 5, L-Met; 6, Ala; 7, Pro; 8, Lys(Z); 9, Gly.

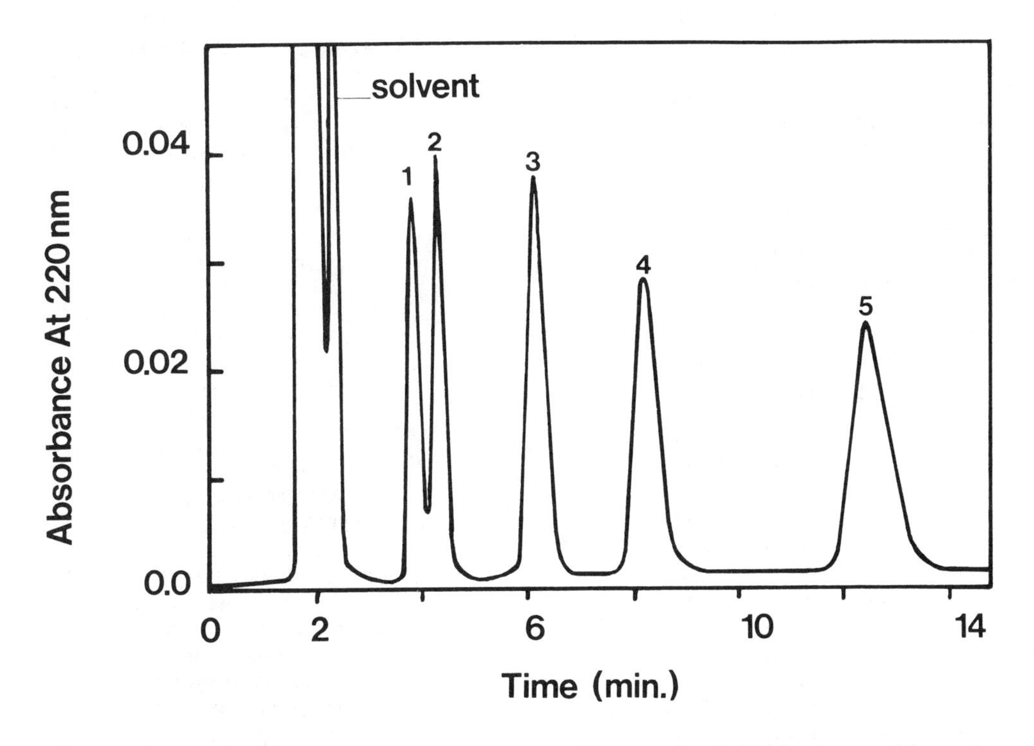

FIGURE 2. Isocratic separation of Boc-Gly-Met_n-OMe peptides; n = 1 to 5. Mobile phase, cyclohexane/isopropanol/methanol (92:6:2) at a flow rate of 2.0 mℓ/min on a μPorasil® column (30 cm × 3.9 mm i.d.) and at ambient temperatures.

$$-CH_2-CH_2-CH_2-CH_2-NH-\overset{O}{\overset{\|}{C}}-O-CH_2-C_6H_5$$

I

$$-CH_2-CH_2-\overset{O}{\overset{\|}{C}}-O-CH_2-C_6H_5$$

II

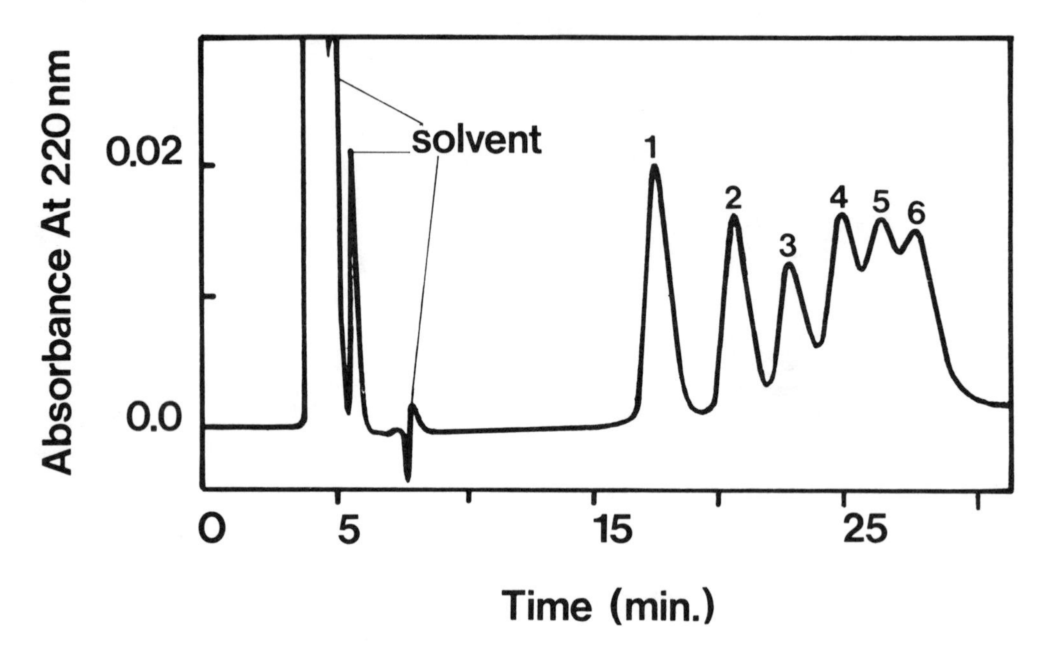

FIGURE 3. Isocratic separation of six isomeric hexapeptides; 1, Boc-Met_3-Gly-Met_2-OMe; 2, Boc-Met_2-Gly-Met_3-OMe; 3, Boc-Met_5-Gly-OMe; 4, Boc-Gly-Met_5-OMe; 5, Boc-Met_4-Gly-Met-OMe; 6, Boc-Met-Gly-Met_4-OMe. Mobile phase, cyclohexane/isopropanol/methanol (92:6:2) at a flow rate of 1.0 mℓ/min on a μPorasil® column (30 cm × 3.9 mm i.d.) and at ambient temperatures.

Except for the NH linkage in side chain I these side chains contain quite similar bonds. Yet the Glu(OBzl)-containing peptides exhibit mobilities similar to the homomethionines whereas those with Lys(Z) appear as much more polar compounds. We conclude that interaction between urethane or amide-like NH groups with silica gel play a dominant role in determining the mobility of peptides.

Separation of Stereoisomers

As seen in Figure 1 Boc-Met_4-OMe is resolved from Boc-L-Met-D-Met-L-Met-L-Met-OMe in cyclohexane/isopropanol 94:6. The L and D diastereomers were also resolved for Boc-Met-X-Met-OMe although significantly smaller differences were observed. It is clear that HPLC on normal-phase silica can be used to distinguish diastereomeric peptides.

Effect of Chain Length on Mobility

The data in Table 3 and Figure 2 clearly illustrate that increasing the chain length leads to retardation on the μPorasil® column. This finding has been observed with Boc-Met_n-OMe (n = 2 to 7),[2] Boc-Val-Met_n-OMe (n = 2 to 6) and Boc-Gly-Met_n-OMe[2] (n = 2 to 6) oligopeptides. Similar trends were observed for Boc-$[Glu(OMe)]_n$-OMe (n = 2 to 7) oligopeptides although significantly more polar mobile phases were required to elute the glutamate oligopeptides.[2] For the methionine-containing peptides a system was developed which gave excellent separation from the dipeptide through to the hexapeptide. Under these conditions the respective heptapeptides required relatively long times to elute from the column and exhibited badly distorted elution profiles. Attempts to develop an isocratic system which would give reasonable k′ values and peak shapes for the heptamer and higher oligopeptides invariably resulted in coelution of the dipeptide and the solvent front. It is possible that the difficulties observed at the heptapeptide level are related to either conformational factors or a drastic decrease in solubility which results in precipitation of the peptide on the column.

A comparison of the data for Boc-Met_7-OMe and Boc-Met-Gly-Met_4-OMe lends support to the latter conclusion. Both of these peptides have very similar k′ values using the same mobile phase. Yet the heptapeptide has a severely distorted peak whereas the hexapeptide exhibited a nearly Gaussian profile. Clearly precipitation of the peptide on the column would result in distorted elution profiles.

Effect of Positional Isomerism on Mobility

The mobility of protected peptides composed of the identical amino acid residues in a different sequence varies significantly. Comparison of oligopeptides of the form Boc-X-Met-OMe and Boc-Met-X-OMe (X = Pro,Ala,Gly) reveals that in every case the peptide with the X residue in the carboxyl terminal position has a lower k′ value.[5] This trend is also observed in tripeptides and hexapeptides composed mostly of methionine. Striking is the fact that six hexapeptides composed of five methionyl and one glycyl residue are completely resolved by a cyclohexane/isopropanol/methanol (92:6:2) mobile phase using a flow rate of 1 to 2 mℓ/min (Figure 3).[2] The k′ values for these hexapeptides vary from a low of 4.43 to a high of 7.71; a marked difference for two compounds composed of the identical number of atoms. Evaluation of six hexapeptides composed of five methionyl and one alanyl residue gave analogous results although smaller differences in k′ values were observed. In both cases the hexamer with the more polar residue at the carboxyl terminus moved more rapidly than the homolog with the more polar residue at the amine terminus. However, whereas Boc-Met_5-Ala-OMe had the lowest k′ value of the Met-Ala hexapeptides, two hexapeptides with Gly at internal positions eluted more rapidly than Boc-Met^5-Gly-OMe. It is likely that the effect of the Ala or Gly residue is a combination of both specific interactions of the polar residue with the silica support and perturbation of the distribution of conformations assumed by the hexapeptide. Since Gly is expected to greatly perturb the conformations assumed by Met peptides, whereas Ala should have less drastic effects, differences in the positional influence of these residues is not unexpected. However, no conformational analysis on these peptides has ever been carried out in cyclohexane/isopropanol/methanol, and it is not possible to ascertain the preferred conformation for hexamethionine in this mobile phase.

Effect of Solvent on Mobility and Peak Shape

The solvents utilized in the studies on protected methionine oligopeptides vary from cyclohexane/isopropanol (96:4) as the least polar to cyclohexane/isopropanol/methanol (92:6:2) as the most polar. Systems containing higher ratios of isopropanol to cyclohexane (25:75) have also been utilized. In general the greatest success in terms of good separation and symmetrical peak shape has been obtained by using systems containing small amounts of methanol rather than using high concentrations of isopropanol. For equivalent values of k′ much more symmetrical peak shapes are found in systems containing a little methanol than for systems composed solely of cyclohexane and isopropanol. Similar improvements in peak shape for octapeptides were also observed when trace amounts of acid (0.01 to 0.2%) are added to a cyclohexane/isopropanol/methanol mobile phase.[2]

MOBILITY OF (γ-METHYL) GLUTAMATE PEPTIDES ON μPORASIL®

Analysis of homo- and co-oligopeptides composed primarily of γ-methyl-L-glutamate revealed trends quite similar to those found for methionine peptides. Since γ-methyl glutamate is less hydrophobic than methionine, stronger mobile phases were necessary to elute the glutamate peptides. Excellent separation was achieved with cyclohexane/isopropanol/methanol, 8:1:1. With this mobile-phase dimer through heptamer of the Boc-$[Glu(OMe)]_n$OMe series were separable under isocratic conditions, although the heptamer exhibited severe

Table 4
MOBILITY OF (γ-METHYL)GLUTAMATE PEPTIDES ON μPORASIL®

Peptide	Mobile phase	k′[a]	Mobile phase	k′[a]
Boc[Glu(OMe)]$_2$OMe	A[b]	1.48	B[c]	0.70
Boc[Glu(OMe)]$_3$OMe	A	3.40	B	1.08
Boc[Glu(OMe)]$_4$OMe	A	7.96	B	1.65
Boc[Glu(OMe)]$_5$OMe	A	20.01	B	2.65
Boc[Glu(OMe)]$_6$OMe	—	—	B	4.26
Boc[Glu(OMe)]$_7$OMe	—	—	B	7.08
BocGly[Glu(OMe)]$_5$OMe	—	—	B	4.65
BocGlu(OMe)Gly[Glu(OMe)]$_4$OMe	—	—	B	4.85
Boc[Glu(OMe)]$_2$Gly[Glu(OMe)]$_3$OMe	—	—	B	4.19
Boc[Glu(OMe)]$_3$Gly[Glu(OMe)]$_2$OMe	—	—	B	3.89
Boc[Glu(OMe)]$_4$GlyGlu(OMe)OMe	—	—	B	4.37

[a] See Table 2, Footnote a.
[b] A, Cyclohexane/isopropanol/methanol 92:6:2.
[c] B, Cyclohexane/isopropanol/methanol 8:1:1.

tailing. The octamer did not elute with the above mobile phase. Insertion of glycine at various positions of a glutamate hexamer resulted in changes in retention analogous to those found for Boc-Met$_6$-OMe (Table 4).[2] As with the Met-Gly series, Gly at internal positions of Boc-[Glu(OMe)]$_6$OMe resulted in least retention on the column. In fact two of the co-oligopeptides elute faster than the homohexaglutamate. Since studies on shorter co-oligopeptides suggest that γ-methyl glutamate is less polar than glycine, the positional influence of the glycine residue appears to be based on conformational perturbations.

Plots of log k′ vs. the number of amino acid residues for the glutamate, and methionine peptides give reasonably good straight lines. For the glutamate oligomers a minor deviation occurs at the heptapeptide. For the Boc-Met$_n$-OMe series the heptapeptide deviates markedly, a finding consistent with the severely distorted peak obtained for this oligomer. The lines obtained for the Boc-Met$_n$-OMe, Boc-Gly-Met$_n$-OMe, Boc-[Glu(OMe)]$_n$OMe and Boc-Gly-[Glu(OMe)]$_n$OMe oligomers were fairly parallel whereas that found for Boc-Val-Met$_n$-OMe series had a significantly greater slope.[2] The differences in slope appear to reflect the solvent system since the similar slopes were attained for cyclohexane/isopropanol/methanol mobile phases whereas the Boc-Val-Met$_n$-OMe series was measured in cyclohexane/isopropanol. The trends obtained suggest that as long as precipitation or severe changes in conformation do not occur at a specific chain length, HPLC mobilities for short oligomers on μPorasil® can be used to predict k′ values for higher homologs. This observation might prove useful in the design of solvent systems for the analysis and purification of these oligomers.

ACKNOWLEDGMENTS

The authors gratefully acknowledge the technical contributions made by James Champi, Michael Huchital, Robert Sipzner, and Dr. P. Shenbagamurthi. This work was supported by grants from The National Institute of Allergy and Infectious Diseases (AI-14387), from The National Institute of General Medical Sciences (GM22086-07 and GM22087-07), and from the PSC-HBE fund of the City University of New York.

REFERENCES

1. **Naider, F. and Goodman, M.,** Conformational analysis of oligopeptides by spectral techniques, in *Bioorganic Chemistry; Macro and Molecular Systems,* Vol. 3, van Tamelin, E. E., Ed., Academic, New York, 1977, 177.
2. **Huchital, M., Becker, J. M., and Naider, F.,** Effect of positional isomerism and conformation on the mobility of oligopeptides in silica gel liquid chromatography, submitted for publication.
3. **Hara, S., Ohsawa, A., and Dobashi, A.,** Design of binary solvent systems for separation of protected oligopeptides in silica gel liquid chromatography, *J. Liq. Chromatogr.*, 4, 409, 1981.
4. **Rekker, R. F.,** *The Hydrophobic Fragmental Constant,* Elsevier, New York, 1977, 301.
5. **Naider, F., Sipzner, R., Steinfeld, A. S., and Becker, J. M.,** Separation of protected oligopeptides by normal-phase high pressure liquid chromatography, *J. Chromatogr.*, 176, 264, 1979.

A MODEL STUDY FOR MONITORING MERRIFIELD SOLID-PHASE PEPTIDE SYNTHESIS THROUGH *N*-2,4-DINITROPHENYL DERIVATIZATION

Stephen I. Sallay and Stephen Oroszlan

The numerous advantages of the Merrifield solid-phase peptide synthesis (SPPS[1]) over the classical solution technique have been clearly demonstrated during the past two decades. One of its inherent disadvantages is that its intermediates cannot be purified during the synthesis. Short of that, there seems to be no method available by which the exact course of SPPS can be monitored. Although there are numerous destructive and nondestructive methods available for determining the completeness of coupling and deblocking steps during the synthesis[1,2] those techniques do not reveal the exact chemistry of the synthesis on a step-by-step basis. Therefore, undesirable side reactions[1] which can take place remain undetected during the course of the synthesis.

We wished to develop a HPLC technique for monitoring the course of the Merrifield synthesis in a stepwise fashion. In the following, our preliminary findings are summarized.[3]

First the free amino termini, remaining after incomplete coupling reactions, were terminated with a UV absorbing 2,4-dinitrophenyl (DNP) group.[4] Then the same reagent was used for tagging a few milligrams of resin-bound peptide following each deblocking step. After the tagged peptides were cleaved from the resin, they were separated by HPLC. Scheme 1 demonstrates the general synthesis of a tetrapeptide. If the coupling reactions are not complete, four DNP derivatives (I to IV) should be observed, compound I being the major component at the end of the synthesis of a tetrapeptide. However, due to incomplete termination reactions, DNP.A_3.A_1, DNP.A_4.A_3.A_1, DNP.A_4.A_2.A_1, and DNP.A_4.A_1 failure sequences could also develop.

First, we explored the usefulness of 2,4-dinitro-1-fluorobenzene (DNFB).[5] A large excess (>10x) of 0.1 *M* DNFB and an equimolar amount of diisopropylethylamine (DIEA) in dichloromethane was used for the termination and tagging reactions.

There are several advantages of this procedure: (1) DNP peptides are permanently blocked; (2) DNP amino acids and peptides are readily separable by HPLC, and (3) the high molar absorbance of DNP derivatives at 254 nm provide sensitive UV detectability. As an example, 10^{-8} g DNP-Ala was readily detected by HPLC. Furthermore, the ready separation of DNP amino acids: DNP-Ala, DNP-Val, DNP-Leu, and DNP-Ile (Figure 1) and DNP peptides (Figure 7) was demonstrated by reversed-phase HPLC (RP-HPLC).

As a model for monitoring the course of SPPS, Ile-Leu-Val-Ala (1) was synthesized. A Vega-Fox peptide synthesizer was charged with 0.5 g Boc-Ala resin (1.4 mmol/g) and the synthesis was carried out according to the procedure of Yamashiro and Li.[6]

The completion of the deblocking and coupling steps were monitored by the picrate method[7] and the uncoupled amino groups of peptide chains were terminated by DNFB, as described above. Then, after each deblocking step a few milligrams of peptide-resin sample were withdrawn and treated with 2 mℓ 0.1 *M* DNFB solution and equivalent amount of DIEA in dichloromethane for 1 hr. Following the tagging reaction, the peptide resin sample was thoroughly washed and the DNP peptides were cleaved by saturated hydrogen bromide in trifluoroacetic acid (TFA) for 40 min. After the usual work-up, the crude mixture of peptides was separated.

The DNP derivatives (3, 5, and 9) (Figure 2) were detected by HPLC as single, major peaks. Combined aliquots of these products were then separated by HPLC (Figure 3). The smaller peaks were identified as DNP.Ala (2) and the DNP-terminated failure sequences: DNP.Leu-Ala (4) and DNP.Ile-Leu-Ala (8). Among the possible failure sequences DNP.Ile-Ala (6) and DNP.Ile-Val-Ala (7) have not been observed.

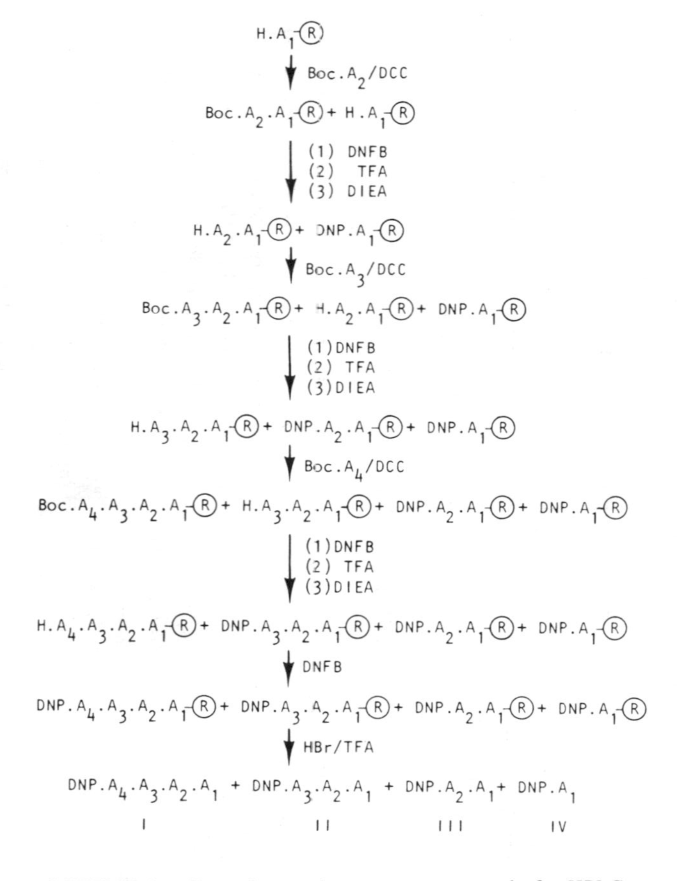

SCHEME 1. General procedure to prepare sample for HPLC analysis.

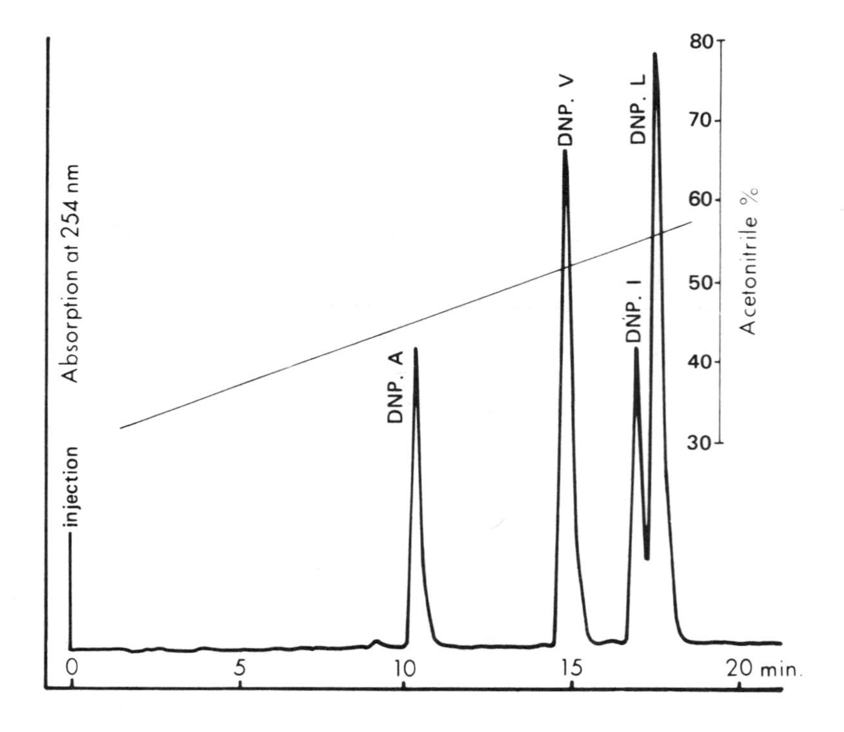

FIGURE 1. Separation of DNP.A, DNP.V, DNP.I, and DNP.L[Ile] (A, Ala; V, Val; I, ILe; L, Leu); LC conditions are given in Reference 11 for all Figures.

DNP. A — 2

DNP. V — A — 3

DNP. L ——— A — 4

DNP. L — V — A — 5

DNP. I ————— A — 6

DNP. I ——— V — A — 7

DNP. I — L ——— A — 8

DNP. I — L — V — A — 9

FIGURE 2. DNP peptides and their failure sequences.

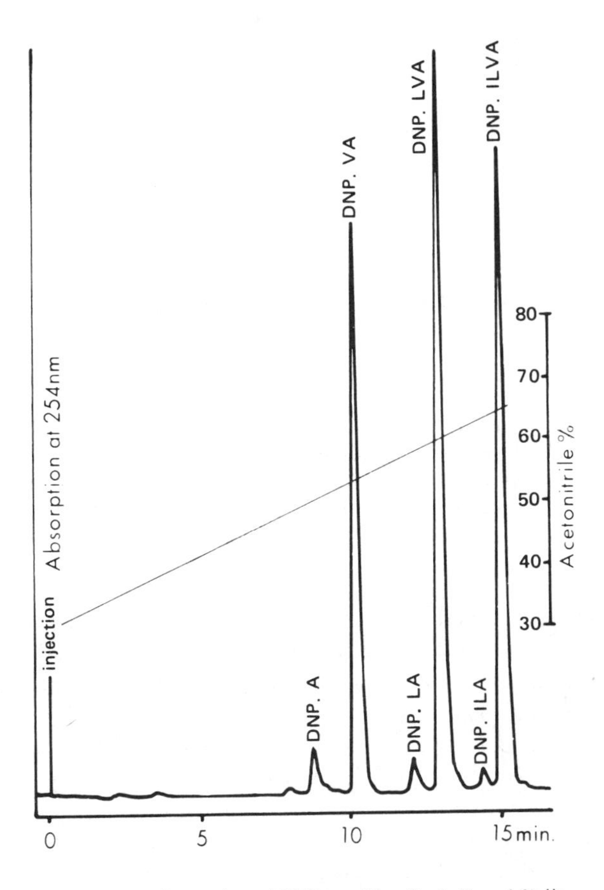

FIGURE 3. Separation of DNP peptides (2, 5, 7, and 9).[11a]

The building blocks of the tetrapeptide (1) were selected from "UV-blind" amino acids. Thus, only the DNP substituents contributed to the absorbance of these products. Consequently, the integral of the peaks quantitated their molecular ratios.

In a parallel synthesis of the tetrapeptide (1), less than the theoretically required amount of Boc-Val anhydride was used during its coupling to the resin-bound alanine residues. In addition, after the coupling reaction the uncoupled alanine was not terminated by DNFB/

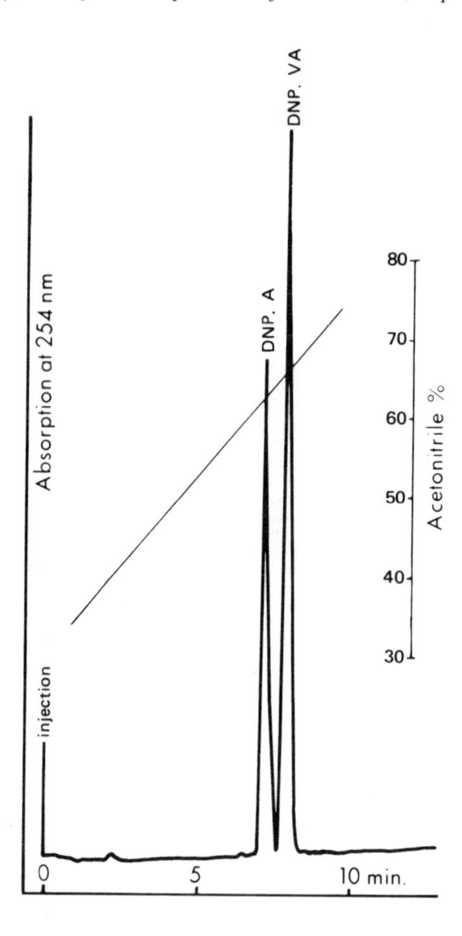

FIGURE 4. Products 2 and 3 from the incomplete coupling reaction between Boc-Val and H.Ala-R.[11c]

DIEA. As a result, DNP derivatization and cleavage of a deblocked sample of the first coupling step showed the formation of 59% of DNP.Val-Ala (3) (Figure 4). After the completion of the synthesis of the tetrapeptide DNP.Ala (2) and the three expected DNP peptides (3, 5, 9) were observed. Furthermore, the H.Ala-R sites (41%) which were purposely left uncoupled after the first cycle gave rise to two failure sequences, DNP.Leu-Ala (4) and DNP.Ile-Leu-Ala (8) during the synthesis (Figures 5, 6). The combined aliquots of the six DNP derivatives, (2 to 5, 8, and 9) were separated by HPLC (Figure 7).

In order to verify our results, unambiguous syntheses of failure sequences DNP.Leu-Ala (4) and DNP.Ile-Leu-Ala (8) were carried out (Figures 8, 9). These authentic specimens exhibited identical retention times with the original failure sequences of Figures 5, 6, and 7, during separate and mixed HPLC analyses.

While studying the usefulness of DNFB in monitoring SPPS, for comparison, another batch of Leu-Val-Ala-R was synthesized. At this time after couplings and deblocking steps, DNFB was used for both the termination and tagging reactions. The HPLC analysis of the HBr/TFA cleavage product (Figure 14) showed the presence of 91.3% DNP.Leu-Val-Ala (5), 3.6% DNP.Ala (2), 1.7% DNP.Val-Ala (3), and 3.6% DNP.Leu-Ala (4). The presence of the latter indicated incomplete termination and coupling reactions of the alanine residue.

The incomplete termination reactions of the resin-bound peptides by the DNFB method

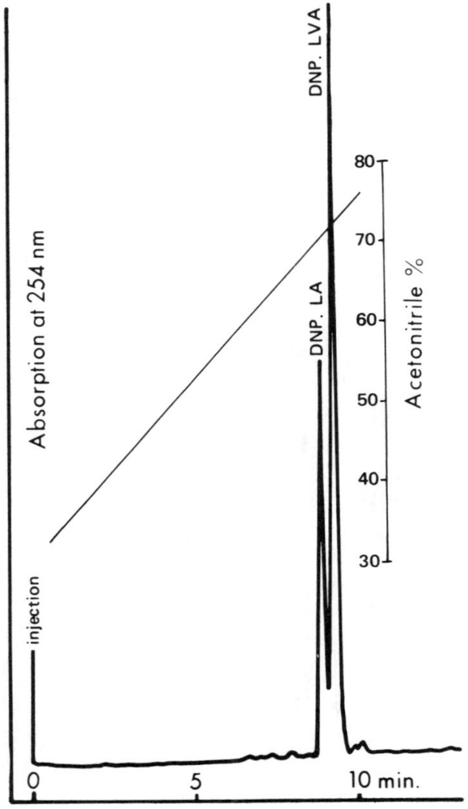

FIGURE 5. Detection of failure sequence DNP.Leu-Ala (4).[11c]

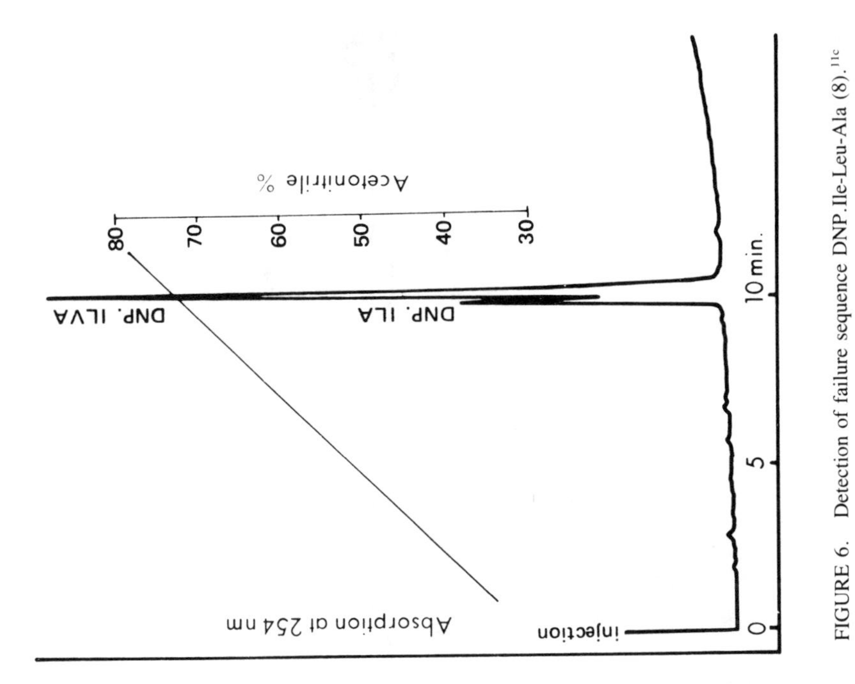

FIGURE 6. Detection of failure sequence DNP.Ile-Leu-Ala (8).[11c]

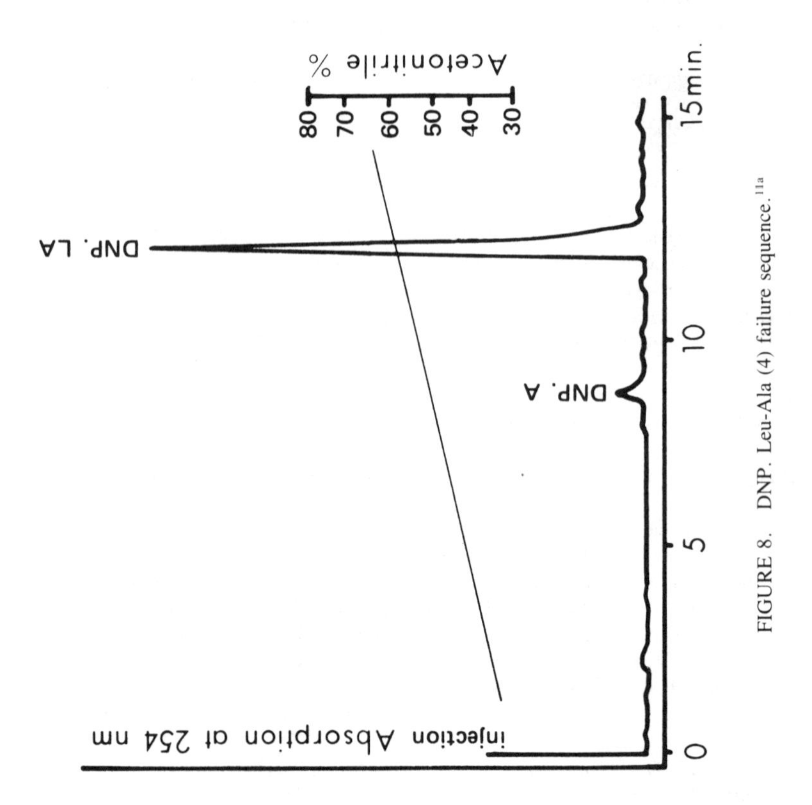

FIGURE 8. DNP. Leu-Ala (4) failure sequence.[11a]

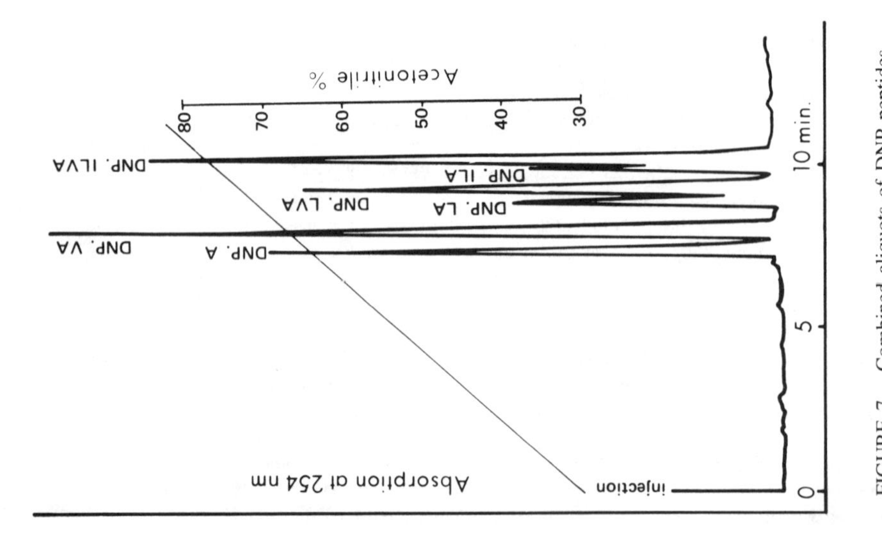

FIGURE 7. Combined aliquots of DNP peptides (2 to 5, 8, and 9).[11a]

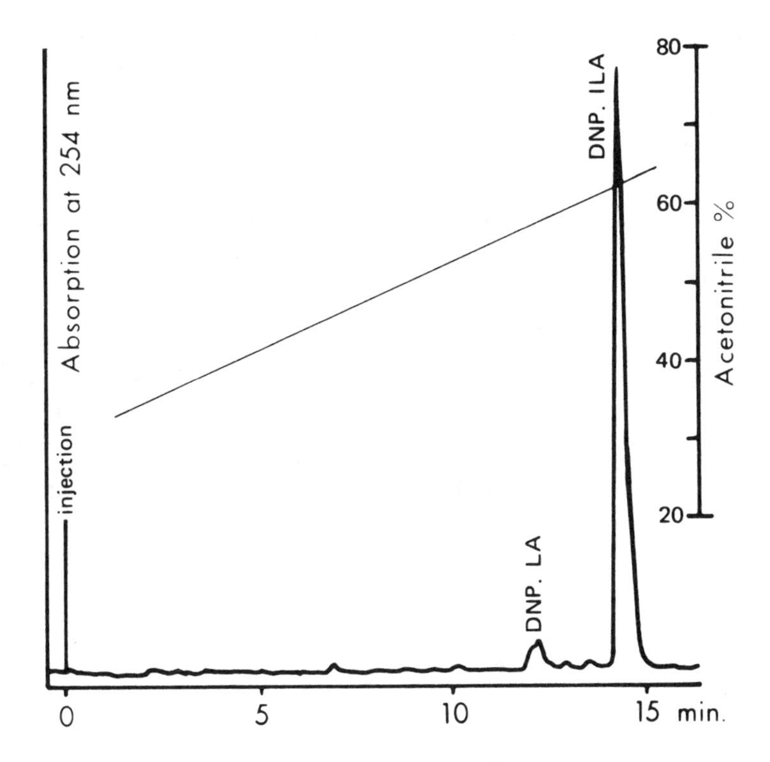

FIGURE 9. DNP.Ile-Leu-Ala (8) failure sequence.[11a]

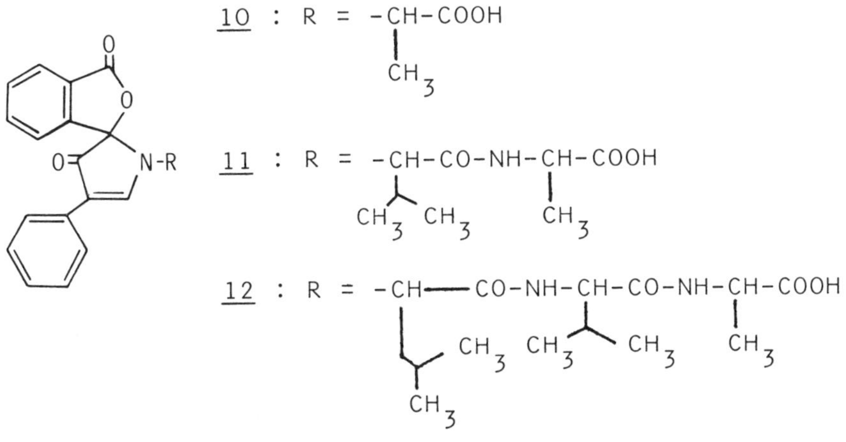

10 : R = -CH(CH3)-COOH

11 : R = -CH(CH(CH3)CH3)-CO-NH-CH(CH3)-COOH

12 : R = -CH(CH2CH(CH3)CH3)-CO-NH-CH(CH(CH3)CH3)-CO-NH-CH(CH3)-COOH

FIGURE 10. Structure of nonfluorophor-Ala (10), Val-Ala (11), and Leu-Val-Ala (12) derivatives.

prompted us to compare our findings with the highly sensitive fluorescamine reagent which was used for terminating in SPPS by Felix and co-workers.[8,9] Accordingly, during the synthesis of Leu-Val-Ala, DIEA and fluorescamine were used for the termination reactions. Furthermore, after every deblocking step, a few milligram peptide-resin was tagged by the same reagent. Thorough washing of the fluorescent samples was then followed by 40-min dry HBr/TFA cleavage. The HPLC analysis of the expected nonfluorophor-Ala (10), Val-Ala (11), and Leu-Val-Ala (12) derivatives (Figure 10) has shown a complex mixture (Figures 11 to 13) indicating extensive transformation of the reported spiro-lactone structures (10 to 12).[8]

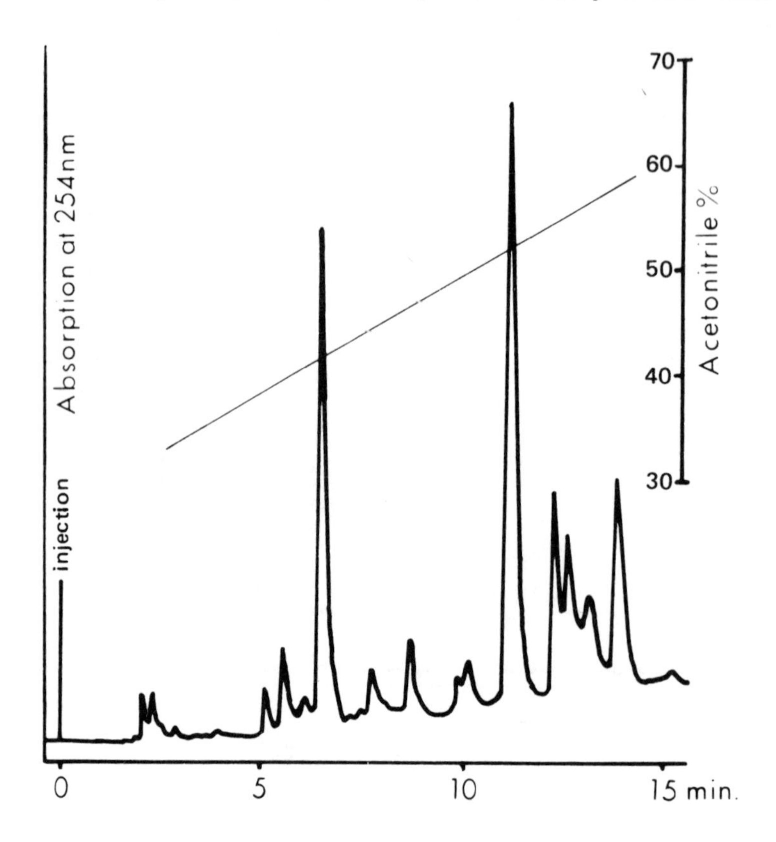

FIGURE 11. HBr/TFA cleavage products of fluorescamine-treated H.Ala-R.[11b]

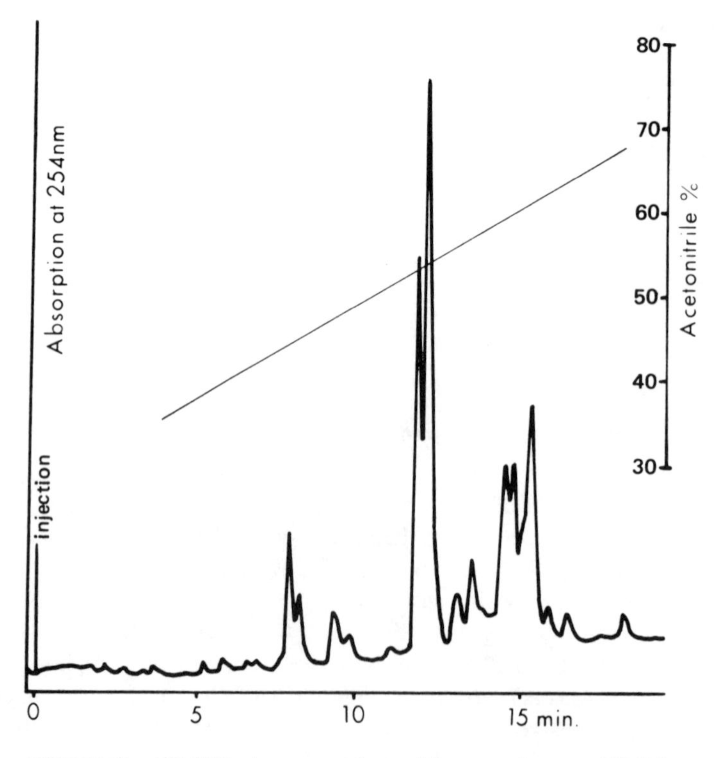

FIGURE 12. HBr/TFA cleavage products of fluorescamine-treated H.Val-Ala-R.[11b]

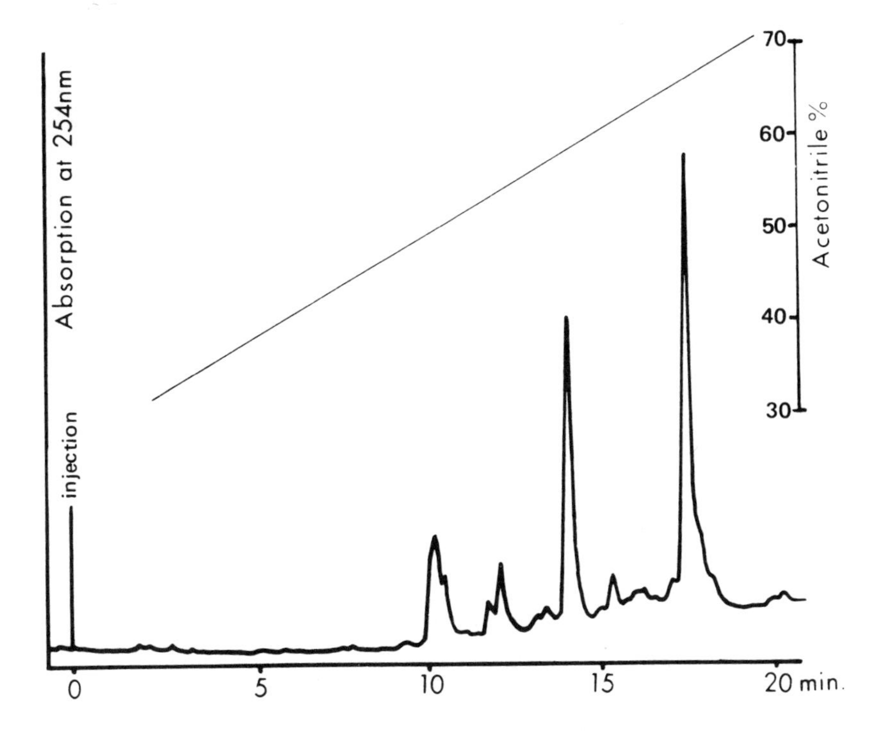

FIGURE 13. HBr/TFA cleavage products of fluorescamine-treated H.Leu-Val-Ala-R.[11b]

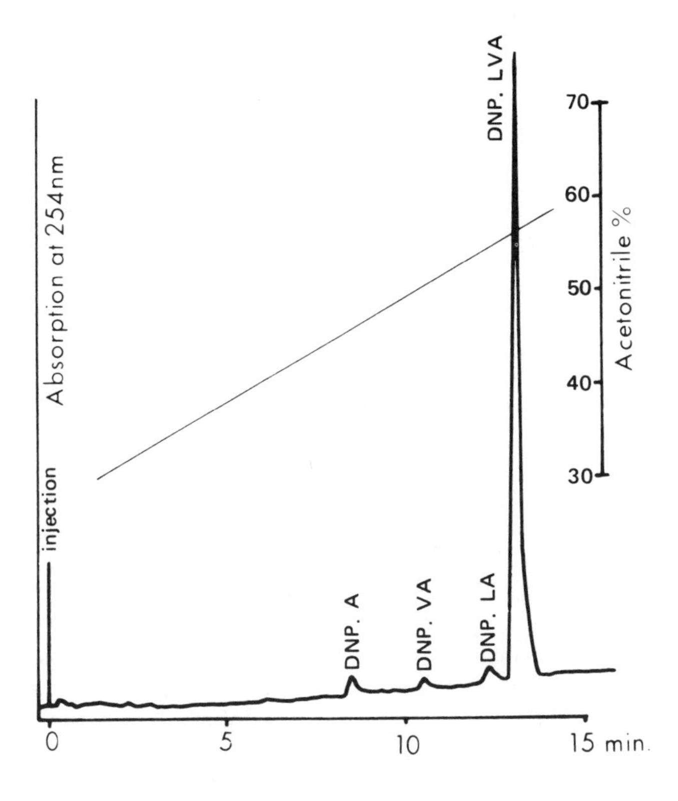

FIGURE 14. DNP.Leu-Val-Ala (5).[11b]

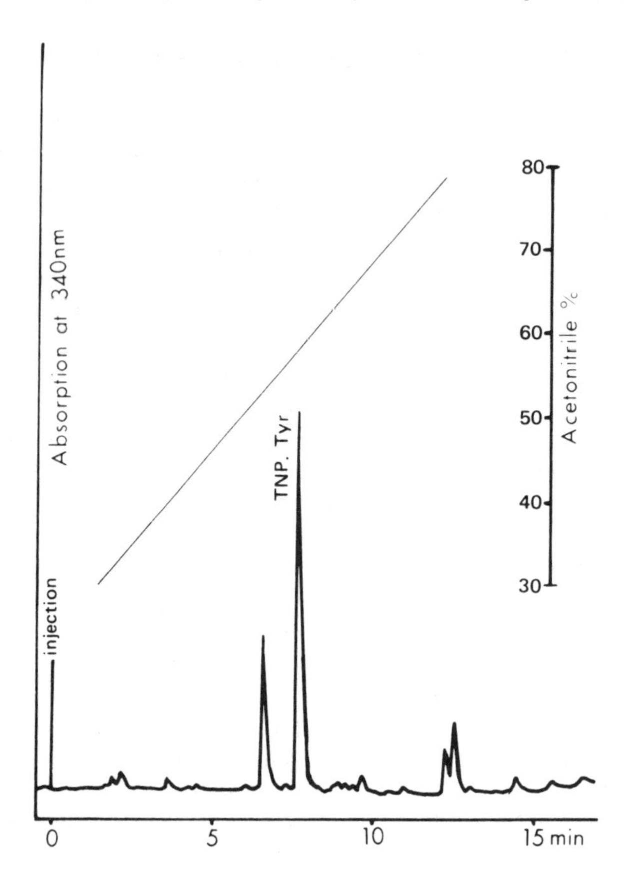

FIGURE 15. HBr/TFA cleavage products of TNP-Tyr-(O-Bzl-2,6-Cl_2)-R, measured at 340 nm.[11d]

As the next model, the synthesis of the 25 to 28 sequence of woolly monkey and gibbon ape type-C RNA virus gs-antigen, H. Ala-Asp-(O-Bzl-4-Cl)-Leu-Tyr-(O-Bzl-2,6-Cl_2)[10] was carried out. During the monitoring of this synthesis at 254 nm, the cleavage products of the side-chain protective groups complicated the HPLC analysis of the DNP peptide derivatives.

In attempting to circumvent the problem, a longer wavelength absorbing terminating reagent was sought. Whereas, DNP alanine exhibits $\epsilon_{238} = 1.58 \times 10^4$ and $\epsilon_{305} = 9.73 \times 10^3$ absorption, 2,4,6-trinitrophenyl (TNP) amino acids and TNP peptides show an $\epsilon_{340} = 1.05 - 1.27 \times 10^4$.[12]

The combination of the high molar absorptivity and the longer wavelength absorption maximum of the TNP derivatives was expected to simplify our chromatograms measured at 340 nm. For the demonstration of this idea, *N*-(2,4,6-trinitrophenyl)-*O*-(2,6-dichlorobenzyl)-tyrosine-resin was cleaved by HBr. The released mixture of products indeed showed a simpler chromatogram when measured at 340 nm (Figure 15) than at 254 nm (Figure 16). No effort was made to identify the peaks of the chromatograms other than the TNP-Tyr.

The TNP tagging and terminating reactions were similar to the DNP derivatization, except that a DMF solution of 2,4,6-trinitrobenzenesulfonic acid (TNBS) was used. Preliminary results indicated that 0.1 *M* dichloromethane solutions of DNFB and TNBS have about the same reactivity during SPPS.

In summary, the Merrifield synthesis of shorter peptides was successfully monitored by HPLC through DNP derivatization. The course of reactions were quantitated and the HPLC separation of the closely related peptide derivatives were demonstrated.

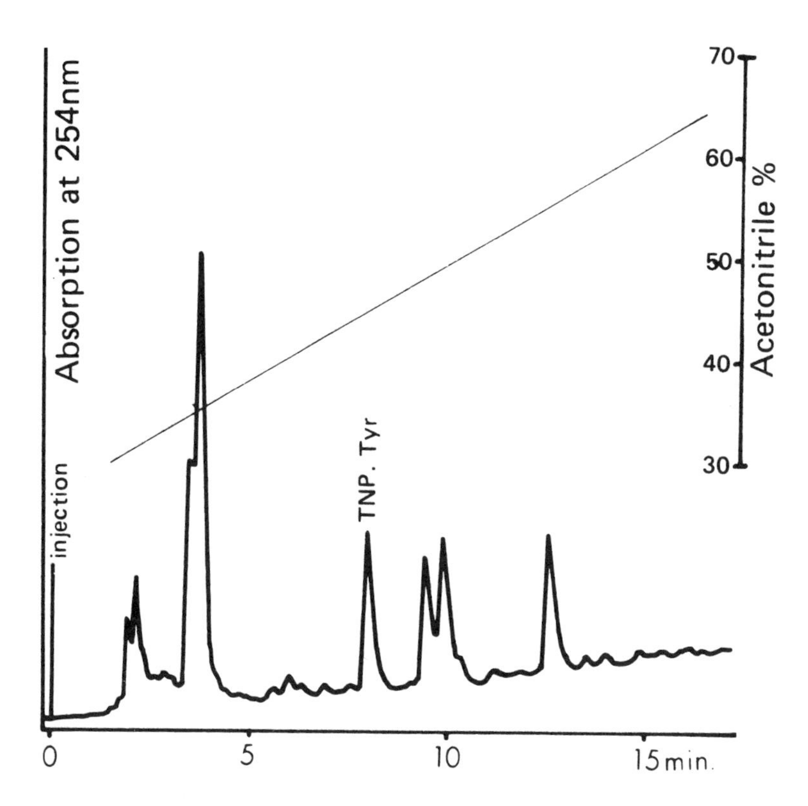

FIGURE 16. HBr/TFA cleavage products of TNP-Tyr-(O-Bzl-2,6-Cl_2)-R, measured at 254 nm.[11d]

REFERENCES AND NOTES

1. **Erickson, B. W. and Merrifield, R. B.,** in *The Proteins,* Vol. 2, Neurath, H. and Hill, R. L., Eds., Academic Press, New York, 1976, 255.
2. **Hirt, J., deLeer, E. W. B., and Beyerman, H. C.,** in *The Chemistry of Polypeptides,* Katsoyannis, P. G., Ed., Plenum Press, New York, 1973, 363.
3. **Sallay, S. I. and Oroszlan, S.,** A model study for monitoring Merrifield solid-phase peptide synthesis by high-pressure liquid chromatography in *Biological/Biomedical Application of Liquid Chromatography,* Hawk, G. L., Ed., Marcel Decker, New York, 1979, 199.
4. **Zhukova, G. F., Ravdel, G. A., and Shchukina, L. A.,** Synthesis of eleodisin analogs on a polymer, *Zh. Obshch. Khim.,* 40, 2753, 1970.
5. **Sanger, F.,** Free amino groups of insulin, *Biochem. J.,* 39, 507, 1945.
6. **Yamashiro, D. and Li, C. H.,** Synthesis of a pentacontapeptide with high lipolytic activity corresponding to carboxyl-terminal fifty amino acids of ovine β-lipotropin, *Proc. Natl. Acad. Sci. U.S.A.,* 71, 4945, 1974.
7. **Gisin, B. F.,** Monitoring of reactions in solid-phase peptide synthesis with picric acid, *Anal. Chim. Acta,* 58, 248, 1972.
8. **Felix, A. M., Jimenez, M. H., Vergona, R., and Cohen, M. R.,** Fluorescamine as a terminating agent in solid-phase peptide synthesis, *Int. J. Peptide Protein Res.,* 7, 11, 1975.
9. **Felix, A. M. and Jimenez, M. H.,** Rapid fluorometric detection for completeness in solid-phase coupling reactions, *Anal. Biochem.,* 52, 377, 1973.
10. **Sallay, S. I., Srivastava, K. S. L., Oroszlan, S., and Gilden, R. V.,** in *Peptides, Structure and Biological Function,* Proc. 6th American Peptide Symp., Gross, E. and Meienhofer, J., Eds., Pierce Chemical Company, Rockford, IL, 1979, 377.

11. A Waters Associates Model ALC/GPC 244 HPLC instrument with two, Model 6000 A solvent delivery systems, a Model 660 solvent programmer and a Model 440 absorbance detector was used throughout this work. LC analysis was performed on a 300 × 7 mm μBondapak® C_{18}-column using linear gradient mixture of acetonitrite/water (+1% AcOH,v/v).

	Flowrate (mℓ/min)	**Programming (% AN/min)**	**AUFS**
a	5	30—100/30	1.0
b	5	30—100/30	0.1
c	5	30—100/15	0.5
d	5	30—100/15	0.1
e	5	30—100/45	0.5

12. **Satake, K., Okuyama, T., Ohashi, M., and Shinoda, T.,** The spectrophotometric determination of amines, amino acids, and peptides with 2,4,6-trinitrobenzene-1-sulfonic acid, *J. Biochem. (Tokyo),* 47, 654, 1960.

SEPARATION OF SYNTHETIC ANALOGS OF PEPTIDE HORMONES

Ken Inouye, Kunio Watanabe, and Ryusei Konaka

INTRODUCTION

Since alkylsilanized silica supports were introduced as stationary phases for peptide separation,[1-4] reversed-phase high performance liquid chromatography (RP-HPLC) has received particular interest in its wider applications to the field of peptide research. At the end of 1978, the increasing numbers of reports were covering from relatively small peptides with less than ten amino acid residues[5-13] to larger peptides including some proteins.[14-18] RP-HPLC has now become a well-established method for the separation and analysis of both native and synthetic peptides.[19,20]

A detailed examination of the behavior of 32 hormonal peptides and 9 proteins on reversed-phase columns was reported by O'Hare and Nice,[21,22] in which they employed as mobile phase one of the phosphate buffer (pH 2.1)-acetonitrile systems, used by Molnár and Horváth[10] for small peptides, to show that the method might be of general use for peptides of a wider range. The use of ion-pairing reagents to control or to modify the retention of peptides was investigated by Hancock et al.[17] We also reported the successful separation of a number of synthetic corticotropin (ACTH) analogs and some other hormonal peptides including insulins, in which an octadecyl silica column was used in combination with tartrate buffer (pH 3.0)-acetonitrile systems containing sodium sulfate and sodium 1-butanesulfonate as mobile phase.[23] The sodium 1-butanesulfonate, as an ion-pairing reagent, and sodium sulfate were added to improve the resolution of closely related peptides and to control their retention. Later the tartrate buffer was replaced by a phosphate buffer and the resulting mobile phase was used for monitoring of the reaction and the analysis of products in the semisynthesis of insulin analogs.[24,25] The low UV absorbance of the phosphate buffer permitted us to use it at much higher concentrations than the carboxylic acid buffer.

In the present report we describe the examination and reexamination of the separation of a variety of synthetic ACTH peptides,[23] native and semisynthetic insulins,[23-25] and solid-phase synthetic human C-peptide[26] and its fragments on a reversed-phase column. RP-HPLC as a powerful tool for the separation of ACTHs,[15,21-23,27] insulins,[15,16,22-24,27,29,30] and related peptides and has been described repeatedly.

The chromatography was performed under isocratic[21,23,24,28,30] or gradient[15,16,21,22,27] elution conditions and in most cases an acetonitrile-water or methanol-water system (pH 2 or 3) containing phosphate, sulfate, acetate, tartrate, or trifluoroacetate was used as mobile phase. The present experiments were carried out by isocratic elution with phosphate buffer (pH 3.0 or 6.5)-acetonitrile systems containing some sodium sulfate. Similar mobile phases have recently been used by Terabe et al.[31] for separation of cytochrome *c* samples from different species.

EXPERIMENTAL

Materials

ACTH peptides — The primary structures of porcine ACTH and human ACTH are shown in Figure 1. All the ACTH peptides including porcine and human ACTHs were synthesized by the conventional solution method in these laboratories except for ACTH-(1-24), which was isolated from a commercial product Cortrosyn® Z (N.V. Organon, The Netherlands). These peptides were purified by repeated carboxymethylcellulose column chromatography

```
1                   5                   10                  15                  20
H-Ser-Tyr-Ser-Met-Glu-His-Phe-Arg-Trp-Gly-Lys-Pro-Val-Gly-Lys-Lys-Arg-Arg-Pro-Val-

  21                  25                  30                  35
  Lys-Val-Tyr-Pro-Asn-Gly-Ala-Glu-Asp-Glu- X -Ala-Glu-Ala-Phe-Pro-Leu-Glu-Phe-OH

                      Human ACTH:     X = Ser
                      Porcine ACTH:   X = Leu

            1                   5                   10                  15                  20
A-Chain     H-Gly-Ile-Val-Glu-Gln-Cys-Cys- X -Ser- Y -Cys-Ser-Leu-Tyr-Gln-Leu-Glu-Asn-Tyr-Cys-Asn-OH

            1                   5                   10                  15                  20
B-Chain     H-Phe-Val-Asn-Gln-His-Leu-Cys-Gly-Ser-His-Leu-Val-Glu-Ala-Leu-Tyr-Leu-Val-Cys-Gly-

              21                  25                  30
              Glu-Arg-Gly-Phe-Phe-Tyr-Thr-Pro-Lys- Z -OH

                      Human insulin:     X = Thr, Y = Ile, Z = Thr
                      Porcine insulin:   X = Thr, Y = Ile, Z = Ala
                      Bovine insulin:    X = Ala, Y = Val, Z = Ala

1                   5                   10                  15                  20
H-Glu-Ala-Glu-Asp-Leu-Gln-Val-Gly-Gln-Val-Glu-Leu-Gly-Gly-Gly-Pro-Gly-Ala-Gly-Ser-

  21                  25                  30
  Leu-Gln-Pro-Leu-Ala-Leu-Glu-Gly-Ser-Leu-Gln-OH

                      Human proinsulin C-peptide [Proinsulin-(33-63)]
```

FIGURE 1. Primary structure of corticotropins (ACTHs), insulins, and human proinsulin C-peptide.

and by partition chromatography on Sephadex® G-25 columns. Their purity was assessed by conventional analytical methods including thin-layer chromatography (TLC). However, it became apparent, when examined by RP-HPLC, that most of them were still contaminated with some minor impurities.[23] In the present work, they were further purified by HPLC as will be described later.

Insulins — The structures of human, porcine, and bovine insulins are shown in Figure 1. Crystalline bovine insulin was purchased from Calbiochem (San Diego, CA) and crystalline porcine insulin was kindly supplied by Lilly Research Laboratories. These two insulins were freed from zinc and then purified on a QAE-Sephadex® A-25 column. Human insulin was derived from porcine insulin by the semisynthetic procedure involving a trypsin-catalyzed coupling of desoctapeptide-(B23-B30)-insulin with a synthetic octapeptide corresponding to positions B23-B30 of human insulin.[32] The [Leu^{B24}]-, [Leu^{B25}]-, and [Leu^{B24}, Leu^{B25}]-analogs of human insulin[24] and the [Thr^{B30}]-, [Ala^{B23}, Thr^{B30}]-, [Ala^{B24}, Thr^{B30}]-, [Ala^{B25}, Thr^{B30}]-, and [Ala^{B26}, Thr^{B30}]-analogs of bovine insulin[25] were also semisynthesized from porcine insulin and bovine insulin, respectively. These insulins were further purified by HPLC if required.

C-Peptide and fragments — Human proinsulin C-peptide (CP), a 31-amino acid peptide corresponding to positions 33-63 of proinsulin (Figure 1), was synthesized by the solid-phase method in these laboratories.[26] The C-peptide fragments CP-(18-31), CP-(10-31), and CP-(7-31) were derived from the corresponding peptide-resin intermediates during the course of the C-peptide synthesis. These peptides were purified to a single component by chromatography on a DEAE-Sephadex® column and by RP-HPLC.

Apparatus and Methods

HPLC was performed by an isocratic elution method using a Waters Associates Model 6000A solvent delivery system, equipped with a Waters U6K injector, and a Japan Spectroscopic UVIDEC-100-II variable wavelength UV detector. A 25 × 0.4 cm i.d. column packed with Nucleosil® $5C_{18}$ (Macherey, Nagel & Co., Düren, Germany) was used throughout. The column was enclosed in a jacket thermostatically maintained at 25°C.

A representative mobile phase was prepared by mixing the following two solutions in a 1:1 ratio: solution A, 100 m*M* sodium phosphate buffer (pH 3.0) containing 300 m*M* sodium sulfate; solution B, (2 × *a*) % acetonitrile in water. The acetonitrile concentration (*a* %) in the mobile phase was so chosen as to optimize retention and separation. This was used as the standard mobile phase in the present work for the separation experiments of most of the peptides except insulins, for which a 50 m*M* sodium phosphate buffer (pH 3.0) containing 50 m*M* sodium sulfate, 5 m*M* sodium 1-butanesulfonate, and 27 to 28% acetonitrile was used.[24] In some experiments the standard mobile phase was modified for tests or to get better resolution by varying the pH or the salt concentration. The flow rate was 1.0 mℓ/min throughout and the eluate was continuously monitored at 220 nm or at 280 nm (for peptides containing Trp and/or Tyr).

These HPLC conditions were also employed for the final purification of the peptide samples when it was required. The sample (50 to 700 μg) was applied to a Nucleosil® $5C_{18}$ column (25 × 0.4 cm) with an appropriate mobile phase selected from those described above. The eluate was monitored at 220 or 280 nm and the fraction corresponding to the main peak was collected. The resulting solution containing the desired peptide in a homogeneous state was stored in a refrigerator. For the separation experiment, a small aliquot (corresponding to 3 to 5 μg peptide) from this solution was used.

RESULTS AND DISCUSSION

ACTH Peptides

Figure 2 illustrates the chromatograms of a partially purified preparation of octadecapeptide [Gly^1]-ACTH-(1-18)-NH_2,[33] in which a 50 m*M* sodium phosphate buffer (pH 3.0) containing 17% acetonitrile and varying concentrations of sodium sulfate was used as mobile phase. At low salt concentrations, an unsymmetrical peak with extensive tailing was produced and the poor resolution could not be improved by increasing the buffer concentration up to 250 m*M* (not shown). However, when the concentration of sodium sulfate was increased to 150 m*M* or higher, the presence of two distinct peaks became apparent. This heterogeneity had not been detected by conventional analytical methods including CM-cellulose column chromatography and TLC. The two peaks were best separated when the mobile phase contained 50 m*M* sodium phosphate (pH 3.0), 150 m*M* sodium sulfate, and 16% acetonitrile. This mobile phase could be applied successfully to most of the separation experiments by varying the concentration of acetonitrile only. This system was therefore used as standard mobile phase in the present work.

The N-terminal half of the ACTH molecule is basic in nature because of the presence of the clustered Lys and Arg residues in positions 15 to 18, while the C-terminal half is very rich in acidic amino acid residues and contains no basic amino acids except for the Lys in position 21. Figure 3 shows the chromatograms of a mixture of ACTH-(1-18)-OH[33] and ACTH-(1-18)-NH_2[33] obtained with different mobile phases. At pH 3.0 with the standard mobile phase these two peptides gave a very sharp peak with good symmetry, but they were not resolved at all (Figure 3A). At a neutral pH the carboxylate group should be dissociated and may be clearly distinguished from the carboxamide group in terms of hydrophobicity. However, the two octadecapeptide analogs were hardly separated also at pH 6.5 (Figure 3B) except for one case, in which some separation was observed (Figure 3C). The poor

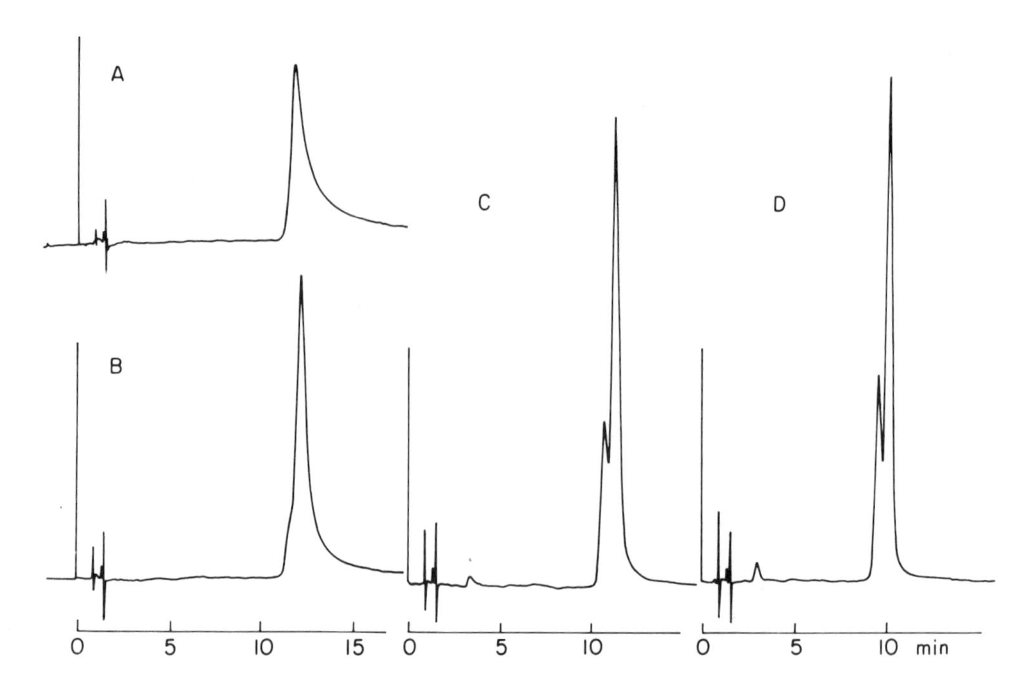

FIGURE 2. Influence of salt concentration in the mobile phase on the chromatography of a preparation of synthetic [Gly[1]]-ACTH-(1-18)-NH_2. Column, Nucleosil® $5C_{18}$, 25 × 0.4 cm; mobile phase, 50 m*M* phosphate buffer (pH 3.0) containing 17% CH_3CN and (A) 25 m*M*, (B) 50 m*M*, (C) 100 m*M*, or (D) 150 m*M* Na_2SO_4; flow rate, 1 mℓ/min; detection, 220 nm (range 0.08); temperature, 25°C.

separation of these peptides may be ascribed to their basic nature. The strong tendency of basic peptides to bind to the column under neutral conditions may have prevented their good separation. Therefore, the HPLC of basic peptides was performed under acidic conditions in the present work.

Figure 4 shows the chromatograms of a partially purified human ACTH-(22-39)[34] at pH 3.0 and pH 6.7. In contrast to ACTH-(1-18), this acidic peptide gave good results with both acidic (Figure 4A) and neutral (Figure 4B) mobile phases, although there is some difference in elution profiles which characterizes the two chromatograms.

Since the ionic nature of peptides varies with pH, the concentration of acetonitrile as organic modifier required for the proper retention should be influenced by the pH of the mobile phase. In fact, as is seen in Figures 3 and 4, the basic peptide ACTH-(1-18) was eluted with mobile phases containing 16.5 and 20.5% acetonitrile under acidic and neutral conditions, respectively, while in the case of the acidic peptide ACTH-(22-39) the acetonitrile concentrations required were 27 and 18.5% under acidic and neutral conditions, respectively.

Figure 5 shows the chromatograms of ACTH peptides with different chain lengths. As seen in Figure 5A, ACTH-(1-18), ACTH-(1-26),[35] ACTH-(1-27),[35] and ACTH-(1-24) were eluted in that order with the standard mobile phase containing 18% acetonitrile. Addition of a fragment Pro-Val-Lys-Val-Tyr-Pro to the C-terminus of ACTH-(1-18) gave ACTH-(1-24) which showed a markedly retarded elution. The further attachment of hydrophilic fragments Asn-Gly and Asn-Gly-Ala to ACTH-(1-24) produces ACTH-(1-26) and ACTH-(1-27), respectively, which were eluted faster than the tetracosapeptide. For elution of natural ACTHs the acetonitrile concentration in the mobile phase had to be elevated to 26.5%. Figure 5B shows an excellent separation of synthetic preparations of human ACTH and porcine ACTH. They differ only by a single amino acid residue in position 31, Ser in human and Leu in porcine hormone, for their 39 amino acid sequences.

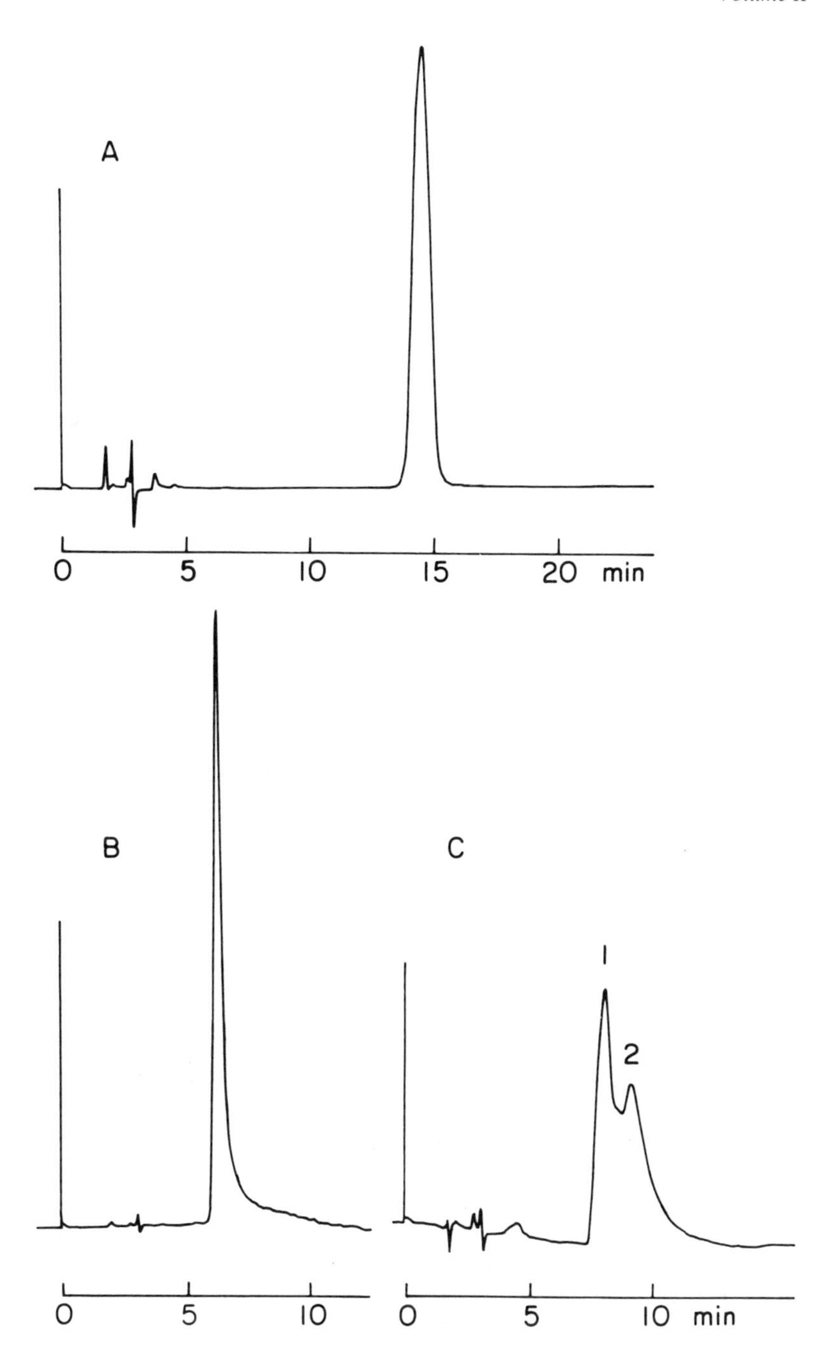

FIGURE 3. Chromatogram of a mixture of ACTH-(1-18)-OH and ACTH-(1-18)-NH_2. Column, Nucleosil® $5C_{18}$, 25 × 0.4 cm; mobile phase, (A) 50 m*M* phosphate buffer (pH 3.0) containing 150 m*M* Na_2SO_4 and 16.5% CH_3CN, (B) 50 m*M* phosphate buffer (pH 6.5) containing 150 m*M* Na_2SO_4 and 20.5% CH_3CN, (C) 250 m*M* phosphate buffer (pH 6.5) containing 20.5% CH_3CN; flow rate, 1 mℓ/min; detection, 220 nm (range 0.08); temperature, 25°C.

Figure 6 shows the separation of ACTH-(1-18)-NH_2 and its [X^1]-analogs,[33] which have a usual or unusual amino acid X substituted for Ser in position 1, and [X^{10}]-analogs[36] having X substituted for Gly in position 10. Among the [X^1]-analogs shown in Figure 6A, [Gly^1]-

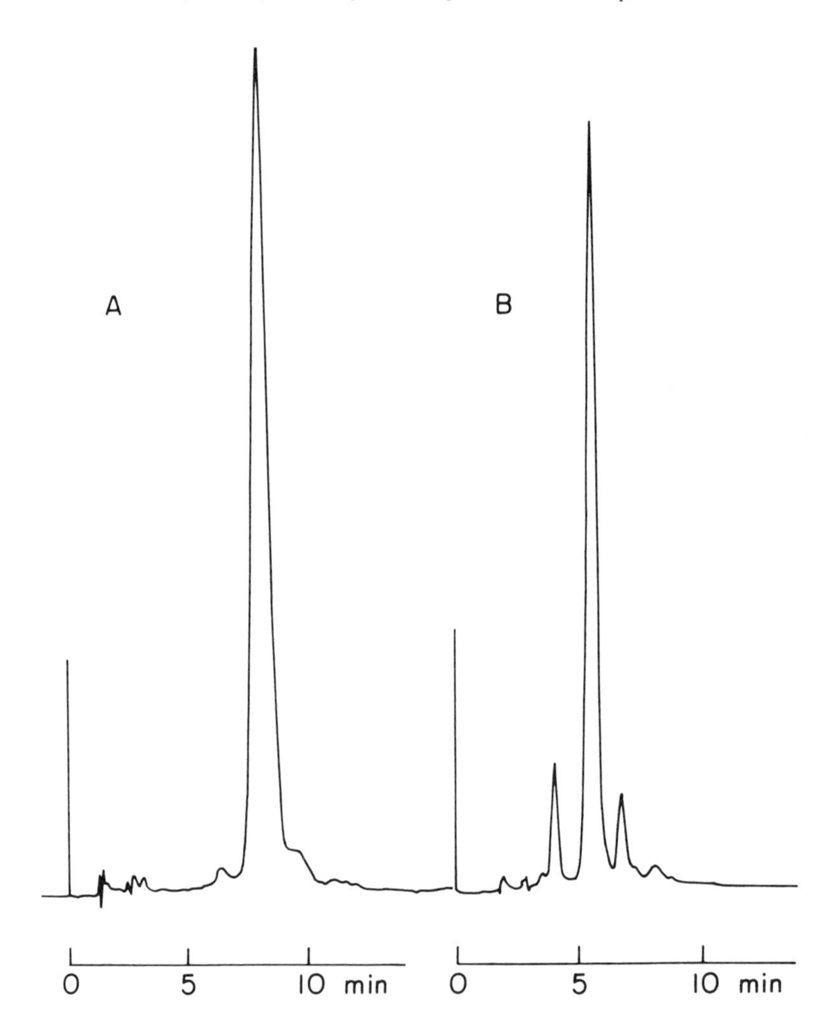

FIGURE 4. Chromatogram of a partially purified preparation of synthetic human ACTH-(22-39). Column, Nucleosil® $5C_{18}$, 25 × 0.4 cm; mobile phase, (A) 50 m*M* phosphate buffer (pH 3.0) containing 150 m*M* Na_2SO_4 and 27% CH_3CN or (B) 100 m*M* phosphate buffer (pH 6.7) containing 18.5% CH_3CN; flow rate, 1 mℓ/min; detection, 220 nm (range 0.08); temperature, 25°C.

ACTH-(1-18)-NH_2 and [βAla^1]-ACTH-(1-18)-NH_2 are hardly resolved, although a slight separation was observed when the acetonitrile concentration was reduced to 16% (not shown). The retention time of the four peptides increases in the order, Ser $<$ Gly $\leq$ βAla $\ll$ Aib (α-aminoisobutyric acid), with respect to the N-terminal residue (X^1) and this order of retention agrees well with the order of hydrophobicity of X.

Figure 6B shows the good separation of ACTH-(1-18)-NH_2 and its four [X^{10}]-analogs except for the pair of ACTH-(1-18)-NH_2 and [Ala^{10}]-ACTH-(1-18)-NH_2, which are only slightly separated from each other. It is worthy of note that [βAla^{10}]-ACTH-(1-18)-NH_2 is well separated from ACTH-(1-18)-NH_2 (with Gly in position 10). This represents a striking contrast to the poor separation between the [Gly^1]- and [βAla^1]-analogs. In addition, βAla^{10}-peptide is eluted faster than Gly^{10}-peptide, despite the fact that the former has one more methylene unit than the latter. The production of different effects by the same kind of substitution may be ascribed to where the substitution occurs. When βAla is substituted for Gly in an *exo* position of the peptide chain, the extra methylene group may not affect the

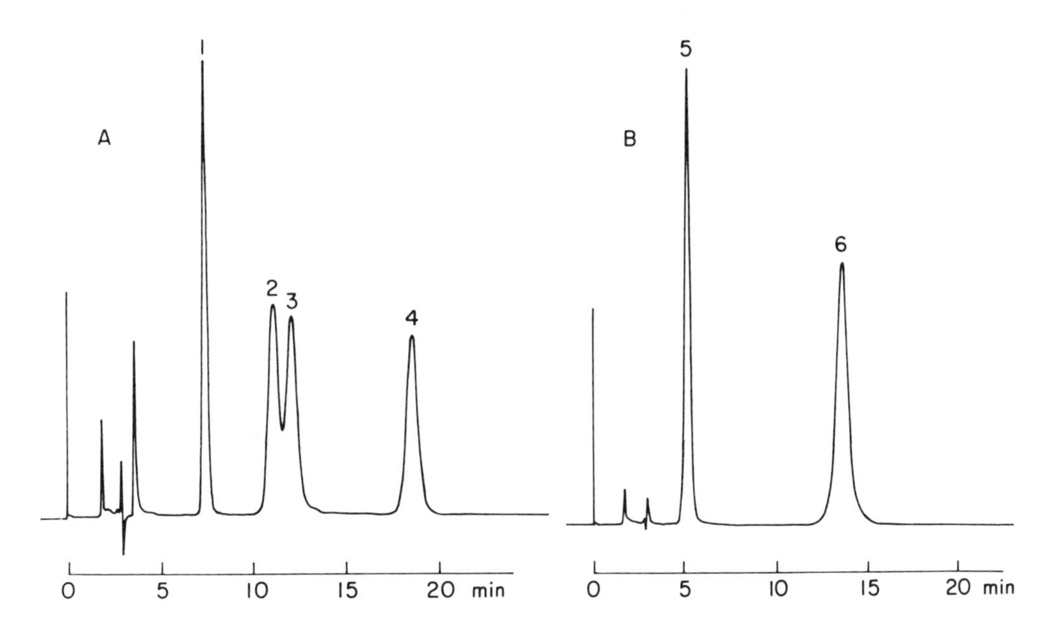

FIGURE 5. Separation of ACTH peptides with different chain lengths. (A) A mixture of ACTH-(1-18) (1), ACTH-(1-26) (2), ACTH-(1-27) (3) and ACTH-(1-24) (4). (B) A mixture of human ACTH (5) and porcine ACTH (6). Column, Nucleosil® $5C_{18}$, 25 × 0.4 cm; mobile phase, 50 m*M* phosphate buffer (pH 3.0) containing 150 m*M* Na_2SO_4 and (A) 18% or (B) 26.5% CH_3CN; flow rate, 1 mℓ/min; detection, 220 nm (range 0.04); temperature, 25°C.

nature of peptide more than slightly. When the same substitution occurs in an *endo* position, however, the extra methylene unit inserted into the peptide backbone will cause a so-called ''frame shift'', which alters the steric relations of side chain groups and probably leads to a significant change in conformation or conformational freedom of the peptide. This must affect the hydrophobicity of the whole molecule and should be a main cause for the unexpected chromatographic behavior of βAla^{10}-peptide. This kind of structural change usually leads to a considerable loss of biological activity, as is observed with [βAla^{10}]-ACTH-(1-18)-NH_2.[36]

As mentioned above ACTH-(1-18)-NH_2 and [Ala^{10}]-ACTH-(1-18)-NH_2 showed poor resolution. The structural difference between these peptides (Gly/Ala at position 10) may be similar to that between [Gly^1]-ACTH-(1-18)-NH_2 and [βAla^1]-ACTH-(1-18)-NH_2 (Gly/βAla at position 1, Figure 6A) and also to that between [βAla^1, Orn^{15}]-ACTH-(1-18)-NH_2[33] (not shown) and [βAla^1]-ACTH-(1-18)-NH_2 (Orn/Lys at position 15); the latter of each pair having an extra methylene unit in the amino acid side chain. The last pair has never been separated at all. Another example of a difficult separation is shown in Figure 7. The figure shows the chromatograms of [Aib^1, Lys^{17}, Lys^{18}]-ACTH-(1-18)-R, where R denotes different amide groups at the C-terminal, in which the two peptides with R = NH_2 and $N(CH_3)_2$, respectively, did not show any separation from each other although they were fairly well separated from the pyrrolidide analog (Figure 7B).

Figure 8 shows two examples of successful separation of diastereoisomeric mixtures. One is [Ala^{10}]-ACTH-(1-18)-NH_2 and [D-Ala^{10}]-ACTH-(1-18)-NH_2[36] (Figure 8A), and the other is [βAla^1, Orn^{15}]-ACTH-(1-18)-NH_2 (with Phe in position 7) and [βAla^1, D-Phe^7, Orn^{15}]-ACTH-(1-18)-NH_2[33] (Figure 8B). The D-Ala-isomer is more retarded than the corresponding L-isomer, while the D-Phe-isomer is eluted faster than the L-isomer. The good separation of these isomers from each other suggests their significant conformational differences. In the strict sense racemization always occurs during the course of peptide synthesis. Therefore,

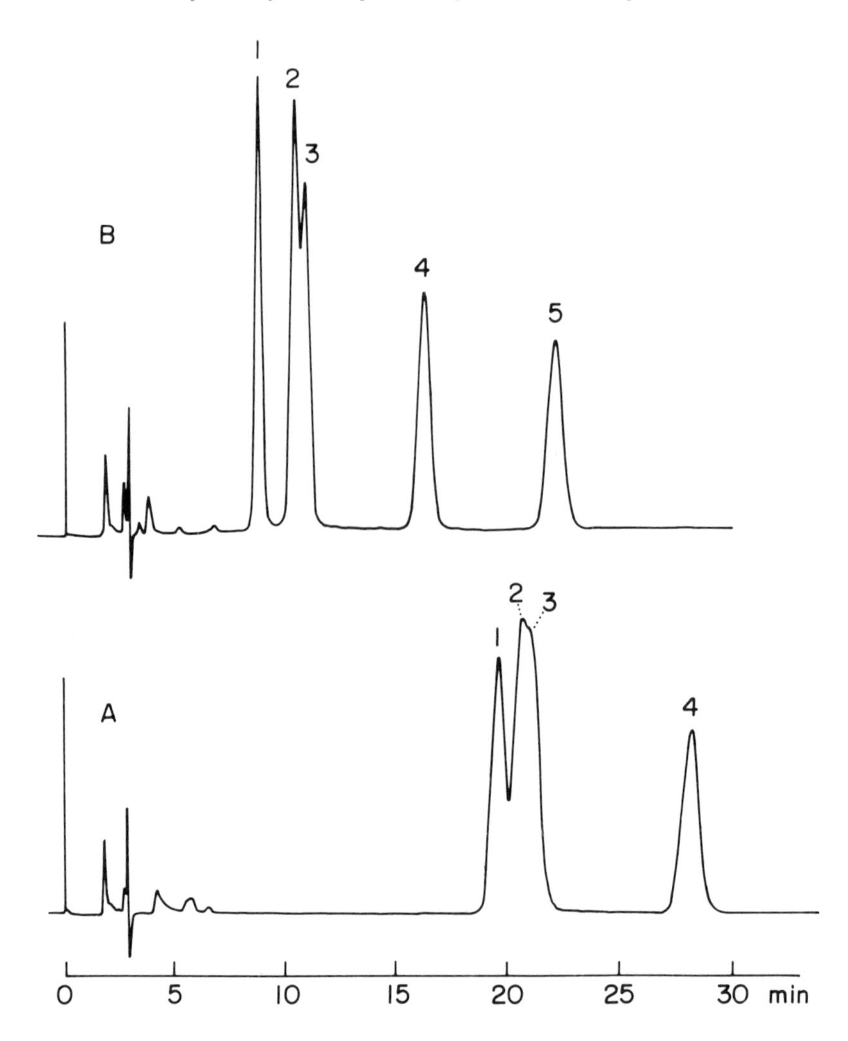

FIGURE 6. Separation of ACTH-(1-18)-NH_2 and its analogs with substitution in position 1 or 10. (A) A mixture of [X^1]-ACTH-(1-18)-NH_2; X = Ser (normal, 1), Gly (2), βAla (3) and Aib (4). (B) A mixture of [X^{10}]-ACTH-(1-18)-NH_2; X = βAla (1), Gly (normal, 2), Ala (3), D-Ala (4) and Aib (5). Column, Nucleosil® $5C_{18}$, 25 × 0.4 cm; mobile phase, 50 m*M* phosphate buffer (pH 3.0) containing 150 m*M* Na_2SO_4 and (A) 16% or (B) 17% CH_3CN; flow rate, 1 mℓ/min; detection, 220 nm (range 0.08); temperature, 25°C.

the detection of minute amounts of diastereoisomers and their separation from the desired product are among the most important problems in peptide synthesis. The present results as well as those described in the literature[11,23,37-40] clearly indicate that RP-HPLC is particularly useful for this purpose.

Figure 9 shows chromatograms for a mixture of ACTH-(1-18)-NH_2 and its positional isomer [Lys^3, Ser^{11}]-ACTH-(1-18)-NH_2. The latter peptide has the structure obtained by exchanging the two amino acid residues Ser in position 3 and Lys in position 11 of the former sequence. The mixture also contains [Lys^3]-ACTH-(1-18)-NH_2 for comparison. The first two peptides were not separated when the acetonitrile concentration was 17% (Figure 9A), but they exhibited some separation as the acetonitrile concentration was reduced to 16% (Figure 9B). This separation cannot be explained in terms of the total hydrophobicity of constituent amino acid residues but in terms of the hydrophobicity of the whole molecule,

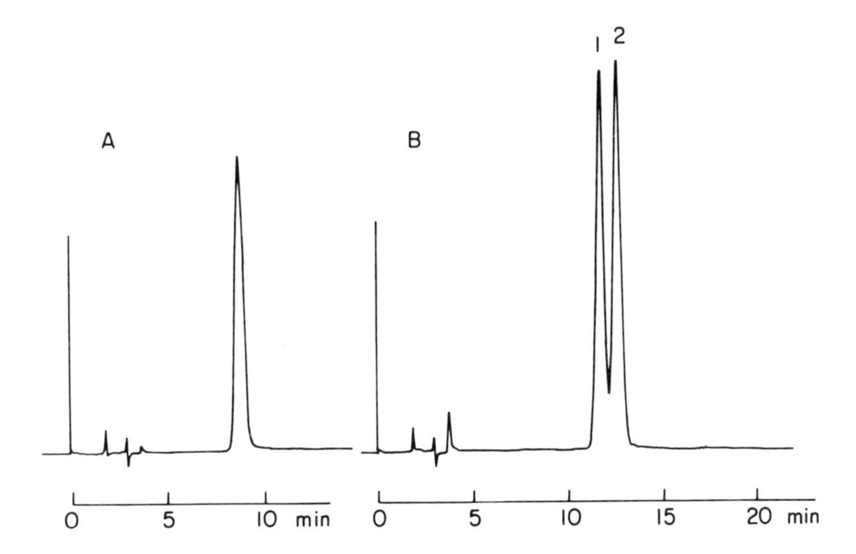

FIGURE 7. Separation of ACTH octadecapeptides with different C-terminal amide groups. (A) A mixture of [Aib^1, Lys^{17}, Lys^{18}]-ACTH-(1-18)-R; R = NH_2 and $N(CH_3)_2$. (B) A mixture of [Aib^1, Lys^{17}, Lys^{18}]-ACTH-(1-18)-R; R = $N(CH_3)_2$ (1) and N☐ (2). Column, Nucleosil® $5C_{18}$, 25 × 0.4 cm; mobile phase, 50 m*M* phosphate buffer (pH 3.0) containing 150 m*M* Na_2SO_4 and (A) 18% or (B) 17.5% CH_3CN; flow rate, 1 mℓ/min; detection, 220 nm (range 0.08); temperature, 25°C.

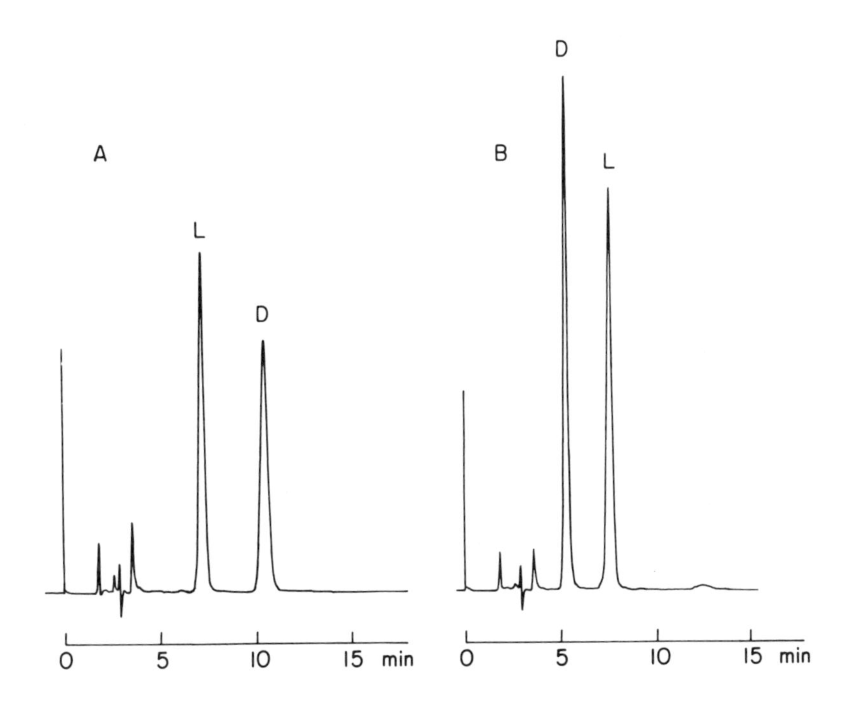

FIGURE 8. Separation of diastereoisomeric mixtures of ACTH octadecapeptides. (A) A mixture of [X^{10}]-ACTH-(1-18)-NH_2; X = Ala (L) and D-Ala (D). (B) A mixture of [βAla^1, X^7, Orn^{15}]-ACTH-(1-18)-NH_2; X = Phe (normal, L) and D-Phe (D). Column, Nucleosil® $5C_{18}$, 25 × 0.4 cm; mobile phase, 50 m*M* phosphate buffer (pH 3.0) containing 150 m*M* Na_2SO_4 and (A) 18% or (B) 17.5% CH_3CN; flow rate, 1 mℓ/min; detection, 220 nm (range 0.08); temperature, 25°C.

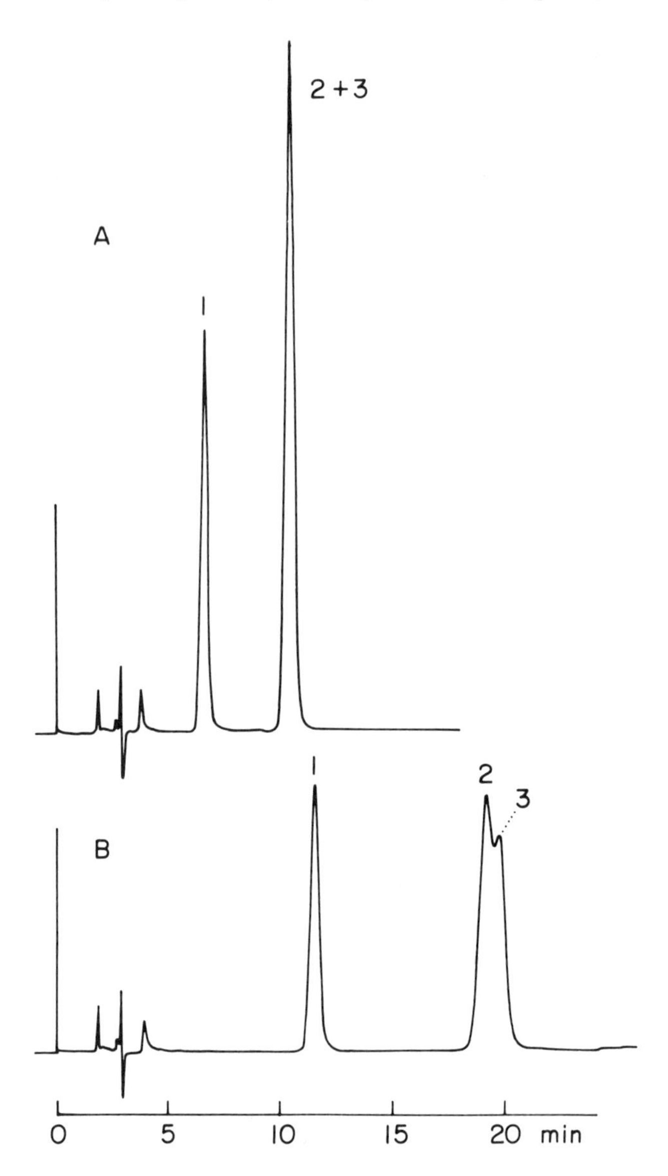

FIGURE 9. Separation of [X^3, Y^{11}]-ACTH-(1-18)-NH_2. 1: X = Y = Lys; 2 (normal): X = Ser, Y = Lys; 3: X = Lys, Y = Ser. Column, Nucleosil® $5C_{18}$, 25 × 0.4 cm; mobile phase, 50 m*M* phosphate buffer (pH 3.0) containing 150 m*M* Na_2SO_4 and (A) 17% or (B) 16% CH_3CN; flow rate, 1 mℓ/min; detection, 220 nm (range 0.08); temperature, 25°C.

which must be slightly different between the two isomers because of their conformational difference.

In Figure 10 a comparison is made for the chromatographic retention of [Aib^1]-ACTH-(1-18)-NH_2[33] (with two arginines in positions 17 and 18), [Aib^1, Lys^{17}, Lys^{18}]-ACTH-(1-18)-NH_2 and their Nle^4-analogs which have a norleucine residue substituted for Met in position 4. The isosteric relationship between Arg and Lys and that between Met and Nle are shown as follows:

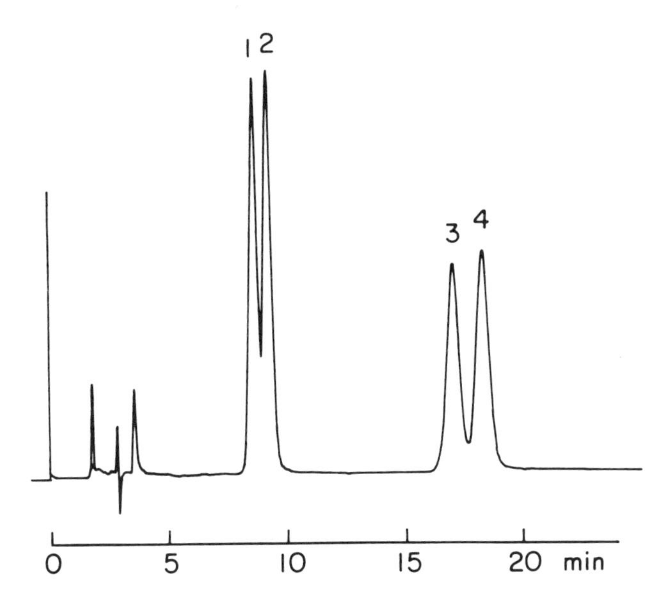

FIGURE 10. Separation of [Aib[1], X[4], Y[17,18]]-ACTH-(1-18)-NH_2. (1) X = Met, Y = Lys; (2) (normal): X = Met, Y = Arg; (3) X = Nle, Y = Lys; (4) X = Nle, Y = Arg. Column, Nucleosil® $5C_{18}$, 25 × 0.4 cm; mobile phase, 50 m*M* phosphate buffer (pH 3.0) containing 150 m*M* Na_2SO_4 and 18% CH_3CN; flow rate, 1 mℓ/min; detection, 220 nm (range 0.08); temperature, 25°C.

Lys	Arg	Met	Nle
NH_2	$NH{=}C{-}NH_2$		
CH_2	NH	CH_3	CH_3
CH_2	CH_2	S	CH_2
CH_2	CH_2	CH_2	CH_2
CH_2	CH_2	CH_2	CH_2
NH—CH—CO	NH—CH—CO	NH—CH—CO	NH—CH—CO

Figure 10 shows that the Lys-analogs are eluted a little faster than the corresponding Arg-peptides and that the Nle-analogs are much more retarded than the corresponding Met-peptides.

Figure 11 shows the separation of [Aib[1]]-ACTH-(1-16)-Lys_n-NH_2 (n = 1, 2, 3, or 4), in which the extra lysines are attached to the C-terminus of the hexadecapeptide. The retention time of these analogs shortened as the number of lysines increased, indicating that the ε-amino groups of the lysines affect the retention of peptide more strongly than their butyl side chains at pH 3.0.

Figure 12 is a chromatogram of a mixture of ACTH heptacosapeptides. The figure shows the separation of ACTH-(1-27)[35] and [Asp[25], Ala[26], Gly[27]]-ACTH-(1-27),[41] the latter containing a wrong sequence Asp-Ala-Gly (proposed for human ACTH in 1961)[42] instead of the correct Asn-Gly-Ala sequence in positions 25 to 27.

The ACTH peptides mentioned above are mostly related to the N-terminal half of the

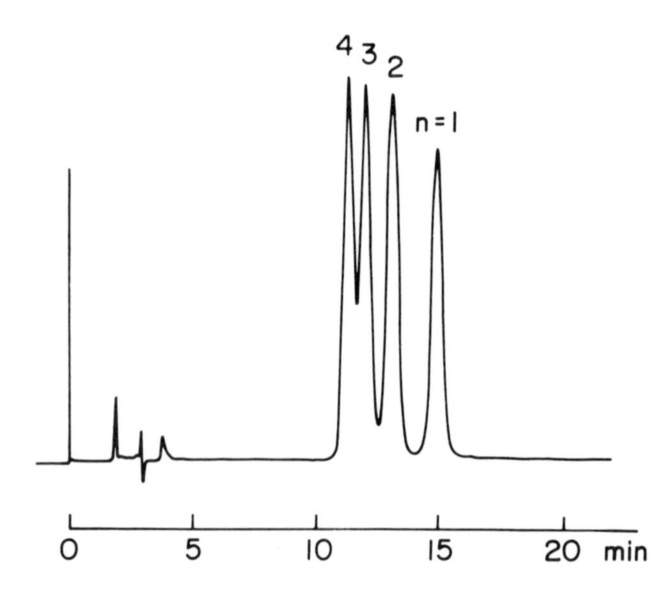

FIGURE 11. Separation of [Aib[1]]-ACTH-(1-16)-Lys_n-NH_2 (n = 1, 2, 3, and 4). Column, Nucleosil® $5C_{18}$, 0.4 × 25 cm; mobile phase, 50 m*M* phosphate buffer (pH 3.0) containing 150 m*M* Na_2SO_4 and 17% CH_3CN; flow rate, 1 mℓ/min; detection, 220 nm (range 0.08); temperature, 25°C.

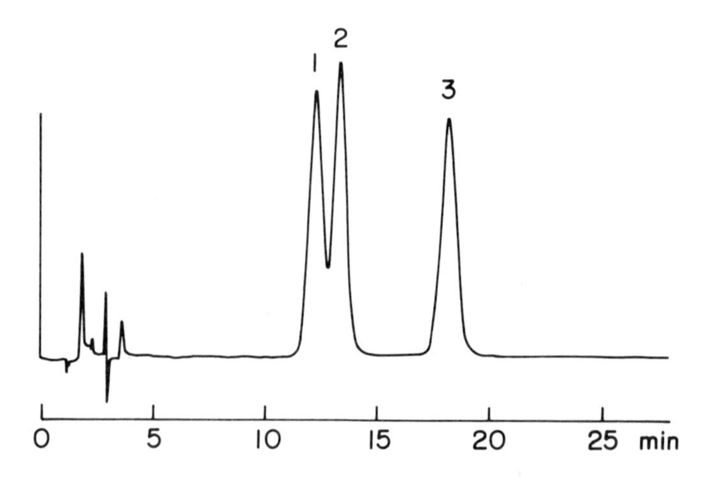

FIGURE 12. Separation of ACTH-(1-27) and its analogs. (1) ACTH-(1-27) with Asn-Gly-Ala in positions 25-27, (2) [Asp^{25}, Ala^{26}, Gly^{27}]-ACTH-(1-27), (3) [Aib^{1}, Asp^{25}, Ala^{26}, Gly^{27}]-ACTH-(1-27). Column, Nucleosil® $5C^{18}$, 25 × 0.4 cm; mobile phase, 50 m*M* phosphate buffer (pH 3.0) containing 150 m*M* Na_2SO_4 and 18% CH_3CN; flow rate, 1 mℓ/min; detection, 220 nm (range 0.08); temperature, 25°C.

hormone molecule. The chromatograms of the C-terminal octadecapeptide, ACTH-(22-39), of human hormone and the corresponding porcine peptide[34] are shown in Figure 13. Although the two peptides differ by only a single amino acid residue at position 31 (Ser in human peptide and Leu in porcine peptide) as mentioned already, they were well separated at both pH 3.0 (Figure 13A) and pH 6.7 (Figure 13B). In Figure 13B a small peak, indicated by the vertical arrow, was increased with time when the porcine peptide was kept in a neutral solution. This peak material is most likely to be the succinimide intermediate,

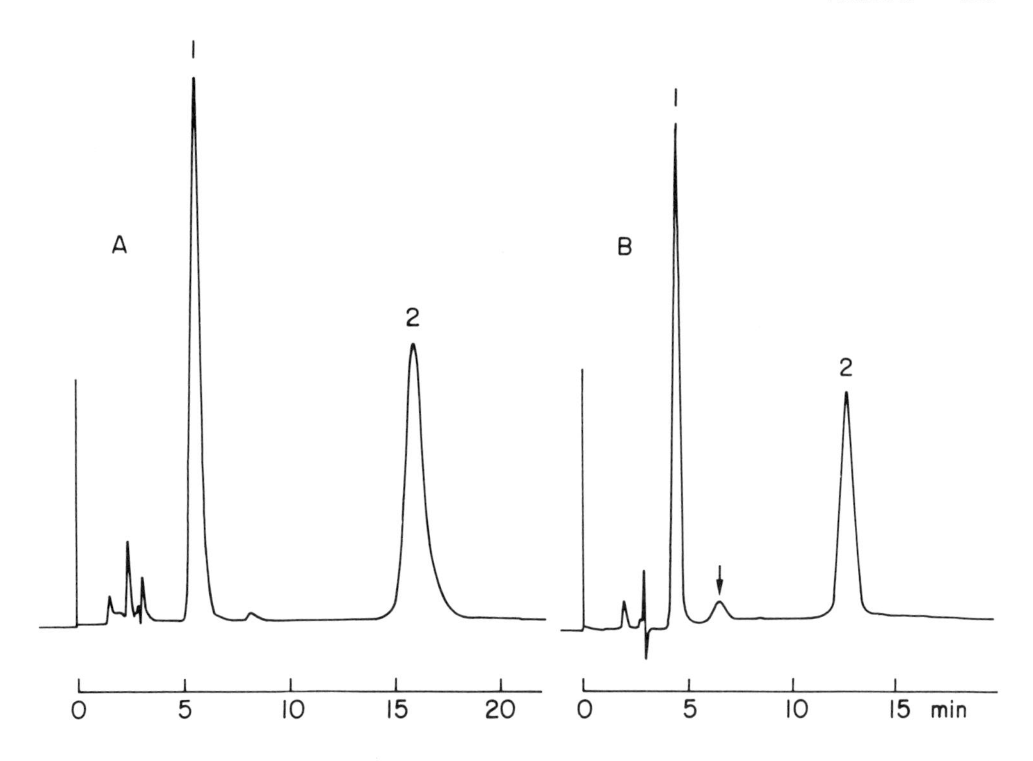

FIGURE 13. Separation of ACTH octadecapeptides. (1) Human ACTH-(22-39), (2) porcine ACTH-(22-39). Column, Nucleosil® $5C^{18}$, 25 × 0.4 cm; mobile phase, (A) 50 m*M* phosphate buffer (pH 3.0) containing 150 m*M* Na_2SO_4 and 28.5% CH_3CN or (B) 100 m*M* phosphate buffer (pH 6.7) containing 19% CH_3CN; flow rate, 1 mℓ/min; detection, 220 nm (range 0.08); temperature, 25°C. The vertical arrow indicates the presence of the possible β-Asp^{25}-peptide derived from porcine ACTH-(22-39) (see text).

$$\begin{array}{l} -\text{NHCH-CO} \diagdown \\ \quad\;\; | \qquad\qquad\;\; \text{N}- \\ \quad \text{CH}_2\text{CO} \diagup \end{array}$$

or the β-Asp-peptide,

$$\begin{array}{l} -\text{NHCH-COOH} \\ \quad\;\; | \\ \quad \text{CH}_2\text{CONH}- \end{array}$$

which could be derived by intramolecular rearrangement at the Asn-Gly bond (positions 25-26), since ACTH is known to undergo deamidation at Asn-25 to form a mixture of α-Asp- and β-Asp-peptides when treated with 0.1 *M* ammonia.[43]

Insulins

Insulin consists of two peptide chains, the A-chain containing 21 amino acid residues and the B-chain containing 30 amino acid residues, connected by two disulfide linkages (Figure 1). In contrast to ACTH, which is a linear and structurally flexible peptide, insulin has a rather rigid tertiary structure. The separation experiments of insulins were all performed with a 50 m*M* sodium phosphate buffer (pH 3.0) containing 50 m*M* sodium sulfate, 5 m*M* sodium 1-butanesulfonate, and 27 to 28% acetonitrile as mobile phase, although in our recent work the standard mobile phase (see Apparatus and Methods) containing 26% acetonitrile and no butanesulfonate as ion-pairing reagent has been found to give almost the same retention and separation.

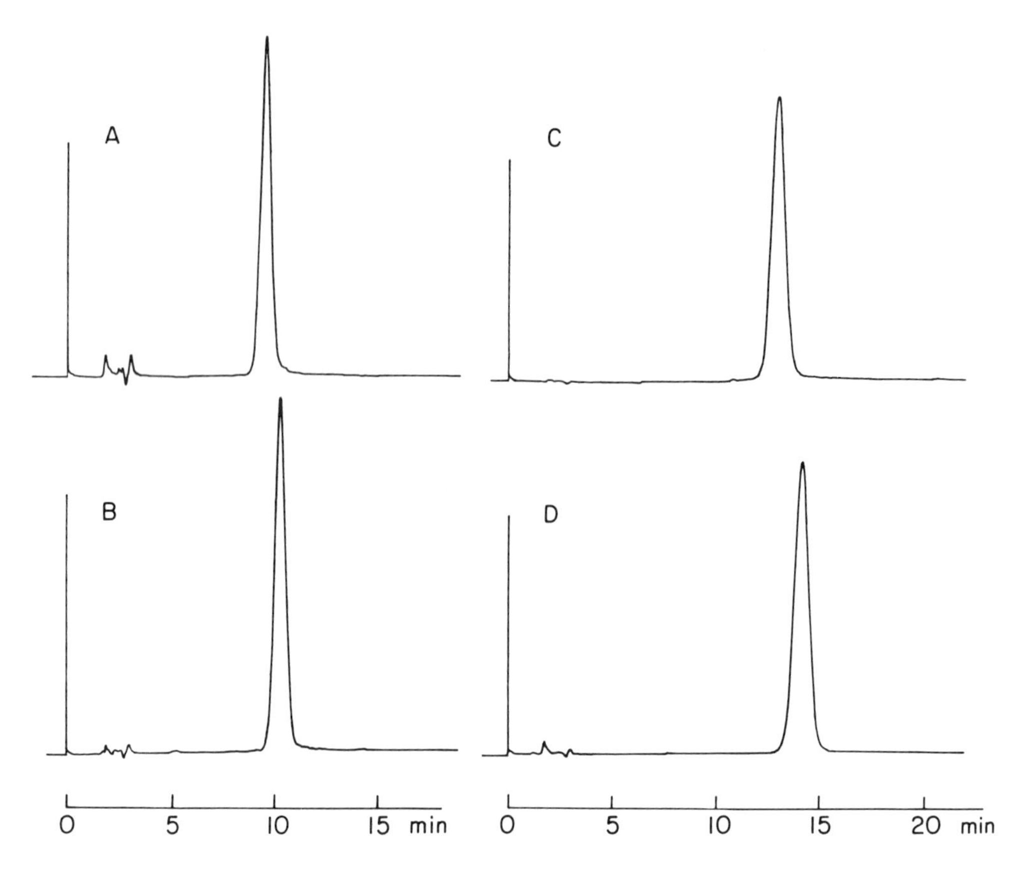

FIGURE 14. Chromatograms of purified insulins. (A) Semisynthetic [Thr^{B30}]-bovine insulin derived from bovine insulin (B). (C) Semisynthetic human insulin derived from porcine insulin (D). Column, Nucleosil® 5C_{18}, 25 × 0.4 cm; mobile phase, 50 m*M* phosphate buffer (pH 3.0) containing 50 m*M* Na_2SO_4, 5 m*M* 1-BuSO_3Na and 28% CH_3CN; flow rate, 1 mℓ/min; detection, 220 nm (range 0.08); temperature, 25°C.

Figure 14 shows chromatograms of bovine and porcine insulins from natural origin and those of [Thr^{B30}]-bovine insulin[25] and [Thr^{B30}]-porcine insulin (= human insulin)[32] prepared semisynthetically from bovine and porcine insulins, respectively. Figure 15 demonstrates good separation of these four insulins. Porcine insulin differs from bovine hormone by two amino acid residues on the A-chain; Thr instead of Ala in position A8 and Ile instead of Val in position A10 (Figure 1). The two semisynthetic insulins differ from their parent insulins only by a single amino acid residue at the C-terminal of the B-chain; Thr instead of Ala in position B30. The elution order of the four insulins was [Thr^{B30}]-bovine, bovine, human, and porcine.

Figure 16 shows excellent separation of semisynthetic preparations of human insulin and its three analogs having one or two Leu residues substituted for the Phe residues in positions B24 and B25. They were eluted in the order, [Leu^{B25}]-insulin, [Leu^{B24}, Leu^{B25}]-insulin, normal insulin, and [Leu^{B24}]-insulin.[24] Leu^{B24}-insulin and Leu^{B25}-insulin have the identical amino acid composition but their retention times were considerably different. We have recently found that the circular dichroism spectrum of Leu^{B25}-insulin is rather similar to, but that of Leu^{B24}-insulin is clearly distinct from, that of normal insulin indicating a substantial conformational difference between the two Leu-substituted insulins.[25] This must have been reflected on the observed difference in their retention times. Leu^{B24}- and Leu^{B25}-insulins exhibited 20 to 30 and 1 to 2%, respectively, of the biological activities of normal insulin in in vitro assays.[24]

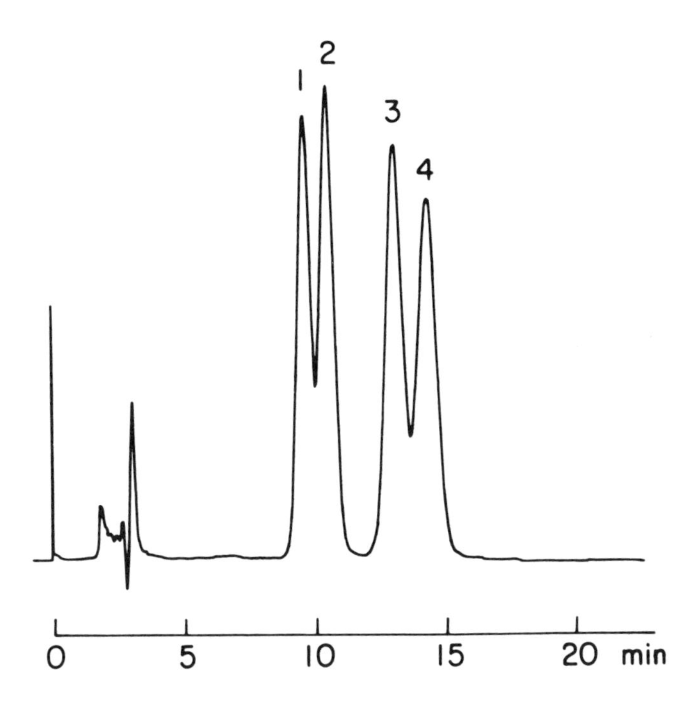

FIGURE 15. Separation of insulins. A mixture of [Thr^{B30}]-bovine insulin (1), bovine insulin (2), human insulin (3), and porcine insulin (4) was chromatographed under the same conditions as in Figure 14. (From Terabe, S., et al., *J. Chromatogr.*, 172, 163, 1979. With permission.)

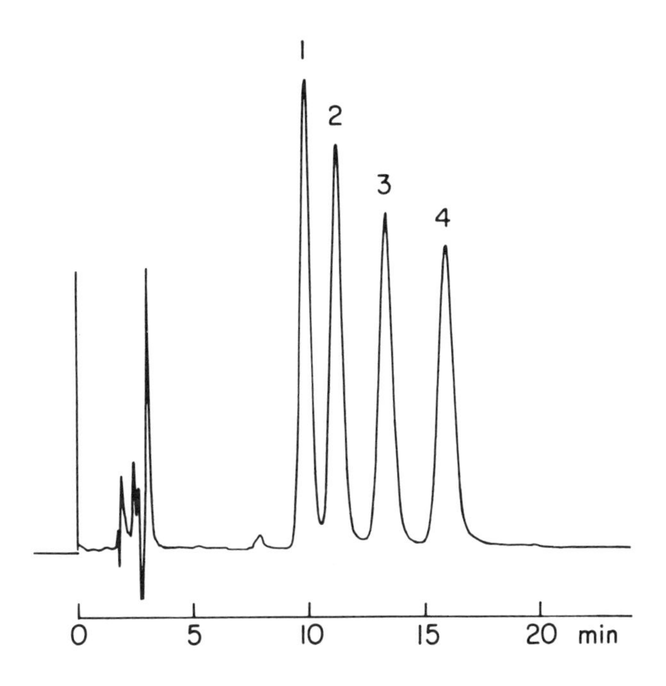

FIGURE 16. Separation of semisynthetic human insulin and its analogs. (1) [Leu^{B25}]-insulin; (2) [Leu^{B24}, Leu^{B25}]-insulin; (3) normal insulin; (4) [Leu^{B24}]-insulin. Column, Nucleosil® 5C$_{18}$, 25 × 0.4 cm; mobile phase, 50 m*M* phosphate buffer (pH 3.0) containing 50 m*M* Na_2SO_4, 5 m*M* 1-$BuSO_3Na$ and 28% CH_3CN; flow rate, 1 mℓ/min; detection, 220 nm (range 0.08); temperature, 25°C. (From Inouye, K., et al., *Experientia*, 37, 811, 1981. With permission.)

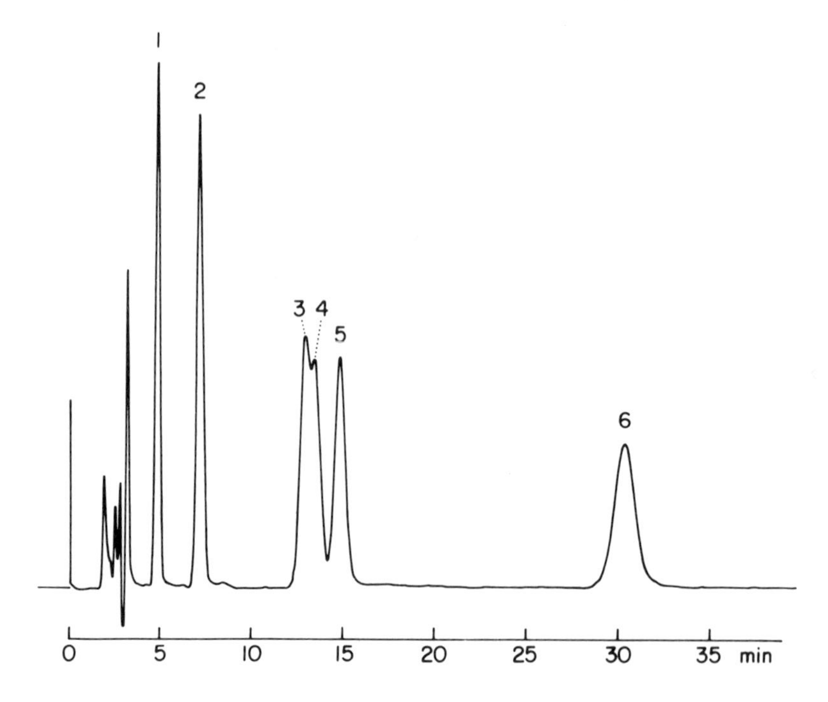

FIGURE 17. Separation of bovine insulin and its semisynthetic analogs. (1) [Ala^{B25}, Thr^{B30}]-insulin; (2) [Ala^{B24}, Thr^{B30}]-insulin; (3) [Ala^{B26}, Thr^{B30}]-insulin; (4) [Thr^{B30}]-insulin; (5) normal insulin; (6) [Ala^{B23}, Thr^{B30}]-insulin. Column, Nucleosil® 5C_{18}, 25 × 0.4 cm; mobile phase, 50 m*M* phosphate buffer (pH 3.0) containing 50 m*M* Na_2SO_4, 5 m*M* 1-$BuSO_3Na$ and 27% CH_3CN; flow rate, 1 mℓ/min; detection, 220 nm (range 0.08); temperature, 25°C.

A mixture of semisynthetic preparations of [Thr^{B30}]-bovine insulin and its four analogs[25] which contain an Ala residue substituted for Gly-B23, Phe-B24, Phe-B25, or Tyr-B26 was also chromatographed and the result is shown in Figure 17. For comparison natural bovine insulin was included in the mixture. The retention times were increased in the order, [Ala^{B25}, Thr^{B30}]-insulin,[Ala^{B24}, Thr^{B30}]-insulin, [Ala^{B26}, Thr^{B30}]-insulin, [Thr^{B30}]-insulin, bovine insulin, and [Ala^{B23}, Thr^{B30}]-insulin. The substitution of Phe by Leu is to change the side chain group from phenyl to isopropyl and may not alter the hydrophobic nature greatly, whereas the substitution of Phe by Ala is to remove the phenyl group from the side chain. The latter substitution in position B24 or B25 causes a remarkable decrease in the hydrophobicity of the insulin molecule as shown in Figure 17 by the fact that Ala^{B24}- and Ala^{B25}-insulins are eluted much faster than normal insulin, although the substitution for Tyr-B26 (removal of a hydroxyphenyl group) appears to cause little change in the molecular hydrophobicity as indicated by the poor separation between [Ala^{B26}, Thr^{B30}]- and [Thr^{B30}]-insulins (Figure 17). The substitution of Gly-B23 by Ala gives Ala^{B23}-insulin which differs from normal insulin only by an extra methyl group. In spite of this structural resemblance, Ala^{B23}-insulin had a much longer retention time than normal insulin (Figure 17), suggesting that a significant conformational difference may exist between the two insulins. Ala^{B26}-insulin was fully active, Ala^{B25}-insulin was partially active, and Ala^{B24}- and Ala^{B23}-insulins were only slightly active in biological assays.[25]

C-Peptide (CP) and Fragments

Human CP is a 31-amino acid peptide and may be characterized by the absence of aromatic, basic, and sulfur-containing amino acid residues in its structure (Figure 1). Figure 18 shows the good separation of CP and its fragments under both acidic and neutral conditions. The

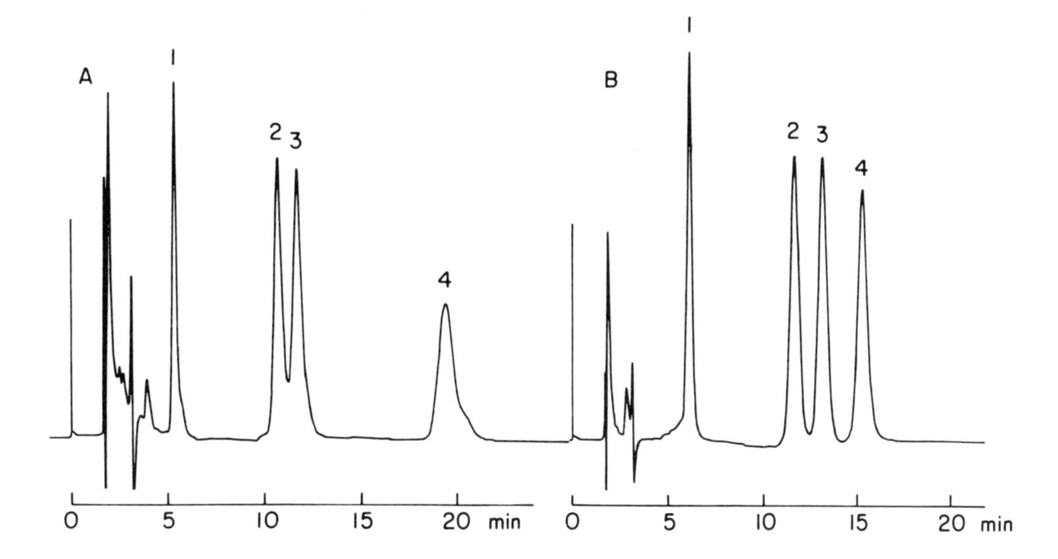

FIGURE 18. Separation of human proinsulin C-peptide (CP) and its fragments. (1) CP-(18-31), (2) CP-(1-31), (3) CP-(7-31), (4) CP-(10-31). Column, Nucleosil® $5C_{18}$, 25 × 0.4 cm; mobile phase, (A) 50 m*M* phosphate buffer (pH 3.0) containing 150 m*M* Na_2SO_4 and 24% CH_3CN or (B) 50 m*M* phosphate buffer (pH 6.5) containing 150 m*M* Na_2SO_4 and 20.5% CH_3CN; flow rate, 1 mℓ/min; detection, 220 nm (range 0.08); temperature, 25°C.

elution order is: CP-(18-31), CP-(1-31), CP-(7-31), and CP-(10-31). This order was not varied with the pH change of the mobile phase, although CP-(10-31) showed a remarkable retardation at pH 3.0 compared to at pH 6.5. These are acidic peptides that contain 1, 5, 2, and 2, acidic amino acid residues, respectively. However, there seems to be little or no apparent correlation between the primary structure of peptides and their retention times as far as the CP fragments are concerned within the narrow range of their single isocratic elution system.

CONCLUSION

The separation experiments were carried out with a variety of 17- to 51-amino acid peptides including ACTHs, insulins, and human C-peptide by RP-HPLC under isocratic conditions, in which an octadecyl silica column was used as stationary phase and phosphate buffer-acetonitrile systems as the mobile phase. The results clearly demonstrate that the HPLC method used is well suited to the rapid and efficient separation and analysis of peptides with molecular weights of up to about 6000.

In the present work isocratic elution was extensively used for known mixtures of closely related peptides and has proved to be useful for the separation of compounds having similar hydrophobicities. However, when the composition of the sample is not known or when the mixture has a wide range of hydrophobicities, the gradient elution method may be more suitable.

REFERENCES

1. **Rzeszotarski, W. J. and Mauger, A. B.,** Reversed-phase high-pressure liquid chromatography of actinomycins, *J. Chromatogr.*, 86, 246, 1973.
2. **Tsuji, K., Robertson, J. H., and Bach, J. A.,** Quantitative high-pressure liquid chromatographic analysis of bacitracin, a polypeptide antibiotic, *J. Chromatogr.*, 99, 597, 1974.
3. **Tsuji, K. and Robertson, J. H.,** Improved high-performance liquid chromatographic method for polypeptide antibiotics and its application to study the effects of treatments to reduce microbial levels in bacitracin powder, *J. Chromatogr.*, 112, 663, 1975.
4. **Hancock, W. S., Bishop, C. A., and Hearn, M. T. W.,** High pressure liquid chromatography in the analysis of underivatised peptides using a sensitive and rapid procedure, *FEBS Lett.*, 72, 139, 1976.
5. **Krummen, K. and Frei, R. W.,** The separation of nonapeptides by reversed-phase high-performance liquid chromatography, *J. Chromatogr.*, 132, 27, 1977.
6. **Krummen, K. and Frei, R. W.,** Quantitative analysis of nonapeptides in pharmaceutical dosage forms by high-performance liquid chromatography, *J. Chromatogr.*, 132, 429, 1977.
7. **Hansen, J. J., Greibrokk, T., Curie, B. L., and Johansson, K. N.-G.,** High-pressure liquid chromatography of peptides, *J. Chromatogr.*, 135, 155, 1977.
8. **Axelsen, K. S. and Vogelsang, S. H.,** High-performance liquid chromatographic analysis of gramicidin, a polypeptide antibiotic, *J. Chromatogr.*, 140, 174, 1977.
9. **Mönch, W. and Dehnen, W.,** High-performance liquid chromatography of peptides, *J. Chromatogr.*, 140, 260, 1977.
10. **Molnár, I. and Horváth, C.,** Separation of amino acids and peptides on non-polar stationary phases by high-performance liquid chromatography, *J. Chromatogr.*, 142, 623, 1977.
11. **Larsen, B., Viswanatha, V., Chang, S. Y., and Hruby, V.,** Reversed phase high pressure liquid chromatography for the separation of peptide hormone diastereomers, *J. Chromatogr. Sci.*, 16, 207, 1978.
12. **Lundanes, E. and Greibrokk, T.,** Reversed-phase chromatography of peptides, *J. Chromatogr.*, 149, 241, 1978.
13. **Hancock, W. S., Bishop, C. A., Prestidge, R. L., Harding, D. R. K., and Hearn, M. T. W.,** High-pressure liquid chromatography of peptides and proteins. II. The use of phosphoric acid in the analysis of underivatised peptides by reversed-phase high pressure liquid chromatography, *J. Chromatogr.*, 153, 391, 1978.
14. **Burgus, R. and Rivier, J.,** Use of high pressure liquid chromatography in the purification of peptides, in *Peptides 1976, Proc. 14th Eur. Peptide Symp.*, Loffet, A., Ed., University Bruxelles, Brussels, Belgium, 1976, 85.
15. **Bennet, H. P. J., Hudson, A. M., McMartin, C., and Purdon, G. E.,** Use of octadecasilyl-silica for extraction and purification of peptides in biological samples. Application to the identification of circulating metabolites of corticotropin-(1-24)-tetracosapeptide and somatostatin *in vivo, Biochem. J.*, 168, 9, 1977.
16. **Mönch, W. and Dehnen, W.,** High-performance liquid chromatography of polypeptides and proteins on a reversed-phase support, *J. Chromatogr.*, 147, 415, 1978.
17. **Hancock, W. S., Bishop, C. A., Prestidge, R. L., Harding, D. R. K., and Hearn, M. T. W.,** Reversed-phase high-pressure liquid chromatography of peptides and proteins with ion-pairing reagents, *Science*, 200, 1168, 1978.
18. **Rivier, J. E.,** Use of trialkylammonium phosphate (TAAP) buffers in reversed phase HPLC for high resolution and high recovery of peptides and proteins, *J. Liq. Chromatogr.*, 1, 343, 1978.
19. **Regnier, F. E. and Gooding, K. M.,** High-performance liquid chromatography of proteins, *Anal. Biochem.*, 103, 1, 1980.
20. **Krummen, K.,** HPLC in the analysis and separation of pharmaceutically important peptides, *J. Liq. Chromatogr.*, 3, 1243, 1980.
21. **Nice, E. C. and O'Hare, M. J.,** Simultaneous separation of β-lipotropin, adrenocorticotropic hormone, endorphins and enkephalins by high-performance liquid chromatography, *J. Chromatogr.*, 162, 401, 1979.
22. **O'Hare, M. J. and Nice, E. C.,** Hydrophobic high-performance liquid chromatography of hormonal polypeptides and proteins on alkylsilane-bonded silica, *J. Chromatogr.*, 171, 209, 1979.
23. **Terabe, S., Konaka, R., and Inouye, K.,** Separation of some polypeptide hormones by high-performance liquid chromatography, *J. Chromatogr.*, 172, 163, 1979.
24. **Inouye, K., Watanabe, K., Tochino, Y., Kanaya, T., Kobayashi, M., and Shigeta, Y.,** Semisynthesis and biological properties of the [B24-leucine]-, [B25-leucine]- and [B24-leucine, B25-leucine]-analogues of human insulin, *Experientia*, 37, 811, 1981.
25. **Inouye, K., Watanabe, K., Tochino, Y., Kobayahsi, M., and Shigeta, Y.,** Semisynthesis and properties of some insulin analogs, *Biopolymers*, 20, 1845, 1981.
26. **Igano, K., Minotani, Y., Yoshida, N., Kono, M., and Inouye, K.,** A synthesis of human proinsulin C-peptide, *Bull. Chem. Soc. Jpn.*, 54, 3088, 1981.

27. **Biemond, M. E. F., Sipman, W. A., and Olivié, J.,** Quantitative determination of polypeptides by gradient elution high-pressure liquid chromatography, *J. Liq. Chromatogr.*, 2, 1407, 1979.
28. **Dinner, A. and Lorenz, L.,** High performance liquid chromatographic determination of bovine insulin, *Anal. Chem.*, 51, 1872, 1979.
29. **Damgaad, U. and Markussen, J.,** Analysis of insulins and related compounds by HPLC, *Horm. Metab. Res.*, 11, 580, 1979.
30. **Lloyd, L. F. and Calam, D. H.,** Separation of human insulin and some structural isomers by high-performance liquid chromatography, *J. Chromatogr.*, 237, 511, 1982.
31. **Terabe, S., Nishi, H., and Ando, T.,** Separation of cytochromes *c* by reversed-phase high-performance liquid chromatography, *J. Chromatogr.*, 212, 295, 1981.
32. **Inouye, K., Watanabe, K., Morihara, K., Tochino, Y., Kanaya, T., Emura, J., and Sakakibara, S.,** Enzyme-assisted semisynthesis of human insulin, *J. Am. Chem. Soc.*, 101, 751, 1979.
33. **Otsuka, H. and Inouye, K.,** Structure-activity relationships of adrenocorticotropin, *Pharmacol. Ther. B.*, 1, 501, 1975.
34. **Watanabe, K. and Inouye, K.,** Synthesis of corticotropin peptides. XIV. The synthesis of two octadecapeptides corresponding to the amino acid sequence 22-39 of porcine and human corticotropins, *Bull. Chem. Soc. Jpn.*, 50, 201, 1977.
35. **Inouye, K., Sumitomo, Y., and Shin, M.,** Synthesis of corticotropin peptides. XIII. The synthesis of a hexacosapeptide and a heptacosapeptide corresponding to the first twenty-six and twenty-seven amino acid residues of corticotropin (ACTH), *Bull. Chem. Soc. Jpn.*, 49, 3620, 1976.
36. **Inouye, K., Shin, M., Nakamura, M., and Tanaka, A.,** Synthesis and biological properties of the 10-substituted analogues of ACTH-(1-18)-NH_2, *Peptide Chemistry 1977, Proc. 15th Symp. Peptide Chem.*, Shiba, T., Ed., Protein Research Foundation, Osaka, Japan, 1978, 177.
37. **Kroeff, E. P. and Pietrzyk, D. J.,** High performance liquid chromatographic study of the retention and separation of short chain peptide diastereomers on a C_8 bonded phase, *Anal. Chem.*, 50, 1353, 1978.
38. **Larsen, B., Fox, B. L., Burke, M. F., and Hruby, V. J.,** The separation of peptide hormone diastereoisomers by reversed phase high pressure liquid chromatography. Factors affecting separation of oxytocin and its diastereoisomers — Structural implications, *Int. J. Peptide Protein Res.*, 13, 12, 1979.
39. **Blevins, D. D., Burke, M. F., Hruby, V. J., and Larsen, B. R.,** Factors affecting the separation of arginine vasopressin peptide diastereoisomers by HPLC, *J. Liq. Chromatogr.*, 3, 1299, 1980.
40. **Hunter, C., Sugden, K., and Lloyd-Jones, J. G.,** HPLC of peptides and peptide diastereoisomers on ODS- and cyano-propyl-silica gel column packing materials, *J. Liq. Chromatogr.*, 3, 1335, 1980.
41. **Otsuka, H., Watanabe, K., and Inouye, K.,** Synthesis of a heptacosapeptide corresponding to the human corticotropin 1-27 sequence, *Bull. Chem. Soc. Jpn.*, 43, 2278, 1970.
42. **Lee, T. H., Lerner, A. B., and Buettner-Janusch, V.,** On the structure of human corticotropin (adrenocorticotropic hormone), *J. Biol. Chem.*, 236, 2970, 1961.
43. **Gráf, L., Bajusz, S., Patthy, A., Barát, E., and Cseh, G.,** Revised amide location for porcine and human adrenocorticotropic hormone, *Acta Biochim. Biophys. Acad. Sci. Hung.*, 6, 415, 1971.

OXYTOCIN INTERMEDIATES

F. Nachtmann and K. Gstrein

INTRODUCTION

Nowadays oxytocin, a cyclic nonapeptide, is almost exclusively produced by chemical synthesis. The synthesis differs in the protective groups used, the peptide linkage methods employed, and the plan followed in building up the molecule. A well-known method is the use of tosyl and carbobenzoxy residues as protecting groups. The nonapeptide may be constructed on the 6 + 3 or the (5 + 2) +2 plan, these intermediates being synthesized step by step.[1] In the absence of a 100% yield for each synthetic step, peptide impurities similar to the main product will occur. For separation and determination of the impurities chromatographic techniques are necessary. A review of these techniques available up until 1972, such as thin-layer chromatography (TLC), paper chromatography (PC), gas chromatography (GC) or electrophoresis, was published by Rosmus and Deyl.[2] TLC and PC permit rapid qualitative results, but exact quantitative determinations are difficult to obtain. GC can only be used after sample derivatization. Important improvements have been made since the development of chemically modified stationary phases with small particle size for high performance liquid chromatography (HPLC). Reviews for the HPLC separation of peptides and proteins have been given by several authors.[3-5] Recently determinations of protected peptides and amino acids were described.[6,7] A systematic investigation for the separation of protected peptides on silica-gel columns was published by Hara et al.[8]

Of ever increasing importance for peptide separation is the technique of the reverse-phase HPLC (RP-HPLC). The factors which influence the retention time have been studied by numerous authors.[9-12] On the basis of the amino acid composition of the peptides their retention time can be estimated.[13] The great selectivity of RP-HPLC was demonstrated by separating oxytocin from some of its diastereoisomers and analogous compounds.[14-17] A mixed separation mechanism between adsorption and reversed-phase chromatography is achieved on new "radial compression" columns, which possess residual silanol-groups.[18,19]

SEPARATION OF OXYTOCIN INTERMEDIATES BY RP-HPLC

Initial results for the separation and determination of synthetic oxytocin intermediates using RP-HPLC were published by Nachtmann.[20] As the stationary phase a C-8 material, particle size 5 μm, packed in steel columns, 150 × 3.2 mm i.d., was used. After isocratic elution UV detection was employed at 215 nm. In this way, successive lengthening of the peptide chain by single amino acids can be studied (Figure 1).

Free pentapeptideamide (structure see Table 1) is reacted with CbO-glutamine-nitrophenylester to give the CbO-hexapeptideamide. For optimal reaction control it should be possible to monitor the concentration of all reactants in the same run. As can be seen from Figure 1, a separation of the relevant compounds is possible within a few minutes. Additionally, CbO-pentapeptideamide that may be present as an impurity, is detectable. For a quantitative determination the substances must be dissolved in the mobile phase, because the free pentapeptideamide is eluted shortly after the dead volume time.

If oxytocin is synthesized according to a 6 + 3 plan, in the final step a tripeptide and a hexapeptide have to be coupled. This reaction can easily be monitored by RP-HPLC.[21] A typical chromatogram is shown in Figure 2. The reactants, hexapeptideamide and Tos-tripeptide acid, as well as the reaction product Tos-nonapeptideamide, are well separated.

Systematic investigations showed that the best separations on reversed phases are possible

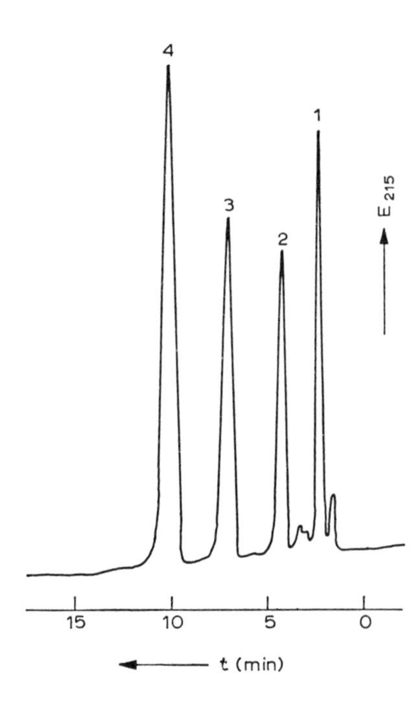

FIGURE 1. Separation of pentapeptideamide (1), CbO-hexapeptideamide (2), CbO-pentapeptideamide (3), and CbO-Gln-ONP (4). Column: LiChrosorb® RP-8, 5 μm, 15 cm × 3.2 mm i.d. Mobile phase: phosphate buffer (0.015 *M*, pH 7.0)-acetonitrile (70:40); pressure, 130 bar; flow rate, 0.6 mℓ/min. Detector: Perkin-Elmer LC 55.

with mixtures of acetonitrile and phosphate buffer (different pH). Retention data for numerous synthetic oxytocin intermediates are given in Table 1. The elution of oxytocin is noteworthy. This substance has a shorter elution time than all other tested peptides. An explanation can be given by considering the conformation of oxytocin. Oxytocin has a cyclic structure. The ring has an antiparallel pleated sheet conformation with intramolecular hydrogen bonds, and the side chain is set back to the ring.[22]

The apolar, protecting groups have a big influence on the chromatographic characteristics of these compounds. Esterification prohibits the ionization of the carboxyl-groups of peptides resulting in a drastic increase in k′ values. This can be seen by comparing the k′ values of Tos-dipeptide acid and Tos-dipeptide ester or Tos-tripeptide acid and Tos-tripeptide ester.

If peptides contain the same protecting groups such as CbO-hexapeptideamide, CbO-pentapeptideamide, and CbO-tetrapeptideamide the elution does not follow the molecular weight, but instead follows the retention contribution of the single amino acids as described in the literature.[13,23]

For the determination of the protected peptides, a pH of 7.0 for the mobile phase is optimal. In the case of free peptides better results were achieved at pH 3.0. The acid pH resulted in an improvement of peak symmetry and selectivity. This was demonstrated for the separation of oxytocin and an intermediate, the reduced nonapeptideamide.[20]

QUANTITATION

Excellent quantitative determinations of synthetic oxytocin intermediates are possible with

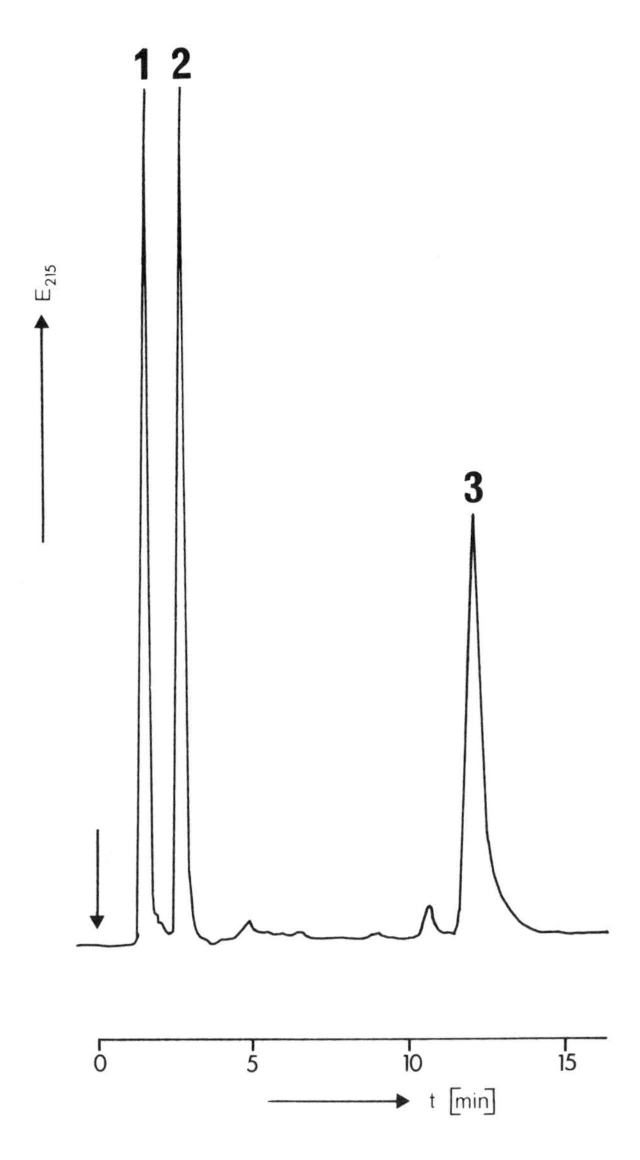

FIGURE 2. Separation of hexapeptideamide (1), Tos-tripeptide acid (2), and Tos-nonapeptideamide (3). Column: LiChrosorb® RP-8, 5 μm, 15 cm × 3.2 mm i.d. Mobile phase: phosphate buffer (0.015 *M*, pH 7.0)-acetonitrile (60:40); pressure, 170 bar; flow rate 0.8 mℓ/min.; Detector: Uvikon LCD 725.

215 nm UV detection. The correlation coefficients for a number of peptides were determined.[20] Linear calibration curves were found in a concentration range of 3 to 300 μg peptide per milliliter. The detection limit was in the nanogram range.

Table 1
CHROMATOGRAPHIC DATA FOR SYNTHETIC OXYTOCIN INTERMEDIATES

Structure	Compound	Phosphate buffer pH 7/CH_3CN	Retention time (min)	k′
Cys-Tyr-Ile-Gln-Asn-Cys-Pro-Leu-Gly-NH_2	Oxytocin	82/18	9.8	6.2
CbO-Leu-OH	CbO-Leucine	82/18	11.9	7.7
H-Tyr-OEt		70/30	2.2	0.6
CbO-Leu-OH		70/30	2.3	0.7
CbO-Leu-DCHA		70/30	2.3	0.7
Tos-S-benzyl-Cys-OH		70/30	4.1	2.0
Tos-S-benzyl-Cys-Tyr-OH	Tos-dipeptide acid	70/30	5.6	3.1
CbO-Pro-Leu-Gly-NH_2		70/30	5.7	3.2
CbO-Gln-Asn-S-benzyl-Cys-Pro-Leu-Gly-NH_2	CbO-hexapeptideamide	70/30	9.7	6.1
CbO-Asn-S-benzyl-Cys-Pro-Leu-Gly-NH_2	CbO-pentapeptideamide	70/30	17.4	11.8
H-Pro-Leu-Gly-NH_2		60/40	1.3	0
Tos-S-benzyl-Cys-Tyr-Ile-OH	Tos-tripeptide acid	60/40	2.7	1.0
CbO-Asn-ONP		60/40	5.4	2.9
CbO-Gln-ONP		60/40	5.3	2.9
CbO-Pro-Leu-Gly-OEt	CbO-tripeptide ester	60/40	6.8	4.0
CbO-Leu-Gly-OEt	CbO-dipeptide ester	60/40	8.0	4.9
CbO-S-benzyl-Cys-Pro-Leu-Gly-NH_2	CbO-tetrapeptideamide	60/40	9.1	5.7
Tos-S-benzyl-Cys-Tyr-Ile-Gln-Asn-Cys-Pro-Leu-Gly-NH_2	Tos-nonapeptideamide	60/40	12.3	8.0
Tos-S-benzyl-Cys-Tyr-OEt	Tos-dipeptide ester	50/50	5.9	3.3
Tos-S-benzyl-Cys- Tyr-Ile-OMe	Tos-tripeptide ester	50/50	7.2	4.3
CbO-S-benzyl-Cys-ONP		50/50	17.8	12.1
Gln-Asn-S-benzyl-Cys-Pro-Leu-Gly-NH_2	Hexapeptideamide	78[a]/22	3.6	1.6
Asn-S-benzyl-Cys-Pro-Leu-Gly-NH_2	Pentapeptideamide	78[a]/22	6.2	3.6

Note: Column: Lichrosorb® RP8, 5 μm, 15 cm × 3.2 mm i.d.; temperature, 20 to 22°C; flow rate, 0.8 mℓ/min; detection wavelength, 215 nm; determination of dead volume, thiourea. CbO, carbobenzoxy; Tos, tosyl; DCHA, dicyclohexylamide; -ONP, -ortho-nitrophenylester; Et, ethyl; Me, methyl.

[a] Phosphate buffer pH 3.0.

REFERENCES

1. **Nachtmann, F., Krummen, K., Maxl, F., and Riemer, E.,** Oxytocin, in *Analytical Profiles of Drug Substances,* Vol. 10, Florey, K., Ed., Academic Press, New York, 1981, 563.
2. **Rosmus, J. and Deyl, Z.,** Chromatographic methods in the analysis of protein structure. Methods for identification of N-terminal amino acids in peptides and proteins, *J. Chromatogr.,* 70, 221, 1972.
3. **Regnier, F. E. and Gooding, K. M.,** High-performance liquid chromatography of proteins, *Anal. Biochem.,* 103, 1, 1980.
4. **Hearn, M. T. W.,** The use of reversed phase high-performance liquid chromatography for the structural mapping of polypeptides and proteins, *J. Liq. Chromatogr.,* 3, 1255, 1980.
5. **Rubinstein, M.,** Preparative high-performance liquid partition chromatography of proteins, *Anal. Biochem.,* 98, 1, 1979.
6. **Naider, F., Sipzner, R., and Steinfeld, A. S.,** Separation of protected hydrophobic oligopeptides by normal phase high-pressure liquid chromatography, *J. Chromatogr.,* 176, 264, 1979.
7. **Kullmann, W.,** Monitoring of protease-catalyzed peptide synthesis by high performance liquid chromatography, *J. Liq. Chromatogr.,* 4, 1121, 1981.
8. **Hara, S., Ohsawa, A., and Dobashi, A.,** Design of binary solvent systems for separation of protected oligopeptides in silica gel liquid chromatography, *J. Liq. Chromatogr.,* 4, 409, 1981.
9. **Rivier, J. and Burgus, R.,** Application of reverse phase high pressure liquid chromatography to peptides, *Chromatogr. Sci. Biol. Biomed. Appl. Liq. Chromatogr.,* 10, 147, 1979.
10. **Hearn, M. T. W. and Grego, B.,** High performance liquid chromatography of amino acids, peptides and proteins. XXXVI. Organic solvent modifier effects in the separation of unprotected peptides by reversed phase liquid chromatography, *J. Chromatogr.,* 218, 497, 1981.
11. **Hearn, M. T. W. and Grego, B.,** High performance liquid chromatography of amino acids, peptides and proteins. XXVII. Solvophobic considerations for the separation of unprotected peptides on chemically bonded hydrocarbonaceous stationary phases, *J. Chromatogr.,* 203, 349, 1981.
12. **Hearn, M. T. W., Su, S. J., and Grego, B.,** Pairing ion effects in the reversed phase high performance liquid chromatography of peptides in the presence of alkylsulphonates, *J. Liq. Chromatogr.,* 4, 1547, 1981.
13. **Meek, J. L. and Rosetti, Z. L.,** Factors affecting retention and resolution of peptides in high performance liquid chromatography, *J. Chromatogr.,* 211, 15, 1981.
14. **Larsen, B., Fox, B. L., Burke, M. F., and Hruby, V. J.,** The separation of peptide hormone diastereoisomers by reverse phase high pressure liquid chromatography, *Int. J. Peptide Protein Res.,* 13, 12, 1979.
15. **Larsen, B., Viswanatha, V., Chang, S. Y., and Hruby, V. J.,** Reverse phase high pressure liquid chromatography for the separation of peptide hormone diastereomers, *J. Chromatogr. Sci.,* 16, 207, 1978.
16. **Krummen, K., Maxl, F., and Nachtmann, F.,** The use of HPLC in the quality control of oxytocin, *Pharm. Technol.,* 3, 77, 1979.
17. **Lindenberg, G.,** Separation of vasopressin analogues by reversed phase high performance liquid chromatography, *J. Chromatogr.,* 193, 427, 1980.
18. **Hancock, W. S., Capra, J. D., Bradley, W. A., and Sparrow, J. T.,** The use of reversed phase high performance liquid chromatography with radial compression for the analysis of peptide and protein mixtures, *J. Chromatogr.,* 206, 59, 1981.
19. **Hancock, W. S., Sparrow, J. T.,** Use of mixed-mode, high performance liquid chromatography for the separation of peptide and protein mixtures, *J. Chromatogr.,* 206, 71, 1981.
20. **Nachtmann, F.,** High performance liquid chromatography of intermediates in the oxytocin synthesis, *J. Chromatogr.,* 176, 391, 1979.
21. **Gstrein, K. and Nachtmann, F.,** unpublished results.
22. **Gross, E. and Meienhofer, J.,** *The Peptides,* Vol. 1, Academic Press, New York, 1979, 5.
23. **Su, S. J., Grego, B., Niven, B., and Hearn, M. T. W.,** Analysis of group retention contributions for peptides separated by reversed phase high performance liquid chromatography, *J. Liq. Chromatogr.,* 4, 1745, 1981.

DIASTEREOISOMER SEPARATIONS

Dennis D. Blevins, Michael F. Burke, and Victor J. Hruby

INTRODUCTION

Purification and/or analysis of the purity of peptides still presents a formidable challenge. A particularly interesting and increasingly important aspect of this area is the separation of peptide diastereoisomers in which one or more of the asymmetric centers in a peptide has been converted from an S (or R) configuration to the opposite configuration. The effect of such configurational changes is to produce diastereoisomers which can have similar or quite different physical, chemical, and biological properties. Therefore, it is critical that analytical and preparative high-performance liquid chromatography (HPLC) methods be found to separate these compounds. Some of the most important and increasingly relevant reasons for developing methods for these separations include: (1) for determination of diastereoisomers in synthetic peptides which have resulted from the synthetic methods employed, or because amino acid derivatives were used in the synthesis which were not pure enantiomers. This is particularly important when examining synthetic peptides in which one or more of the amino acids is D-, since most of the commercially available D-amino acids have small amounts of the L-amino acid. (2) Introduction of isotopes into peptides by total synthesis often is most efficiently accomplished by use of the racemic modification of the amino acid and subsequent separation of the diastereoisomeric peptides. (3) Structure-function studies of biologically active peptides have shown that often diastereoisomeric analogs have interesting biological activities (e.g., antagonist, superagonist, or prolonged activity in vivo). (4) It has been observed that peptide diastereoisomers often have different conformational properties than their all L analogs. It would, therefore, be interesting to develop an understanding of the chromatographic principles which can maximize diastereoisomer separations and provide insight into the effect of configurational change on conformational properties.

Krummen[1] has reviewed the preparation of several pharmaceutically important peptides by HPLC, including a few diastereoisomers of oxytocin, and has summarized some very important parameters to consider when attempting to chromatograph certain peptides. *Analytical Chemistry* biannually publishes a review on ion-exchange and liquid column chromatography including some amino acid and peptide separations.[2]

In this review we will examine the work which has appeared in the literature regarding peptide diastereoisomer separation by HPLC. We will try to emphasize how careful consideration of the properties of the stationary phase and mobile phase, as well as the characteristics of the peptide diastereoisomers, can lead to efficient analytical and preparative separation of these compounds.

SEPARATIONS OF AMINO ACID AND DIPEPTIDE DIASTEREOISOMERS

For HPLC separation of amino acid enantiomers, the enantiomer must be converted to a diastereoisomer either intramolecularly, by reaction with another chiral reagent, or intermolecularly, by the use of chiral stationary or mobile phases. Linder et al.[3] reported on the separation of 5-dimethylaminonapthalene-1-sulfonyl (Dns, dansyl) derivatives of amino acids on a reversed-phase C_8-column with the addition of a chiral metal chelate additive to the mobile phase. The mobile phase consisted of an aqueous ammonium acetate buffer with acetonitrile, tetrahydrofuran (THF), or methanol as the organic modifier. The effects of pH and temperature were also reported. In this work the authors also report the separation of some dipeptide diastereoisomers without a derivatization step, again employing the same metal chelate additive to the mobile phase.

Hare and Gil-Av[4] report the separation of amino acid enantiomers on an ion-exchange column (Table 1). The mobile phase consisted of a sodium acetate buffer containing copper (II) sulfate with proline added as the chiral reagent. They reported that the separation was dependent upon the stereochemistry of the chiral eluent employed and investigated the effects of ionic strength, pH, and temperature on the separations. Detection was via post-column *o*-phthaladehyde (*o*-PTH) derivatization and fluorometry.

Cahill et al.[5] described the determination of DL amino acids after their conversion to dipeptide diastereoisomers utilizing an *N*-hydroxysuccinimide ester of a *t*-butyloxycarbonyl-L-amino acid. The same approach was used for the derivatization and subsequent determination of certain dipeptides. Several reversed-phase columns were employed including a C_{18}- and two different C_8-columns. The mobile phases were an aqueous phosphate buffer, with added sodium chloride to adjust the ionic strength, and varying percentages of acetonitrile.

Lundanes and Greibrokk[6] reported the separation of several dipeptides on two different C_{18}-reversed-phase columns with the mobile phase consisting of an aqueous ammonium acetate buffer and methanol. Their study of varying the percentage methanol in the mobile phase demonstrated that the column employed has a drastic effect upon the retention of diastereoisomers. They also state that other C_{18} columns tested were unable to separate the diastereoisomers.

Kroeff and Pietrzyk[7] separated several dipeptide and tripeptide diastereoisomers on a reversed-phase C_8 column. The mobile phase employed consisted of an aqueous phosphate buffer with controlled pH and ionic strength, and either ethanol or acetonitrile as the organic solvent. They investigated the separation as a function of the peptide structure, eluent pH, solvent composition, and stereochemical properties of the peptides. This report also included a discussion of the influence of the lipophilicities of the amino acids and the relationships of the lipophilic amino acid subunits to the charged sites, on retention.

Iskandarani and Pietrzyk[8] report the separation of several di- and tripeptide diastereoisomers on a porous polystyrene-divinylbenzene copolymer absorbent. One of the advantages of the porous polymer over a silica gel packing is the pH range can be extended past the pH 2 to 7.5 range. The mobile phase employed consisted of an aqueous phosphate buffer, or HCl with NaCl added to adjust the ionic strength, with acetonitrile as the organic modifier. The effects of pH on the separation were investigated (Table 1).

SEPARATION OF BIOLOGICALLY ACTIVE PEPTIDE DIASTEREOISOMERS

The HPLC of selected enkephalin and endorphin diastereoisomer analogs was reported by Currie et al.[9] A phenyl column was employed with the mobile phase consisting of an aqueous ammonium acetate buffer and acetonitrile. The percentage of acetonitrile was varied to enhance the separation of the analogs. The relative retention of the peptides was reported to be sensitive to both the buffer concentration and the percentage of acetonitrile. Mousa et al.[10] also reported the separation of enkephalin diastereoisomers on a C_{18} column using an aqueous phosphate buffer and differing percentages of acetonitrile. In addition, they employed these HPLC systems to separate and quantitate the enkephalins for investigation of pharmacodynamics of enkephalin synthesis and release.

Larsen et al.[11] reported the separation of several peptide diastereoisomers of oxytocin with a reversed-phase C_{18} column using an aqueous ammonium acetate buffer at pH 4.0 and acetonitrile as the organic modifier. They later expanded the study to include the effects of pH, the concentration of the buffer, and the choice of organic modifier: acetonitrile, THF, or dioxane.[12] It is interesting to note that the elution order of the diastereoisomers was dependent upon the organic modifier employed. The optimization of the separation was a function of several parameters. Upon optimization of the separation conditions, these authors reported a preparative separation of the diastereoisomers of a [3-DL-[2-^{13}C]leucine]oxytocin

Table 1
SEPARATIONS OF AMINO ACID DERIVATIVES AND DIPEPTIDE DIASTEREOISOMERS

Diastereoisomer sample	Column[a]	Mobile phase	pH	Detector	Ref.
AA-Dns derivatives	C_8 (made by authors)	NH_4OAc and chiral chelate agent and metal ion	7	UV and fluor.	3
AA-underivatized	Ion exchange (DC-4a)	Chiral copper proline complex NaOAc	5.5	Post-column derivatization, *o*-PTH fluor.	4
Dipeptides-derivatized with Boc-L-AA-OSu	LiChrosorb® RP-8 μBondapak® C_{18}	CH_3CN/phosphate	2.0 7.5	UV UV	5
Dipeptides	Phenyl Sil-X-1 μBondapak® NH_2 Nucleosil® 5 CN Spherisorb S5W-ODS Spherisorb S5-phenyl ODS-Hypersil	MeOH/NH_4OAc	3.47 4.85 7.00	UV and RI	6
Dipeptides	LiChrosorb® RP-8	CH_3CN or EtOH/ phosphate	3.38 5.82 7.91	UV	7
Dipeptides	PRP-1	CH_3CN/phosphate	1.56—1.75 5.00—5.25 11.00	UV	8

[a] Refer to Reference 22 for column identification and manufacture.

Note: AA, amino acid; OSu, *N*-hydroxysuccinimide; RI, refractive index.

mixture and of a [8-DL-[2-^{13}C]leucine]oxytocin mixture by HPLC.[13] The former diastereoisomeric mixture could not be separated by partition chromatography methods.

Blevins et al.[14] reported the separation of several diastereoisomers of arginine vasopressin (AVP) on several C_{18} columns and a single C_8 column. In addition, they varied the concentrations of the aqueous triethylammonium acetate (TEAA) buffer, the organic modifier (either methanol, acetonitrile, or THF), and the percentage of organic modifier in the mobile phase. They determined that the elution order of the peptides was dependent upon which C_{18} column was employed.

Burgus and Rivier[15] report the separation of several luteinizing hormone-releasing hormone (LH-RH) diastereoisomeric analogs on a reversed-phase C_{18} column with an aqueous ammonium acetate buffer, and acetonitrile or ethanol as the organic modifier in the mobile phase. In this article, they also report the separation of several somatostatin analogs, and the use of HPLC to characterize thyrotropin-releasing hormone (TRH). The method is sensitive enough to determine as little as 0.7% [D-His^2]TRH in the presence of all L-TRH. Rivier et al.[16,17] later report the separation of LH-RH diastereoisomers employing different buffer systems. They evaluated an aqueous triethylammonium formate (TEAF) buffer and an aqueous triethylammonium phosphate (TEAP) buffer, and state that even though TEAF is volatile, TEAP is favored for some separations because of improved resolution for the separation of unprotected peptides and higher recovery of the peptides even though desalting of the isolated peptides is necessary. TEAP is also advantageous because it is UV transparent down to 195 nm. Sertl et al.[18] also reported the separation of LH-RH diastereoisomers by HPLC employing a C_{18} column and an aqueous phosphate buffer and acetonitrile mixture as the mobile phase. The effects of the sample matrix on the determination were reported.

Meyers et al.[19] reported the separation of several diastereoisomeric analogs of somatostatin by HPLC. A reversed-phase C_{18} column was employed with an aqueous ammonium acetate (pH 4.1) buffer and acetonitrile as the mobile phase. They used HPLC to check the purity of the final peptides.

Bakkum et al.[20] reported the separation of several protected analogs of secretin diastereoisomers on a C_{18} column with a methanol, water, acetic acid mixture as the mobile phase. They demonstrated that HPLC is an excellent method for the analysis of peptides synthesized by a sequential strategy. Voskamp et al.[21] also reported the reversed-phase separation of several secretin diastereoisomers on a C_{18} column with a methanol, water, trifluoroacetic acid (TFA) mobile-phase eluent.

DISCUSSION

Chromatographic systems which utilize a hydrocarbon bonded to a solid silica support material as an adsorbent are commonly referred to as reversed-phase systems. Reversed phase simply implies that the effective stationary phase is less polar than the mobile phase. In order to discuss the interaction between the solute molecules and the mobile and stationary phases which are responsible for the chromatographic selectivity, it is necessary to recognize the chemical and physical nature of the effective stationary phase. The stationary phase consists of the solid support and the bonded hydrocarbon, both of which are solvated by the components of the mobile phase. Therefore, a change in the chemical composition of the mobile phase not only changes the eluting strength of this phase but also changes the retentive nature of the stationary phase. The chromatographic behavior of a diastereoisomer in different solvent systems must, therefore, be considered in terms of the changes occurring in both phases.

The most common type of bonded species employed for the separation of diastereoisomers has been an octadecyl group chemically bonded to the chromatographic support. Some of the octadecyl modified silica gels are more commonly denoted as ODS, RP-18, or C_{18}.

Other chemically bonded species included an octyl group (C_8 or RP-8), a dimethyl grop (RP-2), a phenyl, a propylcyano, or a propylamino function group.

The separation of peptide diastereoisomers is particularly sensitive to the differences in the composition of the stationary phase. If the individual is not aware of the mechanism which dictates the selectivity for an HPLC separation, frustration can result. Research has shown that not all octadecyl columns have identical separation characteristics. The selectivity of the octadecyl column is a function of the amount of hydrocarbon bonded per unit area of the available surface. In addition, the chemical composition of the stationary phase is a function of unreacted surface available to the mobile phase components. The nature of the stationary phase is also influenced by ''endcapping'', a technique used which attempts to reduce the number of residual silanols on the surface by reacting the chemically modified silica with a smaller organosilane such as trimethylchlorosilane. However, even columns made by the same manufacturers suffer from a lack of reproducibility. This has given HPLC a bad name in some instances which is unjustified.

Despite these problems, HPLC has been widely used for trace determination of specific peptides in physiological fluids and for purification of synthetic peptides for biological, chemical, and biophysical studies. The synthesis of specific diastereoisomer analogs of naturally occurring peptides has proven to be a valuable tool for evaluating the role of specific amino acid residues in determining the biological activity of a variety of peptide hormones. In addition, recent studies in our laboratory indicate that the highly selective interactions of the peptides with the chromatographic stationary phase can be used to obtain information about the topological properties of these molecules. Chromatographic studies of these diastereoisomers have been shown to provide unique insight into the changes in structure which occur upon substitution of a single D amino acid residue into the peptide.

Scientific studies employing peptide diastereoisomers have always been plagued by the problem of ensuring that only the desired diastereoisomer is present. HPLC has provided a highly sensitive technique for checking the purity of specific peptide diastereoisomers and, if necessary, can also provide a method for purification on a preparative scale. A variety of mobile and stationary phases has been reported in the literature (Table 1, Table 2) and it has become clear that there is an interrelationship between the two in determining the selectivity between the diastereoisomers.

Bonded-phase HPLC has been demonstrated to be a valuable technique for the separation of peptide diastereoisomers under a variety of conditions. The applications which have been reported to date represent only the beginning efforts of a limited number of researchers and a broad range of applications can be expected in the near future. In addition to the analytical and preparative experiments which have been reported, bonded phase chromatography offers the potential of providing structure-function information. Correlation of the retention of biologically active compounds on a lipophilic stationary phase with biological activity should be possible given a better understanding of the chemical and physical nature of the adsorbent surface.

ACKNOWLEDGMENTS

This research was supported by grants from the National Science Foundation and the U.S. Public Health Service.

Table 2
SEPARATIONS OF BIOLOGICAL ACTIVE PEPTIDE DIASTEREOISOMERS

Diastereoisomer sample	Column[a]	Mobile phase	pH	Detector	Ref.
Enkephalin endorphin	μBondapak® Phenyl	CH_3CN/NH_4OAc	4.5	UV	9
Met and leu enkephalin	Ultrasphere ODS	CH_3CN/phosphate	2.1	UV	10
Oxytocin	μBondapak® C_{18}	CH_3CN/NH_4OAc	4.0	UV	11
Oxytocin	μBondapak® C_{18}	CH_3CN, THF, or dioxane/NH_4OAc	4.0 6.0	UV	12
Oxytocin	Partisil® 10 ODS	CH_3CN/NH_4OAc	4.0	UV	13
AVP	μBondapak® C_{18} Spherisorb ODS LiChrosorb® RP-18 LiChrosorb® RP-8	CH_3OH, CH_3CN, or THF/TEAA	4.0	UV	14
LH-RH, TRH, somatostatin	μBondapak® C_{18}	CH_3CN/NH_4OAc	4.0	UV	15
LH-RH	μBondapak® C_{18} μBondapak® CN	CH_3CN/TEAF TEAP	3.0	UV	16
LH-RH	μBondapak® CN μBondapak® phenyl μBondapak® C_{18}	CH_3CN/TEAP	3.0	UV	17
LH-RH	ODS-Hypersil	CH_3CN/phosphate	6.5	UV	18
Somatostatin	μBondapak® C_{18}	CH_3CN/NH_4OAc	4.1	UV	19
Secretin	LiChrosorb® RP-18	$MeOH/H_2O/HOAc$	NR	UV, RI	20
Secretin	LiChrosorb® RP-18 Nucleosil® C_{18} Polygosil CN Polygosil phenyl	MeOH or CH_3CN/ TFA or phosphate buffer	NR	UV	21

Note: NR, not reported; RI, refractive index.

[a] Refer to Reference 22 for column identification and manufacturer.

REFERENCES

1. **Krummen, K.,** HPLC in the analysis and separation of pharmaceutically important peptides, *J. Liq. Chromatogr.*, 3, 1243, 1980.
2. **Majors, R. E. and Lochmuller, C.,** Ion exchange and liquid chromatography, *Anal. Chem.*, 54, 1982.
3. **Linder, W., LePage, J. N., Davies, G., Seitz, D. E., and Karger, B. L.,** Reversed-phase separation of optical isomers of Dns-amino acids and peptides using chiral metal chelate additives, *J. Chromatogr.*, 185, 323, 1979.
4. **Hare, P. E. and Gil-Av, E.,** Separation of D and L amino acids by liquid chromatography: use of chiral eluants, *Science*, 204, 1226, 1979.
5. **Cahill, W. R., Jr., Kroeff, E. P., and Pietrzyk, D. J.,** Applications of tert-butyloxycarbonyl-L-amino acid-*N*-hydroxysuccinimide esters in the chromatographic separation and determination of D,L-amino acids and diastereomeric dipeptides, *J. Liq. Chromatogr.*, 3, 1319, 1980.
6. **Lundanes, E. and Greibrokk, T.,** Reversed-phase chromatography of peptides, *J. Chromatogr.*, 149, 241, 1978.

7. **Kroeff, E. P. and Pietrzyk, D. J.,** High performance liquid chromatographic study of the retention and separation of short chain peptide diastereomers on a C_8-bonded phase, *Anal. Chem.*, 50, 1353, 1978.
8. **Iskandarani, Z. and Pietrzyk, D. J.,** Liquid chromatographic separation of amino acids, peptides, and derivatives on a porous polystyrene-divinylbenzene copolymer, *Anal. Chem.*, 53, 489, 1981.
9. **Currie, B. L., Chang, J. K., and Cooley, R.,** High performance liquid liquid chromatography of enkephalin and endorphin peptide analogs, *J. Liq. Chromatogr.*, 3, 513, 1980.
10. **Mousa, S., Mullet, D., and Couri, D.,** Sensitive and specific high performance liquid chromatographic method for methionine and leucine enkephalins, *Life Sci.*, 29, 61, 1981.
11. **Larsen, B., Viswanatha, V., Chang, S. Y., and Hruby, V. J.,** Reversed phase high pressure liquid chromatography for the separation of peptide hormone diastereoisomers, *J. Chromatogr. Sci.*, 16, 207, 1978.
12. **Larsen, B., Fox, B. L., Burke, M. F., and Hruby, V. J.,** The separation of peptide hormone diastereoisomers by reverse phase high pressure liquid chromatography. Factors affecting separation of oxytocin and its diastereoisomers — structural implications, *Int. J. Peptide Protein Res.*, 13, 12, 1979.
13. **Viswanatha, V., Larsen, B., and Hruby, V. J.,** Synthesis of DL-[2-^{13}C] Leucine and its use in the preparation of [3-DL-[2-^{13}C]leucine]oxytocin and [8-DL-[2-^{13}C]leucine]oxytocin. Preparative separation of diastereoisomeric peptides by partition chromatography and high pressure liquid chromatography, *Tetrahedron*, 35, 1575, 1979.
14. **Blevins, D. D., Burke, M. F., Hruby, V. J., and Larsen, B. R.,** Factors affecting the separation of arginine vasopressin peptide diastereoisomers by HPLC, *J. Liq. Chromatogr.*, 3, 1299, 1980.
15. **Burgus, R. and Rivier, J.,** Use of high pressure liquid chromatography in the purification of peptides, in *Peptides 1976*, Loffet, C. A., Ed., Editions de L'Universite de Bruxelles, Belgium, 1976, 85.
16. **Rivier, J., Spiess, J., Perrin, M., and Vale, W.,** Application of HPLC in the isolation of unprotected peptides, in *Biological/Biomedical Applications of Liquid Chromatography II*, Hawk, G. L., Ed., Marcel Dekker, New York, 1979, 223.
17. **Rivier, J. E.,** Use of trialkylammonium phosphate (TAAP) buffers in reverse phase HPLC for high resolution and high recovery of peptides and proteins, *J. Liq. Chromatogr.*, 1, 343, 1978.
18. **Sertl, D. C., Johnson, R. N., and Kho, B. T.,** An accurate, specific HPLC method for the analysis of a decapeptide in a lactose matrix, *J. Liq. Chromatogr.*, 4, 1134, 1981.
19. **Meyers, C. A., Coy, D. H., Huang, W. Y., Schally, A. V., and Redding, T. W.,** Highly active position right analogues of somatostatin and separation of peptide diastereomers by partition chromatography, *Biochemistry*, 17, 2326, 1978.
20. **Bakkum, J. T. M., Beyerman, H. C., Hoogerhout, P., Olieman, C., and Voskamp, D.,** Reverse-phase high performance liquid chromatography of protected peptides in the sequential synthesis of secretin and analogues, *Recueil, J. R. Netherlands Chem. Soc.*, 96, 301, 1977.
21. **Voskamp, D., Olieman, C., and Beyerman, H. C.,** The use of trifluoroacetic acid in the reverse-phase liquid chromatography of peptides including secretin, *Recueil, J. R. Netherlands Chem. Soc.*, 99, 105, 1980.
22. **Majors, R. E.,** Recent advances in HPLC packings and columns, *J. Chromatogr. Sci.*, 18, 488, 1980.

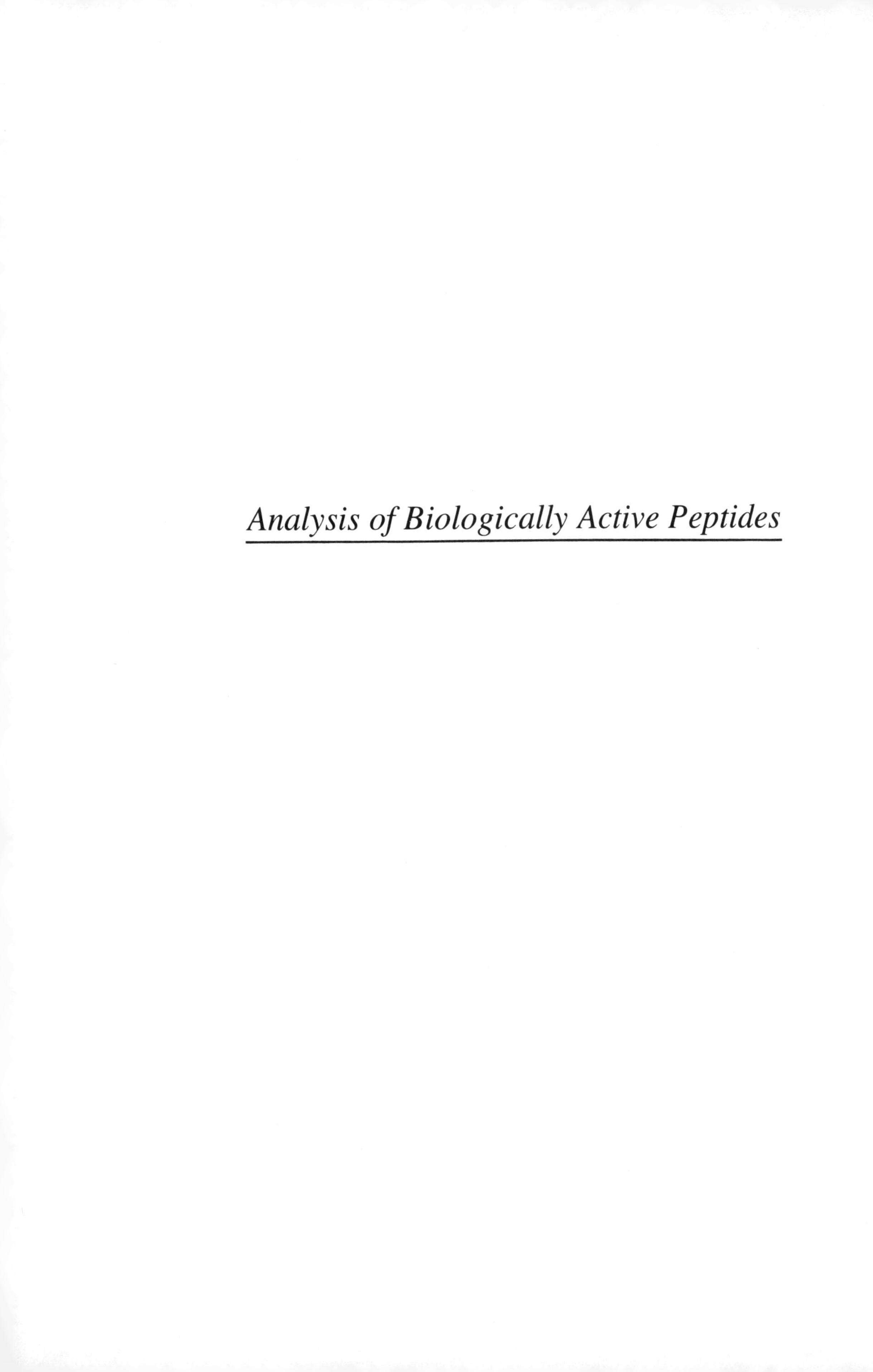

Analysis of Biologically Active Peptides

METABOLISM OF PEPTIDES

Colin McMartin

INTRODUCTION

A detailed study of the catabolism of a peptide is complicated because most peptides can be cleaved at a number of sites along the amino acid chain and consequently give rise to a multiplicity of products. Methods such as radioimmunoassay (RIA) and bioassay yield valuable information about rates of inactivation of a peptide and the sites in the body where inactivation occurs. However, more complete information about the mechanisms of inactivation or other more subtle transformations requires the separation and identification of fragments. For this purpose HPLC methods can be extremely useful. The metabolism of peptide hormones has been recently reviewed.[1] It is only in the last few years that HPLC methods have been developed and applied to peptide catabolism and the number of applications is still small. The present report describes methods which the author or his colleagues have found to be useful.

PRINCIPLES UNDERLYING THE USE OF OCTADECYLSILYL-SILICA FOR EXTRACTIONS AND SEPARATIONS

Peptides which contain aromatic amino acids adsorb strongly to octadecylsilyl-silica (ODS-silica) when applied in aqueous solutions. The pH must be less than 7 because the bonding of the ODS group to the silica is unstable to alkaline conditions. Elution can be accomplished using a water-miscible organic solvent, e.g., acetonitrile or methanol, and an electrolyte. Acetonitrile is a more powerful eluant than methanol and the adsorption of a given peptide is very concentration dependent. This may be due to the fact that a peptide (P) will interact with the same area of ODS-silica as a large number of organic solvent molecules so that the equilibrium will be

$$\mathrm{ODS.(MeOH)_n} + \mathrm{P_{aqueous}} \rightleftharpoons \mathrm{ODS.P} + \mathrm{nMeOH_{aqueous}}$$

The extent of binding of the peptide depends on the nth power of the methanol concentration and if n is large this will make binding highly dependent on concentration.

The presence of an electrolyte in the eluate ensures sharp peaks and good recoveries, probably by suppressing ion-exchange effects which can arise owing to the polyelectrolyte character of most peptides.

Proteins do not bind to any great extent to ODS-silica possibly because their size prevents entry into the gel matrix. In addition most of the hydrophobic residues in a water-soluble protein are buried in the interior so that adsorption is less likely to occur.

If a biological sample is applied to a short extraction column of ODS-silica the peptides are adsorbed while proteins and inorganic electrolytes wash through. The peptides can then be eluted with 80% methanol/1% trifluoroacetic acid (TFA).

The eluted sample can be diluted to 20% methanol to give a solution which, because of the dependency of binding on solvent concentration, can be applied directly to an HPLC column. The volume applied can be several times the column volume because the peptides will concentrate at the head of the column and elute as sharp peaks when a solvent gradient is run. Thus the adsorption of peptides to ODS-silica can be exploited for their extraction from biological fluids and tissues and enable direct loading of extracts on to HPLC columns without prior evaporation which can be time consuming and could result in loss of peptide.

HPLC PROCEDURES

A variety of columns are suitable; 4 mm × 250 mm columns containing Partisil®-10-ODS (Reeve Angel, London) or Nucleosil®-10-C18 (Alltech, Carnforth, Lancashire) have both been used successfully. If fractions from HPLC are to be further purified or subjected to amino acid analysis it is an advantage to use a volatile solvent system. This can be achieved by using 1% TFA as the electrolyte. Good separations are achieved using a linear gradient of increasing concentration of organic solvent. The retention time of a peptide fragment is normally comparable to that of the starting peptide.

For initial experiments, a gradient from water/TFA (99:1 by vol) to methanol/water/TFA (80:19:1 by vol) with a total volume of 100 mℓ is most useful. The total volume and starting and final concentrations can then be adjusted to give a more rapid or more efficient separation of the products actually present.

A linear gradient can be generated using two identically shaped flasks connected by a siphon. Flask A contains final buffer as does the siphon which is a U-shaped piece of glass tubing. Flask B contains a volume of starting buffer equal to the volume in flask A and is stirred. Buffer is pumped from flask B onto the column. A Dosapro Milton Roy minipump (Dosapro Milton Roy (U.K.) Ltd., Chertsey, London) is suitable. In general the column is washed for 5 min with a solution of organic solvent, e.g., methanol/water/TFA (80:19:1) and then for 5 min with starting buffer. The sample is pumped through the pump head and onto the column and the pump inlet is then transferred to flask B to start the gradient. The buffers are either purged with helium or kept slightly warm to prevent the formation of gas bubbles in the pump head.

A flow rate of 0.7 mℓ/min gives reasonably sharp peaks and the eluate can be monitored either by liquid scintillation counting of fractions collected every half minute or by using a Cecil spectrophotometer (Cecil Instruments, Cambridge) with a 10-μℓ cell to detect UV-absorbing compounds. When tryptophan is present in the peptide microgram amounts can be detected at 280 nm (see Figure 1). If tryptophan is not present and radioactively labeled peptide is not available the eluate can be monitored at shorter wavelengths, e.g., 224 nm. At this wavelength a chromatographic grade of acetonitrile or methanol must be used and 1% TFA replaced by a transparent electrolyte, e.g., 0.5% phosphoric acid. If the eluate is to be dried and the peptides are positively charged at low pH values, the phosphoric acid can be selectively exchanged for acetic acid by passing the appropriate portion of the eluate through a bed of Dowex® 8% cross-linked anion-exchange resin in the acetate form. The peptides, being cationic, pass straight through the resin bed.

Small peptides lacking aromatic residues will only bind weakly to the ODS-silica. For such compounds it may nevertheless be possible to accomplish separations using isocratic elution conditions and adjusting the pH and ionic strength (see Applications regarding muramyldipeptide below).

EXTRACTION PROCEDURES

The sample is loaded onto the top of a bed of ODS-silica held between two porous teflon discs in a disposable plastic syringe barrel. The sample is forced through the bed using the syringe plunger, the plunger withdrawn, and the process repeated until the whole sample (10 to 20 bed volumes) has been passed through the extraction column. A similar process is used to wash the extraction column with several bed volumes of 1% TFA and then to elute the peptides with two or three bed volumes of methanol/water/TFA (80:19:1 by volume). The eluate is diluted with three volumes of 1% TFA to obtain a solution containing 20% methanol, suitable for pumping onto an HPLC column.

Gillette (Greenford, London) 1, 2, or 5mℓ syringes are suitable for the extraction column

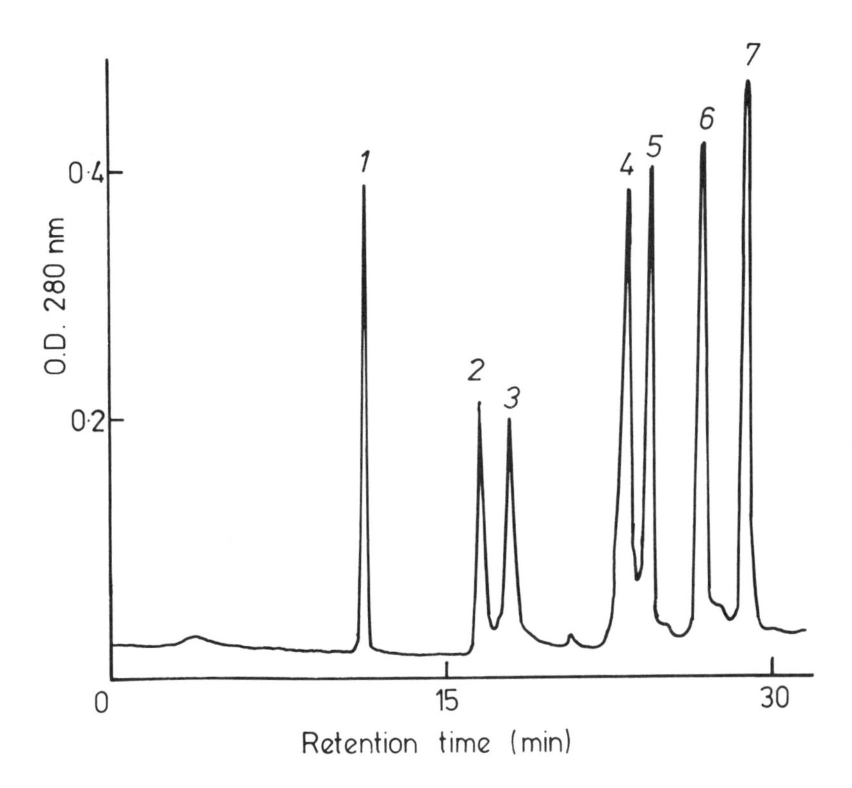

FIGURE 1. HPLC separation of a mixture of peptides. A Partisil®-ODS 4 mm × 250 mm column was eluted at 1 mℓ/min with a linear gradient, total volume 30 mℓ, composition: acetonitrile/water/TFA 0/99/1 to 80/19/1 (v:v:v). (1) D-Ser-Tyr, 20 μg; (2) D-Ser^{1}, Lys^{17,18}, 1-18 ACTH octadecapeptide, 20 μg; (3) 1-24 ACTH tetracosapeptide, 25 μg; (4) Human ACTH, 100 μg; (5) Somatostatin, 25 μg; (6) Porcine insulin, 100 μg; (7) Glucagon, 50 μg.

and porous teflon can be obtained from Pampas fluorplast, Newcastle, Staffordshire. Spherisorb 10-ODS (Phase-Separations Led., Queensferry, Clwyd) can be used as the packing material but if the sample contains protein, or particulate matter, or is large in volume, a larger particle size ODS-silica is preferable. Porasil® A (35- to 70-μm mesh; Waters Associates, Northwick, Cheshire) can be readily bonded with ODS groups in the laboratory[2] to give a suitable product for packing into extraction columns.

Samples can be prepared for extraction in several ways. Plasma can be extracted directly. Alternatively, prior to extraction, it can be either diluted with 1% TFA or deproteinized using a final concentration of 15% trifluoroacetic acid. If the latter procedure is contemplated it is important to check peptide recoveries because those with a high content of hydrophobic residues may coprecipitate with the plasma proteins. It is often possible to increase the recovery of peptide in a protein-free supernatant by increasing the volume of solution added to a fixed volume of plasma, by reducing the TFA concentration or by washing the precipitate with the same concentration of TFA used for the precipitation.

To prepare tissue samples for extraction it is necessary to homogenize them in a medium which precipitates proteins but not peptides and at the same time inactivates the peptidases which are released in large amounts when tissue is damaged. A suitable medium is 50 mℓ formic acid, 150 mℓ of TFA, 10 g of sodium chloride and 91 mℓ of 11 *M* hydrochloric acid made up to 1 ℓ with water. The TFA precipitates proteins and the hydrochloric acid ensures the environment is sufficiently acidic to retard degradation. The medium is used at a rate of 10 mℓ/g of tissue (see Figure 2). Once again the volume of medium, its composition,

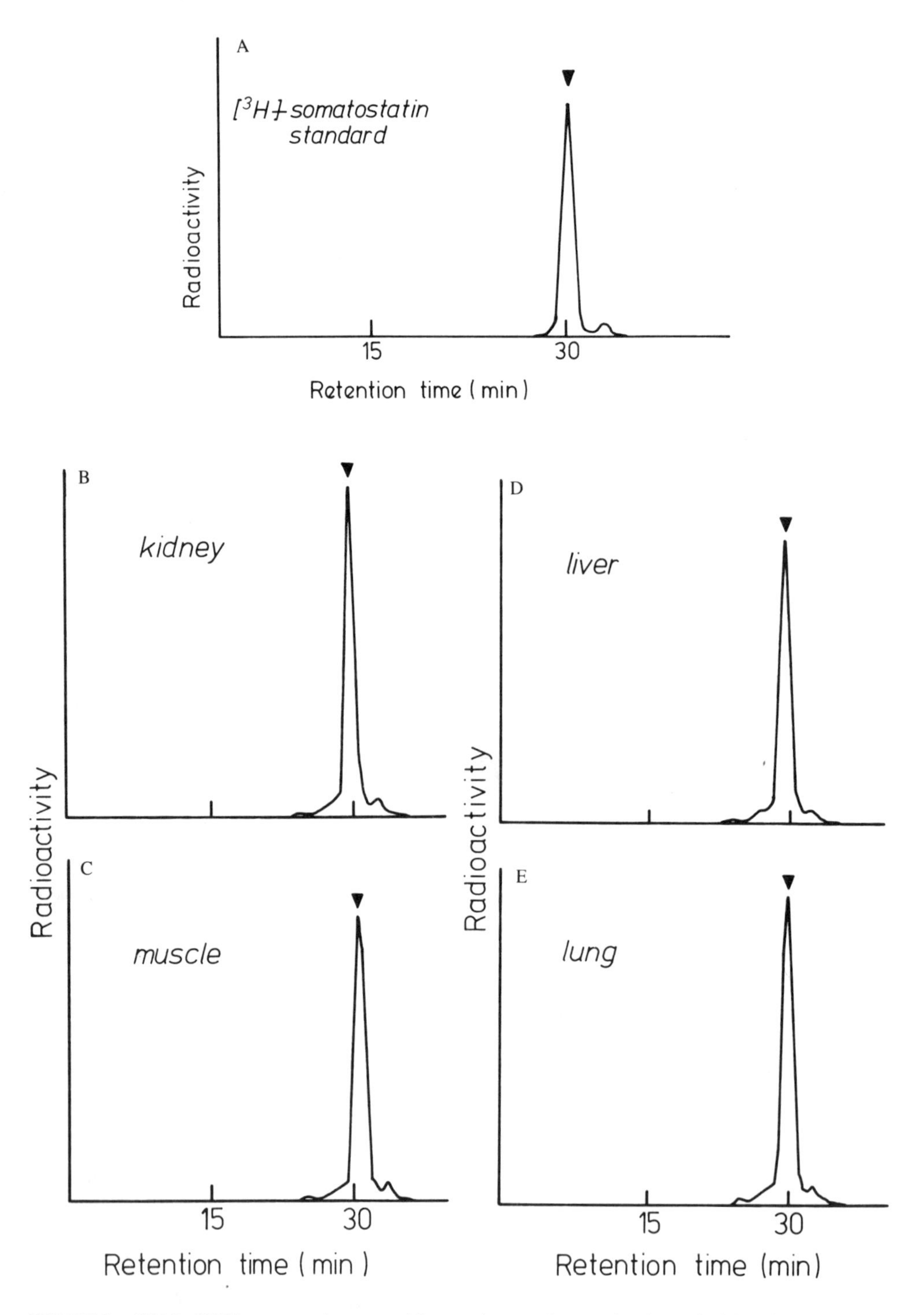

FIGURE 2. HPLC of [^{3}H]-somatostatin extracted from various rat tissues using the method described in the text. Recoveries were 80 to 90% and Figure 2A shows the profile of [^{3}H]-somatostatin applied directly to the column. The column was eluted at 0.7 mℓ/min with a linear gradient, total volume 60 mℓ, composition: acetonitrile/water/TFA 20/79/1 to 80/19/1 (v:v:v).

and whether or not the tissue protein pellet should be washed can be decided by test extractions using the particular peptide for which the method is being developed.

For muramylpeptides a different procedure of sample preparation is required. These compounds only adsorb weakly onto ODS-silica but the hydrophobic bonding forces are enhanced in the presence of concentrated salt solutions. Plasma proteins are precipitated using 1 *M* hydrochloric acid saturated with ammonium sulfate. The supernatant is then extracted using a bed of ODS-silica which is then washed with water. The products are eluted with methanol/water/TFA (80:19:1 by volume) and the eluate is film dried *in vacuo* and redissolved in phosphate buffer prior to loading for HPLC analysis.

Tissue (up to 1 g) is homogenized in 5 mℓ of 3.5% perchloric acid then 2.5 g ammonium sulfate is dissolved in the supernatant prior to extraction. Subsequent procedures for column washing, elution, and preparation for HPLC are the same as for plasma.

APPLICATIONS

The HPLC and extraction methods described above have been used to demonstrate the appearance in rat plasma of fragments of corticotrophin-(1-24)-tetracosapeptide following i.v. administration of tritium-labeled compound.[3] The plasma was carefully prepared by injecting heparin and collecting a large volume of blood through a cannula inserted into the dorsal aorta. After further purification using carboxymethylcellulose chromatography, a number of the fragments could be identified by amino acid analysis. Thus it was possible to obtain information about the sites within the molecule where cleavage had occurred.

The dose used for identification of fragments was quite large (1 mg per animal) but it was possible to show that a small dose of 3 μg per animal gave a very similar radioactivity profile after HPLC. In these experiments tissue extracts were also examined by HPLC. The amounts of peptide were too small to identify unequivocally but by using samples of peptide labeled in different known positions and by comparing tissue and plasma profiles it was possible to obtain information about the role of liver, kidney, muscle, and skin in the catabolism of corticotrophin-(1-24)-tetracosapeptide. A similar though less detailed study has been carried out for human corticotrophin.[4]

Investigation using HPLC of the fate of unlabeled somatostatin in the rat showed that rapid conversion to des-Ala[1]-somatostatin occurs in vivo and in plasma in vitro. Des-Ala[1]-somatostatin did not separate from somatostatin during HPLC and it was necessary to perform amino acid analysis to demonstrate the presence of the metabolite. A more detailed study of the fate of somatostatin has been carried out by applying HPLC to samples obtained after administration of tritium-labeled somatostatin.[5]

HPLC has been used to investigate the catabolism and elimination of muramyldipeptide and nor-muramyldipeptide.[6] These products are immune-modulators and consist of a sugar residue linked to Ala-D-iGln. The binding to ODS-silica is weak and extraction is accomplished in the presence of ammonium sulfate (see Extraction Procedures above). Isocratic elution of HPLC columns (0.05 *M* NaH_2PO_4, pH 4.5) provided evidence that for MDP and nor-MDP most of the product was rapidly eliminated unchanged in the urine. It was clear, however, that small amounts of metabolites were formed.

CONCLUSIONS

The rationale for the use of ODS-silica for extraction and separation of peptides is straightforward and depends mainly on the hydrophobic bonding between sample and solid phase. The methods can usually be applied in a predictable manner and give good yields. Although products will not always be separated it is often possible to achieve separation of a large number of components in a single run and the method can therefore give highly informative results. Recoveries are good and the procedure is a useful tool in the investigation of mechanisms of catabolism of peptides.

In addition to the use of linear solvent gradients described above it is possible to change the column selectivity by using ion-pairing reagents. This method has been exploited by Burbach and colleagues[7] to separate oxytocin fragments.

ACKNOWLEDGMENTS

The author wishes to acknowledge Drs. W. Rittel and B. Riniker, CIBA-GEIGY, Basel, Switzerland for ACTH and ACTH analogs, Drs. D. E. Brundish, R. Wade, and Mr. M. C. Allen for the [^{3}H]-somatostatin, and Mr. S. Metcalfe and Dr. G. E. Purdon for the results shown in Figures 1 and 2.

REFERENCES

1. **Bennett, H. P. J. and McMartin, C.,** Peptide hormones and their analogues: distribution, clearance from the circulation and inactivation *in vivo, Pharmacol. Rev.*, 30, 247, 1979.
2. **McMartin, C. and Purdon, G. E.,** Early fate of somatostatin in the circulation of the rat after intravenous injection, *J. Endocrinol.*, 77, 67, 1978.
3. **Hudson, A. M. and McMartin, C.,** Mechanisms of catabolism of corticotrophin-(1-24)-tetracosapeptide in the rat *in vivo, J. Endocrinol.*, 85, 93, 1980.
4. **Ambler, L., Bennett, H. P. J., Hudson, A. M., and McMartin, C.,** Fate of human corticotrophin immediately after intravenous administration to the rat, *J. Endocrinol.*, 93, in press, 1982.
5. **Peters, G. E.,** Distribution and metabolism of exogenous somatostatin in the rat, *Regul. Peptides,* 3, 361, 1982.
6. **Ambler, L. A. and Hudson, A. M.,** Metabolism of muramyldipeptides in the mouse, in preparation, 1982.
7. **Burbach, J. P. H., De Kloet, E. R., and De Weid, D.,** Oxytocin biotransformation in the rat limbic brain: characterization of peptidase activity and significance in the formation of oxytocin fragments, *Brain Res.*, 202, 401, 1980.

PEPTIDE ANTIBIOTICS

Shumpei Sakakibara and Yukio Kimura

INTRODUCTION

In 1939, Dubos first isolated a peptide antibiotic from a culture broth of *Bacillus brevis* and named it tyrothricin; later, this material was found to be a mixture of two different types of compounds, gramicidins (linear) and tyrocidins. Since then, more than 200 types of peptide antibiotics have been isolated from fermentation mixtures of bacteria, mainly genera *Bacillus* and *Streptomyces*, or fungi.

The characteristics of these antibiotic peptides differ distinctly from those of normal peptides originating from proteins. Although normal peptides only consist of ordinary L-amino acids, the antibiotic peptides frequently contain fatty acyl groups, D-amino acids and/or nonprotein amino acids in addition to the normal L-amino acids. Furthermore, antibiotic peptides are apt to be produced as an intimate mixture of analogous peptides. These phenomena may be explained by the difference in the pathway of their biosynthesis; peptides of protein origin are synthesized by the mechanism involving nucleic acids and ribosomes, but almost all peptide antibiotics are synthesized by the so-called multienzyme thiotemplate mechanism,[1] with which similar but different amino acids or nonprotein amino acids are commonly activated and incorporated into the products. Thus, the isolation of a homogeneous peptide antibiotic is not an easy task. Under these circumstances, the history of peptide antibiotics has been one of repeated detection of contaminants and structure revision. Today, HPLC is probably the best separation technique thus far developed for detecting analogous peptides. Attempts to separate peptide antibiotics by HPLC are now at the beginning stage in the long history of antibiotics, but the importance has been widely recognized. Although the number of published articles in this field is rather small, examples of the separation of peptide antibiotics by HPLC are reviewed in the following sections.

EXAMPLES OF ANTIBIOTIC SEPARATIONS

Actinomycin — This material was first isolated by Waksman et al. in 1940 as a reddish brown chromopeptide. Since then, more than 50 closely related analogs have been isolated from various strains of *Streptomyces*. These analogs only differ in the structure of the peptide moieties. In 1973, Rzeszotarski and Mauger[2] first separated the actinomycin C complex into three components, C_1 (which is identical with D), C_2, and C_3 by reversed-phase (RP) HPLC using a long μBondapak® C_{18}/Corasil column (1.8 m × 2.3 mm i.d.) and a mixture of acetonitrile-water (1:1) as the mobile phase. The structures of these components are shown in Figure 1. They further succeeded in separating actinomycins CP_2, CP_3, and D by a similar technique from a complex produced by *S. parvullus* (actinomycin D producing strain) in the presence of *cis*-4-chloro-L-proline. These separations may be the first examples of HPLC separation of peptide antibiotics using an alkyl-bonded silica column.

Alamethicin — This antibiotic is produced by the fungus *Trichoderma viride* as a mixture of closely related compounds. In the early stage of the structure determination, various structures (Figure 2) were proposed and then revised several times. Gisin et al.[3] synthesized a peptide having one of the proposed structures (compound D with Ala at the 6th position in Figure 2) as the most promising one. They also separated Upjohn's natural alamethicin on HPLC into two major peaks, compared component I with their synthetic product, and observed that these two compounds appeared to have identical properties in physical, chem-

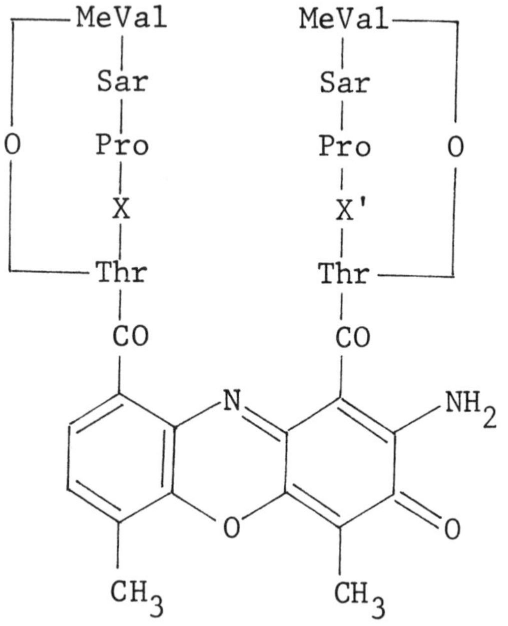

Actinomycin	X	X'
C_1	D-Var	D-Val
C_2	D-Var	D-aIle
C_3	D-aIle	D-aIle

FIGURE 1. Structure of actinomycin C. MeVal, *N*-methyl-valine; aIle, allo-isoleucine.

A. Pro-Aib-Ala-Aib-Ala-Gln-Aib-Val-Aib-Gly-Leu-Aib-Pro-Val-Aib-Aib-Gu-Gln-OH

B. Ac-Aib-Pro-Aib-Ala-Aib-Ala-Gln-Aib-Val-Aib-Gly-Leu-Aib-Pro-Val-Aib-Aib-Glu(Phol)-Gln-OH

C. Ac-Aib-Pro-Aib-Ala-Aib-(Ala/Aib)-Gln-Aib-Val-Aib-Gly-Leu-Aib-Pro-Val-Aib-Aib-Glu-OH (with Glu(Phol)-NH_2 attached to the final Glu)

D. Ac-Aib-Pro-Aib-Ala-Aib-(Ala/Aib)-Gln-Aib-Val-Aib-Gly-Leu-Aib-Pro-Val-Aib-Aib-Glu-Gln-Phol

FIGURE 2. Structures proposed for alamethicin. Aib, 2-aminoisobutyric acid; Phol, phenylalaninol.

ical, and biological analyses. The HPLC system was: column, Partisil® 5 octadecylsilyl (ODS) (250 × 4.6 mm i.d.); mobile phase, methanol-20 m*M* aqueous triethylammonium acetate (TEAA) pH 5 (85:15). Independently, Balasubramanian et al.[4] also synthesized the

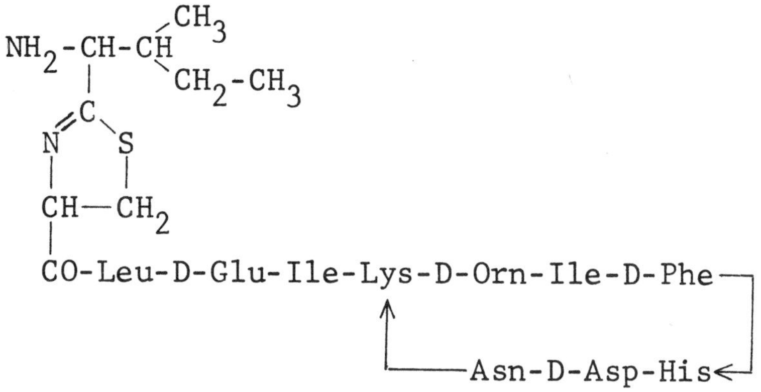

```
             CH3
            /
NH2-CH-CH
       |    \
       |     CH2-CH3
      C
    //  \
   N     S
   |     |
   CH—CH2
   |
   CO-Leu-D-Glu-Ile-Lys-D-Orn-Ile-D-Phe—┐
                           ↑              |
                           └——Asn-D-Asp-His←┘
```

FIGURE 3. Structure of bacitracin A.

same peptide. They applied Upjohn's standard alamethicin to their improved HPLC system and found that this material was composed of at least 12 different components. Their synthetic peptide corresponded closely to the major component by all criteria they examined. HPLC was carried out by isocratic and/or gradient elution on a μBondapak® C_{18}-column and a Spherisorb-ODS column. The mobile phases were: (1) 0.25 *N* triethylammonium phosphate (TEAP, pH 3.5) or 0.05 *N* TEAA (pH 3.5); (2) a mixture of tetrahydrofuran (THF), acetonitrile, and (1) (8:1:2).

Bacitracin — Commercially available bacitracin, produced by strains of *Bacillus licheniformis*, is a mixture of various components. This material was originally separated by counter-current distribution and/or ion-exchange column chromatography into groups A, B, C, D, E, F, and G. HPLC of this material was first carried out by Tsuji et al.[5] in 1974 on a μBondapak® C_{18}/Corasil column. More than 22 components were separated, but the resolution of each was not very high. In 1975, the same group[6] published an improved procedure for their HPLC system to obtain baseline separation of ten major components, for which a μBondapak® C_{18}-column (300 × 4.6mm i.d.) and the following linear gradient systems were used: (1) a mixture of methanol, water and 0.1 *M* phosphate buffer (pH 4.5) (1:17:2) and (2) methanol, acetonitrile, water, and 0.1 *M* phosphate buffer (pH 4.5) (5:2:2:1). The components were eluted in the following sequence: X, D, E, B_1, B_2, A, C, G, F_2, F_1. The HPLC technique was successfully applied to the quantitative determination of these components in commercial products during the process of reducing microbial contamination. The reported structure of the major component A is shown in Figure 3; B differs from A only by having Val instead of Ile, but the position of the replacement is not clear. Structures of the other components have not been elucidated yet.

Bleomycin — This material is produced by *Streptomyces verticillus* as a complex mixture with a heavy metal, and the major components are Bleomycin A_2 (55 to 70%) and B_2 (about 25%); the structures are shown in Figure 4. Since these compounds are highly cationic, paired-ion chromatography using an alkylsulfonic acid is useful for their separation by HPLC. Sakai[7] separated these two major components on a μBondapak® C_{18}-column (300 × 4.6 mm i.d.) using a mobile phase of 5 m*M* heptanesulfonic acid in 50% aqueous methanol at pH 3.5 to 4; better separation was obtained, however, by adjusting the solution with concentrated aqueous ammonia to pH 8.3. Rapid HPLC determination of Bleomycin A_2 in plasma was also demonstrated by Shiu et al.[8] using similar conditions; the standard curve for the assay was linear over the range of 0.5 to 5 μg/mℓ.

Brevistin — This group of antibiotics was isolated from *Bacillus brevis* by Shoji et al. in 1976 as a mixture of components A and B; the structures were determined as shown in

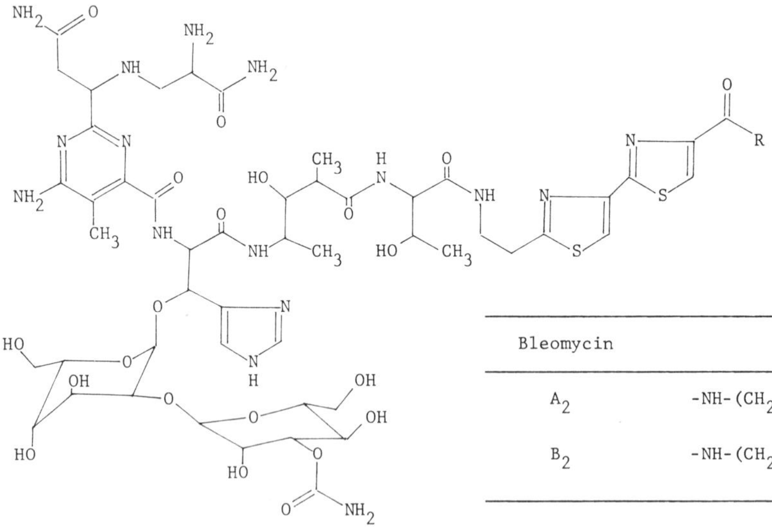

Bleomycin	R
A_2	$-NH-(CH_2)_3-\overset{+}{\underset{CH_3}{S}}-CH_3\ X^-$
B_2	$-NH-(CH_2)_4-NH-\underset{NH}{\overset{\|}{C}}-NH_2$

FIGURE 4. Structure of bleomycin A_2 and B_2.

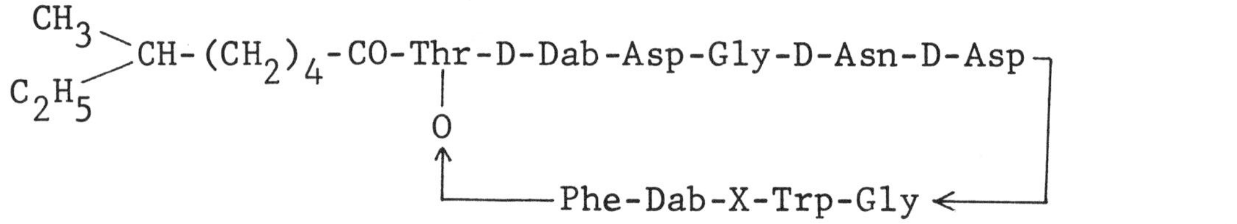

Brevistin	X
A	Ile
B	Val

FIGURE 5. Structure of Brevistin A and B. Dab, 2,4-diaminobutyric acid.

Figure 5. These components were separated by HPLC by the same group[9] as shown in Figure 6. The separation conditions were: column, Nucleosil® 5 C_{18} (200 × 4 mm i.d.); mobile phase, acetonitrile and 10 m*M* ammonium sulfate (pH 7.0) (45:55).

Cerexins — Cerexins A and B were isolated by Shoji et al. from the culture broths of *Bacillus cereus* 60-6 and *B. cereus* Gp-3, respectively. Later, cerexins C and D were also isolated from improved media. Although cerexins A and C were purified as single entities by conventional procedures, B and D each proved to be a complex mixture of four different components, which differ from one another only in the acyl residue. Complete resolution of these complexes by HPLC was achieved by Shoji et al.[10] on a column of Nucleosil® 5 C_{18} (250 × 4 mm i.d.) using: (1) a mixture of acetonitrile and 5 m*M* phosphate buffer

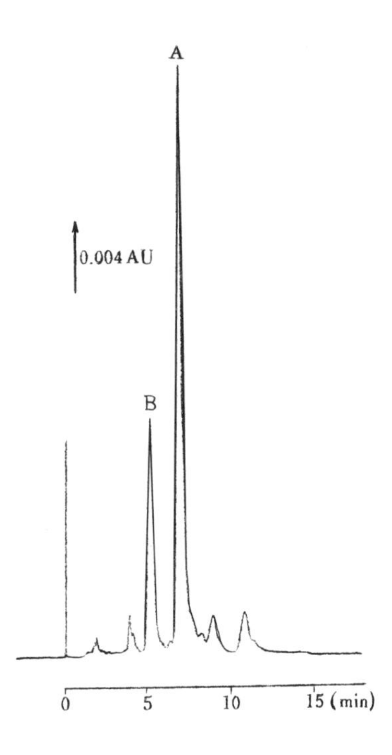

FIGURE 6. HPLC chromatogram of brevistins.

adjusted to pH 7 or (2) a mixture of acetonitrile and 10 m*M* ammonium sulfate adjusted to pH 7; the acetonitrile concentrations were both 35%. The retention times of the separated components increased in the order of B_1, B_2, B_3, B_4, and D_1, D_2, D_3, D_4, respectively. Structures determined for the separated materials are shown in Figure 7. Both cerexins C and D containing a Lys residue have longer retention time than the corresponding components A and B which contain a hydroxylysyl residue in place of the lysyl residue.

Cyclosporin — Cyclosporin A and D are both fungal cyclic peptides (Figure 8) with immunosuppressive activity isolated from *Trichoderama polysporium*. HPLC was first developed for the determination of component A in human plasma and urine by Niederberger et al.[11] in 1980. Chromatography was carried out on a column of 5 μm LiChrosorb® RP-8 (125 × 3 mm i.d.) using a gradient system with two mobile phases: (1) water-acetonitrile-methanol (5:75:20) and (2) water-acetonitrile-methanol (60:20:20). For the quantitation, a known quantity of cyclosporin D was added to the system as the internal standard. To increase the detection sensitivity, Gfeller et al.[12] attempted to convert this material into the β-naphthyl-seleno-substituted THF derivative just before applying it to the column. The derivative was readily separated on a column of LiChrosorb® RP-8 (250 × 4.5 mm i.d.) using a mixture of methanol and water (5:1) as the mobile phase, and the detection sensitivity was enhanced from three- to fourfold that of the original compound.

Enduracidin — Enduracidin is a chlorine-containing depsipeptide which was isolated from the mycelium of *Streptomyces fungicidicus* B-5477. This material was considered to be a single entity from its behavior on paper chromatography and counter-current distribution. Later, it was separated into two components A and B by linear gradient elution from a column of Amberlite® XAD-2 using solvent A (0.5% NaCl in 50% methanol) as the starting solvent and solvent B (3/500 *N* HCl in 50% methanol) as the limiting solvent.[13] This may be the first report of the separation of peptide antibiotics on a macroreticular styrene-divinylbenzene copolymer in a reversed-phase adsorption mode. The structures of enduracidin A and B are shown in Figure 9.

FA-D-Asn-D-Val-X-Asn-D-Asn-Y-D-aThr-Z-D-Trp-D-aIle-OH

Cerexin	X	Y	Z	FA
A	Val	γ-Hyl	Ser	3-Hydroxy-9-methyldecanoyl
B_1	Phe	γ-Hyl	Gly	3-Hydroxy-8-methylnonanoyl
B_2	Phe	γ-Hyl	Gly	3-Hydroxydecanoyl
B_3	Phe	γ-Hyl	Gly	3-Hydroxy-8-methyldecanoyl
B_4	Phe	γ-Hyl	Gly	3-Hydroxy-9-methyldecanoyl
C	Val	Lys	Ser	3-Hydroxy-9-methyldecanoyl
D_1	Phe	Lys	Gly	3-Hydroxy-8-methylnonanoyl
D_2	Phe	Lys	Gly	3-Hydroxydecanoyl
D_3	Phe	Lys	Gly	3-Hydroxy-8-methyldecanoyl
D_4	Phe	Lys	Gly	3-Hydroxy-9-methyldecanoyl

FIGURE 7. Structure of cerexins. γ-Hyl, 4-hydroxy-lysine; aThr, allo-threonine; aIle, allo-isoleucine.

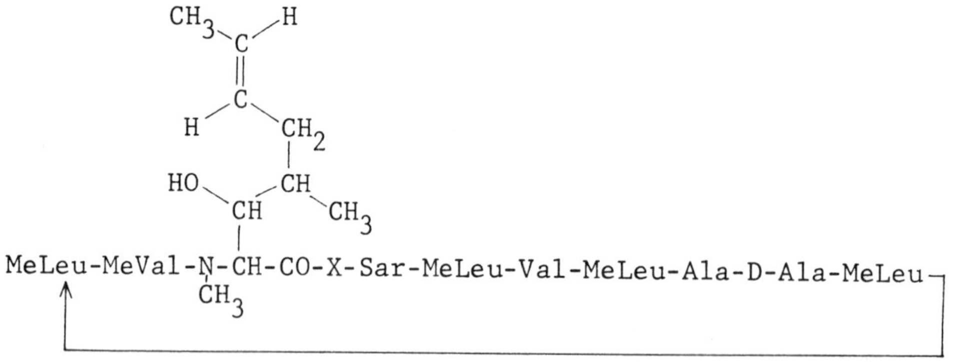

Cyclosporin	X
A	Abu
D	Val

FIGURE 8. Structure of cyclosporin A and D. MeVal, *N*-methyl-valine; MeLeu, *N*-methyl-leucine; Abu, 2-aminobutyric acid.

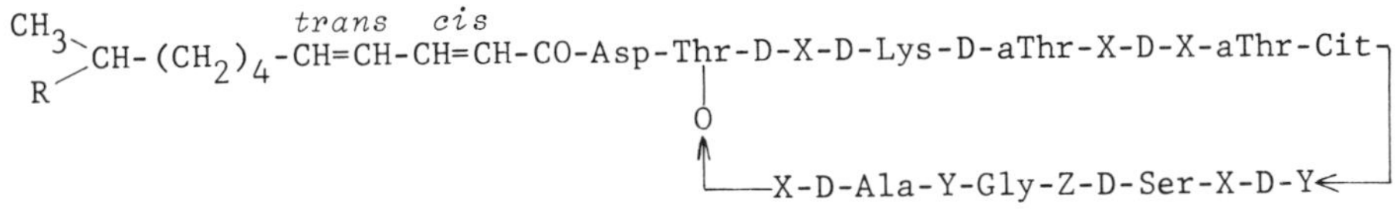

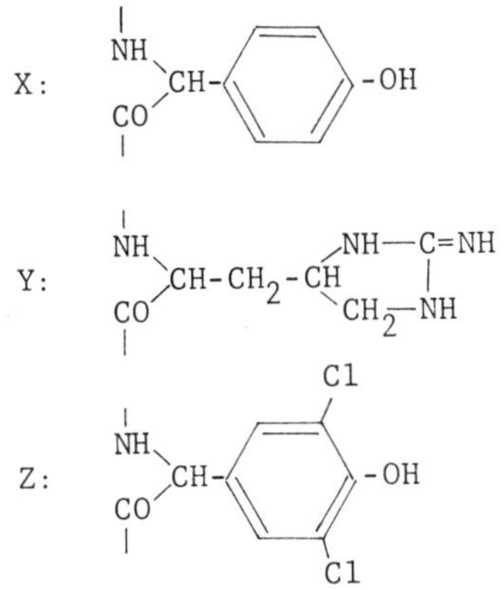

CH_3(R)>CH-$(CH_2)_4$-CH=CH-CH=CH-CO-Asp-Thr-D-X-D-Lys-D-aThr-X-D-X-aThr-Cit (*trans*, *cis*)

O (from Thr) ← X-D-Ala-Y-Gly-Z-D-Ser-X-D-Y ← (from Cit)

X: NH / CH-(4-hydroxyphenyl) / CO

Y: NH / CH-CH_2-CH (ring: NH—C=NH, CH_2—NH) / CO

Z: NH / CH-(3,5-dichloro-4-hydroxyphenyl) / CO

Enduracidin	R
A	CH_3
B	C_2H_5

FIGURE 9. Structure of enduracidin A and B. aThr, allo-threonine; Cit, citrulline.

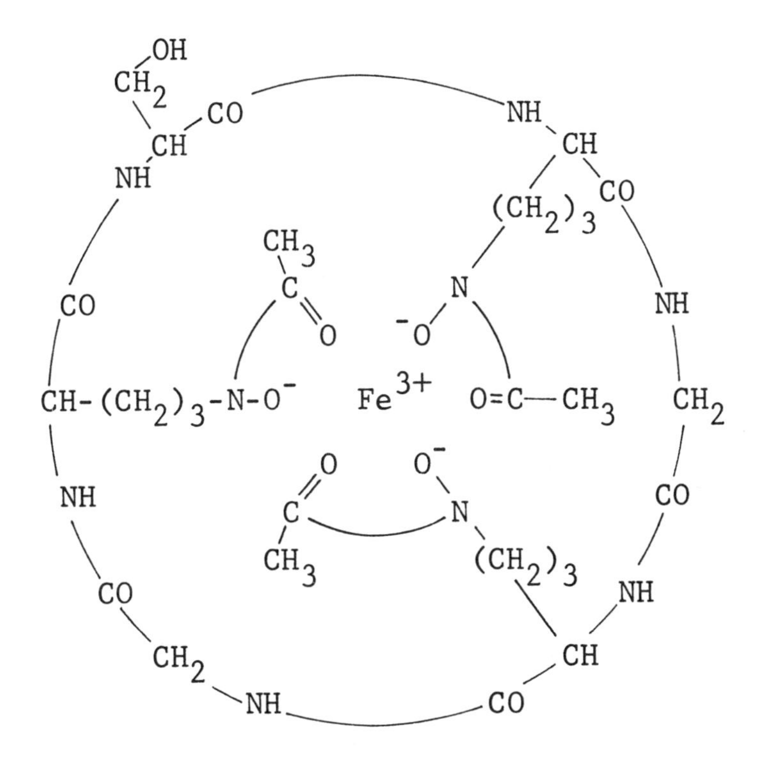

FIGURE 10. Structure of ferricrocin.

Ferricrocin — This material is an iron complex of a cyclic peptide (Figure 10), which was isolated from the fermentation broth of *Aspergillus viridi-nutans* as a metabolic product. HPLC of this material was reported by Fiedler[14] in 1981 on a column of 7 μm LiChrosorb® RP-8 (250 × 4.5 mm i.d.) for analysis and on a column of the same support (250 × 16

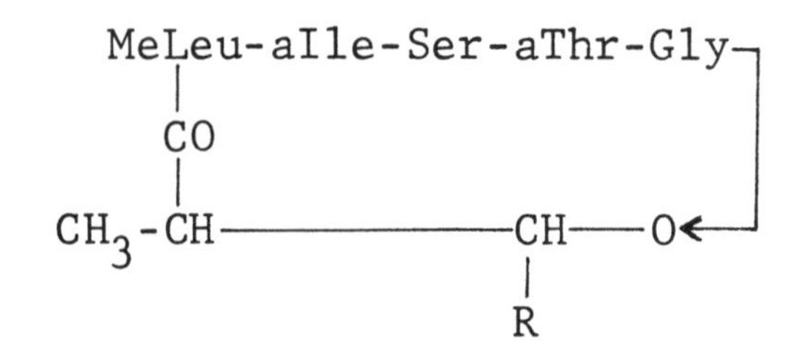

Compound	R
Globomycin	$-(CH_2)_5-CH_3$
SF-1902 A_5	$-(CH_2)_7-CH_3$

FIGURE 11. Structure of globomycins. MeLeu, *N*-methyl-leucine; aIle, allo-isoleucine; aThr, allo-threonine.

mm i.d.) for preparative separation (up to 1 g) using water-acetonitrile (9:1) as the mobile phase.

Globomycin — Globomycin (Figure 11) was first isolated by Inukai et al. from a broth of *Streptomyces globocacience*. Independently, Shomura et al. isolated a similar antibiotic SF-1902 from *S. hygroscopicus* SF-1902. This material seemed to be homogeneous on TLC, but HPLC analysis disclosed it to be a mixture of at least five major components, SF-1902, A_2, A_3, A_4, and A_5.[15] Among them, SF-1902 was identical with globomycin. The conditions employed for the HPLC analysis were: column, μBondapak® C_{18} (300 × 8 mm i.d.); mobile phase, acetonitrile-water (60:40). Component A_5 was isolated as homogeneous crystals, and the determined structure, shown in Figure 11, differed from that of globomycin only in the structure of the fatty acyl residue.

Gramicidin(linear) — This material was originally isolated from a fermentation product, tyrothricin, and has been fractionated into three major components by counter-current distribution. These linear gramicidins(Gdin) A, B and C were each further separated into two closely related components, Val- or Ile-containing components, by continuing the counter-current distribution for a total of 2000 transfers. The structures determined are shown in Figure 12. Separation of each component by HPLC was first reported in 1977 on a column of Zorbax® ODS (250 × 2.1 mm i.d.) using a mixture of methanol and aqueous 5 m*M* ammonium sulfate solution (74:24); the components were eluted in the following sequence: [Val]-Gdin C, [Ile]-Gdin C, [Val]-Gdin A, [Ile]-Gdin A, [Val]-Gdin B, and [Ile]-Gdin B.[16] Similar separation was also achieved in 1981 on a column of 10 μm phenyl-silica (300 × 3.9 mm i.d.) using a mobile phase of methanol-water (75:25).[17] This procedure was applied to the separation of crude gramicidins of up to the 100-mg scale using a larger column (3000 × 7.8 mm i.d.).

Isariin — This material, isolated from *Isaria felina,* was assigned the structure shown in Figure 13. Recently, four similar components, I, II, III, and IV, were isolated and partially purified on TLC. Final purification was completed by HPLC and yielded crystalline compounds.[18] Structural studies revealed that I was identical to the originally characterized isariin, and others were novel analogs, which were named isariin B, C, and D; the structures are

HCO-X-Gly-Ala-D-Leu-Ala-D-Val-Val-D-Val-Trp-D-Leu-
Y-D-Leu-Trp-D-Leu-Trp-NH-$(CH_2)_2$-OH

Gramicidin	X	Y
[Val]-Gdin A	Val	Trp
[Ile]-Gdin A	Ile	Trp
[Val]-Gdin B	Val	Phe
[Ile]-Gdin B	Ile	Phe
[Val]-Gdin C	Val	Tyr
[Ile]-Gdin C	Ile	Tyr

FIGURE 12. Structure of linear gramicidins.

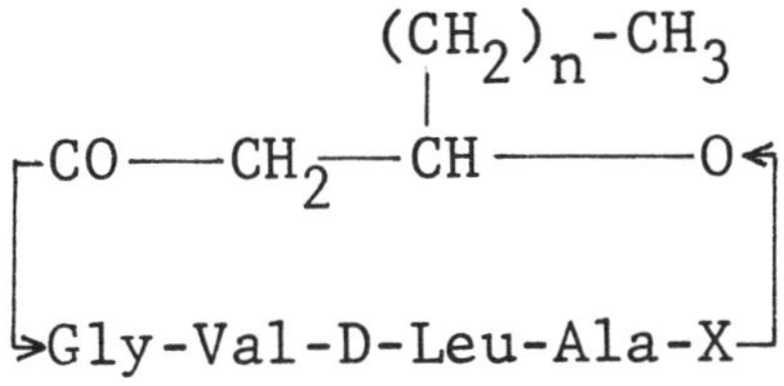

$(CH_2)_n$-CH_3 (on CH)
CO—CH_2—CH———O
Gly-Val-D-Leu-Ala-X

Compoumd	X	n
Isariin	Val	8
Isariin B	Val	5
Isariin C	Ala	5
Isariin D	Ala	3

FIGURE 13. Structure of isariins.

```
FA-X-Dab-Dab-Y-Z-Dab-Dab-Leu┐
       ↑                    |
       └────────────────────┘
```

Octapeptin	X	Y	Z	FA
A_1	D-Dab	D-Leu	Leu	3-Hydroxy-8-methyldecanoyl
A_2	D-Dab	D-Leu	Leu	3-Hydroxy-8-methylnonanoyl
A_3	D-Dab	D-Leu	Leu	3-Hydroxydecanoyl
A_4	D-Dab	D-Leu	Leu	3-Hydroxy-9-methyldecanoyl
B_1	D-Dab	D-Leu	Phe	3-Hydroxy-8-methyldecanoyl
B_2	D-Dab	D-Leu	Phe	3-Hydroxy-8-methylnonanoyl
B_3	D-Dab	D-Leu	Phe	3-Hydroxydecanoyl
B_4	D-Dab	D-Leu	Phe	3-Hydroxy-9-methyldecanoyl
C_1	D-Dab	D-Phe	Leu	3-Hydroxy-6-methyloctanoyl
D_1	D-Ser	D-Leu	Leu	3-Hydroxy-8-methyldecanoyl
D_2	D-Ser	D-Leu	Leu	3-Hydroxy-9-methylnonanoyl
D_3	D-Ser	D-Leu	Leu	3-Hydroxydecanoyl
D_4	D-Ser	D-Leu	Leu	3-Hydroxy-9-methyldecanoyl

FIGURE 14. Structure of octapeptins. Dab, 2,4-diaminobutyric acid.

also given in Figure 13. The HPLC conditions were: column, Porasil® C_{18} (300 × 3.9 mm i.d.); mobile phase, acetonitrile-water (63:37).

Octapeptin — In 1973, a new complex of peptide antibiotics, EM 49, was isolated from a culture broth of *Bacillus circulans ATCC* 21,656. The complex was separated into four components by CM-Cellulose chromatography, but two of them were still considered to be complexes. Independently, two closely related antibiotics were isolated from a different strain of *B. circulans* 333-25. One was identical with EM 49, but the other, 333-25, was a new antibiotic related to EM 49; this group of antibiotics was named octapeptin. The structure is similar to that of polymyxins but the branched group is composed of one amino acid residue acylated by a fatty acid. These complexes were completely separated by HPLC into nine components, A_1, A_2, A_3, A_4, B_1, B_2, B_3, B_4, and C_1, and the structures assigned are shown in Figure 14.[19] The columns employed for the analysis were Nucleosil® 10 C_{18} (250 × 10 mm i.d.) for A and B and Nucleosil® 5 C_{18} (200 × 4 mm i.d.) for C. The mobile phase was a mixture of acetonitrile-5 m*M* sodium tartrate buffer (pH 3) containing 5 m*M* sodium 1-butanesulfonate and 50 m*M* sodium sulfate; the acetonitrile contents were 30% for octapeptin C and 31% for A/B. The last group, octapeptin D, was isolated separately from a *Bacillus* strain, JP-301, and further separated into four different components, D_1, D_2, D_3, and D_4, by HPLC;[20] the separation conditions were the same as those for the separation of A and B except that the acetonitrile content was 35%. The structures are also given in Figure 14.

Polymyxin — Thirteen complexes of the polymyxin antibiotics, A, B, C, D, E, F, K, M, P, S, T, colistin, and circulin, have been isolated from the culture broths of *Bacillus*

```
6-Methyloctanoyl-Dab-Thr-Dab-Dab-Dab-D-Leu-Leu-┐
                             ↑                  │
                             └─Thr-Dab-Dab ←────┘
```

FIGURE 15. A proposed structure of polymyxin E_1. Dab, 2,4-diaminobutyric acid.

Table 1
AMINO ACID AND FATTY ACID COMPOSITION, AND RETENTION TIMES OF POLYMYXINS ANALYZED ON A HITACHI GEL 3011 COLUMN

	Amino acid							
Polymyxin	Dab[a]	Thr	Ser	Leu	Ile	Phe	Fatty acid	Retention time
K_1	6	3	—	1	—	—	Unknown	9.2
A_1	6	3	—	1	—	—	6-MOA[a]	10.8
M_1	6	3	—	1	—	—	6-MOA	11.4
D_1	5	3	1	1	—	—	6-MOA	13.8
P_1	6	3	—	—	—	1	6-MOA	15.8
Circulin A	6	2	—	1	1	—	6-MOA	16.6
Colistin A	6	2	—	2	—	—	6-MOA	20.6
E_1	6	2	—	2	—	—	6-MOA	21.8
B_1	6	2	—	1	—	1	6-MOA	40.4
K_2	6	3	—	1	—	—	Unknown	6.0
A_2	6	3	—	1	—	—	IOA[a]	6.8
M_2	6	3	—	1	—	—	IOA	7.0
D_2	5	3	1	1	—	—	IOA	8.5
P_2	6	3	—	—	—	1	IOA	9.2
Circulin B	6	2	—	1	1	—	IOA	9.8
Colistin B	6	2	—	2	—	—	IOA	11.8
E_2	6	2	—	2	—	—	IOA	12.2
B_2	6	2	—	1	—	1	IOA	21.0

[a] 6-MOA: 6-methyloctanoic acid; IOA, isooctanoic acid; Dab, 2,4-diaminobutyric acid.

polymyxa and related species. They are strongly basic fatty acyl peptides which contain five or six Dab residues in the respective molecules. Each complex was further separated into two or three major components by counter-current distribution techniques, which depend on differences between the fatty acyl residues. Many different structures had been proposed for polymyxins. Among them, the structure of polymyxin E_1 (colistin A) proposed by Suzuki et al. in 1964 (Figure 15) is now widely accepted as the most reasonable one for this antibiotic. However, some of the recently observed data are inconsistent with this structure.[21] In 1971, a colistin complex was first separated by Kimura et al.[22] into three major components, A, B, and C, with reversed-phase adsorption chromatography on macroreticular styrene-divinylbenzene copolymer(Amberlite® XAD-2, 100 mesh) using a mixture of methanol and 0.2 *N* KCl-HCl buffer (pH 2) (1:1). A similar system with Hitachi gel #3011 (10 μm) was much more effective for separating 18 different polymyxin components within 30 min for each complex.[23] The amino acid and fatty acid compositions, and the retention times of these components are listed in Table 1. Recently, Kimura[24] further found that Hamilton PRP-1 resin is the best in terms of resolving power among similar resins thus far examined, as demonstrated by the separation of colistin shown in Figure 16.

In 1975, RP-HPLC was first applied to the separation of polymyxin B, circulin, and colistin on a μBondapak® C_{18}-column (300 × 4 mm i.d.) using a linear gradient of mobile

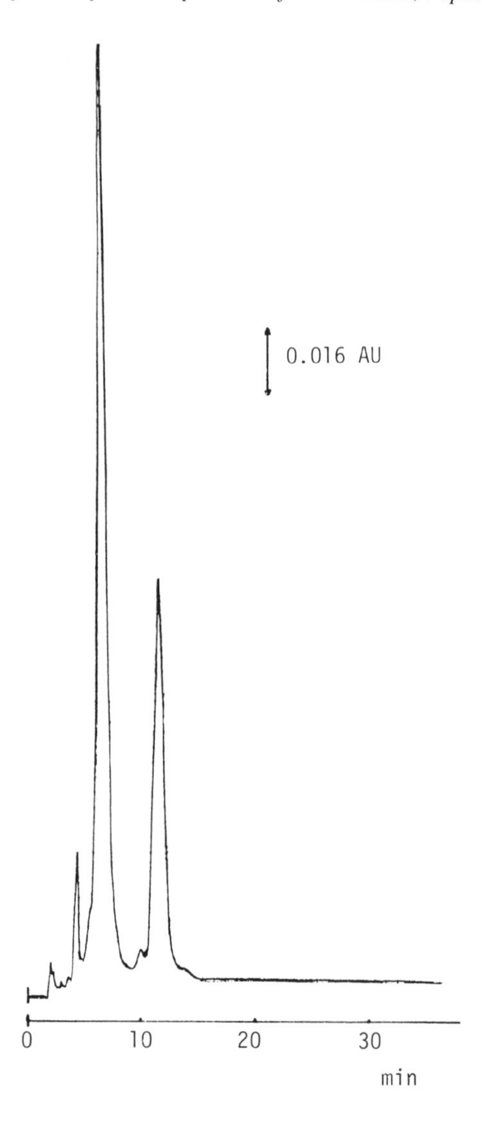

FIGURE 16. Reversed-phase adsorption chromatogram of colistin. Sample: colistin sulfate 50 ng (Banyu Pharm. Co., Ltd.). Column: Hamilton PRP-1 (250 × 4.6 mm i.d.). Mobile phase: 0.005 *M* Gly-H_2SO_4 buffer (pH 2.3) (containing 0.005 *M* sodium 1-butanesulfonate and 0.05 *M* sodium sulfate)-acetonitrile (80:20). Flow rate: 1.0 mℓ/min. Detection: 210 nm.

phase from 20% acetonitrile to 50% acetonitrile-20% methanol, both in phosphate buffer (pH 2.0).[6] In this case, a moving-wire flame ionization detector was used. Later in 1979, polymyxin B was separated into two major components, B_1 and B_2, under isocratic conditions on a 5-μm Hypersil-ODS column (250 × 3 mm i.d.) using a mobile phase containing 22.5% acetonitrile in an aqueous mixture of 0.5% tetramethylammonium chloride and 0.2% sulfuric acid, adjusted to pH 2.7 with 0.5 *M* K_2HPO_4.[25] Successful separation of a series of polymyxins was achieved by Terabe et al.[19] in 1979 under the conditions of isocratic ion-paired RP-HPLC on a Nucleosil® C_{18}-column (200 × 4 mm i.d.). The mobile phase was a mixture of acetonitrile and 5 m*M* sodium tartrate buffer (pH 3.0) containing 5 m*M* sodium

1-butanesulfonate and 50 m*M* sodium sulfate; the acetonitrile content was 21% for polymyxins C, D, M, and S, 22.5% for polymyxins E, B, and S, 30% for polymyxins F and T. Under these conditions, the following polymyxins were separated: M_1, M_2, D_1, D_2, C_1, C_2, S_1, E_1 (colistin A), E_2 (colistin B), B_1, B_2, B_3, F_1, F_2, F_3, T_1, T_2. The above procedure was applied by Thomas et al.[26] to the separation of various commercial samples of polymyxins B and E on a 5-μm Spherisorb ODS column (150 × 4 mm i.d.), and polymyxin B complex was further separated into 11 different components including minor ones and polymyxin E (colistin) into 13 components. Later, a modification of Terabe's procedure was published by Whall,[27] in which he recommended the use of a 5-μm Ultrasphere ion-pair column (250 × 4.5 mm i.d.) and a mobile phase of 0.1 *M* sodium phosphate in acetonitrile-water (23:77) at pH 3.0. Elverdam et al.[28] also separated commercial polymyxins B and E into 10 to 13 components on a Nucleosil® 5 C_{18}-column (150 × 4.6 mm i.d.) by isocratic elution using a mobile phase of 22% acetonitrile in 23 m*M* phosphoric acid, 10 m*M* acetic acid, 50 m*M* sodium sulfate buffer adjusted to pH 2.5 with triethylamine. Under the present conditions, they deduced the presence of three new polymyxins, Ile-polymyxin B_1, Val-polymyxin E_2, and Nva-polymyxin E_1 from the starting materials. Furthermore, separation of more than 1 g of polymyxin B complex was successfully achieved under almost the same conditions using a 10-μm octadecyldimethylsilyl-substituted LiChrosorb® Si 100 column (250 × 40 mm i.d.).[28] Kalasz and Horvath[29] introduced a displacement mode in HPLC using a standard column (250 × 4.6 mm i.d.), which could be used to separate a complex containing polymyxins B_1 and B_2 on a preparative scale. The column was packed with 5-μm LiChrosorb® RP-8, and up to 300 mg of polymyxin was adsorbed onto the resin. The adsorbed material was displaced using a solution of 0.05 *M* octyldodecyldimethylammonium chloride in water containing 10% acetonitrile; resolution of the components B_1 and B_2 was satisfactory.

Quinomycin — Since 1954, several different kinds of peptide antibiotics containing quinoxaline groups have been isolated from *Streptomyces*. HPLC of a quinomycin complex was first carried out by Shoji et al.[30] in 1976, and the five components A, D, B, E, and C were clearly separated. The column was Micropak CH-10 (250 × 2.1 mm i.d.) and the mobile phase was a linear gradient of acetonitrile-water from (35:65) to (7:1000). If the proposed structure (Figure 17) is correct, two positional isomers might be present for each component, but the present HPLC analysis failed to detect any. However, ^{13}C NMR analysis of component E revealed the presence of the isomers as had been expected.[30]

Tridecaptin — In 1978, tridecaptin A, B, and C were isolated from culture broths of *Bacillus polymyxa* AR-110, B-2, and E-23, respectively. These were characterized as complexes of acyl tridecapeptides. HPLC of tridecaptin A revealed that the complex is a mixture of two components, A_α and A_β.[10] Similarly, tridecaptins B and C were further separated into B_α, B_β, B_γ, B_δ, and $C_{\alpha 1}$, $C_{\alpha 2}$, $C_{\beta 1}$.[10] Separation was achieved on a column of Nucleosil® 5 C_{18} (200 × 4 mm i.d.) using a mixture of acetonitrile-10 m*M* ammonium sulfate (pH 7) (33:67) for A and C, and a mixture of acetonitrile-5 m*M* sodium tartrate buffer (pH 3.0) containing 5 m*M* sodium 1-butanesulfonate and 50 m*M* sodium sulfate (30:70) for B. The structures of these tridecaptins are shown in Figure 18.

Zervamicin — Zervamicin (I and II) and emerimicin II complexes, which are isolated from *Emericellopsis salmosynnemata* and *E. microspora*, respectively, are a group of peptaibophol. Crude zervamicin I was purified by RP-HPLC on an Ultrasphere-ODS column using a mixture of methanol-water-2-propanol-acetic acid (50:34:16:0.1) to separate Zervamicin IC, as the major component, and zervamicin IA, IB, IB′ as the minor components.[31] Zervamicin II and emerimicin II were similarly separated into two major components each (IIA and IIB), and five additional neutral zervamicins (II-1, II-2, II-3, II-4, and II-5) were also separated using methanol-water-2-propanol (52:32:16). During the analysis, the major components of zervamicin, IIA and IIB, and emerimicin, IIA and IIB, were found to give identical amino acid compositions and FAB mass spectra; thus, use of the zervamicin

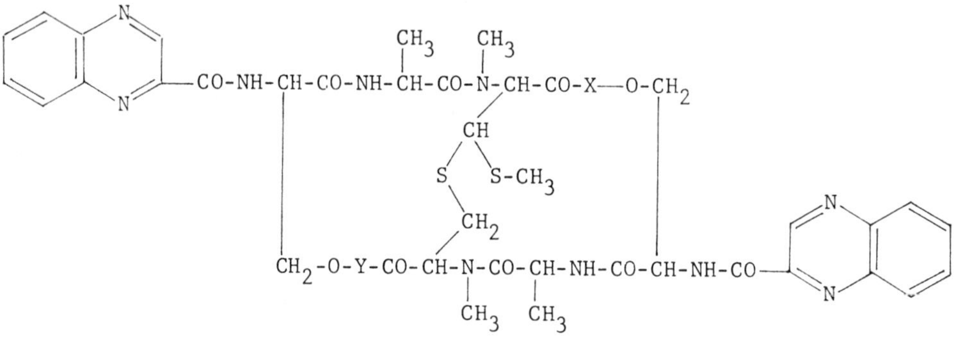

```
                          CH3    CH3
                          |      |
[quinoxaline]-CO-NH-CH-CO-NH-CH-CO-N-CH-CO-X—O-CH2
                    |               /           |
                    |             CH            |
                    |            /  \           |
                    |           S    S-CH3      |
                    |            \              |
                    |             CH2           |
                    |            /              |
                    CH2-O-Y-CO-CH-N-CO-CH-NH-CO-CH-NH-CO-[quinoxaline]
                                  |     |
                                  CH3   CH3
```

Quinomycin	X	Y
A	MeVal	MeVal
B	Me-aIle	Me-aIle
C	DiMe-aIle	DiMe-aIle
D	MeVal	Me-aIle
	or	
	Me-aIle	MeVal
E	1: Me-aIle	DiMe-aIle
	2: DiMe-aIle	Me-aIle

FIGURE 17. Structure of quinomycins. MeVal, *N*-methyl-valine; Me-aIle, *N*-methyl-allo-isoleucine; DiMe-aIle, *N*,4-dimethyl-allo-isoleucine.

FA-W-D-Dab-Gly-D-Ser-D-Trp-Ser-Dab-D-Dab-X-Glu-Val-Y-Z-OH

Tridecaptin	W	X	Y	Z	FA
A_{α}	D-Val	Phe	D-aIle	Ala	3-Hydroxy-6-methyloctanoyl
A_{β}	D-Val	Phe	D-Val	Ala	3-Hydroxy-6-methyloctanoyl
B_{α}	Gly	Ile	D-aIle	Ser	6-Methyloctanoyl
B_{β}	Gly	Ile	D-Val	Ser	6-Methyloctanoyl
B_{γ}	Gly	Val	D-aIle	Ser	6-Methyloctanoyl
B_{δ}	Gly	Val	D-Val	Ser	6-Methyloctanoyl
$C_{\alpha 1}$	D-Val	Phe	D-Val	Ser	3-Hydroxy-8-methyldecanoyl
$C_{\alpha 2}$	D-Val	Phe	D-Val	Ser	3-Hydroxy-8-methylnonanoyl
$C_{\beta 1}$	D-Val	Phe	D-aIle	Ser	3-Hydroxy-8-methyldecanoyl

FIGURE 18. Structure of tridecaptins. Dab, 2,4-diaminobutyric acid; aIle, allo-isoleucine.

nomenclature was proposed for these two compounds. Structures of zervamicins determined by FAB MS, GC/MS, and amino acid analysis are given in Figure 19.

Ac-Trp-W-Glu-X-Y-Thr-Aib-Z-Aib-Hyp-Gln-Aib-Hyp-Aib-Pro-Phol

Zervamicin	W	X	Y	Z
IA	Ile	Iva	Val	Leu
IB	Val	Iva	Ile	Leu
IB'	Ile	Aib	Ile	Leu
IC	Ile	Iva	Ile	Leu
IIA	Ile	Aib	Ile	Leu
IIB	Ile	Iva	Ile	Leu
II-1	Ile	Aib	Val	Leu
II-2	Ile	Aib	Ile	Val
II-3	Val	Aib	Ile	Leu
II-4	Ile	Iva	Val	Leu
II-5	Ile	Iva	Ile	Val

FIGURE 19. Structure of zervamicins. Aib, 2-aminoisobutyric acid; Iva, 2-ethylalanine.

REFERENCES

1. **Kurahashi, K.,** Biosynthesis of peptide antibiotics, in *Antibiotics, Vol. IV, Biosynthesis,* Corcoran, J. W., Ed., Springer-Verlag, Berlin, 1981, 325.
2. **Rzeszotarski, W. J. and Mauger, A. B.,** Reversed-phase high-pressure liquid chromatography of actinomycins, *J. Chromatogr.,* 86, 246, 1973.
3. **Gisin, B. F., Davis, D. G., Borowska, Z. K., Hall, J. E., and Kobayashi, S.,** Synthesis of the major component of alamethicin, *J. Am. Chem. Soc.,* 103, 6373, 1981.
4. **Balasubramanian, T. M., Kendrick, N. C. E., Taylor, M., Marshall, G. R., Hall, J. E., Vodyanoy, I., and Reusser, F.,** Synthesis and characterization of the major component of alamethicin, *J. Am. Chem. Soc.,* 103, 6127, 1981.
5. **Tsuji, K., Robertson, J. H., and Bach, J. A.,** Quantitative high-pressure liquid chromatographic analysis of bacitracin, a polypeptide antibiotic, *J. Chromatogr.,* 99, 597, 1974.
6. **Tsuji, K. and Robertson, J. H.,**Improved high-performance liquid chromatographic method for polypeptide antibiotics and its application to study the effects of treatments to reduce microbial levels in bacitracin powder, *J. Chromatogr.,* 112, 663, 1975.
7. **Sakai, T. T.,** Paired-ion high-performance liquid chromatography of bleomycins, *J. Chromatogr.,* 161, 389, 1978.
8. **Shiu, G. K., Goehl, T. J., and Pitlick, W. H.,** Rapid high-performance liquid chromatographic determination of bleomycin A_2 in plasma, *J. Pharm. Sci.,* 68, 232, 1979.
9. **Konaka, R. and Shoji, J.,** High performance liquid chromatography of peptide antibiotics, in *Biomedical Chromatography,* Vol. 2, Hara, S., Nakajima, T., and Hirobe, M., Eds., Nankodo, Tokyo, 1981, 151.

10. **Shoji, J., Kato, T., Terabe, S., and Konaka, R.,** Resolution of peptide antibiotics, cerexins and tridecaptins, by high performance liquid chromatography (studies on antibiotics from the genus *Bacillus*. XXVI), *J. Antibiot.*, 32, 313, 1979.
11. **Niederberger, W., Schaub, P., and Beveridge, T.,** High-performance liquid chromatographic determination of cyclosporin A in human plasma and urine, *J. Chromatogr.*, 182, 454, 1980.
12. **Gfeller, J.-C., Beck, A. K., and Seebach, D.,** Erhöhung der Nachweisempfindlichkeit von Cyclosporin A durch Derivatisierung mit 2-Naphthylselenylchlorid, *Helv. Chim. Acta*, 63, 728, 1980.
13. **Hori, M., Sugita, N., and Miyazaki, M.,** Enduracidin, a new antibiotic. VI. Separation and determination of enduracidins A and B by column chromatography, *Chem. Pharm. Bull.*, 21, 1171, 1973.
14. **Fiedler, H.-P.,** Preparative-scale high-performance liquid chromatography of ferricrocin, a microbial product, *J. Chromatogr.*, 209, 103, 1981.
15. **Omoto, S., Suzuki, H., and Inouye, S.,** Isolation and structure of SF-1902 A_5, a new globomycin analogue, *J. Antibiot.*, 32, 83, 1979.
16. **Axelsen, K. S. and Vogelsang, S. H.,** High-performance liquid chromatographic analysis of gramicidin, a polypeptide antibiotic, *J. Chromatogr.*, 140, 174, 1977.
17. **Koeppe, R. E., II, and Weiss, L. B.,** Resolution of linear gramicidins by preparative reversed-phase high-performance liquid chromatography, *J. Chromatogr.*, 208, 414, 1981.
18. **Baute, R., Deffieux, G., Merlet, D., Baute, M.-A., and Neveu, A.,** New insecticidal cyclodepsipeptides from the fungus *Isaria felina*. I. Production, isolation and insecticidal properties of isariins B, C, and D, *J. Antibiot.*, 34, 1261, 1981.
19. **Terabe, S., Konaka, R., and Shoji, J.,** Separation of polymyxins and octapeptins by high-performance liquid chromatography, *J. Chromatogr.*, 173, 313, 1979.
20. **Kato, T. and Shoji, J.,** The structure of octapeptin D (studies on antibiotics from the genus *Bacillus*. XXVIII), *J. Antibiot.*, 33, 186, 1980.
21. **Kimura, Y.,** unpublished data, 1982.
22. **Kimura, Y., Araki, T., Murai, E., Ando, K., and Nakamura, K.,** Isolation of new minor components, colistin D and colistin E from commercial colistin, in *Proc. 9th Symp. Peptide Chem. Japan*, Yanaihara, Y., Ed., Protein Research Foundation, Osaka, 1972, 159.
23. **Kimura, Y., Nakamura, K., Araki, T., and Baba, M.,** New identification method of polymyxin group antibiotics by porous polymer liquid chromatography, in *Proc. 11th Symp. Peptide Chem. Japan*, Kotake, H., Ed., Protein Research Foundation, Osaka, 1974, 129; **Kimura, Y., Kitamura, H., Araki, T., Noguchi, K., Baba, M., and Hori, M.,** Analytical and preparative methods for polymyxin antibiotics using high-performance liquid chromatography with a porous styrene-divinylbenzene copolymer packing, *J. Chromatogr.*, 206, 563, 1981.
24. **Kimura, Y.,** unpublished data, 1982.
25. **Fong, G. W. K. and Kho, B. T.,** Improved high performance liquid chromatography of cyclic polypeptide antibiotics — polymyxins B — and its application to assays of pharmaceutical formulations, *J. Liq. Chromatogr.*, 2, 957, 1979.
26. **Thomas, A. H., Thomas, J. M., and Holloway, I.,** Microbiological and chemical analysis of polymyxin B and polymyxin E (colistin) sulfates, *Analyst*, 105, 1068, 1980.
27. **Whall, T. J.,** High-performance liquid chromatography of polymyxin B sulfate and colistin sulfate, *J. Chromatogr.*, 208, 118, 1981.
28. **Elverdam, I., Larsen, P., and Lund, E.,** Isolation and characterization of three new polymyxins in polymyxins B and E by high-performance liquid chromatography, *J. Chromatogr.*, 218, 653, 1981.
29. **Kalasz, H. and Horvath, C.,** Preparative-scale separation of polymyxins with an analytical high-performance liquid chromatography system by using displacement chromatography, *J. Chromatogr.*, 215, 295, 1981.
30. **Shoji, J., Konaka, R., Kawano, K., Higuchi, N., and Kyogoku, Y.,** Presence of isomers in quinomycin E., *J. Antibiot.*, 29, 1246, 1976.
31. **Rinehart, K. L., Jr., Gaudioso, L. A., Moore, M. L., Pandey, R. C., Cook, J. C., Jr., Barber, M., Sedgwick, R. D., Bordoli, R. S., Tyler, A. N., and Green, B. N.,** Structures of eleven zervamicin and two emerimicin peptide antibiotics studied by fast atom bombardment mass spectrometry, *J. Am. Chem. Soc.*, 103, 6517, 1981.

CARBA ANALOGS OF NEUROHYPOPHYSIAL HORMONES

Michal Lebl

Pioneering syntheses[1] of the neurohypophysial hormones oxytocin and vasopressin carried out in the 1950s led to the possibility of their study by synthesis of suitably modified analogs. About 600 have been prepared so far. The preparation of analogs with a modified disulfide bridge (for examples see References 2 to 4 and references given in Table 2; for a survey see Reference 5) which is one of the very important structural features of these compounds, led both to a modification of the views on the mechanism of action of neurohypophysial hormones,[2] and to the attainment of compounds with distinctly increased biological activities.[6,7] For the purification of synthetic compounds we used a number of methods (gels, ion-exchange and partition chromatography, free-flow electrophoresis, and counter-current distribution). We have recently substituted reversed-phase high performance liquid chromatography (RP-HPLC) for these methods. Special stress is laid on the so-called lyophilizable mobile phase (0.1% trifluoroacetic acid (TFA), 0.1 *M* triethylammonium trifluoroacetate at pH 4 to 5, 0.05 *M* ammonium acetate at pH 6 to 7.5, 0.05 *M* triethylammonium carbonate at pH 7 to 8.5). When nonvolatile buffer components are used, the eluates from preparative chromatography should be desalted[8] on Sep-Pak® C_{18} cartridges. (Note: their capacity for deaminooxytocin is at least 3 mg per cartridge.)

Substitution of sulfur by a methylene group increases the lipophilic character[9] of the compound and when chromatographed on a reversed phase column should exhibit a corresponding increase in the capacity factor value in comparison with the nonmodified substance. However, in the case of carba-analogs of oxytocin we observed the opposite effect,[10] which was consistent with the behavior of some cyclic dipeptides[10] which differed in sulfur content. When comparing simple compounds, such as protected methionine and norleucine, it may be observed, however, that they behave as expected (Table 1), i.e., the norleucine derivative has a longer retention time. Therefore, we compared the behavior of some protected and free peptides which were intermediary products of the synthesis of carba-analogs of oxytocin, differing in the content and position of the sulfur atom. A protected tetrapeptide (IIIc) — in agreement with the assumption of the increased lipophilicity of compounds with a CH_2 group replacing the sulfur atom — was eluted later than the peptides containing sulfur. In the case of pentapeptides (IVa-c) the differences in retention times are already minimal or zero. In free heptapeptides (Va-c) the compound without sulfur was eluted between the two substances containing sulfur, while in the case of protected octapeptides (VIa-c) the order was analogous. On passing to cyclic peptides the situation changes dramatically: compounds containing a CH_2 group instead of sulfur (X-XI) are eluted sooner (see Table 2) and the retention times of the analogs with different positions for sulfur differ considerably. The same is true of the analogs containing an α-amino group (VII, VIII), a contracted cyclic structure (XIII, XIV), or a modified amino acid in position 2 (XLIII, XLIV). In analogs of vasopressin (XIX-XXIII) a distinct dependence of the k value on the pH of the mobile phase is evident (see Figure 1). This dependence is of greater difference in the carba-6-analog (XXI) than in the carba-1-analog (XX) which is similar to a compound with a preserved disulfide bridge (XIX). In contrast to this the carba-6-analog containing D-arginine (XXIII) displays a pH-dependence similar to that of the disulfide analog (XXII).

A complete omission of the disulfide bridge, connected with the elimination of the cyclic structure[11] (in compound XVII) leads to a decrease of the retention time, while the substitution of cystine by two *S*-methylcysteines (XVIII) leads to an increased retention; both these facts may be predicted by estimating the lipophilicity change in the analog formed.

The position of the substitution for sulfur in the disulfide bridge can alter the effect of

Table 1
REVERSED-PHASE CHROMATOGRAPHIC K′ VALUES OF SYNTHETIC INTERMEDIATES OF SOME CARBA-ANALOGS

Compound	Structure[a]	$k_S^{b,d}$	$k_{SO}^{c,d}$	Ref.
Ia	Boc-Met-OH	1.96	1.01	28
Ib	Boc-Nle-OH	3.16		28
IIa	Boc-Cys(C_3H_6COOMe)-OH	2.12	1.11	29
IIb	Boc-Hcy(C_2H_4COOMe)-OH	1.92	1.09	23
IIIa	Z-Cys(C_3H_6COOMe)-Pro-Leu-Gly-NH_2	3.06	1.91	3
IIIb	Z-Hcy(C_2H_4COOMe)-Pro-Leu-Gly-NH_2	2.81	1.67	4
IIIc	Z-Asu(OMe)-Pro-Leu-Gly-NH_2	3.30		4
IVa	Nps-Asn-Cys(C_3H_6COOMe)-Pro-Leu-Gly-NH_2	1.99		3
IVb	Nps-Asn-Hcy(C_2H_4COOMe)-Pro-Leu-Gly-NH_2	1.80		4
IVc	Nps-Asn-Asu(OMe)-Pro-Leu-Gly-NH_2	1.97		4
Va	H-Ile-Gln-Asn-Cys(C_3H_6COOH)-Pro-Leu-Gly-NH_2	2.88[e]	2.22[e]	3
Vb	H-Ile-Gln-Asn-Hcy(C_2H_4COOH)-Pro-Leu-Gly-NH_2	2.53[e]	1.81[e]	4
Vc	H-Ile-Gln-Asn-Asu-Pro-Leu-Gly-NH_2	2.61[e]		4
VIa	Boc-Tyr(Bu^t)-Ile-Gln-Asn-Cys(C_3H_6COOH)-Pro-Leu-Gly-NH_2	5.84	4.98	3
VIb	Boc-Tyr(Bu^t)-Ile-Gln-Asn-Hcy(C_2H_4COOH)-Pro-Leu-Gly-NH_2	5.61	4.39	4
VIc	Boc-Tyr(Bu^t)-Ile-Gln-Asn-Asu-Pro-Leu-Gly-NH_2	5.62		4

[a] Asu, α-aminosuberic acid, Hcy, homocysteine, Nle, norleucine.
[b] Values for sulfide form.
[c] Values for corresponding sulfoxide.
[d] Separon SI-C-18 (25 × 0.4 cm), 0.05% TFA-methanol (30:70).
[e] Same as for d but (50:50).

substitution on retention time. Substitution of sulfur in position 1 has a substantially lower effect on the retention characteristics than the same substitution in position 6 (see Table 2). This again confirms the nonequivalence of the two sulfur atoms which had previously been established with the biological activities of carba-analogs[7,12] and from the study of their CD spectra.[13] On the basis of these spectra an interaction of the sulfur atom in position 6 with the aromatic ring of tyrosine (i.e., in carba-1-analogs) was demonstrated and therefore it may be assumed that this sulfur atom is less accessible for other interactions. On the other hand, the elimination of the interaction between sulfur and the aromatic ring by a carba-substitution of this atom may lead to an increase in the freedom of movement of the tyrosine side chain, which is one of the most important structural elements responsible for the interaction of the peptide with the stationary phase. Thus this substitution can lead to a more dramatic change in retention than the substitution of sulfur in position 1, which is oriented to the "other side" of the molecule and is less likely to interact with the hydrophobic stationary phase. The original assumption[10] that the shielding of sulfur at position 1 inside the molecule was shown to be false by a kinetic study of the oxidation of carba-analogs,[14] where the deamino-6-carba-oxytocin (XI) was oxidized substantially more rapidly than the 1-carba-analog (X). Inspection of a model[15,16] (considering an interaction of the aromatic ring with the sulfur in the 6-position) also shows that access for the relatively small and hydrophilic molecules of the oxidant to the sulfur atom is easier in position 1. The slow rate of oxidation of sulfur at position 6 could be due to the sulfur-aromatic ring interaction or to the steric hindrance to the access of the oxidant by the side chain of the amino acid in position 2. We studied the kinetics of the oxidation of analogs containing the sterically demanding *tert*-leucine in position 2. Again a more rapid oxidation of the compound containing sulfur in position 1 was observed,[17] even though the observed difference was slightly smaller than for substances containing tyrosine.[14] Thus, it may be said that the shielding

Table 2
REVERSED-PHASE CHROMATOGRAPHIC K′ VALUES OF SOME CARBA-ANALOGS

Compound	Structure[a]	k[b]	Mobile phase[c]	k[b]	Mobile phase[c]	Ref.
VII	OT	4.18	A 50	2.00	B 50	1
VIII	C^{1}OT	3.97 (3.19)	A 50	1.83 (1.56 + 1.60)	B 50	30
IX	dOT	4.98	A 50	3.22	B 50	31
X	dC^{1}OT	4.32 (3.20)	A 50	2.93 (1.93 + 2.22)	B 50	3
XI	dC^{6}OT	11.7 (8.56)	A 40	2.00 (1.55)	B 50	4
XII	dC^{1}C^{6}OT	3.35	A 50	1.98	B 50	4
XIII	dCH$_2$$^{1\text{-}6}$OT	2.63	A 50	1.62	B 50	32
XIV	dS$^{1\text{-}6}$OT	4.17 (2.83)	A 50	2.54 (1.97)	B 50	32
XV	dCH$_2$SCH$_2$$^{1\text{-}6}$OT	4.26 (2.65)	A 50	2.57 (1.51)	B 50	33
XVI	dC^{1}OT-SO$_2$			4.72	B 50	12
XVII	Ala^1, Ala^{6}OT	2.70	A 50			11
XVIII	Cys(Me)^1, Cys(Me)^{6}OT	6.28 (3.09)	A 50			11
XIX	Arg^{8}dVP	2.70	A 50	8.70	C 40	34
XX	Arg^{8}dC^{1}VP	2.34 (2.14)	A 50	6.48 (4.84)	C 40	35
XXI	Arg^{8}dC^{6}VP	3.13 (2.73)	A 50	5.76 (4.61)	C 40	36
XXII	D-Arg^{8}dVP	2.98	A 50	10.8	C 40	37
XXIII	D-Arg^{8}dC^{6}VP	2.82 (2.63)	A 50	9.84 (7.98)	C 40	36
XXIV	Ile^2-dC^{1}OT	9.81 (7.64)	A 50	6.43 (3.78 + 6.15)	B 50	38
XXV	Tle^{2}dC^{1}OT	3.78 (2.83)	A 55	4.69 (3.19 + 3.63)	E 55	17
XXVI	Phe^{2}dC^{1}OT	13.6 (10.3)	A 50	8.72 (5.52 + 8.11)	B 50	38
XXVII	Tyr(Me)^{2}dC^{1}OT	13.1 (8.29)	A 50	8.42 (5.62 + 6.74)	B 50	38
XXVIII	Phe(F)^{2}dC^{1}OT	3.04 (2.00 + 2.55)	E 65	7.22 (4.11 + 6.16)	B 50	18
XXIX	Tyr(J$_2$)^{2}dC^{1}OT	5.43 (3.62 + 4.64)	A 60			29
XXX	Glu^{4}dC^{1}OT	1.59 (1.20)	A 63	2.33 (1.53 + 1.75)	B 50	39
XXXI	Tyr(J$_2$)^2,Glu^{4}dC^{1}OT	4.52 (3.27 + 3.98)	A 63			29
XXXII	Ile^{2}dC^{6}OT	9.33 (7.96)	A 50	6.18 (4.12)	B 50	40
XXXIII	Tle^{2}dC^{6}OT	3.66 (2.90)	A 55	4.59 (3.65)	E 55	17
XXXIV	Phe(NH$_2$)^{2}dC^{6}OT	1.76	A 55	2.23 (1.65)	D 55	23

Table 2 (continued)
REVERSED-PHASE CHROMATOGRAPHIC K′ VALUES OF SOME CARBA-ANALOGS

Compound	Structure[a]	k[b]	Mobile phase[c]	k[b]	Mobile phase[c]	Ref.
XXXV	Phe(NO_2)[2]dC[6]OT	4.68 (3.72)	A 55	6.41 (5.25)	D 55	23
XXXVI	Tyr(Me)[2]dC[6]OT	5.01 (4.23)	A 55	6.81 (5.56)	D 55	40
XXXVII	Phe[2]dC[6]OT	5.26 (4.39)	A 55	7.27 (5.77)	D 55	23
XXXVIII	Phe(NMe_2)[2]dC[6]OT	2.92 (2.21)	A 55	13.4 (11.2)	D 55	23
XXXIX	Tyr(Et)[2]dC[6]OT	8.01 (6.64)	A 55	11.4 (9.11)	D 55	23
XL	Phe(Me)[2]dC[6]OT	8.79 (7.36)	A 55	12.7 (9.78)	D 55	23
XLI	Phe(Cl)[2]dC[6]OT	5.05 (4.19)	A 60	15.0 (11.7)	D 55	23
XLII	Phe(NHZ)[2]dC[6]OT	14.2 (12.0)	A 55	22.9 (18.6)	D 55	23
XLIII	Phe(Et)[2]dC[6]OT	4.69 (4.02)	A 65	23.4 (18.0)	D 55	23
XLIV	Phe(Et)[2]dOT	5.73	A 65	27.7	D 55	29
XLV	D-Tyr[2]dC[6]OT	11.2 (5.30)	A 40	13.6 (6.52)	C 40	29
XLVI	D-Phe(Cl)[2]dC[6]OT	6.98 (5.91)	A 60	19.8 (14.8)	D 55	29

[a]
```
    CH2 ———————— R2 ———————— CH2
     |                        |
R1–CH–CO–Tyr–R3–Gln–Asn–NH–CH–CO–Pro–R4–Gly–NH2,
    1        2   3   4   5       6       7   8   9
```

OT denotes oxytocin (R^1 = NH_2, R^2 = –S–S–, R^3 = Ile, R^4 = Leu), VP denotes vasopressin (R^1 = NH_2, R^2 = –S–S–, R^3 = Phe, R^4 = Arg or Lys), d means deamino (R^1 = H), C^1 means 1-carba (R^2 = $-CH_2-S-$), C^6 means 6-carba (R^2 = $-S-CH_2$), C^1C^6 means di-carba (R^2 = $-CH_2-CH_2-$), $CH_2^{1\text{-}6}$, $S^{1\text{-}6}$, $CH_2SCH_2^{1\text{-}6}$ means that R^2 = $-CH_2-$, –S–, $-CH_2SCH_2-$.

[b] Value for corresponding sulfoxide is given in parentheses.

[c] A, 0.1% TFA B, 0.08 *M* triethylammonium trifluoroacetate buffer of pH 4.5; C, 0.05 *M* ammonium acetate buffer at pH 7.0, D, 0.01 *M* triethylammonium borate buffer at pH 8.1, E, 0.01 *M* sodium phosphate buffer at pH 4.4, the number indicates the percentage of methanol in the mixture.

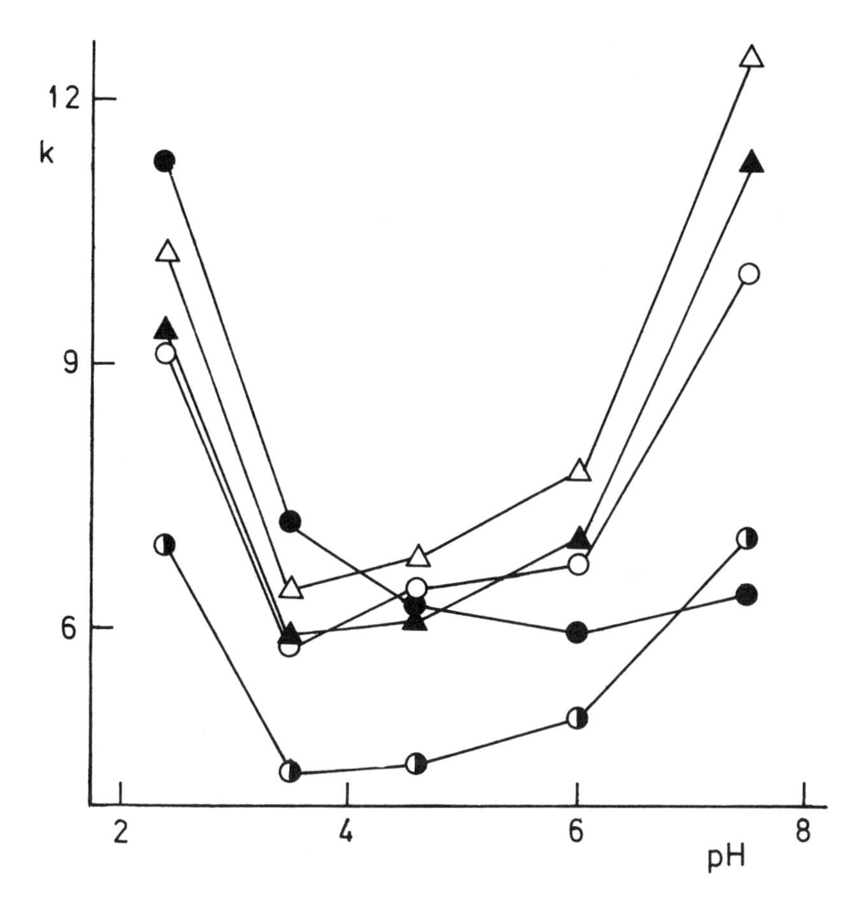

FIGURE 1. Dependence of k values of vasopressin analogs on pH of the mobile phase. ○, [8-Arginine]deamino-vasopressin; ◑ - [8-Arginine]deamino-1-carba-vasopressin; ● - [8-arginine]deamino-6-carba-vasopressin; △ - [8-D-arginine]deamino-vasopressin; ▲ - [8-D-arginine]deamino-6-carba-vasopressin. Conditions: Separon SI-C-18 (25 × 0.4 cm), 0.05 *M* phosphate buffer of different pH – methanol (60:40), flow 1.5 mℓ/min.

effect is at least partly steric in nature. We also compared the oxidation rate of carba-analogs of vasopressin;[14] in these compounds the oxidation is considerably faster than in oxytocin derivatives. No important difference in oxidation rate, dependent on the position of sulfur atom, was observed in this case. For comparison the kinetics of the oxidation of substance P was also studied. Substance P is oxidized several orders of magnitude more rapidly. All kinetic experiments were carried out separately and they were checked using HPLC. To compare the oxidation rates, competition experiments may be used, where both substrates (present in equal concentrations) are oxidized simultaneously (see Figure 2).

With carba-1-analogs of oxytocin a separation of diastereoisomeric sulfoxides may be achieved[10,14] in depending on their structure and the mobile phase used. The most different elution characteristics were observed with sulfoxides of [2-*p*-fluorophenylalanine]deamino-1-carba-oxytocin,[18] which could be separated even with a methanol-water mixture. Proof that the substances were indeed diastereoisomeric sulfoxides was by reducing the isolated sulfoxides, which led to identical sulfides. When these were oxidized, identical mixtures of sulfoxides were formed in both cases.[18] Oxidation with sodium periodate and reduction with hydrogen bromide and acetone (monitored by HPLC) became our routine laboratory method for checking the degree of oxidation of sulfur in synthetic substances.[10,14] Examples of the behavior of sulfides and sulfoxides in reversed-phase chromatography may be found in Table

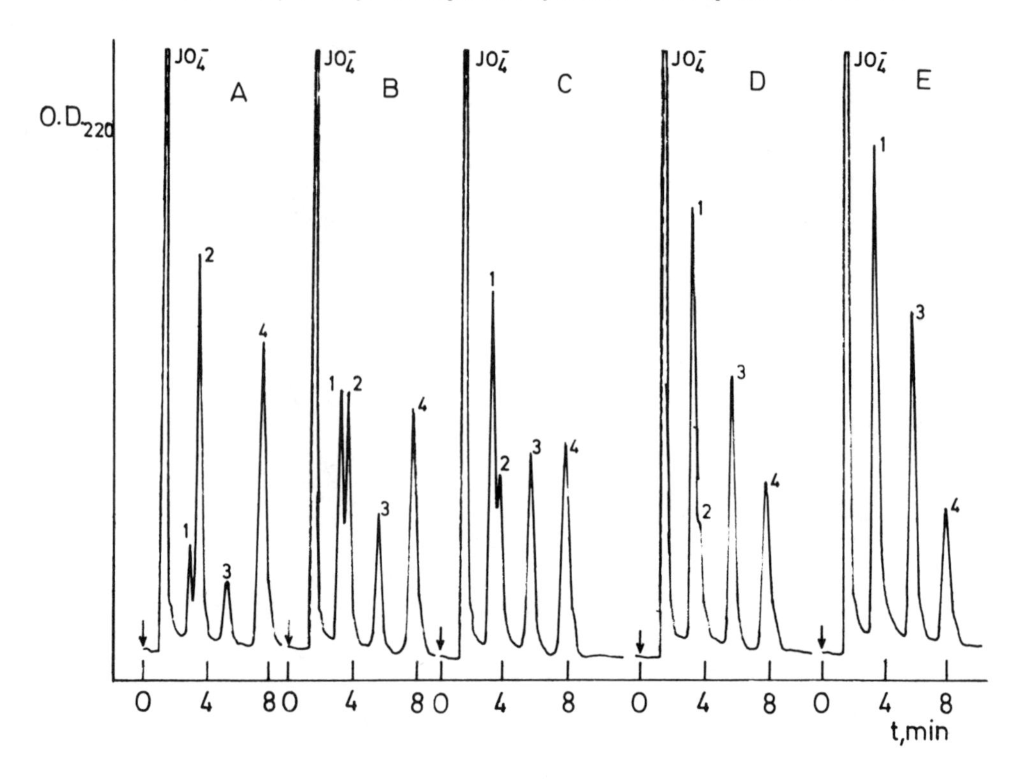

FIGURE 2. Oxidation of a mixture of [8-arginine]deamino-1-carba-vasopressin (2) and deamino-1-carba-oxytocin (4) to corresponding sulfoxides (2 → 1, 4 → 3) by sodium periodate. Conditions: Separon SI-C-18 (25 × 0.4 cm), 0.1% TFA – methanol (55:45), flow 1.5 mℓ/min. (A) Immediately after periodate addition; (B) after 10 min; (C) after 20 min; (D) after 32 min; (E) after 43 min.

1 and 2. The separation of diastereoisomeric sulfoxides could not be achieved with an acidic or neutral mobile phase unless an amine had been added. Triethylammonium trifluoroacetate buffer at pH 4 to 4.5 was the best mobile phase. A phosphate buffer with a similar pH may also be used.[14] Diastereoisomeric sulfoxides of carba-vasopressins were separated with a buffer of pH 6.4. However, in none of the mobile phases tested were we able to achieve the separation of sulfoxides of 6-carba-analogs of oxytocin. The explanation is again based on the assumption that interaction of the sulfur atom in position 6 with the aromatic ring is prevented by oxidation and the configuration of the sulfoxide leads to a different orientation of the tyrosine side chain. (We tried to use this fact — under assumption of the validity of the oxytocin model considered[15,16] — for the prediction of the absolute configuration of individual stereoisomers of sulfoxides.[14,18]) By contrast the oxidation of sulfur in position 1 does not significantly influence the conformation of the tyrosine. However, similar behavior is not limited to analogs containing an aromatic ring in position 2. The analogs containing a *tert*-leucine in this position (XXV, XXXIII) differ only slightly in their capacity factors;[17] nevertheless, the 1-carba-analog affords on oxidation a separable mixture of diastereoisomers, while the 6-carba-analog does not.

Using preparative HPLC we separated the diastereoisomers of some sulfoxides and determined their biological activity.[14,18] The diastereoisomers of 1-carba-analog mostly differ in activity by as much as several orders of magnitude.[14,18]

In the case of deamino-1-carba-oxytocin, the activity of the less active isomer approaches that of the sulfone (XVI) rather than of the other diastereoisomer. In sulfoxides of 6-carba-analogs the activity is decreased relatively little in comparison with the sulfides, which may be explained by the fact that the steric arrangement of the elements important for eliciting receptor response (which are[19] the tyrosine, and asparagine side chains) is not significantly disturbed. In the case of carba-analogs of vasopressin it may be seen[14] again that the oxidation of a 1-carba-analog leads to a much greater decrease in activity and that the difference of k values between diastereoisomers is also larger.

We also investigated the degradation rate of *tert*-butyl-sulfonium salts of deamino-1-carba and 6-carba-oxytocin.[20] The salt derived from 6-carba-oxytocin is much more stable which again shows that the model[16] considering the interaction of the side chain of the amino acid residue in position 2 with the sulfur atom in position 6 is correct, since the model would predict the destabilization of the sulfonium salt of the 1-carba-analog.

If a single structural feature (i.e., substituent) is changed in the whole molecule, then the chromatographic behavior of the substance formed should be predictable on the basis of a change in lipophilicity,[21] unless a change in the conformation of the whole molecule takes place.[22] We prepared[23] a number of analogs of deamino-6-carba-oxytocin differing merely by substitution of the para position of the aromatic nucleus of the amino acid in position 2 (XXXIV-XLIII). Retention characteristics of these substances were well correlated[24] with the π-values of corresponding substituents and thus it is probable that these analogs do not differ in conformation even though they differ substantially in their biological activities.[23]

Substitution of the aromatic nucleus of tyrosine by two iodine atoms in positions 3 and 5 led to a very pronounced increase in the elution time of the analog (X, XXX, and XXIX, XXXI). Using HPLC we observed that when the iodination was carried out in alkaline medium,[25] both for deamino-1-carba-oxytocin (X) and for the analog (XXX), oxidation of sulfur did not take place. The sulfoxides formed by oxidation with periodate were easily separable into individual diastereoisomers.

We also used HPLC to follow the enzymatic cleavage of the carba-analogs. The effect of the post-proline cleaving enzyme (EC 3.4.21.26) was better observed with deamino-1-carba-oxytocin, than with oxytocin as this analog did not inactivate the enzyme; (oxytocin evidently blocks its free SH group.) The action of chymotrypsin on analogs which did not contain an aromatic amino acid in the position 2, gave analogs without glycine amide at the C-terminus of the molecule. These analogs are interesting for their activities in affecting the CNS. For a preparative purification of these compounds it is advantageous to use volatile buffers at neutral pH. At this pH the peptide with a free carboxyl group is eluted earlier than the starting analog (when using an acid mobile phase it is eluted later).

Using RP-HPLC separation of diastereoisomeric peptides may be achieved.[26,27] We made use of this fact in the synthesis of carba-analogs containing unnatural amino acids in position 2. In Figure 3 an elution profile[41] is shown for the preparative chromatography of the crude reaction mixture after cyclization of [2-D,L-*p*-chlorophenylalanine]deamino-6-carba-oxytocin. A similar profile was also observed in the preparation of the analog containing a disulfide bridge, or an amino group, or some other modification in position 1. When the analog XLV containing D-tyrosine in position 2 was prepared, it was essential to eliminate any traces of the substance containing this amino acid of L-configuration (XI). On comparison with a standard it was found, however, that both analogs differ from one another only very slightly (α = 1.05), while their sulfoxides give an α = 1.47. The analog was therefore purified by conversion to the sulfoxide which was purified by preparative chromatography, reduced and rechromatographed. The whole operation was carried out on a 10-mg scale within one working day.

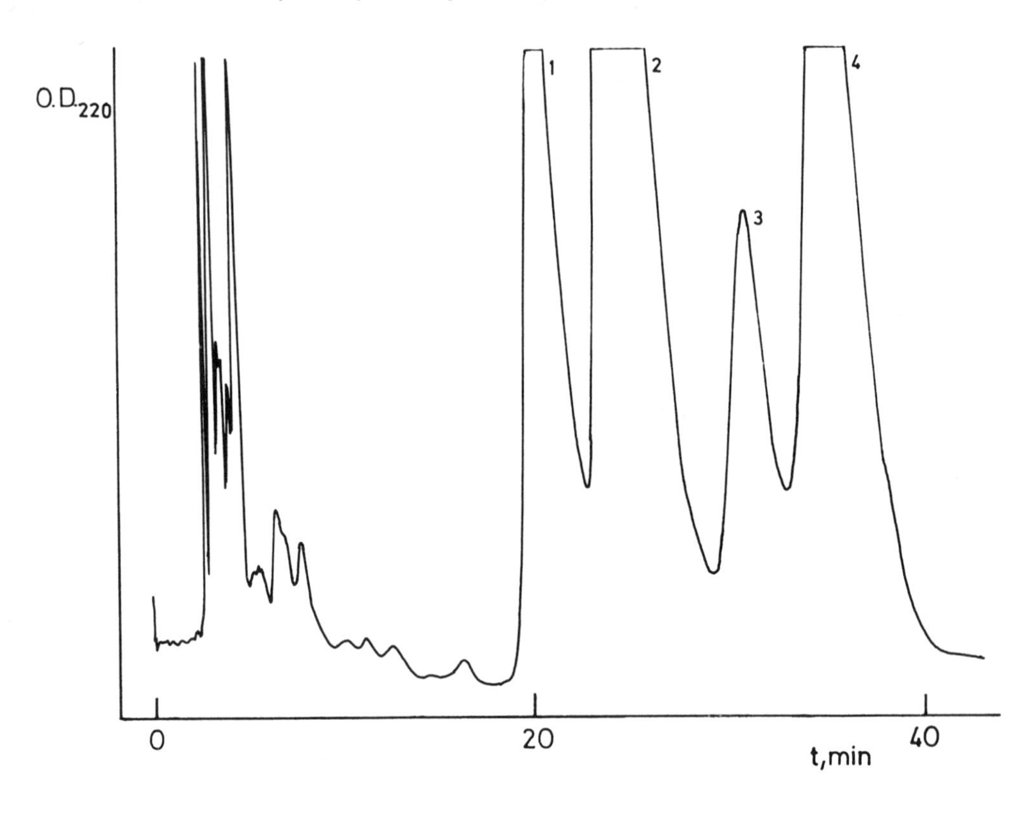

FIGURE 3. Preparative chromatography of the crude mixture after cyclization of carba-analog. [2-L-*p*-chlorophenylalanine]deamino-6-carba-oxytocin (2), its sulfoxide (1), [2-D-*p*-chlorophenylalanine]deamino-6-carba-oxytocin (4) and its sulfoxide (3). Conditions: Partisil® ODS-2 (50 × 0.9 cm), 0.05% TFA – methanol (45:55), flow 5 mℓ/min, load 28 mg.

REFERENCES

1. **du Vigneaud, V., Ressler, Ch., Swan, J. M., Roberts, C. W., Katsoyannis, P. G., and Gordon, S.,** The synthesis of an octapeptide amide with the hormonal activity of oxytocin, *J. Am. Chem. Soc.*, 25, 4879, 1953.
2. **Rudinger, J. and Jošt, K.,** A biologically active analogue of oxytocin not containing disulphide group, *Experientia*, 20, 570, 1964.
3. **Jošt, K.,** An improved synthesis of deamino-carba[1]-oxytocin; comparison of various methods for peptide cyclisation, *Coll. Czech. Chem. Commun.*, 36, 218, 1971.
4. **Jošt, K. and Šorm, F.,** The effect of the presence of sulphur atoms on the biological activity of oxytocin, *Coll. Czech. Chem. Commun.*, 36, 234, 1971.
5. **Smith, G. W., Walter, R., Moore, S., Makofske, R. C., and Meienhofer, J.,** Replacement of the disulfide bond in oxytocin by an amide group, *J. Med. Chem.*, 21, 117, 1978.
6. **Barth, T., Krejči, I., Kupková, B., and Jošt, K.,** Pharmacology of cyclic analogues of deamino-oxytocin not containing a disulphide group (carba-analogues), *Eur. J. Pharmacol.*, 24, 183, 1973.
7. **Barth, T., Jošt, K., and Rychlik, I.,** Milk ejecting and uterotonic activities of oxytocin analogues in rats, *Endocrinol. Exp.*, 9, 35, 1975.
8. **Böhlen, P., Castillo, F., Ling, N., and Guillemin, R.,** Purification of peptides: an efficient procedure for the separation of peptides from amino acids and salt, *Int. J. Peptide Protein Res.*, 16, 306, 1980.
9. **Hansch, C. and Leo, A.,** *Substituent Constants for Correlation Analysis in Chemistry and Biology*, John Wiley & Sons, New York, 1979, 49.

10. **Lebl, M.,** High-performance liquid chromatography of carba-analogues of oxytocin, *Coll. Czech. Chem. Commun.*, 45, 2927, 1980.
11. **Jošt, K., Debabov, V. G., Nesvadba, H., and Rudinger, J.,** A synthesis of dethiooxytocin and other acyclic peptides related to oxytocin, *Coll. Czech. Chem. Commun.*, 29, 419, 1964.
12. **Lebl, M., Barth, T., and Jošt, K.,** Synthesis, reduction, and pharmacological properties of the sulfoxides of some carba-analogues of oxytocin, *Coll. Czech. Chem. Commun.*, 43, 1538, 1978.
13. **Frič, I., Kodiček, M., Jošt, K., and Bláha, K.,** Chiroptical properties of carba-analogues of oxytocin: conformational considerations, *Coll. Czech. Chem. Commun.*, 39, 1271, 1974.
14. **Lebl, M., Barth, T., and Jošt, K.,** Chromatographic and pharmacological properties of carba-analogues of neurohypophyseal hormone sulfoxides. A general method of sulfoxide group determination in peptides, in *Peptides 1980,* Brunfeldt, K., Ed., Scriptor, Copenhagen, 1981, 719.
15. **Urry, D. W. and Walter, R.,** Proposed conformation for oxytocin in solution, *Proc. Natl. Acad. Sci. U.S.A.*, 68, 956, 1971.
16. **Bláha, K. and Frič, I.,** unpublished results, 1980.
17. **Lebl, M., Pospišek, J., Hlaváček, J., Barth, T., Maloň, P., Servitová, L., Hauzer, K., and Jošt, K.,** Analogues of neurohypophysial hormones containing tert. leucine, *Coll. Czech. Chem. Commun.*, 47, 689, 1982.
18. **Procházka, Z., Lebl, M., Servitová, L., Barth, T., and Jošt, K.,** Synthesis and pharmacological properties of [2-*p*-fluorophenylalanine]deamino-1-carba-oxytocin and its diastereoisomeric sulfoxides, *Coll. Czech. Chem. Commun.*, 46, 947, 1981.
19. **Walter, R.,** Identification of sites in oxytocin involved in uterine receptor recognition, *Fed. Proc., Fed. Am. Soc. Exp. Biol.*, 36, 1872, 1977.
20. **Bienert, M., Lebl, M., Mehlis, B., and Niedrich, H.,** Synthesis and application of S-tert.butylsulfonium peptides, in *Peptides 1980,* Brunfeldt, K., Ed., Scriptor, Copenhagen, 1981, 127.
21. **O'Hare, M. J. and Nice, E. C.,** Hydrophobic high-performance liquid chromatography of hormonal polypeptides and proteins on alkylsilane-bonded silica, *J. Chromatogr.*, 171, 209, 1979.
22. **Blevins, D. D., Burke, M. F., and Hruby, V. J.,** Parameters affecting high performance liquid chromatographic separations of neurohypophyseal peptide hormones, *Anal. Chem.*, 52, 420, 1980.
23. **Lebl, M., Hrbas, P., Škopková, J., Slaninová, J., Machová, A., Barth, T., and Jošt, J.,** Synthesis and properties of oxytocin analogues with high and selective natriuretic activity, *Coll. Czech. Chem. Commun.*, in press, 1982.
24. **Lebl, M.,** Correlation between hydrophobicity of substituents in the phenylalanine moiety of oxytocin carba-analogues and reversed-phase-chromatographic k values, *J. Chromatogr.*, in press, 1982.
25. **Flouret, C., Terada, S., Yang, F., Nakagawa, S. H., Nakahara, T., and Hechter, O.,** Iodinated neurohypophyseal hormones as potential ligands for receptor binding and intermediates in synthesis of tritiated hormones, *Biochemistry,* 16, 2119, 1977.
26. **Larsen, B., Fox, B. L., Burke, M. F., and Hruby, V. J.,** The separation of peptide hormone diastereoisomers by reverse phase high pressure liquid chromatography, *Int. J. Peptide Protein Res.*, 13, 12, 1979.
27. **Lindeberg, G.,** Separation of vasopressin analogues by reversed-phase high-performance liquid chromatography, *J. Chromatogr.*, 193, 427, 1980.
28. **Wünsch, E.,** *Synthese von Peptiden,* Thieme, Stuttgart, 1974, 131.
29. **Lebl, M.,** unpublished results, 1982.
30. **Jošt, K., Barth, T., Krejčí, I., Fruhaufová, L., Procházka, Z., and Šorm, F.,** Carba[1]-oxytocin: synthesis and some of its biological properties, *Coll. Czech. Chem. Commun.*, 38, 1073, 1973.
31. **Hope, D. B., Murti, V. V. S., and du Vigneaud, V.,** A highly potent analogue of oxytocin, deamino-oxytocin, *J. Biol. Chem.*, 237, 1563, 1962.
32. **Jošt, K. and Šorm, F.,** Synthesis of two analogues of deamino-oxytocin with a diminished ring not containing a disulphide bond, *Coll. Czech. Chem. Commun.*, 36, 2795, 1971.
33. **Procházka, Z., Jošt, K., and Šorm, F.,** Synthesis of deamino-[1,6-homolanthionine]-oxytocin, *Coll. Czech. Chem. Commun.*, 37, 289, 1972.
34. **Huguenin, R. L. and Boissonnas, R. A.,** Synthèse de la désamino[1]-Arg[8]-vasopressine et de la désamino-Phé[2]-Arg[8]-vasopressine, *Helv. Chim. Acta,* 49, 695, 1966.
35. **Procházka, Z., Barth, T., Cort, J. R., Jošt, K., and Šorm, F.,** Synthesis and some pharmacological properties of [8-L-arginine]deamino-1-carba-vasopressin, *Coll. Czech. Chem. Commun.*, 43, 655, 1978.
36. **Jošt, K., Procházka, Z., Cort, J. H., Barth, T., Škopková, J., Prusík, Z., and Šorm, F.,** Synthesis and some biological activities of analogues of deamino-vasopressin with the disulphide bridge altered to a thioether bridge, *Coll. Czech. Chem. Commun.*, 39, 2835, 1974.
37. **Krchňák, V. and Zaoral, M.,** Synthesis of [1-β-mercaptopropionic acid, 8-D-arginine]vasopressin (DDAVP) in solid phase, *Coll. Czech. Chem. Commun.*, 44, 1174, 1979.
38. **Frič, I., Kodíček, M., Procházka, Z., Jošt, K., and Bláha, K.,** Synthesis and circular dichroism of some analogues of deamino-1-carba-oxytocin with modification of the amino-acid residue in position 2, *Coll. Czech. Chem. Commun.*, 39, 1290, 1974.

39. **Lebl, M. and Jošt, K.,** Synthesis of [4-glutamic acid]-deamino-1-carba-oxytocin, *Coll. Czech. Chem. Commun.*, 43, 523, 1978.
40. **Lebl, M., Barth, T., and Jošt, K.,** Synthesis and pharmacological properties of oxytocin analogues modified simultaneously in position 2 and in the disulfide bridge, *Coll. Czech. Chem. Commun.*, 45, 2855, 1980.
41. **Lebl, M.,** Amino acids and peptides. CLXXXI. Separation of diastereoisomers of oxytocin analogues, *J. Chromatogr.*, 264, 459, 1983.

ANGIOTENSINS

Jeroen A. D. M. Tonnaer

INTRODUCTION

The group of angiotensins (A) comprises a number of closely related biologically active as well as inactive peptides. Figure 1 depicts the amino acid sequence of those angiotensins that possess apparent biological activity, i.e., angiotensin I (A I, A-(1-10)), angiotensin II (A II, A-(1-8)) and angiotensin III (A III, A-(2-8)).

A-(1-10) is released upon cleavage of a precursor molecule by the carboxyl protease renin. This decapeptide can be converted to A-(1-8), the principal bioactive compound of the angiotensin family, by a peptidyl dipeptide carboxypeptidase which is generally referred to as angiotensin-converting enzyme. A-(1-8), in its turn, can be converted to the secondary bioactive molecule A-(2-8) by amino-peptidase activity. Each of the above mentioned peptides is substrate for a group of nonspecific proteolytic enzymes, the angiotensinases. Metabolism of the angiotensins by these enzymes is generally accompanied by a fall in the biological activity of the peptides.

The complex of angiotensins and their metabolizing enzymes is known as the renin-angiotensin system. Due to the prominent role of the angiotensins in the control of blood pressure and body fluid homeostasis, the renin-angiotensin system has been a subject of study throughout the last 50 years. Much attention has been paid to the synthesis of angiotensin agonists and antagonists (see Reference 1 for review) and to the analysis of the function of the components of the renin-angiotensin system and the estimation of their activity in the circulatory system. This field has been comprehensively reviewed by Peach.[2] However, tissue bound components of the renin-angiotensin system might enable the local formation and degradation of angiotensins outside the circulatory system. In this respect the presence of renin-like activity in the brain, and the apparent dipsogenic and pressor effects of intracerebroventricularly administered angiotensins (see Reference 3 for review) are of particular interest.

In former years various separation techniques such as polyacrylamide gel electrophoresis,[4] isoelectric focusing,[4] paper chromatography,[5,6] and thin-layer chromatography[7] have been employed for the analysis of synthetic angiotensin analogs as well as for quantitation of angiotensins and their metabolizing enzymes in body fluids or in tissues. Complete resolution of the closely similar angiotensins, however, was not always attained. The introduction of reversed-phase (RP) HPLC initiated a renewed interest in this field and, as a result, most synthetic angiotensins and related peptides can now be separated successfully. However, systematic research focused on the optimal conditions for resolution of the angiotensins has not been performed. Many literature reports have dealt with the separation of nonangiotensin-like peptides and the systems described have been employed for the angiotensins. As a result, a variety of columns and separation systems, whether or not containing ion-pairing reagents, have been used for these particular peptides.

In the next paragraphs, the different separation systems and their application fields are described. In addition, Table 1 summarizes the literature data over the last 5 years on this topic.

SEPARATION AND ANALYSIS OF SYNTHETIC ANGIOTENSINS

Molnár and Horváth[8] separated synthetic angiotensins and some C-terminal fragments,

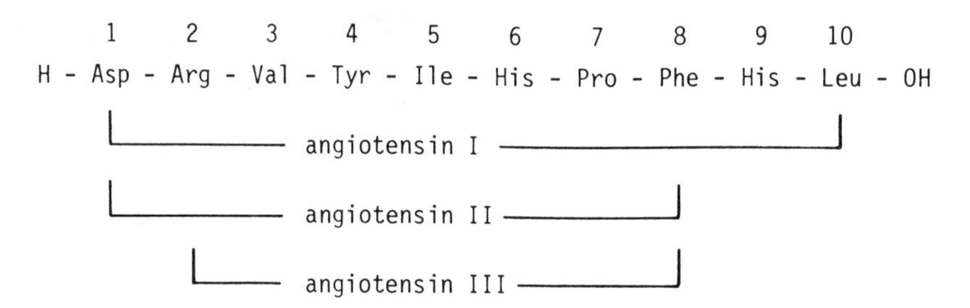

FIGURE 1. Amino acid sequences of angiotensins.

Table 1A
HPLC SEPARATION SYSTEMS FOR ANGIOTENSINS AND ANGIOTENSIN ANALOGS

Peptides studied	Applications	Ref.
Asn[1]-A-(1-10), Asn[1]-A-(1-8), A-(1-8), A-(2-8), A-(3-8)	Separation of synthetic peptides	8
A-(1-10), ^{125}I-A-(1-10), A-(1-8), ^{125}I-A-(1-8), A-(2-8), ^{125}I-A-(2-8), Sar[1]-Ala[8]-A-(1-8), Ile[8]-A-(2-8)	Separation of synthetic (iodinated) peptides	9
Tetradecapeptide (TDP), Leu-Val-Tyr-Ser, A-(1-10), A-(1-8)	Separation of synthetic peptides, TDP-metabolites, estimation of renin activity	10
TDP, A-(1-10), A-(1-8), Val[5]-A-(1-8), A-(2-8), A-(3-8), A-(4-8), Sar[1]-Ala[8]-A-(1-8) and other polypeptides	Identification of angiotensins in rat brain tissue	11
A-(1-10), A-(1-8), A-(2-8), A-(3-8), A-(4-8), A-(5-8), A-(6-8), A-(7-8)	Analysis of angiotensin metabolism in rat brain tissue	12
A-(1-10), A-(1-8), Val[5]-A-(1-8)	Purification of synthetic peptides	13
A-(3-8), Val[4]-A-(4-8), A-(4-8)	Purification of synthetic peptides	14
A-(1-8)	Purification of synthetic peptide	15
Various angiotensin fragments, Ile[5]/Val[5]-angiotensins and optical isomers	Analysis of synthetic peptides	16
A-(1-10), A-(1-8), monoiodinated and diiodinated derivatives	Purification of radiolabeled tracers	17
A-(1-8) and various other polypeptides	Separation of synthetic peptides	18
A-(1-10), A-(1-8), A-(2-8)	Estimation of angiotensins in human plasma	19
TDP and TDP-metabolites	Identification of renin-like enzymes in rat brain tissue	20
A-(1-10), A-(1-8), A-(2-8)	Identification of the renin-angiotensinogen-cleavage product	21
Hip-His-Leu, hippurate	Estimation of serum angiotensin-converting enzyme activity	23
TDP, Leu-Val-Tyr-Ser, A-(1-10)	Determination of renin and cathepsin D	24
TDP, Leu-Val-Tyr-Ser, A-(1-10), A-(1-8), A-(2-8)	Identification of renin-like enzymes in rat brain tissue	25

using a elution system consisting of an acetonitrile gradient in phosphate buffer (pH 2.1). The retention times of the peptides appeared to be dependent on both the hydrophobicity of the side chains and on the size of the molecules. No resolution, however, was obtained between Asn[1]-A-(1-10) and the C-terminal heptapeptide. Complete resolution between the deca-, octa-, and C-terminal heptapeptide was obtained by Guy et al.,[9] using a similar system

Table 1B
HPLC SEPARATION SYSTEMS FOR ANGIOTENSINS AND ANGIOTENSIN ANALOGS

Column packing	Mobile phase	Ref.
Lichrosorb® RP-8	0.1 *M* phosphate buffer (pH 2.1), acetonitrile gradient	8
Biosil® ODS-10	0.05 *M* phosphate buffer (pH 6.0), 25% acetonitrile	9
μBondapak® C_{18}	0.01 *M* ammonium acetate/acetic acid buffer (pH 4.2), methanol gradient	10
μBondapak® C_{18}	0.01 *M* ammonium acetate/acetic acid buffer (pH 4.5/5.4), methanol gradient, or 0.01 *M* triethylammonium phosphate/phosphoric acid buffer (pH 3.0), isopropanol gradient	11
μBondapak® C_{18}	See Reference 10	12
Octadecylsilane	0.1 *M* triethylammonium phosphate buffer (pH 3.5), 19% acetonitrile	13
μBondapak® — fatty acid analysis	75% water, 25% acetonitrile, 0.1% phosphoric acid, or 50% water, 50% methanol, 0.1% phosphoric acid	14
Fatty acid analysis	8% acetic acid, 40% methanol	15
Nucleosil® 5C12	0.1 *M* phosphoric acid, or 0.1 *M* phosphoric acid, acetonitrile, or phosphate buffers, acetonitrile	16
μBondapak® C_{18}	0.02 *M* triethylammonium phosphate buffer (pH 3.0), isopropanol gradient	17
Hypersil-ODS	0.1 *M* sodium phosphate, phosphoric acid buffer (pH 2.1), acetonitrile gradient	18
IEX-530-CMK	0.03 *M* phosphate buffer (pH 6.5), sodium chloride gradient	19
μBondapak® C_{18}	See Reference 10	20
μBondapak® C_{18}	Methanol, methylglycol, water (40:5:55; v/v/v), HCl (pH 3.0)	21
μBondapak® C_{18}	0.2 *M* potassium phosphate buffer (pH 8.0), 40% methanol	23
μBondapak® C_{18}	See Reference 10	24
μBondapak® C_{18}	See Reference 10	25

but at a higher pH (6.0). Similar results have been achieved using ammonium acetate buffer systems in combination with a methanol gradient (Tonnaer et al.,[10] Hermann et al.[11]) or a phosphate buffer with an isopropanol gradient.[11] Tonnaer et al.,[12] in addition, separated a number of C-terminal fragments shorter than the heptapeptide A-(2-8). Figure 2 shows a base-line resolution of these peptides, using a 40-min linear methanol gradient in ammonium acetate buffer. The octapeptide A-(1-8) eluted with a retention time identical to that of its C-terminal hexapeptide fragment, whereas partial overlap was observed with A-(2-8). The latter peptide was completely resolved using an isocratic system, comprising of 41% (v/v) methanol in ammonium acetate buffer (Table 2). The gradient elution time could be shortened to 20 min, and Table 2 summarizes the retention times of A-(1-10), A-(1-8), as well as C- and N-terminal fragments using that system.

Separation of the deca- and octapeptide, resolving the Ile^5 and Val^5 analogs, was performed by Margolis and Longenbach[13] and Hermann et al.[11]

Hancock and co-workers[14] described a combination of RP-HPLC and paired-ion chromatography (using phosphoric acid as counter-ion) for the separation of C-terminal angiotensin fragments and a synthetic deletion product. They concluded that the described system could be applied to the detection of minor synthesis deletion products. Similarly, Feldman et al.[15] analyzed synthetic A-(1-8) by HPLC and found at least seven impurities in a commercially available preparation. Several different HPLC systems have been applied by Takai and co-workers[16] to the separation of C-terminal as well as N-terminal fragments, Ile^5 and Val^5 analogs and, additionally, optical isomers of angiotensins. Their procedures enable the detection of synthesis impurities at levels down to 0.1%.[16]

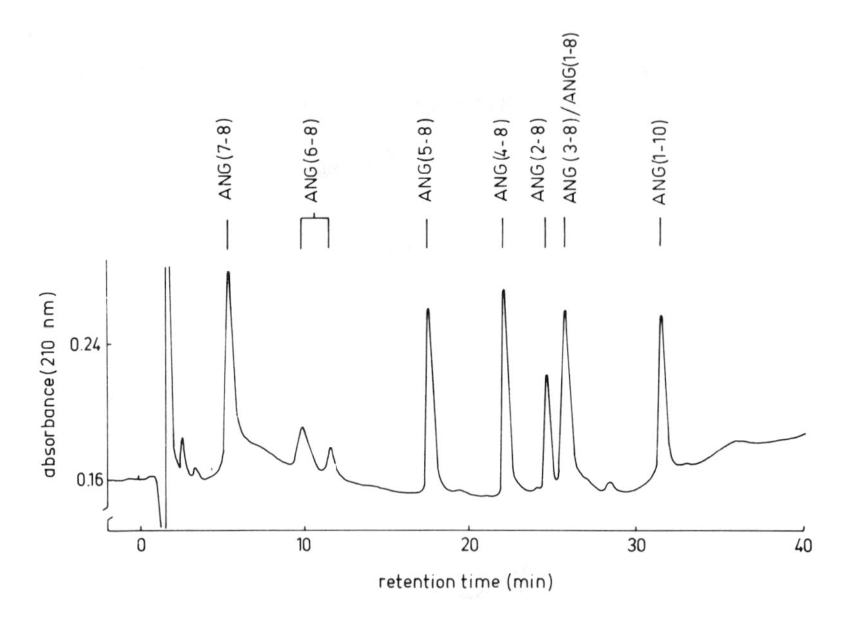

FIGURE 2. Chromatogram of a mixture of synthetic angiotensins. The peptides were separated on a RP μBondapak® C_{18} column (30 cm × 0.39 cm i.d.; 10 μm, Waters Associates) using a 40-min linear methanol gradient from 10 to 75% (v/v) in 0.01 *M* ammonium acetate buffer (pH 4.2). Methanol contained 0.5 mℓ of acetic acid per liter in order to make the solution equitransparent to the ammonium acetate buffer at 210 nm. A flow rate of 2 mℓ/min was used.

IODINATED ANGIOTENSINS

HPLC separation of iodinated angiotensins and unlabeled peptides appears to be a promising method for the purification of tracers to be used in radioimmunoassay (RIA) or radioligand binding assay systems. Introduction of a hydrophobic iodogroup into the molecules should increase their hydrophobicity and, as a consequence, their HPLC retention time. Guy et al.[9] separated commercially available iodinated angiotensins from their unlabeled analogs. Upon chloramine-T labeling Seidah and co-workers[17] achieved a good resolution between monoiodinated angiotensins, their diiodinated derivatives, and the native molecules. In addition, the latter authors were able to separate the native molecules from their oxidized metabolites, which might be generated as a result of the chloramine-T treatment. HPLC purification of radiolabeled angiotensins permits a high titer and consequently a high sensitivity in the RIA.[17]

IDENTIFICATION AND QUANTITATION OF ENDOGENOUS ANGIOTENSINS

The high resolving power of RP-HPLC offers the unique possibility to identify and quantitate the individual, structurally related angiotensins in body fluids as well as in tissues. In this respect it is of particular interest that the angiotensins can be recognized in mixtures among numerous unrelated polypeptides such as bradykinin, neurohypophyseal peptides, pituitary peptides, gut peptides, and others.[11,18]

Following prepurification on Sep-Pak® C_{18} cartridges prior to HPLC separation, Hermann et al.[11] were able to identify and quantitate A-(1-10), A-(1-8), and A-(2-8) in extracts of rat brain tissue. Similarly, Sakurai et al.[19] specifically determined the levels of these three peptides in human plasma, following HPLC fractionation.

Table 2
SEPARATION OF ANGIOTENSINS IN AMMONIUM ACETATE-METHANOL RP-HPLC SYSTEMS

Peptide	Retention time (min)	
A-(1-10)	19.6[a]	
A-(9-10)	2.6	
A-(1-8)	16.9	8.6[b]
A-(2-8)	16.4	7.0
A-(3-8)	16.9	
A-(4-8)	15.1	
A-(5-8)	12.9	
A-(6-8)	8.5	
A-(7-8)	5.3	
A-(1-7)	12.4	
Phe[c]	3.2	

[a] Peptides were separated on a RP μBondapak® C_{18}-column (30 cm × 0.39 cm i.d.; 10 μm, Waters Associates) using a 20-min linear methanol gradient from 10 to 75% (v/v) in 0.01 *M* ammonium acetate buffer (pH 4.2), 2 mℓ/min.
[b] Idem, but an isocratic system of 41% (v/v) methanol in 0.01 *M* ammonium acetate buffer (pH 4.2) was used.
[c] Phe represents residue 8 in the angiotensin molecules.

Quantitation of synthetic angiotensins is normally performed by continuous monitoring of the UV absorbance of the HPLC eluate. Although dependent on the sensitivity of the UV detector used, the sensitivity obtained will be limited to the microgram range[10] and does not allow the detection of endogenous angiotensins. The pronounced pressor activity of A-(1-10), A-(1-8), and A-(2-8) enables characterization and a more sensitive quantitation (nanogram range) of these peptides in collected fractions of the eluate, using bioassay.[20] Moreover, sensitive RIA systems can be used for the specific characterization of angiotensins in HPLC eluates,[20,21] or for detection of picogram amounts of the peptides in tissue extracts or body fluids as has been mentioned above.[11,19]

It should be noted that the specificity of peptide characterization by combined HPLC-RIA systems is not necessarily dependent on the specificity of the antiserum used. In contrast, nonspecific antisera will recognize several cross-reacting peptides specifically, following adequate HPLC separation of the peptides.[22] Thus, using an antiserum raised against A-(1-8), but cross-reacting with A-(2-8) and smaller C-terminal fragments, it is possible to estimate the true levels of these peptides in tissue[11] or in plasma.[19]

ANGIOTENSIN METABOLISM

A special application of HPLC is found in the study of angiotensin metabolism. Upon proteolytic cleavage of the angiotensins the generated metabolites can be analyzed properly following HPLC fractionation of the incubation mixtures. Characterization and quantitation of the accumulating cleavage products provides information on the character and activity of the proteolytic enzymes involved. A similar methodology can be applied to the study of the processing of natural and synthetic angiotensin precursors and for the study of the metabolism of angiotensin analogs.

This latter approach has been followed by Chiknas[23] for the estimation of serum angiotensin-converting enzyme activity. Following incubation of the synthetic substrate Hip-His-Leu with serum samples, the released hippurate was separated from the native molecule by HPLC and quantitated by UV monitoring of the HPLC eluate.

Speck and co-workers[21] measured A-(1-10) release from rat plasma angiotensinogen by brain renin-like activity, using a combined HPLC-RIA system. Tonnaer et al. described the formation of A-(1-10) from a synthetic renin substrate by purified kidney renin[10,24] and used that methodology for the analysis of renin-like enzymes in rat brain tissue.[20,25] In addition, they studied the route of angiotensin metabolism in rat brain tissue using a combined RP-HPLC-amino acid analysis.[12]

CONCLUSION

The introduction of RP-HPLC has enabled the adequate separation and purification of structurally similar angiotensins. It is to be expected that the HPLC methodology will be applied successfully to the study of endogenous angiotensins and their metabolism.

REFERENCES

1. **Regoli, D., Park, W. K., and Rioux, F.,** Pharmacology of angiotensin, *Pharmacol. Rev.*, 26, 69, 1974.
2. **Peach, M. J.,** Renin-angiotensin system: biochemistry and mechanism of action, *Physiol. Rev.*, 57, 313, 1977.
3. **Ganten, D. and Speck, G.,** The brain renin-angiotensin system: a model for the synthesis of peptides in the brain, *Biochem. Pharmacol.*, 27, 2379, 1978.
4. **Corvol, P., Rodbard, D., Drouet, J., Catt, K., and Ménard, J.,** Monoiodinated angiotensins: preparation and characterization by polyacrylamide gel electrophoresis and isoelectric focussing, *Biochem. Biophys. Acta*, 322, 392, 1973.
5. **Mendelsohn, F. A. and Johnston, C. I.,** A radiochemical renin assay, *Biochem. J.*, 121, 241, 1971.
6. **Semple, P. F., Boyd, A. S., Dawes, P. M., and Morton, J. J.,** Angiotensin II and its heptapeptide (2-8), hexapeptide (3-8), and pentapeptide (4-8) metabolites in arterial and venous blood of man, *Circ. Res.*, 39, 671, 1976.
7. **Cohen, S., Taylor, J. M., Murakami, K., Michelakis, A. M., and Inagami, T.,** Isolation and characterization of renin-like enzymes from mouse submaxillary glands, *Biochemistry*, 11, 4286, 1972.
8. **Molnár, I. and Horváth, C.,** Separation of amino acids and peptides on non-polar stationary phases by high-performance liquid chromatography, *J. Chromatogr.*, 142, 623, 1977.
9. **Guy, M. N., Roberson, G. M., and Barnes, L. D.,** Analysis of angiotensins I, II, III and iodinated derivatives by high-performance liquid chromatography, *Anal. Biochem.*, 112, 272, 1981.
10. **Tonnaer, J. A. D. M., Verhoef, J., Wiegant, V. M., and De Jong, W.,** Separation and quantification of angiotensins and some related peptides by high-performance liquid chromatography, *J. Chromatogr.*, 183, 303, 1980.
11. **Hermann, K., Ganten, D., Bayer, C., Unger, T., Lang, R. E., and Rascher, W.,** Definite evidence for the presence of [Ile^5]-angiotensin I and [Ile^5]-angiotensin II in the brain of rats, *Exp. Brain Res.*, Suppl. 4, 192, 1982.
12. **Tonnaer, J. A. D. M., Engels, G. M. H., Wiegant, V. M., Burbach, J. P. H., De Jong, W., and De Wied, D.,** Proteolytic conversion of angiotensins in rat brain tissue, *Eur. J. Biochem.*, 131, 415, 1983.
13. **Margolis, S. A. and Longenbach, P. J.,** Separation of structurally similar, biologically active peptides from their impurities, *J. High Resol. Chromatogr. Commun.*, 2, 255, 1979.

14. **Hancock, W. S., Bishop, C. A., Prestidge, R. L., Harding, D. R. K., and Hearn, M. T. W.,** Reversed-phase, high-pressure liquid chromatography of peptides and proteins with ion-pairing reagents, *Science,* 200, 1168, 1978.
15. **Feldman, J. A., Cohn, M. L., and Blair, D.,** Neuroendocrine peptides, analysis by reversed phase high performance liquid chromatography, *J. Liq. Chromatogr.,* 142, 623, 1977.
16. **Takai, M., Kumagae, S., Kishida, Y., and Sakakibara, S.,** High-performance liquid chromatography of synthetic peptides. I. Purity measurement of synthetic angiotensins, *Peptide Chem.,* 16, 67, 1979.
17. **Seidah, N. G., Dennis, M., Corvol, P., Rochemont, J., and Chrétien, M.,** A rapid high-performance liquid chromatography purification method of iodinated polypeptide hormones, *Anal. Biochem.,* 109, 185, 1980.
18. **O'Hare, M. J. and Nice, E. C.,** Hydrophobic high-performance liquid chromatography of hormonal polypeptides and proteins on alkylsilane-bonded silica, *J. Chromatogr.,* 171, 209, 1979.
19. **Sakurai, H., Hoshino, T., Iseki, M., Sakuma, M., Shimizu, M., and Kurimoto, F.,** A selective determination of angiotensin I, II and III using high-performance liquid chromatography, in Abstr. 9th Sci. Meet. Int. Soc. Hypertension, Mexico City, 1982, Abstract 362.
20. **Tonnaer, J. A. D. M., Wiegant, V. M., and De Jong, W.,** Subcellular localization in rat brain of angiotensin I-generating endopeptidase activity distinct from cathepsin D, *J. Neurochem.,* 38, 1356, 1982.
21. **Speck, G., Poulsen, K., Unger, T., Rettig, R., Bayer, C., Schölkens, B., and Ganten, D.,** In vivo activity of purified mouse brain renin, *Brain Res.,* 219, 371, 1981.
22. **Loeber, J. G., Verhoef, J., Burbach, J. P. H., and Witter, A.,** Combination of high pressure liquid chromatography and radioimmunoassay is a powerful tool for the specific and quantitative determination of endorphins and related peptides, *Biochem. Biophys. Res. Commun.,* 86, 1288, 1979.
23. **Chiknas, S. G.,** A liquid chromatography-assisted assay for angiotensin-converting enzyme (peptidyl dipeptidase) in serum, *Clin. Chem.,* 25, 1259, 1979.
24. **Tonnaer, J. A. D. M., Wiegant, V. M., and De Jong, W.,** Angiotensin generation in the brain and drinking: indications for the involvement of endopeptidase activity distinct from cathepsin D, *Brain Res.,* 223, 343, 1981.
25. **Tonnaer, J. A. D. M., Wiegant, V. M., and De Jong, W.,** In vitro and in vivo evidence for angiotensin generation in the rat brain by endopeptidase activity distinct from cathepsin D, *Exp. Brain Res.,* Suppl. 4, 109, 1982.

ACTH- AND MSH-LIKE NEUROPEPTIDES

J. Verhoef, E. E. Codd, J. P. H. Burbach, and A. Witter

OCCURRENCE, BIOSYNTHESIS AND BIOLOGICAL ACTIVITY

The anterior pituitary hormone adrenocorticotropin (ACTH) is a polypeptide of 39 amino acid residues which regulates the growth and function of the adrenal cortex. It stimulates the production and release of adrenal glucocorticosteroids. This pituitary-adrenal system has been shown to play an important role in the physiological adaptation of an organism to noxious systemic as well as to neurogenic and psychic stimuli.[1] α-Melanocyte-stimulating hormone (α-MSH) is identical to $N^{\alpha 1}$-acetyl, $C^{\alpha 13}$-ACTH 1-13-amide. The sequence of human β-MSH, a peptide of 22 amino acid residues, is present within the pituitary polypeptide β-lipotropin (β-LPH) as residues 37-58. Both MSH peptides have been isolated from the pituitary intermediate lobe of various species. They share a common sequence with ACTH 4-10 and β-LPH 47-53 (Table 1) and can increase skin pigmentation by stimulating the dispersion of melanin granules in melanocytes. Neither MSH peptide exhibits significant adrenocorticotropic activity. Recent evidence exists that pituitary ACTH- and β-LPH-related peptides comprise a family of peptides that are synthesized as parts of a common precursor molecule, pro-opiomelanocortin which has a molecular weight of about 31,000.[2,3] This glycoprotein contains β-LPH in the C-terminal segment and ACTH near the middle of its sequence. The primary structure of the precursor includes several pairs of basic amino acid residues which can be cleaved by proteolytic enzymes to yield smaller, biologically active peptides. The posttranslational processing appears to vary in different regions of the pituitary gland, leading to different patterns of peptides in the anterior and intermediate lobe.[4,5] In the anterior lobe the larger peptides ACTH and β-LPH are the main products, whereas the smaller fragments α-MSH, corticotropin-like intermediate lobe peptide (CLIP; ACTH 18-39), β-MSH and β-endorphin (β-LPH 61-91) are present in minute amounts. The reverse is found in the intermediate lobe, where the smaller peptides predominate and ACTH and β-LPH are only minor components.

Accumulating evidence is available that ACTH and α-MSH — as well as their N-terminal fragments that are virtually devoid of endocrine activities — play a crucial role in the acquisition and maintenance of a variety of behavioral tasks in animals and man through direct action on the central nervous system.[6-8] In this respect it is of particular interest that ACTH- and α-MSH-related peptides have also been found to be endogenous compounds in the brain,[9-13] a putative target tissue for these neuropeptides. The biosynthetic pathway in the brain hypothalamic area seems to be quite similar to that in the pituitary gland.[14]

Within the N-terminal segment of pro-opiomelanocortin a peptide fragment is located containing a melanotropin-like sequence designated as γ-MSH.[2] Recently, γ-MSH-like peptides have been isolated from the neurointermediate lobe of bovine and rat pituitary tissue.[15,16] The biological function of γ-MSH peptides is not yet known. Compared to α-MSH, they possess very low melanotropic activity.[17] On the other hand, it has been suggested that γ-MSH elicits a synergistic effect on ACTH-stimulated steroidogenesis[18] and that it may act as a functional antagonist of β-endorphin.[19]

REVERSED-PHASE HPLC OF ACTH- AND MSH-LIKE PEPTIDES

Reversed-phase (RP) HPLC has been established as a powerful technique for the analysis and separation of complex mixtures of underivatized peptides.[20-24] Retention on a reversed-

Table 1
AMINO ACID SEQUENCES OF HUMAN ACTH, α-MSH, β-MSH, AND OF BOVINE γ_3-MSH[a]

Peptide	Sequence
ACTH	(1, 5, 10, 15, 20) H-Ser-Tyr-Ser-Met-Glu-His-Phe-Arg-Trp-Gly-Lys-Pro-Val-Gly-Lys-Lys-Arg-Arg-Pro-Val-
	-Lys-Val-Tyr-Pro-Asn-Gly-Ala-Glu-Asp-Glu-Ser-Ala-Glu-Ala-Phe-Pro-Leu-Glu-Phe-OH (25, 30, 35, 39)
α-MSH	(1, 5, 10, 13) Ac-Ser-Tyr-Ser-Met-Glu-His-Phe-Arg-Trp-Gly-Lys-Pro-Val-NH_2
β-MSH	(1, 5) H-Ala-Glu-Lys-Lys-Asp-Glu-Gly-
	-Pro-Tyr-Arg-Met-Glu-His-Phe-Arg-Trp-Gly-Ser-Pro-Pro-Lys-Asp-OH (10, 15, 20, 22)
γ_3-MSH[a]	(1, 5, 10, 15) H-Tyr-Val-Met-Gly-His-Phe-Arg-Trp-Asp-Arg-Phe-Gly-Arg-Arg-Asn-Gly-Ser-Ser-Ser-
	-Ser-Gly-Val-Gly-Gly-Ala-Ala-Gln-OH (20, 25, 27)

Note: The common melanotropic sequences are underlined.

[a] Synthetized by Ling et al.[17] and derived from the cDNA sequence encoding for bovine pro-opiomelanocortin mRNA;[2] (γ_1-MSH = γ_3-MSH 1-11; γ_2-MSH = γ_3-MSH 1-12).

phase column, which has apolar characteristics, is highly dependent on both the molecular weight and polarity of the compound applied. Retention times for peptides are usually proportional to their chain lengths and inversely proportional to their polarities (as indicated by the inverse relation between retention time and the solubilities of the constituent amino acids in water).[24] In the literature a large variety of mobile-phase combinations has been described. The effects of organic solvent modifiers as well as the composition of the aqueous buffers, including the addition of hydrophilic or hydrophobic ion-pairing reagents, have received detailed attention.[20-22,25-28] This report presents a survey of reversed-phase liquid chromatography of ACTH- and MSH-related peptides, as summarized in Table 2, and reviews some biological applications, in particular in the isolation of naturally occurring ACTH- and MSH-like (neuro)peptides.

SEPARATION PROCEDURES

In their pioneering paper, Burgus and Rivier[29] introduced ammonium acetate (0.01 *M*, pH 4.2) as an aqueous buffer in the application of RP-HPLC for synthetic and natural underivatized peptides. This buffer allows excellent separation of oligopeptides, is volatile for easy sample recovery by evaporation or lyophilization, and is UV transparent enough to obtain sensitive detection of peptides at low wavelengths (e.g., 210 nm, at which wavelength peptide bonds strongly absorb).[30] Using ammonium acetate (0.01 *M*) and/or acetic acid (1 to 5%) resolution can be obtained between α-MSH and ACTH 18-39,[29] α-MSH and des-$N^{\alpha 1}$-acetyl-α-MSH,[31-36] α-MSH and ACTH 1-24.[23] In addition, satisfactory separation between several N-terminal fragments of ACTH 1-16 has been demonstrated.[36] The retention order of these fragments appears to correlate with their hydrophobic amino acid content. $C^{\alpha 16}$-ACTH 1-16-amide has a shorter retention time than its smaller fragments ACTH 1-10 and ACTH 4-10, probably because the presence of three hydrophilic lysine residues within the C-terminal sequence 11-16 (Table 1) more than compensates the effect of increased

Table 2
CONDITIONS FOR SEPARATION OF ACTH- AND MSH-RELATED PEPTIDES BY RP-HPLC[a]

	Mobile phase						
Column packing	Aqueous buffer (A)	Organic solvent (B)	Gradient (%B)	Flowrate (mℓ/min)	UV[b] (nm)	Retention order of ACTH and MSH-peptides investigated[c]	Ref.
μBondapak® C_{18}	NH_4OAc (0.01 *M*, pH 4.2)	CH_3CN	Isocratic (25)	2.5	210	α-MSH<hACTH 18-39	29
μBondapak® C_{18}	NH_4OAc (0.01 *M*, pH 4.0)	CH_3CN	Isocratic (28)	2	IR	des-Ac-α-MSH<α-MSH	31,33,34
μBondapak® C_{18}	NH_4OAc (0.01 *M*, pH 4.2)	MeOH	Stepwise (20—75)	2	210	des-Ac-α-MSH<α-MSH	35
μBondapak® C_{18}	NH_4OAc (0.01 *M*, pH 4.2)	MeOH	Convex (5—50)	2	210	6-7<5-7<8-10<4-7<6-10<7-10<4-9 = 1-16 amide<4-10<3-10<des-Ac-α-MSH<2-10<1-10<α-MSH	36
μBondapak® C_{18}	HOAc (1%)	MeOH	Linear (16—40)	2	275	4-10<des-Ac-α-MSH<1-10<α-MSH	32
μBondapak® C_{18}	HOAc (5%)	*n*-Propanol	Concave (10—40)	1	280	α-MSH<ACTH 1-24	23
Hypersil-ODS and other C_{18} and C_8 packings	Phosphate (0.1 *M*, pH 2.1)	CH_3CN, dioxane, MeOH, THF	Stepwise (0—60)	0.5—1.5	225	5-10<1-18<4-10<1-24<α-MSH<h(1-39) = h(18-39)<34-39<p(1-39)	22,38
Partisil®-ODS	TFA (1%)	MeOH	Linear (0—80)	0.7	280	4-10<1-24<h(1-39)	20
Partisil®-ODS	TFA (1%)	MeOH	Linear (20—55)	0.7	280 ^3H	18-24 = 17-24<9-24<1-20 = 2-20 = 3-20<1-15<1-24 sulfoxide<1-24	20,42
μBondapak® C_{18}	TFA (0.01 *M*)	CH_3CN	Linear (20—40)	1.5	210	1-24<α-MSH<h(18-39)<h(1-39)	27,28
μBondapak® C_{18}	HFBA (0.01 *M*)	CH_3CN	Linear (20—40)	1.5	210	α-MSH<1-24<h(18-39)<h(1-39)	27,28
μBondapak® C_{18}	TFA (0.08%)	CH_3CN	Linear (30—50)	1	206	1-24<α-MSH<h(11-39)<h(17-39)<h(1-39)	39
Lichrosorb® RP-18	TFA (0.05 *M*, pH 1.3)	CH_3CN	Stepwise (5—80)	1.3	275	5-10<1-18<4-10<α-MSH<1-24<h(1-39)<p(1-39)	40,41

Table 2 (continued)
CONDITIONS FOR SEPARATION OF ACTH- AND MSH-RELATED PEPTIDES BY RP-HPLC[a]

Column packing	Mobile phase: Aqueous buffer (A)	Mobile phase: Organic solvent (B)	Gradient (%B)	Flowrate (mℓ/min)	UV[b] (nm)	Retention order of ACTH and MSH-peptides investigated[c]	Ref.
μBondapak® C_{18}	TFA (1%)	MeOH	Linear (40—70)	1	210	4-7 = 11-24<4-9 = 4-10<1-16 amide<1-10 = des-Ac-α-MSH<αMSH<1-24<h(18-39)<p(25-39)<p(1-39)	36
μBondapak® C_{18}	TEAF (0.25 *M*, pH 3—3.5)	CH_3CN	Linear (12—24)	1.5	210, 254	deAc-α-MSH sulfoxide<α-MSH-sulfoxide<des-Ac-α-MSH<αMSH	44
Zorbax® ODS	TEAP (0.25 *M*, pH 3.0)	CH_3CN	Isocratic (80)	1.5	210	Analysis of γ_1-, Ac-γ_1-, γ_2-, and γ_3-MSH	17
μBondapak® CN	TEAP (0.02 *M*, pH 3.0)	CH_3CN	Linear (0—32)	1	210	ACTH 1-8<γ-MSH 1-13<α-MSH<sACTH 1-39	45
μBondapak® C_{18}	TEAP (0.02 *M*, pH 3.0)	CH_3CN	Linear (0—35)	2	210	Separation of γ-MSH 1-13 and its tryptic fragments	45
Spherisorb 10 ODS	Pyridine-formate (pH 3.0)	*n*-Propanol	Stepwise (0—32)	2	^3H	Separation of α-MSH, des-Ac-α-MSH, and their tryptic fragments	47,48

[a] Abbreviations used: Ac, acetyl; CH_3CN, acetonitrile; des-Ac, des-$N^{\alpha 1}$-acetyl; h, human; ^3H, measurement of tritium; HFBA, heptafluorobutyric acid; HOAc, acetic acid; IR, measurement of immunoreactivity; MeOH, methanol; NH_4OAc, ammonium acetate; p, porcine; s, sheep; TEAF, triethylammonium formate; TEAP, triethylammonium phosphate; TFA, trifluoroacetic acid; THF, tetrahydrofuran.

[b] Detection by monitoring UV absorbance, unless stated otherwise.

[c] Sequences presented refer to those derived from ACTH 1-39, unless mentioned otherwise.

chain length. In the studies mentioned above (Table 2), acetonitrile and methanol as organic modifiers result in quite similar retention orders of the peptides investigated (e.g., compare References 33 and 34 with 35 and 36). However, the concentrations of methanol that are required are generally higher than those of acetonitrile, because of the higher polarity of methanol. An attendant disadvantage of ammonium acetate and/or acetic acid in the mobile phase may be peak broadening and long retention times resulting in decreased resolution of larger peptides such as ACTH 1-24 and ACTH 1-39.[23,36]

Hancock et al.[37] have reported that hydrophilic anionic reagents more polar than acetate, e.g., phosphate, decrease the retention of hydrophobic peptides and improve resolution on reversed-phase column supports when added to the mobile phase, presumably by increasing the polarity of the peptides through formation of hydrophilic ion-paired complexes. For instance, in a mobile phase of aqueous phosphoric acid (0.1%) with 40% methanol ACTH 1-24 elutes rapidly from a μBondapak® C_{18}-Corasil column, whereas the peptide is retained indefinitely in the absence of the ion-pairing reagent.[25] Using a 0.1 *M* phosphate buffer, O'Hare and Nice have carried out a systematic study on the separation of numerous peptides, including ACTH- and MSH-related fragments.[22,38] Sharp peaks without peak broadening or deterioration of peak shape occur even for late-eluting peptides such as ACTH 1-39. In the phosphate system the retention times of the peptides depend on their total content of hydrophobic amino acid residues,[22,38] similar to the situation mentioned for the resolution of N-terminal ACTH fragments in the ammonium acetate system.[36] Nevertheless, anomalies have been observed for the larger peptides ($>$ 15 residues), possibly because secondary and tertiary structure characteristics can modify the number of exposed residues in these larger compounds. The high resolving power of the phosphate system is illustrated by the resolution of human and porcine ACTH 1-39, peptides differing in only one amino acid residue at position 31 (Table 1). The retention times of the peptides investigated appear to be minimally influenced ($<$ 15%) by differences in flow rate (0.5 to 1.5 mℓ/min) or differences in temperature (15 to 70°C). In addition, low pH values ($<$ 4.0) and high buffer molarities ($\geqslant$ 0.1 *M*) are required for reproducible and high resolution, and minor selective effects have been noted with different organic solvent modifiers or with different C_8- and C_{18}-alkylsilane-bonded column packings.[22]

For reversed-phase liquid chromatography of peptides Bennett et al.[20] introduced trifluoroacetic acid (TFA) into the mobile phase to obtain high recoveries. In contrast to phosphate, TFA is a volatile ion-pairing compound allowing good peptide recoveries after sample lyophilization. Replacing phosphoric acid with TFA has been shown not to affect the high resolution of various ACTH-like peptides, but only to increase their retention times.[39] Successful separations of ACTH and α-MSH peptides can be obtained by using aqueous TFA at concentrations ranging from 0.1 to 1% (Table 2). Sharp, symmetric peptide peaks are found even for the larger peptides ACTH 1-24 and ACTH 1-39 which appear to tail in the ammonium acetate systems.[36] In general, the retention orders of peptides investigated with TFA in the mobile phase are comparable to those found with the ammonium acetate or phosphate systems. Different organic solvents such as acetonitrile and methanol do not significantly modify the selectivity (e.g., compare References 27, 39, and 40 with 20 and 36). Separation of human and porcine ACTH 1-39 can be achieved both in phosphate[22,38] and in TFA[40,41] systems. The oxidized derivative of ACTH 1-24 ([Met(O)^4]-ACTH 1-24; ACTH 1-24 sulfoxide) has a lower retention time than ATH 1-24, in accordance with the higher polarity of the oxidized form.[20,42] Replacing TFA by heptafluorobutyric acid (HFBA), a compound that is also volatile and sufficiently transparent at 210 nm to facilitate UV monitoring, enhances retention times and can change the order of elution of the peptides in reversed-phase liquid chromatography.[27,28] For instance, the retention order of ACTH 1-24 and α-MSH in the HFBA system is reversed compared to that in the TFA system. This effect can offer an advantage in the purification and isolation of peptides from biological material.[27,28,43]

The influence of the ion-pairing reagents, triethylammonium formate (TEAF) and triethylammonium phosphate (TEAP), in the mobile phase for RP-HPLC of numerous peptides has been studied extensively by Rivier.[21] These compounds are UV transparent to less than 200 nm. High resolution and improved recoveries can be obtained for closely related peptides and optimal results have been shown at low pH values ($\leqslant 3$), flow rate of 1 to 1.5 mℓ/min and at ambient or even lower temperatures. However, for good sample recoveries TEAF might be preferable because TEAF (in contrast to TEAP) is a lyophilizable reagent. Using TEAF in the mobile phase, excellent resolution on a μBondapak® C_{18}-column has been reported for α-MSH, des-$N^{\alpha 1}$-acetyl-α-MSH and their sulfoxide derivatives, with retention orders proportional to their polarities.[44] In addition, aqueous TEAP buffers have been found suitable for the analysis of γ-MSH peptides on different reversed-phase column packings.[17,45]

Application of pyridine-acetate or pyridine-formate buffers in combination with *n*-propanol as a mobile phase in RP-HPLC has been shown to be successful in the resolution of opioid peptides.[46] Such systems can also result in satisfactory separation between α-MSH, des-$N^{\alpha 1}$-acetyl-α-MSH and their fragments generated by tryptic digestion.[47,48] Pyridine in the mobile phase, however, does not allow detection in the low UV wavelength area and is commonly used in connection with fluorometric assay procedures.[46]

BIOLOGICAL APPLICATIONS

The α-MSH-like peptides in biological material have been widely studied using RP-HPLC with acetate or TEAF buffers in the mobile phase. O'Donohue et al.[49] have shown that the major peak of α-MSH immunoreactive material obtained from the rat pituitary gland, had a retention time similar to that of $O^{\beta 1}$-acetyl-α-MSH, whereas α-MSH and des-$N^{\alpha 1}$-acetyl-α-MSH represented minor components. These results corroborate those found by Browne et al.[28] In contrast, the human fetal pituitary gland appeared to contain primarily des-$N^{\alpha 1}$-acetyl-α-MSH.[34] As reported by various authors using different HPLC systems, rat brain contains approximately equimolar amounts of α-MSH and des-$N^{\alpha 1}$-acetyl-α-MSH.[32,49] In human hypothalamic tissue, however, des-$N^{\alpha 1}$-acetyl-α-MSH was shown to be the predominant form of total α-MSH-like material.[33,49] Some doubt still exists regarding the characteristics of α-MSH peptides in cerebrospinal fluid (CSF). According to O'Donohue et al.,[44] both rat and human CSF is comprised primarily of α-MSH and des-$N^{\alpha 1}$-acetyl-α-MSH. On the other hand, De Rotte et al.[35] detected α-MSH but not des-$N^{\alpha 1}$-acetyl-α-MSH in rat CSF.

Using TFA in the mobile phase, HPLC analysis of rat pituitary pars intermedia extracts has shown three main ACTH-immunoreactive peaks with as yet unknown identities.[39] A comparable HPLC system revealed that only a minor amount of human plasma ACTH immunoreactivity could be attributed to ACTH 1-39 and that the other ACTH-related compounds probably originated from peripheral metabolic processes.[40,41] Following i.v. administration of ACTH 1-24 in rats, extensive degradation in both plasma and peripheral tissue was observed.[20,42] Many degradation products have been identified, originating from enzymatic cleavage of peptide bonds following the amino acid residues 1, 2, 8, 15, 16, 17, 19, 20, and 21 of the ACTH 1-24 molecule.

Alternative TFA and HFBA buffer systems in RP-HPLC have been applied to the isolation of ACTH- and MSH-related peptides from the pituitary gland of various species. In the rat anterior lobe the two major peaks of ACTH-immunoreactivity were determined as ACTH 1-39 and $O^{\beta 31}$-phosphoryl-ACTH 1-39.[27] On the other hand, ACTH-immunoreactivity in the calf anterior pituitary was identified as ACTH 1-39, ACTH 1-38, and CLIP (ACTH 18-39).[43] The main form of CLIP in the rat neurointermediate lobe was found to be $O^{\beta 31}$-phosphoryl-CLIP, whereas CLIP itself was present in low amounts.[28] Moreover, α-MSH appeared to be principally $O^{\beta 1}$-acetyl-α-MSH, and to a much smaller extent α-MSH[28] (see also Reference 49). γ-MSH, purified from the rat neurointermediate lobe, was demonstrated to be Lys-γ_3-MSH, a glycopeptide of 25 amino acid residues. The γ_3-MSH sequence corresponded to that predicted by sequence analysis of cDNA encoding for rat pro-opiomelanocortin mRNA.[16] TEAP buffers in RP-HPLC have shown to be applicable to studies of the biosynthesis of ACTH and MSH peptides.[50] Using a similar system, Seidah et al.[51] have isolated from the human pituitary gland the N-terminal segment of pro-opiomelanocortin, in which the γ-MSH region appeared to be identical to that of bovine γ-MSH.[2]

Using liquid chromatography and a mobile phase consisting of a pyridine buffer and *n*-propanol, Böhlen et al.[15] have purified to homogeneity a γ-MSH-related peptide from bovine pituitary neurointermediate lobes. Sequence analysis has established this peptide to be identical to Lys-γ_1-MSH.[17] Comparable HPLC approaches by Martens et al.[48] have revealed evidence that in the pituitary intermediate lobe of amphibians des-$N^{\alpha 1}$-acetyl-α-MSH functions as the precursor for α-MSH, acetylation occurring just before or during release.

CONCLUSION

It is obvious that the use of HPLC has contributed significantly to the isolation of naturally occurring ACTH- and MSH-(neuro)peptides and in understanding the patterns of their processing. In fact, these developments could possibly only have evolved by virtue of the high resolving power of HPLC systems in separating chemically closely related peptides.

REFERENCES

1. **Seleye, H.**, *Stress. The Physiology and Pathology of Exposure to Stress,* Acta Medical Publication, Montreal, 1950.
2. **Nakanishi, S., Inoue, A., Kita, T., Nakamura, M., Chang, A. C. Y., Cohen, S. N., and Numa, S.**, Nucleotide sequence of cloned cDNA for bovine corticotropin-β-lipotropin precursor, *Nature (London),* 278, 423, 1979.
3. **Eipper, B. A. and Mains, R. E.**, Structure and biosynthesis of pro-adrenocorticotropin/endorphin and related peptides, *Endocrinol. Rev.,* 1, 1, 1980.
4. **Scott, A. P., Ratcliffe, J. G., Rees, L. H., Landon, J., Bennett, H. P. J., Lowry, P. J., and McMartin, C.**, Pituitary peptides, *Nature New Biol.,* 244, 57, 1973.
5. **Jackson, S. and Lowry, P. J.**, Distribution of adrenocorticotrophic and lipotrophic peptides in the rat, *J. Endocrinol.,* 86, 205, 1980.
6. **De Wied, D.**, Pituitary-adrenal system hormones and behavior, in *The Neurosciences Third Study Program,* Schmitt, F. O. and Worden, F. G., Eds., Rockefeller University Press, New York, 1974, 653.
7. **Kastin, A. J., Sandman, C. A., Stratton, L. O., Schally, A. V., and Miller, L. H.**, Behavioral and electrographic changes in rat and man after MSH, in *Progress in Brain Research,* Vol. 42, Gispen, W. H., Van Wimersma Greidanus, Tj. B., Bohus, B., and De Wied, D., Eds., Elsevier, Amsterdam, 1975, 143.
8. **Bohus, B. and De Wied, D.**, Pituitary-adrenal system hormones and adaptive behaviour, in *General, Comparative and Clinical Endocrinology of the Adrenal Cortex,* Jones, I. C. and Henderson, I. W., Eds., Academic Press, London, 1980, 265.
9. **Krieger, D. T., Liotta, A., and Brownstein, M. J.**, Presence of corticotropin in limbic system of normal and hypophysectomized rats, *Brain Res.,* 128, 575, 1977.
10. **Watson, S. J., Richard, R. W., III, and Barchas, J. D.**, Adrenocorticotropin in rat brain: immunocytochemical localization in cells and axons, *Science,* 200, 1180, 1978.
11. **Oliver, C. and Porter, J. C.**, Distribution and characterization of α-melanocyte-stimulating hormone in the rat brain, *Endocrinology,* 102, 697, 1978.
12. **Jacobowitz, D. M. and O'Donohue, T. L.**, α-Melanocyte-stimulating hormone: immunohistochemical identification and mapping in neurons of rat brain, *Proc. Natl. Acad. Sci. U.S.A.,* 75, 6300, 1978.
13. **Kleber, G., Gramsch, C., Höllt, V., Mehraein, P., Pasi, A., and Herz, A.**, Extrahypothalamic corticotropin and α-melanotropin in human brain, *Neuroendocrinology,* 31, 39, 1980.

14. **Liotta, A. S., Gildersleeve, D., Brownstein, M. J., and Krieger, D. T.,** Biosynthesis in vitro of immunoreactive 31,000 dalton corticotropin/β-endorphin-like material by bovine hypothalamus, *Proc. Natl. Acad. Sci. U.S.A.*, 76, 1448, 1979.
15. **Böhlen, P., Esch, F., Shibasaki, T., Baird, A., Ling, N., and Guillemin, R.,** Isolation and characterization of a γ_1-melanotropin-like peptide from bovine neurointermediate pituitary, *FEBS Lett.*, 128, 67, 1981.
16. **Browne, C. A., Bennett, H. P. J., and Solomon, S.,** The isolation and characterization of γ_3-melanotropin from the neurointermediary lobe of the rat pituitary, *Biochem. Biophys. Res. Commun.*, 100, 336, 1981.
17. **Ling, N., Ying, S., Minick, S., and Guillemin, R.,** Synthesis and biological activity of four γ-melanotropins derived from the cryptic region of the adrenocorticotropin/β-lipotropin precursor, *Life Sci.*, 25, 1773, 1979.
18. **Pedersen, R. C., Brownie, A. C., and Ling, N.,** Pro-adrenocorticotropin/endorphin-derived peptides: coordinate action on adrenal steroidogenesis, *Science*, 208, 1044, 1980.
19. **Van Ree, J. M., Bohus, B., Csontos, K. M., Gispen, W. H., Greven, H. M., Nijkamp, F. P., Opmeer, F. A., De Rotte, A. A., Van Wimersma Greidanus, Tj. B., Witter, A., and De Wied, D.,** Behavioral profile of γ-MSH: relationship with ACTH and β-endorphin action, *Life Sci.*, 28, 2875, 1981.
20. **Bennett, H. P. J., Hudson, A. M., McMartin, C., and Purdon, G. E.,** Use of octadecasilyl-silica for the extraction and purification of peptides in biological samples, *Biochem. J.*, 168, 9, 1977.
21. **Rivier, J. E.,** Use of trialkyl ammonium phosphate (TAAP) buffers in reverse phase HPLC for high resolution and high recovery of peptides and proteins, *J. Liq. Chromatogr.*, 1, 343, 1978.
22. **O'Hare, M. J. and Nice, E. C.,** Hydrophobic high-performance liquid chromatography of hormonal polypeptides and proteins on alkylsilane-bonded silica, *J. Chromatogr.*, 171, 209, 1979.
23. **Morris, H. R., Etienne, A. T., Dell, A., and Albuquerque, R.,** A rapid and specific method for the high resolution purification and characterization of neuropeptides, *J. Neurochem.*, 34, 574, 1980.
24. **Verhoef, J. and Witter, A.,** Reversed-phase high performance liquid chromatography of neuropeptides, in *Biological/Biomedical Applications of Liquid Chromatography*, Vol. 4, Hawk, G. L., Ed., Marcel Dekker, New York, 1982, 57.
25. **Hancock, W. S., Bishop, C. A., Prestidge, R. L., Harding, D. R. K., and Hearn, M. T. W.,** Reversed-phase, high-pressure liquid chromatography of peptides and proteins with ion-pairing reagents, *Science*, 200, 1168, 1978.
26. **Mahoney, W. C. and Hermodson, M. A.,** Separation of large denaturated peptides by reversed phase high performance liquid chromatography, *J. Biol. Chem.*, 255, 11199, 1980.
27. **Bennett, H. P. J., Browne, C. A., and Solomon, S.,** Purification of the two major forms of rat pituitary corticotropin using only reversed-phase liquid chromatography, *Biochemistry*, 20, 4530, 1981.
28. **Browne, C. A., Bennett, H. P. J., and Solomon, S.,** Isolation and characterization of corticotropin- and melanotropin-related peptides from the neurointermediary lobe of the rat pituitary by reversed-phase liquid chromatography, *Biochemistry*, 20, 4538, 1981.
29. **Burgus, R. and Rivier, J.,** Use of high pressure liquid chromatography in the purification of peptides, in *Peptides 1976, Proc. 14th Eur. Peptide Symp.*, Loffett, A., Ed., University of Brussels, Brussels, 1976, 85.
30. **Woods, A. H. and O'Bar, P. R.,** Absorption of proteins and peptides in the far ultraviolet, *Science*, 167, 179, 1970.
31. **O'Donohue, T. L., Miller, R., and Jacobowitz, D. M.,** Identification, characterization and stereotaxic mapping of intraneuronal α-melanocyte stimulating hormone-like immunoreactive peptides in discrete regions of the rat brain, *Brain Res.*, 176, 101, 1979.
32. **Loh, Y. P., Eskay, R. L., and Brownstein, M.,** α-MSH-like peptides in rat brain: identification and changes in level during development, *Biochem. Biophys. Res. Commun.*, 94, 916, 1980.
33. **Parker, C. R., Jr., Barnea, A., Tilders, F. J. H., and Porter, J. C.,** Characterization of immunoreactive α-melanocyte stimulating hormone (α-MSH$_i$) in human brain tissue, *Brain Res. Bull.*, 6, 275, 1981.
34. **Tilders, F. J. H., Parker, C. R., Jr., Barnea, A., and Porter, J. C.,** The major immunoreactive α-melanocyte-stimulating hormone (α-MSH)-like substance found in human fetal pituitary tissue is not α-MSH but may be desacetyl-α-MSH (adrenocorticotropin 1-13 NH_2), *J. Clin. Endocrinol. Metab.*, 52, 319, 1981.
35. **De Rotte, A. A., Verhoef, J., Mens, W. B. J., and Van Wimersma Greidanus, Tj. B.,** Characterization and release of α-MSH-like immunoreactivity in blood and cerebrospinal fluid of intact and hypophysectomized rats, *Peptides*, submitted.
36. **Verhoef, J., Codd, E., Burbach, J. P. H., and Witter, A.,** Reversed-phase high pressure liquid chromatography of neuropeptides related to ACTH, including a potent ACTH 4-9 analog (ORG 2766), *J. Chromatogr.*, 233, 317, 1982.
37. **Hancock, W. S., Bishop, C. A., Prestidge, R. L., and Harding, D. R. K.,** High-pressure liquid chromatography of peptides and proteins. II. The use of phosphoric acid in the analysis of underivatised peptides by reversed-phase high-pressure liquid chromatography, *J. Chromatogr.*, 153, 391, 1978.

38. **Nice, E. C. and O'Hare, M. J.,** Simultaneous separation of β-lipotropin, adrenocorticotropic hormone, endorphins and enkephalins by high-performance liquid chromatography, *J. Chromatogr.*, 162, 401, 1979.
39. **McDermott, J. R., Smith, A. I., Biggins, J. A., Chyad Al-Noaemi, M., and Edwardson, J. A.,** Characterization and determination of neuropeptides by high-performance liquid chromatography and radioimmunoassay, *J. Chromatogr.*, 222, 371, 1981.
40. **Schöneshöfer, M. and Fenner, A.,** Hydrophilic ion-pair reversed-phase chromatography of biogenic peptides prior to immunoassay, *J. Chromatogr.*, 224, 472, 1981.
41. **Schöneshöfer, M. and Fenner, A.,** ACTH immunoreactivities predominating in normal human plasma are not attributable to the human ACTH 1-39 molecule, *Biochem. Biophys. Res. Commun.*, 102, 476, 1981.
42. **Hudson, A. M. and McMartin, C.,** Mechanisms of catabolism of corticotrophin-(1-24)-tetracosapeptide in the rat in vivo, *J. Endocrinol.*, 85, 93, 1980.
43. **Brubaker, P. L., Bennett, H. P. J., Baird, A. C., and Solomon, S.,** Isolation of ACTH 1-39, ACTH 1-38 and CLIP from the calf anterior pituitary, *Biochem. Biophys. Res. Commun.*, 96, 1441, 1980.
44. **O'Donohue, T. L., Charlton, C. G., Thoa, N. B., Helke, C. J., Moody, T. W., Pert, A., Williams, A., Miller, R. L., and Jacobowitz, D. M.,** Release of alpha-melanocyte stimulating hormone into rat and human cerebrospinal fluid in vivo and from rat hypothalamus slices in vitro, *Peptides*, 2, 93, 1981.
45. **Seidah, N. G., Routhier, R., Benjannet, S., Larivière, N., Gossard, F., and Chrétien, M.,** Reversed-phase high-performance liquid chromatographic purification and characterization of the adrenocorticotropin/lipotropin precursor and its fragments, *J. Chromatogr.*, 193, 291, 1980.
46. **Lewis, R. V., Stein, S., and Udenfriend, S.,** Separation of opioid peptides utilizing high performance liquid chromatography, *Int. J. Peptide Protein Res.*, 13, 493, 1979.
47. **Martens, G. J. M., Jenks, B. G., and Van Overbeeke, A. P.,** Analysis of peptide biosynthesis in the neurointermediate lobe of xenopus laevis using high-performance liquid chromatography: occurrence of small bioactive products, *Comp. Biochem. Physiol.*, 67, 493, 1980.
48. **Martens, G. J. M., Jenks, B. G., and Van Overbeeke, A. P.,** *N*α-acetylation is linked to α-MSH release from pars intermedia of the amphibian pituitary gland, *Nature (London)*, 294, 558, 1981.
49. **O'Donohue, T. L., Handelmann, G. E., Chaconas, T., Miller, R. L., and Jacobowitz, D. M.,** Evidence that *N*-acetylation regulates the behavioral activity of α-MSH in the rat and human central nervous system, *Peptides*, 2, 333, 1981.
50. **Gianoulakis, C., Seidah, N. G., Routhier, R., and Chrétien, M.,** Biosynthesis and characterization of adrenocorticotropic hormone, α-melanocyte-stimulating hormone, and an NH_2-terminal fragment of the adrenocorticotropic hormone/β-lipotropin precursor from rat pars intermedia, *J. Biol. Chem.*, 254, 11903, 1979.
51. **Seidah, N. G., Benjannet, S., Routhier, R., De Serres, G., Rochemont, J., Lis, M., and Chrétien, M.,** Purification and characterization of the N-terminal fragment of pro-opiomelanocortin from human pituitaries: homology to the bovine sequence, *Biochem. Biophys. Res. Commun.*, 95, 1417, 1980.

HYPOTHALAMIC RELEASING FACTORS

David H. Coy

INTRODUCTION

Although the enormous potential of high-performance liquid chromatography (HPLC) for rapidly analyzing and separating mixtures of organic compounds was recognized in the early 1970s, its perhaps even more dramatic impact on the peptide and protein fields was not felt until several years later when the silica-bonded hydrophobic phase packings became commercially available. Probably the first biologically active peptides to be fully evaluated using the new technique and octadecylsilane (ODS)-bonded columns were the hypothalamic releasing hormones TRH, LH-RH, and somatostatin (Figure 1) and their numerous, readily available analogs.

GENERAL ELUTION CONDITIONS

Simple Acidic Ammonium Acetate Systems

Simple, fully volatile (which is important for preparative uses) elution systems were soon found, such as the popular acetonitrile-0.01 *M* ammonium acetate mixtures,[1] which were very effective in resolving closely related analogs of TRH and LH-RH. The superiority of ODS-phases over C_8, CN, and Phe-phases with this class of peptides was also suggested[1] at this time, although it would appear that Phe-silicas should be potentially very effective with these highly aromatic peptides and would still be worth a fuller examination. The resolution of even diastereoisomeric pairs of peptides such as L- and D-His^2-TRH and -LH-RH[1,2] indicated the tremendous value of the technique in establishing the homogeneity of synthetic peptides.

Also at that time it became apparent that many peptides are surprisingly hydrophobic and have very high affinities for ODS-silica so that they could only be eluted at high concentrations of organic component. The resulting loss of buffering power from the aqueous component resulted in substantial peak broadening and loss of symmetry and resolution. This problem arose particularly with somatostatin and became even worse with certain extremely hydrophobic LH-RH analogs and with most large peptides such as β-endorphin and other neuro-GI peptides.

Several methods have now been developed to overcome this problem. The use of more hydrophobic organic solvents such as 2-propanol or 1-propanol in conjunction with regular ammonium acetate buffers results in the elution of hydrophobic peptides at much lower concentrations of organic solvent.[3] Excellent results have been obtained with this type of system with numerous peptides, including the separation of somatostatin stereoisomers[4] and the larger somatostatin-28 and catfish somatostatin I species.[5]

Buffers Utilizing Triethylamine

We were also able to improve the elution characteristics of the regular acetonitrile-ammonium acetate system by including 0.1% triethylamine in the organic component.[6] This worked very well for somatostatin and its analogs, β-endorphin,[6] and several GI peptides including VIP.[7] Improved buffers employing triethylamine were also developed by Rivier et al.[8,9] The acidic triethylammonium phosphate (TEAP)-acetonitrile system has shown great utility with LH-RH, somatostatin and their analogs,[8] and also with many larger peptides. The one drawback with this, and other phosphate based eluants, is their lack of volatility which can be a nuisance when actual isolation of the emerging peptide is required. Trie-

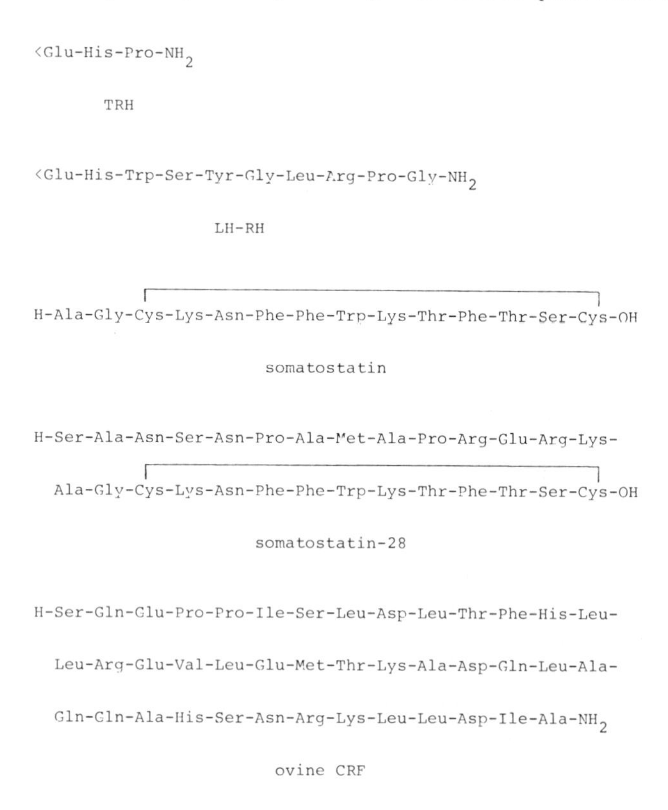

FIGURE 1. Amino acid sequences of TRH, LH-RH, somatostatin, somatostatin-28, and ovine CRF.

thylammonium formate (TEAF) buffers have been proposed[8] to overcome this, but these may not have the resolving power of the TEAP system.

Ion-pairing Trifluoracetic Acid (TFA) Systems

Another aqueous component of considerable value and convenience with many peptides in this field, such as somatostatin-28 and CRF (Figure 1), is the dilute TFA solution of Voskamp et al.[10] used in conjunction with either acetonitrile or one of the propanols. Not only is 0.1% TFA solution fully volatile and quite transparent at low wavelengths, but it often greatly alters elution positions compared to other systems described here. This makes it a useful additional system for homogeneity determinations. It is also particularly applicable to preparative HPLC of large peptides which will be described in a subsequent section.

SPECIAL PROBLEMS EXPERIENCED WITH CRF

Few difficulties are experienced in choosing a suitable solvent-buffer combination for high-efficiency elution of TRH, LH-RH, and somatostatin peptides, including somatostatin-28, on regular 60-Å ODS columns. However, corticotropin-releasing factor (CRF) which is one of the newest of the hypothalamic peptides just identified and isolated from the sheep,[11] contains 41 residues (Figure 1) and approaches the size-exclusion limit for normal ODS columns. During a recent solid-phase synthesis of ovine CRF[12] we were able to utilize a preparative 60-Å ODS column for the major purification step of the crude peptide. It became clear, however, from simultaneous analytical runs on 60-Å, 10-μm ODS columns that major peak broadening and poor resolution of contaminants could not be remedied by

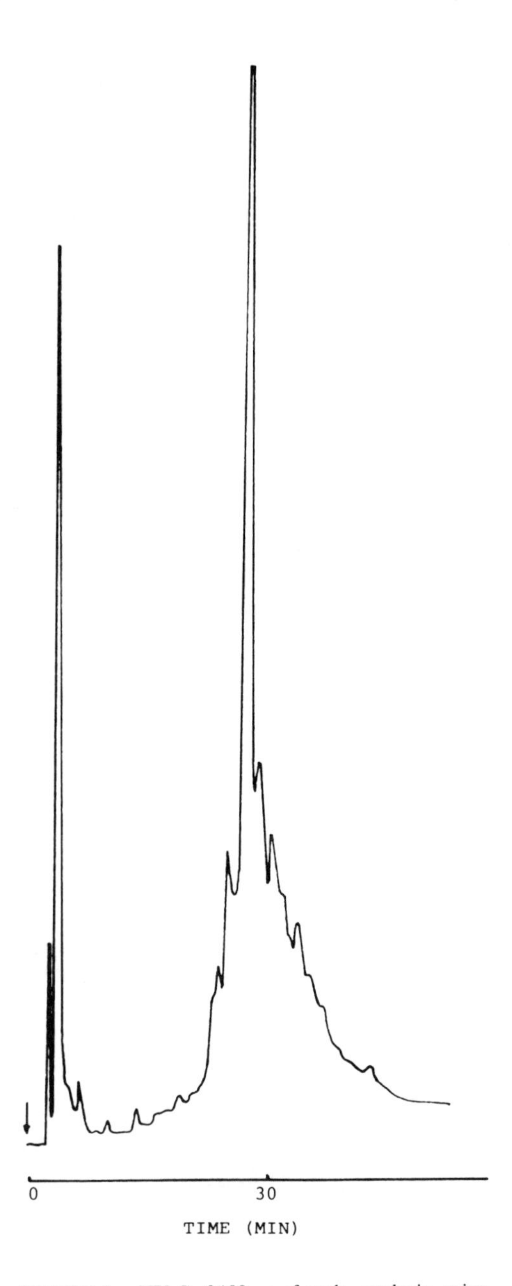

FIGURE 2. HPLC of 100 μg of crude, synthetic, ovine CRF on a column (0.4 × 25 cm) of Synchropack RP-P (10 μm, 300 Å) using linear gradient of 25 to 35% 2-propanol in 0.1% TFA over 30 min. Flow rate 1 mℓ/min, OD at 215 nm.

the choice of eluants. This problem was overcome by changing to a column of 10 μm Synchropak RP-P which is a 300-Å ODS-silica sold by Synchrom. Elution (Figure 2) of crude, synthetic CRF with a 2-propanol gradient and 0.1% TFA clearly revealed the presence of a major component together with numerous contaminants at the leading and trailing edges. On 60-Å columns, these impurities could only be seen as poorly resolved shoulders on the main peak.

We have subsequently examined the behavior of a number of other large peptides of

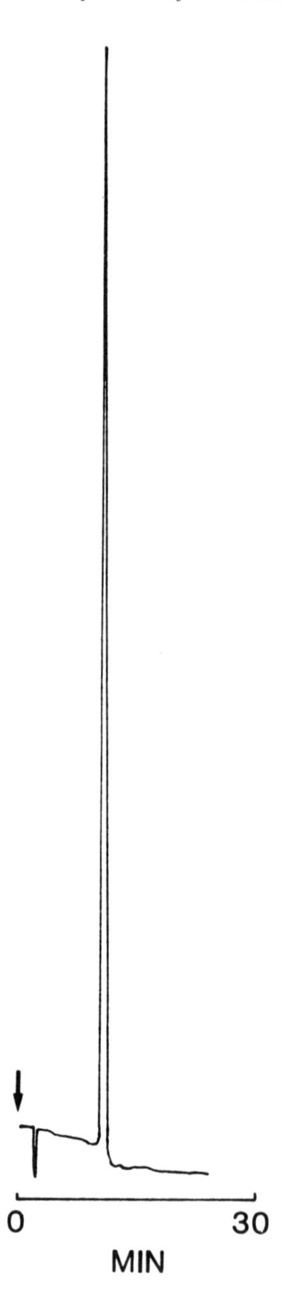

FIGURE 3. HPLC of 10 μg of somatostatin-28 using the same column as in Figure 2 and a linear gradient of 15 to 35% 2-propanol in 0.1% TFA over 15 min. Flow rate 1.5 mℓ/min, OD at 215 nm.

somewhat lower molecular weights than CRF on 300-Å columns. We also appear to obtain superior results with β-endorphin, secretin, VIP, and somatostatin-28, a representative elution profile for which is shown in Figure 3.

PREPARATIVE HPLC OF RELEASING FACTORS AND ANALOGS

Certainly of equal importance to the analytical uses of HPLC is its development as a rapid and more efficient purification method for both naturally isolated and synthetic peptides, particularly those prepared by solid-phase methods where complex mixtures of contaminants can be expected. In our laboratory, virtually all synthetic peptides undergo a simple purification protocol involving an initial clean-up by gel filtration on Sephadex® followed by one or two medium pressure HPLC steps. Complex purification routes which used to take weeks to complete can now be routinely performed in several days.

Rather than occupying expensive analytical HPLC equipment for preparative purposes, it seemed more economical to build our own medium pressure systems from components. Each system which we designed has a choice of 2.5 or 1.5 × 50 cm thick-walled glass columns, such as those supplied by Altex, which can withstand pressures up to 100 psi. They are eluted with inexpensive Fluid Metering Co. chromatography pumps and, with a medium pressure Altex or Rheodyne injection valve, the whole system can be assembled for well under $1000. Several companies now supply ODS-silica in bulk at reasonable prices. We have had good results with Whatman LRP-1 which has a mesh size of 13 to 26 μm and can be packed satisfactorily as a slurry under simple gravity flow. The necessary gradients of organic solvent can be generated using the standard two-reservoir system employed extensively for ion-exchange chromatography. Peptide loads of up to 1 g can frequently be purified in a single run, usually lasting about 3 hr, using flow rates from 2.5 to 10 mℓ/min. The high concentrations of peptide normally encountered in regular scale peptide synthesis precludes the monitoring of column effluent in the super-sensitive 210-nm region. We normally monitor aromatic absorptions at 280 or 254 nm and recheck tubes by analytical HPLC at 210 nm and/or TLC.

Very many LH-RH and somatostatin analogs together with β-endorphin, somatostatin-28, secretin, VIP, glucagon, CRF, and some of their analogs have been purified satisfactorily and quickly in this manner. Purification methods necessary for one class of LH-RH analogs — the LH-RH antagonists — are worthy of special attention. Many of these analogs, particularly the most biologically effective ones (for instance see Reference 13), are very lipophilic and poorly soluble in aqueous or organic media except at high acetic acid concentrations. This severely limits the purification methods that can be employed. Indeed, without the availability of preparative ODS-silica chromatography it would be hard to suggest a suitable alternative approach. To overcome the solubility problem during HPLC we use 20 to 30% acetic acid solutions and elute with 1-propanol gradients. Figure 4 shows a typical preparative elution profile obtained during the purification of about 250 mg of [Ac-D-Leu^{1},D-*p*-Cl-Phe^{2},D-Trp^{3},D-Arg^{6},D-Ala^{10}]-LH-RH on a 2.5-cm column. This analog had previously been partially purified by gel filtration on Sephadex® G-25 in 50% acetic acid.

SUMMARY

As in most other areas of peptide and protein research, HPLC has become an indispensible tool in both the isolation and synthesis of hypothalamic hormones. The technique can no longer be considered a purely analytical one since it is rapidly becoming the preparative purification method of choice in many laboratories. The development of even finer particle sizes, larger pore sizes, and new types of bonded phases can be expected to further increase the speed, power, and flexibility of the technique.

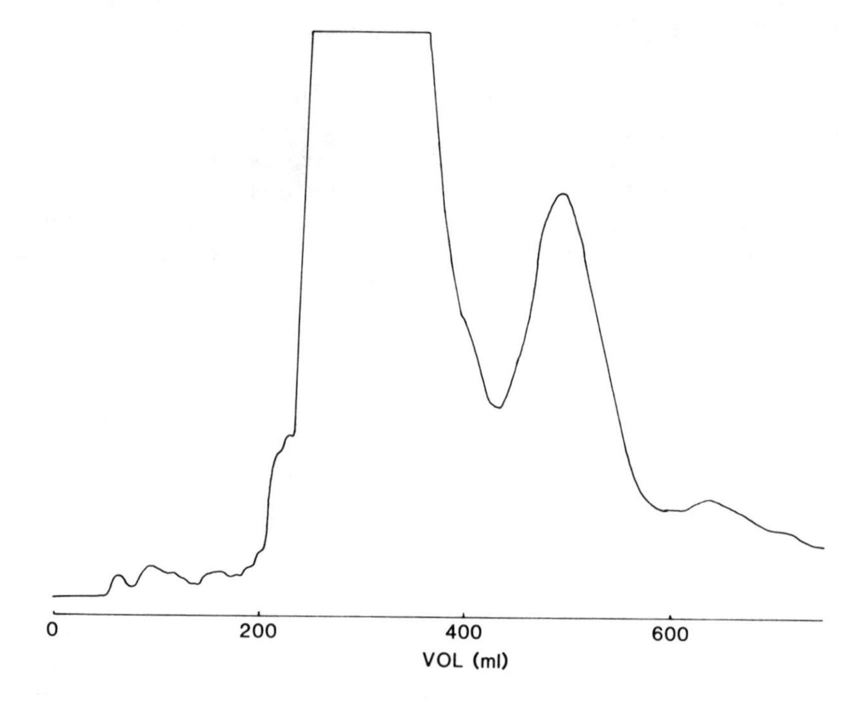

FIGURE 4. Preparative HPLC of 250 mg of partially purified synthetic [Ac-D-Leu^1,D-*p*-Cl-Phe^2,D-Trp^3,D-Arg^6,D-Ala^{10}]-LH-RH on a column (2.5 × 45 cm) of Whatman LRP-1 using a linear gradient of 10-40% 1-propanol in 20% acetic acid. Flow rate 5 mℓ/min, OD at 280 nm.

REFERENCES

1. **Burgus, R. and Rivier, J.,** Use of high pressure liquid chromatography in the purification of peptides, in *Peptides, 1976,* Loffet, A., Ed., Editions de l'Universite de Bruxelles, Brussels, 1976, 85.
2. **Rivier, J., Wolbers, R., and Burgus, R.,** Application of high pressure liquid chromatography to peptides, in *Peptides: Proc. 5th Am. Peptide Symp.*, Goodman, M. and Meienhofer, J., Eds., John Wiley & Sons, New York, 1977, 52.
3. **Coy, D. H.,** Determination of the purity of large synthetic peptides by HPLC, in *Biological-Biomedical Applications of Liquid Chromatography II,* Hawk, G. L., Ed., Marcel Dekker, New York, 1979.
4. **Meyers, C. A., Coy, D. H., Huang, W. Y., Schally, A. V., and Redding, T. W.,** Highly active position 8 analogs of somatostatin and the role of tryptophan in position 8: separation of diastereoisomers by partition chromatography, *Biochemistry,* 17, 2326, 1978.
5. **Coy, D. H., Murphy, W. A., Meyers, C. A., and Fries, J. L.,** Synthesis and some biological properties of somatostatin-28 and catfish somatostatin I, in *Peptides, 1980,* Brunfeldt, K., Ed., Scriptor, Copenhagen, 1981, 308.
6. **Coy, D. H., Gill, P., Kastin, A. J., Dupont, A., Cusan, L., Labrie, F., Britton, D., and Fertel, R.,** Synthetic and biological studies on unmodified and modified fragments of human β-lipotropin with opioid activities, in *Peptides: Proc. 5th Am. Peptide Symp.*, Goodman, M. and Meienhofer, J., Eds., John Wiley & Sons, New York, 1977, 107.
7. **Coy, D. H. and Gardner, J.,** Solid-phase synthesis of porcine vasoactive polypeptide, *Int. J. Peptide Protein Res.*, 15, 73, 1980.
8. **Rivier, J., Desmond, J., Spiess, J., Perrin, M., Vale, W., Eksteen, R., and Karger, B.,** Peptide and amino acid analysis by RP-HPLC, in *Peptides: Proc. 6th Am. Peptide Symp.*, Gross, E. and Meienhofer, J., Eds., Pierce, Rockford, 1979, 125.

9. **Rivier, J., Spiess, J., Perrin, M., and Vale, W.,** Application of HPLC in the isolation of unprotected peptides, in *Biological-Biomedical Applications of Liquid Chromatography II,* Hawk, G. L., Ed., Marcel Dekker, New York, 1978, 223.
10. **Voskamp, D., Olieman, C., and Beyerman, H. C.,** The use of trifluoracetic acid in the reverse-phase liquid chromatography of peptides including secretin, *Recl. Trav. Chim. Pays-Bas,* 99, 105, 1980.
11. **Vale, W., Spiess, J., Rivier, C., and Rivier, J.,** Characterization of a 41-residue peptide that stimulates secretion of corticotrophin and β-endorphin, *Science,* 213, 1394, 1981.
12. **Sueiras-Diaz, J., Coy, D. H., Vigh, S., Redding, T. W., Huang, C.-W., Torres-Aleman, I., and Schally, A. V.,** Synthesis and biological properties of ovine CRF, *Life Sci.,* 31, 429, 1982.
13. **Erchegyi, J., Coy, D. H., Nekola, M. V., Coy, E. J., Schally, A. V., Mezo, I., and Teplan, I.,** LH-RH analogs with increased antiovulatory activity, *Biochem. Biophys. Res. Commun.,* 100, 915, 1981.

EXTRAHYPOTHALAMIC THYROTROPIN-RELEASING HORMONE (TRH)

Olli Vuolteenaho and Juhani Leppäluoto

Thyrotropin-releasing hormone (TRH) was originally identified in hypothalamic extracts on the basis of its ability to release thyrotropin from the pituitary gland.[1] After the structure of TRH was elucidated[2-3] the availability of large quantities of synthetic TRH made possible the development of sensitive and specific radioimmunoassays (RIA) for TRH.[4] The RIAs revealed that TRH is present not only in the hypothalami of several species, but is also detectable in virtually all parts of extrahypothalamic brain.[5-7] It was calculated that quantitatively over 70% of total brain TRH is situated outside the hypothalamus. Later TRH has also been found in several extraneural tissues, including pancreas and intestines.[8-11] Several studies have shown frog skin to be a particularly rich source of TRH.[12-14] Table 1 presents the distribution of TRH in extrahypothalamic locations in several species. For further discussion on the distribution of TRH the reader may consult recent reviews.[31-33,50]

Although TRH has been shown to be distributed rather widely, no physiological function has thus far been attributed to TRH in extrahypothalamic locations. There are, however, interesting results from pharmacological studies which may shed some light on the significance of TRH in normal animals. Very small doses of intracisternally given TRH induce tachycardia and a rise in blood pressure.[34] Intravenously administered TRH inhibits stimulated secretion of gastric acid and pancreatic enzymes.[35,36] TRH improves cardiovascular function in endotoxic and hemorrhagic shock[37] and improves neurological recovery after spinal trauma.[38] More studies are needed to establish whether these pharmacological effects reflect the physiological effects of TRH.

Bioassays and several chromatographic methods have been used to characterize the extrahypothalamic TRH-like material; due to low levels obtainable extrahypothalamic TRH has not as yet been chemically characterized except in the case of frog skin.[12] The most popular chromatographic methods are gel filtration, ion-exchange chromatography, and reversed-phase (RP) HPLC. HPLC has proved to be the most suitable of these methods because of its high resolving power and its easy reproducibility. Gel filtration on Sephadex® or Bio-Gel® columns is not satisfactory as an analytical tool since TRH behaves anomalously in these systems. It coelutes with LRH and somatostatin, for example, molecules that are several times larger than TRH. Ion-exchange chromatography (in practice cation-exchange chromatography) has a very high resolving power. In our hands SP-Sephadex® C-25 resin gives equal, if not better, resolution than HPLC for TRH.[33] Ion-exchange chromatography, however, suffers from the difficulty of reproducibility of the chromatographic runs.

A variety of HPLC chromatography systems have been published for the analysis of extrahypothalamic TRH (Table 2). No systematic study has been performed to establish whether one of the published HPLC methods is superior to others, and one main reason for this is probably the lack of knowledge of the contaminating material from which TRH-like material should be separated by HPLC. One common feature of the extrahypothalamic TRH HPLC systems is the rather low efficiency compared with separations of other similar-sized molecules, e.g., amino acids.[47] The ion-pair system of Spindel and Wurtman is probably the most satisfactory in this respect. It gives a theoretical plate number of about 2500 as calculated from Figure 3 in Reference 41 (μBondapak® C_{18}-column, 3.9 × 300 mm) for immunoreactive TRF in striatal extract from the rat. By comparison we routinely obtain 10,000 theoretical plates for dansyl amino acids on the same column. The reason for the low efficiency is not known. We have studied the possibility that it might result from interaction of the imidazole proton of TRH with the negatively charged unbonded silica in the HPLC columns, but we have found that raising the buffer pH above the isoelectric point

Table 1
DISTRIBUTION OF TRH-LIKE MATERIAL IN EXTRAHYPOTHALAMIC TISSUES

Tissue	TRH (pg/mg wet weight)	Ref.
Extrahypothalamic brain		
Rat extrahypothalamic brain	2—6	5
	1.2—43	6
	2—12	7
Frog brain	55—520	5
Rat brain stem, spinal cord	+[a]	15
Human fetal cerebrum, cerebellum	0.11—18	16
Human cortex, thalamus	1—9	17
Human brain, various areas	26—9900[b]	18
Human fetal cortex	0.28—92.6	19
Human brain, various areas	10—144[b]	20
Rat spinal cord, anterior horn	1010[b]	21
Rat retina	ca.120—300[b]	22
	below 0.1	23
Human retina	0.68—1.02	24
Rat olfactory bulb	60	25
Human fetal cerebellum	8.8—20	26
Rat cerebellum	145.3	27
Gastrointestinal tissues		
Rat pancreas	0.9	8
	3.4	9
	1.0	10
Rat pancreas, isolated islets	76—158	11
Rat gastrointestinal tract	0.3—0.5	8
	0.4—3.3	9
	0.3—0.5	10
Newborn rat pancreas	603	28
Human gastrointestinal tract	0.45—3.27[b]	52
Other tissues		
Rat prostata	0.16	29
Rat testes	44	29
Human placenta	19.8[b]	30
Rat placenta	1.9	28
Frog skin *(Bombina orientalis)*	38,000[d]	12
(Rana pipiens)	3,500	13
Human and bovine milk	0.16—0.34[c]	51

[a] Positive reaction in immunofluorescence.
[b] pg/mg protein.
[c] pg/mℓ
[d] Measured by amino acid analysis.

of the imidazole group has no beneficial effect on the chromatographic behavior of TRH, indicating that other factors are responsible for the inefficiency of TRH HPLC systems. There appear to be no significant differences in the separation efficiencies between the different types of reversed-phase packings used for analysis of extrahypothalamic TRH (C_{18}, C_8, and CN, see Table 2). The cation-exchange HPLC column used by Jackson and Reichlin[14]

Table 2
HPLC METHODS USED FOR THE CHARACTERIZATION OF TRH-LIKE MATERIAL IN EXTRAHYPOTHALAMIC TISSUES

Tissue	Column	Solvent system	Temperature	Flow rate (mℓ/min)	Retention time of TRH (min)	Ref.
Rat pancreas	MicroPak MCH-10 (C_{18}) 2.1 × 250 mm[a]	A: 10 m*M* NH_4OAc, pH 4; B: CH_3CN; 1%/min gradient from A to B	Ambient	1	10—13	10
Frog skin	Partisil® (Cation-exchange) 4.6 × 250 mm[b]	10 % ethanol in 0.2 *M* NH_4OAc, pH 4.6	—[f]	—[f]	6—8	14
Rat pancreas, isolated pancreatic islets, eye, and pineal	ODS Hypersil (C_{18}) 5 × 250 mm[c]	A: 0.1 *M* NaH_2PO_4, pH 2.1; B: CH_3CN; 2 min A then 3%/min gradient from A to 9% B, then 0.8%/min to 21% B	Ambient	1	18—21	39,40
Rat pancreas, brain stem, spinal cord, and cortex	μBondapak® C_{18}, 3.9 × 300 mm[b]	2.75% CH_3CN, 0.1% hexane sulfonic acid in 0.02 *M* HOAc	60°C	2	15—18	41
Extrahypothalamic brain and spinal cord, alfalfa	μBondapak C_{18}, 3.9 × 300 mm[b]	10% ethanol in 0.2 *M* NH_4OAc pH 4.6	—[f]	1	8.5—11.5	42,43
Frog skin	ODS Hypersil (C_{18}) 5 × 250 mm[c]	50% methanol in 0.1% HOAc	—[f]	1	8—10	44
Human pancreas	MicroPak MCH-10 (C_{18}) 4 × 300 mm[a]	A: 10 m*M* NH_4OAc, pH 4.5; B:CH_3CN; 1%/min gradient from A to B	30°C	1	17—21	45
Human cerebellum	μBondapak® C_{18} 3.9 × 300 mm[b]	(1) 28% CH_3CN in 10 m*M* NH_4OAc, pH 4; (2) 0.5% CH_3CN in 10 m*M* NH_4OAc, pH 4	—[f] —[f]	1 2	4—5.2 7.2—8.7	26
Rat cerebellum	RP-10 (C_8)[d]	A: 0.1% H_3PO_4; B: 40% MetOH in A gradient from 99% A to 99% B in 10 min	—[f]	1	—[f]	27
Rat pancreas	MicroPak CN-10 (cyanopropyl) 4 × 300 mm[a]	A: 10 m*M* NH_4OAc, pH 4.5; B: 2-propanol gradient from A to 15% B in 15 min	Ambient	1	7.5—9.5	33

Table 2 (continued)
HPLC METHODS USED FOR THE CHARACTERIZATION OF TRH-LIKE MATERIAL IN EXTRAHYPOTHALAMIC TISSUES

Tissue	Column	Solvent system	Temperature	Flow rate (mℓ/min)	Retention time of TRH (min)	Ref.
Human urine	RP-8 (C_8)[c]	A: 10 m*M* NH_4OAc, pH 4; B: 2-propanol gradient from A to 30% B in 30 min	Ambient	1	13—16	46
Human and bovine milk	RP-18(C18)[g]	A: 50 m*M* NH_4OAc pH 6.5 B: 50% 2-propanol in A 15 min A then gradient to 100% B in 30 min	35°C	1	28—30	51
Various tissues and body fluids	Partisil® 10 (C18)[g]	50% propanol in 6.7 m*M* pyridine/3 m*M* HOAc	—[f]	3	7—8	53

[a] Varian.
[b] Waters-Associates.
[c] Shandon Southern.
[d] Brownlee Labs.
[e] Kontron Analytical.
[f] Not mentioned in the original paper.
[g] Merck.
[h] Whatman.

appears to give a rather similar separation compared with the reversed-phase columns used by others.

HPLC has proved its applicability in the validation of the measurement of TRH in extrahypothalamic locations where the concentrations are so low that there is no possibility for the chemical characterization at the low level amino acids present in the TRH-like material (e.g., urine[46]). It is, however, clear that this type of characterization gives only relative information; at best it only makes it probable that the measured material is authentic TRH, but never proves it. As long as the nature of the substances that may interfere with the measurement of extrahypothalamic TRH is not known there is no reason to believe that RP-HPLC is necessarily better than other chromatographic methods for the characterization of extrahypothalamic TRH.

At present the most important aim of extrahypothalamic TRH research is probably the chemical characterization of the TRH-like material. There are, however, considerable problems related mainly to the isolation of sufficient amounts of material for this purpose. If one aims to get 100% pure TRH from mammalian extrahypothalamic tissues it would require 10 to 100 million times purification. To obtain this, the preparative methods should be highly efficient. On the other hand, to keep the scale of the project within reasonable limits multistep purifications, such as was used in the original isolation[2,3] of TRH from hypothalamic extracts cannot be applicable.

RP-HPLC in conjunction with cation-exchange chromatography has proved to be an efficient purification method for extrahypothalamic TRH.[33] It should, however, be possible to improve further the separation efficiency of the HPLC step of this purification procedure, e.g., by using the ion-pair agents suggested by Spindel and Wurtman[41] (see above). The disadvantage of the use of sulfonic acids in the HPLC solvents is related to their nonvolatility, since the desalting of TRH-sized molecule is often problematic and requires a further chromatographic step. In addition Spindel and Wurtman have reported that the sulfonic acid interferes with TRH RIA. We have recently found that a completely volatile system consisting of linear 0.5%/min gradient of 2-propanol from 0.05% trifluoroacetic acid (TFA) with a flow rate of 1 mℓ/min gives an efficiency comparable with the sulfonic acid systems when used for the analysis of TRH in extrahypothalamic locations. In this system TRH elutes at 11 to 13 min. Examples of the use of the system for extrahypothalamic TRH-research are shown in Figure 1.

Problems in the isolation of small amounts of TRH from extrahypothalamic tissues are encountered also in the detection of the HPLC effluents. Thus far extrahypothalamic TRH has virtually always been detected by RIA, but the purification work requires also a less specific detection method. Since the NH_2-terminus of TRH is blocked, its highly sensitive detection in the chromatographic effluent by post-column derivatization with fluorescamine[48] or *o*-phthaldialdehyde[49] is not suitable. Absorbance at 210 nm is one possibility for the detection, but it is often problematic, since unrelated peptide/amino acid material interferes severely with the detection when high sensitivity is employed. One possibility for the detection would be the pre-column derivatization with dansyl-Cl of the material to be injected on to the HPLC column. The presence of histidine imidazole in TRH makes the molecule reactive with dansyl-Cl in spite of the fact that the NH_2-terminus is blocked. The native molecule can readily be recovered from the dansyl derivative by acid treatment. According to our experience, the fluorescent detection of dansyl derivatives of peptides in HPLC effluents is rather insensitive to interference by impurities in the solvents. Dansyl derivatization and HPLC can easily be used for highly sensitive analysis of the amino acid composition of the isolated peptide.

To conclude, the bulk of the evidence suggests that extrahypothalamic tissues contain significant amounts of TRH. This point will not, however, be absolutely certain until the extrahypothalamic TRH-like material has been thoroughly characterized. HPLC will be most valuable in the characterization work both as a preparative and analytical tool.

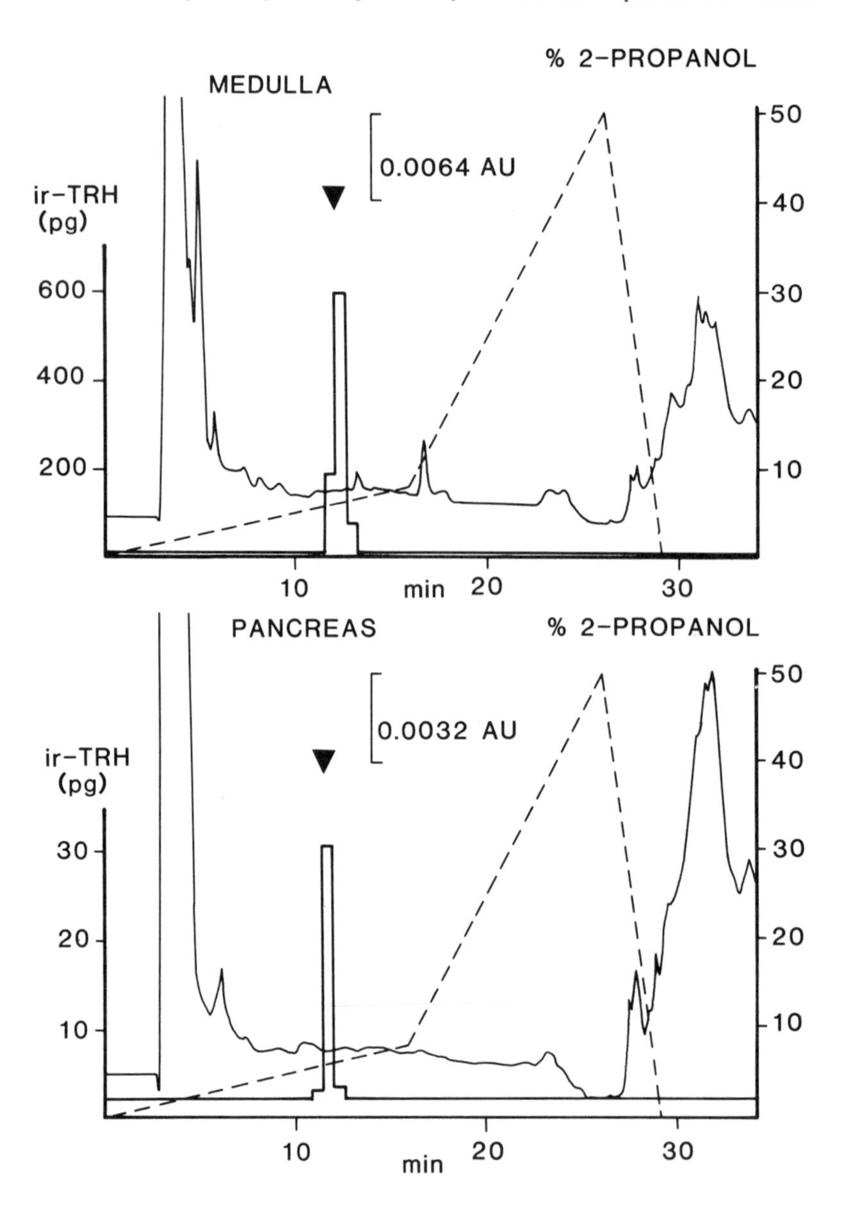

FIGURE 1. HPLC run of porcine medullary and rat pancreatic extract. Medulla and the first few cm of cervical spinal cord (330 g) were extracted with acidified acetone, precipitated with acetone, evaporated, and subjected to gel filtration in Sephadex® G-25 fine in 3% HOAc. TRH-containing fractions K_{av} 0.5 (as measured by radioimmunoassay[33]) were lyophilized and an aliquot was subjected to HPLC in the following conditions: μBondapak® C_{18} 0.4 × 30 cm, buffer A = 0.05% TFA, B = 2-propanol, gradient as shown in figure, flow 1 mℓ/min, temperature 29°C. Absorbance 220 nm. The arrowhead shows the elution position of synthetic TRF (Beckman). Rat pancreata (38 g) were extracted with methanol, centrifuged, evaporated, and subjected to cation-exchange chromatography and HPLC and the TRH-containing fractions were located by radioimmunoassay as described earlier.[33] Part of the dried material was then subjected to HPLC using the conditions described.

REFERENCES

1. **Guillemin, R., Yamazaki, E., Jutisz, M., and Sakiz, E.,** Présence dans un extrait de tissus hypothalamiques d'une substance stimulant la sécrétion de l'hormone hypophysaire thyréotrope (TSH). Premiére purification par filtration sur gel Sephadex, *C.R. Acad. Sci. Paris,* 255, 1018, 1962.
2. **Burgus, R., Dunn, T. F., Desiderio, D., and Guillemin, R.,** Structure moleculaire du facteur hypothalamique hypophysiotrope TRF d'origine ovine: mise en évidence par spectrometrie de masse de la séquence, *C.R. Acad. Sci. Paris,* 269, 1870, 1969.
3. **Bøler, J., Enzmann, F., Folkers, K., Bowers, C. Y., and Schally, A. V.,** The identity of chemical and hormonal properties of the thyrotropin releasing hormone and pyro-glutamyl-histidyl-proline amide, *Biochem. Biophys. Res. Commun.,* 37, 705, 1969.
4. **Bassiri, R. M. and Utiger, R. D.,** The preparation and specificity of antibody to thyrotropin releasing hormone, *Endocrinology,* 90, 722, 1972.
5. **Jackson, I. M. D. and Reichlin, S.,** Thyrotropin-releasing hormone (TRH): distribution in hypothalamic and extrahypothalamic brain tissues of mammalian and submammalian chordates, *Endocrinology,* 95, 854, 1974.
6. **Oliver, C., Eskay, R. L., Ben-Jonathan, N., and Porter, J. C.,** Distribution and concentration of TRH in the rat brain, *Endocrinology,* 95, 540, 1974.
7. **Winokur, A. and Utiger, R. D.,** Thyrotropin-releasing hormone: regional distribution in rat brain, *Science,* 185, 265, 1974.
8. **Leppäluoto, J., Koivusalo, F., and Kraama, R.,** Existence of TRF-like immunoreactivity in neuroectodermal tissues, Proc. Int. Union Physiolog. Sci., XIII, 440, 1977.
9. **Morley, J. E., Todd, J. G., Pekary, A. E., and Hershman, J. M.,** Thyrotropin-releasing hormone in the gastrointestinal tract, *Biochem. Biophys. Res. Commun.,* 79, 314, 1977.
10. **Leppäluoto, J., Koivusalo, F., and Kraama, R.,** Thyrotropin-releasing factor: distribution in neural and gastrointestinal tissues, *Acta Physiol. Scand.,* 104, 175, 1978.
11. **Martino, E., Lernmark, Å., Seo, H., Steiner, D. F., and Refetoff, S.,** High concentration of thyrotropin-releasing hormone in pancreatic islets, *Proc. Natl. Acad. Sci. U.S.A.,* 75, 4265, 1978.
12. **Yajima, H., Kitagawa, K., Segawa, T., Nakano, M., and Kataoka, K.,** Occurrence of Pyr-His-Pro-NH_2 in the frog skin, *Chem. Pharm. Bull.,* 23, 3301, 1975.
13. **Jackson, I. M. D. and Reichlin, S.,** Thyrotropin-releasing hormone: abundance in the skin of the frog, *Rana pipiens, Science,* 198, 414, 1977.
14. **Jackson, I. M. D. and Reichlin, S.,** Thyrotropin-releasing hormone in the blood of the frog, *Rana pipiens:* Its nature and possible derivation from regional locations in the skin, *Endocrinology,* 104, 1814, 1979.
15. **Hökfelt, T., Fuxe, K., Johansson, O., Jeffocate, S., and White, N.,** Thyrotropin releasing hormone (TRF)-containing nerve terminals in certain brain stem nuclei and in the spinal cord, *Neurosci. Lett.,* 1, 133, 1975.
16. **Winters, A. J., Eskay, F. L., and Porter, J. C.,** Concentration and distribution of TRH and LRH in the human fetal brain, *J. Clin. Endocrinol. Metab.,* 3, 960, 1974.
17. **Okon, E. and Koch, Y.,** Localisation of gonadotropin-releasing and thyrotropin-releasing hormones in human brain by radioimmunoassay, *Nature (London),* 263, 345, 1976.
18. **Guansing, A. R. and Murk, L. M.,** Distribution of thyrotropin-releasing hormone in human brain, *Horm. Metab. Res.,* 8, 493, 1976.
19. **Aubert, M. L., Grumbach, M. M., and Kaplan, S. L.,** The ontogenesis of human fetal hormones. IV. Somatostatin, luteinizing hormone releasing factor, and thyrotropin releasing factor in hypothalamus and cerebral cortex of human fetuses 10—22 weeks of age, *J. Clin. Endocrinol. Metab.,* 44, 1130, 1977.
20. **Kubek, M. J., Lorincz, M. A., and Wilber, J. F.,** The identification of thyrotropin releasing hormone (TRH) in hypothalamic and extrahypothalamic loci of the human nervous system, *Brain Res.,* 126, 196, 1977.
21. **Kardon, F. C., Winokur, A., and Utiger, R. D.,** Thyrotropin-releasing hormone (TRH) in rat spinal cord, *Brain Res.,* 122, 578, 1977.
22. **Schaeffer, J. M., Brownstein, M. J., and Axelrod, J.,** Thyrotropin-releasing hormone-like material in rat retina: changes due to environmental lighting, *Proc. Natl. Acad. Sci. U.S.A.,* 74, 3579, 1977.
23. **Eskay, R. L., Long, R. T., and Iuvone, P. M.,** Evidence that TRH, somatostatin and substance P are present in neurosectory elements of the vertebrate retina, *Brain Res.,* 196, 554, 1980.
24. **Martino, F., Nardi, M., Vaudagna, G., Simonetti, S., Cilott, A., Pinchera, A., Venturi, G., Seo, H., and Baschieri,** Thyrotropin-releasing hormone-like material in human retina, *J. Endocrinol. Invest.,* 3, 267, 1980.
25. **Kreider, M. S., Winokur, A., and Krieger, N. R.,** The olfactory bulb is rich in TRH immunoreactivity, *Brain Res.,* 217, 69, 1981.

26. **Parker, C. R., Jr.,** Characterization of immunoreactive thyrotropin releasing hormone in human fetal cerebellum, *J. Neurochem.*, 37, 1266, 1981.
27. **Pacheco, M. F., McKelvy, J. F., Woodward, D. J., Loudes, C., Joseph-Bravo, P., Krulich, L., and Griffin, W. S. T.,** TRH in the rat cerebellum. I. Distribution and concentration, *Peptides*, 2, 277, 1981.
28. **Koivusalo, F. and Leppäluoto, J.,** High TRF immunoreactivity in purified pancreatic extracts of fetal and newborn rats, *Life Sci.*, 24, 1655, 1979.
29. **Pekary, A. E., Meyer, N. V., Vaillant, C., and Hershman, J. M.,** Thyrotropin-releasing hormone and a homologous peptide in the male rat reproductive system, *Biochem. Biophys. Res. Commun.*, 95, 993, 1980.
30. **Shambaugh, G., III, Kubeck, M., and Wilberg, J. F.,** Thyrotropin-releasing hormone activity in the human placenta, *J. Clin. Endocrinol. Metab.*, 48, 483, 1979.
31. **Jackson, I. M. D. and Reichlin, S.,** Distribution and biosynthesis of TRH in the nervous system, in *Central Nervous System Effects of Hypothalamic Hormones and Other Peptides*, Collu, R., Barbeau, A., Ducharme, J. R., and Rochefort, J.-G., Eds., Raven Press, New York, 1979, 3.
32. **Morley, J. E.,** Extrahypothalamic thyrotropin releasing hormone (TRH) — its distribution and its function, *Life Sci.*, 25, 1539, 1979.
33. **Leppäluoto, J., Vuolteenaho, O., and Koivusalo, F.,** Thyrotropin releasing factor: radioimmunoassay and distribution in biological fluids and tissues, *Med. Biol.*, 59, 85, 1981.
34. **Koivusalo, F., Paakkari, I., Leppäluoto, J., and Karppanen, H.,** The effect of centrally administered TRH on blood pressure, heart rate and ventilation in rat, *Acta Physiol. Scand.*, 106, 83, 1979.
35. **Dolva, L. Ö., Hanssen, K. F., and Berstad, A.,** Actions of thyrotropin-releasing hormone on the gastrointestinal function in man, *Scand. J. Gastroenterol.*, 14, 33, 1979.
36. **Gullo, L. and Labo, G.,** Thyrotropin-relasing hormone inhibits pancreatic enzyme secretion in humans, *Gastroenterology*, 80, 735, 1981.
37. **Holaday, J. W., D'Amato, R. J., and Faden, A. I.,** Thyrotropin-releasing hormone improves cardiovascular function in experimental endotoxic and hemorrhagic shock, *Science*, 213, 216, 1981.
38. **Faden, A. I., Jacobs, T. P., and Holaday, J. W.,** Thyrotropin-releasing hormone improves neurologic recovery after spinal trauma in cats, *N. Engl. J. Med.*, 305, 1063, 1981.
39. **Kellokumpu, S., Vuolteenaho, O., and Leppäluoto, J.,** Behavior of rat hypothalamic and extrahypothalamic immunoreactive TRF in thin-layer chromatography (TLC) and high pressure liquid chromatography (HPLC), *Life Sci.*, 26, 475, 1980.
40. **Koivusalo, F., Leppäluoto, J., Knip, M., and Rajaniemi, H.,** Presence of TRF immunoreactivity in marginal islet cells in rat pancreas, *Acta Endocrinol.*, 97, 398, 1981.
41. **Spindel, E. and Wurtman, R. J.,** TRH immunoreactivity in rat brain regions, spinal cord and pancreas: validation by high-pressure liquid chromatography and thin-layer chromatography, *Brain Res.*, 201, 279, 1980.
42. **Jackson, I. M. D.,** TRH in the rat nervous system: identity with synthetic TRH on high performance liquid chromatography following affinity chromatography, *Brain Res.*, 201, 245, 1980.
43. **Jackson, I. M. D.,** Abundance of immunoreactive thyrotropin-releasing hormone-like material in the alfalfa plant, *Endocrinology*, 108, 344, 1981.
44. **Bennett, G. W., Balls, M., Clothier, R. H., Marsden, C. A., Robinson, G., and Wemyss-Holden, G. D.,** Location and release of TRH and 5-HT from amphibian skin, *Cell Biol. Int. Rep.*, 5, 151, 1981.
45. **Koivusalo, F.,** Evidence of thyrotropin-releasing hormone activity in autopsy pancreata from newborns, *J. Clin. Endocrinol. Metab.*, 53, 734, 1981.
46. **Leppäluoto, J. and Suhonen, A.-S.,** High pressure liquid chromatography purification of human urinary samples for thyrotropin-releasing hormone radioimmunoassay, *J. Clin. Endocrinol. Metab.*, 54, 914, 1982.
47. **Zimmerman, C. L., Appella, E., and Pisano, J. J.,** Rapid analysis of amino acid phenylthiohydantoins by high-performance liquid chromatography, *Anal. Biochem.*, 77, 569, 1977.
48. **Böhlen, P., Stein, S., Stone, J., and Udenfriend, S.,** Automatic monitoring of primary amines in preparative column effluents with fluorescamine, *Anal. Biochem.*, 67, 438, 1975.
49. **Lee, K. and Drescher, D. G.,** Fluorometric amino-acid analysis with *o*-phthaldialdehyde (OPA), *Int. J. Biochem.*, 9, 457, 1978.
50. **Dolva, L. Ö. and Hanssen, K. F.,** Thyrotropin-releasing hormone. Distribution and actions in the gastrointestinal tract, *Scand. J. Gastroenterol.*, 17, 705, 1982.
51. **Amarant, T., Fridkin, M., and Koch, Y.,** Luteinizing hormone-releasing hormone and thyrotropin-releasing hormone in human and bovine milk, *Eur. J. Biochem.*, 127, 647, 1982.
52. **Dolva, L. Ö., Hanssen, K. F., Aadland, E., and Sand, T.,** Thyrotropin-releasing hormone immunoreactivity in the gastrointestinal tract of man, *J. Clin. Endocrinol. Metab.*, 56, 524, 1983.
53. **Busby, W. H., Youngblood, W. W., Humm, J., and Kizer, J. S.,** A reliable method for the quantification of thyrotropin-releasing hormone (TRH) in tissue and biological fluids, *J. Neurosci. Meth.*, 4, 315, 1981.

LUTEINIZING HORMONE RELEASING HORMONE (LH-RH) AND ANALOGS

Anne Guyon-Gruaz, Daniel Raulais, and Pierre Rivaille

INTRODUCTION

LH-RH* is a hypothalamic peptide involved in the stimulation of release of luteinizing hormone (LH) and follicle-stimulating hormone (FSH) from the pituitary gland of humans and animals.[1-2] Factors isolated both from porcine[3-5] and ovine hypothalamus[6] possess the same decapeptide structure

$$\text{Pyro-Glu-His-Trp-Ser-Tyr-Gly-Leu-Arg-Pro-Gly-NH}_2$$
$$1 \quad 2 \quad 3 \quad 4 \quad 5 \quad 6 \quad 7 \quad 8 \quad 9 \quad 10$$

Previously, before HPLC became widely used, purification of LH-RH from thousands of hypothalami necessitated 12 successive steps, including gel filtrations, chromatographies on carboxymethyl cellulose, free flow electrophoresis, counter-current distribution, partition chromatography, and high-voltage zone electrophoresis.[7]

CHEMICAL IDENTIFICATION OF LH-RH

LH-RH can be chemically characterized by its amino acid analysis (after acidic or basic hydrolysis), thin-layer chromatography (TLC) on silicagel or cellulose, or paper electrophoresis. LH-RH shows an Erlich and Pauly positive test and a ninhydrin negative test $[\alpha]_D^{20°} = -49$ (C = 1, AcOH 14%).[8]

BIOLOGICAL ESTIMATION OF LH-RH

Both in vitro and in vivo methods involving measurement of LH and FSH are generally used for estimating biological activities of this releasing factor. LH is determined by a double antibody radioimmunoassay (RIA) method for rat LH.[9] FSH is determined by bioassay[10] or a double antibody RIA for rat FSH.[11] The in vitro methods can be performed either through short-term incubation of the anterior pituitaries[12] or by long-term tissue culture method.[13]

The in vivo methods are performed on ovariectomized rats for measurement of LH[14] or FSH[15] released. Another in vivo assay uses induction of ovulation in rats,[16] hamsters,[17] or rabbits.[18] A RIA has also been developed for the evaluation of LH-RH[19] contents.

HPLC OF LH-RH

HPLC began to be widely used when biological or clinical investigations caused a considerable increase in requests for synthetic LH-RH or its analogs. LH-RH was usually analyzed under isocratic conditions (Table 1) or with gradient elution (Table 2) on reverse-phase columns. Synthetic LH-RH was also purified under isocratic elution condition.[20]

The retention time was found to depend on the elution flow rate or the temperature as illustrated in Table 3.[21] Retention time was also influenced by the nature of the stationary

* Other names found in the literature for LH-RH are L.R.F., luteinizing hormone releasing factor, L.R.H., luteinizing hormone-releasing hormone, and also luliberin.

Table 1
HPLC OF LH-RH IN ISOCRATIC CONDITIONS. RETENTION TIME (R.T.) IS CLASSIFIED IN INCREASING ORDER

Column								
ø(mm)	height (cm)	Phase	Solvent	pH	F.R. (mℓ/min.) (t)	UV Detect. nm	R.T. (min.)	Ref.
4	60	µBondapak® C_{18}	CH_3CN, 0.01 *M* NH_4 OAc (25-75, v/v)	4	2.5	210	4.2	3
4	60	µBondapak® C_{18}	CH_3CN, TEAF buffer (30-70, v/v)	3	3	280	5.5	8
4	30	µBondapak® C_{18}	CH_3CN, 0.01 *M* NH_4 OAc (22-78, v/v)	4	2	210	6.2	23
6.3	30	Partisil® 10 ODS	CH_3OH, H_2O, HCOOH (27-73-0.25, v/v)	—	1 (30°)	225	7.2	28
4	30	µCyanopropyl	CH_3CN, TEAP buffer (12-88, v/v)	3	1.5	210	8.3	13
4	60	µBondapak® C_{18}	CH_3CN, 0.01 *M* NH_4 OAc (20-80, v/v)	4	2.5	210	9.8	23
4	30	µBondapak® C_{18}	EtOH, 0.01 *M* NH_4 OAc (22-78, v/v)	4	1.5	210	11.8	23
8	30	µBondapak® C_{18}	CH_3CN, TEAF buffer (19-81, v/v)	3.1	3	254	13	20
4	25	Lichrosorb® RP_{18}	CH_3CN, TEAP buffer (20-80, v/v)	3	—	—	—	29
3.9	60	µBondapak® C_{18}	CH_3CN, TEAF buffer (16.2-83.8, v/v)	3	1.5	280	16.5	24
4	60	µBondapak® C_{18}	CH_3CN, 0.01 *M* NH_4 OAc (17-83, v/v)	4	2.5	210	27.5	23

Note: Stationary phases µBondapak® C_{18} (Waters Associates), Lichrosorb® RP^{18} (Merck). Partisil® 10 ODS (Whatmann), µCyanopropyl (Waters) TEAP buffer = 0.25 *N* phosphoric acid brought to pH 3 with triethylamine. TEAF buffer = 0.25 *N* formic acid brought to pH 3 with triethylamine. — = data not given.

Table 2
RETENTION TIMES OF LH-RH WITH DIFFERENT LINEAR GRADIENT SYSTEMS

Column			Solvents							
ø(mm)	height(cm)	Phase	A	B	Slope (%)	Time (min)	pH	Flow rate (mℓ/min)	Retention time (min)	Ref.
—	—	μBondapak® C_{18}	0.01 *M* aq. TFA	CH_3CN, 0.01 *M*. aq. TFA (80—20, v/v)	B 25—73	90	—	1.5	7.5	29
4	30	μCyanopropyl	Aq. TEAP buffer	CH_3CN, aq. TEAP buffer (60—40, v/v)	B 12—33	30	3	1.5	8	21
—	—	μBondapak® C_{18}	0.01 *M* aq. PFPA	CH_3CN, 0.01 *M* aq. PFPA (80—20, v/v)	B 25—73	90	—	1.5	12.5	30
—	—	μBondapak® C_{18}	0.01 *M* aq. HFBA	CH_3CN, 0.01 *M* aq. HFBA (80—20, v/v)	B 25—73	90	—	1.5	17	30
—	—	μBondapak® C_{18}	0.01 *M* aq. UFCA	CH_3CN, 0.01 *M* aq. UFCA (80—20, v/v)	B 25—73	90	—	1.5	32	30
—	—	Lichrosorb® RP_{18}	0.1 *M* $NaClO_4$ + 0,1% H_3PO_4	CH_3CN, 0.1 *M* $NaClO_4$ + 0,1% H_3PO_4 (60—40, v/v)	B 0—100	80	2.1	1	37	31
—	—	Biosil® ODS	0.1 *M* $NaClO_4$, 5m *M* phosphate buffer, pH7.4	CH_3CN, 0.01 *M* $NaClO_4$ (60—40, v/v)	B 0—100	80	7.4	1	42.8	31
—	—	Biosil® ODS	0.1 *M* $NaClO_4$, 0.1% H_3PO_4	CH_3CN, 0.1 *M* $NaClO_4$— 0,1% H_3PO_4 (60—40, v/v)	B 0—100	80	2.1	1	49,5	31

Note: —, data not given; TFA, trifluoroacetic acid, TEAP buffer : 0.25 *N* phosphoric acid brought to pH 3 with triethylamine. PFPA, pentafluoropropionic acid; HFBA, heptafluorobutyric acid; UFCA, undecafluorocaproic acid. μBondapak® C_{18}, μCyanopropyl (Waters associates), Lichrosorb® RP^{18} (Merck), Biosil ODS (Biorad).

Table 3
INFLUENCE OF THE FLOW RATE (A) AND THE TEMPERATURE (B) ON THE RETENTION TIME

Flow rate (mℓ/min)	Pressure (psi)	Retention time (min)	Temp (°C)	Pressure (psi)	Retention time (min)
1	300	26.6	25	2000	20
1.5	700	17.3	35	1800	15
2	900	12.9	45	1600	12
3	1400	8.5	55	1400	9
	A			B	

	A		B
Column :	μBondapak-alkylphenyl (Waters) ø = 4mm, h = 30 cm.	Column :	μ Bondapak® C_{18} (Waters) ø = 4mm, h = 30 cm.
Solvent :	CH_3CN, TEAP[a] buffer (15—85, v/v) pH 3.	Solvent :	CH_3CN, TEAP[a] buffer (15—85, v/v), pH 3. Flow rate = 1.5 mℓ/min.
U.V. detection :	210nm	U.V. detection :	210 nm

[a] TEAP buffer : 0.25 *N* phosphoric acid brought to pH 3 with triethylamine.

Table 4
INFLUENCE OF THE SUPPORTS AND SOLVENT COMPOSITION ON RETENTION TIME

Phase	Pressure (psi)	Solvent	pH	Retention Time (min)
μCyanopropyl	1000	CH_3CN, TEAP buffer (7.2—92.8, v/v)	3	21
μAlkylphenyl	1300	CH_3CN, TEAP buffer (15—85, v/v)	3	21
μBondapak® C_{18}	1400	CH_3CN, TEAP buffer (15.6—84.4, v/v)	3	21

Note: Column : ϕ = 4 mm, h = 30 cm, Flow rate, 1.5 mℓ/min; UV detection 210 nm. TEAP buffer : 0.25 *N* phosphoric acid brought to pH 3 with triethylamine.

From Rivier, J., *J. Liq. Chromatogr.*, 1(3), 343, 1978. With permission.

phase as described in Table 4.[21] Using different stationary phases the composition of the solvent did, indeed, have to be modified in order to conserve the same retention times.[21]

Structural differences in vertebrate hypothalamic immunoreactive LH-RH were confirmed by HPLC[22] with the use of a gradient elution program (Table 5). HPLC also permitted the analysis of several analogs of LH-RH under isocratic conditions (Table 6).[23-25]

The metabolic breakdown of LH-RH could lead to ten possible pyroglutamyl fragments, which were analyzed by HPLC according to the diagram shown in Figure 1.[26] From reading the different tables in this report, it can be observed that retention time of LH-RH analyzed by HPLC, in fact, depends on the conditions of analysis : flow rate, solvent mixture, temperature, or support. In order to avoid any confusion in the identification of this releasing

Table 5
RETENTION TIME OF LH-RH EXTRACTED FROM HYPOTHALMII OF DIFFERENT SPECIES

LH-RH	Retention time (min)
Synthetic and frog	17.25
Chicken and teleost	14.8
Mammalian	17.1

Note: Column : Phase RP8 (Spectra-Physics), ϕ = 1 mm, h = 22 cm. Solvents : A = CH_3OH, 0.01 *M* NH_4 OAc (90—10, v/v), pH = 4.6. B = CH_3OH, 0.01 *M* NH_4 OAc (10—90, v/v), pH = 4.6. Gradient profile : (1) B 30 to 70% linear in 14.50 min. (2) Isocratic B 70% for 10 min. (3) B 70 to 99% in 10.50 min.

From King, J. A. and Hillar, R. P., *Endocrinology*, 106(3), 707, 1980. With permission.

factor, the peak of the standard or supposed LH-RH must be collected and chemically or biologically authenticated[27] whether previously described conditions are used or a new system of HPLC is established.

Table 6
HPLC OF SOME LH-RH ANALOGS

LH-RH Analogs	Column		Phase	Solvent	pH	Flow rate mℓ/min.	Retention time (min)	UV detect. (nm)	Ref.
	φ (mm)	Height (cm)							
[D-Lys[6]] — LH-RH	4	30	μBondapak® C_{18}	CH_3CN, 0.01 *M* NH_4 OAc (22—78, v/v)	4	2	3.5	210	23
[D-Ala[6]] — LH-RH	4	30	μBondapak® C_{18}	EtOH, 0.01 *M* NH_4 OAc (22—78, v/v)	4	4	5.9	210	23
[D-Ala[6]] — LH-RH	4	30	μBondapak® C_{18}	CH_3CN, 0.01 *M* NH_4 OAc (22—78, v/v)	4	2	6.2	210	23
[D-Ala[6]] — LH-RH	4	30	μBondapak® C_{18}	CH_3CN, 0.01 *M* NH_4 OAc (22—78, v/v)	4	4	6.5	210	23
[L-Hhi[2]] — LH-RH[a]	9	20	Lichrosorb® 5 μ RP_{18}	CH_3CN, 0.01 *M* NH_4 OAc (22—78, v/v)	4	3.2	11	280	25
[Des-Ser[4]] — LH-RH	4	30	μBondapak® C_{18}	EtOH, 0.01 *M* NH_4 OAc (22—78, v/v)	4	1.5	12.5	210	23
[Des-Gly[6]] — LH-RH	4	30	μBondpak® C_{18}	EtOH, 0.01 *M* NH_4 OAc (22—78, v/v)	4	1.5	12.7	210	23
[Sar30] — LH-RH	4		μBondapak® C_{18}	EtOH, 0.01 *M* NH_4 OAc (22—78, v/v)	4	1.5	13	210	23
[D-Tyr[6]] — LH-RH	4	30	μBondapak® C_{18}	CH_3CN, 0.01 *M* NH_4 OAc (22—78, v/v)	4	2	13	210	23
[D-Pro[9]] — LH-RH	4	30	μBondapak® C_{18}	CH_3CN, 0.25 *M* TEAF (16.2—83.8, v/v)	3	1.5	13.5	280	24
[L-Hhi[2]] — LH-RH	9	20	Lichrosorb® 5 μ RP^{18}	CH_3CN, 0.2 *M* NH_4 OAc (27—73, v/v)	4	3	16.5	280	25
[D-His[2]] — LH-RH	4	60	μBondapak® C_{18}	CH_3CN, 0.01 *M* NH_4 OAc (17—83, v/v)	4	2.5	22.5	210	23

Note: TEAF : 0.25 *M* Formic acid brought to pH 3 with triethylamine. μBondapak® C_{18} (Waters), Lichrosorb® 5 μ RP_{18} (Merck)

[a] [L-Hhi[2]] — LH-RH = [L-Homohistidine[2]] — LH-RH.

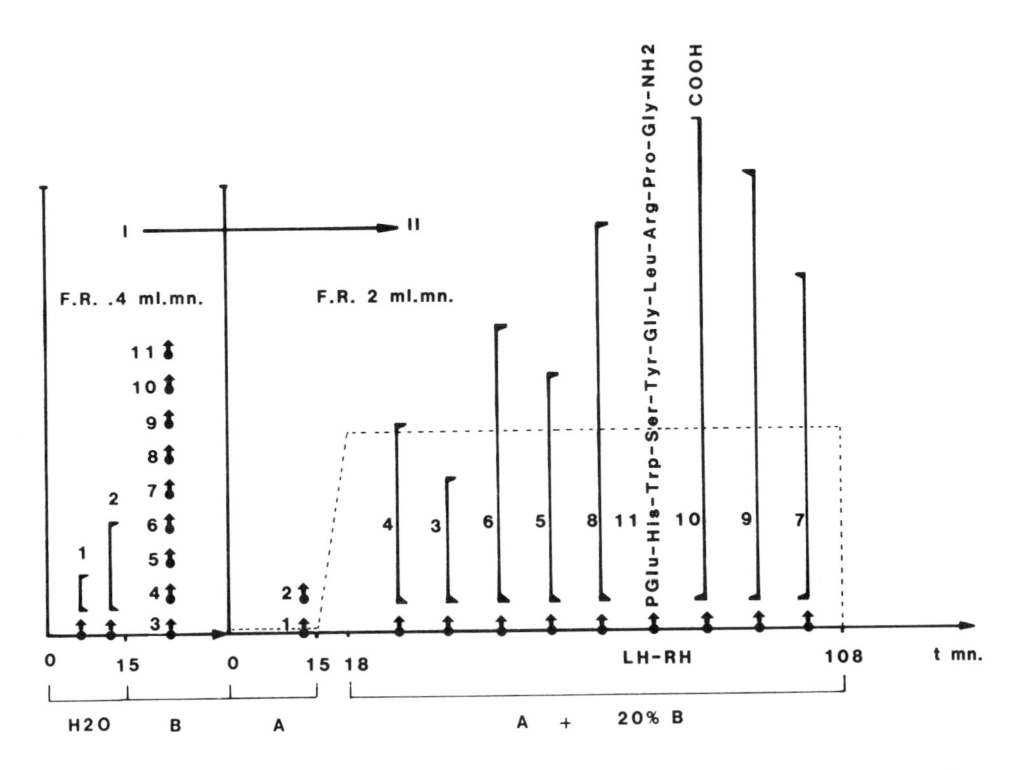

FIGURE 1. Separation of LH-RH metabolites. Only the elution position of the LH-RH metabolites are given. Solvents : (A) CH_3CN, 0,1% H_3PO_4 (1—99 v/v). (B) CH_3CN, 0,1% H_3PO_4 (60—40 v/v), pH 2,4. Column : ϕ, 39 mm, H, 30 cm, μBondapak® C_{18} (Waters); detection uv, 220 nm. (From Stetler-Stevenson, M. A., Yang, D. C., McCartney, L., Peterson, D., Lipkowski, A. W., and Flouret, G., in *Proc. 16th Eur. Peptide Symp.*, Brunfeldt, K., Ed., Scriptor, Copenhagen, 1981, 725. With permission.)

REFERENCES

1. **Mittler, J. C. and Meites, J.,** In vitro stimulation of pituitary follicle-stimulating-hormone release by hypothalamic extract, *Proc. Soc. Exp. Biol. Med.*, 117, 309, 1964.
2. **McCann, S. M., Taleisnik, S., and Friedman, H. M.,** LH releasing activity in hypothalamic extracts, *Proc. Soc. Exp. Biol. Med.*, 104, 432, 1960.
3. **Shally, A. V., Arimura, A., Baba, Y., Nair, R. M. G., Matsuo, H., Redding, T. W., Debeljuk, L., and White, W. F.,** Isolation and properties of the FSH and LH-releasing hormone, *Biochem. Biophys. Res. Commun.*, 43, 393, 1971.
4. **Shally, A. V., Nair, R. M. G., Redding, T. W., and Arimura, A.,** Isolation of the luteinizing hormone and follicle-stimulating hormone-releasing hormone from porcine hypothalami, *J. Biol. Chem.*, 246, 7230, 1971.
5. **Matsuo, H., Baba, Y., Nair, R. M. G., Arimura, A., and Shally, A. V.,** Structure of the porcine LH and FSH-Releasing hormone. I. The proposed amino-acid sequence, *Biochem. Biophys. Res. Commun.*, 43, 1334, 1971.
6. **Burgus, R., Butcher, M., Amoss, M., Ling, N., Monahan, M., Rivier, J., Fellows, R., Blackwell, R., Vale, W., and Guillemin, R.,** Primary structure of the ovine hypothalamic luteinizing-hormone-releasing factor (LRF), *Proc. Natl. Acad. Sci. U.S.A.*, 69, 278, 1972.
7. **Shally, A. V.,** Aspect of hypothalamic regulation of the pituitary gland, *Science*, 202, 18, 1978.
8. **Rivaille, P., Gautron, J. P., Castro, B., and Milhaud, G.,** Synthesis of LH-RH using a new phenolic polymer as solid support and "BOP" reagent for fragment coupling, *Tetahedron*, 36, 3413, 1980.

9. **Niswender, G. D., Midgley, A. R., Monroe, S. E., and Reichert, L. E.,** Radioimmunoassay for rat luteinizing hormone with antiovine LH serum and ovine LH-^{131}I *Proc. Soc. Exp. Biol. Med.,* 128, 807, 1968.
10. **Steelman, S. L. and Pohley, F.,** Assay of the follicle stimulating hormone based on the augmentation with human chorionic gonadotropin, *Endocrinology,* 53, 604, 1953.
11. **Daane, T. A. and Parlow, A. F.,** Periovulatory patterns of rat serum follicle stimulating hormone and luteinizing hormone during the normal estrous cycle : effect of pentobarbital, *Endocrinology,* 88, 653, 1971.
12. **Mittler, J. C. and Meites, J.,** Effects of hypothalamic extract and androgen on pituitary FSH-Release in vitro, *Endocrinology,* 78, 500, 1966.
13. **Reddings, T. W., Schally, A. V., Arimura, A., and Matsuo, H.,** Stimulation of release and synthesis of luteinizing hormone (LH) and follicle stimulating hormone (FSH) in tissue cultures of rat pituitaries in response to natural and synthetic LH and FSH releasing hormone, *Endocrinology,* 90, 764, 1972.
14. **Ramirez, V. D. and McCann, S. M.,** A highly sensitive test for LH-releasing activity : the ovariectomized, estrogen progesterone blocked rats, *Endocrinology,* 73, 193, 1963.
15. **Arimura, A., Debeljuk, L., and Schally, A. V.,** Stimulation of FSH release in vivo by prolonged infusion of synthetic LH-RH, *Endocrinology,* 91, 529, 1972.
16. **Arimura, A., Schally, A. V., Saito, T., Muller, E. E., and Bowers, C. Y.,** Induction of ovulation in rats by highly purified pig LH-Releasing factor (LRF), *Endocrinology,* 80, 515, 1967.
17. **Arimura, A., Matsuo, H., Baba, Y., and Schally, A. V.,** Ovulation induced by synthetic luteinizing hormone-releasing hormone in the hamster, *Science,* 174, 511, 1971.
18. **Hilliard, J., Schally, A. V., and Sawyer, C. S.,** Progesterone blockade of the ovulatory response to intrapituitary infusion of LH-RH in rabbits, *Endocrinology,* 88, 730, 1971.
19. **Deery, D. J.,** Determination by radioimmunoassay of the luteinizing hormone-releasing hormone (LH-RH) content of the hypothalamus of the rat and some lower vertebrates, *Gen. Comp. Endocrinol.,* 24, 280, 1974.
20. **Seyer, R.,** personal communication.
21. **Rivier, J.,** Use of trialkyl ammonium phosphate (TAAP) buffers in reverse phase HPLC for high resolution and high recovery of peptides and proteins, *J. Liq. Chromatogr.,* 1(3), 343, 1978.
22. **King, J. A. and Hillar, R. P.,** Comparative aspects of luteinizing hormone-releasing hormone structure and function in vertebrate phylogeny, *Endocrinology,* 106(3), 707, 1980.
23. **Burgus, R. and Rivier, J.,** Use of high pressure liquid chromatography in the purification of peptides, in *Proc. 15th Eur. Peptide Symp.,* Loffet, A., Ed., Edition de l'Université de Bruxelles, Bruxelles, Belgium, 1976, 85.
24. **Rivier, J., Spiess, J., Perrin, M., and Vale, W.,** Application of HPLC in the isolation of unprotected peptides, in *Biological/Biomedical Applications of Liquid Chromatography II,* Hawk, G. L., Ed., Marcel Dekker, New York, 1979, 223.
25. **Raap, J. and Kerling, K. E. T.,** Studies on polypeptides. XXXIV. Replacement of histidine 2 in luliberin (LH-RH) by L homohistidine, *J. R. Neth. Chem. Soc.,* 100, 2, 62, 1981.
26. **Stetler-Stevenson, M. A., Yang, D. C., McCartney, L., Peterson, D., Lipkowski, A. W., and Flouret, G.,** Synthesis and characterization of the metabolic products of LH-RH breakdown by renal tissue, in *Proc. 16th Eur. Peptide Symp.,* Brunfeldt, K., Ed., Scriptor, Copenhagen, 1981, 725.
27. **Rivaille, P., Raulais, D., and Milhaud, G.,** High performance liquid chromatography analysis of peptides hormones, in *Biological/Biomedical Applications of Liquid Chromatography II,* Hawk, G. L., Ed., Marcel Dekker, New York, 1979, 223.
28. **Sarkiz, G.,** personal communication.
29. **Pedroso, E., Albericio, F., Grandas, A., Giralt, E., Van Rietschoten, J., and Granier, C.,** Two new methods for the solid phase synthesis of protected peptides. Synthesis of apamin and LH-RH protected fragments, in *Proc. 16th Eur. Peptide Symp.,* Brunfeldt, K., Ed., Scriptor, Copenhagen, 1981, 334.
30. **Bennett, H. P. J., Browne, C. A., and Solomon, S.,** The use of perfluorinated carboxylic acids in the reversed phase HPLC of peptides, *J. Liq. Chromatogr.,* 3, 9, 1353, 1980.
31. **Meek, J. L.,** Prediction of peptide retention times in high pressure liquid chromatography on the basis of amino acid composition, *Proc. Natl. Acad. Sci. U.S.A.,* 77, 3, 1632, 1980.

SOMATOSTATIN

Kerstin Gröningsson and Monika Abrahamsson

INTRODUCTION

HPLC has been used for studying the purity (homogeneity) and stability of the cyclic tetradecapeptide somatostatin (Figure 1).[1-3] For purification of somatostatin and its chemical precursors preparative HPLC has been used.[4,5] Lyophilized pharmaceutical preparations have been analyzed for somatostatin in the presence of albumin[1] and lactose. The in vivo degradation of somatostatin in rat after i.v. administration has been followed by analyses of plasma extracts.[6] The biological and immunological properties of synthetic peptides may not always reflect their purity. Thus replacement of one or several amino acid residues in somatostatin by their D-enantiomers enhances its biological potency.[7] Since racemization is one of the side reactions difficult to control in peptide synthesis, the related by-products in a synthetic somatostatin could impart enhanced biological properties even to impure preparations. HPLC is nowadays considered to be a superior technique in purity control as it can separate closely related analogs including diastereoisomers. However, it may be necessary to optimize the chromatographic conditions even for every particular batch, as the impurity pattern (types of analogs) can differ significantly as a result of minor variations in the synthetic procedure.

In this chapter we shall summarize our experiences, and in some cases those of others, in the HPLC separation of somatostatin. The influence of the stationary phase and the composition of the eluent will be discussed.

EXPERIMENTAL

The pump used in our experiments was a Constametric II G pump (LDC, Riviera Beach, FL). A few experiments were performed by using LDC gradient equipment including Constametric I and II G pumps attached to an LDC Gradient Master. The column effluent was generally monitored at 210 nm, though sometimes at 280 nm, with a Spectromonitor III variable UV-detector (LDC, Riviera Beach, FL). The native fluorescence of somatostatin was monitored by a Schoeffel FS 970 Fluorescence Detector (Kratos, Schoeffel Instruments, GFR). The sample was injected with a Rheodyne 7125 loop injector (Berkeley, CA). The columns used were slurry-packed Nucleosil® $5C_8$ and $5C_{18}$, 150 × 4.6 mm (Macherey-Nagel & Co, Düren, GFR) as well as the commercially available columns Biosil® ODS-5S, 150 × 4 mm (Bio-Rad Laboratories, CA); μBondapak® alkylphenyl, 300 × 3.9 mm, 10-μm particles (Waters Associates Inc., Milford, MA); μBondapak® CN, 300 × 3.9 mm, 10-μm particles (Waters Associates Inc., Milford, MA); LiChrosorb® DIOL, 250 × 4.6 mm, 10-μm particles (HIBAR®, E. Merck, Darmstadt, GFR); PRP-1, 150 × 4.1 mm, 10-μm particles (Hamilton Bonaduz AG, Bonaduz, Switzerland). Somastostatin (SRIF) was obtained from Sempa Chemie (Paris, France) while analogs were obtained from UCB (Bruxelles, Belgium).

REVERSED-PHASE CHROMATOGRAPHY

Introduction

All successful HPLC separations of somatostatin have been performed in the reversed-phase mode.[1,2,8,9] The most commonly used stationary phases are chemically bonded hy-

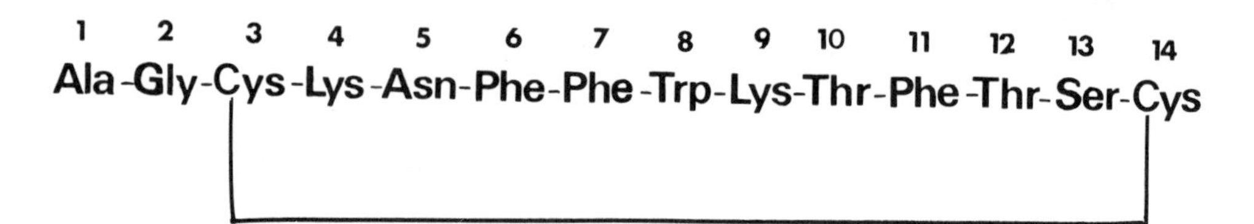

FIGURE 1. Sequence of somatostatin (SRIF).

Table 1
INTERACTION OF SOMATOSTATIN WITH DIFFERENT BONDED PHASES

Bonded phase	%,v/v,1-propanol required for $k'_{somatostatin} = 11$
Nucleosil® $5C_8$	15.5
Nucleosil® $5C_{18}$	16.5
Biosil® ODS-5S	16.5
μBondapak® CN	18.1
μBondapak® alkylphenyl	18.5
Hamilton PRP-1	$\gg$40

Note: Eluent: 1-propanol in phosphate buffer, $\mu = 0.1$. The pH of the mixtures = 4.5 ; the flow rate = 1 mℓ/min.

drocarbonaceous silica (C_8 and C_{18} materials). Bonded phases such as alkylphenyl and 3-cyanopropyl silica gels have been used only occasionally. Our studies have included all these types of stationary phases as well as the recently introduced copolymeric styrene divinylbenzene material PRP-1.

Columns

Solute-Stationary Phase Interactions

A mixture of phenols (phenol, 4-methylphenol, 3,5-dimethylphenol, anisole, and phenetole) used for testing column efficiency was run on each of the columns investigated with an eluent consisting of 60%, v/v, of methanol in water. From these tests the columns could be ranged in the following order of increasing hydrophobicity μBondapak® CN < μBondapak® alkylphenyl < Nucleosil® $5C_8$ < Nucleosil® $5C_{18}$ ≈ Biosil® ODS-5S ≪ Hamilton PRP-1. In order to investigate the interactions of somatostatin with these stationary phases somatostatin was eluted with eluents containing 1-propanol in phosphate buffer (the pH of the mixtures = 4.5). The concentration of 1-propanol was adjusted to give a capacity ratio (k′) of 11. The required concentration of 1-propanol is given in Table 1. The most hydrophobic column, Hamilton PRP-1, strongly retains somatostatin, i.e., 40%, v/v, of 1-propanol does not elute somatostatin within 2 hr. Comparing the two Nucleosil® materials, one finds as expected that the more hydrophobic C_{18} material requires a somewhat higher concentration of 1-propanol. Biosil® ODS, showing the same hydrophobic properties as Nucleosil® C_{18} in the column test, retards somatostatin to the same extent. However, the less hydrophobic materials μBondapak® CN and μBondapak® alkylphenyl require the highest concentrations of 1-propanol indicating stronger interaction with somatostatin. This clearly shows the difficulties in predicting the chromatographic behavior of somatostatin on different stationary phases as their properties, as well as the conformational and hydrophobic properties of the whole somatostatin molecule, have to be taken into consideration. Even changes in the

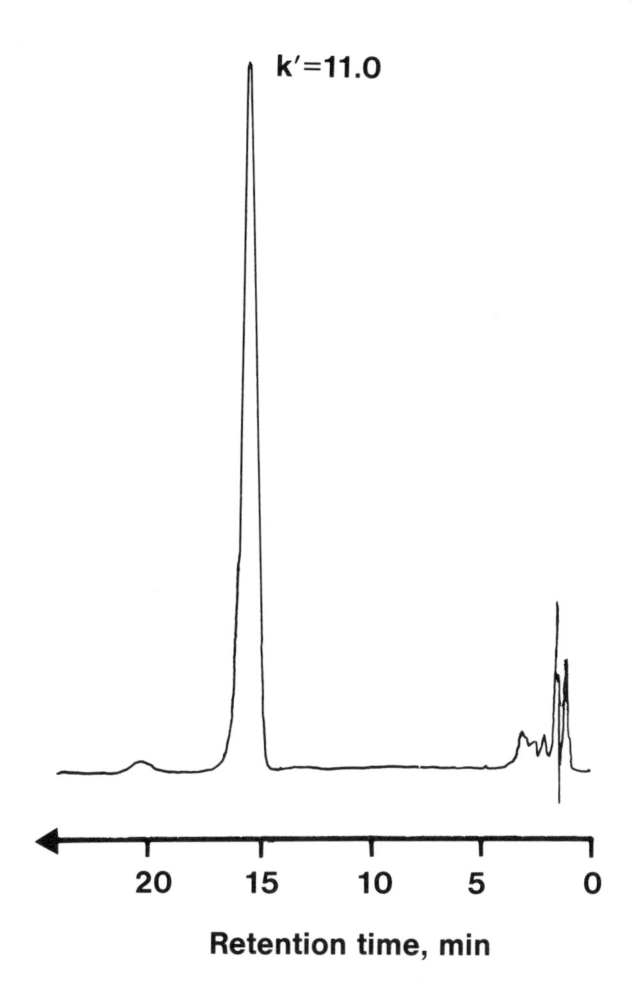

FIGURE 2. Chromatogram of somatostatin. Column: Biosil® ODS-5S, 150 × 4 mm. Eluent: 16.5%, v/v, of 1-propanol in phosphate buffer ($\mu = 0.1$). The pH of the mixture is 4.5. Detection wavelength: 210 nm (0.2 AUFS). Flow rate: 1 mℓ/min. Chart speed: 4 mm/min.

manufacturing procedure of the stationary phases can strongly influence the chromatographic behavior of somatostatin. For instance, the results obtained with the LiChrosorb® RP-8 (E. Merck) material in our previous studies[1] cannot be duplicated with the currently available LiChrosorb® RP-8 material which interacts very strongly with somatostatin. A drastic increase in the concentration of acetonitrile was necessary to elute somastatin, which appeared as a peak exhibiting extreme tailing.

The separating efficiencies (reduced HETP) obtained on the stationary phases given in Table 1 were 7 to 10 for the C_{18} materials and somewhat higher for the other materials. The peak symmetries, A_s, were about 1.0 for μBondapak® CN and μBondapak® alkylphenyl, 1.5 for the C_{18} materials and somewhat higher for the C_8-column. A chromatogram obtained on the Biosil® ODS column is given in Figure 2.

Separating Properties

The selectivity of the different columns (the Hamilton PRP-1 column is omitted) with respect to somatostatin and five analogs was tested with 1-propanol as organic modifier in

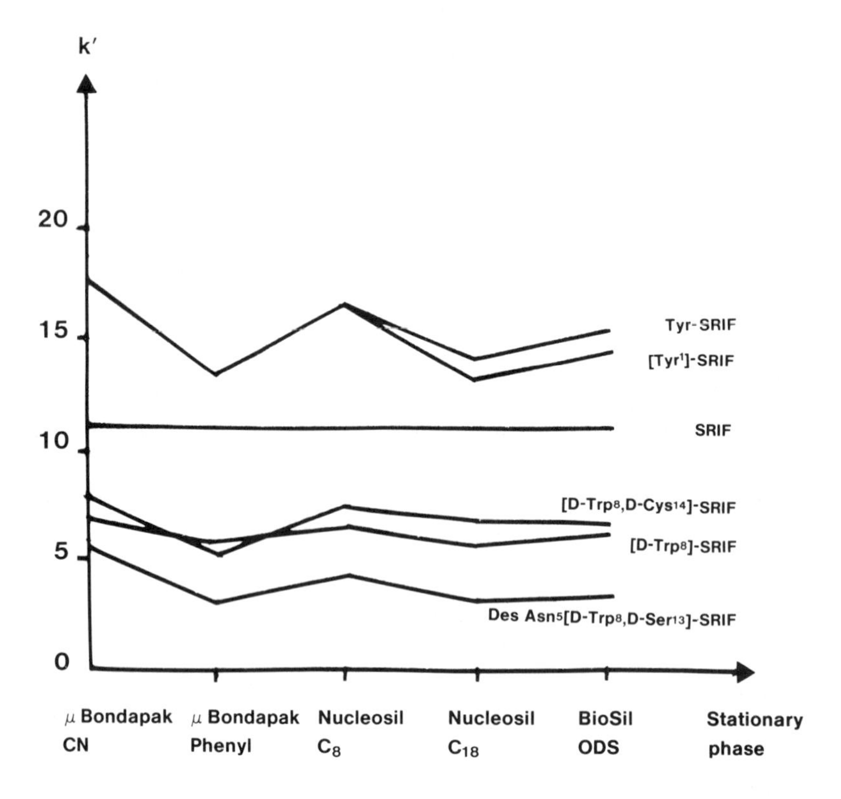

FIGURE 3. Separation of somatostatin and five analogs on different stationary phases. Eluent: 15.5 to 18.5%, v/v, (see Table 1) of 1-propanol in phosphate buffer ($\mu = 0.1$). The pH of the mixture is 4.5.

the eluent. The k' values are given in Figure 3. There are only small differences in the selectivity between the stationary phases, which all separate the analogs from somatostatin. The C_{18} materials are slightly better as there are tendencies to separation between [Tyr¹]-SRIF and Tyr-SRIF on the one hand and [D-Trp⁸]-SRIF and [D-Trp⁸, D-Cys¹⁴]-SRIF on the other hand. The resolution, R_S, is about 0.8 for the latter analog pair which is further resolved by gradient elution (see Figure 4) giving $R_S = 1.3$.

Correspondingly good selectivities have been obtained by using acetonitrile as organic modifier and a μBondapak® C_{18} column.[9] Our earlier studies[1] using LiChrosorb® RP-8 and acetonitrile as organic modifier gave successful isocratic separations between other somatostatin analogs.

In our opinion, isocratic elutions give sufficiently good resolutions between somatostatin and possible closely related peptide impurities, including diastereoisomers. However, gradient elution (with increasing concentrations of organic modifier) may be the method of choice for eluting very hydrophobic impurities such as aggregates (dimers) formed during improper storage of somatostatin.

Eluents

pH

Due to suppression of the ionization of the free amino groups in somatostatin k' increases with increasing pH.[1] A simultaneous increase in HETP is also observed, thus a pH of 2 to 5 is recommended and all our experiments have been performed at pH 4.5.

Buffers

An increase in the buffer concentration (acetate and phosphate buffers) leads to a decrease

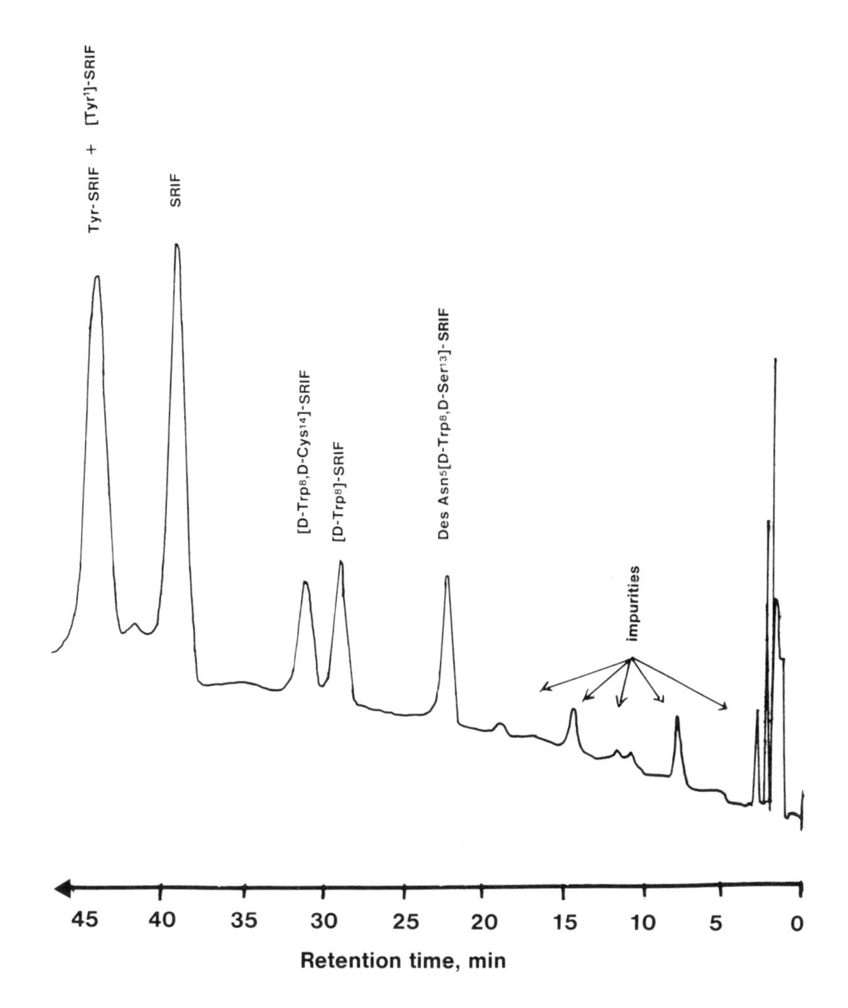

FIGURE 4. Gradient elution of somatostatin and five analogs. Column: Nucleosil® 5 C_{18}, 150 × 4.6 mm. Gradient: 10 to 17%, v/v, of 1-propanol in phosphate buffer, pH 4.5. Flow program: 0.6%/min, exponent = 0.5. Detection wavelength: 210 nm. Flow rate: 1 mℓ/min.

in the k′ values.[1] This effect is very pronounced below 0.1 mol/ℓ of phosphate and 0.2 mol/ℓ of acetate (including acetic acid). For optimum efficiency a salt concentration of at least 0.1 mol/ℓ is needed.

Due to their low UV absorbance, sodium phosphate as well as triethylammonium phosphate (TEAP) buffers have been used most extensively. A volatile buffer such as the recently introduced triethylamine formate (TEAF)[10] has been preferred when the eluted substance is collected for lyophilization or for further investigation, e.g., mass spectrometry.

Ion-Pairing Reagents

For regulation of the retention times, hydrophobic ion-pairing agents such as alkylsulfonates,[1] sodium dodecyl sulphate,[1] D(+)-camphor sulfonic acid,[1] and perfluorinated carboxylic acids[11] have been used. The retention is increased with increasing hydrophobicity and concentration of the ion-pairing anions. However, no effects on the separating efficiency and selectivity have been observed.

Organic Modifiers

Organic modifiers such as methanol, ethanol, acetonitrile, and 1-propanol have been used.

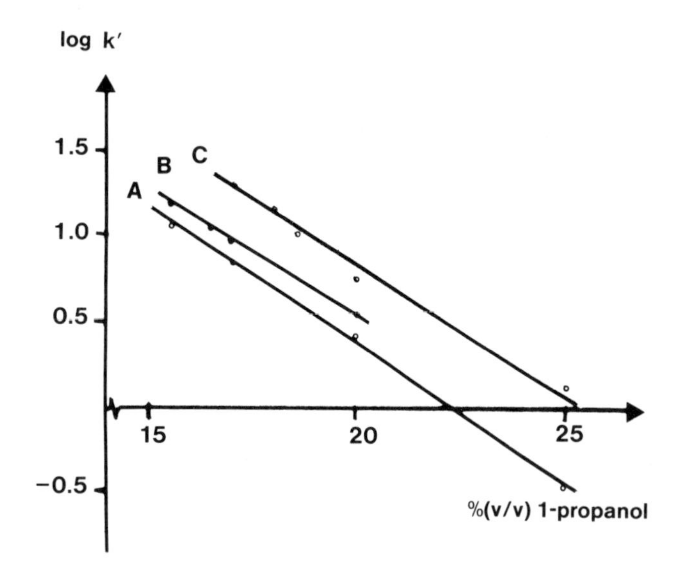

FIGURE 5. Influence of the concentration of 1-propanol on the capacity ratio of somatostatin. Column: (A) Nucleosil® 5 C_8, 150 × 4.6 mm. (B) Nucleosil® 5 C_{18}, 150 × 4.6 mm. (C) μBondapak® alkylphenyl, 300 × 3.9 mm. Eluent: 1-propanol in phosphate buffer ($\mu = 0.1$). The pH of the mixture is 4.5.

No differences regarding peak symmetry or selectivity of the system have been noticed between ethanol and methanol. Acetonitrile and 1-propanol, however, are preferred as they give better separating efficiency and selectivity. Of these two solvents, acetonitrile seems in general to give more efficient separations and should be used for the most difficult separations. However, 1-propanol, being less toxic, less expensive, and having a higher eluotropic strength than acetonitrile is sufficient for most practical separations.

The concentration of organic modifier in the eluent has a drastic effect on the k' value. The influence of the concentration of 1-propanol on different stationary phases is illustrated in Figure 5. An increase in the content of 1-propanol by 5% (e.g., from 20 to 25%) results in an approximately sixfold decrease in the k' value of somatostatin. The slopes of the lines are independent of the type of stationary phase and very similar to earlier published results with acetonitrile as organic modifier and LiChrosorb® RP-8 as stationary phase.[1] The deviations from linearity above a certain concentration of organic modifier observed by, e.g., Hearn et.al.[12] for other polypeptides have not been observed for somatostatin even at such high acetonitrile concentrations as 80%, v/v, on LiChrosorb® RP-8.

Detection

Somatostatin exhibits UV maxima at 210 nm (peptide bonds), 280 nm and 287 nm (tryptophan). The sensitivity is about ten times better at 210 nm ($\epsilon_{210} \approx 7 \times 10^4$) than at 280 nm ($\epsilon_{280} \approx 5 \times 10^3$). The native fluorescence can also be used for detection. When monitoring tryptophan fluorescence at an excitation wavelength of 290 nm and an emission wavelength of 370 nm, 7 ng (4 pmol) can be determined, the same amount as for UV detection at 210 nm. Even higher sensitivity could be achieved by fluorescence derivatization or using a wavelength lower than 210 nm.

SIZE EXCLUSION

For detection of polymeric forms of somatostatin gel filtration (size exclusion) is an

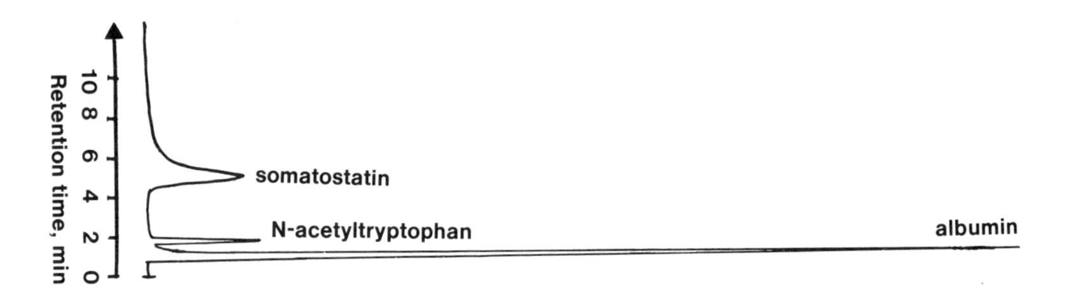

FIGURE 6. Chromatogram of somatostatin in a pharmaceutical preparation wih albumin and *N*-acetyltryptophan. Column: LiChrosorb® DIOL, 10-μm particles. Eluent: Phosphate buffer pH 7.5, $\mu = 0.1$. Flow rate: 1 mℓ/min. Detection wavelength: 210 nm (0.5 AUFS) Amount injected: 2 μg somatostatin, 4 μg of albumin, and 0.08 μg of acetyltryptophan.

alternative to reversed-phase chromatography,[2] where gradient elution is necessary to elute the strongly retained aggregates. Two phases used for size exclusion of proteins have been tried by us, namely LiChrosorb® DIOL (E. Merck) and TSK-gel 3000 SW (Toyo Soda Manufacturing Co. Ltd.). On both phases monomeric somatostatin is adsorbed (tailing peaks with $V_e < V_t$) and thus size exclusion is not the dominant separation mechanism. However, Mabuchi and Nakashi[13] have reported separations on TSK-gel using an eluent containing sodium dodecyl sulfate. In their system V_e for somatostatin is less than V_t.

Most of our studies have been performed on the DIOL-phase and efforts have been made to diminish adsorption by addition of 1-propanol to the eluent, increasing the ionic strength and changing the pH of the eluent. An eluent containing a few percent of 1-propanol and a high concentration of buffer salts (ionic strength > 0.1) is recommended. In our experience the adsorption increases with decreasing particle size of the stationary phase, thus 10-μm particles are preferred to 5-μm particles. The chromatogram given in Figure 6 illustrates how this type of stationary phase can be used for rapid and isocratic separation of compounds of such widely different size and hydrophobicity as somatostatin (1630 mol wt), albumin (66,000 mol wt), and *N*-acetyltryptophan (246 mol wt).

REFERENCES

1. **Abrahamsson, M., and Gröningsson, K.,** High-performance liquid chromatography of the tetradecapeptide somatostatin, *J. Liq. Chromatogr.*, 3, 495, 1980.
2. **Wünsch, E.,** Peptide factors: definition of purity, in *Hormone Receptors in Digestion and Nutrition*, Rosselin, G., Fromageot, P., and Bonfils, S., Eds., Elsevier/North-Holland Biomedical Press, Amsterdam, 1979, 115.
3. **Feldman, J. A., Cohn, M. L., and Blair, D.,** Neuroendocrine peptides — analysis by reversed phase high performance liquid chromatography, *J. Liq. Chromatogr.*, 1, 833, 1978.
4. **Tronquet, C., Guimbard, J-P., and Paolucci, F.,** Application of preparative hplc for the purification of somatostatin and of its chemical precursor, in *Hormone Receptors in Digestion and Nutrition*, Rosselin, G., Fromageot, P., and Bonfils, S., Eds., Elsevier/North-Holland Biomedical Press, Amsterdam, 1979, 89.
5. **Rivier, J., Spiess, J., Perrin, M., and Vale, W.,** Application of hplc in the isolation of unprotected peptides, *Chromatogr. Sci.*, 12, 223, 1979.
6. **Bennett, H. P. J., Hudson, A. M., McMartin, C., and Purdon, G. E.,** Use of octadecyl-silica for the extraction and purification of peptides in biological samples: application to the identification of circulating metabolites of corticotropin - (1-24) - tetracosapeptide and somatostatin in vivo, *Biochem. J.*, 168, 9, 1977.
7. **Rivier, J., Brown, M., and Vale, W.,** D-Trp^8-somatostatin: an analog of somatostatin more potent than the native molecule, *Biochem. Biophys. Res. Commun.*, 65, 746, 1975.

8. **Rivier, J. and Burgus, R.,** Application of reverse phase high pressure liquid chromatography to peptides, in *Biological/Biomedical Applications of Liquid Chromatography,* Vol. 10, Chromatography Science Series, Hawk, G. L., Ed., Marcel Dekker, New York, 1979, 147.
9. **Desiderio, D. M., Sabbatini, J. Z., and Stein, J. L.,** HPLC and field desorption mass spectrometry of hypothalamic oligopeptides, *Adv. Mass Spectrom.,* 8B, 1298, 1980.
10. **Desiderio, D. M. and Cunningham, M. D.,** Triethylamine formate buffer for hplc-field desorption mass spectrometry of oligopeptides, *J. Liq. Chromatogr.,* 4, 721, 1981.
11. **Bennett, H. P. J., Browne, C. A., and Solomon, S.,** The use of perfluorinated carboxylic acids in the reversed-phase hplc of peptides, *J. Liq. Chromatogr.,* 3, 1353, 1980.
12. **Hearn, M. T. W. and Grego, B.,** High performance liquid chromatography of amino acids, peptides and proteins. XXXVI. Organic solvent modifier effects in the separation of unprotected peptides by reversed phase liquid chromatography, *J. Chromatogr.,* 218, 497, 1981.
13. **Mabuchi, H. and Nakahashi, H.,** Systematic separation of medium-sized biologically active peptides by high-performance liquid chromatography, *J. Chromatogr.,* 213, 275, 1981.

SYNTHETIC β-LIPOTROPIN FRAGMENTS

J. W. van Nispen and P. S. L. Janssen

INTRODUCTION

β-Lipotropin (β-lipotropic hormone, β-LPH*) was first discovered and isolated from sheep pituitary glands and later also obtained from bovine, porcine, and human pituitaries.[1] The primary structure was determined and the hormone shown to be a straight-chain peptide of 91 amino acid residues. However, in a recent report Li and Chung provide evidence for revised sequences in the N-terminal parts of the human and bovine β-LPHs.[2] The primary structure of β_h-LPH is now supposed to consist of 89 and β_p-LPH of 93 amino acid residues.[2] Minor changes in the N-terminal part of β_h-LPH were reported later by Hsi et al.[3] although the length of the chain did not change. The latest version of the amino acid sequence of β_h-LPH, is shown in Figure 1. In addition to in vivo and in vitro lipolytic effects of β- and γ-LPH,[1] (γ_h-LPH comprises the 1-56 sequence of β_h-LPH, see Figure 1), melanocyte-stimulating activity was reported[1] for these peptides. The latter finding is not surprising since the sequence 39-56 is identical with that of β-MSH. Bradbury et al.[5] showed that processing of β-LPH in the pituitary gives rise to the fragments 1-56, 1-36, 39-56, 59-85, and 59-89. This latter peptide, i.e., the C-terminal 31-peptide of β-LPH, β-endorphin or C-fragment, was isolated from pituitaries of several species.[6] β-Endorphin was shown to possess very low lipolytic activity but appeared to have a high affinity for opiate-binding sites in the brain. Many studies dealing with the spectrum of activities of β-endorphin, including behavioral properties, have been reported (for reviews see Reference 7). Further processing of β-endorphin by enzymes generates naturally occurring fragments, the N-terminal 16- and 17-peptides, i.e., α- and γ-endorphin, respectively.[8] Figure 2 shows the amino acid sequences of the endorphins and also gives the species differences. Literature data on lipotropin fragments are based on the original numbering system (91 residues); in order to avoid confusion we will use the revised numbering system only for fragments in the N-terminal part. The C-terminal 31-peptide β-endorphin will be the basis for all other (the majority) fragments. [Thus the β_h-endorphin fragment 1-16, i.e., α-endorphin, corresponds to β_h-LPH-(59-74).]

Several applications of the HPLC technique in the β-lipotropin field have been reported:

1. Isolation of β- and γ-LPH and the endorphins from tissue extracts
2. The use of HPLC as an analytical purity criterion for each synthetic (or natural) β-lipotropin fragment
3. Characterization of isolated lipotropin fragments has been carried out via separation by HPLC of tryptic digests (using volatile mobile phases) and the subsequent analysis of the isolated peptide(s)
4. Enzymatic conversion of β-endorphin and the smaller 2-17 fragment using crude membrane preparations of different sources has been followed by HPLC; well-defined synthetic peptides have been used as reference compounds

* Standard abbreviations are used for amino acids [IUPAC-IUB Commission on Biochemical Nomenclature, *Eur. J. Biochem.*, 53, 1, 1975; 27, 201, 1972]. Other abbreviations are: Et_3N, triethylamine; β-LPH, β-lipotropic hormone; lin. grad. linear gradient; SDS, sodium dodecyl sulfate; TBAP, tetrabutylammonium phosphate; TEAA, triethylammonium acetate; TEAF, triethylammonium formate; TEAP triethylammonium phosphate; TMAH, tetramethylammonium hydroxide; TMAP, tetramethylammonium phosphate. The superscripts used to indicate the species are abbreviated as follows: b, bovine; c, camel; e, equine, h, human; p, porcine; r, rat; s, sheep.

H-Glu^{1}-Leu-Thr-Gly-Gln-Arg-Leu-Arg-Glu-Gly^{10}-Asp-Gly-Pro-Asp-Gly-

Pro-Ala-Asp-Asp-Gly^{20}-Ala-Gly-Ala-Gln-Ala-Asp-Leu-Glu-His-Ser^{30}-

Leu-Leu-Val-Ala-Ala-Glu-Lys-Lys-Asp-Glu^{40}-Gly-Pro-Tyr-Arg-Met-Glu-

His-Phe-Arg-Trp^{50}-Gly-Ser-Pro-Pro-Lys-Asp-Lys-Arg-Tyr-Gly^{60}-Gly-

Phe-Met-Thr-Ser-Glu-Lys-Ser-Gln-Thr^{70}-Pro-Leu-Val-Thr-Leu-Phe-Lys-

Asn-Ala-Ile^{80}-Ile-Lys-Asn-Ala-Tyr-Lys-Lys-Gly-Glu^{89}-OH

FIGURE 1. Amino acid sequence of human β-LPH as recently proposed by Hsi et al.[3] γ-LPH comprises the N-terminal 56-peptide while β-endorphin is the C-terminal 31-peptide. Human β-MSH, orginally supposed to consist of 22 amino acid residues is now considered to be of the same length as β-MSH of all mammalians[4]: 18 residues, sequence β_h-LPH-(39-56).

H-Tyr^{1}-Gly-Gly-Phe-Met-Thr-Ser-Glu-Lys-Ser-Gln-Thr-Pro-Leu-Val-

Thr^{16}-Leu^{17}-Phe-Lys-Asn-Ala-Ile-Ile-Lys-Asn-Ala-Tyr-Lys-Lys-Gly-

Glu^{31}-OH

FIGURE 2. Amino acid sequence of human β-endorphin. Species differences occur at the following positions: bovine, camel, rat, and sheep His^{27} and Gln^{31}; porcine Val^{22}, His^{27} and Gln^{31}, equine Ser^{6}, His^{27}, and Gln^{31}. Sequence 1-16 is α-endorphin, 1-17 is γ-endorphin, and 1-19 is δ-endorphin.

In this practical report we will provide data on the HPLC systems that have been used to assess the purity of several *synthetic* β-lipotropin fragments, but in some instances will also give data on systems used for other purposes, e.g., for isolation or reference purposes. The peptides will be divided into four groups namely β-LPH, N-terminal fragments of β-LPH (covering γ-LPH and β-MSH), C-terminal fragment of β-LPH, i.e., β-endorphin and closely related peptides, and finally short peptides within the β-endorphin sequence. Both literature data and unpublished work are summarized in tables; no patent literature has been included. The following aspects are covered in the tables: peptide(s), stationary phase, mobile phase, elution mode (including analysis time, flow rate, and temperature if given), detection method, and a column for remarks. Particle size of the HPLC columns used, if available, is included in the tables except for the μBondapak® columns since these are only available in 10-μm particle size. Dimensions of the columns are not specified in the tables if their i.d. is 3.9 to 4.6 mm and their length 20 to 30 cm; only deviating dimensions will be given.

β-LPH

In 1978 the synthesis of sheep β-LPH was reported but without HPLC analysis of the end-product. Very recently, the same group synthesized the human 89-peptide and did analyze the product by HPLC.[78] HPLC has been used however, for the isolation of β-LPH from natural sources or as a final purification step in such an operation. The same type of stationary and mobile phases as were used for the purity assessment of small peptides have been used (see Table 1). We have also included some data on the separation of (natural) β-

Table 1
HPLC DATA OF β-LIPOTROPINS

Peptide(s)[a]	Stationary phase	Mobile phase	Elution mode	Detection	Remarks	Ref.
β_h-LPH	μBondapak® C_{18}	(A) 0.02 *M* TEAF, pH 3.0 (B) 2-Propanol	Lin. grad. 25—50% B, 75 min. 0.5 mℓ/min.	235 nm	Higher purity than with other techniques	3
β-LPH	μBondapak® CN	(A) 0.1% TFA (B) CH_3CN	Lin. grad. 30—35% B, 20 min	210 nm	Final purification	79
		(A) 0.01 *M* F_3CCOOH (B) CH_3CN	Lin. grad. 42—48% B, 10 min	210 nm	Re-purification of previous product	79
		(A) 0.1% TFA (B) CH_3CN	Lin. grad. 12—80% B, 60 min	210 nm	Main component > 90% retention time 24 min	79
β_h-LPH	μBondapak® C_{18}	(A) 0.1 % TFA (B) 2-Propanol	Lin. grad. 20—60% B, 60 min 1 mℓ/min	235 nm	Final purification	80
Synthetic β_h-LPH	Allteck Vydac 201-TP	(A) 0.1% TFA (B) 2-Propanol	Lin. grad. 10—50% B, 30 min	280 nm	Purity check	78
β_r-LPH	Lichrosorb® RP-18 (3.2 × 250 mm)	(A) 1*M* pyridine-0.5 *M* HOAc (B) 1-propanol	Lin. grad. 0—40% B, 45 min Isocratic 40% B, 15 min, 0.83 mℓ/min; 25°C	Fluorescamine	As final purification step	10
β_h-LPH, synthetic β_h-and α-endorphin	Hypersil ODS (5 μm) (5 × 100 mm)	(A) 0.1 *M* NaH_2PO_4-H_3PO_4, pH 2.1 (B) CH_3CN	Lin. grad. 10—40% B, 45 min, 1.0 mℓ/min; ambient temp.	225 nm	Partial separation of β_h-LPH and β_h-endorphin at room temp.; baseline separation at 45°C	11
β_s-LPH, γ_s-LPH and β_s-endorphin	(I) μBondapak® CN (II) μBondapak® C_{18}	(A) 0.02 *M* TEAP, pH 3.0 (B) CH_3CN-0.2 *N* TEAP (pH 3) (90—10, v/v)	(I) Isocratic 0% CH_3CN, 6 min; lin. grad. 0—31% CH_3CN, 46 min, 1 mℓ/min: room temp. (II) Lin. grad. 25—37% CH_3CN, 30 min, 1 mℓ/min; room temp.	210 nm	(I) No separation of β_s- and γ_s-LPH (II) Good separation of β_s- and γ_s-LPH, partial separation of γ_s-LPH and β_s-endorphin	12

Table 1 (continued)
HPLC DATA OF β-LIPOTROPINS

Peptide(s)[a]	Stationary phase	Mobile phase	Elution mode	Detection	Remarks	Ref.
β-LPH, γ-LPH and β-endorphin	μBondapak® C_{18}	(A) 0.08% TFA (B) CH_3CN	Lin. grad. 3.5—49% B, 20 min, 1 mℓ/min; ambient temp.	206 nm	Retention times 21.4, 19.5, and 22.4 min, respectively; separation?	13
$β_s$-LPH and pGlu[1]-$β_s$-LPH	C_8 support (porosity 330 Å)	(A) 1 *M* pyridine — 0.5 *M* HOAc (B) 1-propanol	Isocratic 15% B, 30 min	Fluorescamine	No separation	14

[a] The β-lipotropins and other peptides mentioned are isolated from pituitary glands from the indicated species (see abbreviations). When no subscript is given, the source of β-LPH is not mentioned.

LPH, γ-LPH, and β-endorphin. It is noteworthy that separation of β_s-LPH and [pGlu[1]]β_s-LPH could not be effected by HPLC (C_8 column) while partition chromatography on agarose results in partial separation of these two closely related peptides.[14]

N-TERMINAL FRAGMENTS OF β-LPH

Only very limited HPLC data are available on γ-LPH and fragments including β-MSH. On the one hand recent interest in β-LPH fragments has been restricted to the C-terminal 31-peptide (β-endorphin) and on the other hand syntheses of the long-known β-MSH were nearly all performed before the introduction of HPLC. In Table 2 we have summarized the limited information from the literature and results from our own, unpublished, work.

C-TERMINAL FRAGMENTS OF β-LPH AND CLOSELY RELATED PEPTIDES

Several syntheses of the C-terminal 31-peptide of β-LPH, β-endorphin, as well as closely related peptides have been reported to date. Available HPLC data are presented in Tables 3 to 5; Table 3 contains information only on synthetic β-endorphin itself, Table 4 on separations of β-endorphin (different species) from acetylated β-endorphin and β-endorphin peptides missing one amino acid residue, and Table 5 on the separation of β-endorphin from γ- and α-endorphin.

Synthetic β-Endorphin

Both the solid-phase technique and the fragment condensation approach in solution have been used for the synthesis of β-endorphin of several species. As stationary phase mainly μ Bondapak® or Lichrosorb® columns have been used. A variety of mobile phases was used containing either an acid (TFA, HCOOH, HCl) or a buffer (NH_4OAc, Et_3N acetate, formate or phosphate, TMAP, pyridine buffers). The addition of a small amount of (an "organic modifier" like) Et_3N to an NH_4OAc-CH_3CN mobile phase system resulted in improvement of the elution pattern.[22] An even larger difference in separation capacity can be found if totally different buffer systems are used. For example, analyzing a synthetic β_h-endorphin preparation, we have compared the widely used NH_4OAc buffer (pH 4.15) in a concave gradient with MeOH (cf. Reference 33) with a linear TMAP-MeOH gradient (both on a μBondapak® C_{18}-column). A considerably better resolution of the peak of the main component from the peaks of the adjacent by-products was found in the latter system. A main component of 91.0% was found in this system, whereas an apparent purity of 97.5% was observed in the NH_4OAc system. In addition, we investigated the effect of column temperature on the resolution capacity by analyzing a solid-phase product (β_p-endorphin, gift to Dr. Rigter) at 20 and 40°C. A marked influence was found: a main component of 86.5% at 20°C and 77.4% at 40°C, while nearly no change in pattern for the β_h-endorphin product, synthesized by the fragment condensation approach, was noticed.[17] Differences in stationary phase can also have a marked influence on the elution profile as was shown with synthetic[19] and natural[31] β-endorphin (Table 3); the cation exchanger Partisil®-SCX was clearly less suitable than the octadecyl columns.

β-Endorphin and Closely Related Peptides

In one of the pioneering papers in the field of HPLC of peptides, Rivier showed the high resolving power of the technique by obtaining a partial separation of human, porcine, and ovine β-endorphin,[34] despite the difference of only one CH_3 group between the latter two peptides. In contrast to this is the inability of Yamashiro and Li[14] to separate β-endorphin from several analogs missing one amino acid residue in the chain (see Table 4). Using Rivier's system but on a C_{18} column, we were unable to separate β_h-endorphin from its des-

Table 2
HPLC DATA OF N-TERMINAL FRAGMENTS OF β-LPH (i.e., γ-LPH AND β-MSH FRAGMENTS)

Peptide(s)	Stationary phase	Mobile phase	Elution mode	Detection	Remarks	Ref.
Natural γ_c-LPH	Partisil®-10-ODS-2	(A) 1 *M* pyridine — 0.5 *M* HOAc (B) 1-propanol	Lin. grad. 10—25% B, 60 min, 0.5 mℓ/min.	Fluorescamine	Final purification step	16
γ_h-LPH[a]	Syn-Chropak® RP-P (30 nm pore size)	(A) 0.1% TFA (B) 2-Propanol	Lin. grad. 0-40% B, 30 min 0.5 mℓ/min	?	Retention time 24.2 min	81
H-Lys-Lys-Asp-Ser-Gly-Pro-Tyr-OH, β_s-LPH-(39—45)	Hypersil-ODS (5 μm) (5 × 100 mm)	(A) 0.1 *M* NaH_2PO_4-H_3PO_4, pH 2.1 (B) CH_3CN	Lin. grad. 10—40% B, 45 min, 1.0 mℓ/min; ambient temp.	225 nm	Retention time ca. 7 min	11
[Tyr[9]]β_p-MSH-(9—18)	μBondapak® C_{18}	(A) 0.01 *M* NH_4OAc, pH 4.0 (B) 2-propanol	Lin. grad. 5—50% B, 25 min, 1.0 mℓ/min; 25°C	220 nm	Retention times of synthetic and natural material the same	15
β_h-LPH-(45—58)[b]	μBondapak® C_{18}	(A) MeOH-H_2O (25—75, v/v) (B) MeOH-H_2O(80—20, v/v) both with 0.05 *M* TMAH and H_3PO_4 till pH 2.8	Lin. grad. 0—70% B, 25 min, isocratic 70% B, 5 min, 2.0 mℓ/min; 20°C	210 nm	Main component 94.0%; bad separation on Nucleosil® 10 C_{18} column	17
β_h-LPH-(45—56)[b]	μBondapak® C_{18}	As for fragment 45—58	As for fragment 45—58	210 nm	Main component 95.8%; bad separation on Nucleosil® 10 C_{18} column	17
β_h-LPH(52—58)[b]	μBondapak® C_{18}	As for fragment 45—58	As for fragment 45—58	210 nm	Main component 94.4%; nearly same elution profile on Nucleosil® 10 C_{18} column	17

[a] The sequence according to the incorrect primary structure (2) was synthesized.
[b] The numbering system for the revised sequence of human β-LPH is used (see Figure 1).

Table 3
HPLC DATA OF SYNTHETIC β-ENDORPHIN

Peptide	Stationary phase	Mobile phase	Elution mode	Detection	Remarks	Ref.
β_h-endorphin	μBondapak® C_{18}	(A) 0.01 *M* NH_4OAc, pH 4.0 (B) CH_3CN containing 1% *N*-ethylmorpholine	Isocratic, 55% B, 20 min	220 nm	Not homogeneous	18
β_h-endorphin	(I) Partisil®-SCX[a]	(A) 0.5 *M* NH_4OAc, pH 6.0 (B) 1.0 *M* NH_4OAc — 10% CH_3CN	(I) Lin. grad.	278 and 230 nm	(I) Single symmetrical peak but in (II) a main component of only 85%	19
	(II) μBondapak® C_{18}	(A) 0.01 *M* NH_4OAc, pH 4.5 (B) CH_3CN	(II) Lin. grad. 5—60% B			
β_c-endorphin	μBondapak® C_{18}	(A) 10 m*M* HCOOH (B) MeOH	Isocratic 50% B, 13 min, 1 mℓ/min	254 nm	Comparison with isolated material; small amounts not recovered (see also Reference 21)	20
β_c-endorphin	μBondapak® C_{18}	(I)(A) 5 m*M* TFA, pH 2.5 (B) CH_3CN (II)(A) 10 m*M* HCOOH (B) CH_3CN	(I) Lin. grad. 30—50% B, 10 min, 2 mℓ/min; 25°C (II) Lin. grad. 30—70% B, 15 min, 2 mℓ/min; 25°C	254 nm 254 nm	TFA more effective to decrease nonspecific adsorption; gives higher recovery and shorter elution time	21
β_h-endorphin	Lichrosorb® RP-18	(I)(A) 0.01 *M* NH_4OAc, pH 4 (B) CH_3CN (II)(A) 0.01 *M* NH_4OAc, pH 4 (B) CH_3CN containing 0.1% Et_3N (III)(A) 0.01 *M* NH_4OAc, pH 4 (B) isopropanol	(I) Lin. grad. 30—50% B, 15 min, 1.5 mℓ/min (II) Isocratic 41% B, 15 min, 1.5 mℓ/min (III) Lin. grad. 25—30% B, 10 min, 1.5 mℓ/min	220 nm 220 nm 220 nm	Broad peak, considerable improvement of elution pattern by 0.1% Et_3N; not homogeneous. In (III) partial separation from [Leu^5]- and [$Phe(Cl)^4$]-β_h-endorphin; coelutes with D-Ala^2-analog	22

Table 3 (continued)
HPLC DATA OF SYNTHETIC β-ENDORPHIN

Peptide	Stationary phase	Mobile phase	Elution mode	Detection	Remarks	Ref.
β_h-endorphin	μBondapak® C_{18}	(A) 0.01 *M* HCl (B) CH_3CN	Isocratic 29% B, 12 min, 2 mℓ/min	210 nm	Analytic run after purification on Lichroprep® RP-8 Lobar using CH_3CN-0.01 *N* HCl (stepwise and lin. grad.; 254 nm)	23
β_h-endorphin	Lichrosorb® RP-8 (5 μm)	(A) 0.1 *M* NH_4OAc (B) CH_3CN	Isocratic 33.3% B, 1.0 mℓ/min	280 nm	Retention time 7.4 min; trace of Met S-oxide analog	24
β_h-endorphin	μBondapak® C_{18}	(A) MeOH-H_2O = 25—75 (B) MeOH-H_2O = 80—20 both with 0.05 *M* TMAH and H_3PO_4 till pH 3.0	Lin. grad. 30—100% B, 45 min, isocratic 100% B, 15 min; 2 mℓ/min; 20°C	210 nm	Main component 91.0%; see also in text	25
β_h-endorphin; β_c-endorphin	μBondapak® Phenyl	(A) 0.01 *M* NH_4OAc, pH 4.5 (B) CH_3CN	Isocratic 40% B or 45% B, 15 min, 1 mℓ/min; ambient temp.	280 and 254 nm	Difference in capacity factors for the two species and for the two mobile phases	26
β_h-endorphin	ES Industries RP-8 (4.6 × 150 mm)	Pyridine -HOAc-CH_3CN-isopropanol-water	Isocratic, ratio 5.9—1.9—13—13—66.2, by volume, in 23 min; 1 mℓ/min	*o*-phthalaldehyde	Symmetrial peak, K′ = 2.8; see also Reference 23	27
β_h-endorphin	Lichrosorb® RP-8 or 18	(A) 0.125 *M* pyridine formate, pH 3.0 (B) 1.0 *M* pyridine acetate (pH 5.5) and 1-propanol (40—60, v/v)	Lin. grad. 0—50% B, 36 min; 0.7 mℓ/min; ambient temp.	Fluorescamine	Comparison actual and calculated elution points	28
β_h-endorphin	Lichroprep® RP-18	(A) 5% Et_3N-acetate, pH 3.9 (B) CH_3CN	Isocratic 30% B	?	Preparative scale	29

β_h-endorphin	Ultrasphere® ODS	(A) 0.12 *M* H_3PO_4 with Et_3N till pH 3.0 (B) 70% CH_3CN-30%A	Isocratic, 15% CH_3CN 10 min, stepwise grad. 15—25% CH_3CN 2 min, 25—40% CH_3CN 30 min; isocratic 40% CH_3CN 10 min, 0.5 mℓ/min	210 nm	Contains Met(O) analog; comparison with isolated radioactive material	30
Natural β_r-endorphin[b]	Lichrosorb® RP-18	(A) 1 *M* pyridine — 0.5 *M* HOAc (B) 1-propanol	Lin. grad. 0—40% B, 120 min, 0.77 mℓ/min; 25°C	Fluorescamine	For final purification; better resolution then on Partisil®-SCX	31

[a] A strong cation (strongly acidic) exchanger, in which benzene sulfonic acid groups are Si-O-Si bonded to the surface hydroxyls of Partisil®.
[b] For comparison, natural rat β-endorphin has also been included in this list. Equine β-endorphin has been purified using the same system.[32]

Table 4
HPLC DATA OF SEPARATIONS OF β-ENDORPHIN AND CLOSELY RELATED PEPTIDES

Peptides	Stationary phase	Mobile phase	Elution mode	Detection	Remarks	Ref.
β_h-, β_p- and β_s-endorphin	μBondapak® CN	(I)(A) 0.05 *M* H_3PO_4 to pH 7.50 with NaOH (B) CH_3CN	(I) Isocratic 32% B, 26 min, 2 mℓ/min.	210 nm	(I) Elution order β_h-, β_p-, β_s- endorphin; β_p- and β_s-endorphin partially separated; conditions detrimental for column	34
		(II)(A) 0.25 *N* H_3PO_4-Et_3N till pH 3 (B) CH_3CN	(II) Isocratic 17.4% B, 14 min, Lin. grad. 17.4—24% B, 3 min, isocratic 24% B 15 min. 1.5 mℓ/min		(II) Slightly better resolution; elution order β_p-, β_s-, β_h-endorphin	
β_h-, $[Tyr(I_2)^1]\beta_h$-, $[Tyr(I_2)^{27}]\beta_h$- and $[Tyr(I_2)^{1,27}]$ β_h-endorphin	Partisil® 10 ODS-2 (10 μm)	(A) 1.0 *M* pyridine — 0.5 *M* HOAc (B) 1-propanol	Lin. grad. 0—40% B, 60 min, 0.5 mℓ/min	Fluorescamine	Rel. fluorescence ca. 1.00, 0.55, 0.50 and 0.20; retention times 58, 62, 62 and 71 min, respectively	35

Table 4 (continued)
HPLC DATA OF SEPARATIONS OF β-ENDORPHIN AND CLOSELY RELATED PEPTIDES

Peptides	Stationary phase	Mobile phase	Elution mode	Detection	Remarks	Ref.
β_h-, [Arg[0]]β_h-, β_c- and [MeAla[2]]β_h-endorphin	μBondapak® Phenyl (10 μm)	(A) 0.01 *M* NH_4OAc, pH 4.5 (B) CH_3CN	Isocratic 40 or 45% B, 15 min, 1 mℓ/min; ambient temp.	280 and 254 nm	Capacity factors increase in order given at both CH_3CN concentrations	26
β_c-endorphin and its des-Gly[2], des-Ser[7]-, des-Gln[11]- and des-Asn[20] analogs	Partisil® ODS-2 (10 μm)	(A) 1 *M* pyridine — 0.5 *M* HOAc (B) 1-propanol	Isocratic 15% B, 30 min	Fluorescamine	No separation of β_c-endorphin and these analogs in contrast to Sephadex® G-50 partition chromatography	14
β_c-, N^α-Ac-β_c-endorphin, β_c-endorphin-(1—27) and its N^α-Ac analog	Ultrasphere® ODS (5 μm)	(A) 50 m*M* Na phosphate, pH 2.7 with 0.05% Et_3N (B) CH_3CN	Lin. grad. 27.5—45% B, 23 min, isocratic 45% B, 3 min; 1 mℓ/min	210 nm	Partial separation of the four compounds	36
β_h-, N^α-Ac-β_h-endorphin, β_h-endorphin-(1—27) and its N^α-Ac-analog	Ultrasphere® ODS (5 μm)	(A) 0.05 *M* HCOOH-Et_3N, pH 3.0 (B) CH_3CN	Stepwise grad., 1 mℓ/min	?	Reference compounds for labeling studies; last compound separable	37
β_h-endorphin-(1-28)	Partisil® ODS-2 (10μm)	(A) 1 *M* pyridine-0.5 *M* HOAc (B) 1-Propanol	Lin. grad. 10-30% B 0.5 mℓ/min	Fluorescamine	Retention time 40.5 min	82

Table 5
HPLC DATA OF SEPARATIONS OF β-, γ-, AND α-ENDORPHIN, AND SMALLER FRAGMENTS

Peptides	Stationary phase	Mobile phase	Elution mode	Detection	Remarks	Ref.
β_h-, γ-, α-endorphin and fragments	μBondapak® C_{18}	(A) 0.01 *M* NH_4OAc, pH 4.15 (B) MeOH	Concave grad. 30—75% B, 45 min, 2 mℓ/min; ambient temperature	210 nm	Separation of 14 peptides; no separation of 1-31 and 2-31, of 1-7 and 1-9, and 2-9 and 10-16, retention times are given	33, 84
β_h-, [Leu^5]β_h-, γ-, α-endorphin, Met- and Leu-enkephalin	Lichrosorb® C_{18} (10μm)	(A) 0.5 *M* HCOOH-pyridine, pH 4.0 (contains 0.001% pentachlorophenol) (B) 1-propanol	Lin. grad. 0—20% B, 36 min, isocratic 20% B, 14 min	Fluorescamine	Baseline separation; EM Lichrosorb® C_{18} and μBondapak® C_{18} were equally effective	39
β-, γ-, α-endorphin and Met- and Leu-enkephalin	μBondapak® C_{18}	(A) 5% HOAc (B) 1-propanol	Lin. grad. 10—40% B, 20 min, 1.0 mℓ/min; room temp.	280 nm	Bad peak performance	40
[^{125}I]β-, γ-, and α-endorphin	Sep-Pak® RP C_{18} cartridge	(A) 0.5 *M* HCOOH-pyridine, pH 4.0 (B) 1-propanol	Stepwise grad. 0-10-20-30% B in 10 min, 2.5 mℓ/min	Fluorescamine and cpm	Baseline separation; no separation of γ- and des-Tyr^1-γ-endorphin nor of β-endorphin and β-LPH with this system	41
β_h-, γ-, α-endorphin and their des-Tyr^1-analogues	Nucleosil® 5 C_{18}	(A) 0.5 *M* Na_2HPO_4 + 0.125 *M* Na_2SO_4, pH to 2.5 with H_3PO_4 (B) H_2O (C) CH_3CN	Lin. grad. 20-65-15 to 20-40-40 in 75 min, 1.0 mℓ/min; 40°C	210 nm	Good separation, see Figure 3; gradient of 20-60-20 to 20-40-40 in 60 min gives similar results	17, 83

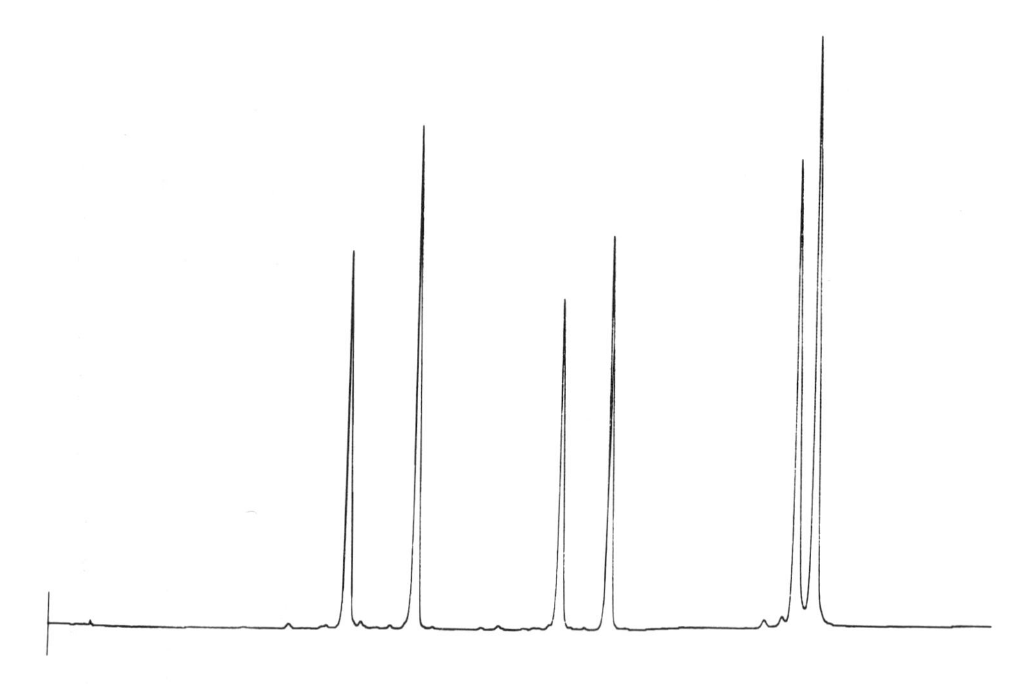

FIGURE 3. Separation of β_h-endorphin and five fragments. From left to right (with increasing retention time): 2-16, 1-16, 2-17, 1-17, 2-31, and 1-31. For conditions see Table 5. (Apparatus: Spectra Physics SP 8000).

Tyr^1-analogue (sequence 2—31); using an NH_4OAc/MeOH gradient no separation was achieved either.[33] With our TMAP system and two Nucleosil® 10 C_{18}-columns in series, temperature 35°C, a reasonable separation could be obtained. Under completely different conditions (Nucleosil® 5 C_{18} and a Na_2HPO_4-Na_2SO_4/ CH_3CN gradient at 40°C; see Figure 3) a clear-cut separation was found. A further, albeit slight improvement was obtained with an Ultrasphere 5 C-18 column as stationary phase.[83]

In Table 4 data are collected from some papers where the (attempted) separation of β-endorphin from closely related substances is described. Separation of synthetic β_h-endorphin from completely unrelated substances has also been described. Here we would only like to mention the paper of Bennett et al.[38] which describes a comparison of the efficacy of several (volatile) perfluorinated carboxylic acids as hydrophobic counter-ions in the mobile phase (CH_3CN gradient in 0.01 *M* aqueous acid; μBondapak® C_{18}-column). Varying retention orders and separation profiles of β_h-endorphin and bovine insulin, human calcitonin, and $ACTH_h$ (1—39) were observed.[38]

Separation of β-Endorphin from γ- and α-Endorphin and Smaller Fragments

β-Endorphin biotransformation in the brain results in the formation of γ- and α-type endorphins. HPLC was used to separate the peptide mixtures obtained after incubation of the 31-peptide with enzymes under different conditions.[42,84] Synthetic peptides were used as reference compounds in these studies.[33] HPLC in combination with specific radioimmunoassay (RIA) systems is also being used to study the levels of β-endorphin and its fragments in pituitary and brain tissue (see References 43 and 85).

The amino acid composition of isolated rat β-endorphin was found to be identical to that of the already known camel β-endorphin.[44] The identity of the two endorphins could be confirmed by comparison of the HPLC patterns of the tryptic digests of both peptides as well as from their mixture.[44] (Lichrosorb® RP-18 column and a pyridine-acetic acid/ 1-propanol mobile phase). Very recently, Liotta et al.[30] showed that a β-endorphin-like peptide synthesized in placental cells was identical with synthetic β_h-endorphin by performing HPLC of the products and their tryptic digests. Two different systems (Ultrasphere ODS, 5 μm, with 0.01 *M* TFA and CH_3CN, 210 nm, and Lichrosorb® RP-18, 10 μm, with a 0.015 *M* NH_4OAc-isopropanol gradient, 220 nm) were used for comparison of the tryptic peptide maps of synthetic β_h-endorphin and purified ^3H-labeled placental β-endorphin-like peptide.[30] The marker peptides, β-endorphin fragments 1-9, 10-19, 20-24, 25-28, and 29-31 were obtained by trypsin treatment and identified by amino acid analysis of the HPLC-resolved fragments.

The separation of the 16-peptide α-endorphin from the 17-peptide γ-endorphin (extended with an apolar leucine residue at the C-terminus) is possible in all systems mentioned in Table 5, the retention time of the latter being increased. The difference in retention time is much smaller if one compares the peptide pairs 1-16 and 2-16 and 1-17 and 2-17.[33] (The Sep-Pak® RP C-18/HCOOH-pyridine-1-propanol system cannot separate the 1-17 and 2-17 peptides[41]). Separation of the longer peptides 1-31 and 2-31 can only be effected using very critical conditions[83] (see also previous section).

SHORT PEPTIDES WITHIN THE β-ENDORPHIN SEQUENCE

Table 6 shows the HPLC data for synthetic α-endorphin; in addition, some separations of α-endorphin from mostly unrelated peptides are given. As one can see, in general the same type of stationary and mobile phases as mentioned before have been used.

Only a few papers reporting HPLC data of synthetic γ-endorphin have been published; these data are collected in Table 7.

Wilson et al.[28] have compared the predicted elution points (based on the calculated hydrophobicity of the peptide according to Rekker[54]) with those measured for 96 peptides among which were β_h-endorphin, α-endorphin, γ-endorphin, and des-Tyr^1-γ-endorphin. A similar study was performed by Meek and co-workers on 100 peptides including α-endorphin.[49] For peptides up to 18 to 20 amino acid residues, both groups found in general a good correlation between the actual and the calculated retention times.

Other sequences within the C-terminal part of β-LPH have been synthesized as well and analyzed by HPLC. Details of these separations are given in Table 8; the numbering system used there is based on β-endorphin. The sequence 1-5, called Met-enkephalin, had been synthesized several times, already before the introduction of HPLC into peptide chemistry. Separation of Met-enkephalin from many other peptides by HPLC has been reported; no separate table with these data will be presented since most of the papers have been referred to in earlier tables (see References 11, 48, 51 to 53).

In addition to the peptides mentioned in Table 8 various other sequences have been synthesized and analyzed, including the fragments 2-19, 2-18, 2-15, 2-13, 2-11, 2-8, 2-7, 2-6, 2-5, 14-16, 4-9, 5-9, and 6-9.[61,86] Their HPLC profile was established using a μBondapak® C_{18} or a Nucleosil® 10 C_{18} column and a mobile phase based on TMAP and MeOH gradients (cf. References 25, 47, 55, 56, and 86). The same holds for several other β_h-endorphin fragments (unpublished results).

The use of HPLC as an attractive technique to use in metabolism studies[33] is once more illustrated in the lower part of Table 8. A mixture of synthetic endorphin fragments has been separated in order to try and identify the peptides obtained after incubation of β-endorphin-(2-17) with brain synaptic membrane associated peptidases,[58] rat brain homogenates,[59,60] or rat striatal slices.[60]

Another application of HPLC might be its use as a method to check for racemization of amino acid residues involved in activation and subsequent coupling reactions. When reference compounds with the corresponding D-amino acids are available (and can be separated) racemization of 0.1% can be detected in the end-products, (e.g., separation of β-endorphin-(6-17) from its D-Ser^7 and D-Lys^9 analogs; unpublished results).

Table 6
HPLC DATA OF SYNTHETIC α-ENDORPHIN AND OF SEPARATIONS OF α-ENDORPHIN FROM OTHER PEPTIDES

Peptide(s)	Stationary phase	Mobile phase	Elution mode	Detection	Remarks	Ref.
α-endorphin	μBondapak® C_{18}	(A) 0.01 *M* NH_4OAc, pH 4 (B) CH_3CN	Isocratic 20% B, 16 min, 2.5 mℓ/min; room temp.	210 nm	Coelutes with natural porcine α-endorphin; not homogeneous	45
α-endorphin	Lichrosorb® RP-18	(A) 1.0 *M* pyridine -0.5 *M* HOAc containing 10^{-2}% (v/v) thiodiglycol and 10^{-4}% (w/v) pentachlorophenol (B) 1-propanol	Lin. grad. 0—11% B	Fluorescamine	Different retention time than γ-endorphin	46
α-endorphin	μBondapak® C_{18}	(A) MeOH-H_2O = 25-75 (B) MeOH-H_2O = 80-20 both with 0.05 *M* TMAH and H_3PO_4 till pH 3.0	Lin. grad. of 0—70% B, 25 min, 2.0 mℓ/min; 20°C	210 nm	Main component 96.1%	47
α-endorphin	Lichrosorb® RP-18 (5 μm)	(A) 0.1 *M* NH_4OAc (B) CH_3CN	Isocratic 20% B, 1.0 mℓ/min; ambient temp.	280 nm	Retention time 5.2 min; trace of Met S-oxide analog	24
α-endorphin	μBondapak® Phenyl	(A) 0.01 *M* NH_4OAc, pH 4.5 (B) CH_3CN	Isocratic 40% B, 1.0 mℓ/min; ambient temp.	280 and 254 nm	Capacity factor 1.94 (vs. 1.15 when 45% B is used)	26
α-endorphin	Bio-Rad® ODS	(I)(A) 0.1 *M* $NaClO_4$ + 5 m*M* phosphate, pH 7.4 (B) CH_3CN (II)(A) 0.1 *M* $NaClO_4$ + 0.1% H_3PO_4, pH 2.1 (B) CH_3CN	Lin. grad. 0—60% B, 1.0 mℓ/min; ambient temp.	200—220 nm and fluorescamine	(I) Retention time 32.3 min (II) Retention time 47.7 min. Comparison with calculated data	48 49
α-endorphin	Lichrosorb® RP-8 or 18	(A) 0.125 *M* pyridine-formate, pH 3.0 (B) 1.0 *M* pyridine-acetate (pH 5.5) and 1-propanol (40-60, v/v)	Lin. grad. 0—50% B, 36 min, 0.7 mℓ/min; ambient temp.	Fluorescamine	Comparison actual and calculated elution points	28

α-endorphin	MicroPak MCH-10	(A) 0.01 *M* KH_2PO_4, pH 4.5 (B) CH_3CN	Lin. grad. 10—50% B, 50 min, 2.0 mℓ/min; 30°C	210 nm	Not homogeneous	50
α- and γ-endorphin	Nucleosil® 10 C_{18}	(A) MeOH-H_2O = 25-75 (B) MeOH-H_2O = 80-20 both with 0.05 *M* TMAH and H_3PO_4 till pH 2.8	Lin. grad. 0—70% B, 25 min, isocratic 70% B, 5 min; 2.0 mℓ/min; 20°C.	210 nm	Clear-cut separation, 1105 vs. 1507 sec	17
α-endorphin and various other peptides	Hypersil-ODS (5 μm) (5 × 100 mm)	(A) 0.1 *M* NaH_2PO_4-H_3PO_4, pH 2.1 (B) CH_3CN	Stepwise grad. 0—10% B, 5 min, 10—40% B 40 min, 40—60% B 5 min, 1.0 mℓ/min; ambient temp.	225 nm	Coelution with Leu-enkephalin	11
α-endorphin,Leu-enkephalin and ACTH-(1—24)	Hypersil-ODS (5 μm) (5 × 100 mm)	(A) 0.1 *M* NaH_2PO_4-H_3PO_4, pH 2.1 (B) CH_3CN	Isocratic 20% B, 5 min, 1.0 mℓ/min; ambient temp.	225 nm	Nearly complete separation	11
α-endorphin and 6 other peptides	Perkin-Elmer 10 C-18	(A) 0.1% H_3PO_4, pH 2.4 (B) CH_3CN	Lin. grad. 15—30% B, 15 min, 40°C	200 nm and native fluorescence	Retention time α-endorphin 9.98 min; no baseline separation from Leu-enkephalin	51
α-endorphin and various other peptides	Cosmosil 5 C-18 (4.6 × 150 mm)	(A) CH_3CN-H_2O(50-50 v/v) with 10 m*M* H_3PO_4 and 15 m*M* SDS (B) CH_3CN-H_2O(60-40) with 15 m*M* SDS (C) CH_3CN-H_2O)(75-25) with 15 m*M* SDS (D) CH_3CN-H_2O(50-50) with 15 m*M* SDS and 1 m*M* TBAP	Stepwise, 10 min of each of A—D 1.5 mℓ/min; room temp.	Fluorescamine	Retention time α-endorphin, 6.4 min	52
α-endorphin and various other peptides	MicroPak AX-10[a]	(A) 0.01 *M* TEAA, pH 6.0 (B) CH_3CN	Lin. grad. 25—100% B, 80 min, 1 mℓ/min; 30°C	225 nm	No baseline separation from EAE-peptide[b]	53

[a] Difunctional weak anion-exchange bonded phase prepared on LiChrosorb® Si-60 silica (10 μm).
[b] Stands for experimental allergic encephalitogenic peptide, i.e., H-Phe-Ser-Trp-Gly-Ala-Glu-Gly-Gln-Arg-OH.

Table 7
HPLC DATA OF SYNTHETIC γ-ENDORPHIN

Peptide	Stationary phase	Mobile phase	Elution mode	Detection	Remarks	Ref.
γ-Endorphin	μBondapak® C_{18}	(A) 0.01 *M* NH_4 OAc, pH 4 (B) CH_3CN	Isocratic 24, 5% B, 16 min, 2.5 mℓ/min; room temp.	210 nm	Coelutes with natural material; not homogeneous	45
γ-Endorphin	μBondapak® C_{18}	(A) MeOH-H_2O = 25-75 (B) MeOH-H_2O = 80-20 both with 0.05 *M* TMAH and H_3PO_4 till pH 3.0	Lin. grad. 0—70% B, 25 min, 2 mℓ/min; 20°C	210 nm	Main component 97.1%	47
γ-Endorphin	Lichrosorb® RP-8 (5 μm)	(A) 0.1 *M* NH_4OAc (B) CH_3CN	Isocratic 20% B, 1.0 mℓ/min; ambient temp.	280 nm	Retention time 12.4 min; trace of Met S-oxide analog	24
γ-Endorphin	μBondapak® Phenyl	(A) 0.01 *M* NH_4OAc, pH 4.5 (B) CH_3CN	Isocratic 40% B, 1.0 mℓ/min; ambient temp.	280 and 254 nm	Capacity factor 2.18 vs. 1.15 when 45% B is used	26
γ-Endorphin	Lichrosorb® RP-8 or 18	(A) 0.125 *M* pyridine-formate, pH 3.0 (B) 1.0 *M* pyridine-acetate (pH 5.5) and 1-propanol (40-60, v/v)	Lin. grad. 0—50% B, 36 min, 0.7 mℓ/min; ambient temp.	Fluorescamine	Comparison actual and calculated elution points	28
γ-Endorphin	Partisil® ODS-2 (10 μm)	(A) 1 *M* pyridine-0.5 *M* HOAc (pH 5.5) (B) 1-Propanol	Lin. grad. 0—30% B, 80 min 0.5 mℓ/min	Fluorescamine	Retention time 42.5 min	87

Table 8
HPLC DATA OF SMALL SYNTHETIC PEPTIDES WITH SEQUENCES IN THE C-TERMINAL PART OF β-LPH, i.e., β-ENDORPHIN

Peptide[a]	Stationary phase	Mobile phase	Elution mode	Detection	Remarks	Ref.
1-19	Lichrosorb® RP-8 (5 μm)	(A) 0.1 *M* NH_4OAc (B) CH_3CN	Isocratic 25% B, 1.0 mℓ/min; ambient temp.	280 nm	Retention time 16.8 min; contains trace of Met S-oxide analog	24
18-31	μBondapak® C_{18}	(A) MeOH-H_2O = 10-90 (B) MeOH-H_2O = 50-50, both with 0.05 *M* TMAH and H_3PO_4 till pH 3.0	Lin. grad. 0—100% B, 15 min, 2.0 mℓ/min; 20°C	210 nm	Main component 99%	25
2-17[b]	μBondapak® C_{18}	(A) MeOH-H_2O = 25-75 (B) MeOH-H_2O = 80-20, both with 0.05 *M* TMAH and H_3PO_4 till pH 3.0	Lin. grad. 0—70% B, 25 min, isocratic 70% B, 5 min, 2.0 mℓ/min; 20°C	210 nm	Main component 96.2%	55
2-17	Lichrosorb® RP-8 or 18	(A) 0.125 *M* pyridine-formate, pH 3.0 (B) 1.0 *M* pyridine-acetate (pH 5.5) and 1-propanol (40-60 v/v)	Lin. grad. 0—50% B, 36 min, 0.7 mℓ/min; ambient temp.	Fluorescamine	Comparison actual and calculated elution points	28
5-17[b,c]	μBondapak® C_{18}	As for 2-17, Reference 55	As for 2-17, Reference 55	210 nm	Main component 95.5%	56
6-17[b]	μBondapak® C_{18}	As for 5-17	As for 5-17	210 nm	Main component 99.0%	56
6-17	μBondapak® C_{18}	(A) 0.1% H_3PO_4 (B) CH_3CN	Isocratic 17% B, 3 mℓ/min	210 nm	Solid-phase product; comparison purification by counter-current chromatography and semiprep. HPLC (0.1% HOAc and MeOH grad.)	57
2-16[d]	μBondapak® C_{18}	As for 5-17	As for 5-17	210 nm	Main component 94.9%	86

Table 8 (continued)
HPLC DATA OF SMALL SYNTHETIC PEPTIDES WITH SEQUENCES IN THE C-TERMINAL PART OF β-LPH, i.e., β-ENDORPHIN

Peptide[a]	Stationary phase	Mobile phase	Elution mode	Detection	Remarks	Ref.
2-9[d]	Nucleosil® 10 C-18	(A) 0.1% H_3PO_4 (pH 2.1) (B) MeOH − H_2O = 50-50, with 0.05 *M* TMAH and H_3PO_4 till pH 2.8	Lin, grad. 20—100% B, 35 min, isocratic 100% B, 5 min, 2 mℓ/min; 20°C	210 nm	Main component 98.1%	86
6-16[d]	μBondapak® C_{18}	(A) MeOH − H_2O = 5-95 (B) MeOH − H_2O = 50-50, both with 0.05 *M* TMAH and H_3PO_4 till pH 3.0	Lin. grad. 0—100% B, 40 min 2 mℓ/min; 20°C	210 nm	Main component 96.9%	86
10-16	μBondapak® C_{18}	As for 5-17	Lin. grad. 0—100% B, 15 min, 2.0 mℓ/min; 20°C	210 nm	Main component 98.5%	47
1-9	μBondapak® C_{18}	As for 5-17	As for 10-16	210 nm	Main component 91%	47
1-9	Partisil®ODS-2 (10 μm)	(A) 1 *M* pyridine-0.5 *M* HOAc (pH 5.5) (B) 1-propanol	Lin. grad. 0—30% B 0.5 mℓ/min	Fluorescamine	Retention time 23 min	88
1-5[e]	μBondapak® C_{18}	As for 5-17	As for 10-16	210 nm	Main component 97.2%	47
28-31 (human sequence?)	μBondapak® C_{18}	(A) 0.08% TFA (B) CH_3CN	Lin. grad. 3.5—49% B, 20 min, 1 mℓ/min; ambient temp.	206 nm	Retention time 3.6 min; separated from several larger β-LPH fragments	13
6-16, 10-16, 8-17, 9-17, 6-17, 7-17, 2-16, 5-17, 10-17, 2-17	μBondapak® C_{18}	(A) 10 m*M* NH_4OAc, pH 4.15 (B) MeOH with 1.5 mℓ HOAc/ℓ	Concave grad. 30—75% B, 45 min, 2 mℓ/min	210 nm	Reference compounds in enzymatic degradation of 2-17; elution in indicated order; coelution of 6-16 and 10-16; 8-17, 9-17, 6-17, and 7-17; 5-17 and 10-17.	58

6-17, 7-17, 10-17, 8-17, 5-17, 2-16, 9-17	μBondapak® C_{18}	(A) 10 m*M* Na phosphate, pH 6.9 (B) MeOH	Concave grad. 30—50% B, 25 min, 2 mℓ/min	210 nm	Elution in indicated order; coelution of 6-17 and 7-17; major column to column differences encountered	58
2-16, 6-17, 5-17, 10-17 and 2-17	Ultrasphere® ODS (5 μm)	(A) 0.1 *M* NaH_2PO_4, pH 2.1 (B) CH_3CN	Curvelinear grad. 18—30% B, 40 min, 2.0 mℓ/min; 40°C	210 nm	Reference compounds in enzymatic digestion of 2-17; baseline separation in order given	59
14-17, 2-16, 6-17, 7-17, 5-17, 1-16, 10-17, 12-17, 4-17, 2-17, 1-17	Ultrasphere® ODS (5 μm)	(A) 0.1 *M* NaH_2PO_4, pH 2.1 (B) CH_3CN	Curvelinear grad. 18—30% B, 40 min, 2.0 mℓ/min; 40°C	210 nm	Baseline separation, except for 10-17 and 12-17 (partial) and 6-17 7-17 (coelution), in order given; calculated detection limit 10 ng for each peptide.	60
6-17; 7-17	HS 5- C-18	(A) 0.1 M NaH_2PO_4, pH 2.1 (B) CH_3CN	Curvelinear grad. 18—30% B, 40 min, 2.0 mℓ/min; 40°C	210 nm	Partial separation	60

[a] Sequence denoted by the number of the first and last residue in β_h-endorphin.
[b] Better peak performance was found with a Nucleosil® 10 C-18 column.
[c] The fragments 7-17, 8-17, 9-17, and 10-17 have been synthesized and analyzed in the same HPLC system.[56,47]
[d] Several fragments of this peptide have been synthesized (see text) and analyzed by essentially the same HPLC system.[86]
[e] I.e. Met-enkephalin; see under D.

DISCUSSION

In the β-lipotropin field, HPLC (predominantly reversed-phase mode) has been used for the isolation of naturally occurring peptides, for the purification and analysis of synthetic peptides, and has proven to be a powerful technique in metabolism studies and for the determination of levels in brain and tissue.

Columns

The commercially available C_{18} material for packing columns consists of a silica matrix chemically modified with octadecyl chains. Due to different manufacturing procedures and column-packing techniques considerable variations can be found in the performance of columns from various manufacturers (see References 62, 63). Also, appreciable column-to-column differences were encountered when using C_{18} columns from the same manufacturer.[58] It is also clear that the nature of the peptide (number and distribution of hydrophobic amino acid residues, basic, acidic or neutral side chains of the amino acids, the size of the peptide, and possibly conformational properties of larger peptides) can result in different HPLC profiles when different columns are used. Some years ago we have evaluated the suitability for peptide analysis of four commercially available reversed-phase columns, μBondapak® C_{18}, Nucleosil® 10 C_{18}, Biosil® ODS-10 and Chromegabond MC-18. We used as one of the test compounds, β-endorphin-(2-17) (both pure and crude batches were checked).[17] As the mobile phase the TMAP-MeOH system was used in all cases. From the results it was clear that the Nucleosil® 10 C_{18} gave the best separation pattern of the crude product; at best, only partial separation of a by-product was obtained using the other columns. The purified peptide[55] gave a nearly identical elution pattern on the four columns. (For the basic peptide ACTH-(1-24) tailing and no resolution of by-products was observed except when the μBondapak® C_{18}-column was used!). A recent extension of this comparative study to include a Lichrosorb® ODS-10, Spherisorb C-18 (7 μm) and a 5-μm Nucleosil® C_{18} column confirmed the earlier results.[17] Of the 10-μm columns, in general the Nucleosil® 10 C_{18} column gives the best results when neutral or acidic peptides were analyzed; for most peptides Lichrosorb® ODS-10 gave also good results. A slight increase of the separation capacity of the latter two columns was obtained by lowering the flow rate from 2 mℓ/min to 1 mℓ/min. The best results however, were obtained with a Nucleosil® C-18 column with 5-μm particles (flow rate of 1 mℓ/min, resulting in a longer analysis time). These results were in line with those of Meek and Rossetti,[49] who found the 5-μm column superior to the 10-μm column (Bio-Rad®) in terms of resolution under each gradient condition. Again, for basic peptides the μBondapak® C_{18} (10-μm, no 5-μm columns available) column clearly gave the best results.

Mobile Phases

The choice of mobile phases used in HPLC of β-LPH peptides has been influenced by pioneering studies of several groups. Examples are: the hydrophilic (e.g., phosphoric acid) or hydrophobic (e.g., hexanesulfonic acid) ion-pairing reagents of the New Zealand groups of Hancock and Hearn,[64,65] the NH_4OAc[66] and TEAP/TEAF buffers[34] introduced by Rivier, the NaH_2PO_4 based system of Molnár and Horváth[67], and the pyridine-based buffers in combination with fluorescence detection, especially used in isolation studies, of Udenfriend and co-workers.[46] The TMAP buffer, originally used in alkaloid-HPLC work,[68] was also introduced into peptide chemistry.[69] We have over the past 4 years very successfully applied this system to the analysis of synthetic lipotropin-derived peptides.

Detection

As to the detection of β-lipotropin peptides, UV detection has been used in most cases.

Although the monitoring wavelength of choice is around 190 to 220 nm[51,70,71] (a great increase in sensitivity relative to 280 or 254 nm and in addition a nearly equivalent molar absorptivity of all peptides is obtained in this region[51]), mobile phases that prohibit detection at these lower wavelengths have been used (formic acid, 1-propanol, and pyridine-containing systems). In addition, detection at 280 and 254 nm misses those by-products that lack aromatic residues and this gives rise to seemingly purer peptides. The use of the very sensitive (in general five to ten times more sensitive than UV detection[72]) but more elaborate[72] (post-column) fluorescence detection is not generally applicable since it is limited to peptides that contain free amino functions.

Temperature

According to Rivier, higher temperatures resulted in lower resolution.[34] This is in contrast to other findings where temperatures of 40°C[17,51,59,60] or 45°C[73] result in better separations, or where the effect of increased temperatures (up to 50°C) is of minor importance.[74] Apparently the effect of the temperature may be dependent of other chromatographic parameters, and the sample to be analyzed (the peptide should be shown to be stable at elevated temperatures).

Purity

In several papers, the (synthesized) compounds were claimed to be "homogeneous" or "very pure". These statements can be misleading in view of the large differences in resolution found in the analysis of β_h-endorphin between two mobile phases (see synthetic β-endorphin) or two stationary phases[19] (see also β-endorphin and closely related peptides) and in view of the inability to separate β-endorphin from analogs missing one residue[14] (but see Table 5) vs. the separation of β-endorphin molecules which differ by only one CH_3 group.[34] Therefore, it is to be preferred to state a degree of purity only in relation to the particular system and conditions used.

Preparative HPLC is gaining interest as the (final) purification step in the synthesis of (lipotropin) peptides. Application of the widely used NH_4OAc buffer in the mobile phase is based on the presumption that it is volatile and can be completely removed by lyophilization. However, it depends on the amino acid composition of the peptide whether this salt can be removed completely.[75] For instance, in order to remove >99% of the NH_4OAc present in a solution of the neutral peptide β-endorphin-(6-17), three successive lyophilizations were necessary.[76] The use of HPLC for desalting and the determination of salt by conductivity measurement has been described by Böhlen et al. for γ-endorphin and insulin.[77] When we applied their procedure and analyzed the isolated peptides (β-endorphin-(2-17) and (6-17), porcine insulin) with isotachophoresis, we still found up to 4% residual salt while the recordings after the desalting experiment led us to believe that all of the salt had been removed (unpublished results).

As a second point to mention here is the use of analytical HPLC to check the purity of a compound after purification on a (semi-)preparative scale using the same system; this of course only proves the purification procedure as such to be successful but does not guarantee that the product really is pure.

We would like to finish this contribution with a quotation from workers in the field: "For determination of homogeneity of synthetic and natural peptides, however, it is not enough to merely separate components, but it is necessary to separate them well enough to determine a small amount or even a trace of some impurities in the presence of a large amount of the major product, a requirement usually much more difficult to meet using chromatographic methods."[66]

REFERENCES

1. **Li, C. H.,** β-Lipotropin: prohormone for β-endorphin with potent analgesic and behavioral activities, in *Versatility of Proteins,* Proc. Int. Symp. on Proteins, Academic Press, New York, 1978, 353; **Chrétien, M. and Lis, M.,** Lipotropins, in *Hormonal Proteins and Peptides,* Vol. 5, Li, C. H., Ed., Academic Press, New York, 1978, 75.
2. **Li, C. H. and Chung, D.,** Isolation, characterization and amino acid sequence of β-lipotropin from human pituitary glands, *Int. J. Peptide Protein Res.,* 17, 131, 1981.
3. **Hsi, K. L., Seidah, N. G., Lu, C. L., and Chrétien, M.,** Reinvestigation of the N-terminal amino acid sequence of β-lipotropin from human pituitary glands, *Biochem. Biophys. Res. Commun.,* 103, 1329, 1981.
4. **Bloomfield, G. A., Scott, A. P., Lowry, P. J., Gilkes, J. J. H., and Rees, L. H.,** A reappraisal of human β MSH, *Nature (London),* 252, 492, 1974.
5. **Bradbury, A. F., Smyth, D. G., and Snell, C. R.,** Biosynthesis of β-MSH and ACTH, in *Peptides, Chemistry, Structure and Biology, Proc. 4th Am. Peptide Symp.,* Walter, R. and Meienhofer, J., Eds., Ann Arbor Science, Ann Arbor, 1975, 609.
6. **Li, C. H.,** β-Endorphin: a new biologically active peptide, in *Hormonal Proteins and Peptides,* Vol. 5, Li, C. H., Ed., Acacemic Press, New York, 1978, 35.

7a. **Li, C. H., Ed.,**, *Hormonal Proteins and Peptides,* Vol. 10, Academic Press, New York, 1981.

7b. **Riley, A. L., Zellner, D. A., and Duncan, H. J.,** The role of endorphins in animal learning and behavior, *Neurosci. Biobehav. Rev.,* 4, 69, 1980.

8. **Ling, N., Burgus, R., and Guillemin, R.,** Isolation, primary structure, and synthesis of α-endorphin and γ-endorphin, two peptides of hypothalamic-hypophysial origin with morphinomimetic activity, *Proc. Natl. Acad. Sci. U.S.A.,* 73, 3942, 1976.
9. **Yamashiro, D. and Li, C. H.,** Total synthesis of ovine β-lipotropin by the solid-phase method, *J. Am. Chem. Soc.,* 100, 5174, 1978.
10. **Rubinstein, M., Stein, S., Gerber, L. D., and Udenfriend, S.,** Isolation and characterization of the opioid peptides from rat pituitary: β-lipotropin, *Proc. Natl. Acad. Sci. U.S.A.,* 74, 3052, 1977.
11. **Nice, E. C. and O'Hare, M. J.,** Simultaneous separation of β-lipotrophin, adrenocorticotropic hormone, endorphins and enkephalins by high performance liquid chromatography, *J. Chromatogr.,* 162, 401, 1979.
12. **Seidah, N. G., Routhier, R., Benjannet, S., Larivière, N., Gossard, F., and Chrétien, M.,** Reversed-phase high-performance liquid chromatographic purification and characterization of the adrenocorticotropin/lipotropin precursor and its fragments, *J. Chromatogr.,* 193, 291, 1980.
13. **McDermott, J. R., Smith, A. I., Biggins, J. A., Chyad Al-Noaemi, M., and Edwardson, J. A.,** Characterization and determination of neuropeptides by high-performance liquid chromatography and radioimmunoassay, *J. Chromatogr.,* 222, 371, 1981.
14. **Yamashiro, D. and Li, C. H.,** Partition and high-performance liquid chromatography of β-lipotropin and synthetic β-endorphin analogues, *J. Chromatogr.,* 215, 255, 1981.
15. **Schally, A. V., Chang, R. C. C., Huang, W.-Y., Coy, D. H., Kastin, A. J., and Redding, T. W.,** Isolation, structure, biological characterization, and synthesis of β-[Tyr^9] melanotropin-(9-18) decapeptide from pig hypothalami, *Proc. Natl. Acad. Sci. U.S.A.,* 77, 3947, 1980.
16. **Ng, T.B., Chung, D., and Li, C. H.,** Isolation and properties of β-endorphin-(1-27), N^{α}-acetyl-β-endorphin, corticotropin, gamma-lipotropin and neurophysin from equine pituitary glands, *Int. J. Peptide Protein Res.,* 18, 443, 1981.
17. **Janssen, P. S. L. and van Nispen, J. W.,** unpublished data.
18. **Coy, D. H., Gill, P., Kastin, A. J., Dupont, A., Cusan, L., Labrie, F., Britton, D., and Fertel, R.,** Synthetic and biological studies on unmodified and modified fragments of human β-lipotropin with opioid activities, in *Peptides, Proc. 5th Am. Peptide Symp.,* Goodman, H. and Meienhofer, J., Eds., John Wiley & Sons, New York, 1977, 107.
19. **Atherton, E., Fox, H., Harkiss, D., and Sheppard, R. C.,** Application of polyamide resins to polypeptide synthesis: an improved synthesis of β-endorphin using fluorenylmethoxycarbonylamino-acids, *J. Chem. Soc. Chem. Commun.,* 539, 1978.
20. **Gentleman, S., Lowney, L. I., Cox, B. M., and Goldstein, A.,** Rapid purification of β-endorphin by high-performance liquid chromatography, *J. Chromatogr.,* 153, 274, 1978.
21. **Dunlap, C. E., III, Gentleman, S., and Lowney, L. I.,** Use of trifluoroacetic acid in the separation of opiates and opioid peptides by reversed-phase high-performance liquid chromatography, *J. Chromatogr.,* 160, 191, 1978.
22. **Coy, D. H.,** Determination of purity of large synthetic peptides using HPLC, in *Biological/Biomedical Applications of Liquid Chromatography II,* Hawk, G. L., Ed., Marcel Dekker, New York, 1979, 283.
23. **Gabriel, T. F., Michalewsky, J. E., and Meienhofer, J.,** Preparative purification of peptides by reversed phase liquid chromatography on inexpensive columns, in *Peptides, Structure and Biological Function, Proc. 6th Am. Peptide Symp.,* Gross, E. and Meienhofer, J., Eds., Pierce Chemical Company, Rockford, Ill., 1979, 105.

24. **Nishimura, O., Shinagawa, S., and Fujino, M.,** Syntheses of human β-endorphin and related peptides, *J. Chem. Res. (S)*, p. 352, 1979.
25. **Van Nispen, J. W., Bijl, W. A. A. J., and Greven, H. M.,** Synthesis of fragments of human β-lipotropin, β_h-LPH. II. The synthesis of β_h-LPH-(61-91), β_h-endorphin, *Recl. Trav. Chim. Pays-Bas*, 99, 57, 1980.
26. **Currie, B. L., Chang, J.-K., and Cooley, R.,** High performance liquid chromatography of enkephalin and endorphin peptide analogs, *J. Liq. Chromatogr.*, 3, 513, 1980.
27. **Tzougrake, C., Makofske, R. C., Gabriel, T. F., Michalewsky, J., Meienhofer, J., and Li, C. H.,** Synthesis of human β-endorphin in solution using benzyl-type side chain protective groups, *Int. J. Peptide Protein Res.*, 15, 377, 1980.
28. **Wilson, K. J., Honegger, A., Stötzel, R. P., and Hughes, G. J.,** The behavior of peptides on reverse-phase supports during high-pressure liquid chromatography, *Biochem. J.*, 199, 31, 1981.
29. **Vilain, E., Brison, J., Loffet, A., and Zanen, J.,** Solid-phase synthesis of human β-endorphin, *Arch. Int. Physiol. Biochim.*, 89, B82, 1981.
30. **Liotta, A. S., Houghten, R., and Krieger, D. T.,** Identification of a β-endorphin-like peptide in cultured human placental cells, *Nature (London)*, 295, 593, 1982.
31. **Rubinstein, M., Stein, S., and Udenfriend, S.,** Isolation and characterization of the opioid peptides from rat pituitary: β-endorphin, *Proc. Natl. Acad. Sci. U.S.A.*, 74, 4969, 1977.
32. **Li, C. H., Ng, T. B., Yamashiro, D., Chung, D., Hammonds, R. G., Jr., and Tseng, L.-F.,** β-Endorphin: isolation, amino acid sequence and synthesis of the hormone from horse pituitary glands, *Int. J. Peptide Protein Res.*, 18, 242, 1981.
33. **Loeber, J. G., Verhoef, J., Burbach, J. P. H., and Witter, A.,** Combination of high pressure liquid chromatography and radioimmunoassay is a powerful tool for the specific and quantitative determination of endorphins and related peptides, *Biochem. Biophys. Res. Commun.*, 86, 1288, 1979.
34. **Rivier, J. E.,** Use of trialkylammonium phosphate (TAAP) buffers in reverse phase HPLC for high resolution and high recovery of peptides and proteins, *J. Liq. Chromatogr.*, 1, 343, 1978.
35. **Houghten, R. A., Chang, W.-C., and Li, C. H.,** Human β-endorphin: synthesis and characterization of analogs iodinated and tritiated at tyrosine residues 1 and 27, *Int. J. Peptide Protein Res.*, 16, 311, 1980.
36. **Akil, H., Ueda, Y., Lin, H. L., and Watson, S. J.,** A sensitive coupled HPLC/RIA technique for separation of endorphins: multiple forms of β-endorphin in rat pituitary intermediate vs. anterior lobe, *Neuropeptides*, 1, 429, 1981.
37. **Liotta, A. S., Yamaguchi, H., and Krieger, D. T.,** Biosynthesis and release of β-endorphin-, *N*-acetyl β-endorphin-, β-endorphin-(1-27)-, and *N*-acetyl β-endorphin-(1-27)-like peptides by rat pituitary neurointermediate lobe: β-endorphin is not further processed by anterior lobe, *J. Neurosci.*, 1, 585, 1981.
38. **Bennett, H. P. J., Browne, C. A., and Solomon, S.,** The use of perfluorinated carboxylic acids in the reversed-phase HPLC of peptides, *J. Liq. Chromatogr.*, 3, 1353, 1980.
39. **Lewis, R. V., Stein, S., and Udenfriend S.,** Separation of opioid peptides utilizing high performance liquid chromatography, *Int. J. Peptide Protein Res.*, 13, 493, 1979.
40. **Morris, H. R., Etienne, A. T., Dell, A., and Albuquerque, R.,** A rapid and specific method for the high resolution purification and characterization of neuropeptides, *J. Neurochem.*, 34, 574, 1980.
41. **Gay, D. D. and Lahti, R. A.,** Rapid separation of enkephalins and endorphins on Sep-Pak reverse phase cartridges, *Int. J. Peptide Protein Res.*, 18, 107, 1981.
42. **Burbach, J. P. H., Loeber, J. G., Verhoef, J., and de Kloet, E. R.,** β-Endorphin biotransformation in brain: formation of γ-endorphin by a synaptosomal plasma membrane associated endopeptidase distinct from cathepsin D, *Biochem. Biophys. Res. Commun.*, 92, 725, 1980.
43. **Verhoef, J., Loeber, J. G., Burbach, J. P. H., Gispen, W. H., Witter, A., and de Wied, D.,** α-Endorphin, γ-endorphin and their des-tyrosine fragments in rat pituitary and brain tissue, *Life Sci.*, 26, 851, 1980.
44. **Rubinstein, M., Chen-Kiang, S., Stein, S., and Udenfriend, S.,** Characterization of proteins and peptides by high-performance liquid chromatography and fluorescence monitoring of their tryptic digests, *Anal. Biochem.*, 95, 117, 1979.
45. **Ling, N.,** Solid phase synthesis of porcine α-endorphin and γ-endorphin, two hypothalamic-pituitary peptides with opiate activity, *Biochem. Biophys. Res. Commun.*, 74, 248, 1977.
46. **Udenfriend, S. and Stein, S.,** Fluorescent techniques for ultramicro peptide and protein chemistry, in *Peptides, Proc. 5th Am. Peptide Symp.*, Goodman, M. and Meienhofer, J., Eds., John Wiley & Sons, New York, 1977, 14.
47. **Bijl, W. A. A. J., van Nispen, J. W. and Greven H. M,.** Synthesis of fragments of human β-lipotropin, β_h-LPH. I. The synthesis of β-LPH-(61-76) and β-LPH-(61-77), i.e., α- and γ-endorphin, respectively, *Recl. Trav. Chim. Pays-Bas*, 98, 571, 1979.
48. **Meek, J. L.,** Prediction of peptide retention times in high-pressure liquid chromatography on the basis of amino acid composition, *Proc. Natl. Acad. Sci. U.S.A.*, 77, 1632, 1980.
49. **Meek, J. L. and Rossetti, Z. L.,** Factors affecting retention and resolution of peptides in high-performance liquid chromatography, *J. Chromatogr.*, 211, 15, 1981.

50. **Wehr, C. T.,** Separation of low molecular weight peptides by reverse phase chromatography, *Liquid Chromatogr. at Work 84 (Varian)*, 1979.
51. **DiCesare, J. L.,** Some aspects in the analysis of peptides by reversed-phase liquid chromatography, *Chromatogr. Newslett.*, 9, 16, 1981.
52. **Mabuchi, H. and Nakahashi, H.,** Systematic separation of medium-sized biologically active peptides by high-performance liquid chromatography, *J. Chromatogr.*, 213, 275, 1981.
53. **Dizdaroglu, M., Krutzsch, H. C., and Simic, M. G.,** Separation of peptides by high-performance liquid chromatography on a weak anion-exchange bonded phase, *J. Chromatogr.*, 237, 417, 1982.
54. **Rekker, R. F.,** *The Hydrophobic Fragmental Constant*, Elsevier, Amsterdam, 1977.
55. **Greven, H. M., Bijl, W. A. A. J., and van Nispen, J. W.,** Synthesis of fragments of human β-lipotropin, β_h-LPH. III. The synthesis of des-1-tyrosine-γ-endorphin, i.e., β-LPH-(62-77), *Recl. Trav. Chim. Pays-Bas*, 99, 63, 1980.
56. **Greven, H. M., van Nispen, J. W., and Bijl, W. A. A. J.,** Synthesis of fragments of human β-lipotropin, β_h-LPH. IV. The synthesis of shortened peptides related to des-1-tyrosine-γ-endorphin [β-LPH-(62-77)], *Recl. Trav. Chim. Pays-Bas*, 99, 284, 1980.
57. **Knight, M., Ito, Y., and Chase, T. N.,** Preparative purification of the peptide des-enkephalin-γ-endorphin. Comparison of high-performance liquid chromatography and counter-current chromatography, *J. Chromatogr.*, 212, 356, 1981.
58. **Burbach, J. P. H., Schotman, P., Verhoef, J., de Kloet, E. R., and de Wied, D.,** Conversion of des-tyrosine-γ-endorphin by brain synaptic membrane associated peptidases: identification of generated peptide fragments, *Biochem. Biophys. Res. Commun.*, 97, 995, 1980.
59. **Davis, T. P., Schoemaker, H., Chen, A., and Yamamura, H. I.,** High performance liquid chromatography of pharmacologically active amines and peptides in biological materials, *Life Sci.*, 30, 971, 1982.
60. **Schoemaker, H., Davis, T.P., Pedigo, N. W., Chen, A., Berens, E. S., Ragan, P., Ling, N. C., and Yamamura, H. I.,** Identification of β-endorphin 6-17 as the principal metabolite of des-tyrosine-γ-endorphin (DT γE) in vitro and assessment of its activity in neurotransmitter receptor binding assays, *Eur. J. Pharmacol.*, 81, 459, 1982.
61. **Van Nispen, J. W. and Greven, H. M.,** Structure-activity relationships of peptides derived from ACTH, β-LPH and MSH with regard to avoidance behavior in rats, *Pharmacol. Ther.*, 16, 67, 1982.
62. **George, R. C. and Patel, C.,** A comparison of HPLC columns, *Pharmac. Technol.*, 6, 88, 1982.
63. **Stoklosa, J. T., Ayi, B. K., Shearer, C. M., and DeAngelis, N. J.,** Separation of minute quantities of impurities in nonapeptides by reverse-phase high-performance liquid chromatography: critical nature of the water/acetonitrile ratio, *Anal. Lett.*, B11, 889, 1978.
64. **Hancock, W. S., Bishop, C. A., Prestidge, R. L., Harding, D. R. K., and Hearn, M. T. W.,** High-pressure liquid chromatography of peptides and proteins. II. The use of phosphoric acid in the analysis of underivatised peptides by reversed-phase high-pressure liquid chromatography, *J. Chromatogr.*, 153, 391, 1978.
65. **Hancock, W. S., Bishop, C. A., Prestidge, R. L., Harding, D. R. K., and Hearn, M. T. W.,** Reversed-phase, high-pressure liquid chromatography of peptides and proteins with ion-pairing reagents, *Science*, 200, 1168, 1978.
66. **Burgus, R. and Rivier, J.,** Use of high pressure liquid chromatography in the purification of peptides, in *Peptides 1976, Proc. 14th Eur. Peptide Symp.*, Loffet, A., Ed., Editions de l'Université de Bruxelles, Brussels, 1976, 85.
67. **Molnár, I. and Horváth, C.,** Separation of amino acids and peptides on non-polar stationary phases by high-performance liquid chromatography, *J. Chromatogr.*, 142, 623, 1977.
68. **Van der Maeden, F. P. B., van Rens, P. T., Buytenhuys, F. A., and Buurman, E.,** Quantitative analysis of *d*-tubocurarine chloride in curare by column liquid chromatography, *J. Chromatogr.*, 142, 715, 1977.
69. **Biemond, M. E. F., Sipman, W. A., and Olivié, J.,** Quantitative determination of polypeptides by gradient elution high pressure liquid chromatography, *J. Liq. Chromatogr.*, 2, 1407, 1979.
70. **Hancock, W. S., Bishop, C. A., and Hearn, M. T. W.,** High pressure liquid chromatography in the analysis of underivatised peptides using a sensitive and rapid procedure, *FEBS Lett.*, 72, 139, 1976.
71. **Wilson, K. J. and Hughes, G. J.,** High-performance liquid chromatography of peptides and proteins. Pharmaceutical and bio-medical applications, *Chimia*, 35, 327, 1981.
72. **Wilson, K. J., Honegger, A., and Hughes, G. J.,** Comparison of buffers and detection systems for high-pressure liquid chromatography of peptide mixtures, *Biochem. J.*, 199, 43, 1981.
73. **Lloyd, L. F. and Corran, P. H.,** Analysis of insulin preparations by reversed-phase high-performance liquid chromatography, *J. Chromatogr.*, 240, 445, 1982.
74. **Abrahamsson, M., and Gröningsson, K.,** High-performance liquid chromatography of the tetradecapeptide somatostatin, *J. Liq. Chromatogr.*, 3, 495, 1980.
75. **Van Nispen, J. W., Janssen, P. S. L., Goverde, B. C., and Greven, H. M.,** Analytical isotachophoresis in peptide chemistry, in *Peptides 1980, Proc. 16th Eur. Peptide Symp.*, Brunfeldt, K., Ed., Scriptor, Copenhagen, 1981, 731.

76. **Janssen, P. S. L. and van Nispen, J. W.,** Isotachophoretic determination of anions and cations in peptides, *J. Chromatogr.*, 287, 166, 1984.
77. **Böhlen, P., Castillo, F., Ling, N., and Guillemin, R.,** Purification of peptides: an efficient procedure for the separation of peptides from amino acids and salt, *Int. J. Peptide Protein Res.*, 16, 306, 1980.
78. **Blake, J. and Li, C. H.,** Total synthesis of human β-lipotropin, *Proc. Natl. Acad. Sci. U.S.A.*, 80, 1556, 1983.
79. **Spiess, J., Mount, C. D., Nicholson, W. E., and Orth, D. N.,** NH_2-terminal amino acid sequence and peptide mapping of purified human β-lipotropin: comparison with previously proposed sequences, *Proc. Natl. Acad. Sci. U.S.A.*, 79, 5071, 1982.
80. **Seidah, N. G., Hsi, K. L., Chrétien, M., Barat, E., Patthy, A., and Graf, L.,** The primary structure of human β-lipotropin, *FEBS Lett.*, 147, 267, 1982.
81. **Izdebski, J., Yamashiro, D., Li, C. H., and Viti, G.,** Synthesis and properties of human γ-lipotropin, *Int. J. Peptide Protein Res.*, 20, 87, 1982.
82. **Garzia, R., Yamashiro, D., Li, C. H., and Nicolas, P.,** β-Endorphin: synthesis and properties of human β-endorphin-(1-28) and its analogs, *Int. J. Peptide Protein Res.*, 20, 194, 1982.
83. **Janssen, P. S. L., van Nispen, J. W., Hamelinck, R. L. A. E., Melgers, P. A. T., and Goverde, B. C.,** Practical applications of HPLC in peptide chemistry, *J. Chromatogr. Sci.*, 1984, in press. Chromatography, Baden-Baden, G.F.R., May, 3-6, 1983.
84. **Burbach, J. P. H., de Kloet, E. R., Schotman, P., and de Wied, D.,** Proteolytic conversion of β-endorphin by brain synaptic membranes, *J. Biol. Chem.*, 256, 12463, 1981.
85. **Verhoef, J. Wiegant, V. M., and de Wied, D.,** Regional distribution of α- and γ-type endorphins in rat brain, *Brain Res.*, 231, 454, 1982.
86. **Bijl, W. A. A. J., van Nispen, J. W., and Greven, H. M.,** Synthesis of fragments of human β-lipotropin, β_h-LPH. VI. The synthesis of des-1-tyrosine-α-endorphin and shortened peptides, *Recl. Trav. Chim. Pays-Bas*, 102, 469, 1983.
87. **Garzia, R., Yamashiro, D., Hammonds, R. G., Jr., and Li, C. H.,** Synthesis and properties of human β-endorphin-(1-17) and its analogs, *Int. J. Peptide Protein Res.*, 19, 432, 1982.
88. **Yamashiro, D., Garzia, R., Hammonds, R. G., Jr., and Li, C. H.,**Synthesis and properties of human β-endorphin-(1-9) and its analogs, *Int. J. Peptide Protein Res.*, 19, 284, 1982.

PROCTOLIN

Alvin N. Starratt and Mary E. Stevens

INTRODUCTION

Proctolin, initially isolated from the cockroach *Periplaneta americana*, was identified as the pentapeptide Arg-Tyr-Leu-Pro-Thr.[1,2] Subsequent studies have indicated that it is widely distributed among insects[3-5] and probably other arthropods[6,7] and that it has potent activity on a number of invertebrate muscle preparations.[6-13]

SEPARATION OF PROCTOLIN BY HPLC

Holman and Cook[4,5] first reported the application of HPLC to studies of proctolin. Following ion exchange, insect extracts were chromatographed on a Poragel PN column and the proctolin content of fractions with myogenic activity was determined by analysis of the phenylthiocarbamyl derivative on a μPorasil® column. They found proctolin in whole body extracts of the stable fly *Stomoxys calcitrans* (12.4 ng/g insect)[4] and reported its presence in foreguts (0.83 ng/gut) and hindguts (3.3 ng/gut) but not in heads of the cockroach *Leucophaea maderae*.[5] Recently, O'Shea and Adams[14] described a method of determining the level of proctolin in insect tissue which employed reversed-phase (RP) HPLC. Using a bioassay system employing a locust leg extensor muscle, which is sensitive to low levels of proctolin, they were able to measure this neuropeptide in an extract of an identified neuron from the abdominal ganglion of the cockroach *Periplaneta americana*. Proctolin was separated from other components in the extract by HPLC on a μBondapak® C_{18}-column with 12% acetonitrile in 50 m*M* ammonium acetate, pH 4.5, as liquid phase.

Starratt and Stevens[15] have also reported the use of reversed-phase columns for HPLC of underivatized proctolin. Using several ion-pairing reagents, proctolin could be separated on either μBondapak® fatty acid analysis or μBondapak® C_{18}-columns from pentapeptide analogs differing by only one amino acid. The five possible diastereoisomers of proctolin resulting from replacement of a single amino acid with the respective D-amino acid could easily be distinguished from proctolin. These diastereoisomers had longer retention times than proctolin indicating that they possess conformations which permit them to interact more readily with the hydrocarbon matrix of the reversed-phase column.

MOBILE-PHASE EFFECTS

Two of the ion-pairing reagents investigated, trifluoroacetic acid (TFA) and heptafluorobutyric acid (HFBA), have the advantage that they can be removed during lyophilization making systems utilizing them useful for preparative chromatography and subsequent direct bioassay. Recently, these studies with HFBA have been extended and other systems with volatile acids and buffers have been compared for the separation of proctolin and its analogs.

Table 1 illustrates the effect of the organic modifier on the elution times of proctolin and several related peptides. Small changes in the separation factors were observed and the capacity factors increased in the order 2-propanol (2-PrOH) $<$ acetonitrile (CH_3CN) $\ll$ methanol (MeOH). Figure 1 shows a chromatograph of five peptides using CH_3CN-water (20:80) containing 5 m*M* HFBA. The increase in peak widths for the more hydrophobic peptides could be avoided by the use of gradients (Figure 2). Such systems have proved useful for the simultaneous analysis of proctolin and compounds differing substantially in hydrophobicity.

Table 1
EFFECT OF ORGANIC MODIFIER ON THE RETENTION TIMES OF PROCTOLIN AND SOME RELATED PENTAPEPTIDES ON A REVERSED-PHASE COLUMN[a]

Peptide	Retention time (min)		
	CH_3CN-H_2O (20:80)	2-PrOH-H_2O (20:80)	MeOH-H_2O (40:60)
Proctolin (Arg-Tyr-Leu-Pro-Thr)	15.7	9.6	15.1
[Ala[1]]-Proctolin	9.5	5.4	10.2
[Ala[3]]-Proctolin	6.7	5.6	6.2
[Ala[4]]-Proctolin	12.0	8.3	11.0
[Ala[5]]-Proctolin	17.9	10.0	15.1
[Orn[1]]-Proctolin	15.3	9.6	17.1
[Ser[5]]-Proctolin	13.1	8.4	12.4
[Phe[2]]-Proctolin	36.7	18.3	35.3

[a] 30 cm × 3.9 mm i.d. μBondapak® C_{18}-column. Solvent systems contained 5 m*M* HFBA; flow rate 1.5 mℓ/min.

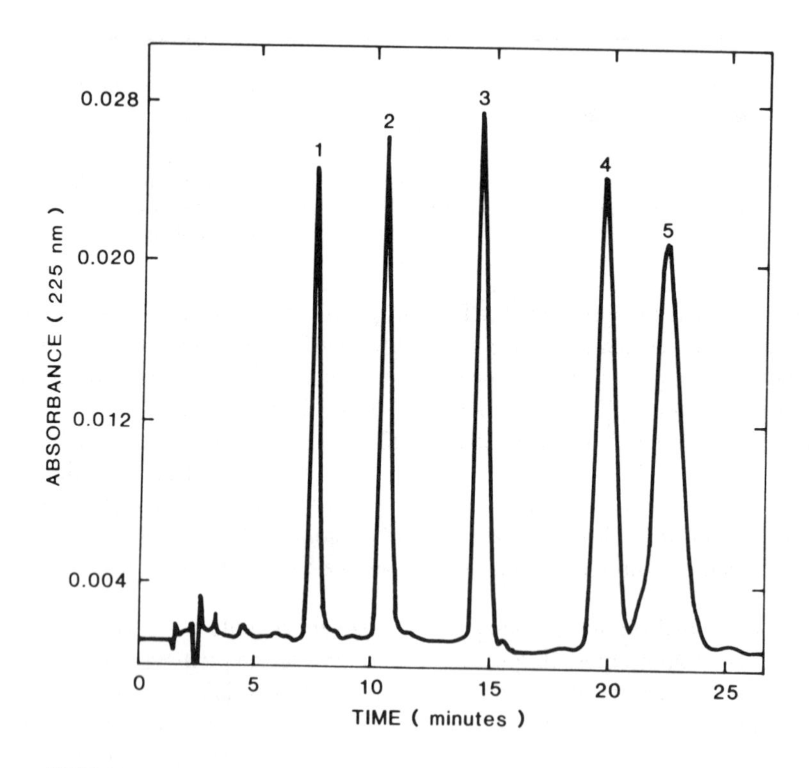

FIGURE 1. Chromatography of proctolin and four analogs on a 30 cm × 3.9 mm i.d. μBondapak® C_{18}-column with CH_3CN-water (1:4) containing 5 m*M* HFBA as eluent at a flow rate of 1.5 mℓ/min. Peaks: 1 = [Ala[3]]-proctolin; 2 = [Ala[1]]-proctolin; 3 = [Ala[4]]-proctolin; 4 = proctolin; 5 = [Ala[5]]-proctolin.

Peptide separation was also achieved with CH_3CN-water systems containing formic and acetic acids as ion-pairing reagents (Table 2A and B) but peaks were unsymmetrical and broader than observed with TFA and HFBA. Good peak shape, comparable to that illustrated

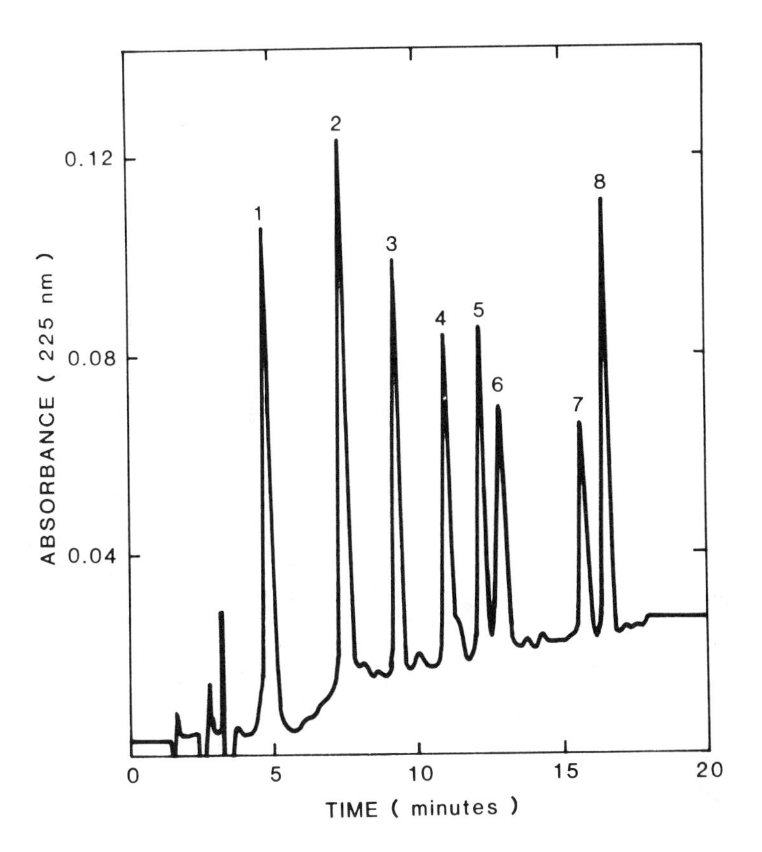

FIGURE 2. Gradient elution profile of proctolin and several related peptides on a 30 cm × 3.9 mm i.d. μBondapak® C_{18}-column. A 20-min linear gradient from 20 to 40% CH_3CN-water containing 5 m*M* HFBA was employed at a flow rate of 1.5 mℓ/min. Peaks: 1 = phenylalanine; 2 = [Ala^3]-proctolin; 3 = [Ala^1]-proctolin; 4 = [Ala^4]-proctolin; 5 = proctolin; 6 = [Ala^5]-proctolin; 7 = [Phe^2]-proctolin; 8 = s36[(*p*-OMe)Phe^2]-proctolin.

(Figure 1) or that obtained with ammonium formate or ammonium acetate, was observed when these systems were buffered by the addition of triethylamine (TEA). Retention times of the basic peptides were similar with triethylammonium formate (TEAF) pH 3.15, and ammonium formate, pH 4.5, but [Ala^1]-proctolin eluted significantly faster at the higher pH (Table 2A). With systems utilizing HFBA-TEA, pH 3.15, an increase in retention times was observed as the amount of HFBA was increased from 5 to 40 m*M* (Table 2C).

RECOVERY OF PROCTOLIN AFTER HPLC SEPARATION

In this laboratory, HPLC with HFBA and TFA as ion-pairing reagents has been used to purify both natural and synthetic proctolin and related peptides. As indicated by rechromatography of collected and lyophilized samples, good recovery of microgram quantities of proctolin is obtained. Although variable losses have been observed at lower levels, this technique has also proved to be useful for the purification and characterization of nanogram amounts of neuropeptide.[16]

ADDENDUM

Ion-pair reversed-phase HPLC on a μBondapak® C_{18} column with CH_3CN-water gradients

Table 2
EFFECT OF VARIOUS VOLATILE ACIDS AND BUFFERS ON THE RETENTION TIMES OF PROCTOLIN AND SOME ANALOGS ON A REVERSED-PHASE COLUMN[a]

	Retention time (min)							
		Proctolin analogs						
Mobile phase	Proctolin	[Ala¹]-	[Ala³]-	[Ala⁴]-	[Ala⁵]-	[Orn¹]-	[Ser⁵]-	[Phe²]-
A. CH_3CN-20 m*M* formic acid (10:90)	13.7	33.1	3.5	7.4	13.9	13.5	10.2	29.2
CH_3CN-20 m*M* formic acid + TEA, pH 3.15 (10:90)	10.2	25.3	2.8	5.7	10.3	10.6	7.8	23.7
CH_3CN-40 m*M* formic acid + TEA, pH 3.15 (10:90)	10.7	26.8	2.9	6.0	10.8	11.2	8.2	26.7
CH_3CN-20 m*M* ammonium formate + formic acid, pH 4.5 (10:90)	13.2	17.4	3.4	7.3	14.2	15.3	10.2	
B. CH_3CN-20 m*M* acetic acid (15:85)	8.5	11.7	3.8	6.0	9.7	8.2	6.9	18.7
CH_3CN-20 m*M* acetic acid + TEA, pH 4.5 (10:90)	11.2	17.8	3.2	6.3	11.7	13.0	8.8	29.8
CH_3CN-20 m*M* ammonium acetate + acetic acid, pH 4.5 (10:90)	15.8	19.7	3.7	8.4	16.8	18.3	12.2	43.8
C. CH_3CN-5 m*M* HFBA + TEA, pH 3.15 (20:80)	9.6	6.7	4.7	7.7	11.2	9.2	7.9	21.7
CH_3CN-20 m*M* HFBA + TEA, pH 3.15 (20:80)	16.0	7.7	5.7	11.8	18.7	15.3	12.8	43.2
CH_3CN-40 m*M* HFBA + TEA, pH 3.15 (20:80)	20.0	8.0	6.2	14.4	23.5	18.8	15.7	58.3

[a] 30 cm × 3.9 mm i.d. μBondapak® C_{18}-column. Flow rate 1.5 mℓ/min.

was used in the study of the in vivo inactivation of [^{14}C-Tyr^2]-proctolin in *P. americana.*[16] Taking advantage of the difference in the separation factors obtained with the ion-pairing reagents TFA and HFBA, proctolin could be separated from tyrosine and all possible tyrosine-containing di-, tri, and tetrapeptides which might result from hydrolysis of proctolin. The separation of proctolin from three other invertebrate neuropeptides, locust adipokinetic hormone, crustacean erythrophore concentrating hormone, and molluscan cardioexcitatory neuropeptide, on Supelcosil® LC-18DB and Zorbax® C-8 columns with CH_3CN-water gradients containing either 0.25 *N* triethylammonium phosphate (pH 2.2) or 0.1% TFA has been described by Jaffe et al.[17] O'Shea and Bishop[18] have recently reported the use of reversed-phase HPLC (μBondapak® C_{18} column; 15% CH_3CN in 50 m*M* ammonium acetate, pH 4.5) in the identification of proctolin associated with a skeletal motoneuron of *P. americana.* The elution time of proctolin was determined by chromatography of [^{3}H-Tyr^2]-proctolin. A report by Dizdaroglu et al.[19] shows that proctolin can be separated from several other naturally occurring peptides including somatostatin, neurotensin, met-enkephalin, and bradykinin on a MicroPak AX-10 column, a silica-based bonded-phase weak anion exchanger, using a gradient of 10 m*M* triethylammonium acetate buffer (pH 6.0) and CH_3CN for elution.

REFERENCES

1. **Brown, B. E. and Starratt, A. N.,** Isolation of proctolin, a myogenic peptide, from *Periplaneta americana, J. Insect Physiol.,* 21, 1879, 1975.
2. **Starratt, A. N. and Brown, B. E.,** Structure of the pentapeptide proctolin, a proposed neurotransmitter in insects, *Life Sci.,* 17, 1253, 1975.
3. **Brown, B. E.,** Occurrence of proctolin in six orders of insects, *J. Insect Physiol.,* 23, 861, 1977.
4. **Holman, G. M. and Cook, B. J.,** The analytical determination of proctolin by HPLC and its pharmacological action in the stable fly, *Comp. Biochem. Physiol.,* 62C, 231, 1979.
5. **Holman, G. M. and Cook, B. J.,** Evidence for proctolin and a second myotropic peptide in the cockroach, *Leucophaea maderae,* determined by bioassay and HPLC analysis, *Insect Biochem.,* 9, 149, 1979.
6. **Sullivan, R. E.,** A proctolin-like peptide in crab pericardial organs, *J. Exp. Biol.,* 210, 543, 1979.
7. **Benson, J. A., Sullivan, R. E., Watson, W. H., and Augustine, G. J.,** The neuropeptide proctolin acts directly on *Limulus* cardiac muscle to increase the amplitude of contraction, *Brain Res.,* 213, 449, 1981.
8. **Brown, B. E.,** Proctolin, a peptide transmitter candidate in insects, *Life Sci.,* 17, 1241, 1975.
9. **Piek, T. and Mantel, P.,** Myogenic contractions in locust muscle induced by proctolin and by wasp, *Philanthus triangulum,* venom, *J. Insect Physiol.,* 23, 321, 1977.
10. **Miller, T.,** Nervous *versus* neurohormonal control of insect heartbeat, *Am. Zool.,* 19, 77, 1979.
11. **Cook, B. J. and Holman, G. M.,** The action of proctolin and L-glutamic acid on the visceral muscles of the hindgut of the cockroach, *Leucophaea maderae, Comp. Biochem. Physiol.,* 64C, 21, 1979.
12. **Cook, B. J. and Meola, S.,** The oviduct musculature of the horsefly, *Tabanus sulcifrons,* and its response to 5-hydroxytryptamine and proctolin, *Physiol. Entomol.,* 3, 273, 1978.
13. **Schwarz, T. L., Harris-Warrick, R. M., Glusman, S., and Kravitz, E. A.,** A peptide action in a lobster neuromuscular preparation, *J. Neurobiol.,* 11, 623, 1980.
14. **O'Shea, M. and Adams, M. E.,** Pentapeptide (proctolin) associated with an identified neuron, *Science,* 213, 567, 1981.
15. **Starratt, A. N. and Stevens, M. E.,** Ion-pair high-performance liquid chromatography of the insect neuropeptide proctolin and some analogs, *J. Chromatogr.,* 194, 421, 1980.
16. **Starratt, A. N. and Steele, R. W.,** *In vivo* inactivation of the insect neuropeptide proctolin in *Periplaneta americana, Insect. Biochem.,* in press.
17. **Jaffe, H., Loeb, M., Hayes, D. K., and Holston, N.,** Rapid isolation of nanogram amounts of crustacean erythrophore concentrating hormone from invertebrate nerve tissue by RP-HPLC, *J. Liquid Chromatogr.,* 5, 1375, 1982.
18. **O'Shea, M. and Bishop, C. A.,** Neuropeptide proctolin associated with an identified skeletal motoneuron, *J. Neurosci.,* 2, 1242, 1982.
19. **Dizdaroglu, M., Krutzsch, H. C., and Simic, M. G.,** Separation of peptides by high-performance liquid chromatography on a weak anion-exchange bonded phase, *J. Chromatogr.,* 237, 417, 1982.

INTESTINAL PEPTIDES

L. Pradayrol, D. Fourmy, and A. Ribet

INTRODUCTION

Principally three areas of research in the field of gastrointestinal (GI) polypeptides biochemistry have seen the use of HPLC developed to its present level.

These branches involve the purification of naturally occurring polypeptides, the estimation of the purity of their preparation, and the characterization of the diverse molecular forms present in tissues and in biological fluids.

Our objective here is not to give an exhaustive review of HPLC applications to intestinal peptides but to illustrate successively the different areas of use, with emphasis on our recent results on the one hand, and by relating our work to the similar findings of other researchers on the other hand.

As in many specialized fields, the choice of type of chromatography for the separation of intestinal peptides is that of reversed-phase polarity, and for this reason the report will deal only with this type of chromatography.

PURIFICATION OF NATURAL POLYPEPTIDES

The use of "reversed-phase separations" is restricted to the final steps in purification of the natural intestinal peptides and complements fractionation with solvents, gel filtration chromatography, and conventional ion-exchange chromatography.

In this application, regarding the small quantities of available material (in the order of a few milligrams), the use of analytical HPLC equipment is perfectly adequate. Our experiences show that for the preparation of somatostatin-28 from porcine small intestine, with this technique (Figure 1), we were able to obtain sufficient pure material for the determination of its structure.[1] Amounts of up to 10 mg were applied to a μBondapak® C_{18}-column (0.39 × 30 cm) without a notable loss in resolution when the eluent consisted of triethylammonium phosphate (TEAP), 0.25 *M*, pH : 3.5 / CH_3CN, 74/26.

Similar work was carried out in the laboratory of Mutt (Karolinska Institute, Stockholm, Sweden), which led McDonald et al. to the characterization of GRP (gastrin-releasing peptide).[2] This work also permitted Tatemoto and Mutt to purify new peptides : PHI (porcine intestinal peptide with N-terminal histidine and C-terminal isoleucine-amide).[3] With a similar approach, Carlquist and Mutt purified VIP (vaso-intestinal peptide) from bovine intestine[4] and showed the presence of a peptide having a "secretin-like" bioactivity from bovine brain.[5] Using a different system, Bataille et al.[6] was able to obtain an enteroglucagon from the distal portion of porcine large intestine.

We have used this technique as well for the characterization of the two molecular forms, 14 and 28 of somatostatin from a tumor of human pancreatic origin.

Although reverse-phase (RP) HPLC consists of a fundamental new method for the characterization of intestinal polypeptides, there still exists a number of limiting factors. Among them we will give examples of the difficulty of obtaining a good resolution in solvents and solutes that are transparent in the far UV (< 210 nm) and yet are completely removed during lyophilization.

At the present time few systems overcome simultaneously these two disadvantages, and consequently a choice is made between detection sensitivity and the necessity to perform a desalting step.

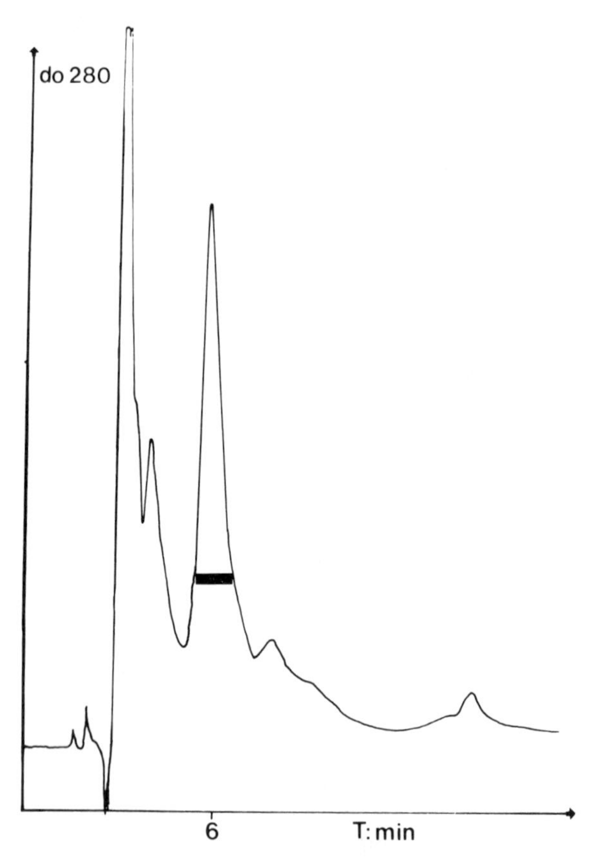

FIGURE 1. HPLC of 10 mg lyophilisate corresponding to the purified somatostatin 28 extract obtained by conventional chromatographies. Column μBondapak® C_{18} (0.39 × 30 cm) ; eluent buffer TEAP 0.25 *N* pH 3.5/acetonitrile 76/24 ; flow rate 2 mℓ/min ; back pressure 1500 psi ; chart speed 1 cm/min ; UV detection 0 → 1 F.S., OD at 280 nm. ■ Somatostatin-28 preparation used for structure elucidation.

PURITY CONTROL

The purity of GI polypeptide preparations can be rapidly evaluated by HPLC. When dealing with polypeptides, no one method alone is enough to determine the purity of a preparation; nevertheless HPLC brings forth important information in this area. Here, we will look at the two principle polypeptide families successively, one being secretin and the other being cholecystokinin-pancreozymin (CCK-PZ).

Shown in Figure 2 is the elution profile obtained after the injection of a VIP, secretin (GIH), and glucagon (Novo) mixture. We ascertain that in addition to the main peaks (t_R 2.48 - 32 - 50.24 min) impurities, in small quantities are present. By individually running each sample separately the contaminants with $t_R < 2$ min proved to be coming from VIP; contaminants with a $t_R = 17.5$ and 38.5 min come from secretin. The remaining impurities with a t_R 10 to 11 min were attributed to glucagon. In the case of secretin, since the impurities which gave peaks at 17.5 and 38.5 min were found both in natural and synthetic preparations it was assumed that they are degradation products.

The behavior of the three basic compounds of CCK-PZ (CCK-39, CCK-33, CCK-8) is illustrated in Figure 3. The polypeptide sequence includes a methionine residue which must not be oxidized in order to ensure the integrity of the biological activity. We were able to ascertain that this oxidation state influenced considerably the retention time of the tetra- and octapeptides and we were thus able to confirm the results obtained by Beinfeld et al.[8] In

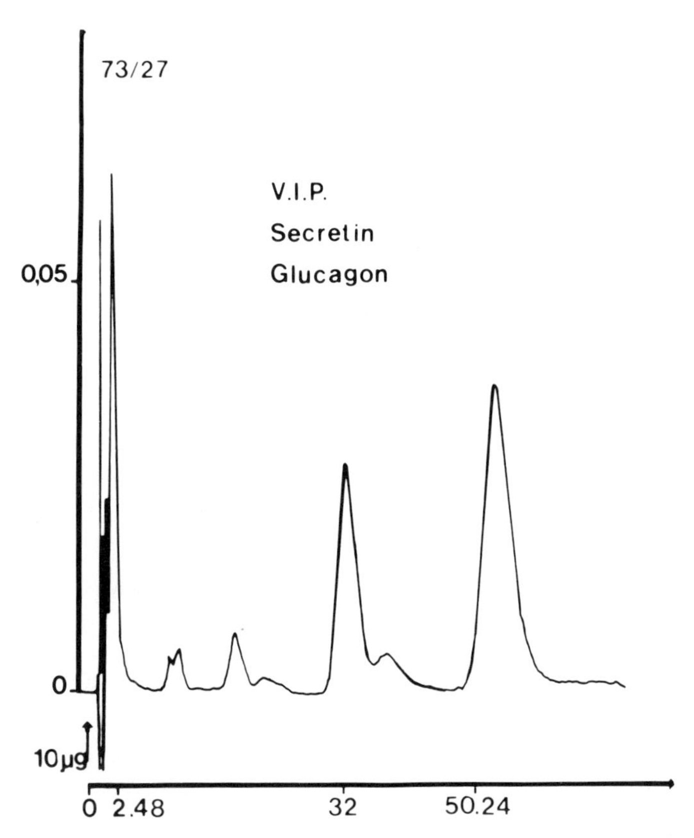

FIGURE 2. RP-HPLC of VIP, secretin, and glucagon. Column: μBondapak® C_{18} (0.39 × 30 cm) ; eluent buffer TEAP 0.25 *N* pH 3.5/ acetonitrile (73/27); flow rate 2 mℓ/min ; Back pressure 1500 psi ; chart speed 1 ch/min ; UV detection 0 → 0.1 F.S., OD at 210 nm; 10 μg each peptide injected. t_R : VIP = 2.48, secretin 32, glucagon 50.2 min.

this particular case the method permits the rapid correlation of the degree of oxidation with any loss in biological activity.

In this same area, a comparative study of different ceruleins permitted Di Castiglione and Salle[9] to show the excellent relationship between the peak area of the corresponding peptide and its biological potency and allowed comparisons of preparations with different origins.

Recently Jörnvall et al.[10] reexamined the preparation of GIP and was able to show by HPLC that it contained a minor component corresponding to another molecular form of this peptide.

ANALYSIS OF DIFFERENT MOLECULAR FORMS OF A POLYPEPTIDE PRESENT IN TISSUES AND BIOLOGICAL FLUIDS

Coupled to a system for specific detection, HPLC should find evergrowing applications in the rapid characterization of molecular forms present in tissues and in biological fluids. This last point will be illustrated by two examples.

In a collaborative work with Uvnas (Stockholm) and Chayvialle (Lyon)[12] we were able to show the concomitant secretion of molecular forms 14 and 28 of somatostatin from antral and duodenal luminal perfusates by coupling HPLC with radioimmunoassay.

In a pilot experiment Carlquist was able to show that the "secretin-like" molecule, carrying the bioactivity present in bovine brain extracts, showed the same tryptic fragments (separated by HPLC) as did secretin isolated from porcine intestine. The purification was monitored by measuring the enrichment of the C-terminal amides according to the method of Tatemoto[3] and by measuring the biological activity.

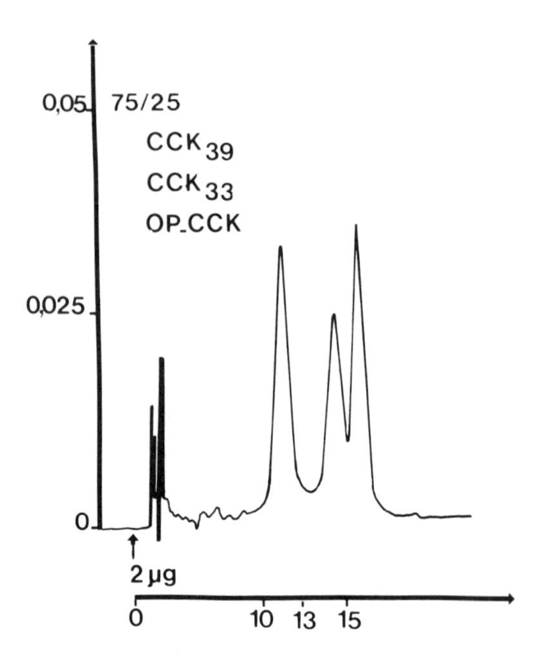

FIGURE 3. HPLC of CCK-39, CCK-33, CCK-8, Column: µBondapak® C_{18} (0.39 × 30 cm) ; eluent buffer TEAP 0.25 *N* pH 3.5 / acetonitrile (75/25) ; Flow rate 2 mℓ/min ; Back pressure 1500 psi ; UV detection 0 → 0.1 F.S., OD at 210 nm. t_R: CCK-39 = 13, CCK-8, 15 min.

REFERENCES

1. **Pradayrol, L., Jörnvall, H., Mutt, V., and Ribet, A.,** N-terminally extended somatostatin : the primary structure of somatostatin-28, *FEBS Lett.*, 109, 55, 1980.
2. **McDonald, T. J., Jörnvall, H., Nilsson, G., Vagne, M., Ghatei, M., Bloom, S. R., and Mutt, V.,** Characterization of a gastrin releasing peptides from porcine non antral gastric tissue, *Biochim. Biophys. Res. Commun.*, 90, 227, 1979.
3. **Tatemoto, K. and Mutt, V.,** Isolation of two novel candidate hormones using a chemical method for feeding naturally occuring polypeptides, *Nature (London)*, 285, 417, 1980.
4. **Carlquist, M., Mutt, V., and Jörnvall, H.,** Isolation and characterization of bovine vasoactive intestinal peptide (VIP), *FEBS Lett.*, 108, 457, 1979.
5. **Mutt, V., Carlquist, M., and Tatemoto, K.,** Secretin-like bioactivity in extracts of porcine brain, *Life Sci.*, 25, 1703, 1979.
6. **Bataille, D., Gespach, C., Coudray, A. M., and Rosselin, G.,** "Enteroglucagon". A specific effect on gastric glands isolated from the rat fundus evidence for an "oxynto modulin" action, *Bio. Sci. Rep.*, 1, 151, 1981.
7. **Pradayrol, L., Chayvialle, J. A., Descos, F., Fagot-Revurat, P., and Galmiche, J. P.,** Multiple forms of somatostatin-like activity in a human pancreatic tumour and its hepatic metastasis : purification and characterization of S28 and S14, in 2nd Int. Symp. on Somatostatin, Athens, June 1 to 3, 1981.
8. **Beinfeld, M. C., Jensen, R. T., and Brownstein, M. J.,** H.P.L.C. separation of cholecystokinin peptides. Two systems, *J. Liq. Chromatogr.*, 3, 1367, 1980.
9. **Di Castiglione, R. and Salle, E.,** Is commercial "cerulein" always true cerulein ?, *Gastroenterology*, 78, 1113, 1980.
10. **Jörnvall, H., Carlquist, M., Brown, J. C., and Mutt, V.,** Amino-acid sequence and heterogeneity of gastric inhibitory polypeptides (GIP), *FEBS Lett.*, 123, 205, 1981.
11. **Carlquist, M. and Mutt, V.,** Isolation of bovine secretin, *Regul. Peptides*, Suppl. 1, S16, 1980.
12. **Uvnas, X. and Chayvialle, J. A.,** unpublished data.

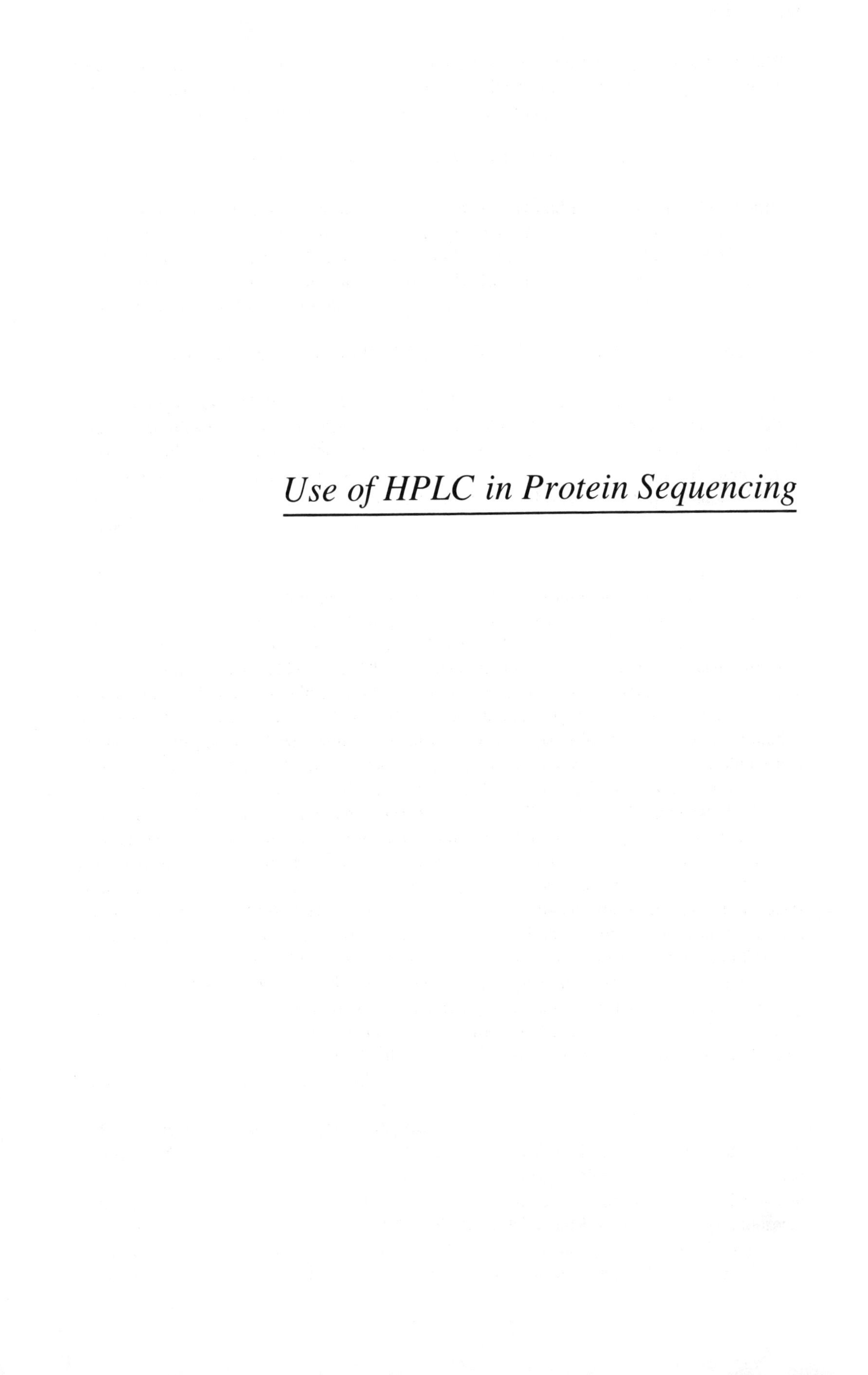

Use of HPLC in Protein Sequencing

THE USE OF HPLC IN PROTEIN SEQUENCING

Ajit S. Bhown and J. Claude Bennett

INTRODUCTION

The study of protein structure presents unique problems that require innovative solutions. The potential for sequence variation in proteins, which makes them biologically so important, inevitably imposes restrictions on their structural analysis. Nevertheless, the primary structure of many proteins has now been successfully determined. These studies have helped in establishing that (1) complex biological functions of proteins can be described in terms of simple chemical concepts, (2) many proteins contain domains and subunits that specify discrete subfunctions, (3) proteins can be grouped into homologous evolutionarily related families, and (4) the structure of a protein can be directly related to its encoding gene.[1]

The elucidation of the primary structure of protein molecules basically involves: (1) purification and isolation of the protein molecule and (2) amino acid sequence analysis of the purified protein and its chemically and enzymatically generated fragments. No one technique has done more than high-pressure/performance liquid chromatography (HPLC) to revolutionize both aspects of structural analysis in terms of speed, reproducibility, sensitivity, predictability, and reliability. Over the period of the last 5 years, methods have emerged which have made possible (1) isolation of large and small complex intracellular, extracellular and membrane proteins to homogeneity with high yields and (2) identification of amino acids and their various derivatives at the level of 10^{-9} to 10^{-12} mol in a relatively short period of time. The recent developments in HPLC relating to the structural studies of protein molecules as developed in this and other laboratories will be described in this section.

PURIFICATION AND ISOLATION OF PROTEINS AND PEPTIDES

Proteins

The limiting factor in achieving the total primary structure of any protein molecule is the availability of sufficient amounts of pure homogeneous protein. The classical methods of protein purification involving open columns packed with different supports to allow size exclusion, ion exchange, or affinity separation suffer numerous major drawbacks. These include sample losses, large volumes, and sample size not to mention the time of separation which frequently extends into days. These drawbacks are more severe when the molecule of interest is a cell surface antigen or one which is available in very minute quantities. On the other hand, HPLC utilizing a support medium of small rigid particles of uniform size has emerged as an excellent alternative[2] to conventional open column techniques of protein and peptide separation.

Numerous reports are now available describing methods for separating proteins by HPLC. Although ion-exchange methods have been described, the most commonly used techniques are gel filtration and/or reverse-phase (RP) HPLC.

Gel Filtration/Size Separation by HPLC

Columns

There is a large selection of columns available to achieve separation on the basis of molecular weight. Most commonly employed are (1) I-250; I-125 and I-60 columns manufactured by Waters Associates and (2) TSK SW 4000; SW 3000 and SW 2000 manufactured by Toya Soda.

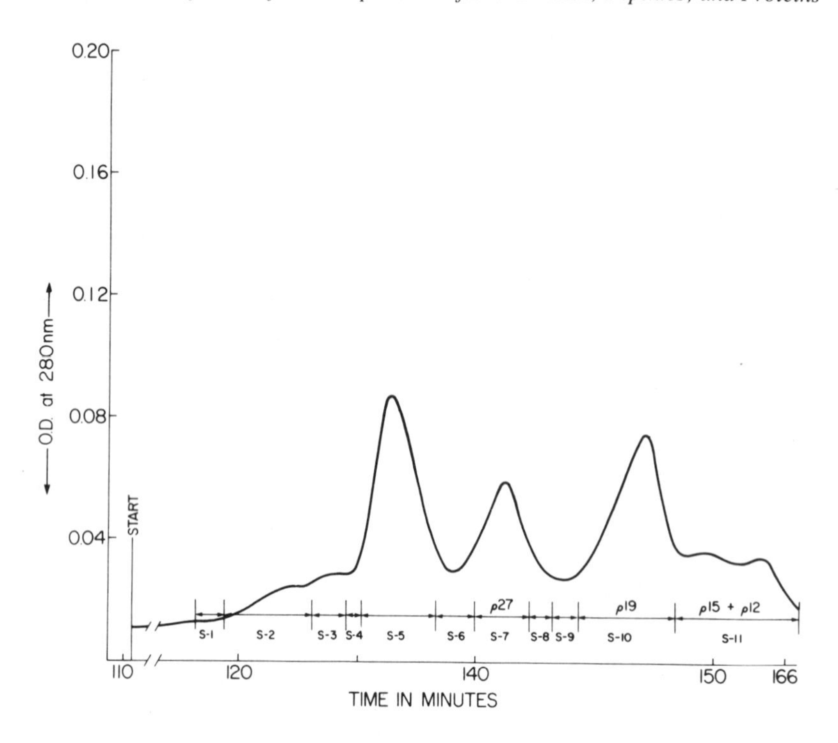

FIGURE 1. Gel permeation HPLC of viral structural proteins. (From Bhown, A. S., Bennett, J. C., Mole, J. E., and Hunter, E., *Anal. Biochem.*, 112, 128, 1981. With permission.)

Mobile Phase

Conventionally, a single solvent system is employed when size separation is to be achieved. Although any buffer system with UV transparent properties can be used it is advantageous to use volatile solvents especially if determination of the primary structure is the major objective. Both inorganic[3,4] and organic[5] solvent systems have been utilized to purify proteins by gel permeation HPLC.

Detection

Column effluents are monitored at 206, 254, or 280 nm in UV detectors equipped with continuous flow, low volume (10 μℓ or less), high-pressure cells with a larger light path to achieve maximum sensitivity.

Viral structural components p27 and p19 have been purified[5] by molecular exclusion HPLC using four I-125 columns (0.78 × 30 cm each; Waters Associates) in series. A mixture of acetic acid:propanol:highly purified water (20:15:65) is used as the mobile phase. The columns are developed at a flow rate of 0.2 mℓ/min and the effluent monitored at 280 nm. A separation is shown in Figure 1 and the purity of the preparation in Figure 2.

Jenix and Porter[3] have compared I-125 and TSK 4000 SW columns using 0.1 *M* potassium phosphate, pH 7.0, buffer as the mobile phase. These authors have concluded that gel permeation columns do not effect separation on size exclusion only, but they also propose some nonspecific solute-column matrix interaction. On the Waters I-125 columns there are negatively charged silanol groups on the silica support. This will cause a repulsion of negatively charged protein molecules resulting in an early exclusion from the column and thus an overestimation of molecular weights. Conversely, basic proteins will tend to adsorb and will be retarded unduly. This tends to cause lower recovery yields. Under certain conditions[6,8] an increase in the ionic concentration of the mobile phase may be helpful in suppressing this phenomenon but will not eliminate it altogether.[9,10]

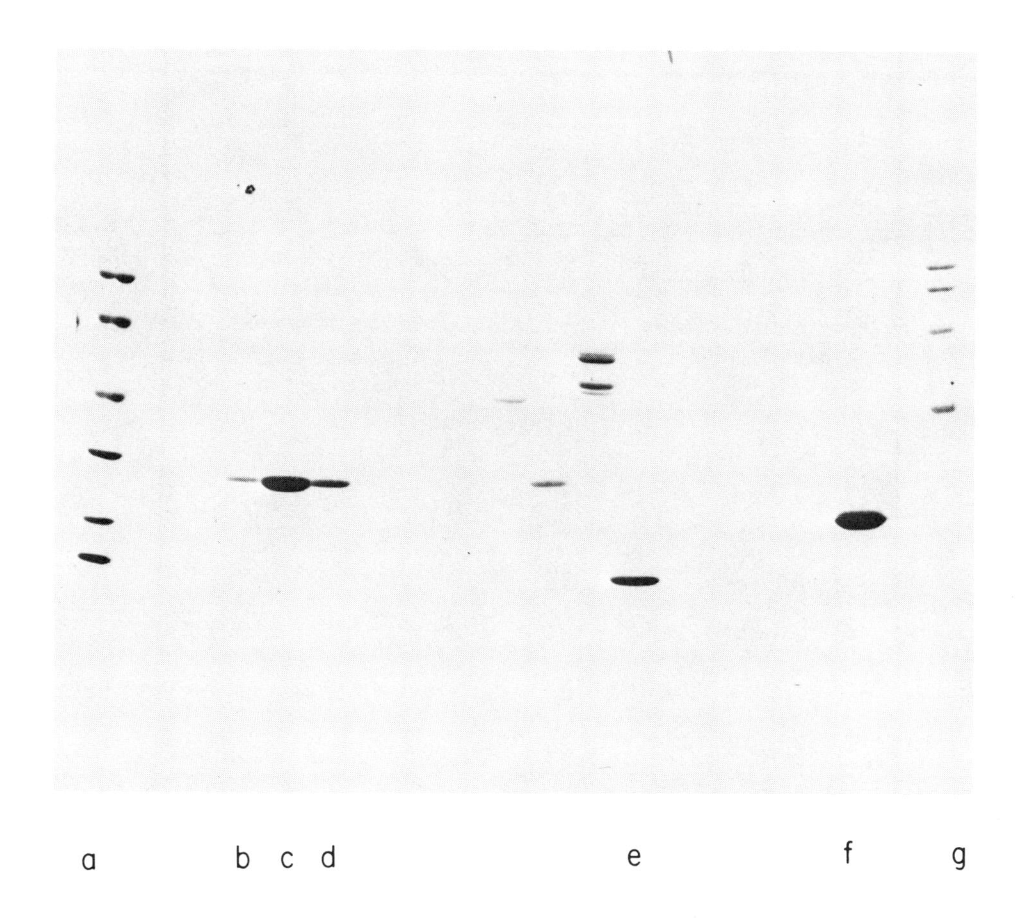

FIGURE 2. SDS-electrophoresis of the products of HPLC separation in Figure 1. (a) standard, (c) S-7, (f) S-10, and (g) standard. (From Bhown, A. S., Bennett, J. C., Mole, J. E., and Hunter, E., *Anal. Biochem.*, 112, 128, 1981. With permission.)

Reverse-Phase HPLC (RP-HPLC)

RP-HPLC is generally used in the final purification step following ion-exchange and/or gel filtration chromatographic procedures. However, recent publications[11-16] indicate that RP-HPLC alone can be effectively employed for the purification of large and small proteins including those of integral cell membrane and of viral origin with and without detergents.

Columns

Various types and sizes of reverse-phase columns depending upon the size of the silica particle (10 μm; 5 μm; 2 μm) and the type of side chain (C_{18}, C_8) are commercially available. Although details of all possible types of columns cannot be given here, examples of protein separations achieved on μBondapak® C_{18} (Waters Associates) and HC-ODS/Sil-X (0.26 × 25 cm; Perkin Elmer) columns will be described.

Solvents

Reverse-phase separation has been achieved in many cases by employing an increasing gradient of acetonitrile as a nonpolar solvent, while polar solvents such as sodium or ammonium acetate, or phosphate, (pH 3 to 5) trifluoroacetic acid (TFA) or heptafluorobutyric acid (HFBA) at low pH (pH 2 to 3) and low ionic concentration (0.01 to 0.5 *M*) have been employed as starting buffer.

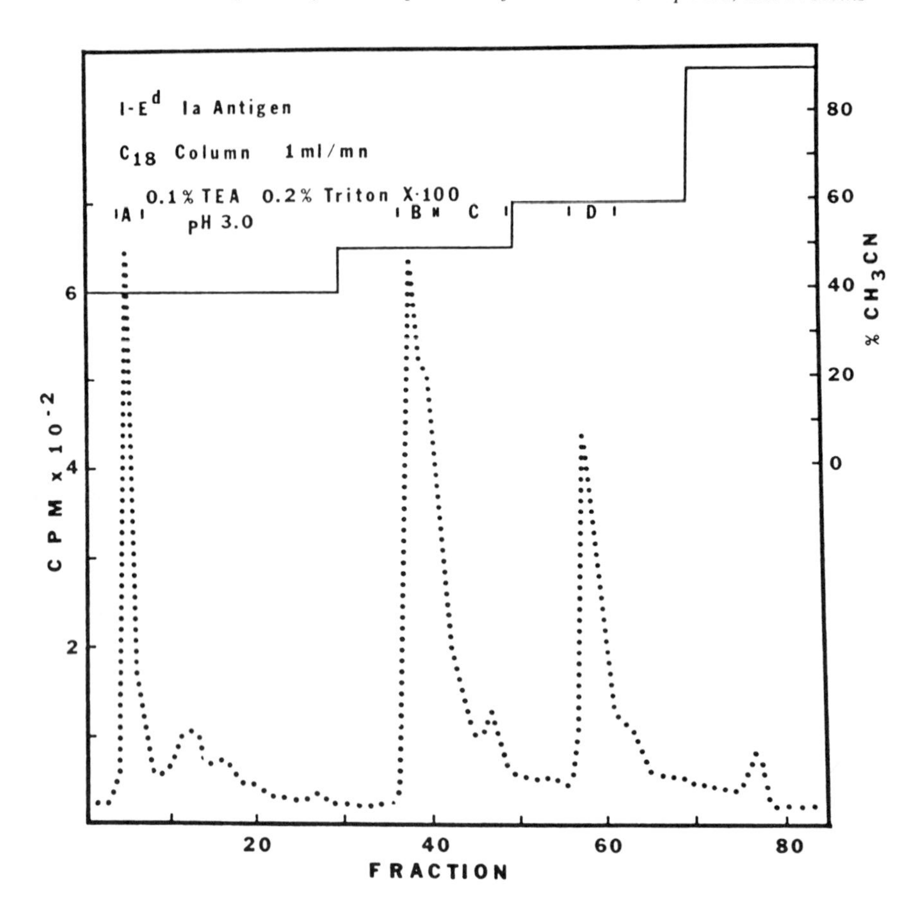

FIGURE 3. Preparative separation of the [^{3}H]-Leu-I-E^d Ia antigen component polypeptide on a C_{18} HPLC column. (From McKean, D. J. and Bell, M., in *Methods in Proteins Sequence Analysis IV*, Elzinga, M., Ed., The Humana Press.

McKean and Bell[11] have reported the isolation of Ia antigen on a μBondapak® C_{18} column. They have employed a gradient system consisting of a 0.1% triethylamine, 0.2% Triton® X-100, pH adjusted to 3.0 with TFA as buffer A and acetonitrile as buffer B. The column is eluted from 0% B to 100% B over a period of 60 min at a flow rate of 1 mℓ/min. In another experiment using a step gradient these authors have achieved the separation of heavy and light chains of Ia antigen (Figure 3).

The system demonstrates the successful use of μBondapak® C_{18} column for a membrane protein component separation in high yields; however, it is limited to radiolabeled polypeptides since the mobile phase is not transparent at the wavelengths which are absorbed by protein.

Separation of a 30,000 mol wt protein from reticuloendotheliosis virus (REV) has been achieved in this laboratory[16] on a HC-ODS/Sil-X column (0.26 × 25 cm; Perkin Elmer). Initial buffer A is composed of 0.1% TFA and buffer B is 100% acetonitrile containing 0.1% TFA. The column is eluted at a flow rate of 1.0 mℓ/min with 0% B for 10 min after injection followed by a linear gradient of 0 to 60% B over a period of 60 min. The separation is shown in Figure 4 and protein purity on SDS-PAGE in Figure 5.

Congote et al.[15] have demonstrated the separation of α, β, ^{G}γ, and ^{A}γ chains from human cord blood hemoglobin on μBondapak® C_{18}.

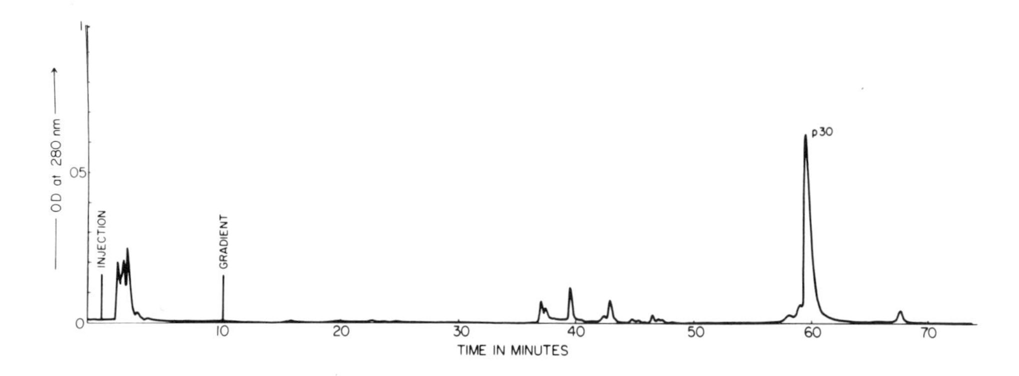

FIGURE 4. RP-HPLC separation of reticuloendotheliosis virus structural protein p30.

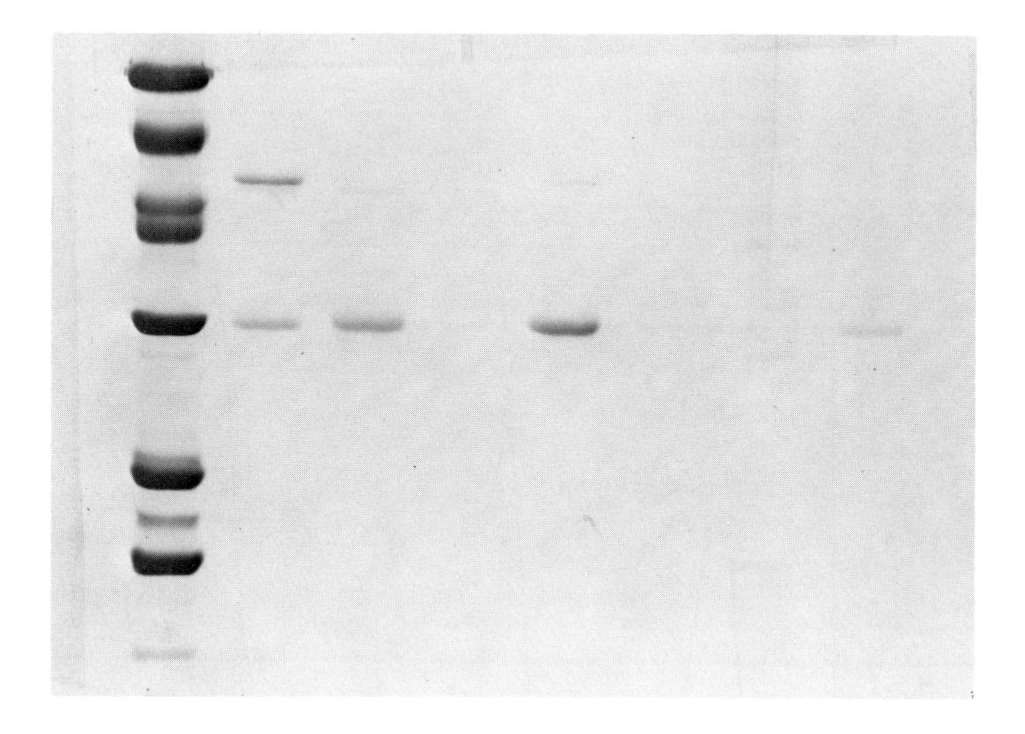

FIGURE 5. SDS-electrophoresis of the products of HPLC separation in Figure 4. Lane 1, standard; Lane 3, p30.

Peptides

The isolation of pure peptides from mixtures generated by enzymatic/chemical cleavage of proteins has always been the most tedious and time-consuming procedure preparatory to amino acid sequence analysis often resulting in low yields of eluted peptides. Utilization of HPLC has enabled investigators to separate protein and peptide mixtures[5,9,11,17-19] and to overcome the previous limitations.

Peptides generated by chemical cleavages are generally large while enzymatic digestion often results in smaller peptides. Employing size separation on gel filtration columns, such as I-125 and I-60, Bhown et al. have separated the mild acid hydrolysis and cyanogen bromide digestion products of the viral structural protein p27 (Figure 6A and B).[20] The protocol was the same as described earlier for the purification of p27. Mahoney and

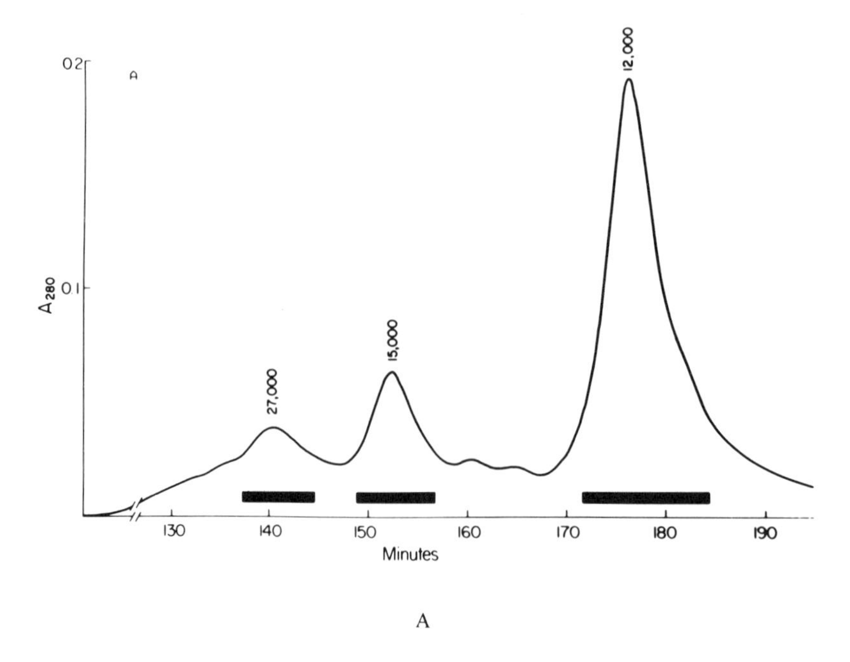

A

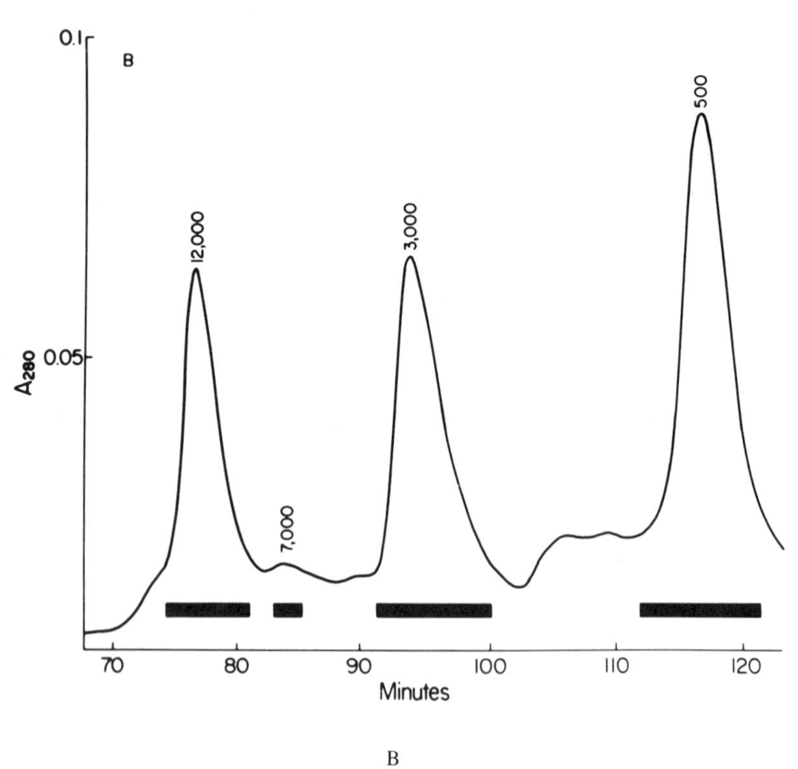

B

FIGURE 6. Gel permeation HPLC of the products of chemical cleavage of the p27 and 12K fragment. (A) Cyanogen bromide digest of the p27; (B) mild acid catalyzed hydrolysis of the 12K fragment.

Hermodson[13] have successfully separated cyanogen bromide peptides of α and β chains of human globin employing a Lichrosorb® RP-8 column (0.46 × 25.0 cm; Brownlee Labs), and a volatile mobile phase consisting of TFA and organic solvent (acetonitrile or 1-propanol).

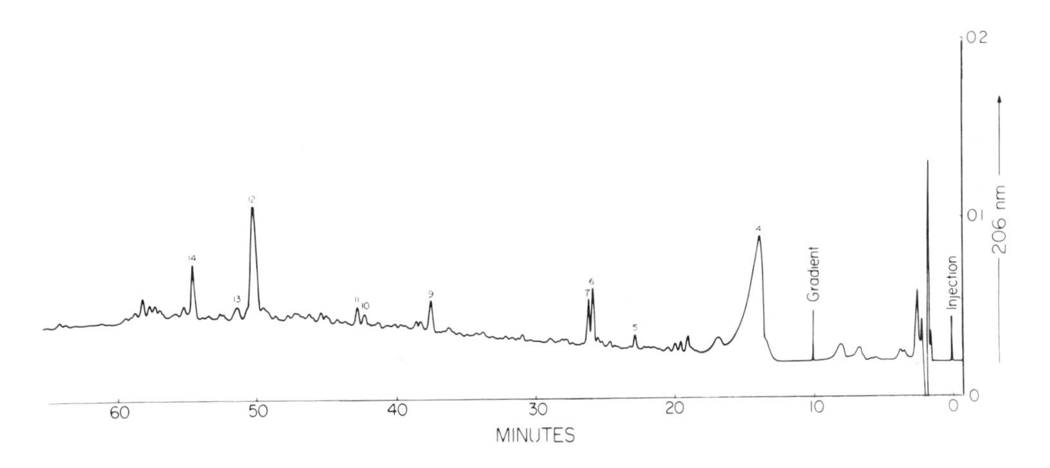

FIGURE 7. RP-HPLC of tryptic digest of the 12K fragment.

Niemann et al.[21] have reported separation of the products of a cyanogen bromide digestion of factor D by molecular exclusion HPLC.

All of these reports have described the use of volatile buffers thus making the separated products suitable for direct amino acid sequence analysis.

Products of enzymatic digestion, which results in a mixture of small peptides, have been effectively and quantitatively separated by ion-pair RP-HPLC. TFA has gained a wide application as an ion-pairing reagent while acetonitrile is the most commonly employed nonpolar limiting solvent. Figure 7 shows the separation of the products of tryptic digestion of a 1200-dalton peptide (Figure 6A) obtained by mild acid hydrolysis of viral structural protein p27 (Figure 1). The separation is achieved on a μBondapak® C_{18} column (0.4 × 25.0 cm; Waters Associates) employing 0.1% TFA as an ion-pairing reagent and acetonitrile containing 0.1% TFA as limiting buffer. The peptides obtained were lyophilized and directly sequenced in an automated amino acid sequencer.

Bennett et al.[12] have reported the use of HFBA in lieu of TFA as an ion-pairing reagent. However, HFBA causes a base line shift far greater than when TFA is employed. Yang et al.[22] have described ammonium acetate pH 6.0 as an alternative ion-pairing system for RP-HPLC. These three systems of ion-paired mobile phases are extremely helpful in conjunction with acetonitrile to affect separation of the mixture of peptides suitable for amino acid sequence analysis.

AMINO ACID SEQUENCE ANALYSIS

Amino acid sequence determination as developed by Edman[23,24] is based on the stepwise degradation of a polypeptide chain releasing one amino acid at a time. This must be followed by an efficient identification of the liberated amino acid derivative. Recent technical developments have greatly improved sequencing capabilities both in terms of automation and sensitivity. Instrument modifications, refinements in the Edman chemistry, and progress in the methods for aminc acid identification have all contributed towards attainment of sequencing capabilities at the pico and femto mole levels.

The major change which has led to microsequencing is the development of detection systems for phenylthiohydantoin (PTH) derivatives of amino acids at picomole levels. This has been made possible by the development of a HPLC system equipped with an autoinjector and a data reduction system. The most sensitive of the HPLC systems when operating optimally can easily identify <10 pmol of material.

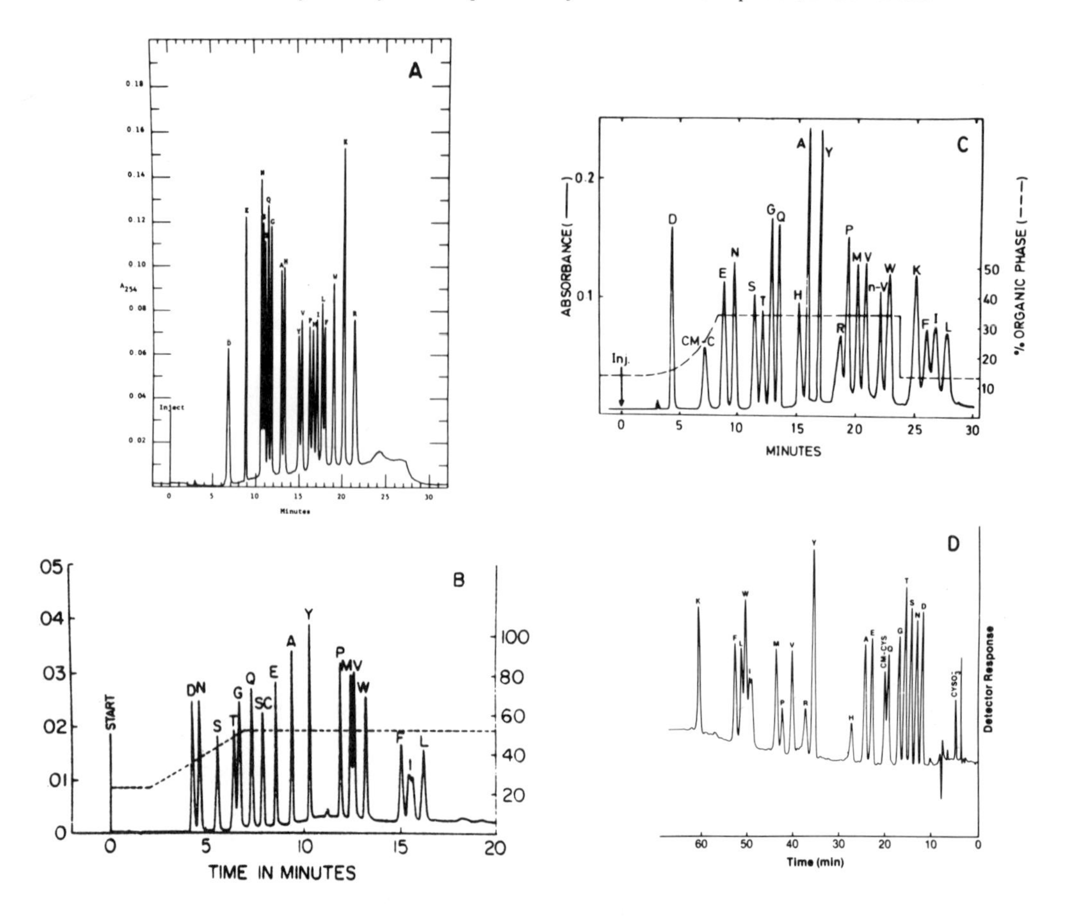

FIGURE 8. HPLC separation of standard PTH amino acid mixture by different methods (Courtesy of Marcel Dekker Inc.)

A number of reports have appeared from this and other laboratories[25-28] in the recent past demonstrating improvements in PTH amino acid separation and time of analysis. Figure 8 shows some of the separations achieved by HPLC using different solvents, columns, and programs. Use of a computer for base line correction followed by peak height expansion has enabled Hunkapiller and Hood[29] to separate 5 pmol of 19 standard PTH amino acids. A computer-controlled system has been developed in this laboratory to separate and collect each peak of a PTH amino acid as it elutes from the HPLC column directly into the scintillation vial for the purpose of radioactive microsequencing.[30]

RP-HPLC has offered promising results for quantitative separation of all the PTH amino acid derivatives within less than 30 min.[25-28] A further advancement with this technique is the capability of simultaneously monitoring the effluent at two different wavelengths (254 and 313 nm). This has helped in the identification and quantitation of PTH-Thr and PTH-Ser in their dehydro forms. PTH-Arg and PTH-His can be separated and quantitatively identified by RP-HPLC.

Figure 9 shows the separation of PTH amino acids with a Waters HPLC system equipped with Model 440 dual-channel (254 nm; 313 nm) UV detector; data module 720; system controller 730 and an Altex 5-μm Ultrasphere ODS column (0.46 × 15.0 cm) marketed by Beckman Instruments (Part #235330). Solvents A and B were constituted as follows: solvent A — 0.04 *M* sodium acetate, pH 3.6 to 3.7, containing 50 μℓ of acetone per liter and filtered through 0.2-μm unipore polycarbonate membranes (Bio-Rad Laboratories);

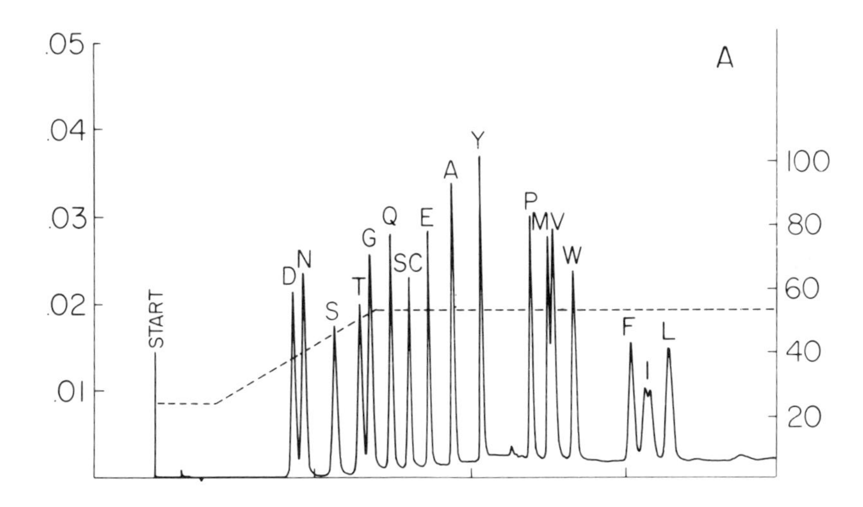

FIGURE 9. Separation of PTH amino acids on Ultrasphere ODS 5-μm column.

solvent B — 100% methanol (Ominsolv-MSB) containing 250 μℓ of glacial acetic acid per liter. The column is developed for 2 min at initial conditions (77% A + 23% B) followed by a 5-min linear gradient to final conditions (47% A + 53% B) at a flow rate of 1.5 mℓ/min.

It is generally preferable that PTH amino acids be identified by two independent methods. Classically the second identification is achieved by thin-layer chromatography (TLC) or acid hydrolysis of PTH amino acids to their parent amino acid followed by analysis on an amino acid analyzer. Both of these systems suffer in being nonquantitative and time consuming. Furthermore, PTH-Ser and PTH-Thr on hydrolysis convert to glycine:alanine and glycine:α-aminobutyric acid, and PTH-Asn and PTH-Gln are hydrolyzed to their respective acids. Thus these residues cannot be assigned unambiguously.

In order to circumvent this problem we have developed a second method of identifying PTH amino acids by HPLC, using a different column in which the elution sequence of PTH amino acids is altered. Thus the same sample is analyzed by two different independent HPLC methods for precise residue identification.

The second HPLC method involves the use of an Altex 5-μm Ultrasphere-CN column (0.46 × 15 cm) marketed by Beckman Instruments Inc. (Part 244070). Solvent A is composed of 0.02 *M* sodium acetate pH 5.4 containing 50 μℓ of acetone per liter and filtered through a 0.2-μm filter. Solvent B is 100% methanol containing 250 μℓ of acetic acid. The column is developed for 2 min at initial conditions (77% A + 23% B) after injection followed by a 5-min linear gradient to final condition (50% A + 50% B) at a flow rate of 1.5 mℓ/min. The separation is shown in Figure 10.

SUMMARY

Protein sequence analysis although automated by Edman in 1967[31] has had to depend on the slow and time-consuming process of protein and peptide purification with poor recoveries. The development of HPLC and the availability of various columns packed with rigid matrices have completely revolutionized the protein-peptide isolation procedures. Furthermore, availability of preparative columns for HPLC has now made it feasible to purify enough protein or peptides in a day or two to keep the automated sequencer committed for rapid analysis.

The second limiting factor for a protein sequencing laboratory has been the slow pace of PTH amino acid identification. Developments such as new HPLC methods, automated

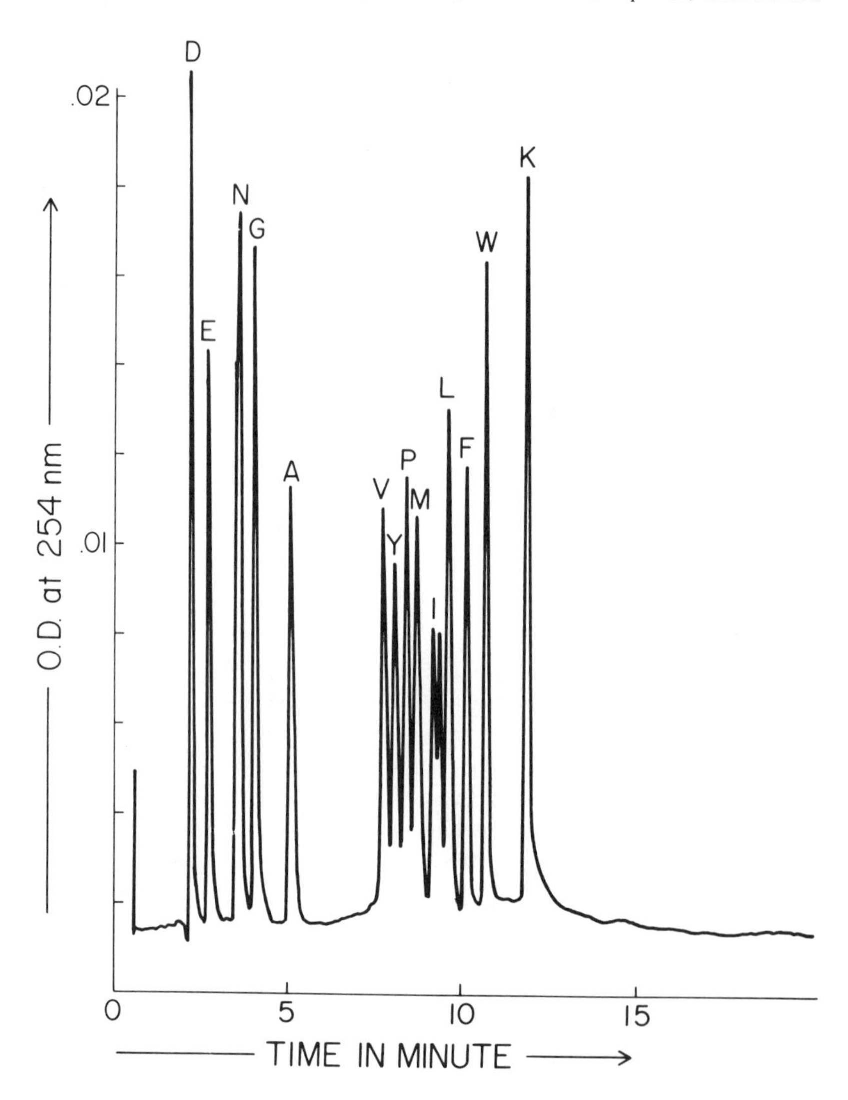

FIGURE 10. Separation of PTH amino acid on Ultrasphere CN 5-μm column.

injection devices, and computer-controlled HPLC for the base line subtraction has overcome this factor. These automated techniques have filled the significant gaps needed for the realization of a fully automated system with a direct read out of amino acid sequence data.

ACKNOWLEDGMENT

The authors wish to acknowledge the help of Mr. Thomas W. Cornelius, Ms. Karen Buttler, and Ms. Mary Elisa Kallman. We also express our thanks to Ms. Brenda Gosnell and Mrs. Melissa Ham for typing this manuscript. Permission by the Humana Press, Academic Press, and Marcel Dekker, and the authors for reproduction of their work is gratefully acknowledged. The work was supported by grant CA-13148 AMO-3555.

REFERENCES

1. **Walsh, K. A., Ericsson, L. H., Parmelee, D. C., and Titani, K.,** Advances in protein sequencing, *Ann. Rev. Biochem,* 50, 261, 1981.
2. **Brown, R. P.,** in *High Pressure Liquid Chromatography,* Academic Press, New York, 1973, 128.
3. **Jenix, R. A. and Porter, J. W.,** High performance liquid chromatography of proteins by gel permeation chromatography, *Anal. Biochem,* 111, 184, 1981.
4. **Chang, S. H., Gooding, K. M., and Reignier, F. E.,** High performance liquid chromatography of proteins, *J. Chromatogr.,* 125, 103, 1976.
5. **Bhown, A. S., Bennett, J. C., Mole, J. E., and Hunter, E.,** Purification and characterization of the "gag" gene products of avian type C retroviruses by high pressure liquid chromatography, *Anal. Biochem.,* 112, 128, 1981.
6. **Stenlund, B.,** Polyelectrolyte effects in gel chromatography, *Adv. Chromatogr.,* 14, 37, 1976.
7. **Crone, H. D. and Dawson, R. M.,** Residual anionic properties of a covalently substituted controlled pore glass glyceryl-CPG, *J. Chromatogr.,* 129, 91, 1976.
8. **Crone, H. D.,** Ion exclusion effects on the chromatography of acetylcholinesterase and other proteins on agarose columns at low ionic strength, *J. Chromatogr.,* 92, 127, 1974.
9. **Reignier, F. E. and Gooding, K. M.,** High performance liquid chromatography of proteins, *Anal. Biochem.,* 103, 1, 1980.
10. **Rokushika, S., Ohkawa, T., and Hatano, H.,** High speed aqueous gel permeation chromatography of proteins, *J. Chromatogr.,* 176, 456, 1979.
11. **McKean, D. J. and Bell, M.,** Preparative isolation of Ia antigen membrane protein component polypeptides on C_{18} reverse phase HPLC, in *Methods in Protein Sequence Analysis IV,* Elzinga, M., Ed., The Humana Press, Clifton, N.J., in press.
12. **Bennett, H. P. J., Browne, C. A., and Soloman, S.,** Purification of the two major forms of rat pituitary corticotropin using only reversed phase liquid chromatography, *Biochemistry,* 20, 4530, 1981.
13. **Mahoney, W. C. and Hermodson, M. A.,** Separation of large denatured peptides by reverse phase high performance liquid chromatography, *J. Biol. Chem.,* 255, 11199, 1980.
14. **Klee, C. B., Oldewurtel, M. D., William, J. F., and Lee, J. W.,** Analysis of Ca^{2+}-binding proteins by high performance liquid chromatography, *Biochem. Int.,* 2, 485, 1981.
15. **Congote, L. F., Bennett, H. J. P., and Soloman, S.,** Rapid separation of the α, β, γ and α human globin chains by reversed phase high pressure liquid chromatography, *Biochem. Biophys, Res. Commun.,* 89, 851, 1979.
16. **Bhown, A. S. and Bennett, J. C.,** unpublished observation.
17. **Kemp, M. C., Hollaway, W. L., Bennett, J. C., and Compans, R. C.,** Reverse phase ion pair high pressure liquid chromatography of viral tryptic glycopeptides, *J. Liq. Chromatogr.,* 4, 587, 1981.
18. **Hollaway, W. L., Prestidge, R. L., Bhown, A. S., Mole, J. E., and Bennett, J. C.,** Hydrophilic ion-pair reversed phase high performance liquid chromatography of peptides and proteins, in *Recent Developments in Chromatography and Electrophresis,* Vol. 10, Frigerio, A. and McCamish, M., Eds., Elsevier, Amsterdam, 1980, 131.
19. **Tomlinson, E., Jefferies, T. M., and Riley, C. M.,** Ion pair high performance liquid chromatography, *J. Chromatogr.,* 159, 315, 1978.
20. **Bhown, A. S., Bennett, J. C., Cornelius, T. W., and Hunter, E.,** Exclusive use of high pressure liquid chromatography for the determination of complete amino acid sequence of the 12K fragment of avian sarcoma virus structural protein p27, *J. Chromatogr.,* submitted.
21. **Niemann, M. A., Bhown, A. S., Bennett, J. C., and Volanakis, J. E.,** Partial amino acid sequence of Human D protein: purification by HPLC and analysis of the CNBr, *o*-Iodosobenzoic acid and trypsin peptides, in Proc. Int. Conf. HPLC, Washington, D.C., in press.
22. **Yang, Y. C., Kratzin, H., and Hilschmann, N.,** Chromatography and rechromatography in HPLC-separation of peptides:poster presentation, in 4th Int. Conf. Methods in Protein Sequence Analysis, Brookhaven National Laboratory, Upton, New York, 1981.
23. **Edman, P.,** A method for the determination of the amino acid sequence in peptides, *Arch. Biochem. Biophys.,* 22, 475, 1949.
24. **Edman, P.,** Method for determination of the amino acid sequence in peptides, *Acta. Chem. Scand.,* 4, 283, 1950.
25. **Bhown, A. S., Mole, J. E., and Bennett, J. C.,** An improved procedure for high sensitivity microsequencing: use of aminoethyl aminopropyl glass beads in the Beckman sequencer and the Ultrasphere ODS column for PTH amino acid identification, *Anal. Biochem.,* 110, 355, 1981.
26. **Johnson, N. D., Hunkapiller, M. W., and Hood, L. E.,** Analysis of phenylthiohydantoin amino acids by high performance liquid chromatography on DuPont Zorbax cyanopropylsilane columns, *Anal. Biochem.,* 100, 335, 1979.

27. **Fohlman, J., Rask, L., and Peterson, P. A.,** High pressure liquid chromatographic identification of phenythyiohydantoin derivatives of all twenty common amino acids, *Anal. Biochem.*, 106, 22, 1980.
28. **Henderson, L. E., Copeland, T. D., and Oroszlan, S.,** Separation of amino acid phenylthiohydantoins by high performance liquid chromatography on phenylalkyl support, *Anal. Biochem.*, 102, 1, 1980.
29. **Hunkapiller, M. W. and Hood, L. E.,** New protein sequenator with increased sensitivity, *Science,* 207, 523, 1980.
30. **Bown, A. S., Mole, J. E., Hollaway, W. L., and Bennett, J. C.,** Computer assisted high pressure liquid chromatography of radiolabelled phenylthiohydantoin amino acids, *J. Chromatogr.*, 156, 35, 1978.
31. **Edman, P. and Begg, G.,** A protein sequenator, *Eur. J. Biochem.*, 1, 80, 1967.

PROTEIN FRAGMENT SEPARATION BY HPLC

R. L. Prestidge

INTRODUCTION

High-performance liquid chromatography (HPLC) continues to become more widely used as a technique for the separation of protein and peptide fragments. An indication of the increasing maturity of the method is that many publications have recently appeared which do not mention HPLC in their titles, but which make extensive use of HPLC for peptide separations.[1-13] Obviously the time is not too far distant when HPLC will not rate even a mention in an abstract, but will be relegated to a mere sentence in the Methods section. This progress towards invisibility demonstrates that HPLC is no longer a trendy new technique, but is becoming just another indispensible tool for the biochemist, like the centrifuge or the pH meter.

Proteins are fragmented, and the fragments separated for a variety of reasons. Analytical separation of protein fragments, or peptide mapping, is used to characterize the parent protein and to answer questions about precursor relationships and identity of closely related proteins, including modified proteins and genetic variants. Preparative separation of protein fragments is essential for protein sequence determination and for the characterization of specific regions of the molecule such as glycosylation sites. Current HPLC procedures are suitable for both analytical and milligram-preparative applications. Examples of both will be reviewed in this report.

Reversed-phase HPLC (RP-HPLC), because of its high resolution and applicability to a wide variety of samples, has become the most widely used separation mode, and it is often stated that 60 to 70% of all recent HPLC applications involve the use of reversed-phase columns.[14-16] For peptide and protein fragments, domination by the reversed-phase column is complete, and alternative separation modes are seldom mentioned. Some laboratories are making extensive use of gel permeation columns in the purification of protein fragments for sequencing studies,[17] but the high cost of these columns and early difficulties in choosing buffer systems to minimize nonspecific column-sample interactions appear to have dissuaded many researchers from making full use of this current generation of support materials.

TECHNIQUES

The development of techniques for the separation of protein fragments by HPLC was aided by the realization[18] that the addition of ionic modifiers to the mobile phase gave much improved resolution of peptide mixtures. The solvent system of Hancock et al.[18] has found widespread use. A typical recent publication on the separation of protein fragments by RP-HPLC would involve the use of an octadecyl-silica column such as the μBondapak® C_{18} with an acetonitrile-water solvent system containing 0.1% phosphoric acid as an ionic modifier, and with low wavelength UV detection.[2,6,11,19-24] This system would appear to be the most suitable starting point for those using HPLC for peptide separations for the first time. Octyl[3,8,10,25,26] and cyano[4,12,13,27-29] columns are finding increased use and appear to offer important variations in selectivity from the octadecyl columns. A report on the use of trimethylsilyl-silica columns[30] suggests that they also may offer a useful alternative to octadecyl-silica. The use of large-pore silica[31,32] appears to improve resolution and column capacity for larger peptides and should be investigated further. The recent development of radially compressed flexible-walled columns[33-35] not only lowers the cost of column replacement but also appears to give the possibility of packing one's own columns with commercially available or homemade stationary phases without requiring special column-packing equipment.

Acetonitrile is by far the most widely used organic solvent for RP-HPLC. Other popular solvents include 1-propanol[3,8-10,36-39] and 2-propanol.[17,31,33,34] Acetone[12,13,29] gives very good resolution of tryptic peptides, but because of its high UV absorption it has only been used with radiolabeled samples. Ethanol[1,26] also appears to give very high resolution of the peptides in various proteolytic digests, and because of its low cost and high purity should be investigated further. A wide variety of ionic modifiers have been investigated in attempts to combine the desirable characteristics of suppression of silanol interactions, UV transparency, and volatility. Volatile acids used include formic,[1,40] trifluoroacetic,[10,31,36,41] heptafluorobutyric,[28] acetic,[42] and hydrochloric;[26] volatile amine salts include ammonium acetate and chloride,[30,43-46] pyridinium acetate and formate,[3,8,9,17,37,39,42] and ammonium bicarbonate.[33] Amine phosphates are not volatile; however, their excellent chromatographic properties have led to wide use.[4,5,27,34,48] An earlier report on the use of *Tris*-acetate buffers[49] does not appear to have been followed up.

The post-column fluorescence detection system of Stein et al. has given excellent results with a wide variety of peptide mixtures,[3,8,9,36,38,39,47] giving not only high sensitivity but also a wide choice of mobile phases to the chromatographer. However, because of the added complexity of such systems, few other workers have used this method of detection.[37] Most continue to use low wavelength UV detection, although the use of internally radioactively labeled peptides, where these are available[10,12,13,29,41,42,50] also gives extreme sensitivity and the ability to use a wide range of UV-absorbing mobile phases.

The evaluation of a new column, mobile phase, or detector and comparison with published systems is still very much an *ad hoc* procedure carried out with whatever peptides are at hand. Several laboratories routinely evaluate all new chromatographic systems with a standard peptide mixture obtained by tryptic digestion of reduced and alkylated lysozyme.[59] The recent publication of several separations of this and related peptide mixtures[23,26,33,36,43] suggests that the separation of tryptic lysozyme peptides could become a generally used standard of performance for HPLC systems. A better standard, however, would be the tryptic peptides of rabbit hemoglobin, because it is readily available internally labeled with ^{3}H or ^{14}C using the reticulocyte lysate system.[51] Internally labeled peptides would make it easy to quantitate the yields obtained with a variety of peptides of different hydrophobicity and charge, and consequently give a more objective basis of comparison for columns, solvents, and ionic modifiers. Whichever test mixture of peptides is chosen, it is important that the peaks obtained are characterized, for example, by amino acid analysis and comparison with published sequence data. Figure 1 compares two published separations[26,43] where this procedure was followed and shows that although the overall trends in hydrophobicity are similar, marked variations in selectivity are obtained from different chromatographic systems. These variations are not obviously related to such factors as the presence of charged groups in the peptides. Three other published separations of these peptides[23,33,36] where the peaks were not characterized cannot be meaningfully compared with other workers' data.

APPLICATIONS

A very large number of proteins have been studied by HPLC mapping of their tryptic peptides. These include ACTH,[7] actinidin,[52] apolipoproteins,[34] bovine serum albumin,[36] calcium binding proteins,[21] CRP,[47] cytochrome b_5,[1] endorphins,[42] enkephalin precursors,[6] human growth hormone,[6,33] Ia antigens,[10-13,29,34] IgG light chain,[34] IgM heavy chain,[23] lysozyme,[23,26,33,36,43] myelin basic protein,[36,48] myoglobin,[17] neurophysin,[27] ovalbumin,[36,47] prolyl hydroxylase,[47] proopiocortin,[9] a protein kinase,[2] relaxin,[53] ribonuclease S-peptide,[25] thyroglobulin,[33] and viral proteins.[50] The study of hemoglobin variants by HPLC tryptic mapping[20,30,37,44,45] is a major area of research, which will be covered elsewhere in this volume. Cyanogen bromide peptides, although more difficult to handle because of their

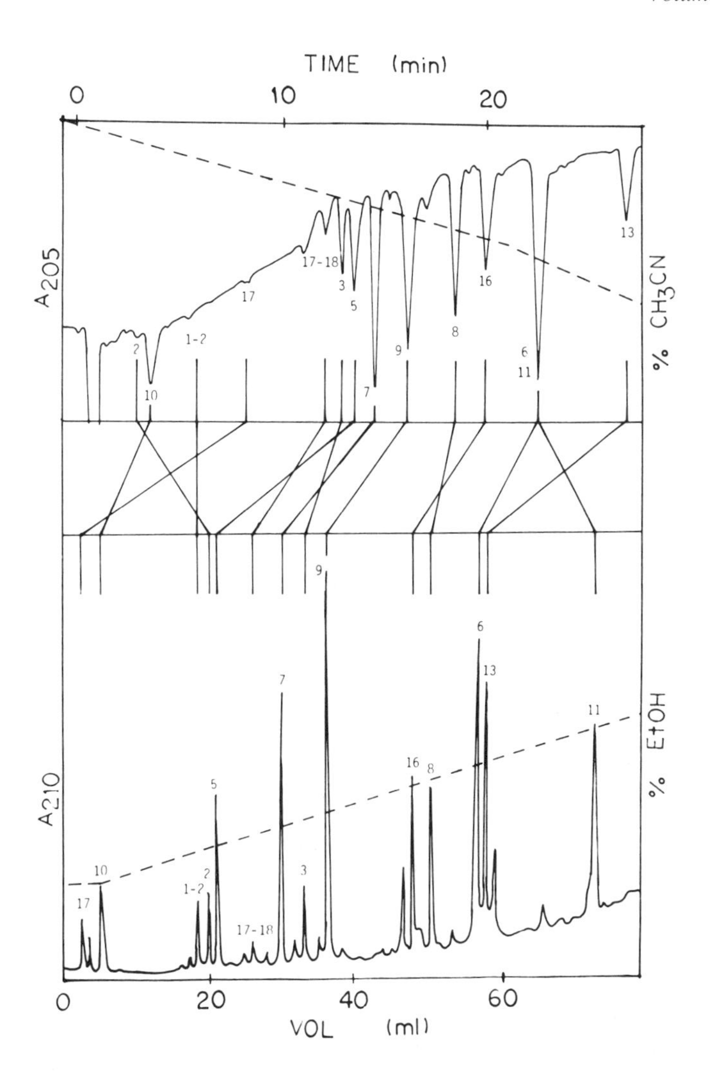

FIGURE 1. Selectivity changes in RP-HPLC of the tryptic peptides of lysozyme. Upper curve: data adapted from Reference 43. Column: Varian Micropak MCH-10 (C_{18} uncapped). Ionic modifier: 100 m*M* ammonium chloride, pH 4.1. Organic solvent: acetonitrile to 40%. Lower curve: data adapted from Reference 26. Column: Merck Lichrosorb® RP-8 (C_8). Ionic modifier: 0.1% HCl. Organic solvent: ethanol to 50%. Peptide numbers after Canfield.[58]

generally hydrophobic nature, have been successfully separated by HPLC in the cases of bacteriorhodopsin,[40] collagen,[24,28] cytochrome b_5,[1] Factor B,[23] globin,[31,37,46] myoglobin,[36] and virus M protein.[23] Other proteolytic digests fractionated by HPLC include the thermolysin digest of acyl carrier protein[52] and corticotropin-releasing factor,[4] the mild acid cleavage of viral p27 protein,[23] the peptic digest of ribonuclease S-protein,[25] the *N*-bromosuccinimide and *Staphlococcus aureus* V8 protease cleavage products of cytochrome b_5,[1] the tonin peptides of ACTH,[5] and the pronase glycopeptides of various viral proteins.[42]

CONCLUSIONS

In the future, it seems likely that RP-HPLC will be used for almost all separations of protein fragments whether for peptide mapping or for sequencing. Already several major sequencing groups have found that they are able to perform all necessary separations by this technique alone or in combination with HPLC gel filtration.[1,3,4] Present HPLC pumps and detectors appear to be adequate for the demands made on them, and because each protein fragment mixture is different, automation will probably not be a major benefit to all users. Consequently the major hardware advances in the near future should be sought in better column technology.

The theoretical basis for reversed-phase separations has been extensively studied,[35,54] and hydrophobicity functions have been defined[55-57] which can be used to predict peptide elution behavior. It is important to note that if the only interactions between the sample and the column are hydrophobic in nature, then changing from one octadecyl-silica column to another, or from octadecyl- to octyl-silica should not affect the order of elution of sample peaks. In fact, it is the rule rather than the exception to find major selectivity changes from one reversed-phase column to another. It appears that many of the selectivity differences between different column packings are related to nonhydrophobic interactions. A major deficiency in our present approach to reversed-phase columns is the failure to understand, control, and utilize silanol and other nonhydrophobic interactions in the separation of peptide mixtures. Free silanol groups have been thought to be completely disadvantageous on a reversed-phase column, and much effort has been devoted to minimizing silanol interactions by column end-capping and the addition of ionic modifiers to the solvent mixture. However, a recent paper[35] has suggested that the presence of a limited number of free silanols may confer desirable properties such as increased column efficiency and useful selectivity variations. For this reason, a systematic study of the variation of column properties with residual silanol content needs to be carried out, and also studies of the effects of end-capping a reversed-phase column with a polar reagent such as a diol rather than the usual trimethylsilane. By these approaches the presence of nonhydrophobic interactions between solutes and the reversed-phase column may be changed from liabilities to assets.

The reversed-phase column has been extensively studied, and much has been learned about the separation mechanism. However many selectivity differences such as those shown in Figure 1 remain to be understood. Only with careful study of defined peptides and column packings in association with an extension of our present theoretical treatments will the current semi-empirical approach be replaced by a more rigorously defined and controlled use of this valuable tool.

REFERENCES

1. **Takagaki, Y., Gerber, G., Nihei, K., and Khorana, H. G.,** Amino acid sequence of the membranous segment of rabbit liver cytochrome b_5. Methodology for separation of hydrophobic peptides, *J. Biol. Chem.*, 255, 1536, 1980.
2. **Nelson, N. C. and Taylor, S. S.,** Differential labeling and identification of the cysteine-containing tryptic peptides of catalytic sub-unit from porcine heart cAMP-dependent protein kinase, *J. Biol. Chem.*, 256, 3743, 1981.
3. **Levy, W. P., Rubenstein, M., Shively, J., Del Valle, U., Lai, C-Y., Moschera, J., Brink, L., Gerber, L., Stein, S., and Pestka, S.,** Amino acid sequence of a human leukocyte interferon, *Proc. Natl. Acad. Sci. U.S.A.*, 78, 6186, 1981.
4. **Spiess, J., Rivier, J., Rivier, C., and Vale, W.,** Primary structure of corticotropin-releasing factor from ovine hypothalamus, *Proc. Natl. Acad. Sci. U.S.A.*, 78, 6517, 1981.
5. **Seidah, N. G., Chan, J. S. D., Mardini, G., Benjannet, S., Chretien, M., Boucher, R., and Genest, J.,** Specific cleavage of beta-LPH and ACTH by tonin: release of an opiate-like peptide beta-LPH (61-78), *Biochem. Biophys. Res. Commun.*, 86, 1002, 1979.

6. **Olson, K. C., Fenno, J., Lin, N., Harkins, R. N., Snider, C., Kohr, W. H., Ross, M. J., Fodge, D., Prender, G., and Stebbing, N.,** Purified human growth hormone from *E. coli* is biologically active, *Nature (London),* 293, 408, 1981.
7. **Brubaker, P. L., Bennett, H. P. J., Baird, A. C., and Solomon, S.,** Isolation of $ACTH_{1-39}$, $ACTH_{1-38}$ and CLIP from the calf anterior pituitary, *Biochem. Biophys. Res. Commun.*, 96, 1441, 1980.
8. **Kimura, S., Lewis, R. V., Stern, A. S., Rossier, J., Stein, S., and Udenfriend, S.,** Probable precursors of (Leu) enkephalin and (Met) enkephalin in adrenal medulla: peptides of 3-5 kilodaltons, *Proc. Natl. Acad. Sci. U.S.A.*, 77, 1681, 1980.
9. **Rubenstein, M., Stein, S., and Udenfriend, S.,** Characterisation of pro-opiocortin, a precursor to opioid peptides and corticotropin, *Proc. Natl. Acad. Sci. U.S.A.*, 75, 669, 1978.
10. **Kimball, E. S., Nathenson, S. G., and Coligan, J. E.,** Amino acid sequence of residues 1-98 of the H-$2K^d$ murine major histocompatibility antigen: comparison with H-$2K^b$ and H-$2D^b$ reveals extensive localised differences, *Biochemistry,* 20, 3301, 1981.
11. **Rose, S. M., Hansen, T. H., and Cullen, S. E.,** Structural relation of murine "third locus" (H-2L) major histocompatibility antigens to the products of H-2K and H-2D loci, *J. Immunol.*, 125, 2044, 1980.
12. **Walker, L. E., Ferrone, S., Pellegrino, M. A., and Reisfeld, R. A.,** Structural polymorphism of the B chain of human HLA-DR antigens, *Mol. Immunol.*, 17, 1443, 1980.
13. **McMillan, M., Cecka, J. M., Hood, L., Murphy, D. B., and McDevitt, H. O.,** Peptide map analyses of murine Ia antigens of the I-E subregion using HPLC, *Nature (London),* 277, 663, 1979.
14. **Deelder, R. S., Linssen, H. A. J., Konijnendijk, A. P., and van de Venne, J. L. M.,** Retention mechanism in reversed-phase ion-pair chromatography of amines and amino acids on bonded phases, *J. Chromatogr.*, 185, 241, 1979.
15. **Colin, H. and Guiochon, G.,** Introduction to reversed-phase high-performance liquid chromatography, *J. Chromatogr.*, 141, 289, 1977.
16. **Karger, B. L. and Giese, R. W.,** Reversed phase liquid chromatography and its application to biochemistry, *Anal. Chem.*, 50, 1048A, 1978.
17. **Holloway, W. L., Bhown, A. S., Mole, J. E., and Bennett, J. C.,** HPLC in the structural studies of proteins, in *Chromatographic Science Series X,* Marcel Dekker, New York, 1978, 163.
18. **Hancock, W. S., Bishop, C. A., Prestidge, R. L., Harding, D. R. K., and Hearn, M. T. W.,** High pressure liquid chromatography of peptides and proteins. II. The use of phosphoric acid in the analysis of underivatised peptides by reversed-phase high-pressure liquid chromatography, *J. Chromatogr.*, 153, 391, 1978.
19. **Shelton, J. B., Shelton, J. R., and Schroeder, W. A.,** Preliminary experiments in the separation of globin chains by high performance liquid chromatography, *Hemoglobin,* 3, 353, 1979.
20. **Johnson, C. S., Moyes, D., Schroeder, W. A., Shelton, J. B., Shelton, J. R., and Beutler, E.,** Hemoglobin Pasadena; identification by high performance liquid chromatography of a new unstable variant with increased oxygen affinity, *Biochem. Biophys. Acta,* 623, 360, 1980.
21. **Fullmer, C. S. and Wasserman, R. H.,** Analytical peptide mapping by high performance liquid chromatography. Application to intestinal calcium-binding proteins, *J. Biol. Chem.*, 254, 7208, 1979.
22. **Hearn, M. T. W., Hancock, W. S., Hurrell, J. G. R., Fleming, R. J., and Kemp, B.,** The analysis of insulin-related peptides by reversed-phase high-performance liquid chromatography, *J. Liq. Chromatogr.*, 2, 919, 1979.
23. **Hollaway, W. L., Prestidge, R. L., Bhown, A. S., Mole, J. E., and Bennett, J. C.,** Hydrophilic ion-pair reversed-phase high-performance liquid chromatography of peptides and proteins, in *Recent Developments in Chromatography and Electrophoresis,* Vol. 10, Frigerio, X. and McCamish, M., Eds., Elsevier, Amsterdam, 1980, 131.
24. **Black, C., Douglas, D. M., and Tanzer, M. L.,** Separation of cyanogen bromide peptides of collagen by means of high-performance liquid chromatography, *J. Chromatogr.*, 190, 393, 1980.
25. **Molnar, I. and Horvath, C.,** Separation of amino acids and peptides on non-polar stationary phases by high-performance liquid chromatography, *J. Chromatogr.*, 142, 623, 1977.
26. **Imoto, T. and Okazaki, K.,** A simple peptide fractionation by hydrophobic chromatography with a prepacked reversed-phase column, *J. Biochem. (Tokyo),* 89, 437, 1981.
27. **Chaiken, I. M. and Hough, C. J.,** Mapping and isolation of large peptide fragments from bovine neurophysins and biosynthetic neurophysin-containing species by high-performance liquid chromatography, *Anal. Biochem.*, 107, 11, 1980.
28. **Van der Rest, M., Bennett, H. P. J., Solomon, S., and Glorieux, F. H.,** Separation of collagen cyanogen bromide-derived peptides by reversed-phase high performance liquid chromatography, *Biochem. J.*, 191, 253, 1980.
29. **McMillan, M., Frelinger, J. A., Jones, P. P., Murphy, D. B., McDevitt, H. O., and Hood, L.,** Structure of murine Ia antigens. Two-dimensional electrophoretic analyses and high-pressure liquid chromatography peptide maps of products of the I-A and I-E subregions and of an associated invariant polypeptide, *J. Exp. Med.*, 153, 936, 1981.

30. **Schroeder, W. A., Shelton, J. B., and Shelton, J. R.,** High performance liquid chromatography in the identification of human hemoglobin variants, in *Advances in Hemoglobin Analysis*, Alan R. Liss, New York, 1981, 1.
31. **Pearson, J. D., Mahoney, M. C., Hermodson, M. A., and Regnier, F. E.,** Reversed-phase supports for the resolution of large denatured protein fragments, *J. Chromatogr.*, 207, 325, 1981.
32. **Lewis, R. V., Fallon, A., Stein, S., Gibson, K. D., and Udenfriend, S.,** Supports for reverse-phase high-performance liquid chromatography of large proteins, *Anal. Biochem.*, 104, 153, 1980.
33. **Hearn, M. T. W., Grego, B., and Bishop, C. A.,** The semi-preparative separation of peptides on reversed-phase silica packed into radially compressed flexible-walled columns, *J. Liq. Chromatogr.*, 4, 1725, 1981.
34. **Hancock, W. S., Capra, J. D., Bradley, W. A., and Sparrow, J. T.,** The use of reversed-phase high-performance liquid chromatography for the analysis of peptide and protein mixtures, *J. Chromatogr.*, 206, 59, 1981.
35. **Hancock, W. S. and Sparrow, J. T.,** Use of mixed-mode, high-performance liquid chromatography for the separation of peptide and protein mixtures, *J. Chromatogr.*, 206, 71, 1981.
36. **Bohlen, P. and Kleeman, G.,** Analytical and preparative mapping of complex peptide mixtures by reversed-phase high performance liquid chromatography, *J. Chromatogr.*, 205, 65, 1981.
37. **Hughes, G. J., Winterhalter, K. H., and Wilson, K. J.,** Microsequence analysis. I. Peptide isolation using high-performance liquid chromatography, *FEBS Lett.*, 108, 81, 1979.
38. **Rubenstein, M., Levy, W. P. Moschera, J. A., Lai, C-Y, Hershberg, R. D., Bartlett, R. T., and Pestka, S.,** Human leukocyte interferon:isolation and characterisation of several molecular forms, *Arch. Biochem. Biophys.*, 210, 307, 1981.
39. **Rubenstein, M., Stein, S., and Udenfriend, S.,** Isolation and characterization of the opioid peptides from rat pituitary: β-endorphin, *Proc. Natl. Acad. Sci. U.S.A.*, 74, 4969, 1977.
40. **Gerber, G. E., Anderegg, R. J., Herlihy, W. C., Gray, C. P., Biemann, K., and Khorana, H. G.,** Partial primary structure of bacteriorhodopsin: sequencing methods for membrane proteins, *Proc. Natl. Acad. Sci. U.S.A.*, 76, 227, 1979.
41. **Congote, L. F., Bennett, H. P. J., and Solomon, S.,** Rapid separation of the α, β, $^{G}\gamma$ and $^{A}\gamma$ human globin chains by reversed-phase high pressure liquid chromatography, *Biochem. Biophy. Res. Commun.*, 89, 851, 1979.
42. **Hollaway, W. L., Kemp, M. C., Bhown, A. S., Compans, R. W., and Bennett, J. C.,** Reversed-phase ion-pair chromatography of viral glycopeptides, *J. High Resolution Chromatogr.*, 2, 149, 1979.
43. **Haeffner-Gormley, L., Poludniak, N. H., and Wetlaufer, D. B.,** Separation of the tryptic peptides from reduced, alkylated hen egg white lysozyme by high-performance liquid chromatography, *J. Chromatogr.*, 214, 185, 1981.
44. **Schroeder, W. A., Shelton, J. B., Shelton, J. R., and Powars, D.,** Separation of peptides by high-pressure liquid chromatography for the identification of a hemoglobin variant, *J. Chromatogr.*, 174, 385, 1979.
45. **Wilson, J. B., Lam, H., Pravatmuang, P., and Huisman, T. H. J.,** Separation of tryptic peptides of normal and abnormal α, β, γ, and δ hemoglobin chains by high performance liquid chromatography, *J. Chromatogr.*, 179, 271, 1979.
46. **Stoming, T. A., Garver, F. A., Gangarosa, M. A., Harrison, J. M., and Huisman, T. H. J.,** Separation of the $^{A}\gamma$ and $^{G}\gamma$ cyanogen bromide peptides of human fetal hemoglobin by high-pressure liquid chromatography, *Anal. Biochem.*, 96, 113, 1979.
47. **Rubenstein, M., Chen-kiang, S., Stein, S., and Udenfriend, S.,** Characterization of proteins and peptides by high-performance liquid chromatography and fluorescence monitoring of their tryptic digests, *Anal. Biochem.*, 95, 117, 1979.
48. **Rivier, J. E.,** Use of trialkyl ammonium phosphate (TAAP) buffers in reverse phase HPLC for high resolution and high recovery of peptides and proteins, *J. Liq. Chromatogr.*, 1, 343, 1978.
49. **Yang, H. S., Studebaker, J. F., and Parravano, C.,** The study of disulfide bond pairing in proteins and protein fragments with HPLC, in *Chromatographic Science Series X*, Marcel Dekker, New York, 1978, 247.
50. **Kemp, M. C., Hollaway, W. L., Prestidge, R. L., Bennett, J. C., and Compans, R. W.,** Reverse-phase ion pair high performance liquid chromatography of viral tryptic glycopeptides, *J. Liq. Chromatogr.*, 4, 587, 1981.
51. **Allen, E. H. and Schweet, R. S.,** Synthesis of hemoglobin in a cell-free system. I. Properties of the complete system, *J. Biol. Chem.*, 237, 760, 1962.
52. **Hancock, W. S., Bishop, C. A., Prestidge, R. L., and Hearn, M. T. W.,** The use of high pressure liquid chromatography (HPLC) for peptide mapping of proteins. IV, *Anal. Biochem.*, 89, 203, 1978.
53. **Schwabe, C. and McDonald, J. K.,** Demonstration of a pyroglutamyl residue at the N terminus of the B-chain of porcine relaxin, *Biochem. Biophys. Res. Commun.*, 74, 1501, 1977.

54. **Hearn, M. T. W. and Grego, B.,** High-performance liquid chromatography of amino acids, peptides and proteins. XXVII. Solvophobic considerations for the separation of unprotected peptides on chemically bonded hydrocarbonaceous stationary phases, *J. Chromatogr.*, 203, 349, 1981.
55. **Leo, A., Hansch, C., and Elkins, D.,** Partition coefficients and their uses, *Chem. Rev.*, 71, 525, 1971.
56. **Rekker, R. F.,** *The Hydrophobic Fragmental Constant,* Elsevier, Amsterdam, 1977, 301.
57. **Nozaki, Y. and Tanford, C.,** The solubility of amino acids and two glycine peptides in aqueous ethanol and dioxane solutions. Establishment of a hydrophobicity scale, *J. Biol. Chem.*, 246, 2211, 1971.
58. **Canfield, R. E.,** Peptides derived from tryptic digestion of egg white lysozyme, *J. Biol. Chem.*, 238, 2691, 1963.
59. **Bhown, A. S. and Bennett, J. C.,** personal communication.

PROTEIN IDENTIFICATION BY PEPTIDE MAPPING

Walter A. Schroeder

INTRODUCTION

Peptide mapping, or "fingerprinting" as it was called by Ingram,[1] its originator, has been of major importance in the determination of protein sequences and is of special usefulness when the point of variation in a mutant protein is under investigation. Although paper, which was Ingram's original support, still has its devotees, peptide mapping has also been done on various thin-layer supports[2] or ion-exchange column chromatograms.[3] Because of the almost universal applicability of HPLC, it is only natural that HPLC should be used for peptide mapping in the identification of proteins.

Peptide mapping on paper or like support relies on two-dimensional electrophoresis and chromatography; an initial separation by cation- or anion-exchange column chromatography usually requires rechromatography on the opposite type of ion exchanger. Paper or similar peptide maps separate peptides rather quickly in small amounts with subsequent rather "dirty" amino acid analyses; ion-exchange chromatography provides almost any desired quantity and clean analyses but is time consuming. Peptide mapping by HPLC has many advantages: speed, sensitivity, nondestructive detection, automation, many parameters (column packings, developers, gradients), etc.

The history of the application of HPLC to peptides is brief, but the number of publications on the topic is growing rapidly as a perusal of other chapters of this handbook will amply prove. Our application of HPLC to peptides has largely involved peptide mapping to identify hemoglobin variants. Our approach has not made wide use of theoretical considerations but has attempted to gain extensive experience with many column packings and solvent systems. Consequently, we have data for that final asymptotic approach to the best conditions for separation which is the inevitable process for each chromatographer.

As we discuss ramifications, problems, and successes of HPLC peptide mapping, each subsection below will have within it the components of procedure, results, and discussion.

HEMOGLOBIN

Because this review deals mainly with the application of HPLC methods to hemoglobin, this brief resumé of hemoglobin structure and nomenclature is provided.

In late fetal and in adult life, the several hemoglobins which occur are composed of α, β, γ, and δ chains in various combinations. In late fetal life, the major hemoglobin is Hb-F but about 20% is Hb-A and traces of Hb-A_2 are present. In the adult (age 6 months or more), Hb-A content will be greater than 90%, Hb-A_2 approximately 2.5% (normal), and Hb-F a few percent (after age 3 or 4 years usually $< 1\%$). Hb-A has two α and two β chains, thus $\alpha_2\beta_2$, Hb-F is $\alpha_2\gamma_2$, and Hb-A_2 is $\alpha_2\delta_2$. The two types of γ chain which are normally produced differ in the presence of glycine ($^G\gamma$ chain) or alanine ($^A\gamma$) in position 136. The $^A\gamma$ usually is $^A\gamma^I$ with isoleucine in position 75, but a variant in considerable frequency has threonine there ($^A\gamma^T$). Mutants of all of these chains are known, and the position and type of mutation may be specified by the following nomenclature; thus, sickle-cell hemoglobin or Hb-S is $\alpha_2\beta_2^S$ or $\alpha_2\beta_2^{6Glu\rightarrow Val}$ where β^S shows the mutation to be in the β chain and $\beta^{6Glu\rightarrow Val}$ defines the replacement of glutamic acid in the 6th position of the normal β (that is, β^A) chain by valine.

Tryptic peptides are denoted by a nomenclature such as α^IT-3,4 in which "α" refers to the chain, the superscript "I" to Hb-I, "T" to tryptic peptide, and "3" and "4" to the

Table 1
RP-HPLC COLUMNS AVAILABLE FOR USE IN THIS LABORATORY

Type	Supplier	Designation	Size (mm)	Particle size (μm)
C_{18}	Altex	Ultrasphere® ODS	4.6 × 250	5
	Altex	Ultrasphere® ODS	10.0 × 250	5
	Bio-Rad	Biosil® ODS-10	4.0 × 250	10
	DuPont	Zorbax® ODS	4.6 × 250	≈5
	IBM	ODS	4.5 × 250	5
	Regis	HZ-Chrom Reversible	4.6 × 250	5
	Waters	μBondapak® C_{18}	3.9 × 300	10
	Waters	μBondapak® C_{18}	7.8 × 300	10
C_8	DuPont	Zorbax® C_8	4.6 × 250	≈5
CN	Altex	Ultrasphere CN	4.6 × 250	5
	DuPont	Zorbax® CN	4.6 × 250	≈5
Phenyl	Waters	μBondapak® Phenyl	3.9 × 300	10
TMS	DuPont	Zorbax® TMS	4.6 × 250	≈5

third and fourth tryptic peptides from the N-terminus. In Hb-I, the mutation is Lys→Glu and the α^{I}T-3,4 peptide results from the loss of a tryptic cleavage point.

EQUIPMENT

Two HPLC systems are in use in this laboratory. The first consists of two Altex Model 110A Metering Pumps (Altex Scientific, Inc., Berkeley, CA), an Altex 420 Microprocessor, a Rheodyne Model 7125 Sample Injector, an Altex/Hitachi Model 155-10 UV-Vis Variable Wavelength Detector, and a single channel recorder (Linear Instruments, Irvine, CA). The second has two Waters 6000A solvent delivery systems (Waters Associates, Milford, MA), a Waters U6K universal injector, an Altex/Hitachi Model 155-10 UV-Vis Variable Wavelength Detector, a Cole Scientific Model 711 HPLC System Controller (Cole Scientific, Calabasas, CA), and a Watanabe Model SR6252 Single Pen Chart Recorder.

In our early experiments, when a single Waters 6000A pump was used, linear gradients were formed with a two-vessel system as described by Bock and Ling.[4] Although less convenient than electronically controlled gradients with two pumps, satisfactory chromatograms were produced.

The equipment has generally behaved satisfactorily although our Waters 6000A pumps frequently stopped pumping because of cavitation or bubble formation, a problem that is not solved by degassing. It is claimed by the manufacturer that this problem no longer occurs in newer models of Waters pumps and can be eliminated from older models. Pumps which are used with phosphate solutions (see below) may require additional maintenance (new seals, etc.) more frequently than those which are used with acetonitrile.

COLUMNS

The utility of a variety of column packings has been explored. The list in Table 1 includes not only columns with different types of packing but also those with the same nominal type but from different manufacturers. Not unexpectedly, C_{18} columns from different manufacturers have significantly different properties under the same chromatographic conditions, a fact that may be used to advantage in achieving separations. Product control appears to be good. In our experience, successive purchases of the same type of column from the same manufacturer have resulted in reproducible patterns.

Because not all columns that are listed in Table 1 have been examined to the same degree, the evaluations below are somewhat subjective. Some packings were investigated because of their chemical differences and others because a manufacturer's statement suggested an advantage. If the initial chromatogram showed no marked superiority over others, further experiments were limited. With one exception, all packings yielded essentially symmetrical unskewed peaks; the Regis reversible column had skewed peaks in both directions even with the test sample provided. In general, separations have been satisfactory on all column types. Peaks were sharper on 5- than on 10-μm packings. In overall performance for tryptic peptide separations, the Altex Ultrasphere® ODS has been excellent, the IBM® ODS with only limited study seemed comparable, the Dupont Zorbax® ODS was very good, and all others were good. The rate of deterioration varied with use. Guard columns have not usually been used. Many chromatograms may be run on Altex Ultrasphere® ODS or Waters μBondapak® C_{18} columns (although the latter is more useful for globin chain (see below) than for tryptic peptide separation). On the other hand, the quality of the chromatograms deteriorated rather rapidly with continued use of Biosil® ODS-10, Zorbax® TMS, and μBondapak® Phenyl columns. Data on other columns that are listed in Table 1 are not available.

DEVELOPERS

In our experience, the most effective developers have used a gradient between a phosphate buffer and acetonitrile. The phosphate buffer is 49 m*M* KH_2PO_4 (6.66 g/ℓ) and 5.4 m*M* H_3PO_4 (0.37 mℓ of 85% H_3PO_4/ℓ); the pH is approximately 2.9. A disadvantage of this system is the involatility of phosphate. If effluent is collected and the solvents are evaporated, the residue of phosphate does not interfere with amino acid analysis, but is undesirable for a subsequent Edman degradation. To circumvent this problem, other authors have used trifluoroacetic acid (TFA)-acetonitrile mixtures. Although our initial usage of TFA developers gave badly skewed peaks on a Waters μBondapak® C_{18}-column, more recent experiments with 0.1% TFA-0.1% TFA in acetonitrile gradients on an Altex Ultrasphere® ODS column gave chromatograms that approached phosphate-acetonitrile gradients in efficacy. The effect of type of developer may be dramatically illustrated by comparing the data of Mahoney and Hermodsen[5] with those of Stoming et al.[6] for the separation of $^{G}\gamma$CB-3 and $^{A}\gamma$CB-3. The TFA-acetonitrile gradient of Mahoney and Hermodsen only partially separated them whereas the ammonium acetate-acetonitrile gradient of Stoming et al.[6] not only separated them completely but permitted their quantitative evaluation. Ammonium acetate-acetonitrile mixtures, therefore, provide another volatile system and were first used here as a result of the experiences of Stoming et al.[6] The ammonium acetate was 10 m*M* (0.77 g/ℓ) and was brought to pH 6.07 with dilute acetic acid. Phosphate may be removed by rechromatographing in ammonium acetate or TFA systems. One drawback of the ammonium acetate system is its lower eluting power. As a result, material tends to build up on the column although the resolving power may not be unduly influenced. This problem has been discussed by Johnson et al.[7] If a column is to be used with phosphate-acetonitrile developers after use with ammonium acetate-acetonitrile, a phosphate-acetonitrile gradient should be passed through it once or twice before a sample is applied.

Although other organic solvents have been repeatedly studied in the HPLC of peptides, the general concensus seems to be that acetonitrile usually is the most effective.

The difference of three units in the pH of the phosphate and ammonium acetate buffers will influence the ionization of carboxyl and imidazole groups of any peptide. A further illustration of the effect of changing the type of developer is presented in Figure 1. The shifts in position, however, are not obviously correlated with differences in ionization. Thus, αT-9, a 29-residue peptide with 4 aspartyl and 3 histidyl residues, does not shift, whereas βT-14 which has a single histidine in a rather hydrophobic peptide shows a pronounced

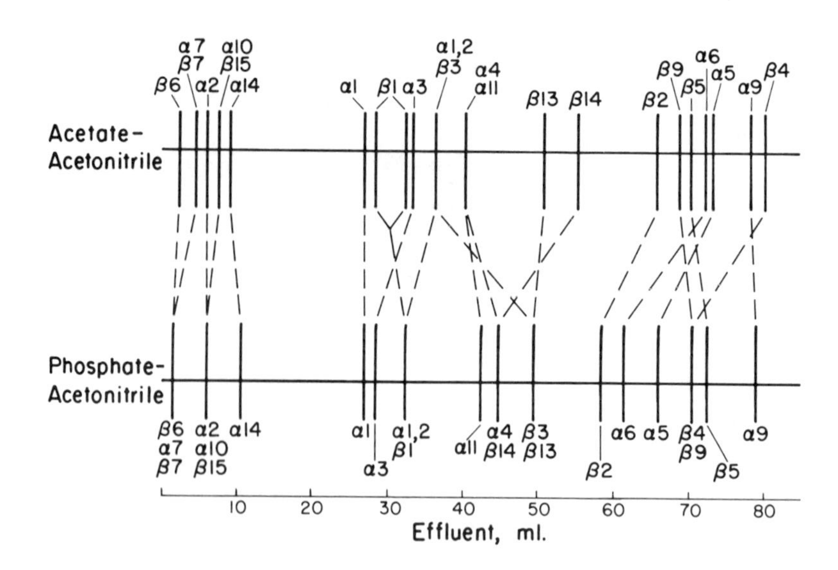

FIGURE 1. Comparison of the elution volumes of tryptic peptides of Hb-A in two developer systems on a 3.9 × 300-mm Waters μBondapak® C_{18}-column. Gradients were linear from buffer to 40% acetonitrile in a total of 100 mℓ at a flow rate of 1 mℓ/min. (From Schroeder, W. A., Shelton, J. B., and Shelton, J. R., *Hemoglobin*, 4, 551, 1980. With permission.)

change. Although the mechanisms that produce these differences in position are obscure and probably multifactorial the practical utility is clear: poor separations in one system may be excellent in the other (although a different mixture may be produced). Although Figure 1 provides examples of the effect of changing the developer, the gradient used was not optimal. The packing for the chromatograms shown in Figure 1 was Waters μBondapak® C_{18}, but similar behavior was observed when an Altex Ultrasphere® ODS column was used.

SAMPLES

Although we have chromatographed chymotryptic peptides as well as those cleaved with cyanogen bromide or acetic acid, tryptic peptides have been used in the majority of our HPLC separations. A typical preparation of a tryptic hydrolysate of hemoglobin as described below no doubt would have to be modified for other proteins because of their different properties.

The sample may be isolated chains or globin, but is is simplest to use hemoglobin itself because heme does not interfere in HPLC analysis. Thus, the hemoglobin (perhaps a chromatographically isolated mutant[9]) dissolved in water is denatured in a boiling water bath for 4 min. Solid ammonium bicarbonate is added to the suspension to a concentration of 25 m*M*, and the pH is adjusted to 8 with NaOH. At 37°C, trypsin (1% by weight of the hemoglobin) is added at 0 and 5 hr, and the digestion is continued for 24 hr. After the pH has been reduced to 2 with HCl, the solution is passed through a 0.5-μm cellulosic filter (Rainin Instrument Co., Woburn, MA) in a Swinney filter holder and then lyophilized.

At the end of a tryptic digestion of hemoglobin, insoluble material (the "core") is present at pH 8 and increases as the pH is reduced to 6.5. However, the precipitate dissolves at pH 2. No attempt has been made to examine soluble and insoluble material separately by HPLC.

After an appropriate weight of lyophilized digest has been dissolved in 100 μℓ of buffer for the HPLC procedure, the solution is filtered using a microfilter system MF-1® with a 0.45-μm BA-85 nitrocellulose filter (Bio-Analytical Systems, Inc., West Lafayette, IN).

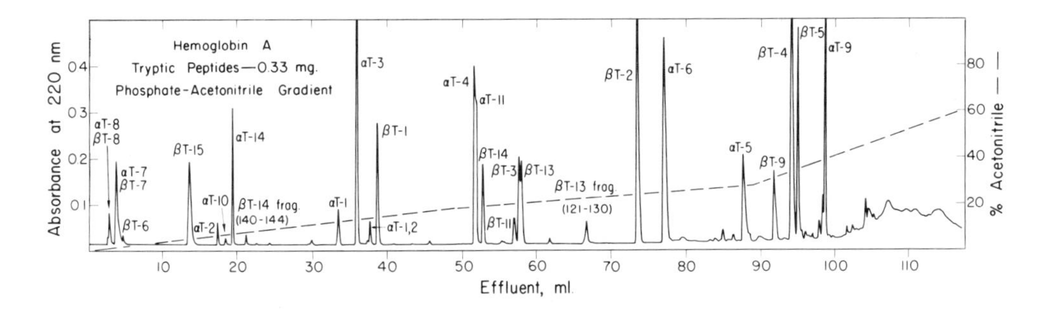

FIGURE 2. Separation of tryptic peptides of Hb-A on a 4.6- × 250-mm Altex Ultrasphere® ODS column with a phosphate-acetonitrile gradient (0 to 18% acetonitrile in 48 mℓ, to 28% in 40 mℓ, to 40% in 10 mℓ, and to 62% in 20 mℓ) at room temperature and a flow rate of one mℓ/min. (From Schroeder, W. A., Shelton, J. B., and Shelton, J. R., in *Advances in Hemoglobin Analysis,* Hanash, S. M. and Brewer, G. J., Eds., Alan R. Liss, New York, 1981, 1. With permission.)

THE HPLC PEPTIDE MAPPING OF HEMOGLOBINS

Although various packings of the same or different type from different manufacturers may be used for the separation of tryptic peptides of hemoglobin with appropriate modifications of the gradient program, the quality of the chromatograms varies in peak width, completeness of separation, etc. In our experience with the packings listed in Table 1, the best performance has come from a phosphate-acetonitrile gradient on an Altex Ultrasphere ODS column which is a spherical 5-μm packing. The HPLC peptide map of any mutant adult hemoglobin is compared with the map of normal adult human Hb-A which is depicted in Figure 2. Similar maps are available for Hb-F and Hb-A_2.

A distinct advantage of using HPLC for peptides is the availability of nondestructive detection at 220 nm (or shorter wavelengths if the solvent does not interfere). Absorption at this wavelength comes from the peptide bond and is somewhat augmented in tyrosyl or tryptophanyl peptides.

Two hours may seem long for an HPLC run when some authors show peptide separations by HPLC in 30 to 40 min. However, the separations shown in Figure 2 are superior to those achieved in 30 to 40 min with such a complex mixture. In our opinion, the extra time is a small price to pay for separations that by two-dimensional fingerprinting or ion-exchange chromatography would require a day or week.

Some of the separations shown in Figure 2 are excellent while others, for example, αT-8 and βT-8 are poor or nonexistent. Actually, αT-8 and βT-8 are both free lysine (traces of some unidentified material in this peak, rather than lysine, probably cause the obvious absorbance). The use of any peptide mapping procedure in the identification of a variant protein involves a search for the change in elution position of a peptide, or the loss or gain of peptides as points of cleavage are lost or gained. The long and highly resolved chromatogram provides ample open spaces for such shifts in retention time or appearances of the aberrant peptide(s) so that modifications in peptides of close pairs such as αT-4 and αT-11 or βT-3 and βT-13 should be readily apparent.

If, for some reason, the aberrant peptide is not obvious, another developer system may be desirable (Figure 1). Likewise, a change in the type of column packing with the same developer program can produce a marked alteration in behavior as can be seen in Figure 3, where the different sample size used for the two chromatograms is at least in part responsible for the broader peaks on the TMS column.

It has been convenient, as the identification of a mutant hemoglobin is begun, to chromatograph a 0.2 to 0.3-mg sample in order to detect a changed pattern. Often, an amino

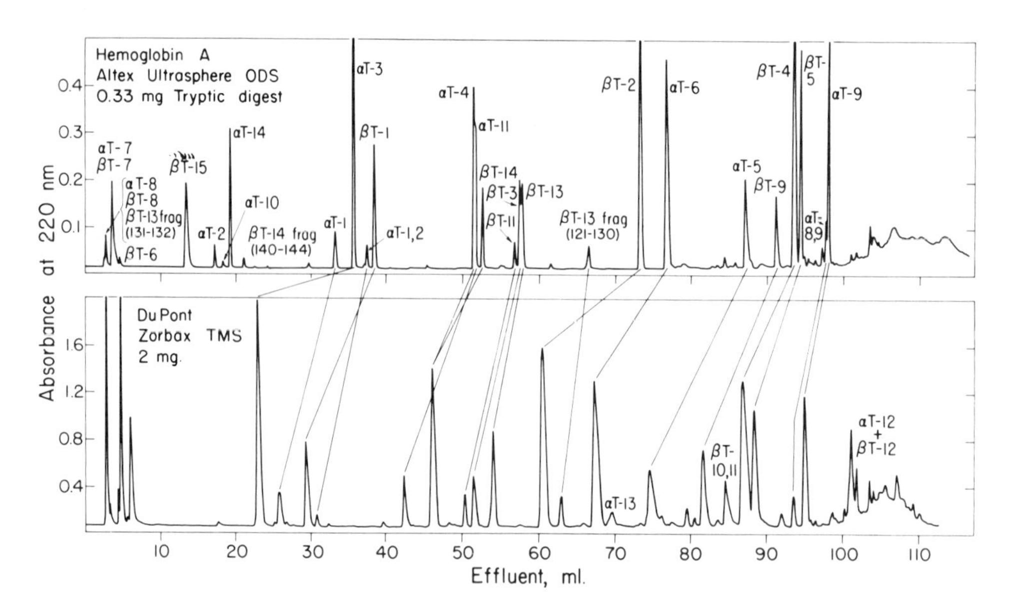

FIGURE 3. Comparison of the separation of tryptic peptides of Hb-A on Altex Ultrasphere® ODS and DuPont Zorbax® TMS columns with the gradient as given for Figure 2. (From Schroeder, W. A., Shelton, J. B., and Shelton, J. R., in *Advances in Hemoglobin Analysis,* Hanash, S. M. and Brewer, G. J., Eds., Alan R. Liss, New York, 1981, 1. With permission.)

acid analysis is sufficient to determine the type and position of the substitution in a hemoglobin variant. Thus, if a sensitive enough amino acid analyzer is available, a 0.2 to 0.3-mg sample provides adequate material. Insensitivity of our amino acid analyzers requires a 2-mg sample in order to provide sufficient material for analysis. In earlier work, although the peptides from as much as 5 to 8 mg of hemoglobin were chromatographed,[11,12] so large a sample was unnecessary and tended to overload the column.

Numerous variants have been identified by these procedures.[13] Figure 4 provides information about the behavior of variant peptides against the background of normal peptides of Hb-A as given in Figure 2. The positions of the aberrant peptides of uncommon variants are shown by the solid bars. The arrows connect the bar and the peptide(s) that has been altered. Thus, in Hb-I, substitution of Lys→Glu in αT-3 prevents the normal cleavage so that α^{I}T-3,4 results. On the other hand, the Gly→Arg substitution in αT-4 of Hb-Handsworth produces two peptides with very altered chromatographic behavior. The dashed bars at the top of the figure show the positions of the aberrant peptides of the common mutants, hemoglobins S, C, E, and G-Philadelphia; for example, Hb-S has Glu→Val in βT-1 and Hb-E has Glu→Lys in βT-3.

If no abnormality is apparent in a chromatogram such as in Figure 2, the mutation may be in αT-12, αT-13, βT-10, or βT-12 which are not detected under these conditions. These are the "core" peptides which normally are insoluble at pH 6.5 and, except for αT-13, contain a cysteinyl residue. That these peptides may be present in the nondescript region between 100 and 120 mℓ of effluent volume is seen from the chromatogram on Zorbax® TMS in Figure 3, where these peptides separate and are identified. Although aminoethylation has often been used in the study of hemoglobin cores, restrictions on the use of ethyleneimine have turned us to oxidation. If, then, no variant peptide is found under the conditions of Figure 2, a sample is oxidized, digested, and chromatographed as in Figure 5. In this case, most of the peptides are without interest and are purposely removed rapidly with the isocratic development before the gradient is used to separate the core peptides.

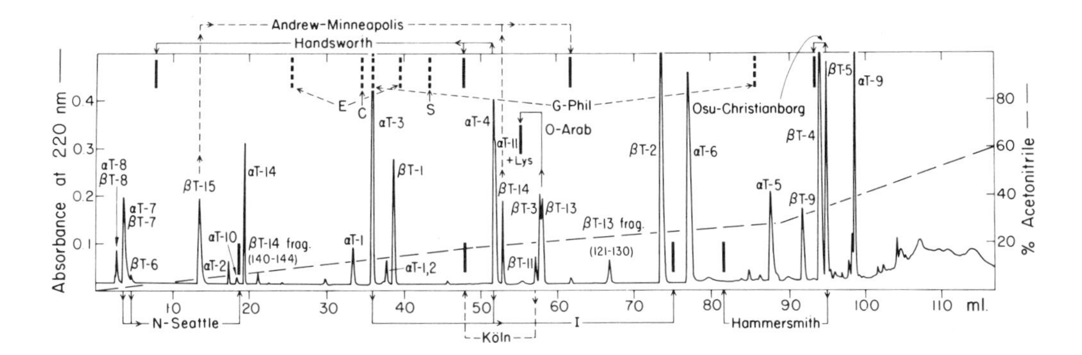

FIGURE 4. Separation of tryptic peptides of Hb-A as in Figure 2. The positions of aberrant peptide(s) of mutant hemoglobins are defined by the vertical solid bars. The dashed bars refer to the peptide(s) from common variants. See text for details. (From Schroeder, W. A., Shelton, J. B., Shelton, J. R., Powars, D., Friedman, S., Baker, J., Finklestein, J. Z., Miller, B., Johnson, C. S., Sharpsteen, J. R., Sieger, L., and Kawaoka, E., *Biochem. Genet.*, 20, 133, 1982. With permission.)

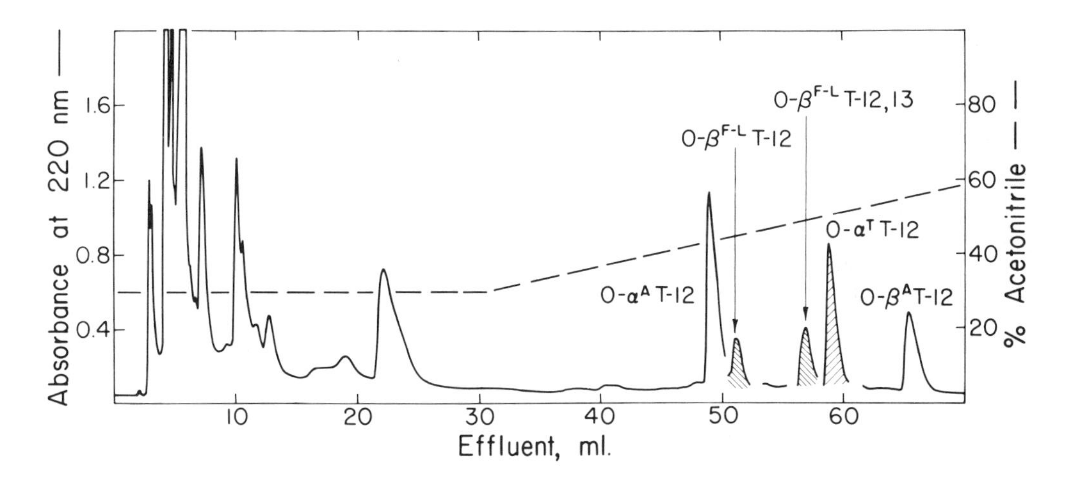

FIGURE 5. Separation of tryptic peptides from oxidized hemoglobins on a DuPont Zorbax® TMS column at one mℓ/min flow rate with 30 mℓ of 70% 10 m*M* ammonium acetate at pH 6.07-30% acetonitrile isocratically and then a 45-mℓ linear gradient from 30 to 62% acetonitrile. Superscripts to Greek letters refer to the hemoglobin from which the peptide derived: A, Hb-A; F-L, Hb-Fannin-Lubbock; T, Hb-Tarrant. (From Schroeder, W. A., Shelton, J. B., Shelton, J. R., Powars, D., Friedman, S., Baker, J., Finklestein, J. Z., Miller, B., Johnson, C. S., Sharpsteen, J. R., Sieger, L., and Kawaoka, E., *Biochem. Genet.*, 20, 133, 1982. With permission.)

HPLC PEPTIDE MAPPING OF HEMOGLOBINS BY OTHER AUTHORS

More and more authors now use HPLC methods to identify abnormal hemoglobins.[14-18] It may be anticipated that the numbers of such papers will increase rapidly. (Since this chapter was written, 13 papers in Volume 6, 1982 of *Hemoglobin* have described the identification of abnormal hemoglobins by HPLC). For the most part, there is an individualistic approach to the use of HPLC. Some authors not only use globin but even separate the chains before preparing the tryptic peptides. Different packings, different solvents, and modified gradient programs are reported. The quality of the separations seems to depend much on the investigator's desire to produce instant HPLC analysis.

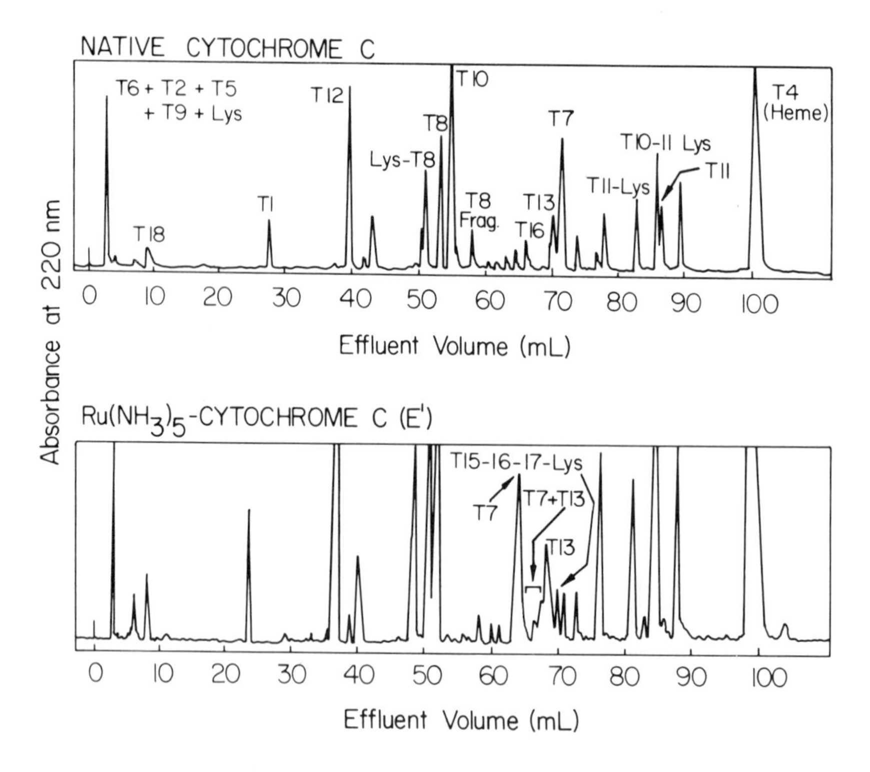

FIGURE 6. Comparison of the tryptic digests of native cytochrome *c* and $Ru(NH_3)_5$-cytochrome *c*. The column was an Altex Ultrasphere® ODS. Development used a linear gradient in 120 mℓ between 49 m*M* KH_2PO_4 and 5.4 m*M* H_3PO_4 at pH 2.85 and 0 to 45% acetonitrile at one mℓ/min flow rate. (From Yocom, K. M., Ph.D. thesis, California Institute of Technology, Pasadena, Calif., 1981; Yocum, K. M., et al., *Proc. Natl. Acad. Sci. U.S.A.*, 79, 7052, 1981. With permission.)

HPLC PEPTIDE MAPPING OF OTHER PROTEINS

As is evident from other chapters in this volume, many peptides have been examined by HPLC with a variety of objectives; for example, synthetic peptides alone or in mixture may have been chromatographed as an illustration of the usefulness of HPLC or, perhaps, to ascertain some aspect of the mechanism of HPLC separation. More and more, however, HPLC is being applied to the identification of mutants as has been discussed in this chapter or for the separation of peptides prior to sequence determination. The following paragraphs summarize data from selected papers which describe the application of HPLC to proteins other than hemoglobin.

Cytochrome *c*

Peptide mapping by HPLC has been used by Yocom[19] to determine the point of reaction of the aquopentaamine ruthenium II ion ($Ru(NH_3)_5H_2O^{2+}$) with horse heart cytochrome *c*. The reaction product was digested with trypsin and compared with the map for underivatized cytochrome *c*. The results are presented in Figure 6 where peptide T-7 can be seen to have changed position. It was anticipated that the $Ru(NH_3)_5H_2O^{2+}$ would react with a histidyl or a methionyl residue. Peptide T-7 contains one histidyl residue in residues 28 to 38 of cytochrome *c*. The ruthenium complex of histidine itself has an absorption maximum at 300 nm. Figure 7 duplicates Figure 6 with recording at 300 nm. In the digest of native protein, the recording at 300 nm detects only T-10, which contains tryptophan, and T-4,

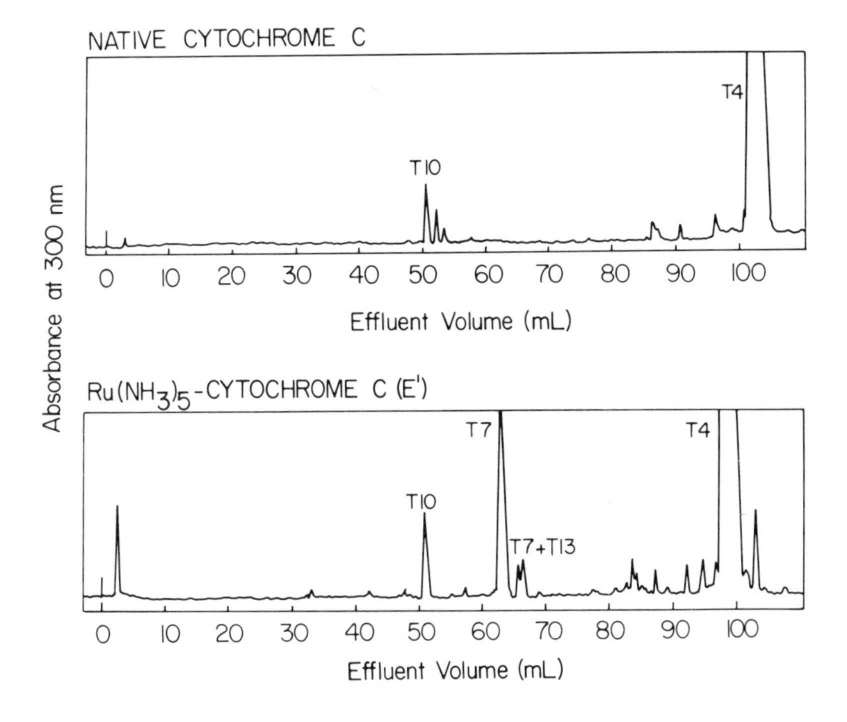

FIGURE 7. Duplicate chromatograms under the conditions of Figure 6 except that the recording was done at 300 nm. (From Yocum, K. M., Ph.D. thesis, California Instiute of Technology, Pasadena, Calif., 1981; Yocum, K. M., et al., *Proc. Natl. Acad. Sci. U.S.A.*, 79, 7052, 1981. With permission.)

the heme peptide. However, T-7 is prominent in the digest of $Ru(NH_3)_5$-cytochrome *c* (Figure 7).

Yeast Alcohol Dehydrogenase

Wills et al.[20] have applied HPLC to identify a mutant of yeast alcohol dehydrogenase (ADH). The sequence of ADH is known from the work of Jörnvall.[21] Figure 8 depicts the peptide maps of wild-type ADH and of a mutant. These chromatograms were obtained from a Waters μBondapak® C_{18}-column with a series of linear gradients between 0.1% trifluoroacetic acid (TFA) and acetonitrile (0 to 18% in 48 mℓ, 18 to 25% in 30 mℓ, 25 to 40% in 30 mℓ, and 40 to 62% in 20 mℓ) at 1 mℓ/min flow rate. The absorbance was monitored at 220 nm. Those of the 32 tryptic peptides that result from the cleavage of the 347 residues of wild-type ADH and have been identified are labeled in Figure 8. In the mutant, T-7 has disappeared and peaks T-7A and T-7B appear. The mutation is Trp→Arg in position 54.

Immunoglobulins

HPLC has been put to good use in the determination of the primary structure of immunoglobulins.[22,23] A variety of column packings and developer systems were employed, and the authors succeeded in determining the sequence of more than 200 residues with only a few milligrams of sample.

Lysozyme

Haeffner-Gormley et al.[24] have examined the tryptic peptides of lysozyme by HPLC on C_{18} columns with ammonium acetate or ammonium chloride-acetonitrile gradients. They were able to separate most of the 14 anticipated peptides. Their use of very small samples

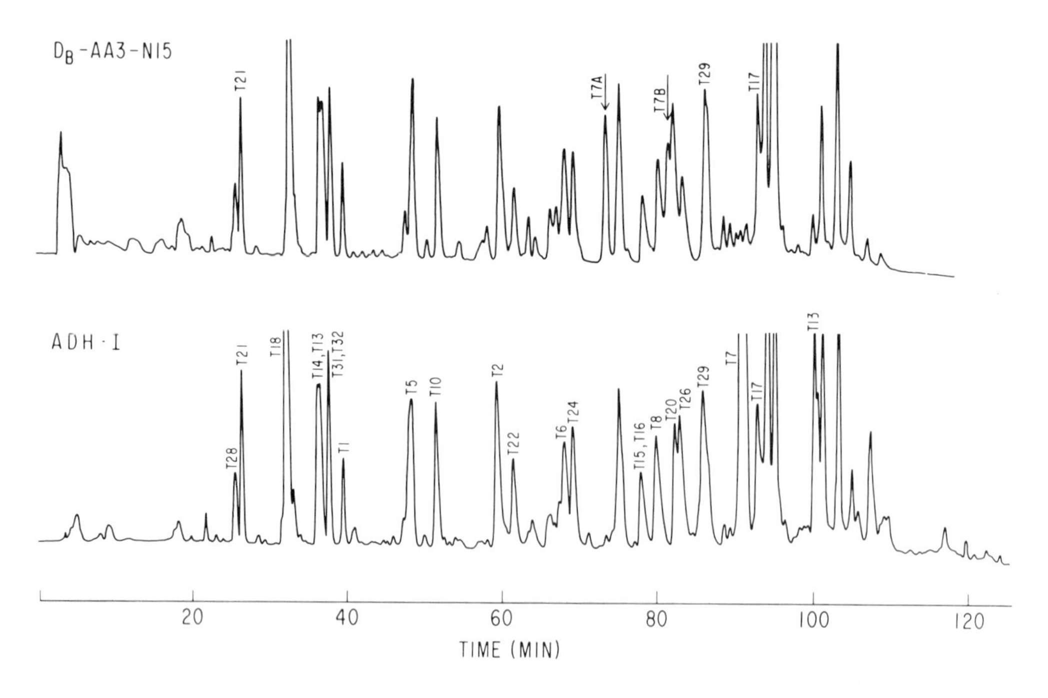

FIGURE 8. Peptide map of wild-type yeast ADH (ADH-I) and of a mutant (D_B-AA3-N15). (From Wills, C., Kratofil, P., and Martin, T., in *Genetic Engineering of Microorganisms for Chemicals*, Hollaender, A., Ed., Plenum Press, New York, in press.)

(a few micrograms) and sensitive recorder settings produced recordings with extremely sloping baselines.

Calcium-Binding Proteins

Fullmer and Wasserman[25] have followed the time course of tryptic digestion of bovine and chick calcium-binding proteins with HPLC and have also isolated, characterized, and compared the peptides. A 0.1% phosphoric acid-acetonitrile gradient on a Waters μBondapak® C_{18}-column provided separation of the peptides.

CHAIN SEPARATIONS OF WHOLE HEMOGLOBINS

In the study of hemoglobins, separation of the chains is an important requirement whether it be as a preliminary to the identification of a variant, the determination of the ratios of biosynthesis, or some other aspect. Such a separation almost universally is done by the method of Clegg et al.[26] on CM-cellulose in 8 *M* urea with a phosphate gradient at pH 6.8.

Despite the fact that the α chain has 141 residues and the β and γ chains 146, their separation by HPLC is readily achieved. Congote et al.[27] and Shelton et al.[28] employed rather different developers for the separations (TFA-acetonitrile and phosphate-acetonitrile) on a Waters μBondapak® C_{18}-column. Huisman and Wilson[29] altered the gradients of Shelton et al.[28] to separate not only the α, β, $^G\gamma$, and $^A\gamma^I$ chains but also the $^A\gamma^T$ chain. Petrides et al.[30] achieved separations on either a Supelcosil® C_{18} or a LiChrosorb® RP-18 column with a gradient between pyridinium formate and propanol which required fluorometric monitoring after reaction with fluorescamine. They succeeded in separating several mutant chains from the normal α or β chain.

In experiments with developers that were based on the perchlorate solutions of Meek,[31] Shelton et al.[32] devised a new procedure which required less than half the time of Huisman

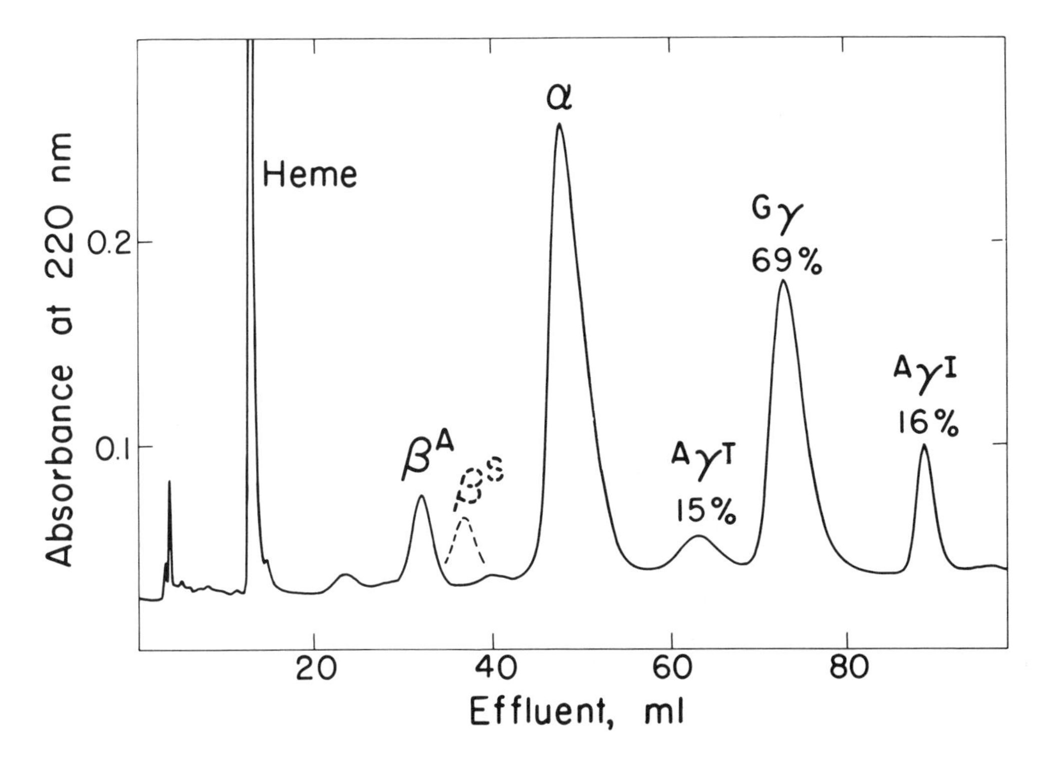

FIGURE 9. Separation of globin chains in a sample of cord blood on a Waters μBondapak® C_{18}-column. Conditions are given in the text.

and Wilson.[29] In more recent experiments, the addition of nonylamine to the developers has produced sharper, well-separated peaks (Figure 9). Developer A was 80:5:15:0.1:0.05 and B was 20:5:75:0.1:0.05 by volume of 0.15 *M* $NaClO_4$:methanol:acetonitrile:85% H_3PO_4:nonylamine. For 57 mℓ, development was isocratic at 62.5% B and was followed by a 37.5-mℓ gradient to 68.5% B. The flow rate was 1.5 mℓ/min. Although the separation may be done isocratically, the gradient is added to elute the $^{A}\gamma^{I}$ chains more rapidly. This modification now produces an excellent separation of the $^{A}\gamma^{T}$ chain, with sharper peaks, and valleys that come close to a slightly rising baseline. In a cord blood with sickle cell trait, the β^S chain falls at the position of the dashed peak (Figure 9). The percentages of the γ chains were determined by planimetry.

The ability to isolate hemoglobin chains by HPLC provides easy insight into genetic aspects of the production, for example, of the γ chains as examined in detail by Huisman et al.[33,34] HPLC should be able to replace the cumbersome Clegg-Naughton-Weatherall method[26] for separating chains to study the biosynthetic ratio; Congote[35] has applied HPLC in this way.

DISCUSSION

Our philosophy of approach to devising methods of peptide mapping of hemoglobin has been an empirical one which chooses an initial, hopefully reasonable, set of variables and refines them with repeated chromatographic runs until a satisfactory chromatogram does (or does not) result. Because of the speed of the HPLC method, not many trials are usually necessary. Thus, it has been of little concern whether ion pairing or some other mechanism brought about the desired separation as long as it was attained. Many papers on the HPLC of peptides, whatever their objective, report the effects of different solutions for development, of flowrate, of the slope of gradient, etc. Any one of these variables will obviously influence the final outcome. Several pertinent references are typical.[36-38] Most papers examine the effect of the variable(s) on a single type of column packing, but few have compared in detail the behavior of a variety of column packings. Our experience suggests that a change in column packing whether of similar or different chemical composition can be of great utility in producing the desired separation. As already mentioned, packings of the same type, for example, C_{18}, from different manufacturers may have very different properties. Thus, tryptic peptides are more effectively separated on an Altex Ultrasphere® ODS than on a Waters μBondapak® C_{18}-column whereas the reverse is true for globin chains. Although the cost of columns may deter comparisons, the choice of a different packing should not be overlooked as an approach to a desired separation.

Whatever may be the exact mechanism that determines the chromatographic behavior of a peptide, it is largely determined by the hydrophobic character of the peptide and is mainly a function of the number and kind of individual amino acid residues in a peptide. Thus, Meek[31] and Meek and Rossetti[36] have determined "retention coefficients" for each type of amino acid residue and for N- and C-termini from a study of 100 peptides. Su et al.[39] have made a similar study. When these coefficients were summed for a given peptide, excellent correlation was observed by Meek and Rossetti between observed and calculated retention times. Because of the effect of developer or packing (as exemplified by Figures 1 and 3) on the chromatographic behavior of a peptide, the quantitative use of "Meek numbers" probably requires that the same packing and developer program be applied. On the other hand, if the chromatographic conditions are similar to Meek's and the peptides are closely related, qualitative data can be meaningful. Thus, if a mutant hemoglobin is being studied, the variant peptide can be readily detected and the sequence of the related normal peptide will be known. The electrophoretic properties of the hemoglobin will tell the relationship of charge between the normal and variant peptide. Consequently, on the basis of single point mutations of the genetic code, certain possibilities of substitution will exist. The "Meek numbers" will tell whether the variant peptide should be more or less rapidly eluted. Consequently, from the known behavior, some possibilities can be eliminated even before the final identification is made. Although these calculations have always correctly ascertained the relative behavior of the variant peptide, the quantitative effect has often been much greater or smaller than anticipated.

ACKNOWLEDGMENTS

Experiments with HPLC in this laboratory as well as the writing of this chapter were supported in part by grant HL-02558 from the National Institutes of Health, U.S. Public Health Service. Joan B. Shelton and J. Roger Shelton were active in experimental design and experimentation. This is Contribution No. 6619 from the Division of Chemistry and Chemical Engineering, California Institute of Technology.

REFERENCES

1. **Ingram, V. M.,** Abnormal human haemoglobins. I. The comparison of normal human and sickle-cell haemoglobins by "fingerprinting", *Biochim. Biophys. Acta,* 28, 539, 1958.
2. **Wajcman, H., Elion, J., Boissel, J. P., Labie, D., Jos, J., and Girot, R.,** A silent hemoglobin variant: Hemoglobin Necker Enfants-Malades α20(B1)His→Tyr, *Hemoglobin,* 4, 177, 1980.
3. **Jones, R. T.,** Structural studies of aminoethylated hemoglobins by automatic peptide chromatography, *Cold Spring Harbor Symp. Quant. Biol.,* 29, 297, 1964.

4. **Bock, R. M. and Ling, N.-S.**, Devices for gradient elution in chromatography, *Anal. Chem.*, 26, 1543, 1954.
5. **Mahoney, W. C. and Hermodsen, M. A.**, Separation of large denatured peptides by reverse phase high performance liquid chromatography. Trifluoroacetic acid as a peptide solvent, *J. Biol. Chem.*, 255, 11199, 1980.
6. **Stoming, T. A., Garver, F. A., Gangarosa, M. A., Harrison, J. M., and Huisman, T. H. J.**, Separation of $^{A}\gamma$ and $^{G}\gamma$ cyanogen bromide peptides of human fetal hemoglobin by high-pressure liquid chromatography, *Anal. Biochem.*, 96, 113, 1979.
7. **Johnson, C. S., Moyes, D., Schroeder, W. A., Shelton, J. B., Shelton, J. R., and Beutler, E.**, Hemoglobin Pasadena, $\alpha_2\beta^{75(E19)Leu\rightarrow Arg}$. Identification by high performance liquid chromatography of a new unstable variant with increased oxygen affinity, *Biochim. Biophys. Acta*, 623, 360, 1980.
8. **Schroeder, W. A., Shelton, J. B., and Shelton, J. R.**, Separation of hemoglobin peptides by high performance liquid chromatography (HPLC), *Hemoglobin*, 4, 551, 1980.
9. **Schroeder, W. A. and Huisman, T. H. J.**, *The Chromatography of Hemoglobin*, Marcel Dekker, New York, 1980.
10. **Schroeder, W. A., Shelton, J. B., and Shelton, J. R.**, High performance liquid chromatography in the identification of human hemoglobin variants, in *Advances in Hemoglobin Analysis*, Hanash, S. M. and Brewer, G. J., Eds., Alan R. Liss, New York, 1981, 1.
11. **Schroeder, W. A., Shelton, J. B., Shelton, J. R., and Powars, D.**, Separation of peptides by high-pressure liquid chromatography for the identification of a hemoglobin variant, *J. Chromatog.*, 174, 385, 1979.
12. **Schroeder, W. A., Shelton, J. B., Shelton, J. R., and Powars, D.**, Hemoglobin Sunshine Seth — α_2(94(G1)Asp→His)β_2, *Hemoglobin*, 3, 145, 1979.
13. **Schroeder, W. A., Shelton, J. B., Shelton, J. R., Powars, D., Friedman, S., Baker, J., Finklestein, J. Z., Miller, B., Johnson, C. S., Sharpsteen, J. R., Sieger, L., and Kawaoka, E.**, Identification of eleven human hemoglobin variants by high-performance liquid chromatography: additional data on functional properties and clinical expression, *Biochem. Genet.*, 20, 133, 1982.
14. **Wilson, J. B., Lam, H., Pravatmuang, P., and Huisman, T. H. J.**, Separation of tryptic peptides of normal and abnormal α, β, γ, and δ hemoglobin chains by high-performance liquid chromatography, *J. Chromatogr.*, 179, 271, 1979.
15. **Sugihara, J., Imamura, T., Imoto, T., and Yanase, T.**, Identification of an abnormal hemoglobin with reduced oxygen affinity by high-performance liquid chromatography, *Biochim. Biophys. Acta*, 669, 105, 1981.
16. **Boissel, J. P., Wajcman, H., Fabritius, H., Cabannes, R., and Labie, D.**, Application of high-performance liquid chromatography to abnormal hemoglobin studies. Characterization of Hemoglobins D in Ivory Coast and description of a new variant Hb Cocody (β21(B3)Asp→Asn), *Biochim. Biophys. Acta*, 670, 203, 1981.
17. **Bishop, C. A., Hancock, W. S., Brennan, S. O., Carrell, R. W., and Hearn, M. T. W.**, High performance chromatography of amino acids, peptides, and proteins. XXIII. Peptide mapping by hydrophilic ion-paired reversed-phase high performance liquid chromatography for the characterization of the tryptic digest of haemoglobin variants, *J. Liq. Chromatogr.*, 4, 599, 1981.
18. **Ibarra, B., Franco-Gamboa, E., Ramirez, M. L., Cantú, J. M., Wilson, J. B., Lam, H., and Huisman, T. H. J.**, Hb Chiapas $\alpha_2$114Pro→Argβ_2: Identification by high pressure liquid chromatography, *Hemoglobin*, 5, 605, 1981.
19. **Yocum, K. M.**, The Synthesis and Characterization of Inorganic Redox Reagent-Modified Cytochromes *c*, Ph.D. thesis, California Institute of Technology, Pasadena, 1981; Yocum, K. M., Shelton, J. B., Shelton, J. R., Schroeder, W. A., Worosila, G., Isied, S. S., Bordignon, E., and Gray, H. B., Preparation and characterization of a pentaamine-ruthenium (III) derivative of horse heart cytochrome c, *Proc. Natl. Acad. Sci. U.S.A.*, 79, 7052, 1981.
20. **Wills, C., Kratofil, P., and Martin, T.**, Functional mutants of yeast alcohol dehydrogenase, in *Genetic Engineering of Microorganisms for Chemicals*, Hollaender, A., Ed., Plenum Press, New York, in press.
21. **Jörnvall, H.**, The primary structure of yeast alcohol dehydrogenase, *Eur. J. Biochem.*, 72, 425, 1977.
22. **Kratzin, H., Yang, C. Y., Krusche, J. U., and Hilschmann, N.**, Präparative Auftrennung des tryptischen Hydrolysats eines Proteins mit Hilfe der Hochdruck-Flüssigkeitschromatographie. Die Primärstruktur eines monoklonalen L-Kette vom K-Typ, Subgruppe I (Bence-Jones-Protein Wes), *Z. Physiol. Chem.*, 361, 1591, 1980.
23. **Yang, C. Y., Pauly, E., Kratzin, H., and Hilschmann, N.**, Chromatographie und Rechromatographie in der Hochdruckflüssigkeitschromatographie von Peptidgemischen. Die vollständige Primärstruktur einer Immunoglobulin L-Kette vom κ-Typ, Subgruppe I (Bence-Jones-Protein Den), *Z. Physiol. Chem.*, 362, 1131, 1981.

24. **Haeffner-Gormley, L., Poludniak, N. H., and Wetlaufer, D. B.,** Separation of the tryptic peptides from reduced, alkylated hen egg white lysozyme by high-performance liquid chromatography, *J. Chromatog.*, 214, 185, 1981.
25. **Fullmer, C. S. and Wasserman, R. H.,** Analytical peptide mapping by high performance liquid chromatography. Application to intestinal calcium-binding proteins, *J. Biol. Chem.*, 254, 7208, 1979.
26. **Clegg, J. B., Naughton, M. A., and Weatherall, D. J.,** Abnormal human haemoglobins. Separation and characterization of the α and β chains by chromatography, and the determination of two new variants, Hb-Chesapeake and Hb J(Bangkok), *J. Mol. Biol.*, 19, 91, 1966.
27. **Congote, L. F., Bennett, H. P. J., and Solomon, S.,** Rapid separation of the α, β, $^{G}\gamma$, and $^{A}\gamma$ human globin chains by reversed-phase high pressure liquid chromatography, *Biochem. Biophys. Res. Commun.*, 89, 851, 1979.
28. **Shelton, J. B., Shelton, J. R., and Schroeder, W. A.,** Preliminary experiments in the separation of globin chains by high performance liquid chromatography, *Hemoglobin*, 3, 353, 1979.
29. **Huisman, T. H. J. and Wilson, J. B.,** Recent advances in the quantitation of human fetal hemoglobins with different gamma chains, *Am. J. Hematol.*, 9, 225, 1980.
30. **Petrides, P. E., Jones, R. T., and Böhlen, P.,** Reverse-phase high performance liquid chromatography of proteins: the separation of hemoglobin chain variants, *Anal. Biochem.*, 105, 383, 1980.
31. **Meek, J. L.,** Prediction of peptide retention times in high-pressure liquid chromatography on the basis of amino acid composition, *Proc. Natl. Acad. Sci. U.S.A.*, 77, 1632, 1980.
32. **Shelton, J. B., Shelton, J. R., and Schroeder, W. A.,** Further experiments in the separation of globin chains by high performance liquid chromatography, *J. Liq. Chromatogr.*, 4, 1381, 1981.
33. **Huisman, T. H. J., Altay, C., Webber, B., Reese, A. L., Gravely, M. E., Okonjo, K., and Wilson, J. B.,** Quantitation of three types of γ chain of Hb F by high pressure liquid chromatography; application of this method to the Hb F of patients with sickle cell anemia or the S-HPFH condition, *Blood*, 57, 75, 1981.
34. **Huisman, T. H. J. and Altay, C.,** The chemical heterogeneity of the fetal hemoglobins of Black newborn babies and adults: a reevaluation, *Blood*, 58, 491, 1981.
35. **Congote, L. F.,** Rapid procedure for globin chain analysis in blood samples of normal and β-thalassemia fetuses, *Blood*, 57, 353, 1981.
36. **Meek, J. L. and Rossetti, Z. L.,** Factors affecting retention and resolution of peptides in high-performance liquid chromatography, *J. Chromatogr.*, 211, 15, 1981.
37. **Wilson, K. J., Honegger, A., and Hughes, G. J.,** Comparison of buffers and detection systems for high-pressure liquid chromatography of peptide mixtures, *Biochem. J.*, 199, 43, 1981.
38. **Hearn, M. T. W. and Grego, B.,** High-performance liquid chromatography of amino acids, peptides and proteins. XXXVI. Organic solvent modifier effects in the separation of unprotected peptides by reversed-phase liquid chromatography, *J. Chromatogr.*, 218, 497, 1981.
39. **Su, S.-J., Grego, B., Niven, B., and Hearn, M. T. W.,** Analysis of group retention contributions for peptides separated by reversed phase high performance liquid chromatography, *J. Liq. Chromatogr.*, 4, 1745, 1981.

Protein Separations

REVIEW OF SEPARATION CONDITIONS

William S. Hancock and David R. K. Harding

INTRODUCTION

While the separation of amino acids and small peptides by reversed-phase high-performance liquid chromatography (RP-HPLC) is becoming an accepted procedure, the separation of proteins by this technique still requires careful evaluation. The biological activities of many proteins are sensitive to denaturation by extremes in pH, by contact with organic solvents or high salt concentrations, by adsorption onto glass or hydrophobic moieties, or at an air-water interface.[1] Since RP-HPLC involves the use of silica particles with chemically bonded hydrophobic groups and with mobile phases which usually contain salts and organic solvents, it can be readily seen that protein separations require careful optimization (see the chapter by Strickler and Gemski).

Hydrophobic chromatography has become a popular technique for the purification of proteins on alkyl- or aryl-substituted agarose and a number of excellent separations have been published.[2,3] Thus the separation of proteins by RP-HPLC is technically feasible. In fact a judicious choice of reversed-phase columns of medium polarity, e.g., C_3H_7-phenyl, when combined with a suitable choice of organic modifier, e.g., isopropanol and optimal salt concentrations, e.g., 0.1 *M* phosphate at pH values of 3 to 7, has allowed the separation of proteins with retention of biological activity. Figure 1 shows a typical protein separation, that of bovine proinsulin on a μBondapak® C_{18}-column.

The fluid mosaic model for membrane structure proposed that membrane proteins float in a sea of lipid. The nonpolar regions of the protein are in contact with the nonpolar lipids present in the membrane, while the polar surface is exposed to the aqueous environment. As is common for many biological systems, the hydrophobic effect plays an important role in this model. The effect attributes a favorable energetic state to the close association of nonpolar (hydrophobic) surfaces in aqueous environments. In the absence of such an association, the nonpolar surfaces must be exposed to an aqueous environment and considerable energy is required to form a solvent cavity for the molecules. There may well be a useful parallel between this model and the mechanism of separation of proteins by RP-HPLC.

In Figure 2 a model is shown for the interaction between protein molecules and a nonpolar support. When a protein is introduced onto the reversed-phase column, the nonpolar regions of the molecule can displace solvent molecules from the surface of the reversed phase. At this stage the protein molecule will probably be adsorbed at the surface of the reversed-phase "bristle". It is unlikely that the protein will penetrate far into the reversed phase unless it contains very large nonpolar surface areas, and thus adsorption rather than partition is the preferred mechanism of interaction.[5] Pearson et al.[6] have suggested that as many as 44 sites of interactions can occur between a typical protein (for example albumin) and a typical reversed-phase column.

This proposed mechanism for protein separations is supported by the theoretical studies of Horvath et al.[7] In these studies, the hydrophobic effect in aqueous-organic systems (termed the Solvophobic Theory) was used to predict the retention of peptides on a nonpolar column. These authors found that the dominant interactions were between the mobile and stationary phase and between the mobile phase and the sample molecules. The driving force in both interactions was the shielding of a nonpolar region of either the column or sample molecule from the polar aqueous phase.

Based on this theory, one can explain the effect of an organic solvent gradient in displacing nonpolar samples from the column during a reversed-phase analysis. The solvent gradient

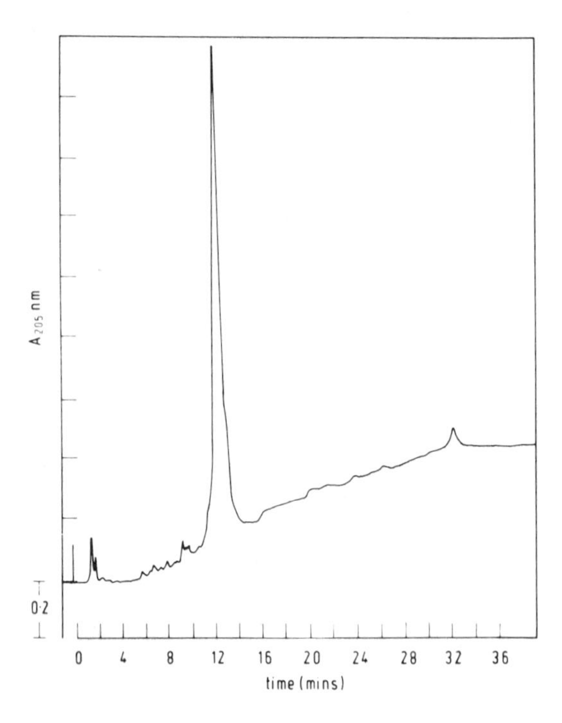

FIGURE 1. Chromatogram obtained with bovine proinsulin using a 30-min linear gradient generated from acetonitrile-water-phosphoric acid (10:90:0.1%) to acetonitrile-water-phosphoric acid (75:25:0.1%). Column, μBondapak® C_{18}, flow rate 2 mℓ/min. The gradient was commenced at the time of loading the sample.

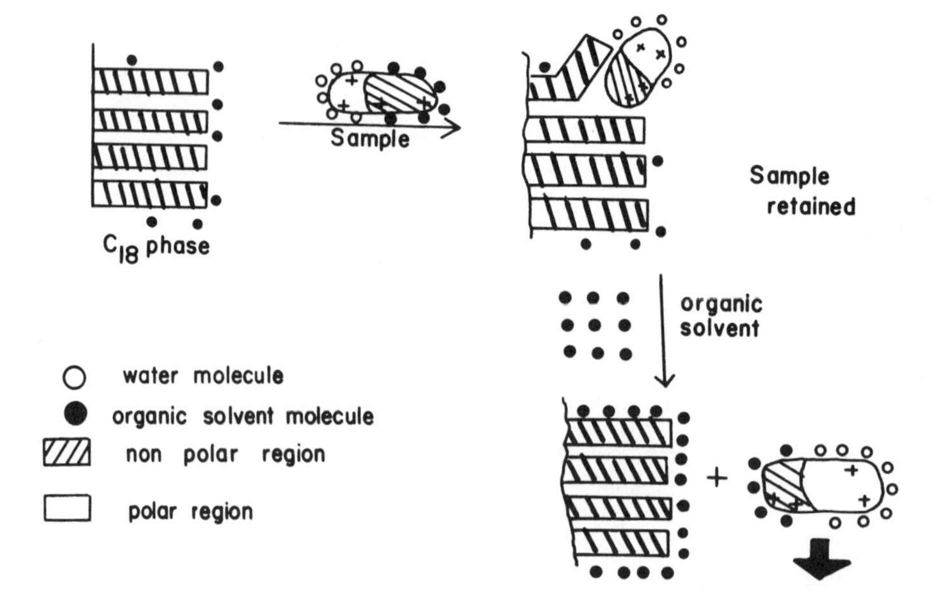

FIGURE 2. A proposed model for the interaction of a protein molecule with the surface of a C_{18}-reversed-phase column. The cross-hatched areas represent the nonpolar regions of the protein and C_{18} molecules. (From Singer, S. J., *J. Colloid Interface Sci.*, 58, 452, 1977. With permission.)

Table 1
EXAMPLES OF RP-HPLC ANALYSIS OF SEVERAL PEPTIDES AND PROTEINS

Sample	Eluant[a]	Type of ion pairing with amino groups	Ion-pairing reagent	Plate counts (plates per meter)	Retention time (min)
Leu-Trp-Met-Arg-Phe	50% CH_3OH				40.6
Leu-Trp-Met-Arg-Phe	50% CH_3OH	Hydrophilic	0.1% H_3PO_4	2700	7.3
Leu-Trp-Met-Arg-Phe	50% CH_3OH	Hydrophobic	5 m*M* sodium hexane sulfonate, pH 6.5	6100	13.6
Linear antamanid	60% CH_3OH				100
Linear antamanid	60% CH_3OH	Hydrophilic	0.1% H_3PO_4 + 0.1 *M* KH_2PO_4	4500	2.4
Porcine insulin	60% CH_3OH	Hydrophilic	0.1% H_3PO_4	1100	6.0[b]
ACTH 1-24 pentaacetates	40% CH_3OH	Hydrophilic	0.1% H_3PO_4	800	2.1[b]
Glucagon	40% CH_3OH	Hydrophilic	0.1% H_3PO_4	2000	4.6[b]
Acyl carrier protein	5% CH_3CN	Hydrophilic	0.1% H_3PO_4	1300	1.8[b]
Acyl carrier protein	30% CH_3CN	Hydrophobic	5 m*M* sodium hexane sulfonate, pH 6.5	3400	4.2[b]

[a] Expressed as a percentage of the organic component: in all cases the other solvent was water.
[b] In the absence of an ion-pairing reagent, the sample was retained indefinitely.

From Hancock, W. S., Bishop, C. A., Prestidge, R. L., Harding, D. R. K., and Hearn, M. T. W., *Science*, 200, 1168, 1978. With permission.

reduces the surface tension of the water molecules and thus decreases the energy required to form a cavity of water molecules surrounding nonpolar molecules. With a sufficient decrease in surface tension, the sample molecule is no longer adsorbed to the reversed phase and is eluted in the mobile phase.

As was mentioned for the section on peptide separations, the addition of an ionic modifier is essential for the separation of protein samples by RP-HPLC. Typically a protein molecule is retained indefinitely on the column in the absence of an ionic modifier added to the mobile phase. A major effect of an added electrolyte is to suppress deleterious interactions between the silanol groups present in all reversed-phase packings and the protein molecule. We have found however, that the addition of ionic modifiers to the mobile phase can also give useful differences in the resolution of complex protein mixtures. Table 1 shows examples of the use of ion-pairing reagents for protein analysis by RP-HPLC and contrasts the results with those obtained for peptides. In general the effect of an ion-pairing reagent is magnified for a protein relative to a peptide sample. This magnification can be attributed to the presence of a much larger number of ionizable groups in the protein sample. For example in Table 1 it is shown that the replacement of 0.1% phosphoric acid with sodium hexanesulfonate (hydrophilic to hydrophobic ion-pairing reagent) resulted in a dramatic increase in retention of acyl carrier protein on a reversed phase column.

EFFECT OF NATURE OF REVERSED-PHASE COLUMN ON THE SEPARATION

In contrast to classical liquid chromatography which involves the separation of relatively simple organic molecules such as pharmaceuticals, the separation of proteins involves high-molecular-weight solutes which are sensitive to denaturation. For example the optimal choice of ion-pairing reagent can be more crucial in a protein separation. Apolipoprotein A-I, which has a molecular weight of 28,000, can be readily eluted from a μBondapak®-alkylphenyl column with a mobile phase that contains 1% triethylammonium phosphate (TEAP) and a gradient of 0 to 40% isopropanol. The substitution of TEAP with 5 m*M* butanesulfonate results in the protein sample being retained indefinitely on the phenyl column. In a similar manner protein separations usually require careful selection of the type of column, as well as the nature of the organic solvent and the precise details of the gradient program.

The chapter written by Lewis shows that the pore size of the silica used in the manufacture of the reversed-phase packing material is crucial for the separation of high-molecular-weight proteins. A recent comparison of LiChrosorb® RP-18 (100Å pores) with Aquapore RP-300 (300-Å pores) showed that the latter column was notably superior for the chromatography of proteins with a molecular weight exceeding 15,000 daltons, e.g., the separation of [^{14}C]-methylated bovine pancreatic trypsin inhibitor, myoglobin, and ovalbumin.[10] Fortunately manufacturers have recognized the importance of pore size in protein separations and recently several wide pore silicas have become commercially available (see the chapters by Wehr and Halaz).

In addition, the type of silica used in the preparation is just as important as the pore size; in fact it may be of crucial importance.[6] Therefore it is difficult to precisely assess the effect of pore size in a study such as that of Wilson et al.[10] where two different types of silicas were compared. We described the use of the Radial Pak-C_{18}-column for the high efficiency separation of complex peptide and protein mixtures. Figure 3 shows the successful chromatography of a mixture of C-apolipoproteins from human very low density lipoproteins (VLDL) using two different gradient programs. In this separation the Radial Pak C_{18}-column clearly separated apolipoprotein C-I from C-$III_{1,2}$ and C-II. With the same chromatographic conditions, a μBondapak® C_{18}-column gave a broad peak with no separation of the protein mixture. An important difference between the two columns is that the μBondapak® C_{18}-column, but not the Radial Pak C_{18}-column, has unreacted silanol groups blocked by "end-capping" with trichloromethylsilane (see the chapter by Wehr). Thus the Radial Pak C_{18}-column contains a significant concentration of free silanol groups which can interact with the sample molecules as well as with the mobile phase. We have shown that such a packing material can give a mixed retention mechanism, where the sample molecule can interact either with silanol or reversed-phase groups depending on the separation conditions. Also it has been observed that reversed-phase columns with a high loading of C_{18} groups are often unsuitable for protein separations.[5] The relatively high column efficiencies obtained with nonpolar stationary phases with low surface coverages can be attributed to a higher concentration of accessible silanol groups. Such groups, which are likely to be hydrated, may reduce mass transfer resistance for a polar solute from the aqueous eluent to the surface of the stationary phase. In addition silanol groups, in the presence of low pH mobile phases, may introduce specific hydrogen bonding interactions with polar solutes. Pearson et al. found that the silica present in Vydak columns was also suitable for protein separations.[6]

The nature of the nonpolar group attached to the silica can be important in protein separations. Nice et al. found that elastase could only be recovered from columns containing C_6 or shorter-chain alkyl groups and that columns which were more nonpolar did not allow elution of the protein.[12] In the same study it was noted that the best recovery (78%) of ovalbumin was obtained from an Ultrasphere SAC (C_3) column.[12] In the separation of the C-apolipoproteins isolated from VLDL, a C_{18} column gave very little separation of the protein mixture and required high levels of acetonitrile to elute the proteins. It can be seen from Figure 4 that the less nonpolar alklylphenyl and nitrile columns gave a good separation of the apolipoproteins. The alkylphenyl column was chosen for subsequent studies because of the somewhat greater selectivity achieved (see Figure 4).

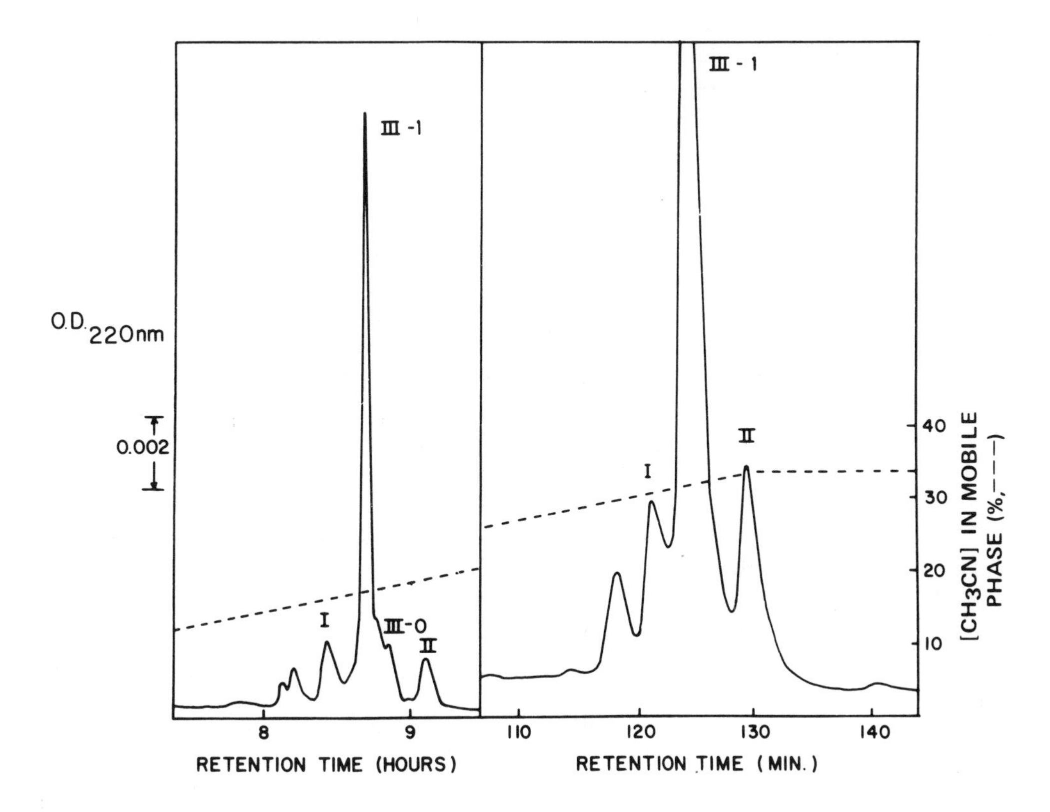

FIGURE 3. The elution profiles of a mixture of C-apolipoproteins obtained from human VLDL when chromatographed on a Radial Pak-C_{18}-column and with different acetonitrile gradients (see dashed lines). A 0.1-mg sample of the protein mixture was dissolved in 0.1 mℓ of 1% TEAP, 6 *M* guanidine hydrochloride and then chromatographed on the Radial Pak-C_{18}-column with an initial mobile phase of 1% TEAP, pH 3.2. (From Hancock, W. S. and Sparrow, J. T., *J. Chromatogr.*, 206, 71, 1981. With permission.)

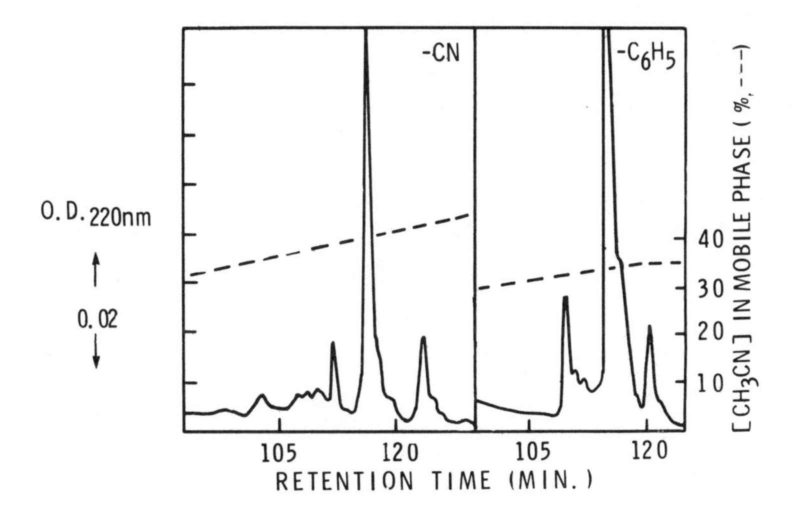

FIGURE 4. Use of two different columns (μBondapak®-CN or μBondapak®-alkylphenyl) in analysis of the C-apolipoprotein mixture. A 100-μg sample was chromatographed on either column with mobile phase of 1% TEAP and flow rate of 1.5 mℓ/min. The dashed line represents the gradient of acetonitrile.

EFFECT OF ORGANIC MODIFIER ON THE SEPARATION

When a protein molecule is introduced on the reversed-phase column, the nonpolar regions of the molecule can displace some of the solvent molecules from the surface of the reversed phase. At this stage, the protein molecule will probably be adsorbed at the surface of the reversed-phase bristle. A protein molecule with more than one nonpolar region can exhibit cooperative binding to the reversed phase.[5] Such binding can explain the reasonable capacities of small pore size reversed-phase packings and the small effect that column length has on separation efficiency, for moderate loadings of some protein samples.

With protein separations it is necessary that the sample does not interact too strongly with the reversed phase by either hydrophobic or silanol interactions. Therefore, many protein separations are better carried out with shallow gradients rather than with an isocratic (i.e., fixed) level of solvent. The gradual increase in organic solvent concentration would serve to continually displace the protein molecules from adsorption sites before irreversible multipoint binding occurs.

In several laboratories, it has been found that the isocratic chromatography of proteins could be achieved, if at all, only over a very narrow range of organic modifier concentration.[9] Slightly above this concentration, the sample is very rapidly eluted; below, the sample is strongly bound to the column. This problem can be overcome by the use of a shallow gradient of organic solvent (see Figure 6) so that the slow increase in organic solvent concentration results in sharp peaks and small elution volumes. In this separation, five different proteins were eluted by a 5% increase in the acetonitrile concentration. In Figure 4, the improved resolution that can be achieved with long gradients at low flowrates is shown. Clearly, the 10-hr gradient allows better resolution of minor components such as C-III_0 than the 2-hr gradient. However, initial optimization of a separation often can be achieved with short gradients as is shown in Figure 5 for the separation of the C-apolipoproteins. The time of the gradient program used in chromatogram 5D was then increased to 2 hr to give the optimized separation shown in Figure 6. It should be remembered that steep gradients of organic solvent can lead to precipitation of protein samples. To check against this problem the recovery of the protein sample should always be determined. Also monitoring of the reverse gradient to reequilibrate the column can be useful to see if any precipitated material has redissolved and subsequently eluted (see Reference 5 for other practical details).

The best organic modifier for the separation of small proteins is often acetonitrile (see the chapter by Welinder on the chromatography of insulin). In the chapter by Lewis and Stern on high-molecular-weight protein separations, it can be seen that *n*-propanol gives excellent results and is to be preferred to acetonitrile. Isopropanol usually gives similar results to *n*-propanol and it was used in the apolipoprotein separation as the proteins exhibited greater solubility in isopropanol.[11]

CHARACTERIZATION OF OPTICAL DENSITY PEAKS OBTAINED IN A PROTEIN ANALYSIS

It is most important that the optical density peaks observed in an HPLC analysis be subjected to an identification procedure because impurities present in the mobile phase can often give rise to spurious peaks. Amino acid analysis of an acid hydrolysate of the isolated peaks after HPLC separation is a crucial technique for identification, quantitation, and determination of purity. Unfortunately, an HPLC separation without this (or equivalent) characterization means little. In a mixture of proteins or peptides, some components may be recovered in almost quantitative yields while others may be isolated with very low yields. This is often observed for samples which aggregate readily or are extremely basic and interact strongly with the silanol groups present in the stationary phase. In the separation of the C-

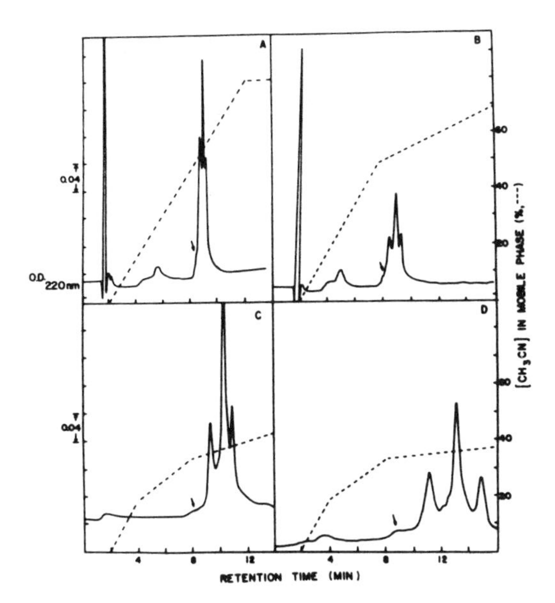

FIGURE 5. Elution profiles obtained for a crude mixture of C-apolipoproteins (50 μg) on a μBondapak®-alkylphenyl column with mobile phase of 1% TEAP, pH 3:2, and flow rate of 1.5 mℓ/min. Chromatograms A to D demonstrate effect of different acetonitrile gradients on separation. The apo-C-I peak is indicated by an arrow because it is present in low concentrations in this serum sample. Dashed line represents the program of the gradient former. (From Hancock, W. S., Bishop, C. A., Gotto, A. M., Harding, D. R. K., Lamplugh, S. M., and Sparrow, J. T., *Lipids,* 16, 250, 1981. With permission.)

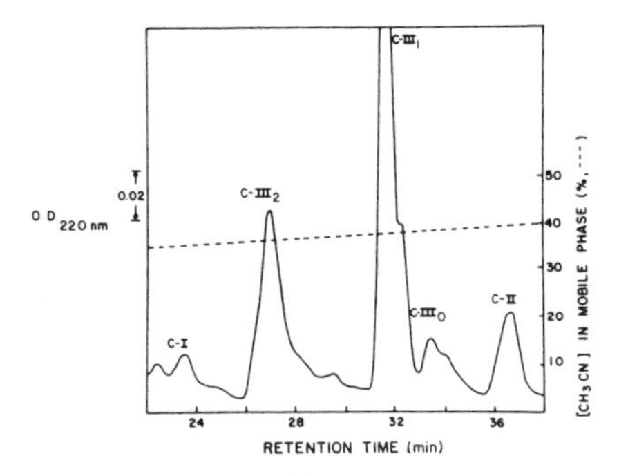

FIGURE 6. Optimized elution profile for separation of C-apolipoprotein mixture (100 μg) using the same conditions as in Figure 5 except that a very shallow gradient of acetonitrile was used at the organic solvent level necessary to elute these proteins (a combination of concave and convex gradient). (From Hancock, W. S., Bishop, C. A., Gotto, A. M., Harding, D. R. K., Lamplugh, S. M., and Sparrow, J. T., *Lipids,* 16, 250, 1981. With permission.)

Table 2
RECOVERIES OF PROTEIN SAMPLES AFTER SEPARATION BY REVERSED PHASE HPLC[a]

Protein sample	Recovery (%)
Tyrosinase, collagen α_1, bovine serum albumin	>80[b]
ACTH	>90[c]
Calcitonin-like materials	80
Insulin, β-lipotropin, albumin, neurotoxin, cytochrome *c*	50—80[d]
Ribonuclease, lysozyme, myoglobin, globin chains	70—100
Somatomedin *c*	>80
Apolipoproteins C-I, C-II, C-III	97,85.5,83
Cytochromes *c*	>80
Growth hormone	61
Parathyroid hormone	79
Proenkephalin	90
Collagen	85[b]

[a] As measured by mass recovery.
[b] For injection of 10-μg samples.
[c] Measured by radioimmunoassay and corticosteroid release form a suspension of isolated cells.
[d] Lower recoveries were observed for small sample loading (1 to 5 μg).

apolipoproteins shown in Figure 6, the recoveries of the different proteins were estimated by amino acid analysis to range from 83 to 97%.[13] With these proteins certain diagnostic amino acid values allowed a good estimation of purity; for example, apolipoprotein C-I does not contain histidine or tyrosine, while apolipoprotein C-II does not contain histidine, and apolipoprotein C-III does not contain isoleucine.

Another simple aid to identification and quantitation is the determination of the UV spectra of the pooled peaks from the separation. Many modern UV detectors allow scanning of the UV spectrum while the protein is contained in the sample cell. Often, proteins have a characteristic UV spectrum which can aid identification (primarily due to the absorption of tryptophan). Quantitation can also be achieved by measurement of the spectrum of the pooled peak (if the extinction coefficient of the protein at a given wavelength is known). Other techniques such as fluorescence measurements and sulfydryl determinations can be used.

THE POTENTIAL AND SCOPE OF HPLC SEPARATION METHODS

Provided the separation parameters are carefully optimized, it would appear that many protein samples can be successfully separated by RP-HPLC. In the chapter by Strickler and Gemski it was shown that enzymes could be separated by this technique with an excellent recovery of enzymatic activity. As is shown in Table 2 a range of proteins have been separated by RP-HPLC with a high recovery (mass measurement). Table 3 shows that the molecular weight of the protein sample is no limitation to the separation, as very high-molecular-weight samples have been chromatographed.

In conclusion, RP-HPLC has rapidly become an important new technique for the separation of proteins as can be judged from the following chapters on the separation of insulin, somatomedins, lipotrophins, endorphins, neurophysins, growth hormone, and milk proteins. In addition the technique will undoubtedly yield much information about the nature of proteins as well as allowing their separation, as can be judged by the chapters on the separation of multiple enzyme forms (Aoshima) and on the measurement of drug-protein interactions (Sebille and Thuand).

Table 3
MOLECULAR WEIGHTS OF PROTEINS CHROMATOGRAPHED SUCCESSFULLY BY RP-HPLC

Protein	Mol wt	Protein	Mol wt
Ferritin	450,000	Ribonuclease	13,700
Catalase	240,000	Cytochrome *c*	11,700
Collagen	200,000	β-Lipotropin	11,600
Somatomedins	140,000	Neurophysin	10,000
Tyrosinase	128,000	Apolipoprotein C-II	10,000
Collagen α_1	95,000	Apolipoprotein C-III	8,800
Serum albumin	65,000	Acyl carrier protein	8,800
Ovalbumin	43,000	Neurotoxin 3	7,800
Thyroid stimulating hormone	30,000	Apolipoprotein C-I	6,600
Proenkephalin	30,000	Insulin	5,400
Apolipoprotein A-I	28,000		
Chymotrypsinogen A	25,000		
Growth hormone	19,000		
Interferon	18,000		
Troponin C	18,000		
Apolipoprotein A-II	17,400		
Lysozyme	14,300		

One area where HPLC separations is having a dramatic impact is in the field of sequence determination of proteins. HPLC has been used in all phases of sequence determination, ranging from isolation of the protein sample, separation of protein fragments after chemical or enzymatic cleavage and separation of PTH derivatives. The overall application of HPLC to this important area of protein chemistry is reviewed in the chapter by Bhown and Bennett, while Prestidge has contributed a chapter on the separation of protein fragments. In another chapter Desiderio has demonstrated the exciting potential of field desorption mass spectroscopy to peptide sequencing. Schroeder has documented the application of RP-HPLC to the identification of related proteins by mapping techniques. Schlesinger has reviewed the separation of PTH amino acid derivatives.

The introduction of a glyceryl coating onto a silica particle has allowed the production of a pressure-resistant support suitable for high-performance gel permeation chromatography (HPGPC). The introduction of the glyceryl group reduces reversed-phase interactions to a sufficient extent that separations based on molecular weight differences can be achieved. It must be emphasized, however, that true molecular weight separations can only be achieved if the composition of the mobile phase is carefully optimized. The potential of the technique and operating conditions are defined in the chapters by Kato (TSK column) and Pollack (Protein-column), while the chapters by Okazaki and Hara, Welinder, and Humphries apply the technique to lipoproteins, insulin, and milk proteins, respectively.

REFERENCES

1. **Haschemeyer, R. H. and Haschemeyer, A. E.,** in *Proteins a Guide to Physical and Chemical Methods,* John Wiley & Sons, New York, 1973, 352.
2. **Hjerten, S.,** Fractionation of membrane proteins by hydrophobic interaction chromatography and by chromatography on agarose equilibrated with a water-alcohol mixture at low or high pH, *J. Chromatogr.,* 159, 85, 1978.
3. **Hofstee, B. H. J. and Otillio, N. F.,** Non-ionic adsorption chromatography of proteins, *J. Chromatogr.,* 159, 57, 1978.
4. **Singer, S. J.,** The proteins of membranes, *J. Colloid Interface Sci.,* 58, 452, 1977.
5. **Hancock, W. S. and Sparrow, J. T.,** *A Laboratory Manual for the Separation of Biological Materials by HPLC,* Marcel Dekker, New York, in press.
6. **Pearson, J. D., Lin, N. T., and Regnier, F. E.,** The importance of silica type for reversed phase protein separations, *Anal. Biochem.,* 124, 217, 1982.
7. **Horvath, C., Melander, W., and Molnar, I.,** Solvophobic interactions in liquid chromatography with non polar stationary phases, *J. Chromatogr.,* 125, 129, 1976.
8. **Hancock, W. S., Bishop, C. A., Prestidge, R. L., Harding, D. R. K., and Hearn, M. T. W.,** Reversed-phase HPLC of peptides and proteins with ion-pairing reagents, *Science,* 200, 1168, 1978.
9. **O'Hare, M. J. and Nice, E. C.,** Hydrophobic high-performance liquid chromatography of hormonal polypeptides and proteins on alkylsilane-bonded silica, *J. Chromatogr.,* 171, 209, 1979.
10. **Wilson, K. J., Van Wieringen, E., Klauser, S., Berchtold, M. W., and Hughes, G. J.,** Comparison of the high-performance liquid chromatography of peptides and proteins on 100- and 300-Å reversed phase supports, *J. Chromatogr.,* 237, 407, 1982.
11. **Hancock, W. S. and Sparrow, J. T.,** Use of mixed-mode high-performance liquid chromatography for the separation of peptide and protein mixtures, *J. Chromatogr.,* 206, 71, 1981.
12. **Nice, E. C., Capp, M. W., Cooke, N., and O'Hare, M. J.,** Comparison of short and ultrashort-chain alkylsilane-bonded silicas for the high-performance liquid chromatography of proteins by hydrophobic interaction methods, *J. Chromatogr.,* 218, 569, 1981.
13. **Hancock, W. S., Bishop, C. A., Gotto, A. M., Harding, D. R. K., Lamplugh, S. M., and Sparrow, J. T.,** Separation of the apoprotein components of human very low density lipoproteins by ion-paired, reversed-phase high performance liquid chromatography, *Lipids,* 16, 250, 1981.

SEPARATION OF LARGE PROTEINS

Randolph V. Lewis and Alvin S. Stern

INTRODUCTION

There is a greatly increased interest in the separation methods for large-molecular-weight biopolymers. Although over the years, these separations have been carried out using a variety of solid supports, the advent of high-performance liquid chromatography (HPLC) has brought not only increased speed of separation, but more importantly, newer, more efficient, and more selective supports for such chromatography.

The technique of reverse-phase (RP) HPLC has been used almost exclusively in our laboratories for the purification of polypeptides (< 50,000 mol wt).[1-5] This technique has also been used in other laboratories for peptides smaller than 13,000 mol wt.[6-8] During these studies a number of important factors have been found that can affect resolution and selectivity in separations involving large proteins. These effects can be categorized as stationary-phase or mobile-phase effects.

STATIONARY-PHASE CONSIDERATIONS

Most bonded-phase HPLC supports have been designed for the separation of small molecules. These supports have generally been based on the silica particle because its rigid nature allows it to withstand the packing and operating pressures needed for HPLC. The silica particles used were 5 to 10 μm in diameter and permeated by pores of varying dimensions typically 10 nm. Attempts in our laboratories to use such supports for larger proteins (> 40,000 mol wt) were unsuccessful due to: (1) poor recovery at low sample loading, (2) low sample capacity, and (3) poor resolution. These problems arise because high-molecular-weight proteins are too large to freely diffuse into the pores of the silica particles. This is demonstrated in Figure 1, which depicts the general phenomenon of decreasing column efficiency with increasing eluant flow rate. The height equivalent to a theoretical plate (HETP) is used here to express column efficiency and is obtained by dividing the length of the column by the theoretical plate count. At all flow rates examined on this 10-nm pore C_8-reverse-phase support, the large protein, bovine serum albumin (BSA; 68,000 mol wt), did not chromatograph as efficiently as the small protein, ribonuclease (13,700 mol wt), and neither protein chromatographed as efficiently as the amino acids, aspartic acid (133 mol wt) and phenylalanine (165 mol wt). This decreased efficiency for proteins is due in part to the fact that large molecules have much lower diffusion rates than do low-molecular-weight compounds and thus equilibrate much more slowly with the stationary phase.

However, the very poor column efficiency obtained on chromatography of BSA (Figure 1) is also the result of its poor penetration into the 10-nm pores of the column support. Therefore, by using a 50-nm pore C_8-support (10-μm particle size), greatly increased efficiency can be obtained for large proteins.[9] As seen in Figure 2, proteins from 12,500 to 95,000 mol wt exhibit similar behavior on the 50-nm pore reverse-phase support and as in Figure 1, approach a limiting value at the lower flow rates. A comparative chromatogram with 10- and 33-nm pore columns using several proteins is shown in Figure 3. The advantage of the larger pores can be seen in this figure.

Therefore, in order to take advantage of reverse-phase chromatography with larger proteins, it is necessary to use silica particles with pore sizes larger than currently available bonded supports. We have used 33- and 50-nm pore-sized silica particles to prepare supports

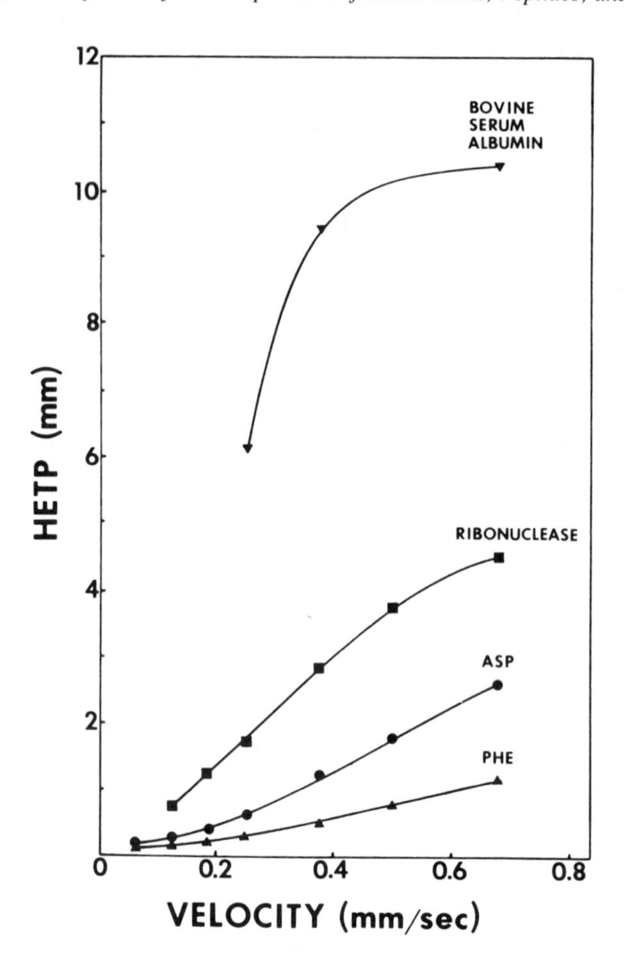

FIGURE 1. Effect of mobile-phase velocity on column efficiency under isocratic conditions. Aspartic acid and phenylalanine were eluted from a 10-nm pore C_8-column (10-μm particle size) with 1 *M* pyridine-0.5 *M* acetic acid (pH 5.5). Ribonuclease was eluted with the same buffer but containing 2% 1-propanol while BSA was eluted with this buffer containing 20% 1-propanol. Each point represents the average of three determinations. With this system a linear velocity of 0.5 mm/sec corresponds to a flow rate of approximately 20 mℓ/hr. (From Jones, B. N., Lewis, R. V., Paabo, S., Kajima, K., Kimura, S., and Stein, S., *J. Liq. Chromatogr.*, 3, 1373, 1980. With permission.)

bonded with octyl, octadecyl, cyanopropyl, or diphenyl groups. These supports with larger pore sizes are now becoming commercially available.

In addition to the pore size, the other stationary-phase factor that must be considered in order to obtain the maximum resolution of large proteins is the type of bonded phase. Our studies have clearly shown that the type of organic phase bound to the silica can greatly affect the retention times of proteins. The bonded phases examined have included octyl, octadecyl, diphenyl, and cyanopropyl phases. The differences in retention time of four proteins chromatographed on these bonded phases are shown in Figure 4. There is almost no difference between the octyl (Figure 4A) and octadecyl (Figure 4B) bonded phases. This suggests the possibility that the octadecyl chain is folding back on itself and thus resembles the octyl chain.

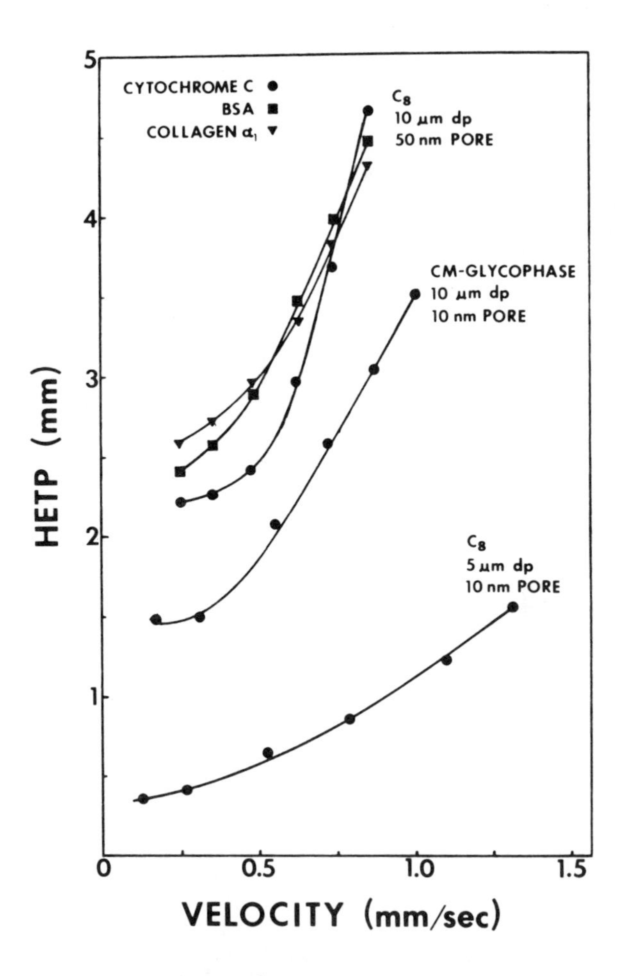

FIGURE 2. Effect of mobile-phase velocity on column efficiency under isocratic conditions. Proteins were eluted from two reverse-phase C_8-columns that differed in both particle size (dp) and pore diameter. The buffer system was 0.5 *M* formic acid-0.4 *M* pyridine (pH 4.0) using 1-propanol concentrations of 20% (collagen α_1), 24% (cytochrome *c*), and 26% (BSA). Cytochrome *c* was also eluted isocratically at various flowrates from a carboxymethyl (CM)-Glycophase ion-exchange column using 0.36 *M* acetic acid-0.09 *M* pyridine (pH 4.0). For the above systems, a flow rate of 20 mℓ/hr resulted in a linear velocity of approximately 0.5 mm/sec. (From Jones, B. N., Lewis, R. V., Paabo, S., Kajima, K., Kimura, S., and Stein, S., *J. Liq. Chromatogr.*, 3, 1373, 1980. With permission.)

The retention times of these same proteins are significantly different on the diphenyl bonded phase (Figure 4C and Table 1). These proteins are retained longer but by varying amounts. It is clear from these data and data obtained with small peptides[4] that polypeptides interact with the diphenyl column in a different manner than they do with alkyl chain-bonded phases. This interaction is probably a combination of hydrophobic and aromatic stacking interactions. Since increased retention times were not observed on a phenyl column, it appears that two phenyl groups are required to observe stacking with aromatic residues on proteins. The smaller effect seen with collagen is probably due to the very low content of aromatic amino acids. The selectivity differences between the alkyl chain- and diphenyl-bonded phases can be exploited to separate closely related proteins and peptides.[3,4]

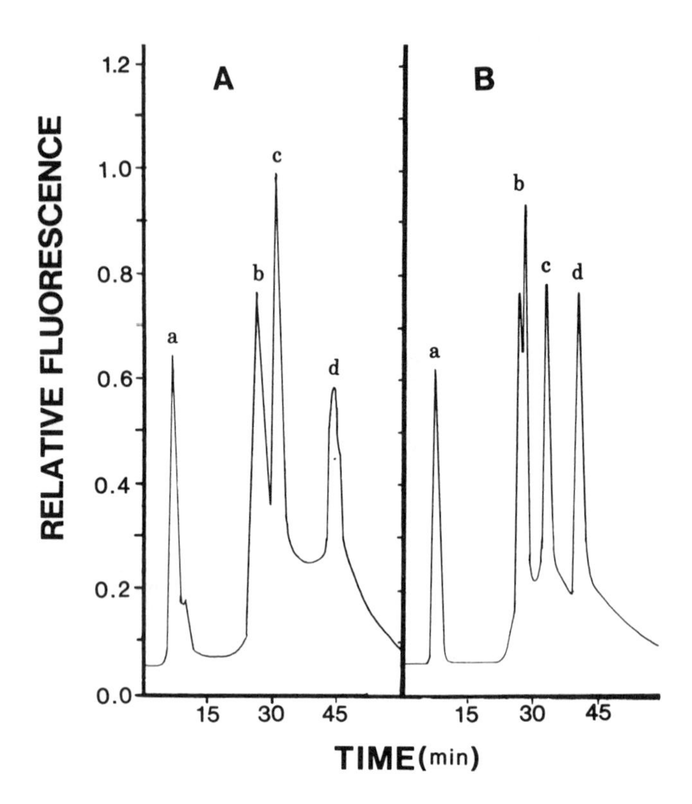

FIGURE 3. Comparison of 10- and 33-nm pore octyl columns. Phosphorylase B (10 μg), collagen (35 μg), human serum albumin (7 μg) and lactoglobulin B (10 μg) were dissolved in 1 mℓ of starting buffer for injection. This is the elution order of these proteins (identified a to d in the figure). The gradient was 1-propanol, 0 to 24% in 15 min, 24 to 48% in 55 min, in 0.5 *M* formic acid/0.4 *M* pyridine pH 4.0 with a flow rate of 0.75 mℓ/min. Panel A is a Lichrosorb® RP-8 column (10-μm particles with 10-nm pores). Panel B is a "Bakerbond" Wide Pore Octyl column (10-μm particles with 33-nm pores). The fluorescamine postcolumn detection system described elsewhere in this volume was used for detection.

The cyanopropyl-bonded phase shows slightly different selectivity from that of the alkyl chain-bonded phases. (Figure 4D and Table 1). This difference in selectivity is especially useful for the separation of proteins that differ in their level of glycosylation.[2]

From these results it is evident that the type of bonded phase can have a profound effect on the retention characteristics of large proteins. As with peptides, these differences can and have been exploited for protein purification. Since gradient conditions are similar for all of these bonded phases, it is possible to take advantage of these selectivity differences without the time-consuming determination of the proper elution conditions.

MOBILE-PHASE CONSIDERATIONS

There are several different factors that must be considered with regard to the mobile phase. These include: (1) flow rate, (2) pH, (3) ionic strength, (4) buffer composition, and (5) organic eluant. Although the studies of these factors are incomplete, several important points have emerged which are discussed below.

The effect that flow rate can have on the resolution of large proteins is clearly seen in

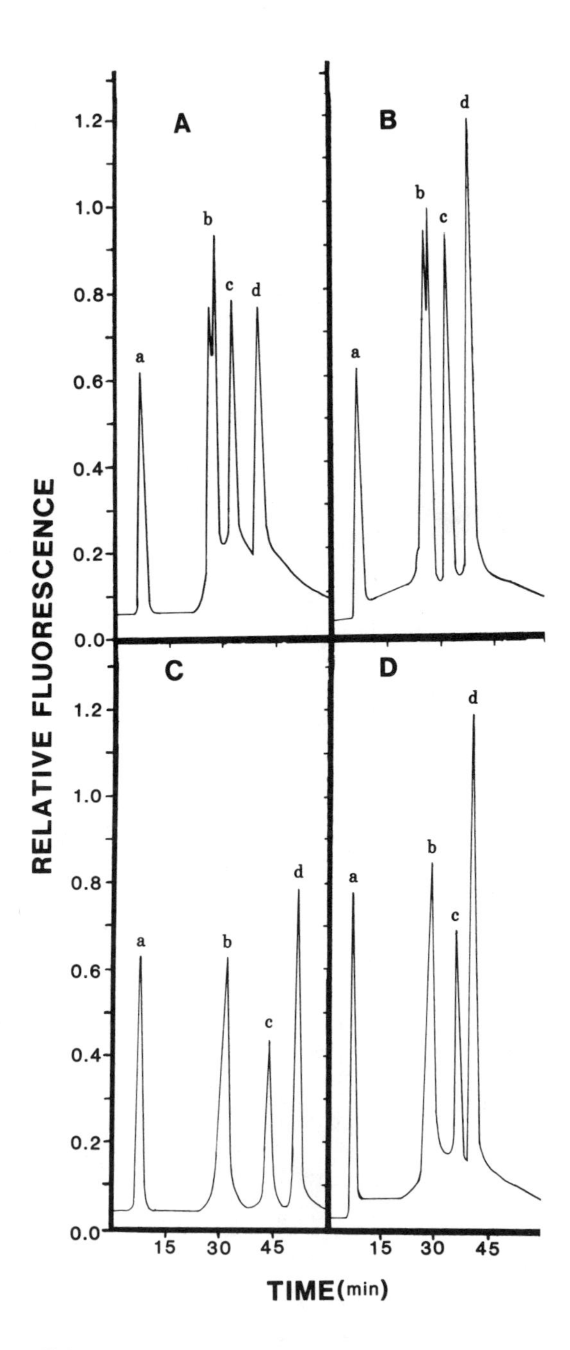

FIGURE 4. Elution from different bonded phases. The proteins and gradient used were the same as in Figure 2. The columns were all "Bakerbond" Wide Pore columns. (A) Octyl, (B) octadecyl, (C) diphenyl, and (D) cyanopropyl. Protein elution order is the same as Figure 3.

Figures 1 and 2. A close examination of Figure 2 shows that significantly improved resolution is obtained for all three of the proteins at linear flow rates of less than 0.5 mm/min, corresponding to about 0.5 mℓ/min for standard size columns. Therefore, the use of a slow flow rate is imperative for the best resolution of large proteins.

The use of elution buffers with pH values below 6 for improved resolution of peptides

Table 1

Column	Elution time (min)[a]			
	PB[b]	CL[b]	HSA[b]	LGB[b]
Octyl (10 nm)[c]	7	26.5 (0.5)	31.5 (0.3)	45 (0.3)
Octyl[d]	7	27.28 (0.3)	34 (0.4)	14 (0.4)
Octadecyl[d]	7	27.28 (0.3)	33 (0.3)	38.5 (0.4)
Cyanopropyl[d]	7	29.5 (0.2)	36.5 (0.4)	41 (0.4)
Diphenyl[d]	7	32.5 (0.1)	45.3 (0.3)	53.5 (0.4)

[a] The times are averages of at least three chromatographic runs with the SEM in parentheses.
[b] The peaks are phosphorylase B (PB), collagen (CL), human serum albumin (HSA), and β-lactoglobulin B (LGB).
[c] E. M. Merck Lichrosorb® octyl column (10-μm particles).
[d] "Bakerbond" Wide Pore column (33-mm pores and 10-μm particles). All columns were 4.6 × 250 nm.

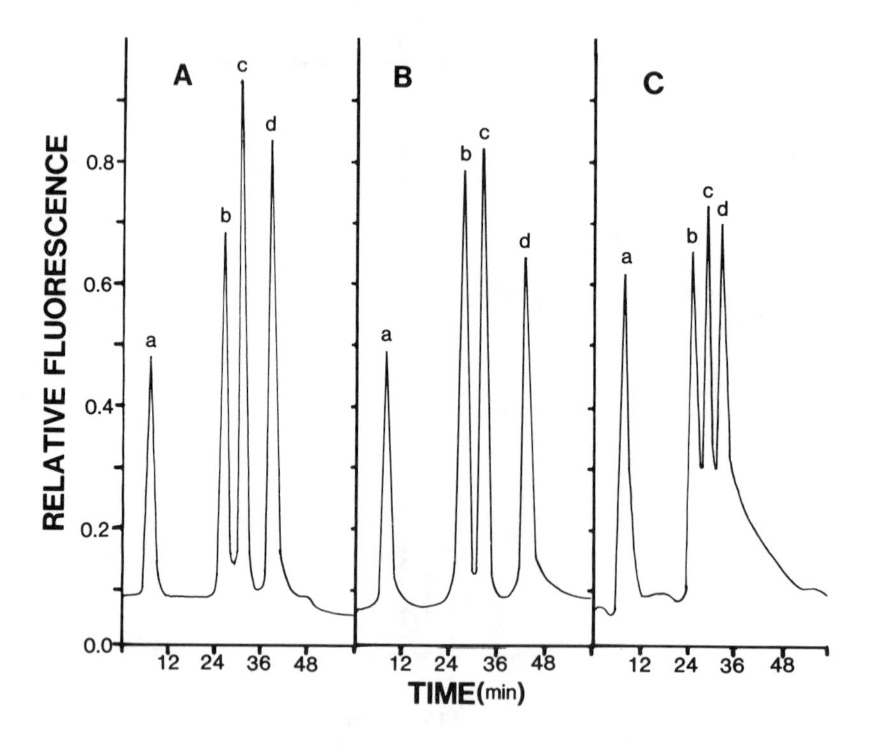

FIGURE 5. Effect of pH. The same proteins were chromatographed as in Figure 3 except the pH of the formate/pyridine buffer was varied: (A) pH 3, (B) pH 4, and (C) pH 5. The 33-nm pore octyl column was used.

has been demonstrated. This was attributed to the protonation of peptide carboxyl groups. Our studies demonstrate that for several buffer systems, lower pH values also give much better resolution with large proteins (0.5 *M* acetate/pyridine buffer is shown in Figure 5). Although the differences between pHs below 5 are small, the effects of pHs above 5 are significant. Resolution is greatly decreased at these higher pH values. This is probably due to a number of effects but does not appear to be due to substantial protein chain unfolding at lower pHs. We have studied the proteins in these experiments using UV and fluorescence

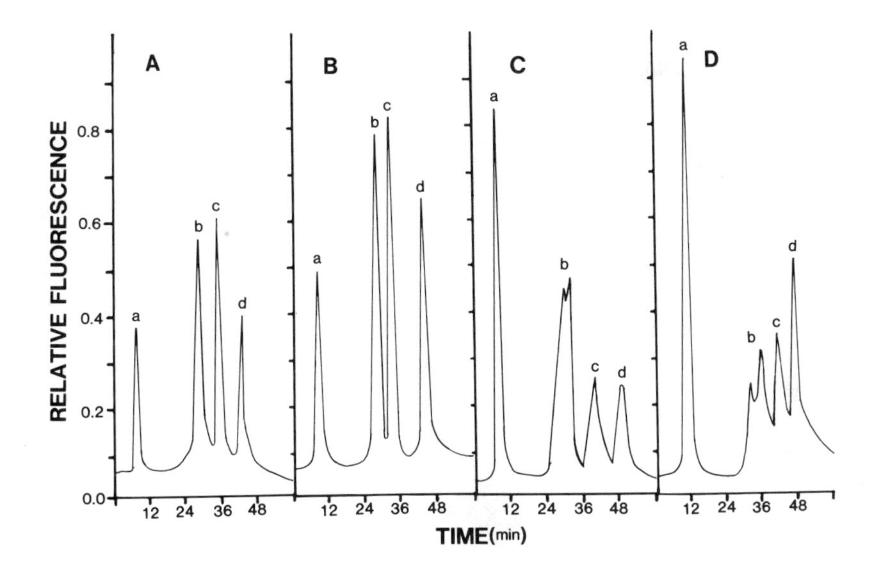

FIGURE 6. Effect of ionic strength. The same proteins were chromatographed as in Figure 3 except the ionic strength of the formate/pyridine buffer (pH 4.0) was varied: (A) 0.5, (B) 0.25, (C) 0.1, and (D) 0.05. The 33-nm pore octyl column was used.

spectroscopy and find very little detectable protein unfolding even down to pH 3 (unpublished results). The effects which are probably responsible are the change in surface charge on the protein and "ion pairing" of the buffer to the protein. Regardless of the reason for this effect, the lower the pH of the buffer (down to pH 3) the better the resolution obtained with large proteins.

The ionic strength of the buffer also has a significant effect on resolution (Figure 6), i.e., ionic strengths greater than 0.2 provide significantly improved resolution of polypeptides. This finding held at all pHs tested and with two other buffer systems. It is probable that this effect is also due to several factors including "ion pairing" and an increased protein hydrophobic interaction with the stationary phase. The effect of ionic strength on elution time is not as predictable. As seen in Figure 6, varying the ionic strength affects the various proteins differently. It may be possible to take advantage of this factor in the purification of some proteins as has been done with collagen.[11]

The composition of the buffer has also been shown to affect the resolution of large proteins.[12] Buffers that are transparent to UV light have been used extensively with peptides and small proteins. These buffers, usually containing phosphoric acid or trifloroacetic acid, give column performance that is significantly worse than with the pyridine-acetate buffers used in our laboratories. If UV absorption is to be used for detection, the pyridine can be replaced with triethylamine or *N*-ethylmorpholine. Other possible counterions that we are investigating are *N*-ethylpiperidine and *N*-methylpyrollidine, both of which resemble pyridine but have saturated carbon rings. The other factor we have considered in our buffer choice is the ability to remove the buffer by lyophylization or vacuum evaporation. This provides an advantage in purification procedures because dialysis and the time it takes coupled with the possibility of protein loss are avoided. For these reasons buffers composed of volatile organic acids and bases are best suited for the RP-HPLC of large proteins.

The final factor to be considered is the organic eluant and the gradient of that eluant used. Methanol and acetonitrile have been used routinely for peptide elution. With large proteins neither of these has sufficient hydrophobicity to be generally useful. However, 1-propanol which has roughly three times the elution power of methanol or acetonitrile, is of much greater utility. The choice of 1-propanol is made due to its increased hydrophobicity and its solubility in water. Gradient elution is usually necessary due to the wide range of hydrophobicities of proteins found in biological samples. Excellent resolution is still achieved due to the very narrow range of organic eluant that will elute a protein.[9] This narrow range makes isocratic elution of very similar proteins difficult. In our experience the vast majority of proteins are eluted between zero and 40% 1-propanol. A few proteins have required a gradient using 60% 1-propanol. Initial studies have utilized 2-hr linear gradients to determine the region of interest. Subsequent gradients are 2- to 4-hr shallow linear gradients with the estimated elution percentage at approximately the two thirds point of the gradient to obtain maximum separation from other components. In summary our findings on the optimization of resolution for large proteins with mobile-phase composition are as follows: (1) low flow rates (less than 0.5 mℓ/min), (2) buffer pHs below 5, (3) ionic strengths greater than 0.2, (4) volatile organic acid/base buffers, and (5) 1-propanol used as organic eluant with a shallow gradient.

EXAMPLES

Proenkephalin

[Met]enkephalin and [Leu]enkephalin are endogenous pentapeptides with opiate-like activity which were originally discovered in the brain.[13] Although β-endorphin, a 31-amino acid opioid peptide which contains the [Met]enkephalin sequence, has been shown to be produced from a common ACTH/β-lipotropin precursor (known as pro-opiocortin) by post-translational processing, accumulating evidence indicates that [Met]enkephalin itself is not derived from the same precursor.[14] Adrenal medulla possesses enkephalin-containing polypeptides in addition to free enkephalin.[15] These molecules range in molecular weight from 500 to 18,200 and have been purified to homogeneity and sequenced.[16] It has been suggested that these polypeptides are intermediates in a biosynthetic pathway starting with "proenkephalin".[17] In an attempt to elucidate the primary structure of this proenkephalin, DNA sequences complementary to mRNA coding for the precursor protein have recently been cloned.[18,19] Nucleotide sequence analysis of the cDNA has revealed that the proenkephalin contains six [Met]enkephalin sequences and one of [Leu]enkephalin. In order to confirm the existence of a primary translation product from the mRNA we attempted the purification of the 30,000-mol wt proenkephalin.

Chromaffin granules (in which the enkephalin-containing polypeptides are contained) from bovine adrenal medulla were isolated, extracted with neutral pH buffer and chromatographed on Sephadex® G-150 as described. The region corresponding to a molecular weight of 30,000 to 40,000 was applied directly onto a Lichrosorb® RP-8 HPLC column (10-μm, 10-nm pore size, 4.6 × 250 mm). Total recovery of biological activity was extremely low (less than 15%) and the chromatography yielded poorly resolved polypeptides. However when chromatographed on a column packed with C_8-bonded silica particles which contained a 50-nm pore size (10 μm, 4.5 × 250 mm), almost total recovery of biological activity was obtained (~ 90%) and resolution of the polypeptides was excellent (Figure 7A). Chromatography of the most active fractions on a large-pore cyanopropyl column yielded a peak of fluorescence which coincided with opioid activity (Figure 7B). The fraction having the highest specific activity was rechromatographed under similar conditions, except for the collection of smaller fractions and the use of a lower flow rate. A single symmetrical peak was obtained. The fraction with the highest specific activity was found to contain a polypeptide of 30,000 mol wt by polyacrylamide gel electrophoresis in SDS. Tryptic mapping of the polypeptide by HPLC[16] revealed the presence of six [Met]enkephalin sequences and one [Leu]enkephalin sequence.

The study of the proenkephalin molecule illustrates the use of RP-HPLC as a powerful

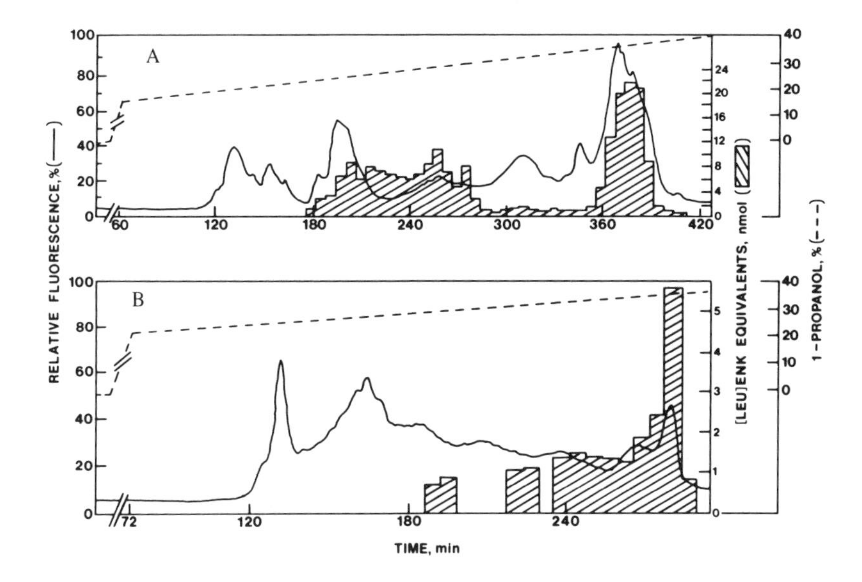

FIGURE 7. Purification of proenkephalin (A) Pooled fractions corresponding to 30,000 to 40,000 mol wt from Sephadex® G-150 chromatography of an extract of chromaffin granules were applied to a large-pore RP-8 column (10-μm, 50-nm pore size, 4.6 × 250 mm) at a rate of 80 mℓ/hr. The column was washed with 20 mℓ of starting buffer (0.5 *M* formic acid/ 0.4 *M* pyridine, pH 4.0) and proteins were eluted with a gradient of 1-propanol (— — —) in the same buffer at a rate of 15 mℓ/hr, 4% of the column effluent was diverted to the post-column fluorescamine detection system. Aliquots of each fraction (3 min) were digested with trypsin and carboxypeptidase B and assayed for opioid activity as previously described.[17] (B) The fraction containing the highest specific activity (276 to 282 min) was lyophilized. The proteins were redissolved in the above starting buffer and then injected onto a large-pore CN-propyl column (10-μm, 33-nm pore size, 4.6 × 250 mm). The proteins were eluted at a flow rate of 15 mℓ/hr and assayed as in A.

tool in the purification of larger proteins. The 10-nm pore reverse-phase supports contained pore sizes too small to be optimally utilized with this protein. Even though the 50-nm C_8 support (and the 33-nm cyanopropyl support) had elution characteristics similar to those of the smaller pore C_8 support with respect to peptides, it proved particularly useful for the purification of proenkephalin.

Collagen Separations

There is evidence that connective tissue contains a number of different collagen α chains.[21] The present methods for separating and identifying these different collagen types is slow and provides poor recoveries. These methods involve salt fractionation, ion-exchange chromatography, or separation of cyanogen bromide fragments and require large amounts of material which can negate many interesting studies. This problem appeared to be an excellent test for the large-pore RP-HPLC supports.

The separation of the α_1 and α_2 chains of Type I collagen (95,000 mol wt) was undertaken first.[9] The collagen was from lathrytic chicks and was denatured by heating to 60°C for 20 min in 0.5 *M* acetate/pyridine buffer at pH 4.0. The two chains were then separated with relative ease on an octyl (C_8) column (Figure 8). SDS-PAGE confirmed the separation and identified Peak A as the α_1 chain, and B as the α_2 chain. Human collagen in addition to these components has β and γ components comprised of chain multimers. Efforts to separate all of these components on C_8 and cyanopropyl columns were unsuccessful. However, the diphenyl column provides a substantial separation of all components (Figure 9).

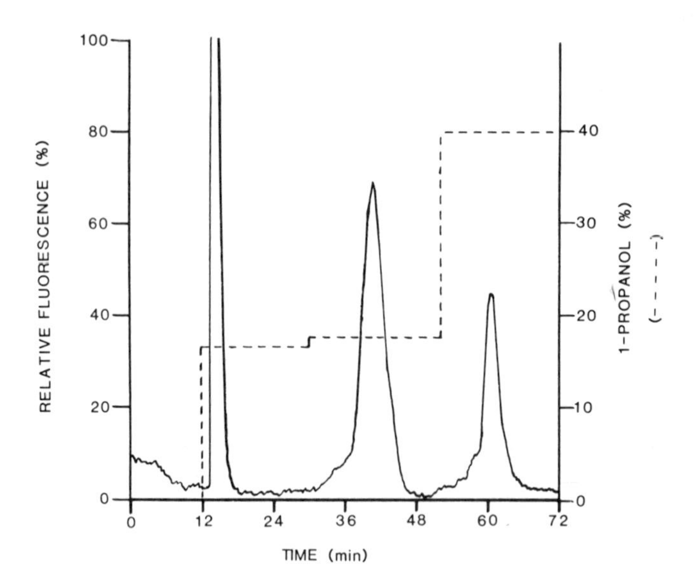

FIGURE 8. Elution of denatured chick Type I collagen (100 μg) from a C_8-column (50-nm pore size). The sample was applied to the column in the starting buffer and eluted with a 1-propanol gradient (as indicated) in 0.5 *M* acetate/pyridine (pH 5.0) at room temperature; the flow rate was 20 mℓ/hr. The earliest peak represents free amino acids, the middle peak the α_1 chain, and the final peak the α_2 chain. (From Lewis, R. V., Fallon, A., Stein, S., Gibson, K. D., and Udenfriend, S., *Anal. Biochem.*, 104, 153, 1980. With permission.)

The ability to separate the different chains suggested it might be possible to separate the intact triple helix collagen chains (300,000 mol wt). These separations required a number of modifications to the buffers used for the collagen chain separations.[11] It was found that the native collagen triple helix of Type I collagen $[(\alpha_1)_2\ \alpha_2]$ behaved like the α_1 chain on the cyanopropyl column. In addition, there was a strong pH dependence on the binding of Type I collagen to this column. It bound to the column only at pHs above 4.5. Thus, using the proper conditions (as described in Figure 10), it was possible to separate native Types I, II, and III collagen (Figure 10). Type III collagen does not bind to the column under these conditions and Types I and II were well separated. SDS-PAGE confirmed the identity of the peaks and in addition, gels of material between the peaks indicated that there was virtually no contamination of the peaks.

Both of these separations were possible using as little as 10 μg of protein with recoveries greater than 85%. These recoveries were determined by measuring protein recovered, and in the case of radiolabeled collagen, the recovery of radioactivity. With the use of these low levels of protein it should be possible to determine changes in the types of collagen in a variety of disease states and possibly in tissue culture as well.

SUMMARY

Due to the size and slow diffusion rates of larger proteins, HPLC supports with 6- to 10-nm pores give poor resolution. To overcome this problem, supports with 33- and 50-nm pore sizes were used successfully. These supports have been shown to provide excellent

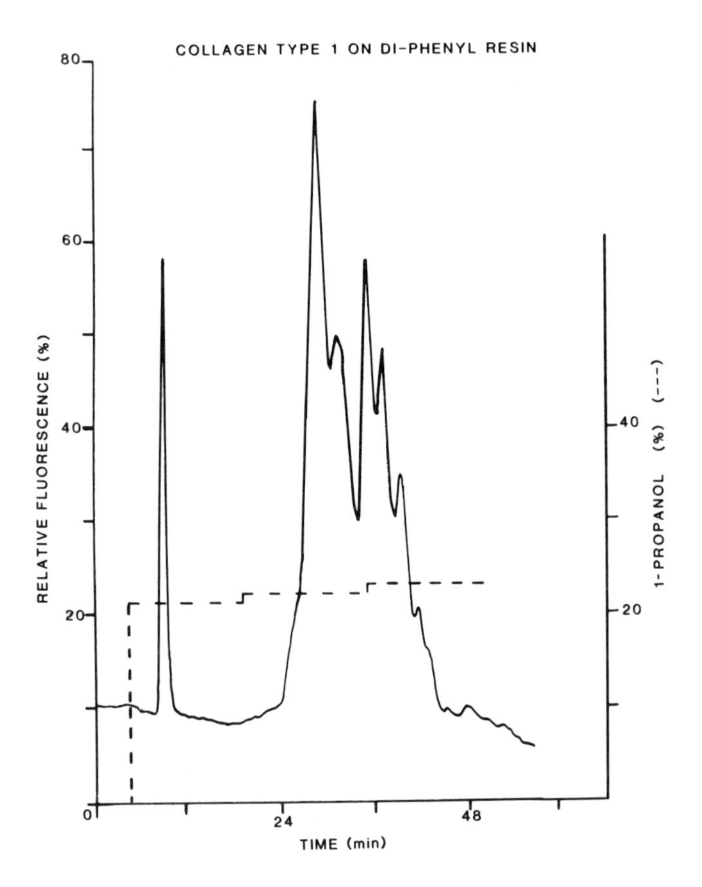

FIGURE 9. Separation of human Type I collagen subunits on a diphenyl column. Type I collagen from human skin was denatured and applied to a 33-nm pore diphenyl-column. Elution was carried out with a 1-propnaol gradient (as indicated) in 0.5 *M* formate/pyridine (pH 4.0) at room temperature; the flow rate was 30 mℓ/hr. The positions of the α_1 and α_2 chains are at 26 and 34 min. (From Lewis, R. V., Fallon, A., Stein, S., Gibson, K. D., and Udenfriend, S., *Anal. Biochem.*, 104, 153, 1980. With permission.)

resolution for proteins from 12,000 to 300,000 mol wt. There are a number of factors that must be considered when designing the mobile phase for these protein separations. By optimizing these various factors, greatly improved resolution of proteins can be obtained.

The examples described provide indications about the advantages of these supports for the isolation of large proteins. Several other applications are presently under study and all the preliminary results support the conclusion that the large pore supports provide excellent resolution of proteins.

ACKNOWLEDGMENTS

We would like to thank Drs. Sidney Udenfriend and Stanley Stein for their interest and help in initiating this work. We would also like to thank E. Gene Yundt for the preparation of this manuscript. This work was supported by grants from Research Corporation and Baker Chemical Company to Randolph V. Lewis.

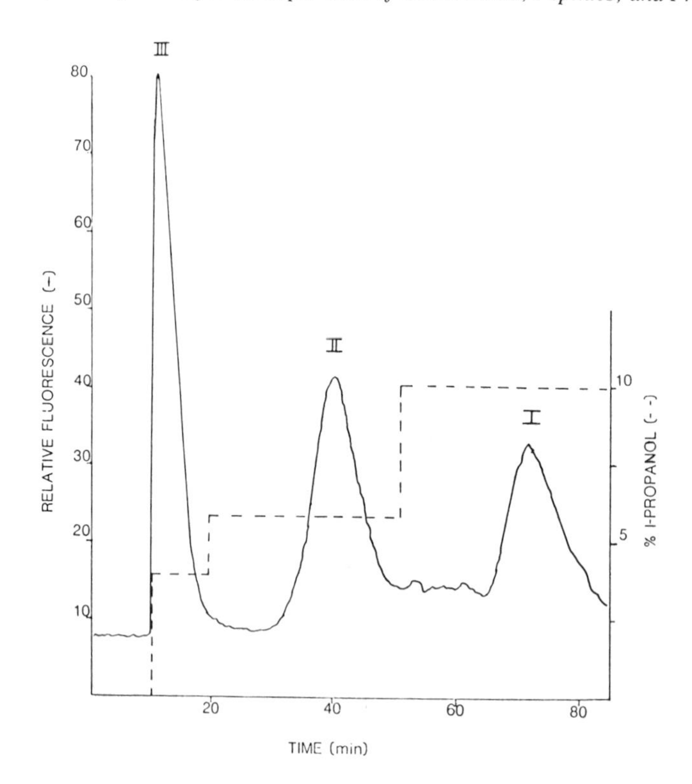

FIGURE 10. Elution of native Types I, II, and III collagen on a cyanopropyl column equilibrated in 1.5 *M* pyridine acetate (pH 4.6); 50-μg amounts of each type of collagen were made up to 0.5 mℓ in starting buffer and gently vortexed before application. The three major peaks (I, II, and III) were eluted using a 1-propanol gradient as shown at a flow rate of 18 mℓ/hr. (From Fallon, A., Lewis, R. V., and Gibson, K. D., *Anal. Biochem.*, 110, 318, 1981. With permission.)

REFERENCES

1. **Rubinstein, M., Rubinstein, S., Familletti, P., Miller, R. S., Waldman, A. A., and Pestka, S.,** Human leukocyte interferon: production, purification to homogeneity, and initial characterization, *Proc. Natl. Acad. Sci. U.S.A.*, 76, 640, 1979.
2. **Kimura, S., Lewis, R. V., Gerber, L. D., Brink, L., Rubinstein, M., Stein, S., and Udenfriend, S.,** Purification to homogeneity of camel pituitary proopiocortin, the common precursor of opioid peptides and corticotropin, *Proc. Natl. Acad. Sci. U.S.A.*, 76, 1756, 1979.
3. **Lewis, R. V., Stern, A. S., Kimura, S., Stein, S., and Udenfriend, S.,** Enkephalin biosynthetic pathway: proteins of 8,000 and 14,000 daltons in bovine adrenal medulla, *Proc. Natl. Acad. Sci. U.S.A.*, 77, 5018, 1980.
4. **Stern, A. S., Lewis, R. V., Kimura, S., Rossier, J., Stein, S., and Udenfriend, S.,** Opioid hexapeptides and heptapeptides in adrenal medulla and brain. Possible implications on the biosynethsis of enkephalins, *Arch. Biochem. Biophys.*, 205, 606, 1980.
5. **Jones, B. N., Shively, J. E., Kilpatrick, D. L., Stern, A. S., Lewis, R. V., Kojima, K., and Udenfriend, S.,** Two adrenal opioid proteins of 8,600 and 12,600 daltons: intermediates in proenkephalin processing, *Proc. Natl. Acad. Sci. U.S.A.*, 79, 2096, 1982.
6. **Glasel, J. A.,** Separation of neurophysal proteins by reversed-phase high-pressure liquid chromatography, *J. Chromatogr.*, 145, 469, 1978.

7. **Hancock, W. S., Bishop, C. A., Prestidge, R. L., Harding, D. R. K., and Hearn, M. T. W.,** High-pressure liquid chromatography of peptides and proteins. II. The use of phosphoric acid in the analysis of underivatised peptides by reversed-phase high-pressure liquid chromatography, *J. Chromatogr.*, 153, 391, 1978.
8. **Rivier, J. E.,** Use of trialkyl ammonium phosphate (TAAP) buffers in reverse phase HPLC for high resolution and high recovery of peptides and proteins, *J. Liq. Chromatogr.*, 1, 343, 1978.
9. **Lewis, R. V., Fallon, A., Stein, S., Gibson, K. D., and Udenfriend, S.,** Reverse-phase HPLC resin for large proteins, *Anal. Biochem.*, 104, 153, 1980.
10. **Lewis, R. V., Stein, S., and Udenfriend, S.,** Separation of all opioid peptides by HPLC, *Int. J. Peptide Protein Res.*, 13, 393, 1979.
11. **Fallon, A., Lewis, R. V., and Gibson, K. D.,** Separation of the major species of interstitial collagen by reverse-phase high-performance liquid chromatography, *Anal. Biochem.*, 110, 318, 1981.
12. **Jones, B. N., Lewis, R. V., Paabo, S., Kajima, K., Kimura, S., and Stein, S.,** Effects of flow rate and eluant composition on the high performance liquid chromatography of proteins, *J. Liq. Chromatogr.*, 3, 1373, 1980.
13. **Hughes, J., Smith, T. W., Kosterlitz, H. W., Fothergill, L., Morgan, B. A., and Morris, H. R.,** Identification of two related pentapeptides from the brain with potent opiate agonist activity, *Nature (London)*, 258, 577, 1975.
14. **Brownstein, M. J.,** Opioid peptides: search for the precursors, *Nature (London)*, 287, 678, 1980.
15. **Lewis, R. V., Stern, A. S., Rossier, J., Stein, S., and Udenfriend, S.,** Putative enkephalin precursors in bovine adrenal medulla, *Biochem. Biophys. Res. Commun.*, 89, 822, 1979.
16. **Stern, A. S. and Lewis, R. V.,** Methods for isolation, characterization, and sequence analysis of enkephalin precursors, *Res. Methods Neurochem.*, 6, in press, 1982.
17. **Lewis, R. V., Stern, A. S., Kimura, S., Rossier, J., Stein, S., and Udenfriend, S.,** A 50,000 Dalton protein in adrenal medulla that may be a common precursor of Met and Leu-enkephalin, *Science*, 208, 1459, 1980.
18. **Gubler, U., Seeburg, P., Hoffman, B. J., Gage, L. P., and Udenfriend, S.,** Molecular cloning establishes proenkephalin as precursor of enkephalin-containing peptides, *Nature (London)*, 295, 206, 1982.
19. **Noda, M., Furutani, Y., Takahashi, H., Toyosato, M., Hirose, T., Inayana, S., Nakarishi, S., and Numa, S.,** Cloning and sequence analysis of cDNA for bovine adrenal pre-proenkephalin, *Nature (London)*, 205, 202, 1982.
20. **Stern, A. S., Jones, B. N., Shively, J. E., Stein, S., and Udenfriend, S.,** Two adrenal opioid polypeptides: proposed intermediates in the processing of proenkephalin, *Proc. Natl. Acad. Sci. U.S.A.*, 78, 1962, 1981.
21. **Burgeson, R. E. and Hollister, D. W.,** Collagen heterogeneity in human cartilages identification of several new collagen chains, *Biochem. Biophys. Res. Commun.*, 87, 1124, 1979.

SEPARATION OF SMALL PROTEINS BY RPLC

Petro E. Petrides

PHYSIOLOGICAL RELEVANCE OF SMALL PROTEINS

Over the last years it has become increasingly clear that small proteins (here operationally defined as possessing a molecular weight between 6000 and 20,000 daltons) play an important role in intercellular communication. From a large number of in vitro and in vivo studies it is apparent that cellular proliferation and differentiation in practically all tissues is under the control of regulatory polypeptides, which are produced by certain cells, secreted into the pericellular fluid and subsequently diffuse to their specific target cells. These factors control a variety of fundamental physiological processes, e.g., the differentiation of pluripotent stem cells of the bone marrow into lymphocytes, erythrocytes, granulocytes, macrophages, megakaryocytes or eosinophiles. In the development of the nervous system regulatory polypeptides are involved in the survival, fiber outgrowth, adhesion to substrata, and differentiation of neurons. In tissue regeneration, these polypeptides are postulated to govern the migration and proliferation of cells (growth factors), the formation of new blood vessels (angiogenesis factors), the differentiation of mesenchymal cells into specialized cells participating in tissue repair, the production of extracellular matrix, or changes in microvascular permeability.

Such activities with peptidic chemical nature have been described to be present in tissue extracts, in media conditioned by normal or malignant cells in culture, or in physiological fluids such as serum, urine, milk, pancreatic juice, seminal fluid, and others. In contrast to classical regulatory polypeptides such as insulin, glucagon, murine epidermal, or nerve growth factor which occur in relatively high abundance in the tissues of origin, the isolation and characterization of the intercellular messengers aforementioned has proven difficult because of their low abundance.

Consequently, the availability of methods for the isolation of these substances which offer high resolving power, excellent recovery, and rapid analysis, is of utmost importance for the progress of our understanding their physiological role.

The extensive purification or at least partial purification is mandatory for several reasons. (1) Tissue extracts or cell conditioned media contain a myriad of polypeptides, some of them with possibly inhibitory or toxic activity which might interfere with the activity of the component of interest. (2) In some instances where partially purified material has been found to act upon several types of cells further purification might reveal a mixture of active components each with a unique target cell specificity. This makes interpretations questionable when studies are carried out with impure components. (3) With partial sequence information it is possible to clone the gene for the component which then can be used to probe tissues to determine where and when the gene is expressed and how the expression is regulated. (4) Large-scale production of the regulatory peptide by genetic engineering techniques will provide sufficient quantities for in vivo and in vitro studies. (5) It is likely that at least some diseases are caused by disturbances in the expression of genes for regulatory polypeptides. Consequently, the availability of gene probes should enable us to test their possible participation. (6) Finally, structural information about the polypeptides and their genes should provide some insight into the evolution of these systems of intercellular communication.

RPLC OF SMALL PROTEINS

Development of RPLC for Small Proteins

Whereas reverse phase peptide separation methods were initiated in the mid 1970s and already well-established techniques only 5 years later, the application of RPLC to small proteins developed at a much slower pace (Table 1). Rubinstein et al.[1] pioneered in 1977 protein isolation (in this case β-lipotropin, a 10,000-dalton polypeptide from rat anterior pituitaries) with RPLC by using a combination of classical gel filtration, ion-exchange HPLC, and final purification on RPLC employing a relatively steep gradient elution. Utilizing a similar procedure the same author[2] was able in 1979 to report the successful purification to homogeneity of human interferon (17,500 daltons) from leukocytes or fibroblasts. This polypeptide was subsequently partially sequenced.[3]

At the same time several studies were published which dealt with the isolation or separation of polypeptides such as 10kd-neurophysins[4] 8- and 14kd-enkephalin containing polypeptides[5] from adrenal medulla as well as the 14kd-proteins cytochrome *c*,[6-8] lysozyme,[7] and lactalbumin.[7]

Investigating the possible participation of a contaminant in the mitogenic activity of insulin (6000 daltons) in tissue culture Petrides and Böhlen[9] described a procedure which allows the separation of insulin from derivatives differing by one amino acid (such as B-31-monoarginine-insulin) or an amidogroup (such as A21-monodesamidoinsulin) by isocratic chromatography and from the larger precursor proinsulin (9000 daltons) by subsequent gradient elution. This method revealed the presence of at least ten compounds in porcine "single component" insulin, which at that time was considered to be a highly purified preparation (Figure 1). To avoid possible buffer salt interference in the biological assay used to test the activity of the purified insulin, the RPLC fractions had to be pumped onto a second ODS column (of 3-cm length), freed of salt — in this case triethylammonium phosphate (TEAP) — and subsequently eluted with *n*-propanol in 0.2 *M* acetic acid. The utilization of volatile solvent systems such as pyridine formate or trifluoroacetic acid (TFA) gives similar separation results, but does not require desalting prior to biological testing.

The evaluation of the purity of hormonal preparations by RPLC is now the method of choice used for proving the homogeneity of human insulin made by recombinant DNA technology and used for the treatment of diabetic patients.[10]

In 1979 and 1980 the potential of RPLC for the separation of proteins larger than 10,000 daltons was demonstrated: Congote et al.[11] reported the separation of various chains of hemoglobin (17,000 daltons) by using the combination of a TFA-gradient (from 0.3 to 0%) to separate alpha- and beta-chains and an organic modifier gradient (to separate alpha-, gammaG-, and gammaA- chains). Petrides et al.[12] systematically investigated the potential of RPLC for the separation of small proteins using mutants of globin chains (which differ by a single amino acid residue only) as model proteins and came to the following conclusions: usually separations of the various mutants can be achieved by isocratic chromatography in less than 1 hr (for a typical separation see Figure 2). The concentration of organic modifier strongly influences retention time and separation. Under isocratic conditions rapid elution of proteins is achieved with relatively high concentration of organic modifier and resolution may be unsatisfactory. By lowering the concentration of the organic modifier, retention time is prolonged and resolution is improved. If the concentration chosen is too low, then the proteins are strongly retarded and may not elute at all. Thus, proper adjustment of the propanol concentration is extremely critical for the resolution of closely related proteins because even slight alterations of the organic solvent can produce drastic changes in retention times and resolution. In order to compromise between the two extremes of a too rapid elution and of a strong retardation it is therefore preferable to use very shallow gradients for elution.

The recovery of the protein also depends upon the organic modifier concentration. With lower concentrations of organic solvents, which elute proteins relatively late and give optimal separation, proteins elute as broad peaks and recoveries are the lowest. This is another reason why shallow gradients are preferred over isocratic chromatography.

The observation that the concentration of organic modifier is critical can be utilized in

Table 1
EXAMPLES FOR SMALL PROTEINS SEPARATED ON OR PURIFIED BY RPLC

Protein	Mol wt (daltons)	Stationary phase (Col. size)	Mobile phase	Flowrate (mℓ/min)	Ref.
β-lipotropin	10,000	ODS (A)[a]	Pyridine acetate, *n*-propanol	0.83	1
Neurophysins	10,000	ODS (A)	0.01 *M* Acetate, pH 5.7, methanol	1.0	4
		ODS (SP)[b]	0.01 *M* Na phosphate, pH 6.0, methanol	2.5	31
Enkephalin-containing peptides	8,000 14,000	ODS, ODS/CN	Pyridine formate, pH 4.0, *n*-propanol	0.33	5
Insulin, Proinsulin	6,000 9,000	ODS (A)	0.25 *M* Triethylammonium, pH 3.0, acetonitrile	1.5	9
Multiplication stimulating activity (MSA)	7,500	ODS (A)	0.05% TFA, acetonitrile	1.0	13
Parathyroid hormone	9,500	ODS, Octyl (A),	0.155 *M* Sodium chloride, pH 2.1, acetonitrile	1.0	20
		ODS (A)	0.13 % TFA or HFBA, acetonitrile	1.5	21
Epidermal growth factor (EGF)	6,000	Octyl (A)	Pyridine formate, pH 3.0, *n*-propanol	0.22	15
		ODS (A)	0.05 *M* triethylamine acetate pH 5.6, acetonitrile	0.5	16
		ODS (A)	0.2% TFA or HFBA pH 2.3, acetonitrile	1.0	17
Transforming growth factors (TGFs)					
	10,000 –12,000	ODS (SP)	0.1% TFA, pH 2.0, acetronitrile	0.8	19
	6,000 –10,000	CN (A)	0.1% TFA pH 2.0, *n*-propanol	1.0	18
	7,400	ODS (A)	0.045% TFA, acetonitrile 0.035% TFA, *n*-propanol	1.0	14
	6,000	Octyl (A, SP, PC)	0.1% TFA, 0.05% HFBA acetonitrile; pyridine formate pH 3.0, h-propanol	0.5 (1.0; 2.5)	25

Table 1 (continued)
EXAMPLES FOR SMALL PROTEINS SEPARATED ON OR PURIFIED BY RPLC

Protein	Mol wt (daltons)	Stationary phase (Col. size)	Mobile phase	Flowrate (mℓ/min)	Ref.
Lysozyme, Lactalbumin	14,300 14,200	ODS (A)	0.1 *M* NaH_2PO_4/H_3PO_4 pH 2.1, acetonitrile	1.0	7
Cytochrome *c*	11,700	ODS (A)	0.05 *M* KH_2PO_4/H_3PO_4 pH 2.0, 2-methoxyethanol/isopropanol	2.0	6
		ODS (A)	0.1 *M* NaH_2PO_4/H_3PO_4 pH 2.1, acetonitrile	1.0	7
		Octyl (A)	Pyridine formate, pH 4.0,	0.5	8
Myoglobin	17,000	ODS (A)	0.1 *M* Na PO /H Po pH 2.1, acetonitrile	1.0	7
Hemoglobin	17,000	ODS (A)	0.5% TFA, acetonitrile	1.5	11
		ODS (A)	pyridine formate, pH 3.0, *n*-propanol	0.6	12
Interferon	17,500	Octyl (A)	1 *M* sodium acetate pH 7.5, pyridine formate pH 4.0, *n*-propanol	0.25	2

[a] Analytical columns with about 0.46-cm diameter.
[b] Semipreparative columns with diameters of about 0.7 to 1.0 cm.
[c] Preparative columns with diameters of 2.5 cm.

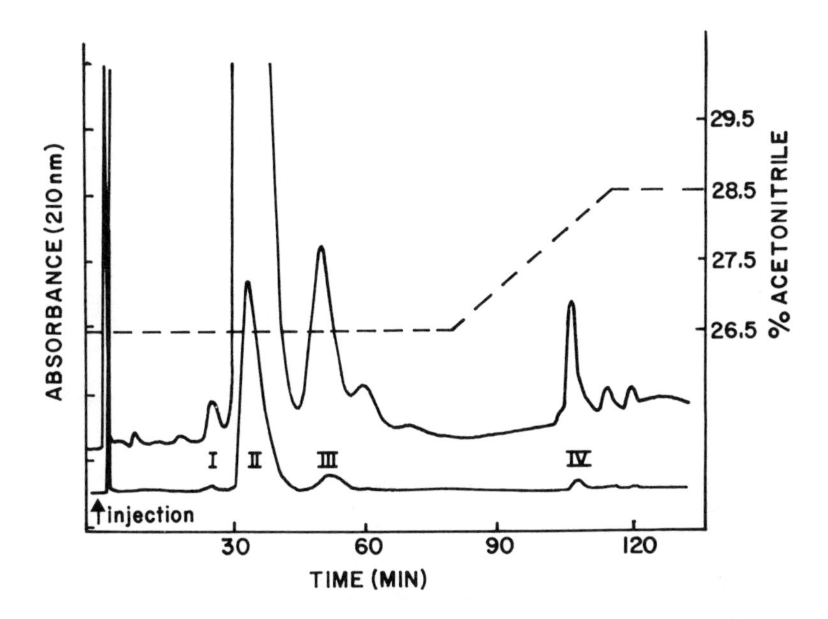

FIGURE 1. RPLC of 500 μg porcine single component insulin. Sensitivity: upper trace 0.02 AUFS, lower trace 0.2 AUFS. Conditions: Supelcosil® C_{18}-column (15 × 0.4 cm, 5-μm particle size), 0.25 *M* TEAP, pH 3.0, acetonitrile; Flow rate 1.5 mℓ/min. (I) B-31 Monoarginine-Insulin; (II) Insulin; (III) A-21-Monodesamido-Insulin; (IV) Proinsulin. (From Petrides, P. E. and Böhlen, P., *Biochem. Biophys. Res. Commun.*, 95, 1138, 1980. With permission.)

batch extractions on reverse phases: if a certain polypeptide elutes at a concentration of 40% organic modifier, for instance, a tissue homogenate supernatant can be loaded in the presence of 20% organic solvent (i.e., below the critical concentration) which will cause salts, amino acids, and hydrophilic polypeptides to pass through the system without interacting with the stationary phase. By increasing the concentration of the organic solvent above the critical modifier concentration for the protein of interest, i.e., 50%, the material can be eluted as an enriched fraction and all other polypeptides which elute at a higher organic solvent concentration are subsequently eluted and discarded (see below).

Results obtained with the separation of hemoglobin variants also further supported the general concept that hydrophobicity primarily dictates the retention behavior of polypeptides in RPLC. Generally, mutants in which hydrophilic amino acids have been substituted with more hydrophobic ones, elute later in solvent systems with acidic pH.

In the beginning of the 1980s Marquardt et al.[13,14] reported the purification of two growth factors (multiplication stimulating activity and transforming growth factor, 7500 daltons) from conditioned media of tumor-derived cells (liver and melanoma cells, respectively), by using a combination of conventional gel filtration and RPLC. At the same time Petrides et al.[15] applied RPLC to the final purification of murine epidermal growth factor (6000 daltons) obtained by conventional gel chromatography and described the presence of two EGF-related molecules which could not be resolved by conventional ion-exchange chromatography. Similar observations were made by Matrisian et al.[16] and Burgess et al.[17] Recently, RPLC was also utilized for the partial purification of other transforming growth factors with molecular weights between 6000 and 12,000 daltons (Roberts et al.[18,19]), identified in sarcoma virus transformed nonneoplastic cells and for the complete purification of transforming growth factors from physiological fluids.[25]

In 1981, Zanelli et al.[20] and Bennett et al.[21] reported the purification of bovine and human parathyroid hormone (9500 daltons) from parathyroid glands and blood plasma.

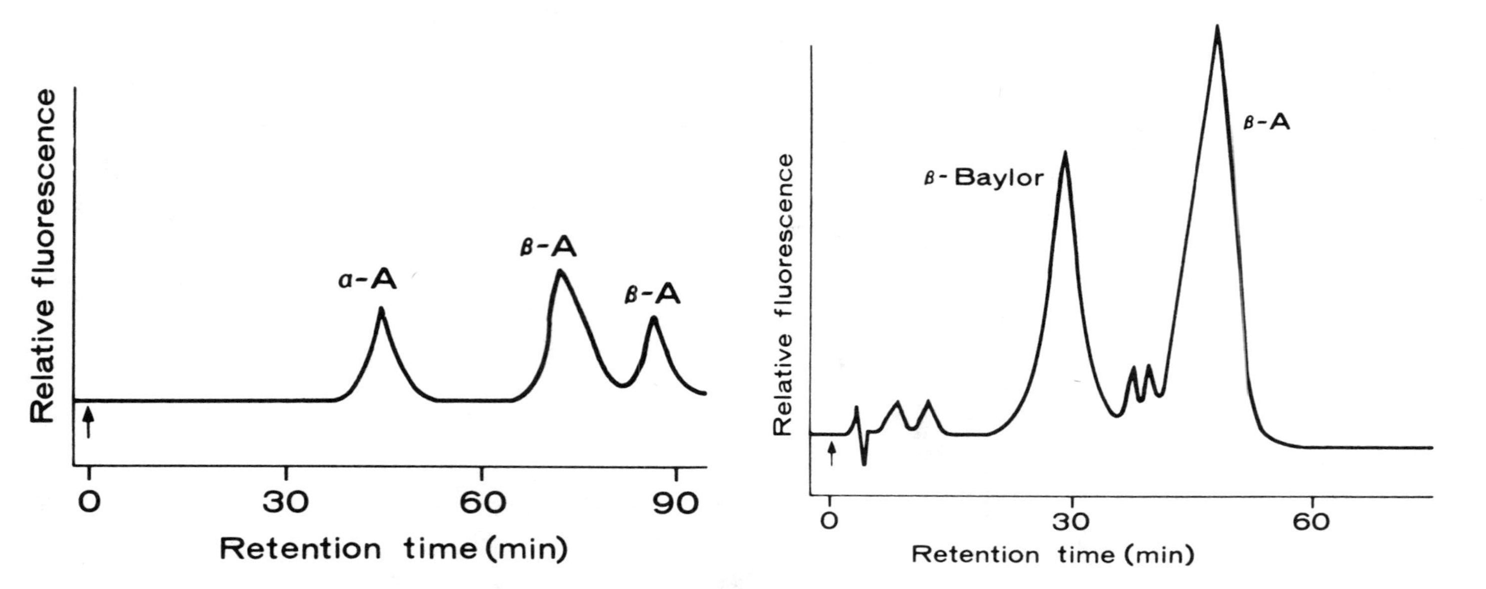

FIGURE 2. (A) RPLC of normal α-1 (20 μg) and β- (50 μg) human globin chains. Conditions: Isocratic chromatography at 26.6% *n*-propanol in pyridine formate pH 3.0 (column: Supelcosil® C_{18} 15 × 0.4 cm, 5-μm particle size). (B) Separation of normal β- (30 μg) and β- Baylor (40 μg) chains at 26.9% *n*-propanol in pyridine formate pH 3.0 (these two polypeptides differ by the substituion of a leucine with an arginine residue). Flow rate: 0.6 mℓ/min. (From Petrides, P. E., Jones, R. T., and Böhlen, P., *Anal. Biochem.*, 105, 383, 1980. With permission.)

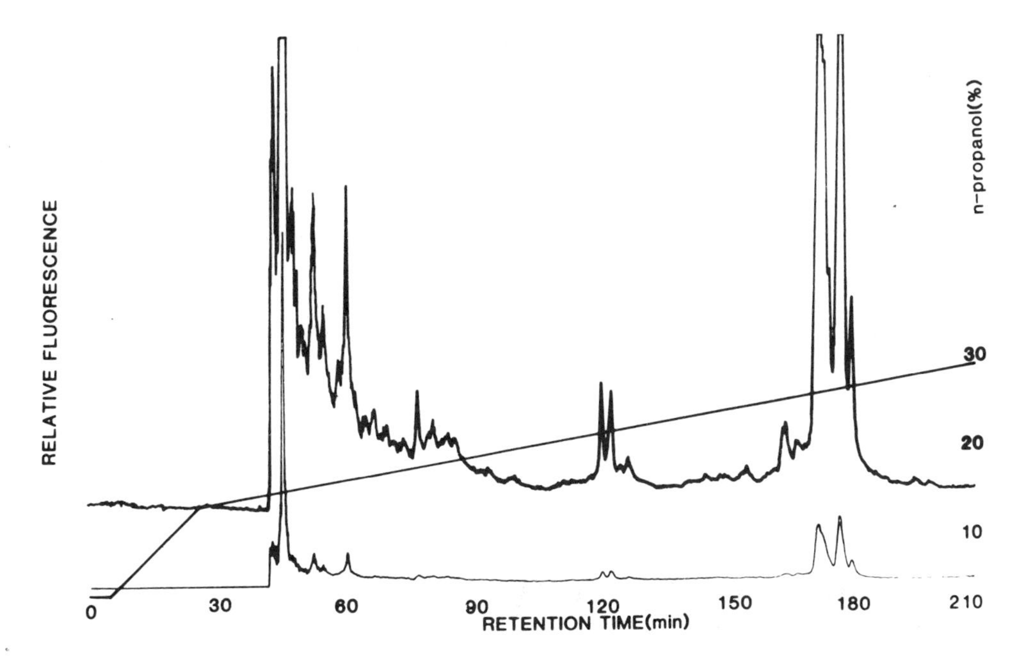

FIGURE 3. RPLC of 4.38 mg murine epidermal growth factor. Conditions: Altex RP8 column (25 × 0.46 cm, 5-μm particle size), pyridine formate, pH 3.0, *n*-propanol; Flow rate 0.22 mℓ/min. The major peaks are subsequently further purified in solvent systems with different selectivity. (From Petrides, P. E., Levine, A. E., and Shooter, E. M., *Peptides: Synthesis-Structure-Function,* Pierce Chemical, Rockford, Ill., 1981, 781. With permission.)

Aspects of Preparative Separations

Two promising approaches should be followed up in the future: (1) the further scaling-up of procedures established on the analytical and semipreparative level and (2) the further development of batch extractions for the rapid production of enriched fractions from tissue homogenate supernatants or physiological fluids.

Whereas numerous methods on the analytical level are now available for RPLC of small proteins, only a few studies have been published which deal with the exploration of the potential of reverse-phase (RP) HPLC for semipreparative and preparative applications (e.g., Petrides et al.,[15] Zanelli et al.,[20] Richter and Schwandt[31]). Up to 5 mg of prepurified epidermal growth factor can be applied to analytical columns for instance (Figure 3) and up to 100 mg can be loaded onto semipreparative columns. Activity containing fractions obtained by conventional gel chromatography can be applied onto semipreparative columns in amounts of 20 mg[19] or more. We have loaded up to 100 mg of prefractionated material onto semipreparative columns without any major loss in resolution.

Over the last years the extraction of small proteins from both tissue homogenates and plasma in a batch procedure — initially introduced for small peptides by Bennett and colleagues[22] and subsequently also applied to small proteins such as parathyroid hormone by the same author[21] — is gaining more popularity.

Although conventional chromatography on dextran or polyacrylamide gels is suitable for the gross fractionation of tissue homogenate supernatants or physiological fluids, its use is limited especially when large quantities of protein have been chromatographed. Physiological fluids have to be reduced in volume either by lyophilization (which is often accompanied by losses due to irreversible adsorption to glass surfaces, loss into the vacuum system, or possibly formation of aggregates) or ultrafiltration which yields viscous protein solutions. In addition, conventional chromatography is time consuming and can give rise to the production of transformation products. Because of the high resolving power of RPLC for closely related molecules which differ from each other, e.g., by the oxidation of methionine residues or deamidation of glutamine or asparagine moieties, it is very much desirable to prevent chemical alterations which may occur during lengthy isolation programs. If this cannot be achieved the investigator might encounter at the final stage of a purification — where separations are done with very shallow gradients for optimal resolution — the presence of several active molecular species which all have to be purified to homogeneity. Only later, when amino acid analysis data are obtained it may turn out that one component represents a transformation product of another. In this regard, Walsh and Niall[23] have shown that the use of C_{18} cartridges (such as Sep-PakC_{18}® from Water Associates) for batch extraction minimizes the proteolysis during the isolation of relaxins (6000 daltons). Because of the fact that adsorption phenomena are often involved in protein chromatography, low abundant peptides and proteins may be spread over a relatively large elution volume and hence may be difficult to detect and to recover.

We have therefore explored the potential use of RPLC for the gross fractionation of complex protein mixtures by combining batch extraction with subsequent elution with organic solvent. For this purpose a preparative glass column — slurry packed with RP 8 stationary phase — was loaded with milk which is known to contain epidermal growth factor activity. 30 mℓ of milk were acidified with acetic acid, mixed with an equal amount of Buffer A (pyridine formate, pH 3.0) pumped onto the column, freed of salt and other hydrophilic components by washing with several column volumes of pyridine formate, and subsequently eluted with a two-step gradient (Figure 4). Figure 5 and 6 demonstrate chromatography of the same material on a six-step gradient or elution with a linear gradient from 0 to 50% Buffer B over 180 minutes, respectively. Reproducibility is, as shown in Figure 6B, excellent.

Amounts of up to 200 mℓ of milk (containing 2 g of protein) can be applied to such a system and excellent fractionation can be achieved in 3 to 4 hr. This approach is particularly suitable when several different molecules have to be isolated from a complex mixture. If the material which passed through the column without interacting with the stationary phase is reloaded and subsequently eluted under identical conditions the chromatogram (Figure 7) reveals that the majority of the material passed through again, but in addition peptides and proteins of various hydrophobicities are retained and eluted with the following organic modifier gradient. This observation corroborates earlier findings that the loading capacity for larger molecular weight proteins is low which seems to be a function of the pore size of the stationary phase[8] and molecular nature of the polypeptide.[24] It is therefore necessary to test in each application whether the protein to be isolated by using such an approach is retained with high affinity or passes through without interacting with the stationary phase.

Whereas good separation is obtained with small- to medium-sized proteins, larger proteins are insufficiently resolved; this observation is consistent with what is seen on analytical and semipreparative systems.

This approach offers a combination of freeing the solution of small molecules such as amino acids, salts, hydrophilic carbohydrates, possibly small and hydrophilic peptides and fractionating peptides and proteins (and probably other biomolecules which interact with the stationary phase) and has provided the basis for the purification of several low abundant α-transforming growth factors from human milk.[25] It has to be emphasized that it is necessary to mix the protein-containing solution with an appropriate amount of RPLC buffer to ascertain solubility of the polypeptides to be chromatographed and optimize interaction with the stationary phase by providing a constant pH.

Bennett et al.[21] and Burgess et al.[16] used similar batch extractions for the isolation of high abundance human parathyroid hormone (PTH) and murine epidermal growth factor (EGF) from adenoma tissue or salivary glands, respectively; for PTH an extract from tumor tissue containing TFA (1%) was loaded onto Sep-Pak® cartridges (coupled in series), washed with 0.1% TFA to remove homogenization salts and residual protein, and subsequently eluted

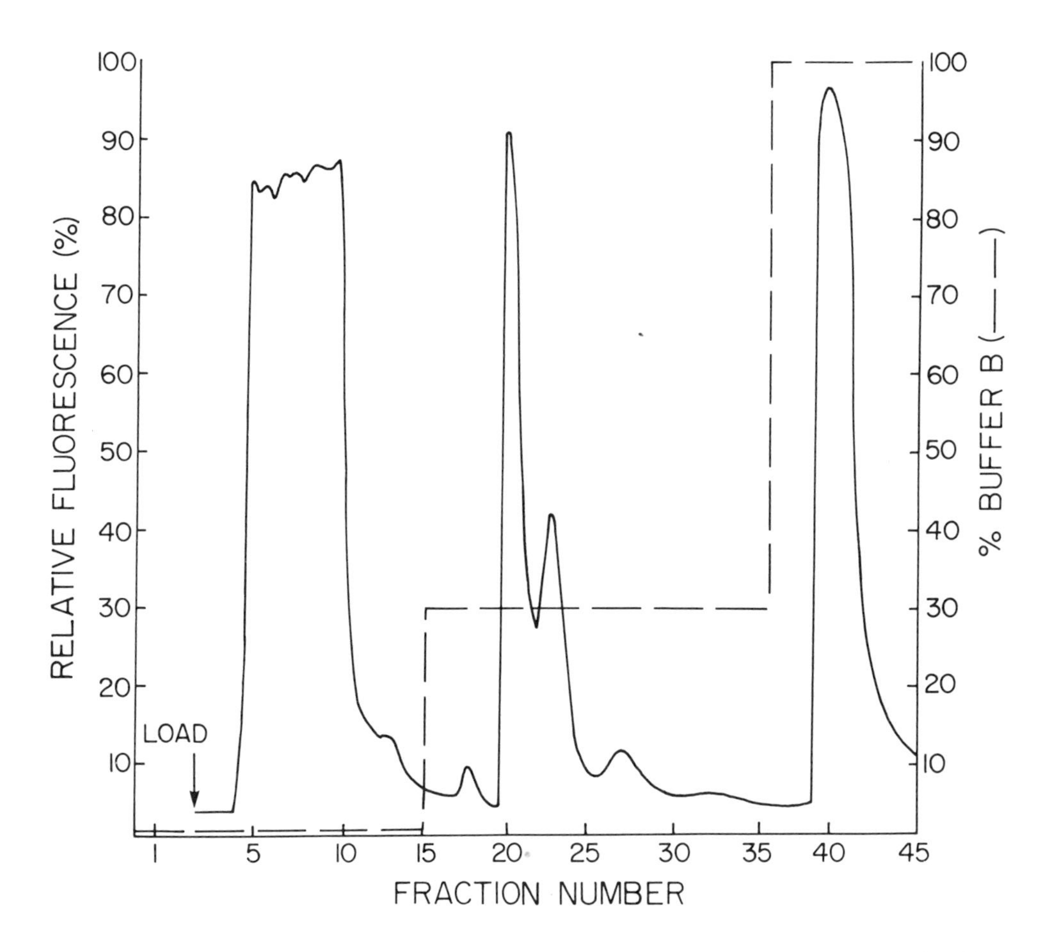

FIGURE 4. Batch extraction with subsequent step-wise elution of acidified human milk (30 mℓ). Conditions: Merck RP8 (7.5 × 2.5 cm, 25 to 40-μm particle size). Buffer A: pyridine formate pH 3.0; Buffer B: pyridine formate pH 3.0, 60% *n*-propanol. Flow rate 2.5 mℓ/min.

with 80% acetonitrile/0.1% TFA. For murine EGF a 1.0 × 9.0 cm column was used onto which the acidic supernatant was loaded in 20% acetonitrile, washed and an EGF-enriched fraction obtained by elution with 50% acetonitrile/0.1% TFA.

Aspects of Analytical Separations

To isolate, and purify a small regulatory protein present in low abundance to apparent homogeneity from an enriched fraction obtained in the manner described above or by more conventional means, is in many cases difficult even with the high resolution of liquid partition chromatography.

This is not only due to the fact that the molecule of interest is present in minute quantities, but also because tissue homogenate supernatants, serum, urine, or other physiological fluids are extremely complex polypeptide mixtures which contain a large number of low abundance molecules.

We use the following guidelines for our work on small proteins with molecular weights between 6000 and 10,000 daltons (insulin, proinsulin, insulin like growth factors, epidermal and transforming growth factors):

Application of sample — The fraction containing the activity of interest — whether it is obtained by batch extraction technique on RP8- or RP18-silica or by conventional chro-

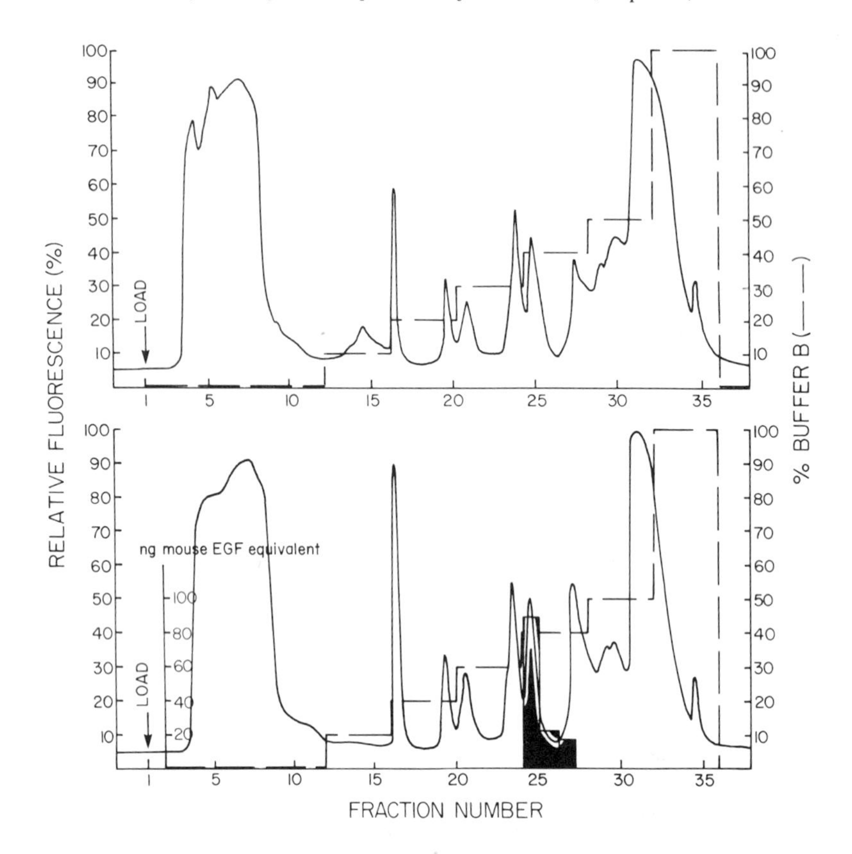

FIGURE 5. Batch extraction with subsequent step-wise elution (two consecutive experiments for comparison). Conditions, see Figure 4.

matography — is diluted with the starting buffer of the first semipreparative or analytical RPLC system in order to reduce the organic modifier concentration below the "critical" concentration (see above) or to adjust the pH to a value compatible with this RPLC system. This allows the direct application of the sample, without prior lyophilization, to the RPLC system.

Selection of solvent systems — Volatile buffer systems with different selectivity such as pyridine formate/ *n*-propanol[1] or TFA (0.1%) or HFBA (0.05% v/v)[21] are used in combination with acetonitrile or *n*-propanol. These solvents provide pH values of 3.0, 2.1, and 2.48, respectively. Chromatography with complex protein mixtures at pH 7 is not recommended because protein-protein interactions are likely to take place. Very shallow organic solvent gradients are employed to achieve optimal resolution (e.g., 1.5% *n*-propanol per hour). Rechromatography in solvent systems with different selectivity is essential to test homogeneity of a small protein sample; Figure 8 illustrates the separation of two small proteins under isocratic conditions; rechromatography (Figure 9) of the second peak in another solvent system reveals the presence of numerous polypeptides in this peak. It has to be emphasized that generally it is not possible to draw conclusions about the approximate molecular weight of a small protein from its retention behavior on RPLC. For instance, limited proteolysis of murine epidermal growth factor yields two fragments; the smaller ($EGF_{49\text{-}53}$) of the two elutes later than the larger ($EGF_{1\text{-}48}$) (Figure 10).

Use of detection system — For the detection of polypeptides chromatographed in the pyridine formate/*n*-propanol system post-column derivatization of an aliquot with subsequent fluorometric detection is used. This system offers the advantage of high specificity (for N-

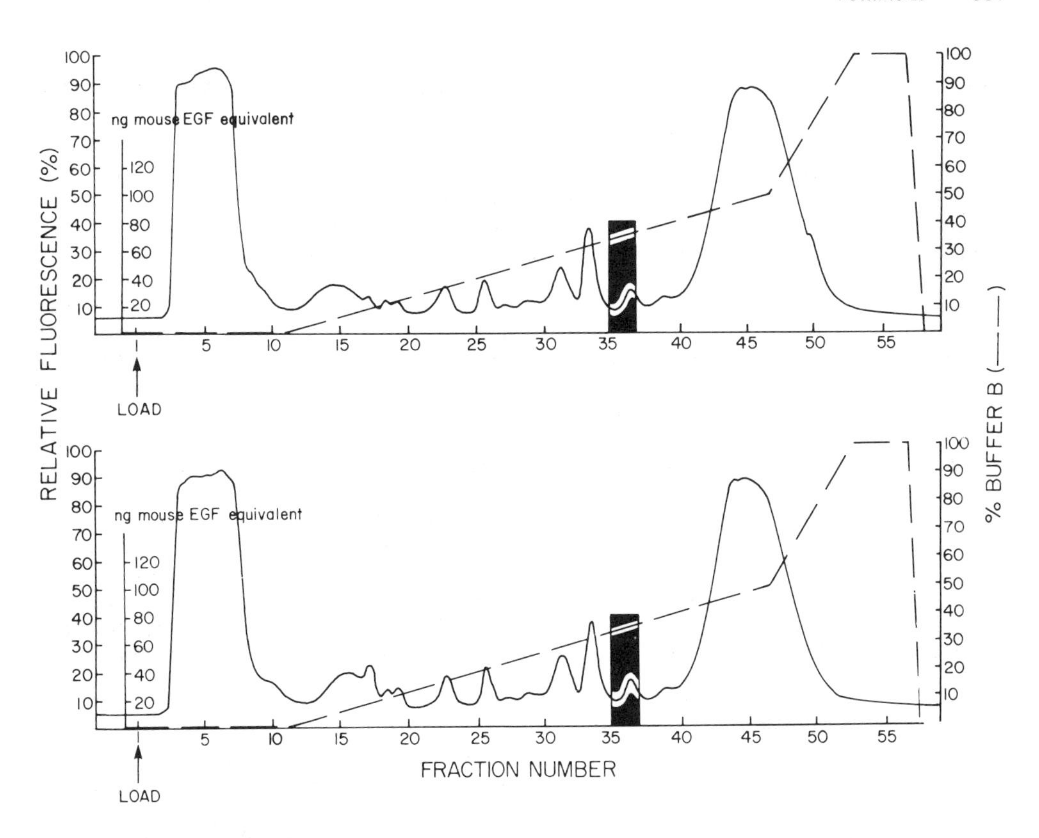

FIGURE 6. Batch extraction and subsequent gradient elution (two subsequent runs for comparison). Conditions, see Figure 4.

terminal residues of polypeptides), but is disadvantageous in situations where the N-terminus is blocked (e.g., by *N*-acetylation) or when the polypeptide starts with a prolyl residue. In addition, as any other postcolumn derivatization system it causes peak broadening. However, since only the portion of the effluent removed for detection is affected, separations are usually better than indicated by the chromatograms. When using perfluoroalkanoic acids at a concentration of 0.1% (v/v), the eluted proteins are first monitored by UV detection at 280 nm and subsequently by automatic post-column derivatization of an aliquot with the fluorometric system. This permits the simultaneous monitoring of the column effluent with two different systems and allows in some cases the detection of substances by UV absorption which are not detected by fluorometric derivatization (Figure 11) and vice versa.

Choice of flow rate — Separations are done with flow rates of 0.5 mℓ or less per minute on analytical or 1.0 mℓ/min on semipreparative columns (which is equivalent to a flowrate of 0.25 mℓ/min in an analytical system). Besides better resolution, lower flow rates offer other advantages such as less consumption of solvents and smaller elution volumes which is preferable for final sample processing (dilution for consecutive RP steps, amino acid analysis, and sequencing) and increases stability of polypeptides.

Choice of stationary phase — For small proteins of 6000 to 10,000 columns from various manufacturers with pore sizes of 100 and 300 Å have been used with similar results. In this regard recovery of the sample is the most important parameter. Recovery is a function of the amount of protein applied to the column, i.e., at early steps of isolations recoveries are usually between 90 and 100%, but may decline at later stages when the material is highly purified to 50%. Because recovery of polypeptides is also influenced by the compatibility of mobile and stationary phase (some columns do not tolerate high perfluoroalkanoic acid

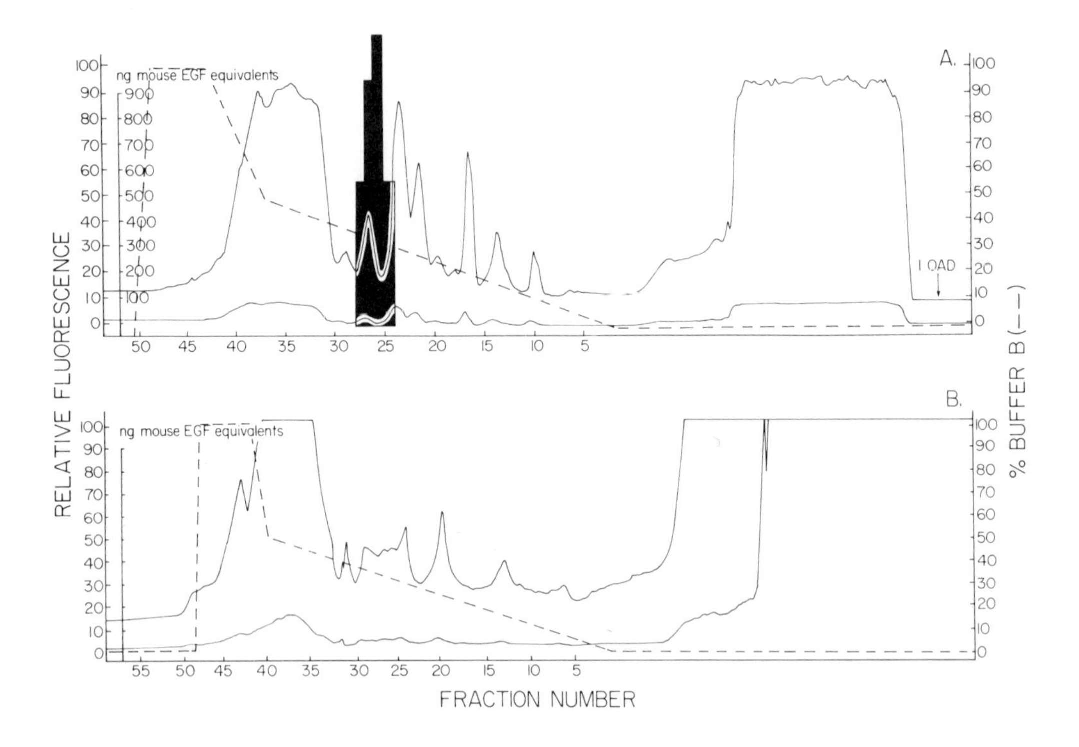

FIGURE 7. (A) Batch extraction with subsequent linear gradient elution. (B) Rechromatography of material which had not been adsorbed to the column during the first chromatography ("flow through peak"). Note that no activity of the polypeptide of interest is observed in the second run indicating that 100% of the activity has been adsorbed and eluted in the initial (conditions see Figures 4 and 5).

concentrations), this parameter has to be checked in pilot experiments when recovery becomes critical. Decreasing recovery over the time of use of a particular column might also indicate column deterioration.

CONCLUSIONS

In general, small protein RPLC does not differ substantially in its basic requirements for optimal separation from the conditions required for small peptides. The major difference seen is that these larger molecules elute in some cases — especially under very shallow gradient conditions or isocratic chromatography — as broad peaks[3,12] which may be due to a slow mass transfer of the proteins between the stationary and the mobile phase. In this regard, acetonitrile seems to give elution of small proteins as sharper peaks than does *n*-propanol. The latter, however, offers the advantage that lower concentrations are required for elution which might prevent certain proteins from becoming irreversibly denatured.[26]

Elution as broad peaks presents two disadvantages; first, chances are increased that another molecule with similar hydrophobicity is coeluting with the protein of interest and second, the molecule is eluted in a relatively larger volume which is less desirable for subsequent analyses such as micro-amino acid analysis or microsequencing. This makes rechromatography of the sample in another solvent with different selectivity even more important, a necessity which might be difficult to follow when only minute quantities of material are available.

It has to be emphasized again that hydrophobicity and not molecular size determines the behavior of polypeptides on reverse phases. We have observed the presence of molecules

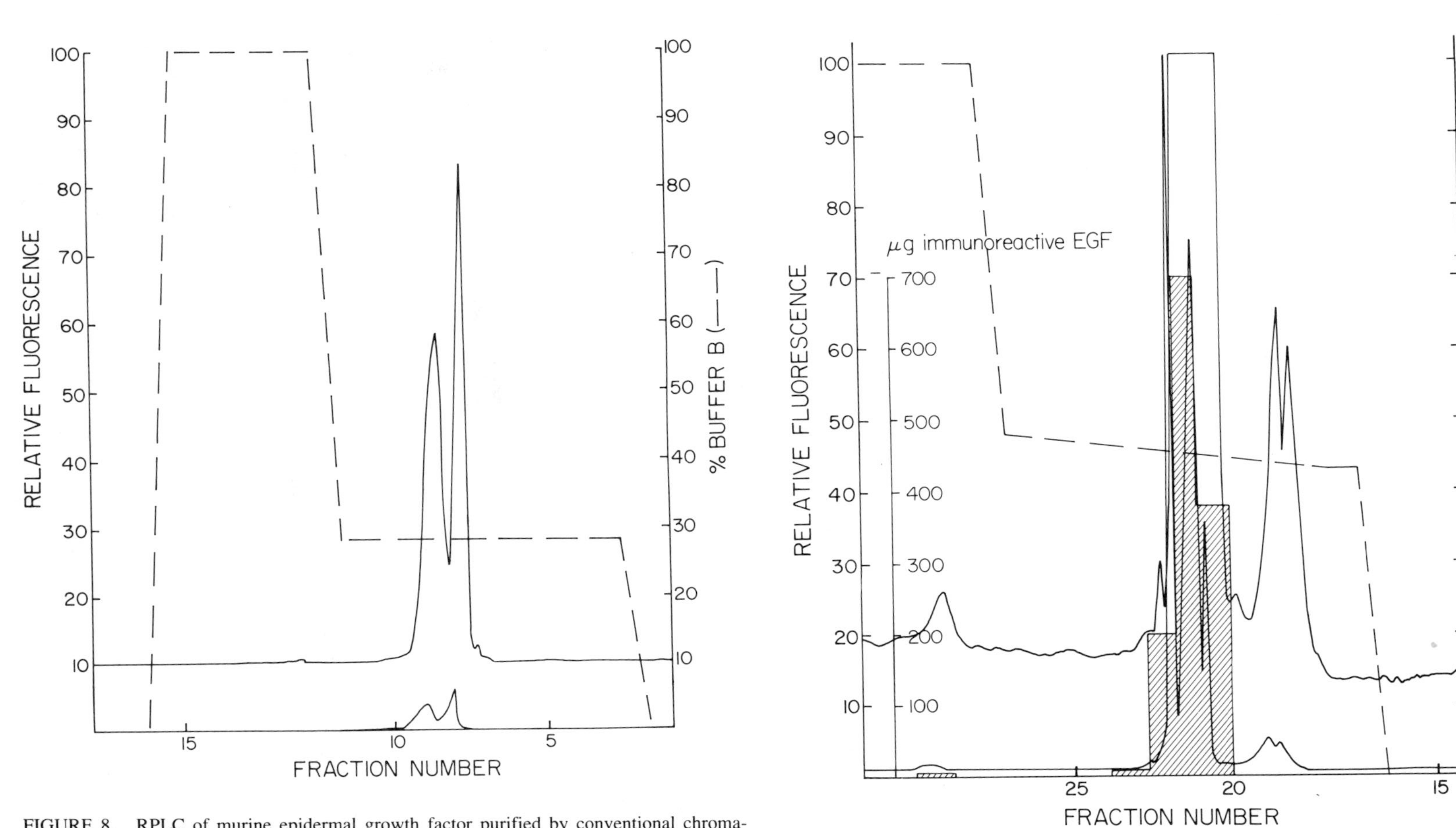

FIGURE 8. RPLC of murine epidermal growth factor purified by conventional chromatography on Bio-Gel® P 10. Conditions: Altex RP8 column (25 × 0.46 cm). Buffer A: pyridine formate pH 3.0; Buffer B: pyridine formate pH 3.0, 60% *n*-propanol. Flow rate: 0.5 mℓ/minute. Both peaks show immunoreactivity.

FIGURE 9. RPLC of the second peak of Figure 8. Conditions: Altex RP 8 column (25 × 0.46 cm). Buffer A: 0.1% TFA. Buffer B: 0.1% TFA, 80% acetonitrile. Flow rate: 0.5 mℓ/minute. Note that the first and second peak do not possess immunoreactivity.

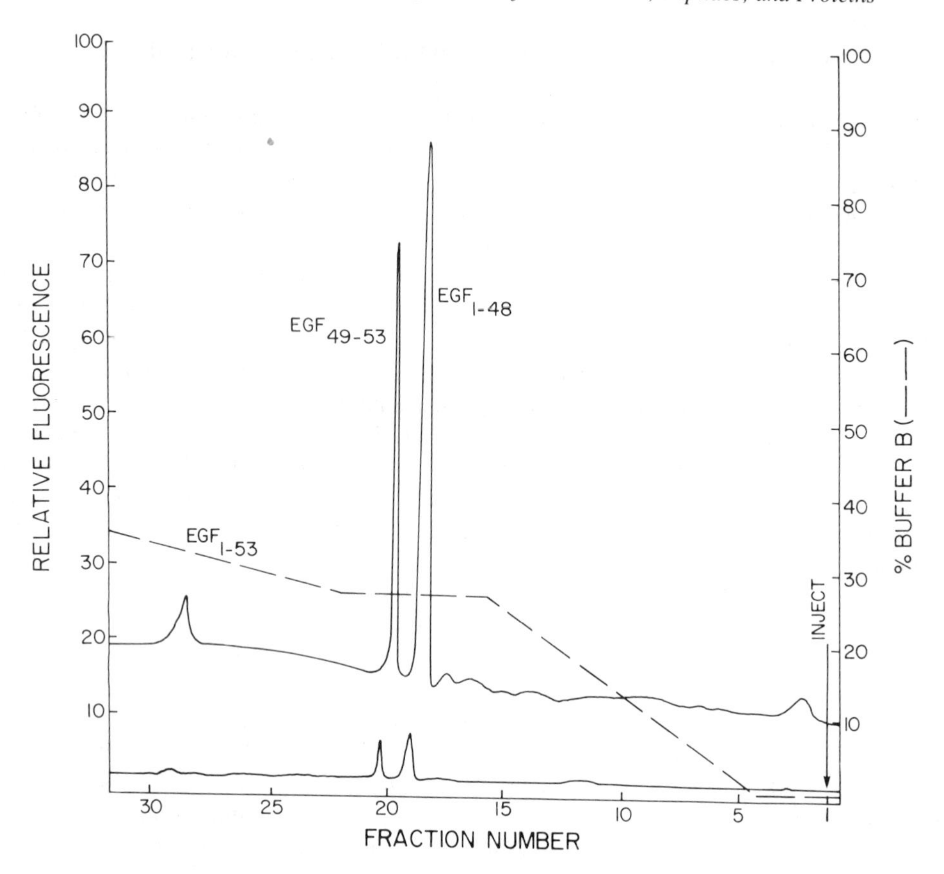

FIGURE 10. RPLC of $EGF_{1\text{-}48}$ and $EGF_{49\text{-}53}$ produced by limited proteolysis of murine EGF with trypsin. Conditions: Altex RP8 column (25 × 0.46 cm). Buffer A: pyridine formate pH 3.0; Buffer B; pyridine formate pH 3.0, 60% *n*-propanol. Flow rate: 0.5 mℓ/minute.

considerably larger (45,000 daltons) than epidermal growth factor (6000 daltons) with similar elution behavior. On the other hand, very hydrophobic small peptides such as the cyanogen bromide fragments of bacteriorhodopsin seem to chromatograph less well than small proteins.[27]

For small protein chromatography two solvent systems have gained wide acceptance: (1) the pyridine formate/acetate/*n*-propanol system introduced by Rubinstein et al.[1] and (2) the use of perfluoroalkanoic acid such as TFA or HFBA as counter-ions[21] in combination with acetonitrile or propane. Flow rates lower than 0.5 mℓ/min (in analytical systems) have been reported to give superior results by several authors;[2,5,8,15] other studies show, however, that satisfactory results can also be obtained with higher flow rates.[13,14,21]

Hence, as with many other parameters in liquid partition chromatography optimal chromatography conditions might differ from protein to protein, which is probably also true for the answer to the question whether octadecasilyl- or octylsilyl- stationary phases are better suited for small protein chromatography.

The fact that even larger proteins than interferon, namely proopiocortin with a molecular weight of 28,000 daltons[28] and T-cell derived suppressor factor with a molecular weight of 24,000 daltons[29] have been purified with RPLC indicates that the potential of this methodology has not been fully explored yet. The further development of stationary phases (RP-3 and 4 columns) and the combination of RPLC with high-performance ion-exchange chromatography[1,30] will in the future simplify the purification of regulatory proteins and spark important advances in our understanding of the the role of these molecules in cell-cell communication.

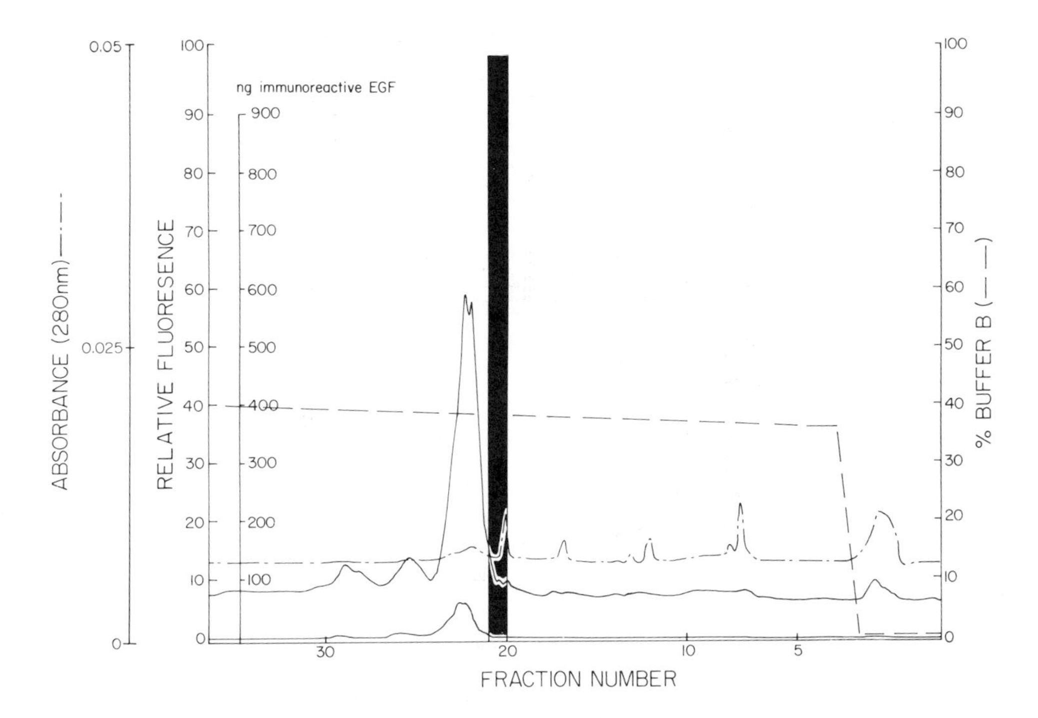

FIGURE 11. RPLC of EGF containing partially purified tissue extract. Conditions: Brownlee RP 300 column (25 × 0.46 cm, 10-μm particle size). Buffer A: 0.1% TFA; Buffer B: 0.1% TFA, 80% acetonitrile. Flow rate: 0.5 mℓ/min. Detection through UV absorption at 280 nm and subsequent automatic post-column derivatization of aliquots of the eluant in regular intervals.

REFERENCES

1. **Rubinstein, M., Stein, S., Gerber, L. D., and Udenfriend, S.,** Isolation and characterization of the opioid peptides from rat pituitary: β-lipotropin, *Proc. Natl. Acad. Sci. U.S.A.*, 74, 3052, 1977.
2. **Rubinstein, M., Rubinstein, S., Familetti, P. C., Miller, R. S., Waldman, A. A., and Pestka, S.,** Human leukocyte interferon: production, purification to homogeneity, and initial characterization, *Proc. Natl. Acad. Sci. U.S.A.*, 76, 640,
3. **Stein, S., Kenny, C., Friesen, H. J., Shively, J., Del Valle, U., and Pestka, S.,** NH_2-terminal amino acid sequence of human fibroblast interferon, *Proc. Natl. Acad. Sci. U.S.A.*, 77, 5716, 1980.
4. **Glasel, J. A.,** Separation of neurophyseal proteins by reversed-phase high pressure liquid chromatography, *J. Chromatogr.*, 145, 469, 1978.
5. **Lewis, R. V., Stern, A. S., Kimura, S., Stein, S., and Udenfriend, S.,** Enkephalin biosynthetic pathway: proteins of 8000 and 14,000 daltons in bovine adrenal medulla, *Proc. Natl. Acad. Sci. U.S.A.*, 77, 5018, 1980.
6. **Mönch, W. and Dehnen, W.,** High performance liquid chromatography of polypeptides and proteins on a reversed phase support, *J. Chromatogr.*, 147, 415, 1978.
7. **O'Hare, M. J. and Nice, E.,** Hydrophobic high performance liquid chromatography of hormonal polypeptides and proteins on alkylsilane-bonded silica, *J. Chromatogr.*, 171, 209, 1979.
8. **Lewis, R. V., Fallon, A., Stein, S., Gibson, K. D., and Udenfriend, S.,** Supports for reverse phase high performance liquid chromatography of large proteins, *Anal. Biochem.*, 104, 153, 1980.
9. **Petrides, P. E. and Böhlen, P.,** The mitogenic activity of insulin: an intrinsic activity of the molecule, *Biochem. Biophys. Res. Commun.*, 95, 1138, 1980.
10. **Johnson, I. S.,** Human insulin from recombinant DNA-technology, *Science*, 219, 632, 1983.

11. **Congote, L. F., Bennett, H. P. J., and Solomon, S.,** Rapid separation of the alpha, beta, G-gamma and A-gamma human globin chains by reversed-phase high pressure liquid chromatography, *Biochem. Biophys. Res. Commun.*, 89, 851, 1979.
12. **Petrides, P. E., Jones, R. T., and Böhlen, P.,** Reverse-phase high performance liquid chromatography of proteins: the separation of hemoglobin chain variants, *Anal. Biochem.*, 105, 383, 1980.
13. **Marquardt, H., Todaro, G. J., Henderson, L. E., and Orozlan, S.,** Purification and primary structure of a polypeptide with multiplication stimulating activity from rat liver cell cultures. Homology with human insulin like growth factor. II, *J. Biol. Chem.*, 256, 6859, 1981.
14. **Marquardt, H. and Todaro, G. J.,** Human transforming growth factor. Production by a melanoma cell line, purification and initial characterization, *J. Biol. Chem.*, 257, 5220, 1982.
15. **Petrides, P. E., Levine, A. E., and Shooter, E. M.,** Preparative reverse phase HPLC: an efficient procedure for the rapid purification of large amounts of biologically active proteins, in *Peptides: Synthesis-Structure-Function*, Rich, D. H., Gross, E., Eds., Pierce Chemical, Rockford, Il., 1981, 781.
16. **Matrisian, L. M., Larsen, B. R., Finch, J. S., and Magun, B. E.,** Further purification of epidermal growth factor by high performance liquid chromatography, *Anal. Biochem.*, 125, 339, 1982.
17. **Burgess, A. W., Knesel, J., Sparrow, L. G., Nicola, N. A., and Nice, E. C.,** Two forms of murine epidermal growth factor: rapid separation by using reverse phase HPLC, *Proc. Natl. Acad. Sci. U.S.A.*, 79, 5753, 1982.
18. **Roberts, A., Anzano, M. A., Lamb, L. C., Smith, J. M., and Sporn, M. B.,** New class of transforming growth factors potentiated by epidermal growth factor: isolation from non-neoplastic tissues, *Proc. Natl. Acad. Sci. U.S.A.*, 78, 5339, 1981.
19. **Roberts, A. B., Anzano, M. A., Lamb, L. C., Smith, J. M., Frolick, C. A., Marquardt, H., Todaro, G. J., and Sporn, M., B.,** Isolation from murine sarcoma cells of novel transforming growth factors potentiated by EGF, *Nature (London)*, 295, 417, 1982.
20. **Zanelli, J. M., O'Hare, M. J., Nice, E. C., and Corran, P. H.,** Purification and assay of bovine parathyroid hormone by reversed-phase high-performance liquid chromatography, *J. Chromatogr.*, 223, 59, 1981.
21. **Bennett, H. P. J., Solomon, S., and Goltzman, D.,** Isolation and analysis of human parathyrin in parathyroid tissue and plasma, *Biochem. J.*, 197, 391, 1981.
22. **Bennett, H. P. J., Hudson, A. M., McMartin, C., and Purdon, G. E.,** Use of octadecasilylsilica for the extraction and purification of peptides in biological samples: application to the identification of circulating metabolites of corticotropin (1-24) tetracosapeptide and somatostatin in vivo, *Biochem. J.*, 168, 9, 1977.
23. **Walsh, J. R. and Niall, H. D.,** Use of an octadecasilica purification method minimizes proteolysis during isolation of porcine and rat relaxins, *Endocrinology*, 107, 1258, 1980.
24. **Böhlen, P., Castillo, F., Ling, N., and Guillemin, R.,** Purification of peptides: an efficient procedure for the separation of peptides from amino acids and salt, *Int. J. Peptide Protein Res.*, 16, 306, 1980.
25. **Petrides, P. E., Hosang, M., Shooter, E. M., and Böhlen, P.,** Transforming growth factors in human milk: isolation and partial biological and chemical characterized, submitted, 1983.
26. **Rubinstein, M.,** Preparative high performance liquid partition chromatography of proteins, *Anal. Biochem.*, 98, 1, 1979.
27. **Gerber, G. E., Anderegg, R. J., Herlihy, W. C., Gray, C. P., Biermann, K., and Khorana, H. G.,** Partial primary structure of bacteriorhodopsin: sequencing methods for membrane proteins, *Proc. Natl. Acad. Sci. U.S.A.*, 76, 227, 1979.
28. **Kimura, S., Lewis, R. V., Gerber, L. D., Brink, L., Rubinstein, M., Stein, S., and Udenfriend, S.,** Purification to homogeneity of camel pituitary pro-opiocortin, the common precursor of opioid peptides and corticotropin, *Proc. Natl. Acad. Sci. U.S.A.*, 76, 1756, 1979.
29. **Krupen, K., Araneo, B. A., Brink, L., Kapp, J. A., Stein, S., Wieder, K. J., and Webb, D. R.,** Purification and characterization of a monoclonal T-cell suppressor factor specific for poly (LGlu60LAla30LTyr10), *Proc. Natl. Acad. Sci. U.S.A.*, 79, 1254, 1982.
30. **Green, M. and Brackmann, K. H.,** The application of high performance liquid chromatography for the resolution of proteins encoded by the human adenovirus type 2 cell transformation region, *Anal. Biochem.*, 124, 209, 1982.
31. **Richter, W. and Schwandt, P.,** Preparation of porcine neurophysin proteins by high performance liquid chromatography, *J. Neurochem.*, 36, 1279, 1981.

SEPARATION OF MULTIPLE PROTEIN FORMS

Hitoshi Aoshima

In their natural state, proteins can exist in multiple forms, differing from one another in their primary structure or in their conformation. These various protein forms may be classified as shown in Table 1.

Proteins have changed their primary structure gradually while maintaining the same function during evolution. Even a single organ sometimes produces several isoenzymes, which are different proteins that catalyze the same reaction. Proteins can change their form during the biosynthetic process. After synthesis, some proteins are cleaved by proteases or converted to conjugated proteins, such as glycoproteins or lipoproteins, and thus assume their functions. Enzymes are sometimes modified chemically in the laboratory to study the mechanism of their catalysis. Such modifications usually produce several enzyme species, the separation of which becomes essential to study the effect of a specific modification on enzyme activity. Some amino acids in proteins may be modified in vivo by reactive chemicals.

These separate protein forms usually differ in some amino acids in their primary structure, which may cause a difference in molecular weight, size, and/or charge. Such proteins have classically been separated and measured by electrophoresis or chromatography on columns packed with compressible ion-exchange resins such as DEAE Sephadex® or CM cellulose. Electrophoresis suffers primarily from being, at best, semiquantitative and normal chromatographic separations are time consuming. Several examples of separation of these proteins by HPLC are tabulated in Table 2.[1-13] HPLC has proved superior not only in speed and convenience but also in resolution, in comparison with classical methods,[1] even though HPLC is limited to smaller sample sizes.[14,15] One can reduce this limitation to some degree by using a large-scale column for preparative separations. The larger column needs a larger quantity of packing material and thus results in a corresponding increase in cost.[14] It is wise to use HPLC for the separaton of a protein already partially purified, or for the purification of a valuable protein which is present in trace amounts, such as a receptor protein or interferon.[10] If the sample protein has already been separated by one of the classical modes of chromatography, then one can choose a comparable column for HPLC. Usually, commercially available ion-exchange or reversed-phase chromatographic columns can be used for the separations,[15] though the latter ones are not applicable to proteins that are irreversibly denatured by organic solvents. In HPLC, a gradient system of an organic mobile phase often becomes necessary for rapid and accurate separations. For special cases, bioaffinity supports for HPLC can be prepared.[5] The eluted proteins can be detected easily by the absorbance measurement at 280 nm or by fluorescence techniques. A post-column reaction system can be made, or enzymic activity can be employed.[2,3] HPLC separations are completed in times ranging from a few minutes to several hours. As equilibration of the protein molecule between mobile and stationary phases needs adequate time depending on the molecular weight, too high a flow rate sacrifices resolution, especially in gel permeation chromatography.[14] Examples of ion-exchange and reversed-phase chromatography are shown in Figures 1 and 2, respectively.

Proteins change their conformation in response to their environment. Heat, pH change, urea, guanidine hydrochloride, or a surfactant such as sodium dodecyl sulfate (SDS) cause a reversible or irreversible denaturation and change the secondary and tertiary structure of the protein. Some proteins have quaternary structures, with association and dissociation of subunits, which also depend on their concentration or the mobile-phase conditions. Specific ligand binding can induce a change in protein conformation, such as binding of substrate to an enzyme, agonist to receptor, or calcium to calmodulin.

Table 1
A CLASSIFICATION OF MULTIPLE PROTEIN FORMS

Primary Structural Variation in Proteins

Isoenzymes
Proteins from different origins but with similar function
Proteins and their precursors
Native and chemically modified proteins

Conformational Variation in Proteins

Native and denatured proteins
Associated or dissociated subunits of protein (differences in the quaternary structure)
Conformational differences in proteins induced by ligand binding

Up to now HPLC has seen limited use in the separation of these multiple forms of proteins that differ only in their conformation. However, native and irreversibly denatured proteins can be separated easily by HPLC using absorption, affinity or gel permeation chromatography. When proteins change their conformation reversibly and rapidly, it is impossible to separate the different species clearly by HPLC. The proteins will be eluted in an average position as a broad peak dependent on the proportion and the eluting position of each species. The equilibrium position for association and dissociation of protein subunits has already been studied by the classical mode of gel permeation chromatography.[16] HPLC with a gel permeation column will enable such measurements to be made accurately and rapidly on small samples. It is to be hoped that further improvements in HPLC sensitivity and resolution will bring about the separations of slightly different forms of proteins, such as intact and degraded receptors, in the future.

Table 2
SEPARATION OF PROTEINS WITH A SLIGHTLY DIFFERENT PRIMARY STRUCTURE BY HPLC

Compound	Column	Mobile Phase	Detection	Ref.
Lactate dehydrogenase				
Five isoenzymes	Anion-exchange (DEAE-Glycophase G)	20 m*M* *Tris*-HCl, pH 7.8 150 m*M* NaCl gradient	Post-column reactor (absorbance or fluorescence)	1—4
Pig heart, rabbit muscle	AMP-silica	0.1 *M* sodium phosphate, pH 7.5 NADH gradient	LDH activity after fractionation	5
Hemoglobin variants				
hemoglobin A_2, S, A_1, F	Anion-exchange (DEAE-Glycophase G)	0.0125 *M* *Tris*-HCl, pH 8 0.15 *M* NaCl gradient	Absorbance at 410 nm	1
hemoglobin A_2, A_0	Anion-exchange (Synchropak AX 300)	0.02 *M* *Tris*-acetic acid, pH 8 0.1 *M* sodium acetate gradient	Absorbance at 410 nm	6
8 β-chain variants	Reversed-phase (Supercosil® C18)	Pyridine formate, pH 3 *n*-propanol gradient	Fluorescence	7
Creatine phosphatase (3 isoenzymes)	Anion-exchange (DEAE-Glycophase G)	50 m*M* *Tris*-HC1, 1 m*M* mercaptoethanol pH 7.5, 0.05—0.3 *M* NaCl gradient	Post-column reactor (absorbance)	1, 4
Alkaline phosphatase (intestine, liver)	Anion-exchange (DEAE-Glycophase G)	0.15 *M* NaCl gradient 1 *M* urea	Post-column reactor (absorbance)	4
Hexokinase (liver, testis)	Anion-exchange (DEAE-Glycophase G)	0.4 *M* NaCl gradient	Post-column reactor (fluorescence, activity)	4
Serum albumin (bovine, human)	Affinity (anti-HSA-silica)	0.1 *M* sodium phosphate, pH 7.5 0.2 *M* glycine-HCl, pH 2.2	Absorbance at 280 nm	5

Table 2 (continued)
SEPARATION OF PROTEINS WITH A SLIGHTLY DIFFERENT PRIMARY STRUCTURE BY HPLC

Compound	Column	Mobile Phase	Detection	Ref.
Arylsulfatase (A, B)	Anion-exchange (DEAE-Glycophase G)	0.5 *M* NaCl gradient	4-methylumbelliferone fluorescence	8
Soybean lipoxygenase-1 (5 components)	Cation-exchange (TSK Gel LS-212)	0.1 *M* sodium phosphate, pH 6.8	Abosrbance at 215 nm	9
Leukocyte interferon (3 components)	Reversed-phase (Lichrosorb® RP-8)	1 *M* sodium acetate, pH 7.5 40% *n*-propanol gradient	Fluorescence fluoroscamine interferon activity, Absorbance at 417 nm	10
Hepatic microsomal cytochrome p-450 (12 components)	Anion-exchange (Anpac)	20 m*M* *Tris*-acetae, 0.2% Emulgen 20% glycerol, pH 7.2 0.8 *M* sodium acetate gradient		11
Cytochrome c (bovine, chicken, dog, horse, rabbit, tuna)	Reversed-phase (Nucleosil® C18,7 C8, 5 CN)	5 or 10 m*M* phosphate, pH 2 or 3 0.1 *M* sodium phosphate, acetonitrile	Absorbance at 210, 220, or 400 nm	12
Trypsin, trypsinogen α-chymotrypsin, α-chymotrypsinogen	Reversed-phase (Nucleosil® CN)	Na ethanesulfonate Na pentanesulfonate Octanesulfonate	Absorbance	13

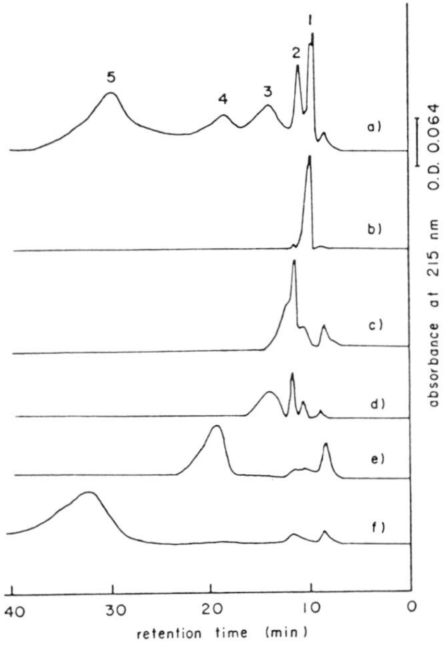

FIGURE 1. High-performance liquid chromatograms of lipoxygenase-1 and its components on cation-exchange resin, TSK Gel LS-212 column eluted with 0.1 *M* sodium phosphate buffer at pH 6.8. One hundred microliters of lipoxygenase-1 solution (1.2 mg/mℓ) purified on DEAE Sephadex® A-50 column was injected for chromatography of (a). One milliliter of each component (peaks 1—5) was injected and analyzed (b—f). Peaks were designated as shown in (a). The effluent was monitored at 215 nm. Flow rate: 1.03 mℓ/min (10 to 20 kg/cm^2); (a) sample lipoxygenase-1, (b) peak 1, (c) peak 2, (d) peak 3, (e) peak 4, (f) peak 5. (From Aoshima, H., *Anal. Biochem.*, 9, 371, 1979. With permission.)

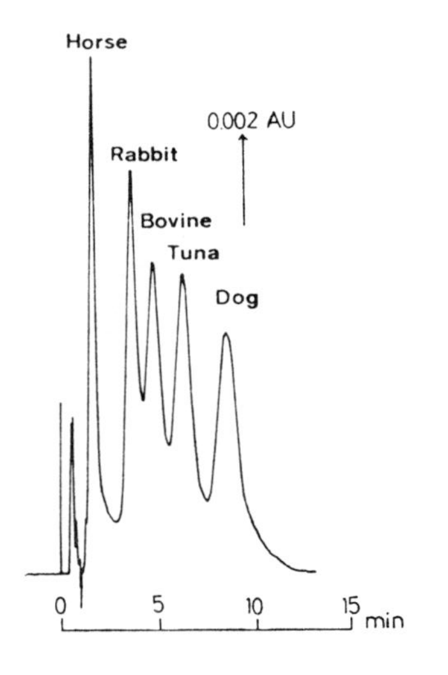

FIGURE 2. Separation of cytochromes *c* on C_8 column. Conditions: mobile phase, mixture of 72.5% of 0.005 *M* phosphate buffer (pH 3.0) containing 0.1 *M* sulphate, and 27.5% of acetonitrile; flow rate, 2.0 mℓ/min; detection, absorption at 220 nm; sample size, total 11 μg (2:2:2:2:3). (From Terabe, S., Nishi, H., and Ando, T., *J. Chromatogr.*, 212, 295, 1981. With permission.)

REFERENCES

1. **Chang, S. H., Gooding, K. M., and Regnier, F. E.,** High-performance liquid chromatography of proteins, *J. Chromatogr.*, 125, 103, 1976.
2. **Fulton, J. A., Schlabach, T. D., Kerl, J. E., and Toren, E. C.,** Dual-detector-post-column reactor system for the detection of isoenzymes separated by high-performance liquid chromatography. II. Evaluation and application to lactate dehydrogenase isoenzymes, *J. Chromatogr.*, 175, 283, 1979.
3. **Schroeder, R. R., Kudirka, P. J., and Toren, E. C., Jr.,** Enzyme-selective detector systems for high-pressure liquid chromatography, *J. Chromatogr.*, 134, 83, 1977.
4. **Schlabach, T. D. and Regnier, F. E.,** Techniques for detecting enzymes in high-performance liquid chromatography, *J. Chromatogr.*, 158, 349, 1978.
5. **Ohlson, S., Hansson, L., Larsson, P. O. and Mosbach, K.,** High performance liquid affinity chromatography (HPLAC) and its application to the separation of enzymes and antigens, *FEBS Lett.*, 93, 5, 1978.
6. **Gooding, K. M., Lu, K. C., and Regnier, F. E.,** High-performance liquid chromatography of hemoglobins. I. Determination of hemoglobin A_2, *J. Chromatogr.*, 164, 506, 1979.
7. **Petrides, P. E., Jones, R. T., and Böhlen, P.,** Reverse-phase high-performance liquid chromatography of proteins: the separation of hemoglobin chain variants, *Anal. Biochem.*, 105, 383, 1980.
8. **Bostick, W. D., Dinsmore, S. R., Mrochek, J. E., and Waalkes, T. P.,** Separation and analysis of arylsulfatase isoenzymes in body fluids of man, *Clin. Chem.*, 24, 1305, 1978.
9. **Aoshima, H.,** High performance liquid chromatography studies on protein: multiple forms of soybean lipoxygenase-1, *Anal. Biochem.*, 95, 371, 1979.
10. **Rubinstein, M., Rubinstein, S., Familletti, P. C., Miller, R. S., Waldan, A. A., and Pestka, S.,** Human leukocyte interferon: production, purification to homogeneity, and initial characterization, *Proc. Natl. Acad. Sci. U.S.A.*, 76, 640, 1979.
11. **Kotake, A. N. and Funae, Y.,** High-performance liquid chromatography technique for resolving multiple forms of hepatic membrane-bound cytochrome P-450, *Proc. Natl. Acad. Sci. U.S.A.*, 77, 6473, 1980.

ENZYME SEPARATION BY REVERSED PHASE HPLC

M. Patricia Strickler and M. Judith Gemski

The full potential of reversed-phase high-performance liquid chromatography (RP-HPLC) for application to the purification and recovery of biologically active enzymes has not yet been realized. A major problem limiting utilization of RP-HPLC for enzyme purification is the apparent poor recovery of active material when eluted from reversed-phase packings. This poor recovery is thought to be due to nonspecific adsorption to the alkylsilane bonded phases or denaturation of the enzyme in the aqueous/organic mobile phases necessary for elution. Another explanation, however, is the improper handling of the fractionated samples. The selection of appropriate postfractionation conditions can result in the recovery of highly purified active enzymes from complex mixtures.

The preparation of chromatographically pure and active trypsin from commercially available sources was accomplished using RP-HPLC.[1] The chromatographic conditions consisted of a 30-min linear gradient run at a flow rate of 2 mℓ/min on a μBondapak® C_{18}-column (3.9 mm × 30 cm). The mobile phase consisted of Solution A, aqueous 0.1% trifluoroacetic acid (TFA) and Solution B, acetonitrile-0.1% TFA, run from 0 to 45% B. The column effluent was monitored at 215 and 280 nm. The fractionation of 1 mg of commercially available trypsin is shown in Figure 1. Four milliliter fractions were collected for 28 min followed by 2-mℓ fractions during the next 6 min. For consistent recovery of enzymatic activity, it was necessary to follow a specific protocol after collecting the fractions. The acetonitrile present in each fraction was removed by evaporation under N_2 at room temperature and then aqueous TFA was removed immediately by lyophilization. The samples were reconstituted in 1.2 mℓ of 0.001 *N* HCl and the amount of protein in each was calculated using the absorbance at 280 nm. All fractions were assayed for enzymatic activity using the TAME procedure, which is specific for trypsin.[2] The results are summarized in Table 1. Fraction 9, which contains the major peak, showed a higher specific activity than the original sample as indicated in Figure 2. A 57% recovery of total protein injection onto the column (calculated on the basis of the absorbance at 280 nm) was obtained from the chromatographic separation. An aliquot of fraction 9 was rechromatographed to confirm its purity (Figure 3).

This technique also provides a convenient purification of small amounts of trypsin directly from biological samples. For example, it was used to quantitate trypsin from the stomach and intestine of a single tsetse fly. The two organs were removed and macerated in 0.001 *N* HCl, and the filtrate injected into the chromatographic system. The resulting chromatogram is shown in Figure 4. The retention times of two additional peaks corresponded to carboxypeptidase and α-chymotrypsin standards.[3] Recovery of active chymotrypsin required a post-collection procedure similar to that for trypsin, however, the carboxypeptidase was inactivated under these conditions.

The purification and recovery of the enzyme prostatic acid phosphatase (PAP) further exemplifies the importance of adequate post-column collection protocols.[4] PAP is a glycoprotein with a molecular weight of approximately 102,000. It has been purified to near homogeneity from human seminal plasma by a combination of techniques such as gel filtration, affinity chromatography, and preparative gel electrophoresis.[5-7] PAP, purified by gel filtration,[5] was chromatographed on a μBondapak® C_{18}-column to assess its purity. A 30-min linear gradient was run at a flow rate of 1.5 mℓ/min. The aqueous 0.1% TFA/acetonitrile mobile phase system was employed at conditions of 12 to 70% B. Only minor impurities were revealed (Figure 5), which corresponded, by retention times, to the peaks present in the RP-HPLC profile of dialyzed human seminal plasma (Figure 6).

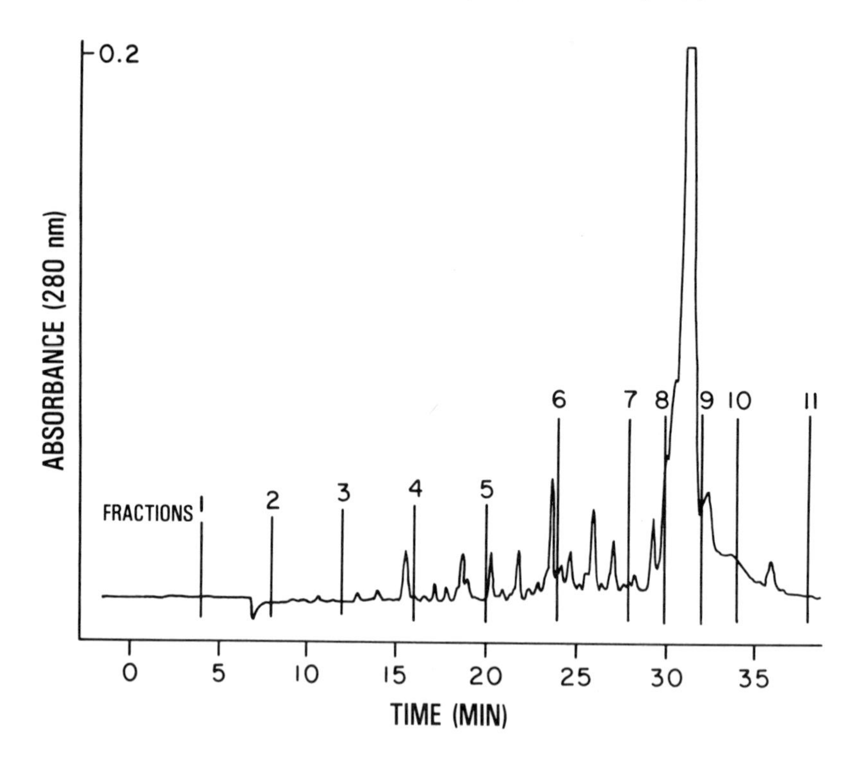

FIGURE 1. Separation of 1 mg of bovine trypsin (Sigma Chemical Co.) collected in fractions indicated by numbers 1 to 11.

Table 1
PROTEIN CONCENTRATION AND ENZYMATIC ACTIVITIES OF THE FRACTIONATED TRYPSIN

Fraction number	Time (min)	Concentration (mg/mℓ calc)	Activity (units/mg protein)
1	4	0.0168	0.0
2	8	0.0133	0.0
3	12	0.0091	0.0
4	16	0.0175	3.7
5	20	0.0231	8.3
6	24	0.0245	0.0
7	28	0.0280	0.0
8	30	0.0637	53.1
9	32	0.2541	151.5
10	34	0.0203	45.1
11	38	0.0147	47.8
Starting sample	—	0.8484	135.8

The chromatographic profile of seminal plasma was reproducible. Injection on column of 100 to 500 μℓ of seminal plasma gave a linear response in terms of the area of integration of selected peaks, including the peak at 22 to 23 min which corresponded to PAP. In contrast to our previous experience with trypsin, however, recovery of active enzyme by lyophilization of the isolated fractions was nonreproducible. An investigation was therefore undertaken to

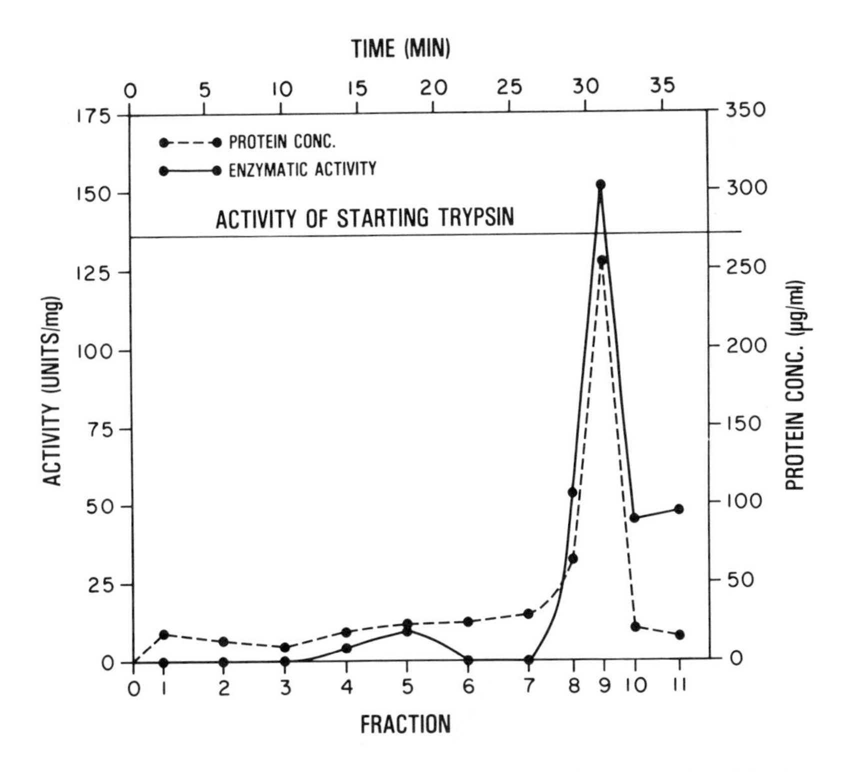

FIGURE 2. Graph showing the enzymatic activity and protein concentration of fractions recovered from the RP-HPLC of trypsin.

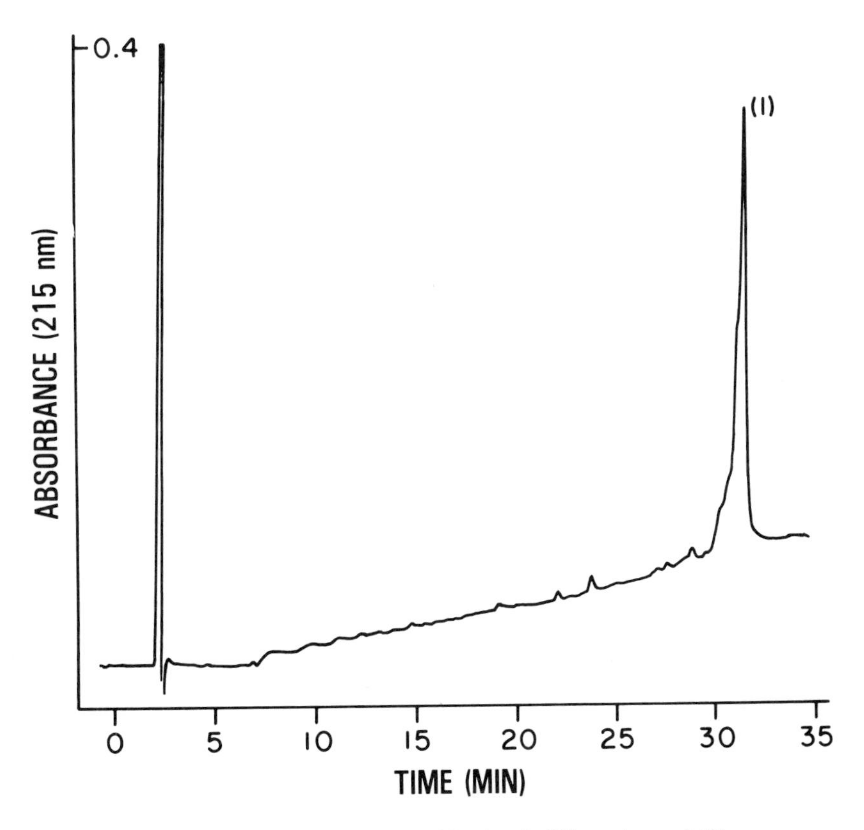

FIGURE 3. Rechromatography of fraction 9 of Figure 1, trypsin(1).

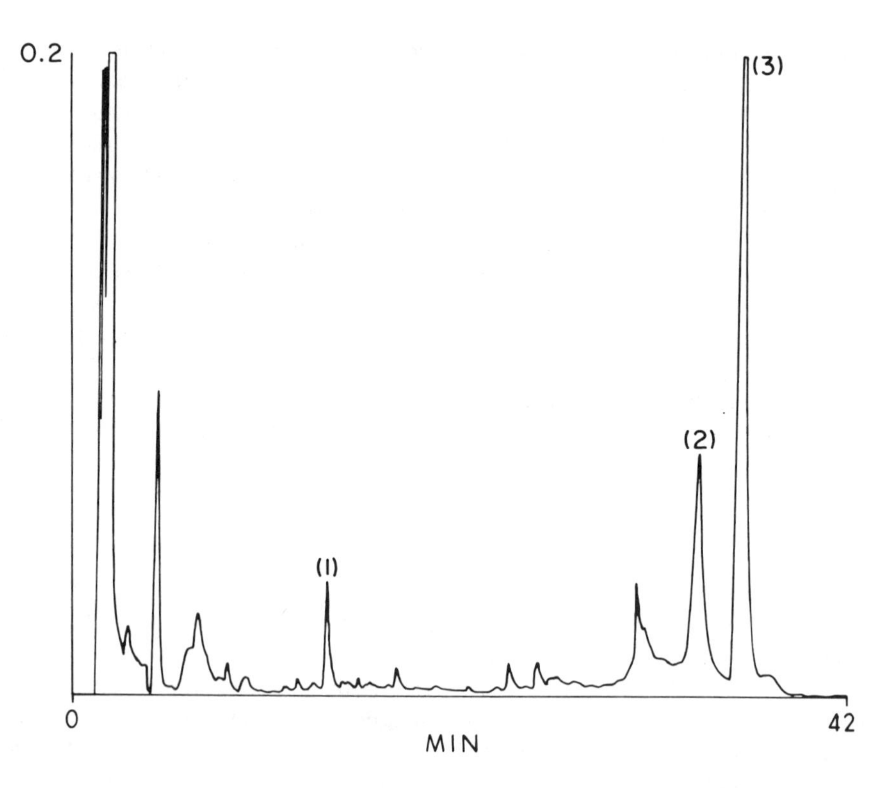

FIGURE 4. Chromatograph of the extract from a single tsetse fly's stomach and intestine. Peaks are identified by comparision with standards or by enzymatic activity, (1) carboxypeptidase, (2) trypsin, and (3) chymotrypsin.

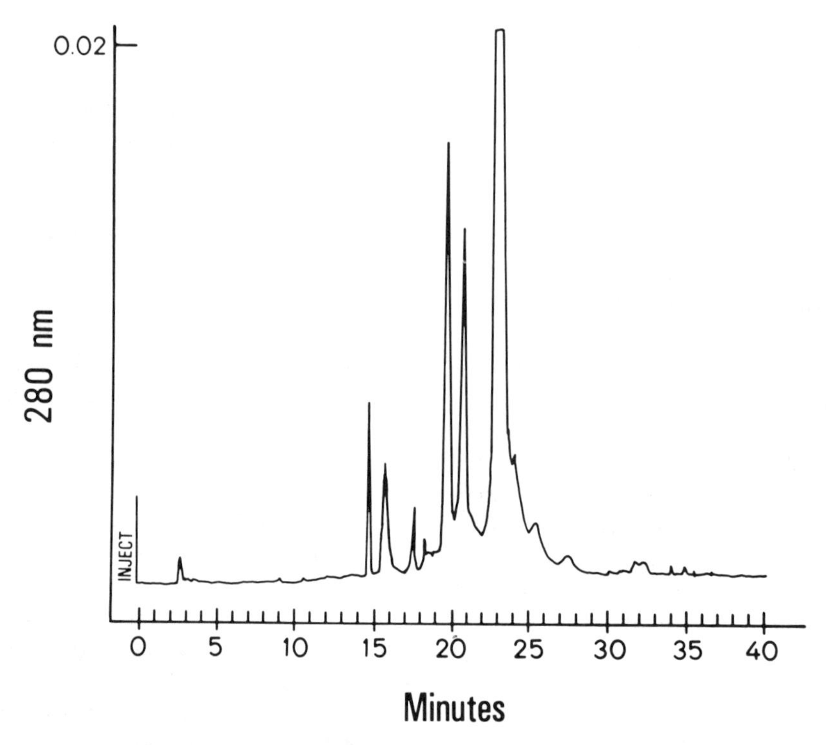

FIGURE 5. Chromatograph of purified prostatic acid phosphatase (PAP), Active fraction occurs at 23 min.

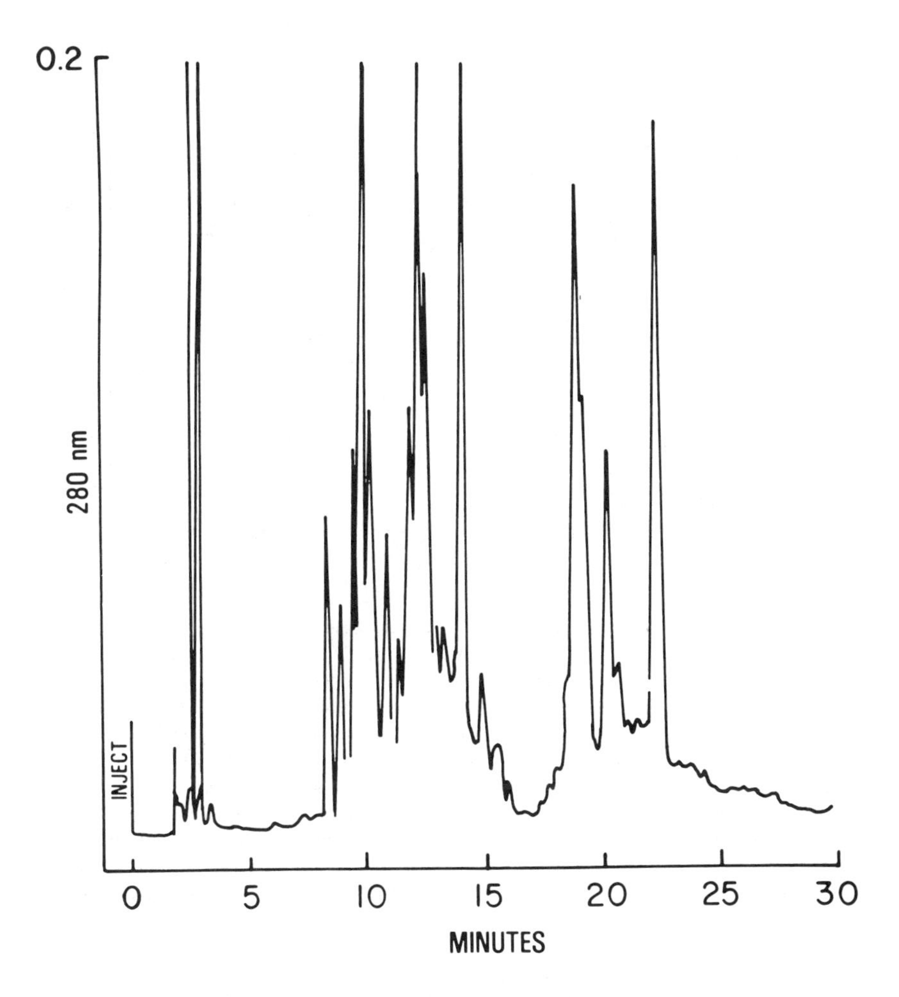

FIGURE 6. Chromatograph of dialyzed human seminal plasma. Active PAP was recovered from the peak at 23 min.

determine whether the poor recovery was attributable to loss of enzyme activity on the column or enzyme denaturation by the mobile phase after fractionation. The chromatographic fractions were collected into a series of buffers and solvents known to stabilize the activity of a variety of enzymes. These substances and their effectiveness in protecting PAP activity are summarized in Table 2. Glycerol was found to be the best protector of PAP activity; Hanks' Balanced Salt Solution (BSS) and Dulbecco's Phosphate-Buffered Saline (PBS), though not nearly as good as glycerol, did significantly increase the recovery of enzymatic activity. The dramatic change in recovery of PAP activity seen in this study, as in the case of trypsin, shows that collection and processing after fractionation is crucial to obtaining active enzyme. The homogeneity of the fractions containing PAP was assessed by rechromatography on RP-HPLC (Figure 7) and by sodium dodecyl sulfate (SDS) gels. The electrophoretic pattern of the SDS gels shows one major band corresponding in molecular weight to the standard PAP and to the active enzyme in seminal plasma. The enzymatic activity of the fractionated PAP was surveyed by the reaction to the substrate *o*-carboxyphenyl phosphate[8] and with a Mallinckrodt RIA kit. The latter also confirms the immunological integrity of the purified PAP. The recovery of immunologically reactive PAP from 100 $\mu\ell$ of human seminal plasma, injected on column, was 77.5%.

Table 2
PROTECTION OF PAP ACTIVITY

Stabilizing agent	Relative activity
No protective agent[d]	Marginal
Hanks' BSS[a,c]	Moderate
Dulbecco's PBS[a,c]	Moderate
Potassium phosphate, 0.1 *M*, pH 7[7 a,c]	None
Sodium acetate, 0.05 *M*, pH 5[a,c]	None
Ammonium bicarbonate, 0.01 *M* pH 8[a,b,d]	None
DMSO[b,c]	None
PEG[b,c]	None
Glycerol (45% v/v)[b,c]	Very active

[a] Aliquots of 0.5 mℓ of the buffered solution were added to each 1.5 mℓ fractions.
[c] None of the fractions were lyophilized prior to testing the activity. The acetonitrile, however, was evaporated under N_2.
[b] Aliquots of 0.1 mℓ of solvents were added to each 1.5 mℓ fractions.
[d] Lyophilization of any of these fractions gave ambigious results, due to the instability of PAP under freezing and thawing conditions. The glycerol was found to be the most effective protector of prostatic acid phosphatase activity.

These findings suggest that loss of material does not necessarily occur during the chromatographic process and that denaturation by the aqueous/organic mobile phase may be prevented or reversed. RP-HPLC when used with the proper postfractionation conditions allows the rapid separation and recovery of nanogram to milligram amounts of highly purified enzyme from complex biological mixtures.

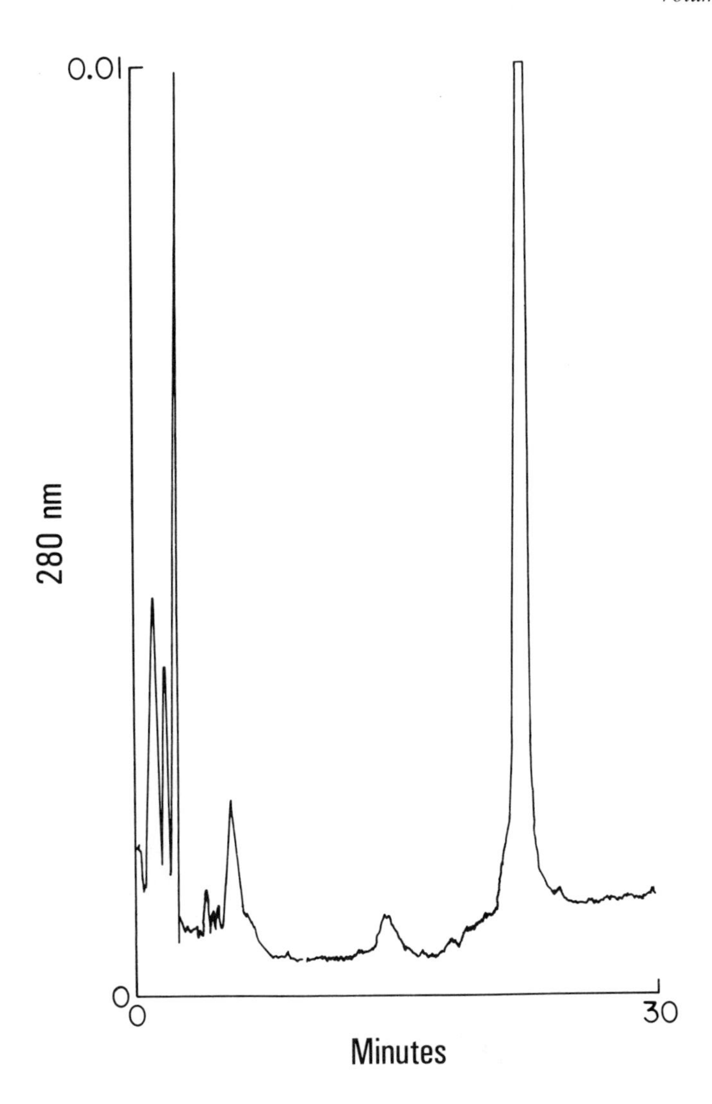

FIGURE 7. Rechromatography of fractions containing PAP activity, isolated from human seminal plasma.

REFERENCES

1. **Strickler, M. P., Gemski, M. J., and Doctor, B. P.,** Purification of commercially prepared bovine trypsin by reverse phase high performance liquid chromatography, *J. Liq. Chromatogr.*, 4, 1765, 1981.
2. **Decker, L. A., Ed.,** Worthington Enzyme Manual, Worthington Biochemical Corporation, Freehold, NJ, 1978, 221.
3. **Gemski, M. J. and Stermer-Cox, M. G.,** unpublished data, 1982.
4. **Strickler, M. P., Kintzios, J., and Gemski, M. J.,** The purification of prostatic acid phosphatase from seminal plasma by reverse phase high performance liquid chromatography, *J. Liq. Chromatogr.*, 5, 1921, 1982.
5. **Mahan, D. E. and Doctor, B. P.,** A radioimmuno assay for human prostatic acid phosphatase-levels in prostatic disease, *Clin. Biochem.*, 12, 10, 1979.
6. **Vihka, P. Kontturi, M., and Korhonew, L. K.,** Purification of human prostatic acid phosphatase by affinity chromatography and isoelectric focusing, I, *Clin. Chem.*, 24, 466, 1978.
7. **Lam, W. K. W., Yam, L. T., Wilbur, H. J., Taft, E., and Li, C. Y.,** Comparision of acid phosphatase isoenzymes of human seminal fluid, prostate and leukocytes, *Clin. Chem.*, 25(F), 1285, 1979.
8. **Decker, L. A., Ed.,** Worthington Enzyme Manual, Worthington Biochemical Corporation, Freehold, NJ, 1977, 145.

HIGH PERFORMANCE ION-EXCHANGE CHROMATOGRAPHY OF INSULIN AND INSULIN DERIVATIVES

B. S. Welinder and S. Linde

INTRODUCTION

Recently a number of ion-exchange supports have become commercially available with characteristics which should allow high-performance ion-exchange chromatography (HPIEC) of peptides and proteins: high mechanical and chemical stability, spherical particles (5 to 10 μm), high ion-exchange capacity, hydrophilic nature, and variable pore size. In addition reports have claimed to show more than 95% recovery of most water-soluble proteins. The supports are based on either silica (Toyo Soda, Synchron) or organic polymers (Pharmacia).[1]

Since low-pressure ion-exchange chromatography of insulin is a well-established process in analytical as well as preparative scale it would be of interest to examine the efficiency of the HPIEC in the purification of insulin. The present communication compares the chromatographic separations of insulin and insulin derivatives on two anion-exchange columns to that obtained by reversed-phase high-performance liquid chromatography (RP-HPLC).

MATERIALS AND METHODS

Insulin — Crystalline porcine Na-insulin (batch No. G-63), desamidoinsulin and b-component were obtained from Nordisk Gentofte. Iodinated insulin with an average iodine content of 1.0 atom of iodine (^{127}I) per molecule of insulin was prepared using the lacto-peroxidase method as previously described.[2] Trace-labeled iodinated insulin labeled with an average iodine content of 0.01 atom of iodine (^{127}I + ^{125}I) per molecule of insulin was prepared using the lactoperoxidase method in buffer containing 6 *M* urea essentially as described previously.[3] Carbamylated insulin was prepared by incubation of insulin in 7 *M* urea at neutral pH for 24 hr.

Columns — Pharmacia Mono Q (5 × 50 mm); Toyo Soda IEX 545 DEAE Sil (6 × 150 mm), HPLC-equipment: 2 P-500 pumps, a GP-250 gradient programmer, a UV-1 photometer, and a FRAC-100 fraction collector (all from Pharmacia) or a Spectra Physics SP-8700 chromatograph, Waters WISP 910B sample processor, and a Pye Unicam UV-photometer. The radioactivity in the collected fractions was counted in a 16-channel gamma counter (Hydrogamma 16). All chemicals were analytical grade. Water was drawn from a Milli-Q system. All buffers were filtered (Millipore 0.22 and 0.45 μm) and degassed before use.

RESULTS

Figure 1 shows the elution of crystalline insulin from a Mono Q column using two different NaCl gradients. The main peak is insulin peptide, the largest minor peak is desamidoinsulin. The separation pattern is similar to that obtained by RP-HPLC (see Figure 2) but much less detailed with respect to minor contaminants such as proinsulin, insulin dimer, arginine insulin, etc.

The separation of insulin iodinated to one I per mole with nonradioactive iodine using the same column buffer and gradients are shown in Figure 3. The main fractions probably correspond to insulin, monoiodoinsulin, and diiodoinsulin, since this separation is similar to separations on AE-cellulose[4] or DEAE-Sephadex®.[5]

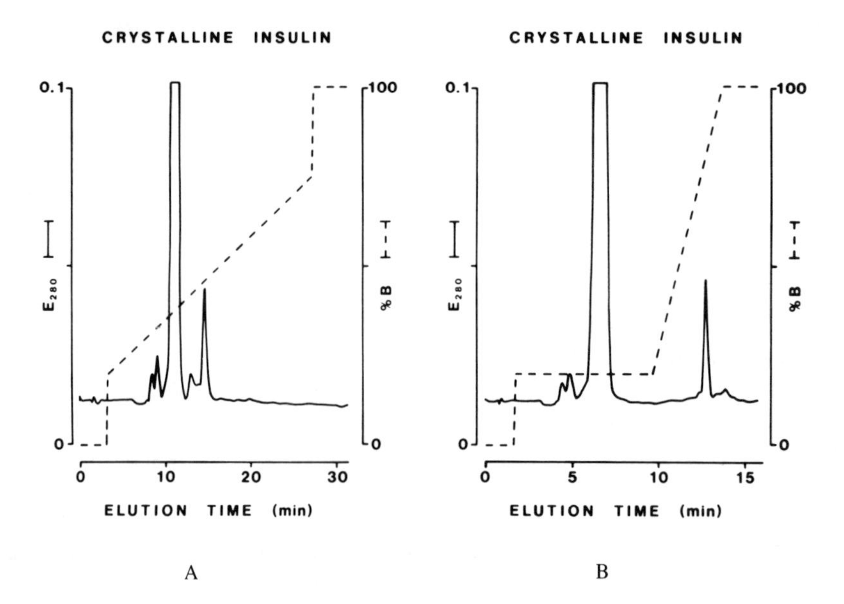

FIGURE 1. Separation of 300 μg crystalline porcine insulin on a Mono Q column. Buffer A: 0.01 *M tris*-HCl/7 *M* urea, pH 8.60. Buffer B: A + 0.30 *M* NaCl. Flow rate: 0.5 mℓ/min (A), 1.0 mℓ/min (B).

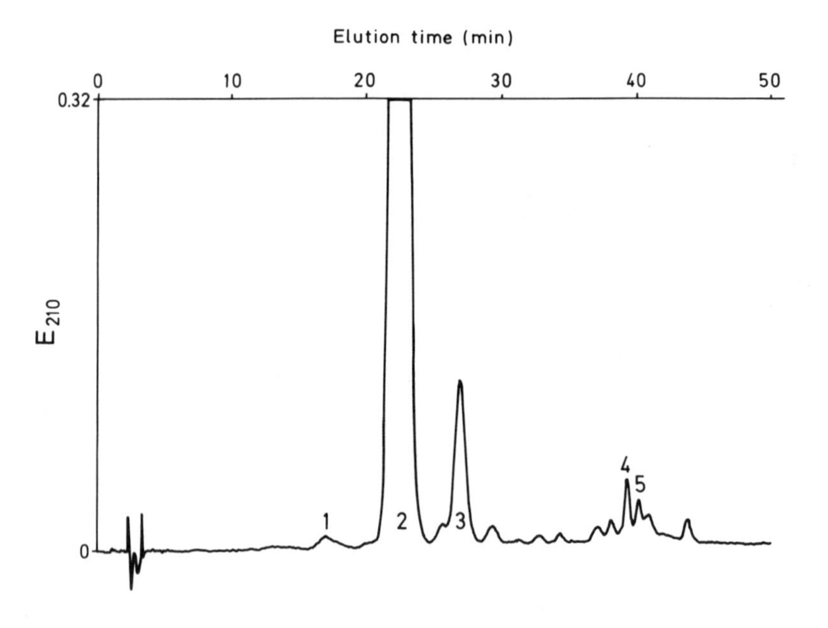

FIGURE 2. Separation of 100 μg crystalline porcine insulin on a Lichrosorb® RP18 (7 μm), 250 × 4.0 mm column. Buffer A: 0.25 *M* TEAP, pH 3.00. Buffer B: 50% A, 50% acetonitrile. Gradient: A linear gradient from 53 to 59% B during 40 min followed by 59% isocratically for 10 min. Flow rate: 1 mℓ/min. The numbers represent the following components: arginine insulin (1), insulin peptide (2), desamidoinsulin (3), proinsulin (4), and insulin dimer (5).

Figure 4 shows the separation of insulin iodinated to 0.01 I per mole (containing trace amounts of ^{125}I). The radioactivity is eluted as one peak although the iodinated insulin is a

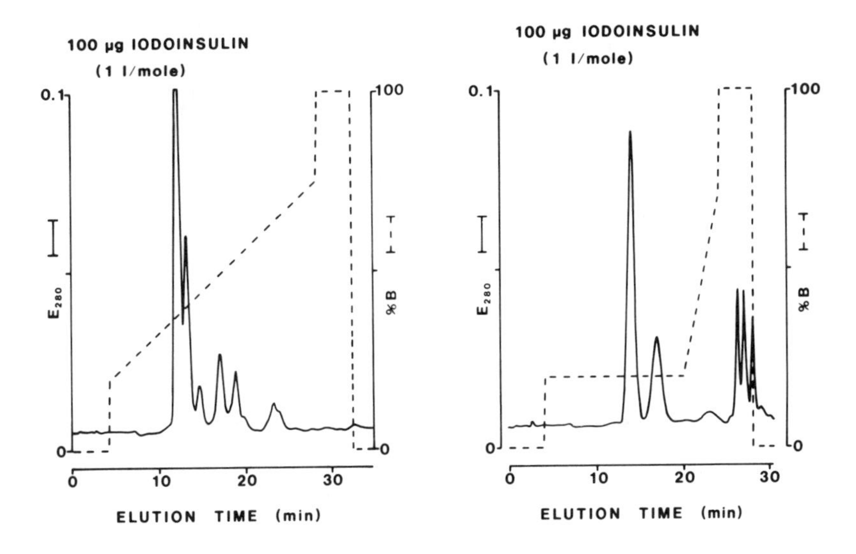

FIGURE 3. Separation of 100 μg iodinated porcine insulin on a Mono Q column. Buffers as in Figure 1. Flow rate: 0.5 mℓ/min.

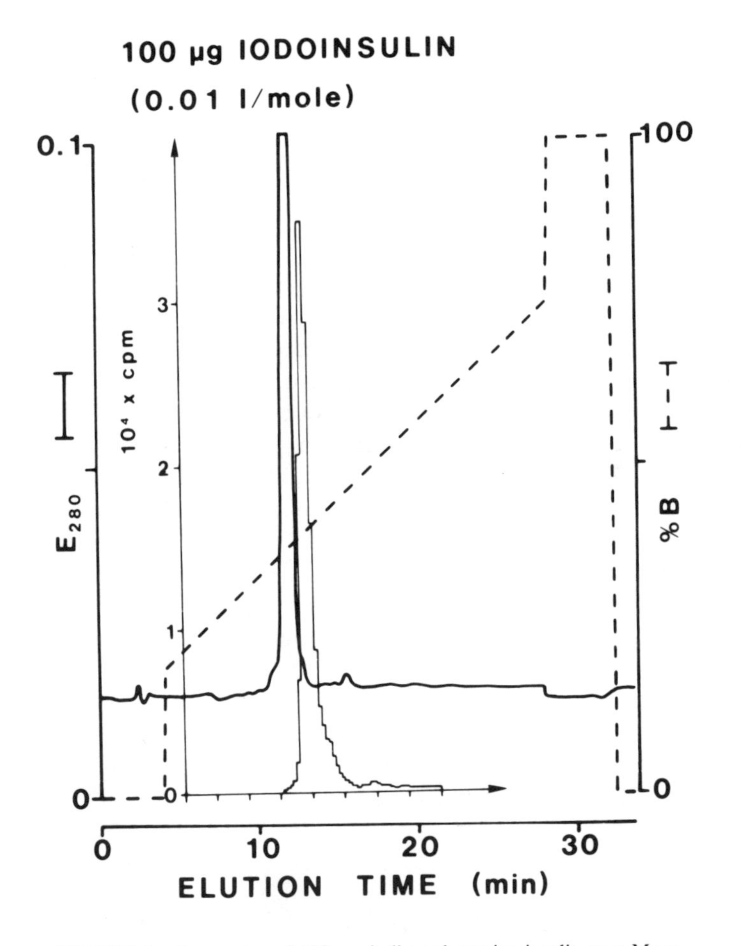

FIGURE 4. Separation of 100 μg iodinated porcine insulin on a Mono Q column. Buffers as in Figure 1. Flow rate: 0.5 mℓ/min. Fractions of 0.2 min were collected and the histogram represents the radioactivity in each fraction.

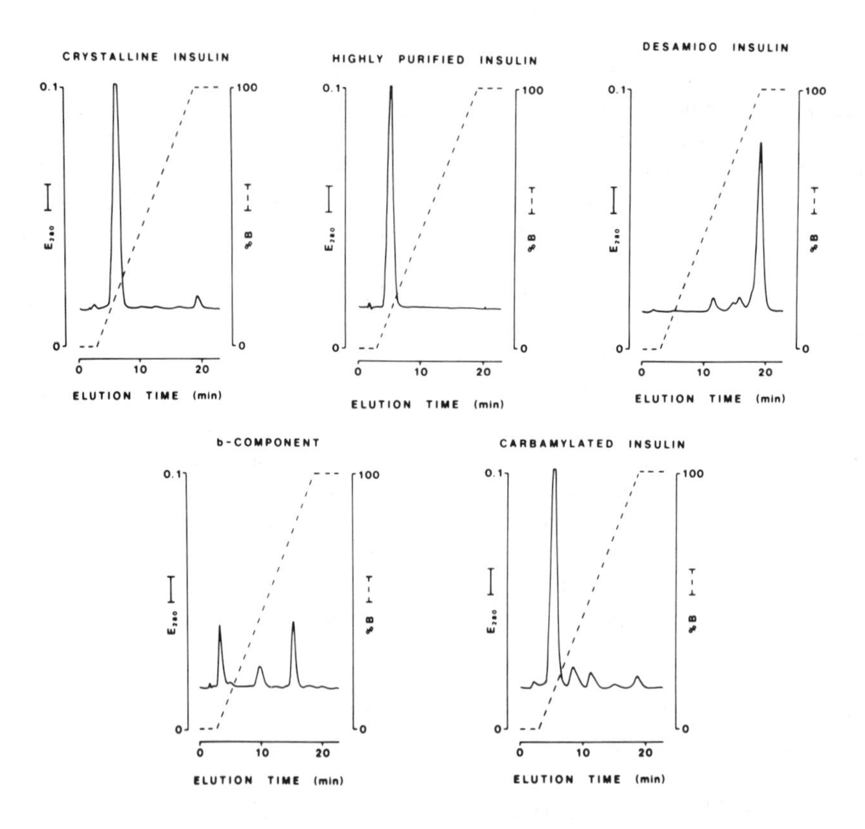

FIGURE 5. Separation of 100 μg porcine insulin and insulin derivatives on a Mono Q column. Flow rate: 0.5 mℓ/min.

mixture of insulins monoiodinated in A14, A19, B16, and B26. The position of the peak of radioactivity corresponds to the second peak in Figure 3 assumed to be monoiodoinsulin.

Instead of NaCl gradient an increasing concentration of a buffer substance and/or a pH gradient can be used for the elution of insulin. Figure 5 shows the fractionation obtained for insulin and insulin derivatives using a Mono Q column where the gradient is formed from two urea/phosphate buffers of varying ion strength and pH (to be published). As can be seen a very good separation between insulin peptide and desamidoinsulin is obtained. Carbamylated insulins and b-fraction (mainly containing proinsulin, insulin dimer, and intermediary insulin) are also well separated from insulin peptide. A highly purified porcine insulin is eluted as a single peak.

This phosphate/pH gradient can also be used for the fractionation of insulin on the silica-based Toyo Soda DEAE-column, but the optimum gradient shape is somewhat different (Figure 6).

The separation pattern for insulin and insulin derivatives on the diethylaminoethyl column is very similar to that obtained on the Mono Q column where the exchange group is $-CH_2-N^+(CH_3)_3$. The only exception is the very close elution of insulin peptide and one of the three main components in the b-fraction.

DISCUSSION

From the experiments described here, it is clearly seen that HPIEC permits rapid fractionation of crystalline insulin. It should be emphasized that a satisfactory separation can be obtained without the use of NaCl gradients, since a number of potential users of this

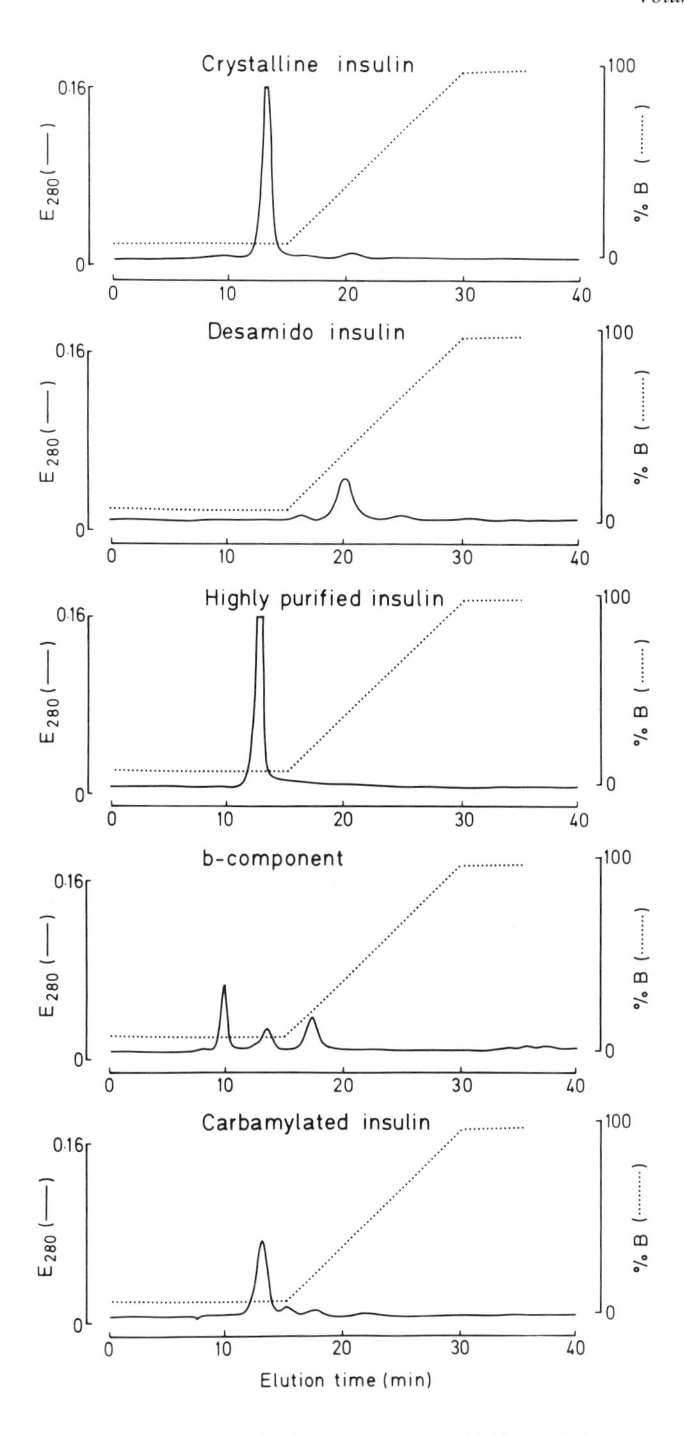

FIGURE 6. Separation of 400 μg crystalline and highly purified porcine insulin and 200 μg desamidoinsulin,b-component and carbamylated insulin on a IEX 545 DEAE Sil column. Buffers as in Figure 5. Flow rate: 0.5 mℓ/min.

technique probably will develop their separation parameters in existing HPLC equipment. Most commercial HPLC equipment is constructed partly in stainless steel and extensive use of halides will damage this material.

Recently a separation between bovine insulin peptide and carbamylated insulin using microparticulate silica coated with a cross-linked polyethylenimine was published.[6] The separation was performed in 20 min using a NaCl gradient in 8 *M* urea per 0.05 *M* triethylamine pH 7.2, and the fractionation pattern was very similar to those for carbamylated insulins described in this communication.

It seems reasonable to compare the separation of crystalline insulin obtained by HPIEC to that found in RP-HPLC. The number of contaminants observed in HPIEC is clearly reduced compared to RP-HPLC. Due to strong tendency to self-aggregation ion-exchange chromatography of insulin must be performed under disaggregating conditions, and the use of 7 *M* urea makes low UV detection at 200 to 220 nm impossible. The reduced sensitivity at 280 nm compared to low UV detection probably explains in part the less detailed ion-exchange chromatograms.

The labeled heterogenous iodoinsulin preparation known to contain a mixture of the four insulins monoiodinated in A14, A19, B16, and B26 was eluted as one peak in HPIEC (the histogram in Figure 4), whereas these four isomers can be separated using RP-HPLC.[7] HPIEC separates according to charge and if the sample components show no (or extremely small) differences in charge (as in the case of the four monoiodo-insulins) the potential in ion-exchange chromatography is limited. Since the hydrophobic interaction between a reversed-phased material and sample components are extensive and can be subjected to considerable changes by varying (especially) the ion-pairing substances in the buffer, RP-HPLC seems to offer greater separation possibilities for insulin and similar peptides than HPIEC.

However, the possibility of performing rapid separations of peptides under mild conditions based upon the electrical charge (without addition of organic modifiers) makes HPIEC an interesting supplement to RP-HPLC. The most obvious application of HPIEC seems to be for the fractionation of proteins, when the organic solvents commonly used in RP-HPLC at present create some difficulties for satisfactory separations.

Really widespread use of HPIEC probably depends upon the availability of the chromatographic support as a bulk material, since a rather expensive prepacked column limits the number and nature of experiments that can be used in search for the optimum separation.

ACKNOWLEDGMENTS

The technical assistance of Ingelise Fabrin and Linda Larsø is gratefully acknowledged. Meda, Denmark, is thanked for the loan of the Pharmacia apparatus.

REFERENCES

1. **Hearn, T. W., Regnier, F. E., and Wehr, C. T.,** HPLC of peptides and proteins, *Int. Lab.*, 13, 16, 1983.
2. **Linde, S. and Hansen, B.,** Monoiodoinsulin specifically substituted in Tyr A14 or Tyr A19, *Int. J. Peptide Protein Res.*, 15, 495, 1980.
3. **Linde, S., Sonne, O., Hansen, B., and Gliemann, J.,** Monoiodoinsulin labelled in tyrosine residue 16 or 26 of the insulin B-chain. Preparation and characterization of some binding properties, *Hoppe-Zeyler's Z. Physiol. Chem.*, 362, 573, 1981.
4. **Linde, S. and Hansen, B.,** Preparation and characterization of monoiodoinsulin, *Int. J. Peptide Protein Res.*, 6, 157, 1974.
5. **Hamlin, J. L. and Arquilla, E. R.,** Monoiodoinsulin. Preparation, purification, and characterization of a biologically active derivative substituted predominantly on tyrosine A14, *J. Biol. Chem.*, 249, 21, 1974.
6. **Richey, F.,** FPLC: a comprehensive separation technique for biopolymers, *Int. Lab.*, 13, 50, 1983.
7. **Welinder, B. S., Linde, S., and Brush, J. S.,** Separation of the monoiodinated isomers of insulin by high-performance liquid chromatography, *J. Chromatogr.*, 257, 162, 1983.

TOYO SODA HIGH PERFORMANCE GEL FILTRATION COLUMNS

Yoshio Kato

THE TOYO SODA TSK-SW COLUMN

The Toyo Soda TSK-SW column was developed for high-speed gel filtration of proteins. The support is a spherical porous silica bead chemically bonded to a hydrophilic compound containing hydroxyl groups. Its main features are low adsorptivity and high resolution.[1] Therefore, the TSK-SW column can separate proteins under the mild elution conditions employed in ordinary gel filtration on soft gels like Sephadex® and yet achieve higher resolution.

The TSK-SW column is available in three grades. Columns of different pore size are listed in Table 1. The analytical column has a 0.75-cm internal diameter and is 30 or 60 cm in length. The preparative column has a 2.15-cm internal diameter and is 30 or 60 cm in length. The support for the preparative column has a slightly larger particle size than that for analytical column.

Gel Filtration in Common Buffers

Separation ranges for globular proteins are shown in Table 2.[2] Proteins over the molecular weight range from several thousands to several millions can be separated by using a TSK-SW column. One of the three columns will provide the best separation based on the molecular weight ranges as shown in Table 3, i.e., the best column should be selected according to the molecular weight of the protein to be separated. In the molecular weight range of 30,000 to 500,000, G3000SW should be selected because G3000SW provides much better separation than G2000SW or G4000SW. In the molecular weight range below 30,000, however, the separations attainable on G2000SW and on G3000SW differ only slightly. However, only G4000SW separates proteins with molecular weights higher than 500,000.

Buffers of pH 6.5 to 7.5 containing 0.1 to 0.4 *M* salts are usually employed as an eluent. At low salt concentrations proteins elute earlier or later than expected from their molecular weights due to ionic interactions with the silanol groups remaining on the surface of the support, while higher salt concentrations cause retardation due to hydrophobic or hydrogen bonding interactions.[3,4] Neutral to slightly basic pH generally provides the highest recovery of proteins.[3] The TSK-SW column must be operated within the pH range 2.5 to 8.0(G2000SW, G3000SW) or 2.5 to 7.5(G4000SW) since the support is silica based.

The TSK-SW column is usually operated at flow rates of 0.5 to 1.0 mℓ/min, which seems the best compromise between separation efficiency, separation time, and column lifetime, although proteins are better separated at flow rates below 0.1 mℓ/min. The operating flow rate is limited up to 1.2 mℓ/min for analytical columns and 8 mℓ/min for preparative columns in order to ensure maximum column lifetime. At higher flow rates, the column performance tends to decrease because voids are created at the column head.

Samples are usually injected at a concentration of 0.01 to 0.5% and a volume not exceeding 1% of the column volume in analytical separations. Sample concentrations higher than 1% or injection volumes larger than 1% of column volume result in a decrease in separation efficiency.[3] On the other hand, extremely low sample loading such as below 10 μg sometimes gives rise to skewed peaks or poor protein recoveries. In the preparative separations, it is better not to inject samples in too small a volume when attempting to achieve a high sample loading. An injection volume of 1 to 3% of the column volume may be a good compromise between separation efficiency and sample loading.

Table 1
PROPERTIES OF THE TSK-SW COLUMNS

Grade	Mean pore diameter (ÅÅ)	Particle size (μm)
G2000SW[a]	125	10 ± 2
G3000SW[a]	250	10 ± 2
G4000SW[a]	500	13 ± 2
G2000SWG[b]	125	13 ± 2
G3000SWG[b]	250	13 ± 2
G4000SWG[b]	500	17 ± 2

[a] Analytical column: size 30 × 0.75 cm i.d. or 60 × 0.75 cm i.d.
[b] Preparative column: size 30 × 2.15 cm i.d. or 60 × 2.15 cm i.d.

Table 2
SEPARATION RANGES OF TSK-SW COLUMNS FOR GLOBULAR PROTEINS IN COMMON BUFFERS[a]

Column	Molecular weight range
G2000SW	5,000—100,000
G3000SW	10,000—500,000
G4000SW	20,000—7,000,000

[a] Data from Reference 2.

Table 3
BEST COLUMN FOR THE SEPARATION OF GLOBULAR PROTEINS IN COMMON BUFFERS[a]

Molecular weight range	Best column
Below 30,000	G2000SW
30,000—500,000	G3000SW
Above 500,000	G4000SW

[a] Data from Reference 2.

Gel Filtration in 6 *M* Guanidine Hydrochloride

Denaturing solvents, such as 6 *M* guanidine hydrochloride, 8 *M* urea, and 0.1% SDS, have been employed in gel filtration for estimating molecular weights of proteins. TSK-SW columns can also be operated using these denaturating solvents.[5-7]

Separation ranges for polypeptides are shown in Table 4.[8] It is possible to separate

Table 4
SEPARATION RANGES OF THE TSK-SW COLUMNS FOR POLYPEPTIDES IN 6 *M* GUANIDINE HYDROCHLORIDE CONTAINING 0.1 *M* SODIUM PHOSPHATE (pH 6.0)[a]

Column	Molecular weight range
G2000SW	1,000—25,000
G3000SW	2,000—70,000
G4000SW	3,000—400,000

[a] Data from Reference 8.

Table 5
BEST COLUMN FOR THE SEPARATION OF POLYPEPTIDES IN 6 *M* GUANIDINE HYDROCHLORIDE CONTAINING 0.1 *M* SODIUM PHOSPHATE (pH 6.0)[a]

Molecular weight range	Best column
Below 10,000	G2000SW
10,000—70,000	G3000SW
Above 70,000	G4000SW

[a] Data from Reference 8.

polypeptides of molecular weights of 1000 to 400,000 by using a TSK-SW column. Since the best separation can be obtained on one of the three columns in respective molecular weight ranges summarized in Table 5,[8] the best column should be selected according to the molecular weight of the polypeptide to be separated. Although G4000SW covers a wide molecular weight range, it is far less effective than G3000SW within the separation range of the G3000SW column.

A TSK-SW column is usually operated at flow rates of 0.3 to 0.5 mℓ/min. Flow rates lower than 0.5 mℓ/min are preferable for both column lifetime and separation efficiency because 6 *M* guanidine hydrochloride has a high viscosity. The higher the viscosity of the eluent, the greater becomes the dependence of separation efficiency on flow rate.

When exchanging a solvent in a column, the first column volume of the new solvent must be introduced at a low flow rate. Especially in the case of an exchange to a highly viscous solvent such as 6 *M* guanidine hydrochloride, it is important to keep the flow rate as low as 0.2 to 0.3 mℓ/min in order to prevent column deterioration.

When a series of measurements has been completed and the next measurements are not scheduled for a while, it is better to return the solvent in the column to water containing an antibacterial agent such as 0.05% sodium azide.

Table 6
SEPARATION RANGES OF THE TSK-SW COLUMNS FOR POLYPEPTIDES IN 0.1% SDS AQUEOUS SOLUTION CONTAINING 0.1 *M* SODIUM PHOSPHATE (pH 7.0)[a]

Column	Molecular weight range
G2000SW	15,000—25,000
G3000SW	10,000—100,000
G4000SW	15,000—300,000

[a] Data from Reference 9.

Table 7
BEST COLUMN FOR THE SEPARATION OF POLYPEPTIDES IN 0.1% SDS AQUEOUS SOLUTION CONTAINING 0.1 *M* SODIUM PHOSPHATE (pH 7.0)[a]

Molecular weight ranges	Best column
Below 60,000	G3000SW
Above 60,000	G4000SW

[a] Data from Reference 9.

Gel Filtration in SDS Aqueous Solution

A 0.1% SDS aqueous solution containing sodium phosphate is usually employed as an eluent. In this case, the sodium phosphate concentration greatly influences the elution behavior of protein-SDS complexes.[7,9,10] As the sodium phosphate concentration increases, the protein-SDS complexes elute later and are better separated. However, an extreme increase in sodium phosphate concentration causes the adsorption of protein-SDS complexes onto the support, resulting in broadening of the peaks and scatter in the plots of molecular weight against elution volume. A 0.1% SDS aqueous solution containing 0.1 *M* sodium phosphate at pH 7 is normally employed.

Table 6 shows the separation ranges for polypeptides in 0.1% SDS aqueous solution containing 0.1 *M* sodium phosphate of pH 7.[9] Polypeptides of molecular weights from 10,000 to 300,000 can be separated by using the TSK-SW column. The separation range is rather limited compared to the case in 6 *M* guanidine hydrochloride, especially at the low-molecular-weight end. The reason is presumably that the size of protein-SDS complexes becomes independent of the molecular weights of proteins in the molecular weight range below 10,000 under these conditions. G3000SW or G4000SW should be used according to the molecular weight of polypeptide to be separated as denoted in Table 7.[9] G2000 SW does not provide the best separation in any molecular weight range.

The SDS concentration also influences the elution behavior of protein-SDS complexes.

Table 8
PROPERTIES OF THE TSK-PW COLUMNS

Grade	Particle size (μm)	Molecular weight range for polyethylene glycol
G1000PW	10 ± 2	— 1,000
G2000PW	10 ± 2	— 5,000
G3000PW	13 ± 2	— 50,000
G4000PW	13 ± 2	2,000—300,000
G5000PW	17 ± 2	4,000—1,000,000
G6000PW	25 ± 5	40,000—8,000,000

The separation range of G3000SW can be extended down to around 1000 by using a 0.2% SDS aqueous solution containing 0.2*M* sodium phosphate at pH 7 as the eluent.[4]

THE TSK-PW COLUMN

The TSK-PW column was also developed for use in high-speed gel filtration. The support is based on a hydrophilic organic copolymer. The TSK-PW column support is available in six grades based on pore size, as listed in Table 8. A column can have a 0.75 (analytical column) or 2.15-cm (preparative column) internal diameter and can be 30 or 60 cm in length. Although the separation range has not been determined for proteins, it may be estimated by multiplying the range for polyethylene glycol by 10. A TSK-PW column can cover a much wider molecular weight range than the TSK-SW column. However, better separation can be achieved on the TSK-SW column within its limited separation range,[11] in which molecular weights of most proteins are included. Consequently, proteins are usually separated on a TSK-SW column, and the TSK-PW column is used for the separation of low-molecular-weight compounds such as oligosaccharides and in the measurement of molecular weight distributions of polymers such as polysaccharides. However, very high-molecular-weight proteins such as serum lipoproteins have been separated on a TSK-PW column.[12,13]

The TSK-PW support contains a small quantity of carboxyl groups. Therefore, the eluent should contain 0.1 to 0.3 *M* salt in order to eliminate any ionic interactions between proteins and support.

The maximum operating flow rate is 1.2 mℓ/min.

THE TSK-TOYOPEARL COLUMN

The TSK-Toyopearl was developed for preparative gel filtration of medium performance. The Toyopearl packing is also based on a hydrophilic organic copolymer and is available in five different grades based on pore size, as listed in Table 9. HW55 is most commonly used in the separation of proteins and has a separation range of 10,000 to 1,500,000 molecular weight for proteins and therefore can be applied to most proteins.[14] However, HW50 provides a slightly better separation for proteins of molecular weights less than several tens of thousands. In the molecular weight range less than several thousands, HW40 provides the best separation. TSK-Toyopearl is also available in three different particle sizes, superfine grade(S, 20 to 40 μm), fine grade (F, 30 to 60 μm), and coarse grade (C, 50 to 100 μm). The fine grade is most commonly used. The superfine grade is best in separations where high resolution is required, and the coarse grade is most convenient in industrial-scale separations.

TSK-Toyopearl can withstand pressures up to several atmospheres and can therefore be packed with a pump into columns easily and rapidly.[15-17] Rapid separations are also possible

Table 9
PROPERTIES OF THE TSK-TOYOPEARL COLUMNS

Grade	Molecular weight range for polyethylene glycol
HW40	— 3,000
HW50	— 18,000
HW55	200—150,000
HW65	10,000—1,000,000
HW75	100,000—5,000,000

due to its high mechanical stability. However, since the separation efficiency depends on the flow rate, separations are normally carried out in 1 to 3 hr. In addition, TSK-Toyopearl is chemically stable and can be used over the pH range of 1 to 14. Organic solvents and denaturing solvents such as 6 *M* guanidine hydrochloride, 8 *M* urea, and SDS can be used. Proteins are usually separated in buffers containing 0.1 to 0.3 *M* salt. The salt is necessary to suppress the ionic interaction between the proteins and carboxyl groups present in a small quantity on the surface of the support.

In general for most proteins better separations can be achieved on TSK-SW than on TSK-Toyopearl. The TSK-SW preparative column is superior in the separation of proteins for sample loading up to about 100 mg per injection.[18]

REFERENCES

1. **Pfannkoch, E., Lu, K. C., Regnier, F. E., and Barth, H. G.,** Characterization of some commercial high performance size-exclusion chromatography columns for water soluble polymers, *J. Chromatogr. Sci.*, 18, 430, 1980.
2. **Kato, Y., Komiya, K., Sasaki, H., and Hashimoto, T.,** Separation range and separation efficiency in high-speed gel filtration on TSK-GEL SW columns, *J. Chromatogr.*, 190, 297, 1980.
3. **Kakuno, T., Hiura, H., Umino, M., Ishikawa, O., Kato, Y., Yamashita, J., and Horio, T.,** High-speed gel filtration chromatography of proteins, in *Gel-Permeation Chromatography of Macromolecules*, Ohsawa, K. and Tanaka, Y., Eds., Kitami, Tokyo, 1980, 137.
4. **Imamura, T., Konishi, K., Yokoyama, M., and Konishi, K.,** High-speed gel filtration of polypeptides in some denaturants, *J. Liq. Chromatogr.*, 4, 613, 1981.
5. **Ui, N.,** Rapid estimation of the molecular weights of protein polypeptide chains using high-pressure liquid chromatography in 6 *M* guanidine hydrochloride, *Anal. Biochem.*, 97, 65, 1979.
6. **Ui, N.,** High-speed gel filtration of glycopolypeptides in 6 *M* guanidine hydrochloride, *J. Chromatogr.*, 215, 289, 1981.
7. **Imamura, T., Konishi, K., Yokoyama, M., and Konishi, K.,** High-speed gel filtration of polypeptides in sodium dodecyl sulfate, *J. Biochem.*, 86, 639, 1979.
8. **Kato, Y., Komiya, K., Sasaki, H., and Hashimoto, T.,** High-speed gel filtration of proteins in 6 *M* guanidine hydrochloride on TSK-GEL SW columns, *J. Chromatogr.*, 193, 458, 1980.
9. **Kato, Y., Komiya, K., Sasaki, H., and Hashimoto, T.,** High-speed gel filtration of proteins in sodium dodecyl sulfate aqueous solution on TSK-GEL SW type, *J. Chromatogr.*, 193, 29, 1980.
10. **Takagi, T., Takeda, K., and Okuno, T.,** Effect of salt concentration on the elution properties of complexes formed between sodium dodecylsulfate and protein polypeptides in high-performance silica gel chromatography, *J. Chromatogr.*, 208, 201, 1981.
11. **Kato, Y., Komiya, K., Sasaki, H., and Hashimoto, T.,** Comparison of TSK-GEL PW type and SW type in high-speed aqueous gel-permeation chromatography, *J. Chromatogr.*, 193, 311, 1980.

12. **Okazaki, M., Ohno, Y., and Hara, I.,** High-performance aqueous gel permeation chromatography of human serum lipoproteins, *J. Chromatogr.*, 221, 257, 1980.
13. **Okazaki, M., Shiraishi, K., Ohno, Y., and Hara, I.,** High-performance aqueous gel permeation chromatography of serum lipoproteins: selective detection of cholesterol by enzymatic reaction, *J. Chromatogr.*, 223, 285, 1981.
14. **Germershausen, J. and Karkas, J. D.,** Preparative high speed gel permeation chromatography of proteins on Toyopearl HW55F, *Biochem. Biophys. Res. Commun.*, 99, 1020, 1981.
15. **Kato, Y., Komiya, K., Iwaeda, T., Sasaki, H., and Hashimoto, T.,** Packing of Toyopearl column for gel filtration. I. Influence of packing velocity on column performance, *J. Chromatogr.*, 205, 185, 1981.
16. **Kato, Y., Komiya, K., Iwaeda, T., Sasaki, H., and Hashimoto, T.,** Packing of Toyopearl columns for gel filtration II. Dependence of optimal packing velocity on column size, *J. Chromatogr.*, 206, 135, 1981.
17. **Kato, Y., Komiya, K., Iwaeda, T., Sasaki, H., and Hashimoto, T.,** Packing of Toyopearl columns for gel filtration. III. Semi-constant pressure packing, *J. Chromatogr.*, 208, 71, 1981.
18. **Kato, Y., Komiya K., Sawada, Y., Sasaki, H., and Hashimoto, T.,** Purification of enzymes by high-speed gel filtration on TSK-GEL SW columns, *J. Chromatogr.*, 190, 305, 1980.

WATERS PROTEIN COLUMNS

J. K. Pollak and M. T. Campbell

INTRODUCTION

The use of high-pressure liquid chromatography (HPLC) for the isolation and separation of proteins provides a rapid and convenient method for protein purification and/or characterization. HPLC is particularly useful for the isolation of unstable proteins, since milligram quantities of proteins may be separated and isolated within minutes and hours rather than hours and days.

The recent advances in the rigid microparticulate HPLC packings and columns for aqueous exclusion chromatography suitable for the separation of proteins have been reviewed by Majors[1] and also by Barth.[2] Majors[1] lists the available packings and columns as provided by 53 different suppliers, while Barth[2] lists packings available by 20 different suppliers. Most of the columns suitable for the separation of proteins by aqueous eluants are silica based, though other packings have also been used, such as sulfonated polystyrene, hydroxylated polyesters, and hydroxylated polyethers.

THE PHYSICAL AND CHEMICAL PROPERTIES OF WATERS PROTEIN COLUMNS

At the time of writing three different Waters protein columns are commercially available, I-60, I-125, and I-250. The columns are made of stainless steel, are 300-mm long, and have an internal diameter of 7.8 mm. Larger columns with a diameter of 25 mm are available on demand. All Waters protein columns are packed with the same rigid, hydrophilic gel which is chemically modified to be effective within the specific limits of molecular weight, shown in Table 1. This gel is composed of silica particles 10 μm in diameter, which have a bonded diol phase:

$$\mathrm{Si{-}O{-}Si(CH_2)_3{-}OCH_2\underset{\displaystyle OH}{\underset{|}{CH}}{-}\underset{\displaystyle OH}{\underset{|}{CH_2}}}$$

On this gel filtration matrix, acidic, neutral, and mildly basic (pI = 8.5) proteins may be partitioned between the silica-bonded surface and the aqueous eluant. However the elution of proteins and polypeptides from such columns cannot be regarded as being solely dependent on gel filtration, because of the presence of unreacted anionic silanol groups on the gel surface. Attraction and repulsion between these silanol moieties and the ionizable sidechains of polypeptides and proteins will occur, as well as hydrogen bond formation between proteins and the particle surface. Therefore on Waters columns the separation of proteins is affected not only by their effective size in solution, but also by conditions such as the ionic strength of the eluting buffer, interaction with buffer salts, pH, and general hydrophobic interactions between the proteins and the stationary phase.

In the following sections a number of examples of protein separations will be described for the different column types. The advantages as well as the limitations of HPLC using Waters protein columns will become apparent from these examples.

Table 1
MOLECULAR WEIGHT RANGES OF WATERS HPLC COLUMNS

Column	Native globular protein (M_r)	Random coil configuration (M_r)
I—60	1,000—20,000	600—8,000
I—125	2,000—80,000	1,000—30,000
I—250	10,000—500,000	2,000—150,000

All columns are packed with particles having an average diameter of 10 μm.

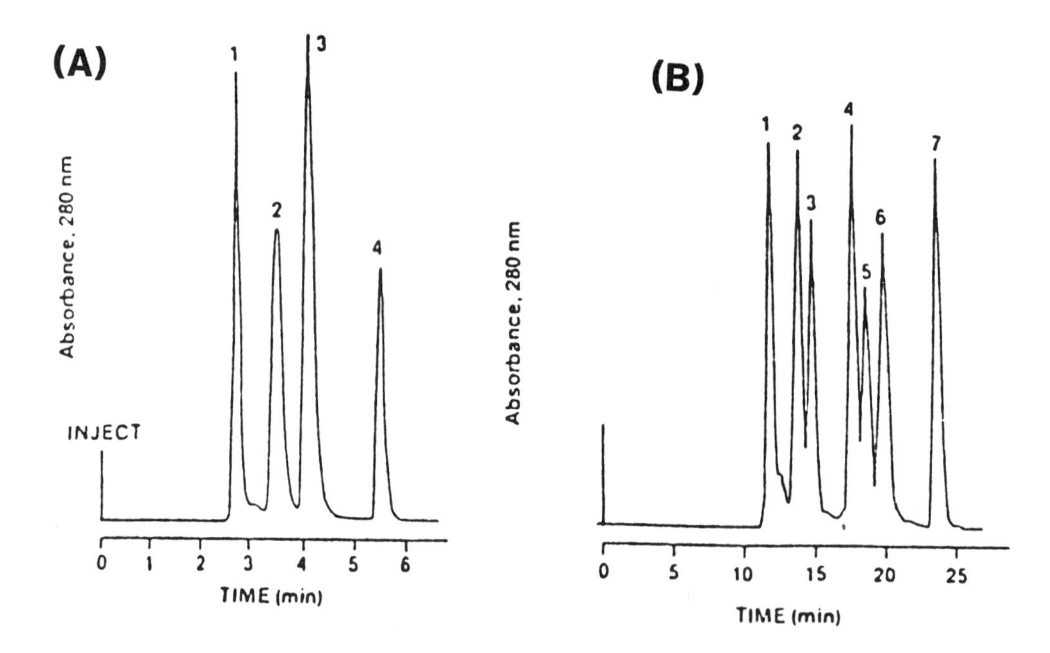

FIGURE 1. (A) Separation of proteins on an I-125 column. 1, Ferritin; 2, ovalbumin; 3, myoglobin; 4, guanosine. (B) Separation of proteins using a dual protein column I-125 kit. 1, Ferritin, 2 BSA; 3, ovalbumin; 4, myoglobin; 5, ribonuclease; 6, cytochrome *c*; 7, guanosine. (From Rittinghaus, K. and Franzen, K. H., *Fresenius Z. Anal. Chem.*, 301, 144, 1980. With permission.)

SELECTED EXAMPLES OF PROTEIN SEPARATIONS ON WATERS PROTEIN COLUMNS

The I-60 Column

This column has only recently been placed on the market and therefore no citations in the literature were cited with respect to the use of this column. The reviewers also have no expertise with this column.

The I-125 Column

An example of the kind of separation of proteins that can be obtained on an I-125 column is shown in Figure 1. Three proteins were separated with baseline resolution; the proteins were chosen so that one was completely excluded (ferritin, M_r = 540,000) and the other two had greatly differing molecular weights (ovalbumin, M_r = 45,000, myoglobin, M_r =

17,000).[3] The fourth peak was that of guanosine which was used as a total inclusion marker (Figure 1A).[3] When two I-125 columns were used in tandem, near baseline separation of six selected protein standards was obtained (Figure 1B).[3]

In a considerably more detailed study the separation of a mixture of five proteins (again selected for the even distribution of their M_r) was investigated.[4] This systematic study evaluated the effects of column length, buffer strength, as well as the dielectric constant of the mobile phase and the effects of the addition of detergents to the mobile phase.[4] Doubling the column length by running two columns in tandem resulted in a dramatic increase in resolution between bovine serum albumin (BSA) and ovalbumin. Further improvement of separation by the addition of a third and fourth column was only marginal and was not commensurate with the additional time taken for the separation of the proteins. Pfeifer et al.[4] suggest a flow rate of 1 mℓ/min; the present reviewers have used 0.05 *M* phosphate buffer pH 6.86 for elution and found that satisfactory separations of proteins could be obtained at a flow rate of 2 mℓ/min, even when only one column length (30 cm) was used. When comparing the elution of mixtures of ferritin, BSA, ovalbumin, trypsin inhibitor, and ribonuclease A in (4-(2-hydroxyethyl)-1-piperazineethanesulfonate) (Hepes) buffer, *Tris*-Na_2SO_4 or *Tris*-NaCl buffer, both the *Tris* buffers allowed elution of the proteins in order of their molecular weight. The addition of the neutral salt NaCl resulted in a better separation between trypsin inhibitor and ribonuclease A.[4] (However Cl^- salts are to be avoided as they corrode stainless steel columns; note added by the reviewers). When Hepes buffer was used all the proteins eluted together in one peak in the excluded volume. Increasing the ionic strength of phosphate buffer at pH 7.25 (0.01 *M* phosphate increasing to 0.5 *M* phosphate) resulted in a marked decrease in the distribution coefficient (K_d) of the two neutral and basic proteins (PI = 7.8 and 9.3), while a similar change in ionic strength increased the K_d of the acidic protein (PI = 4.5).[4]

$$K_d = (V_e - V_o)/(V_t - V_o) \quad (1)$$

where V_e = volume of solvent corresponding to peak concentration of the eluting protein; V_o = excluded volume, as determined by a high-molecular-weight marker; and V_t = total included volume, as determined by a low-molecular-weight marker.

Two I-125 columns in tandem have also been used to study the effect of ionic strength on the distribution coefficient (K_d) of trypsin inhibitor, ovalbumin, and BSA by adding Na_2SO_4 to phosphate buffer to increase the ionic strength; Na_2SO_4 concentration was varied from 0.01 to 0.5 *M*.[5] These authors also showed that immunoglobulin G could be separated by HPLC on a I-125 column into heavy and light chains by using a mobile phase consisting of 8% acetic acid and either 5 or 2.5% (v/v) 1-propanol. The first two peaks contained the heavy chains, while the third peak contained the light chains.[5]

In an earlier investigation, partially purified factor D (fraction of the human complement system) which could not be further purified by conventional gel chromatography, could be further purified by HPLC on an I-125 column. The mobile phase consisted of 20% acetic acid (v/v), eluted at the rate of 1 mℓ/min. Several peaks were obtained during elution of the partially purified component D.[6] The peak eluting between 13.5 to 14.5 min was shown by SDS-polyacrylamide gel electrophoresis (PAGE) to contain a single protein of M_r = 24,000 to 25,000. This protein was subsequently confirmed to be factor D, since it cleaved factor B in the presence of C3b and also caused the lysis of guinea pig erythrocytes in an agarose hemolytic diffusion plate assay.[6]

The present reviewers consider it essential that protein peaks obtained by HPLC be identified and confirmed by SDS-PAGE. This is an essential adjunct for the interpretation of the separations obtained by HPLC.

Nieman et al.[6] also purified serum factor Ba from factor Bb and other cleavage products, using again 20% acetic acid as the mobile phase. This time the flow rate was kept at 0.5 mℓ/min, Factor B having a M_r = 33,000 to 34,000 eluted between 29 to 31 min and was of sufficient purity to be used for amino acid sequence determinations.[6] At the time when this study was carried out the I-250 column was not available and therefore the Bb fragment with a M_r = 60,000 to 65,000 could not be resolved or further purified from the residual factor B, M_r = 90,000 to 100,000, since both proteins were excluded from the I-125 column.

Three I-125 columns in tandem were used for the separation of two enzymes involved in the metabolism of cyclic AMP. The two enzymes were cyclic AMP phosphodiesterase (EC 3.1.4.17, c-AMP diesterase) and ATP pyrophosphorylase (EC 3.6.1.8,ATP hydrolase) which were present in an extract of the slime mold *Dictyostelium discoideum*. The starting material was a 50 to 75% saturated ammonium sulfate fraction, which was dissolved in 40 m*M Tris*-acetate buffer, pH=7.3. The two enzymic activities were eluted using either 0.1 *M Tris* acetate buffer, pH=7.3 or *Tris*-sulfate buffer, pH=7.0 from either a 6% Bio-Gel® column (2.6 × 50 cm) or three I-125 columns run in tandem. The flow rate was 0.5 mℓ/min for the 6% Bio-Gel® column and 1 mℓ/min for the I-125 column. Two main protein peaks and three minor protein peaks were obtained on the I-125 column when monitored at 280 nm. Peak I contained 66% of the c-AMP diesterase activity, with no detectable ATP hydrolase activity, while Peak II contained 31% of the cAMP diesterase activity and 86% of the ATP hydrolase activity.[7]

The Use of the I-125 and I-250 Columns in Tandem

HPLC was used for the rapid purification of the microsomal cytochrome P-450 by Kohli and associates.[8] Under some circumstances it is difficult to remove traces of hemoglobin from the microsomal fraction: the presence of hemoglobin does not interfere with the determination of cytochrome P-450 by difference spectroscopy, but it does not allow the determination of cytochrome P-420, which is the denatured form of cytochrome P-450. Since cytochrome P-450 is a labile protein which is easily converted to cytochrome P-420 it is desirable to remove the last traces of hemoglobin so that both cytochrome P-450 and cytochrome P-420 may be quantitated.

When microsomes were solubilized and then applied onto either a Spherogel TSK-3000SW (0.75 × 60 cm, Altex) column or onto a tandem arrangement of an I-125 and an I-250 column, cytochrome P-450 and cytochrome P-420 were obtained free from hemoglobin.[8] The proteins were eluted with the solubilizing medium, which contained 100 m*M* phosphate (pH 7.4), 0.2% Lubrol WX, 0.5% sodium cholate, 0.1 m*M* ethylenediaminetetraacetic acid, 0.1 m*M* dithiothreitol, and 20% glycerol.[8] When monitored either at 280 nm or at 405 nm, two major and one minor peak were obtained (Figure 2).[8] The first major peak contained cytochrome P-450 and was rich in cytochrome P-420 (peak A), the second major peak contained most of the cytochrome P-450 and no cytochrome P-420 (peak C), while the minor peak contained all the hemoglobin (peak B). In this investigation[8] the authors have presented the results obtained with the Spherogel TSK column. However they indicate that the combined I-125 and I-250 columns produced similar results.[8] In an unpublished observation Pollak and Dalley have recently found that a single I-125 column conveniently separated the combined cytochromes P-450 and P-420 from hemoglobin, so that the difference spectra of both cytochromes could be determined without interference from hemoglobin. After applying the solubilized protein sample onto a single I-125 column, elution was effected with 0.05 *M* phosphate buffer, pH 6.85, containing 1% glycerol at a flow rate of 1 mℓ/min. Cytochrome P-450 and cytochrome P-420 eluted in the first major peak, while hemoglobin was retained and eluted later (Figure 3).

A rapid purification of cytochrome c_1 from the bovine cytochrome bc_1 complex or from

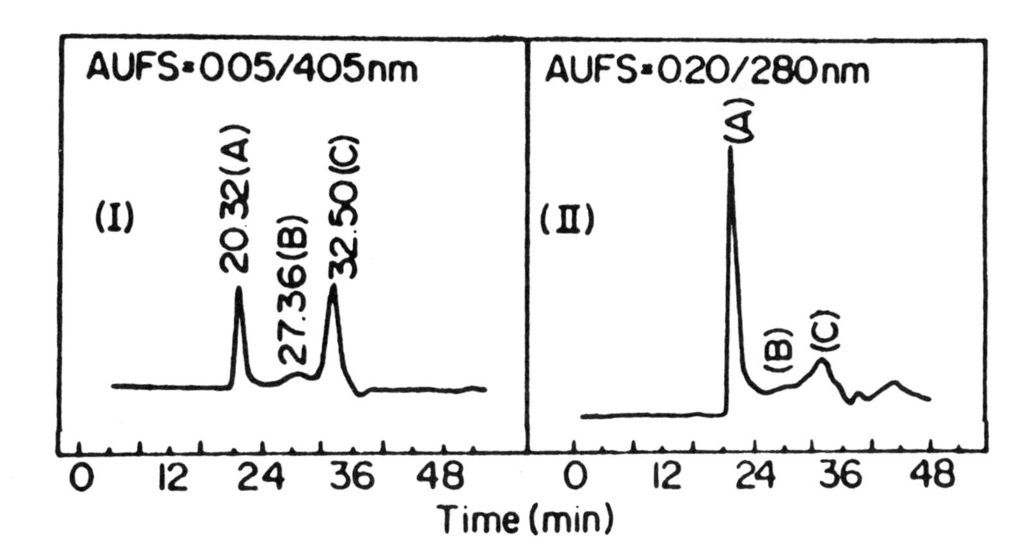

FIGURE 2. HPLC separation of cytochromes P-450 and P-420 from hemoglobin. Peak A is a mixture of cytochrome P-450 and P-420, Peak B is hemoglobin and Peak C is cytochrome P-450. Other details are given in text. (From Kohli, K. K., Hernandez, O., and McKinney, J. D., *J. Liq. Chromatogr.*, 5, 367, 1982. With permission.)

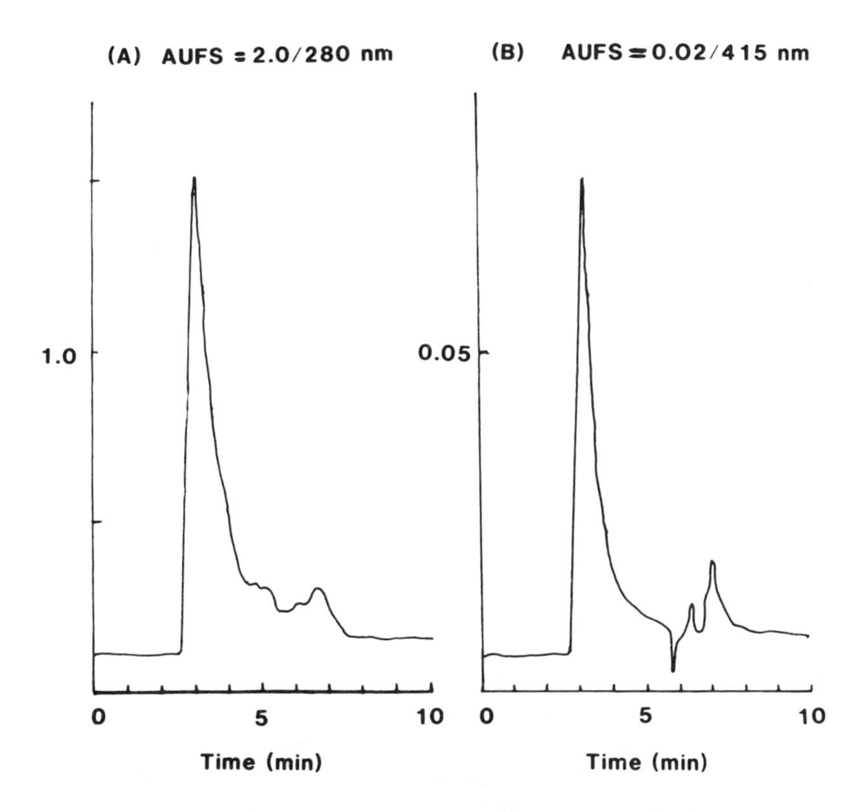

FIGURE 3. Purification of cytochromes P-450 and P-420 free from hemoglobin. HPLC was carried out with an ALC/GPC 204 Liquid Chromatograph (Waters Associates, Milford, Ma.), using a Model 440 wavelength detector with a 280-nm filter and a model 450 variable wavelength detector at 415 nm. A single I-125 column was used, and was kept at a temperature of 5 to 6°C. The elution buffer was 0.05 *M* sodium phosphate pH 6.85, containing 1% glycerol. A flow rate of 2.0 mℓ/min was used, 200 μℓ of solubilized microsomes being injected for each run.

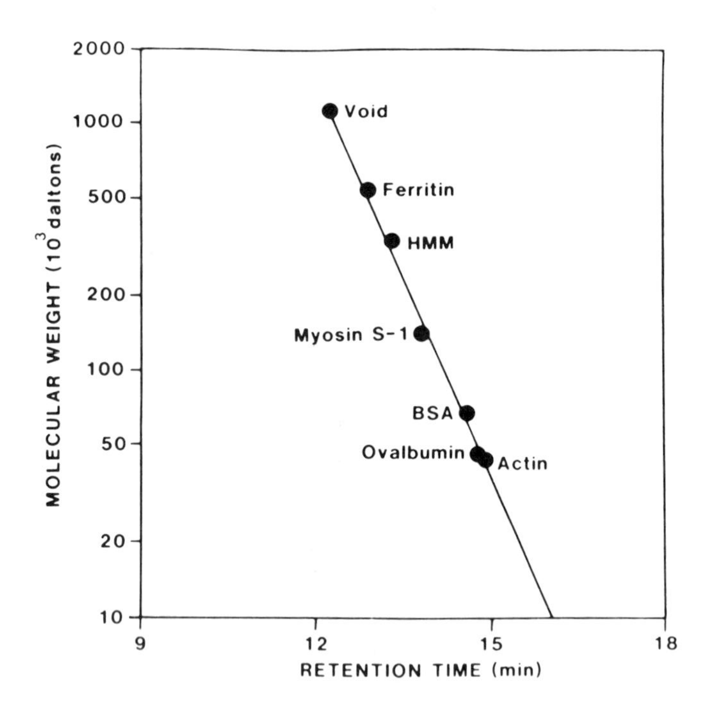

FIGURE 4. A standard curve (linear regression) relating the known molecular weights of six proteins to their separate retention times on tandem HPLC columns (Waters Associates I-250 + I-125). The filled circles represent: (1) the void volumn of the HPLC columns which was determined from the elution time of F-actin and which closely agreed with the manufacturers estimates for the size of excluded molecules, 1,130,000 daltons; (2) ferritin, 540,000 daltons; (3) heavy meromyosin, 350,000 daltons; (4) myosin subfragment-1, 130,000 daltons; (5) BSA, 68,000 daltons; (6) ovalbumin, 45,000 daltons; (7) G-actin, 42,500 daltons; The mobile phase consisted of 20 m*M* sodium acetate at pH 6.5; 0.5 m*M* DDT; and 0.22 *M* glycerol. (From Barden, J. A., Grant, N. J., and dos Remedios, C., *Biochem. Int.*, 5, 685, 1982. With permission.)

partially purified cytochrome c_1 has been described, using an I-300 column (not commercially available) followed by an I-125 column.[9] The sample was applied in 5% sodium deoxycholate and eluted with 0.02 *M Tris*-C1 buffer, pH 8.1, containing 0.09 *M* NaCl, 0.1 m*M* ethylenediaminetetraacetic acid and 1% deoxycholate. However the use of alkaline buffers is not recommended since they tend to remove the bonded diol phase and therefore shorten the lifetime of the column (note added by the reviewers). The isolation of cytochrome c_1 by HPLC took only 20 min compared to 10 hr on a Sephadex® G-200 column, though the cytochrome c_1 isolated by HPLC contained somewhat more impurities than the cytochrome c_1 obtained by conventional Sephadex® column chromatography.[9]

When the tandem arrangement of the I-250 — I-125 columns was eluted with 20 m*M* sodium acetate, pH 6.5, containing 0.5 m*M* dithiothreitol, polypeptides ranging from 42,500 to 1,130,000 daltons showed a linear relationship between retention time and the log of the molecular weight (Figure 4). At a flow rate of 1 mℓ/min the complete separation was obtained within 15 min.[10] This linear relationship between molecular weight and retention time does not depend on the shape of the protein molecules, since proteins which differed greatly in their axial ratios would elute with retention times according to their molecular weights and not on the basis of their Stokes' radii.

The use of an I-250 column followed by an I-125 column enabled Barden and associates to evaluate the extent of polymerization of muscle actin.[10] Using HPLC these workers established that the nucleus for polymerization is a trimer.[10] The advantage of HPLC as compared to conventional column chromatography is found in the speed of the separation, since polymers formed during the early stages of F-actin formation may be separated and defined. Using sodium acetate buffer at 4°C and thermostatting the tandem I-250 — I-125 columns also at 4°C, these authors established that in the absence of KCl, 85% of the protein was present as the monomer.[10] They suggest that the remainder of the proteins are a mixture of poorly resolved oligomers, which may have been formed as they passed through the guard column which remained at room temperature (approximately 23°C) and had a void volume of 0.26 mℓ.[10] The authors state that removal of the guard column restricts polymerization.[13] It should be pointed out that for routine use it is not advisable to remove the guard column (note added by the reviewers).

The I-250 Column

The considerable advantage for the separation or isolation of proteins by HPLC over conventional chromatography is exemplified by the isolation of the enzyme carbamoyl phosphate synthase I (EC. 2.7.2.5), which is extremely prone to proteolytic degradation.[11,12] Carbamoyl phosphate synthase (CPS) catalyses the first step in urea synthesis and is localized in the matrix of mitochondria of ureotelic animals. It is a polypeptide with a M_r 165,000, but even when purified, it degrades rapidly to an enzymically inactive form with a M_r = 155,000.[12] This enzyme has been successfully purified by HPLC using a single I-250 column eluted with 0.05 *M* sodium phosphate buffer, pH 6.85 or 0.1 *M* sodium acetate buffer, pH 7.15 at a flow rate of either 1 mℓ/min or 2 mℓ/min.[13] The column was jacketed so that the buffer and the column were maintained at a temperature of 5 to 6°C. The column was monitored at 280 nm and the eluate was collected in 1-mℓ fractions which were assayed for CPS activity. The fraction that rechromatographed with aldolase (EC 4.1.2.13) which was used as a molecular weight marker, M_r = 150,000 showed the highest specific activity of CPS.[13] Subsequent electrophoresis of purified CPS on SDS-polyacrylamide indicated that more than 88% of the Coomassie Blue® stain was associated with one slow moving band (M_r = 165,000).[13]

In a subsequent investigation the precursor of CPS which is synthesized on cytoplasmic ribosomes was isolated by HPLC on a single I-250 column and then used for studies following the import of the CPS-precursor into the mitochondrial matrix.[14]

The methods used for the purification of the CPS precursor on the I-250 column were essentially the same as those used for the purification and isolation of CPS.[13,14]

Orosomucoid, a glycoprotein from human plasma, which had been purified by ion-exchange chromatography to give one homogeneous peak, was shown by HPLC on an I-250 column to contain several high- and low-molecular-weight protein contaminants. The mobile phase used was 0.1 *M Tris* pH 7.1, containing 0.05 *M* Na_2SO_4, at a flow rate of 1 mℓ/min; good resolution of orosomucoid was achieved within 9 min.[5] These authors also showed that glycoproteins obtained from submaxillary mucin by ion-exchange chromatography were resolvable into several well-defined peaks by HPLC on two I-250 columns run in tandem.[5]

SUMMARY

HPLC columns, such as the series of the three Waters columns, have only recently become commercially available and hence have not made their full impact on the isolation and purification of proteins. It is therefore difficult to make a critical appraisal of the literature, since the technique is still in its infancy. At the present state of the art, the limited number of buffer ions that are of use for the separation of proteins due to the presence of silanol groups in the gel filtration columns, as well as the fact that alkaline pHs severely diminish their lifespan, restrict the wider use of the Waters columns. Varying the ionic strength of the elution buffer was originally thought to be a powerful tool to enhance the separation of proteins of similar molecular weight with different charges by changing their K_ds. However this did not prove to be the case, since the changes in K_d of specific proteins were not sufficiently large to effect significant improvements in their separation and purification.

It is however obvious that HPLC provides the means to isolate proteins rapidly, and therefore this technique will become the method of choice whenever speed is essential for the isolation of active enzymes and undenatured proteins.

ADDENDUM

Since this review went to press, the Waters Protein Pak 250 has been discontinued. This column has been replaced by Protein Pak 300, which is equivalent to the Spherogel TSK 3000 SW column.

REFERENCES

1. **Majors, R. E.,** Recent advances in HPLC packings and columns, *J. Chromatogr. Sci.,* 18, 488, 1980.
2. **Barth, H. G.,** A practical approach to steric exclusion chromatography of water-soluble polymers, *J. Chromatogr. Sci.,* 18, 409, 1980.
3. **Rittinghaus, K. and Franzen, K.-H.,** HPLC in protein analysis: An alternative to gel-filtration and gel-electrophoresis, *Fresenius Z. Anal. Chem.,* 301, 144, 1980.
4. **Pfeifer, R., Skea, W., Waraska, J., Cohen, C., and Burnworth, L.,** Protein purification and analysis by HPLC, in *Biological and Biomedical Applications of Liquid Chromatography,* Vol. 4., Hawke, G., Ed. Marcel Dekker, New York, 1982, 43.
5. **Rittinghaus, K. and Franzen, K.-H.,** Separation of some biologically significant proteins (10 to 500 K_D) by molecular exclusion high performance liquid chromatography, in *High Performance Liquid Chromatography in Protein and Peptide Chemistry,* Lottspeich, F., Henschen, A., and Hupe, K. P., Eds., Walter de Gruyter, Berlin, 1981, 83.
6. **Nieman, M. A., Hollaway, W. L., and Mole, J. E.,** Purification of some biologically significant proteins by molecular exclusion high pressure liquid chromatography, *J. High Resol. Chromatogr. Chromatogr. Commun.,* 2, 743, 1979.
7. **Hodge, J. and Rossomondo, E. F.,** Separation of enzymic activities in *Dictyostelium discoideum* by high-pressure gel permeation chromatography, *Anal. Biochem.,* 100, 174, 1979.
8. **Kohli, K. K., Hernandez, O., and McKinney, J. D.,** Fractionation by high performance liquid chromatography of microsomal cytochrome P-450 induced by hexachlorobiphenyl isomers, *J. Liq. Chromatogr.,* 5, 367, 1982.
9. **Robinson, N. C. and Talbert, L.,** Isolation of bovine cytochrome c_1 as a single non-denatured subunit using gel filtration or high pressure liquid chromatography in deoxycholate, *Biochem. Biophys. Res. Commun.,* 95, 90, 1980.
10. **Barden, J. A., Grant, N. J., and dos Remedios, C.,** Identification of the nucleus of actin polymerization, *Biochem. Int.,* 5, 685, 1982.
11. **Guthohrlein, G. and Knappe, J.,** Structure and function of carbamoyl phosphate synthatase, *Eur. J. Biochem.,* 7, 119, 1968.
12. **Clarke, S.,** A major polypeptide component of rat liver mitochondria: carbamoyl phosphate synthatase, *J. Biol. Chem.,* 251, 950, 1976.
13. **Pollak, J. K, and Campbell, M. T.,** Rapid purification of the unstable enzyme carbamoyl phosphate synthatase by high pressure liquid chromatography, *J. Liq. Chromatogr.,* 4, 629, 1981.
14. **Campbell, M. T., Sutton, R., and Pollak, J. K.,** The import of carbamoyl-phosphate synthase into mitochondria from foetal rat liver, *Eur. J. Biochem.,* 125, 401, 1982.

DRUG-PROTEIN BINDING DETERMINATION BY CHROMATOGRAPHY

Bernard Sebille and Nicole Thuaud

INTRODUCTION

The problem of drug-protein binding, which is of importance in the field of pharmacology and therapeutics, has been the subject of numerous attempts to determine the binding capacity, the number of sites, and the affinity constants for several types of protein. Due to their abundance in blood plasma, human serum albumin (HSA) and bovine serum albumin (BSA) can serve as transport protein and for exogeneous or endogeneous ligands and have been largely studied from this point of view. More generally, the problem of ligand protein binding is of importance in the field of biochemistry and enzymology where determination of the reaction steps for the enzyme substrate or effector affinity is sought.

Two distinct techniques have been used. The first is based on spectroscopic methods in which observations are made on the whole mixture whereas the second uses separation methods: dialysis, ultrafiltration, ultracentrifugation, electrophoresis, and chromatography.

This latter technique has given rise to numerous results since the introduction of gel filtration by Lathe and Ruthven[1] and Porath and Flodin.[2] This topic has already been discussed by Wood and Cooper[3] who used soft gels, the only ones available at that time.

The tremendous development of high-performance liquid chromatography (HPLC) suggests a way of applying this very accurate and fast technique to the study of ligand protein interactions. Moreover, significant progress has been made in the preparation of size-exclusion columns compatible with proteins since the pioneering work of Regnier and Noel[4] and that of Engelhardt and Mathes[5] and Becker and Unger.[6]

It is the purpose of the present review to describe and compare the various ways in which chromatographic techniques have been used in protein-binding studies. Two distinct categories of experimental analysis may be distinguished according to whether it does or does not require the grafting of one of the equilibrium components on to the stationary phase. For a more comprehensive approach we have preferred to begin with a description of the most typical experiments using soft gels and then to discuss the recent works performed with rigid beads and HPLC equipment.

ZONAL CHROMATOGRAPHY

Zonal chromatography has been widely used in the measurements of protein binding. It has been applied to several studies concerning drug or hormone-protein interactions in the field of clinical chemistry. Wood and Cooper[3] reviewed this subject in 1970. At this time, the process used was gel filtration on soft gels (Sephadex® for example).

A small volume of the sample to be analyzed (mixture of protein and ligand) is introduced at the top of the column which is then eluted by an aqueous buffered solution. The complex is separated from the excess ligand by means of the size exclusion effect and emerges first at the bottom of the column (Figure 1). Quantitative evaluation of the complex or free ligand concentration makes it possible to calculate the affinity constant. Validity of the results depends on interaction kinetics since the complex can be dissociated during its passage through the column if the speed of dissociation is too high. Theoretical aspects of the process have been analyzed by Nimmo and Bauermeister[7] and Melenevski et al.[8] In Nimmo's semiquantitative model, axial diffusion is considered as negligible and the calculation leads to theoretical chromatographic profiles with similar curves to experimental ones.[9]

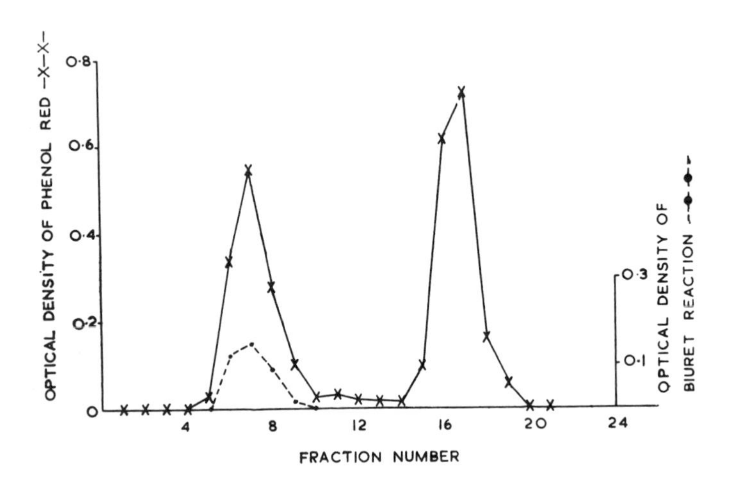

FIGURE 1. Zonal gel filtration chromatography pattern. Column: 1.1 × 12.5 cm Sephadex® G-25. Sample: 0.1 mℓ of a mixture of HSA and Phenol Red. Eluent: distilled water. (From Wood, G. C. and Cooper; P. F., *Chromatogr. Rev.*, 12, 88, 1970. With permission.)

Different approaches to this problem have been considered. By assuming a rapid equilibrium in comparison with the process of gel filtration, Dixon[10] studied the removal of a bound ligand from a macromolecule by gel filtration. This author established a very simple equation linking protein saturation, concentration of the complex, eluting volume, and the affinity constants of the equilibrium. This calculation was analyzed by Zeeberg,[11,12] who proposed a more complete derivation taking into account sample dilution during gel filtration and assuming rapid equilibration of protein and ligand. The determination of the dissociation constant for a tubuline-GDP complex was obtained by this method[11] and was in good agreement with previously determined values.

Zonal elution seems to be convenient when there is a moderate dilution of the eluted sample during chromatography and if the protein-ligand complex dissociation is slower than the gel filtration process.

HUMMEL AND DREYER'S METHOD

One of the most popular methods of studying macromolecule-ligand binding was described by Hummel and Dreyer.[13] in 1962. In this technique, the eluent contains the dissolved ligand. A small volume of protein is injected onto the column, which separates the different species by a size-exclusion mechanism. The chromatogram profile shows a leading peak corresponding to the ligated protein, followed by a trough emerging at the elution volume of the ligand as depicted in Figure 2. The area of the trough depends directly on the amount of bound ligand. In this process, the protein is always in equilibrium with free ligand at a constant concentration level which is that of the eluent solution. The degree of protein saturation is only dependent on the free ligand concentration level. More precise measurements can be obtained by varying the ligand concentration in the injected sample. As pointed out by Hummel and Dreyer, when the concentration of the free ligand in the sample solution is equal to the concentration of ligand in the equilibrium solution, the trough disappears from the chromatogram. Thus, by injecting several sample solutions with different ligand concentrations, it is easy to determine, by interpolation, the mole excess of ligand required to make the trough vanish. This excess is the exact amount of ligand bound to the protein.

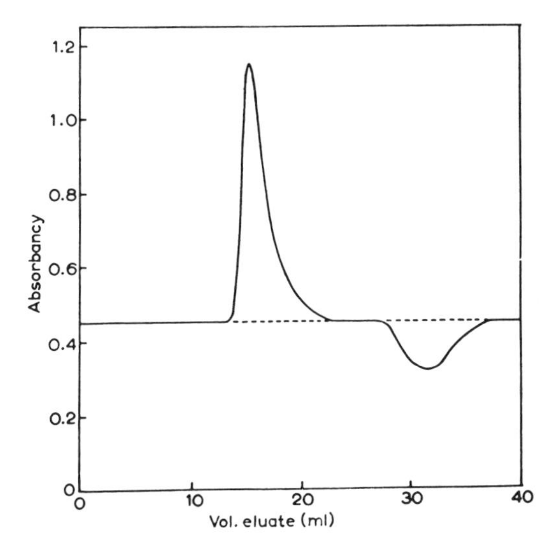

FIGURE 2. Hummel and Dreyer method elution profile. Column: 0.4 × 100 cm, Sephadex® G-25 column sample: 2 mg of pancreatic RNAase in 0.25 mℓ of eluent solution. Eluent: a 9.10^{-5} *M* solution of 2-cytidylic acid in 0.1 *M* acetate at pH 5.3. Monitoring absorbance: 285 nm. (From Hummel, J. P. and Dreyer, W. J., *Biochim. Biophys. Acta,* 63, 530, 1962. With permission.)

This method has been largely applied using soft gels like Sephadex® gels. In the field of drug-protein binding, the binding of sulfonamide,[14] peptides,[15] warfarin and other drugs,[16] furosemide,[17] diazepam,[18] with different kinds of protein, can be cited as examples.

It should be noticed that binding measurements are made possible by means of an appropriate evaluation of the ligand included in the positive protein peak. This has been done with radioactive-labeled ligand such as dodecylsulfate, whose binding to BSA was measured by Allen.[19] The binding of furosemide to human serum protein has also been measured by this method.[17]

Hummel and Dreyer's method is based on the assumption that protein ligand interaction and gel filtration equilibration are fast enough to be completely realized during the sample's passage through the column. This is generally true if the association-dissociation kinetics of the complex is high.

This method offers the major advantage of requiring only a small amount of protein. The determination of all the binding parameters requires several experiments with different ligand concentrations in the eluent. In the case of a rare ligand, for example radioactively labeled material, the cost of the experiment may pose a problem.

A limitation of Hummel and Dreyer's method is observed when systems with low binding properties are used. In this case concentrated ligand solutions are needed and detection of the trough in the chromatogram is difficult due to low sensitivity of the detector when saturation is approached especially with photometric detectors.

Hummel and Dreyer's technique is useful in the study of the simultaneous binding of two (or more) ligands on the same protein in one experiment. In this way, Fairclough and Fruton[15] have measured the competition of acetyl-L-tryptophan and acetyl-L-tryptophanamide on the BSA using Sephadex® gels.

One of the criticisms of Hummel and Dreyer's method is the fact that the protein concentration is not constant during the chromatographic process due to spreading and dilution of the sample.[20] Multiple equilibria theory[21] concludes that the binding ratio is independent of polymer concentration. However, an inverse dependance of several ligand-binding constants upon albumin concentration reflects the existence of a protein self-association, either preexistent or ligand induced.[22,23] In order to overcome this problem, Brumbaugh and Ackers[24] suggested injecting high-volume samples into the column, not to obtain peaks, but rather plateaus with a constant and known protein concentration. The measurements were obtained from the gel absorbance (see below) and made it possible to determine the methyl orange-BSA binding isotherm.

The use of HPLC is very useful in the study of drug-protein binding. Hummel and Dreyer's method can be applied if the column gives no protein adsorption. This is now the case with TSK SW (Toyo Soda), Lichrosorb® diol (Merck), or Synchropak (Synchrom) columns. We have measured[25] the parameters of warfarin-HSA binding with short (10 to 15 cm) Lichrosorb® diol columns (10-μm bead diameter, 100-Å pore diameter). In this case, the separation mechanism was based on exclusion of the protein and partition of the drug between mobile and stationary phases. If the drug is strongly but reversibly retained by the chromatographic support, very short columns can be used for the binding measurement. This new HPLC application has been used to study the competition of warfarin and furosemide on HSA[25] and to determine the affinity of human immunoglobulins to estradiol derivatives.[26]

The concentration dependance of affinity constants is small in a low concentration range and depends on the drug under study. For moderate dilution of the protein solution, as is the case with short chromatographic columns and especially with HPLC equipment, the results of Hummel and Dreyer's method obtained with small volume samples can be considered as valid.[27]

Owing to the speed and accuracy of measurements with HPLC, the study of drug-protein binding with Hummel and Dreyer's method is now easier and can be applied to numerous cases of ligand-macromolecule interaction.

FRONTAL ANALYSIS AND SATURATION METHODS

The use of frontal chromatographic analysis to study the interactions between dissimilar molecules was first introduced by Nichol and Winzor[28] with the lysozyme-ovalbumin association. This general method was then applied to the measurement of the binding of small molecules (sulfonamides) to proteins by Cooper and Wood[29] with Sephadex® gels.

Figure 3 shows the theoretical curve obtained by injecting a large volume of a protein A-ligand B mixture interacting by the reaction $A + B \leftrightarrow C$ (in which equilibrium is rapidly obtained compared with the rate of migration of solutes). The curve reveals three plateaus whose fronts and heights depend on association. As the separation is effected by a size-exclusion mechanism, the first plateau corresponds to the free protein, the second to the complex in equilibrium with the dissociated components, and the third to the free ligand. The height of this last plateau gives the free ligand concentration. During elution, equilibrium conditions are maintained, since the protein is always in contact with unbound drug at the concentration present in the sample. The association parameters can be determined from a Scatchard plot[30] obtained by repeating experiments with a varying protein-ligand composition.

Several protein-drug bindings have been studied using this technique. For example warfarin,[31] salycilate[32] on HSA, and cortisol on whole serum.[33]

Columns containing rigid beads have been used in the same way by Morris and Brown[34] to measure methyl orange-binding to albumin. The column (20-cm length, 4-mm i.d.) is filled with hydrophilic modified glass beads (37 μm diameter, 40-Å pore diameter) and medium pressure (150 lb/in = 1.03 MPa at 2.7 mℓ/mn^{-1} flow rate). A complete chromatogram was obtained in 5 min.

Sebille et al.[35] have studied warfarin-HSA with a μBondagel® column E-125 (Waters

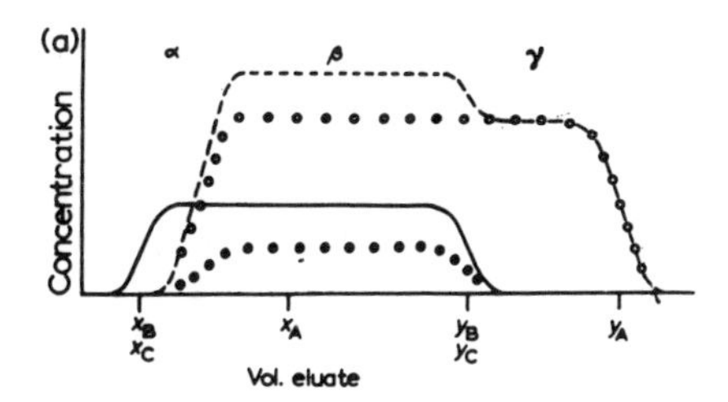

FIGURE 3. Frontal analysis of protein-ligand mixtures: diagrammatic gel filtration chromatograms for a rapidly reversible equilibrium A + B ↔ C. Elution volumes $V_B = V_c < V_A$. —, B + C ≡ total protein; — — — —, [A] + [C] ≡ total ligand (free and bound): O—O, [A] ≡ free ligand; ●—●, [C] ≡ ligand-protein complex. x_A, x_B, x_C denote the elution volumes of pure components. (From Wood, G.C. and Cooper, P. F., *Chromatogr. Rev.*, 12, 88, 1970. With permission.)

Associates) under higher pressure. They noticed some adsorption of ligand and protein to the stationary phase, which reduces accuracy of the assay and the life of the column. Recently, fair chromatograms with good behavior of diazepam and human serum albumin have been observed on a Lichrosorb® diol, column even at a 10 g/ℓ protein concentration.[36]

The application of frontal analysis to binding determination is based on the independence of the observed retention of each component on the support. In other words, the migration rate of each species must be independent of the presence of others. With relatively highly concentrated solutions, a shrinkage of soft gels due to osmotic pressure complicates the analysis of association equilibria because under these conditions the partition coefficients of each compound cannot be considered invariant. Baghurst et al.[37] have rigorously treated this point for a single solute. The problem of osmotic shrinkage may be overcome by choosing rigid beads such as modified glass. It remains, however, necessary to consider the self-interaction of a given solute and possible interaction with a ligand which can induce non-ideality of the mixture and modify penetration into the bead pores. This point was not considered with low ligand concentration when noncompressible beads were used.[35,36]

Another saturation method was introduced by Brumbaugh and Ackers.[24] This method is based on the direct optical scanning of a small chromatographic size-exclusion column which has been saturated with a solution containing a mixture of ligand and macromolecule. As the polymer is excluded from the gel, and remains in the mobile phase, ligand penetration into the gel depends on its affinity for the polymer, and its partition is modified.

Measurement of the optical density of a saturated column zone provides a means of evaluating the extent of binding. This method must take into account ligand fixation, which could occur on the beads due to adsorption phenomena in the absence of polymer. In this case corrections must be made.

The use of this technique in Hummel and Dreyer's method, in which large protein samples are injected, has been mentioned above. Optical density scanning of the different column zones reveals a large trough with a plateau making it possible to assay the bound ligand. The binding of methyl orange to BSA has been extensively studied using these saturation methods.[24]

Improvements in detection was proposed by Ackers et al.[38] They developed a computer-controlled single photon counting spectrophotometer which enhances accuracy in the measurement of a large range of absorbance. This instrument measures the optical densities obtained from small flow cells packed with gel particles.

To avoid the need for a sophisticated monitoring device when scanning, Ford and Winzor[39] have recently described a recycling partition equilibrium technique. The apparatus consists of column filled with gel and connected to a recycle system maintained by a peristaltic pump. The whole system is saturated with a protein-ligand mixture. Optical scanning of the mobile phase permits measurement of the partition coefficient of free ligand and the extent of ligand binding. Successive injections of ligand solutions provide a complete set of results needed to draw a Klotz[21] plot. The high speed attainment of equilibrium allows a moderate experimental time.

We have recently developed[40] the use of a standard HPLC apparatus for drug-binding studies by using the equilibrium saturation method. In a typical experiment a size-exclusion chromatographic support (Lichrosorb® diol from Merck) is eluted with a buffered solution of ligand (warfarin) and macromolecule (HSA). The eluting solution is monitored at the outlet of the column by a photometric detector. The injection of a few microliters of buffered water into the column leads to the appearance of two negative peaks on the chromatogram. The first peak emerges near the void volume of the column and expresses the deficiency of protein and bound ligand caused by the injection of the solvent alone. The second peak, due to the deficiency of the free ligand, provides a means of determining the free ligand concentration from an evaluation of the peak area.

As in Hummel and Dreyer's method, free ligand calibration is obtained by injecting ligand samples with increasing, known concentrations. This calibration is necessary since the response of the detector is not strictly linear over a wide range of absorbances. As the eluent solution contains both macromolecules and ligand, it is generally of a high optical density. The calibration curve obtained from this solution provides greater precision for the free ligand assay. An advantage of this calibration method is that it gives the value of free drug concentration regardless of drug adsorption to the support and any nonideal properties of the mixture.

This method has been applied to the study of the influence of free fatty acids (FFA) on the warfarin-HSA binding.[40] A depicted in Figure 4, increasing concentration of palmitic acid leads to a fall in the free warfarin peak, indicating a binding enhancement. The inverse is observed with octanoic acid. As in the other saturation methods, this technique is of interest when the solubility of the studied compounds is low in the absence of protein.

A similar way to show the influence of one ligand on the protein binding of a second one was introduced with the study of the influence of sodium dodecyl sulfate (SDS) on warfarin-HSA binding.[41] In this experiment (Figure 5) the injection of a small volume of a SDS solution in a column eluted by a buffered HSA-warfarin mixture gives a chromatogram in which a positive peak appears at the SDS retention volume, indicating binding enhancement, while a negative peak appears at the warfarin retention volume due to the deficit of this latter ligand. The reverse was observed with octanoic acid.[42]

All these saturation methods have the advantage of maintaining the constituent concentrations at a constant level and allows measurements under experimental conditions which can be close to physiological ones.

AFFINITY CHROMATOGRAPHY

The use of affinity chromatography has been largely developed for the isolation of natural macromolecules having a biological role. This technique has also been suggested for the measurement of the interaction between enzymes and substrates or effectors.[43] In the field of drug-protein binding, different techniques make it possible to measure the binding parameters. The grafting on to the chromatographic support of the macromolecule (or the

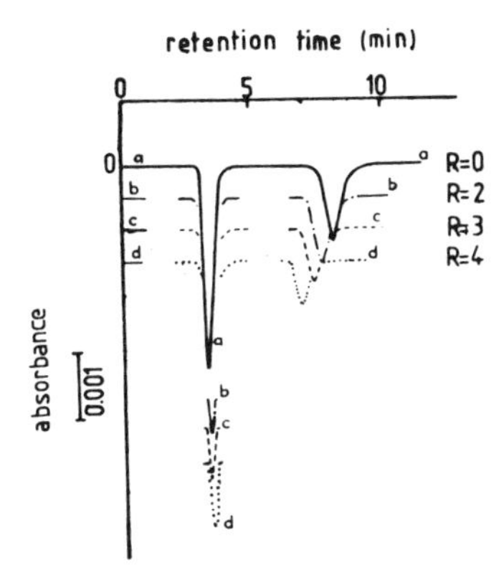

FIGURE 4. HPLC equilibrium saturation chromatograms for warfarin-HSA in the presence of various amounts of palmitic acid. Column: a 4.7-mm i.d. × 15-cm column filled with 10 μm Lichrosorb® diol support. Eluents: 2.5 *M* warfarin, 5.8 *M* serum albumin, palmitic acid solutions in pH 7.4, 0.067 *M* phosphate buffer (R is the molar ratio palmitic acid/serum albumin). Samples: 25 μℓ of pH 7.4 phosphate buffer. (From Sebille, B., Thuaud, N., and Tillement, J. P., *J. Chromatogr.*, 180, 103, 1979. With permission.)

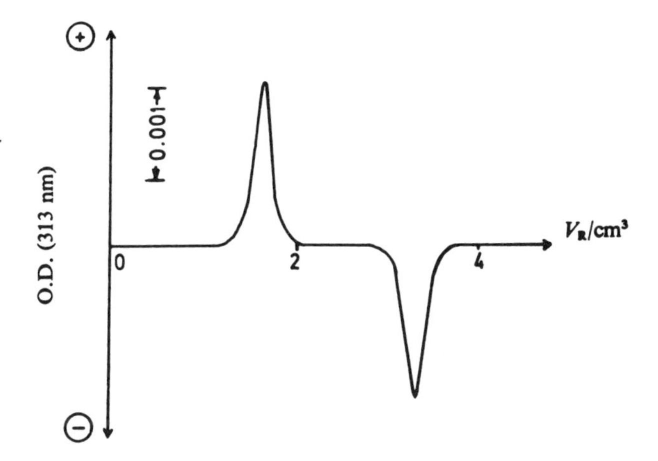

FIGURE 5. Qualitative evaluation of SDS influence on warfarin-serum albumin binding. Column: the same as Figure 4. Eluent: a 5 μ*M* warfarin, 5.8 μ*M* serum albumin solution in 0.067 *M*, pH 7.4 phosphate buffer. Sample: 10 μℓ of a 2.10^{-4} *M* SDS solution in the eluent. (From Sebille, B., Thuaud, N., and Tillement, J. P., *Faraday Discuss. R. Soc. Chlem. Faraday Symp.*, 15, 133, 1980. With permission.)

ligand), or the use of a mobile phase containing the protein, leads to a change in the retention of volume of one of the drugs under study.

Lagercrantz et al.[44] have used columns of serum albumin immobilized by coupling to agarose to measure the fixation of several natural or exogeneous ligands: drugs, fatty acids, and steroids.

The results are obtained from a mathematical expression which derives from those of Nichol et al.[45] and shows a simple relationship between the retention volume of the ligand and binding parameters:

$$V_A - V_A^0 = V_S |X| \sum_{i=1}^{i=m} \frac{n_i k_i}{1 + k_i |A|}$$

where V_A is the ligand elution volume on the affinity column, V_A^0 is the ligand elution volume on unmodified column, V_S is the volume of stationary gel, $|X|$ is the protein concentration on the gel, n_i is the number of binding sites in class i, m is the number of binding classes, and $|A|$ is the concentration of free ligand.

The injection of a very small amount of ligand (radioactively labeled) permits the assumption that $k_i|A| \ll 1$ so that the above equation is simplified to

$$V_A - V_A^0 = V_S |X| \sum_{i=1}^{i=m} n_i k_i$$

In this way, the first association constant k_1 of drugs and steroids to albumin has been measured.[44] Due to the influence of the binding of a component to its retention volume, affinity chromatography makes it possible to separate drugs which have different affinities for a given protein. For example, Lagercrantz et al.,[46] have obtained the complete separation of R and S warfarin by injecting the racemic mixture into a gel filtration column modified by BSA attachment. A complete study dealing with the binding of warfarin or tryptophan to serum albumin from different species demonstrates the selective stereo binding of these ligands.

In order to investigate the validity of frontal affinity chromatography, the same experiments have been performed by injecting continuously different ligand solutions onto albumin grafted columns.[46] In this case, the use of the above equations makes it possible to determine the binding ratio $\bar{r}$ between bound ligand and albumin, since:

$$\bar{r} = \frac{V_A - V_A^0 |A|}{V_S |X|}$$

Then a Scatchard plot[30] can be drawn as usual and gives the binding parameters. Salicylic acid -BSA binding was studied in this way, both by Lagercrantz et al.[44] and Nakano et al.[47] whose results were in good agreement. The mutual displacement interactions in the binding of the two drugs to HSA was studied by Nakano et al.[47] using frontal affinity chromatography. The measurement requires the use of solutions of mixed drugs or the continuous injection of a solution of one drug followed by the injection of the solution of a second one. By this process, it was shown in the case of sulfamethizole and salicyclic acid drugs that the mutual displacement interaction is attributable to the competitive binding of the drugs to the same primary site. The study of the binding capacities of HSA monomer or dimer with several drugs was studied by this method.[48] and reveals a slightly smaller decrease of dimer albumin capacity than in the case of monomer, the reduction being associated with the intrinsic binding constant.

Although there are no experimental results concerning the linkage of a drug to a chromatographic support, in order to study its binding properties, we can mention the work of

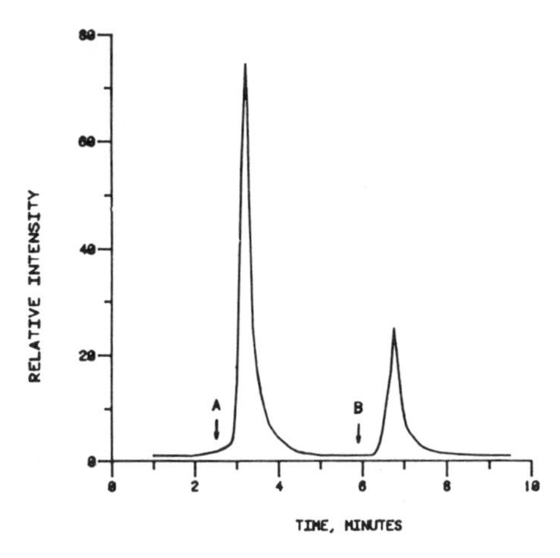

FIGURE 6. Chromatographic method for studying immunosorbent-antigen interaction. Immunosorbent, antihuman-IgG attached to Lichrosphere® Si 1000; column 2 mm i.d. × 4 cm; flow rate 0.5 mℓ/min; mobile phase "A" PBS, pH 7.4; mobile phase "B" 0.01 *M* phosphate buffer, pH 2.2; antigen (IgG) concentration 1.4 mg/mℓ; volume injected 10.0 μℓ; fluorometric detection, excitation 283 nm, emission 335 nm. (From Sportsman, J. R. and Wilson, G. S., *Anal. Chem.*, 52, 2013, 1980. With permission.)

Dunn and Chaiken[49] on the binding of staphylococcal nuclease to thymidine 5′ phosphate. In this method a column of ligand -(TP)-coupled Sepharose serves as a stationary phase for measuring the protein retention volume, from which the binding parameters are calculated.

All the above results were obtained with soft gels. The use of HPLC with rigid gels for the measurement of binding parameters was introduced by Sportsman and Wilson.[50] This technique is termed "high-performance immuno-affinity chromatography" and is applied to the determination of antigen (insulin or human IgG) binding to an antibody. The antibody is covalently immobilized on modified hydrophilic silica. In the first step, a small amount of antigen is injected into the column, which binds a part of the sample, while the unretained antigen appears at the void volume. In the second step the mobile phase is changed to another pH value or modified with acetonitrile. A peak corresponding to the dissociated antigen then appears (Figure 6). The ratio of the amount which elutes immediately upon injection to that which elutes upon changing the mobile phase is a hyperbolic function of concentration, as is the case in ordinary immunoassay procedures. Using the measurements of the mole of antigen bound $|B|$ by the immobilized antibody AB and the mole of antigen $|F|$ which passes unretained through the column, the apparent association constant k′ is expressed as:

$$k' = \frac{|B|}{|AB|\,|F|}$$

and is obtained from the plot of $|B|/|F|$ as a function of $|B|$. It should be noticed that constant values depend on the flow rate and are more satisfactorily obtained by extrapolation when the flow rate approaches zero. This process appears very convenient for investigating the kinetic and thermodynamic parameters of highly associating systems such as antigen-antibody complexes.

Another method has been introduced[51] for drug-protein binding studies, based on an approach avoiding covalent protein linkage to the chromatographic support. This process relies on the use of rigid size exclusion beads eluted with the protein solution. Considering the case where a protein is excluded from the pores, the partition coefficient of small sample of injected drug depends both on its affinity to the stationary phase and to the protein present in the mobile phase.

The drug retention volume in the absence of protein is given in the classical chromatographic relation:

$$V_A = V_0 + KV_s$$

where V_0 is the void volume, V_s is the stationary phase volume, and K is the partition coefficient.

$$K = \frac{|A|_s}{|A|_m}$$

where $|A|_s$ is the ligand concentration in the stationary phase and $|A|_m$ is the ligand concentration in mobile phase.

When the eluent contains a protein, the retention volume becomes:

$$V'_A = V_0 + K'V_s$$

$$\text{with} \quad K' = \frac{|A'|_s}{|A'|_m} = \frac{|A'|_s}{|A|_f + |A|_b}$$

since the ligand concentration in the mobile phase $|A'|_m$ is the sum of free $|A|_f$ and protein-bound $|A|_b$ ligand concentrations

$$|A'|_m = |A|_f + |A|_b$$

It has been demonstrated[51] that

$$\frac{V'_A - V_0}{V_A - V_0} = \frac{1}{a + \sum_{i=1}^{i=m} \frac{n_i k_i |P|}{1 + k_i|A|}} \qquad (1)$$

When the the experiment is performed with an amount of injected ligand small enough to neglect $k_i |A|_f \ll 1$ this equation becomes:

$$\frac{V'_{A1} - V_0}{V_A - V_0} = \frac{1}{1 + \sum_{i=1}^{i=m} n_i k_i |P|} = 1 - \alpha_1$$

V'_{A1} and α_1 are the limit values of V'_A and α ligand binding extent:

$$\alpha = \frac{|A|_b}{|A'|_m}$$

when the injected ligand amount approaches zero. By plotting α_i as a function of log $|P|$, an S-shape curve is obtained from which $\sum_{=1}^{=m} n_i k_i$ can be calculated for every α_i value.

By using this process, the affinity of warfarin, phenylbutazone, furosemide, L-tryptophan to the HSA has been measured.[51] The measured values agree with those obtained using Hummel and Dreyer's method.[25]

A similar HPLC method was previously introduced by Uekama et al.[52,53] for the determination of the stability of inclusion compounds of drugs and prostaglandin with cyclodextrin. In these cases, the chromatographic column was filled with an ion-exchange pellicular support and ligand retention was based on an ion-exchange mechanism.

The measurement of drug retention volumes can afford a means of determining the binding parameters by using the equilibrium saturation method already mentioned, when the eluent contains both protein and ligand. In this case the concentration of total ligand $|A'|_m$ remains constant, and the measurement of V'_A gives $\bar{r}$ and α. We have shown in the case of warfarin that the results obtained from the retention data are equivalent to those obtained by surface measurements.[41]

It must be stressed that all these measurements are only valid if the attainment of an equilibrium between the ligand and the stationary phase on the one hand, and compounds in the mobile phase on the other, is fast enough.

CONCLUSION

Chromatographic techniques provide an extremely precise means of studying the interactions of drugs with proteins. The introduction of HPLC opens new possibilities because of its increased speed and accuracy.

The choice of the method of study must be made according to the kinetics of the equilibria. Thus, classic zone elution chromatography is only suitable for the study of a complex whose dissociation speed is low enough for the complex to pass through the column without being changed. With drugs, this condition is rarely fulfilled and it is usually necessary to employ other chromatographic techniques that keep the concentration of the complex at a constant value. When there is only a small quantity of protein available, Hummel and Dreyer's method is especially recommended. However, with this method it is not possible to maintain the protein concentration at a constant level during elution. Conversely, frontal analysis and saturation methods make it possible to determine the binding ratio under experimental conditions where the concentration of a species is constant and known. These methods require large quantities of the compounds to be studied, a situation that is sometimes difficult to obtain.

In some cases, affinity techniques offer another solution. One component of the equilibrium can be grafted onto the column and, by measuring the retention volume of the other compound eluted on this modified column, the binding parameters can be determined. Another interesting affinity method avoiding all grafting uses a solution of the polymer to be studied.

HPLC techniques represent a means of measuring molecular interactions and should open new horizons in the field of pharmacology for the study of drug transport and activities.

REFERENCES

1. **Lathe, G. H. and Ruthven, C. R. J.,** The separation of substances and estimation of their relative molecular sizes by the use of columns of starch in water, *Biochem. J.*, 62, 665, 1956.
2. **Porath, J. and Flodin, P.,** Gel filtration: a method for desalting and group separation, *Nature (London),* 183, 1657, 1959.
3. **Wood, G. C. and Cooper, P. F.,** The application of gel filtration to the study of protein-binding of small molecules, *Chromatogr. Rev.*, 12, 88, 1970.
4. **Regnier, F. E. and Noel, R.,** Glycerolpropylsilane bonded phases in the steric exclusion chromatography of biological macromolecules, *J. Chromatogr., Sci.*, 14, 316, 1976.
5. **Engelhardt, H. and Mathes, D.,** Chemically bonded stationary phases for aqueous high performance exclusion chromatography, *J. Chromatogr.*, 142, 311, 1977.
6. **Becker, N. and Unger, K. K.,** Separation studies of amino acids, proteins and enzymes on bonded 1,2-dihydroxy-, 1,2-hydroxylamino and amino silica packings, *Chromatographia*, 12, 539, 1979.
7. **Nimmo, I. A. and Bauermeister, A.,** A theoretical analysis of the use of zonal gel filtration in the detection and purification of protein-ligand complexes, *Biochem. J.*, 169, 437, 1978.
8. **Melenevskii, A. T., El'kin, G. E., and Samsonov, G. V.,** Chromatographic determination of rate constants of a reversible reaction involving two components, *Izv. Akad. Nauk S.S.S.R. Ser. Khimi.*, 2, 448, 1979.
9. **Lee, M. and Debro, J. R.,** The application of gel filtration to the measurement of the binding of phenol red by human serum proteins, *J. Chromatogr.*, 10, 68, 1963.
10. **Dixon, H. B. F.,** Removal of a bound ligand from a macromolecule by gel filtration, *Biochem. J.*, 159, 161, 1976.
11. **Zeeberg, B. and Caplow, M.,** Determination of free and bound microtubular protein and guanine nucleotide under equilibrium conditions, *Biochemistry,* 18, 3880, 1979.
12. **Zeeberg, B.,** New approach for studying macromolecular-ligand binding, *J. Biol. Chem.*, 255, 3062, 1980.
13. **Hummel, J. P. and Dreyer, W. J.,** Measurement of protein-binding phenomena by gel filtration, *Biochim. Biophys. Acta.*, 63, 530, 1962.
14. **Clausen, J.,** Binding of sulfonamides to serum proteins, physicochemical and immunochemical studies, *J. Pharmacol. Exp. Ther.*, 153, 167, 1966.
15. **Fairclough, G. F.and Fruton, J. S.,** Peptide-protein interaction as studied by gel filtration, *Biochemistry,* 5, 673, 1966.
16. **Manzini, G., Ciana, A., and Crescenzi, V.,** The interaction of serum albumins with various drugs in aqueous solution. Gel permeation, calorimetric, fluorescence data, *Biophys. Chem.*, 10, 389, 1969.
17. **Forrey, A. W., Kimpel, B., Blair, A. D., and Cutler, R. E.,** Furosemide concentrations in serum and urine, and its binding by serum proteins as measured fluorometrically, *Clin. Chem.*, 20, 152, 1974.
18. **Müller, W. E. and Wollert, U.,** Characterization of the binding of benzodiazepines to human serum albumin, *Naunyn-Schmiedeberg's Arch. Pharmacol.*, 280, 229, 1973.
19. **Allen, G.,** The binding of sodium dodecyl sulphate to bovine serum albumin at high binding ratio,. *Biochem. J.*, 137, 575, 1974.
20. **Parsons, D. L.,** Determination of ligand-macromolecule binding parameters and the method of Hummel and Dreyer, *J. Chromatogr.*, 193, 520, 1980.
21. **Klotz, I. M.,** The application of the law of mass action to binding by proteins. Interactions with calcium, *Arch. Biochem.*, 9, 109, 1946.
22. **Bowmer, C. J. and Lindup, W. E.,** Inverse dependence of binding constants upon albumin concentration, *Biochim. Biophys. Acta,* 624, 260, 1980.
23. **Boobis, S. W. and Chignell, C. F.,** Effect of protein concentration on the binding of drugs of human serum albumin. I. Sulfadiazine, salicylate and phenylbutazone, *Biochem. Pharmacol.*, 28, 751, 1978.
24. **Brumbaugh, E. E. and Ackers, G. K.,** Molecular sieve studies of interacting protein systems. VII. Direct optical scanning method for ligand-macromolecule binding studies, *Anal. Biochem.*, 41, 543, 1971.
25. **Sebille, B., Thuaud, N., and Tillement, J. P.,** Study of binding of low-molecular-weight ligand to biological macromolecules by high performance liquid chromatography. Evaluation of binding parameters for two drugs bound to human serum albumin, *J. Chromatogr.*, 167, 159, 1978.
26. **Pepin, O., Thuaud, N., Sebille, B., Edouard, L., Lemort, N., and Beaumont, J. L.,** Extraction and characterization of antiestradiol immunoglobulins from human serum by high performance liquid chromatography, *Chromatographia,* 16, 267, 1982.
27. **Sebille, B., Thuaud, N., Tillement, J. P.,** Usefulness of chromatographic methods for the determination of drug-protein binding parameters, *J. Chromatogr.*, 193, 522, 1980.
28. **Nichol, L. W. and Winzor, D. J.,** The determination of equilibrium constants from transport data on rapidly reacting systems of the type A + B ↔ C., *J. Phys. Chem.*, 68, 2455, 1964.
29. **Cooper, P. F. and Wood, G. C.,** Protein-binding of small molecules: new gel filtration method, *J. Pharm. Pharmacol.*, 20, (Suppl.,), 150S, 1968.

30. **Scatchard, G.,** The attractions of proteins for small molecules and ions, *Ann. N. Y. Acad. Sci.*, 51, 660, 1949.
31. **Oester, Y. T., Keresztes-Nagy, S., Mais, F. T., Becktel, J., and Zaroslinski, J. F.,** Effect of temperature on binding of warfarin by human serum albumin, *J. Pharm. Sci.*, 65, 1673, 1976.
32. **Zarolinski, J. F., Keresztes-Nagy, S., Mais, R. F., and Oester, Y. T.,** Effect of temperature on the binding of salicylate by human serum-albumin, *Biochem. Pharmacol.*, 23, 1767, 1974.
33. **Burke, C. W.,** Accurate measurement of steroid-protein binding by steady-state gel filtration, *Biochim. Biophys. Acta*, 176, 403, 1969.
34. **Morris, M. J. and Brown, J. R.,** Estimation of the plasma protein binding of drugs by size exclusion chromatography at medium pressure, *J. Pharm. Pharmacol.*, 29, 642, 1977.
35. **Sebille, B., Thuaud, N., and Tillement, J. P.,** Etude de l'intéreaction de macromolecules biologiques et d'un coordinate par chromatographie liquide à haute performance, *C. R. Acad. Sci., Ser. C*, 285, 535, 1977.
36. **Thuaud, N., Sebille, B., Livertoux, M. H., and Bessiere, J.,** Determination of human serum albumin-diazepam binding by polarography and HPLC at different protein concentrations, *J. Chromatogr.*, in press.
37. **Baghurst, P. A., Nichol, L. W., Ogston, A. G., and Winzor, D. J.,** Quantitative interpretation of concentration-dependent migration in gel chromatography of reversibly polymerising solutes, *Biochem. J.*, 147, 575, 1975.
38. **Ackers, G. K., Brumbaugh, E. E., Ip, S. H. C., and Halvorson, H. P.,** Equilibrium gel permeation: a single-photon counting spectrophotometer for studies of protein interaction, *Biophys. Chem.*, 4, 171, 1976.
39. **Ford, C. L. and Winzor, D. J.,** A recycling gel partition technology for the study of protein ligand interactions: the binding of methyl orange to bovine serum albumin, *Anal. Biochem.*, 114, 146, 1981.
40. **Sebille, B., Thuaud, N., and Tillement, J. P.,** Equilibrium saturation chromatographic method for studying the binding of ligands to human serum albumin by high performance liquid chromatography. Influence of fatty acids and sodium dodecylsulphate on warfarin human serum albumin binding, *J. Chromatogr.*, 180, 103, 1979.
41. **Sebille, B., Thuaud, N., and Tillement, J. P.,** Study of the influence of sodium dodecyl-sulphate on warfarin-human serum albumin binding using chromatographic retention data in *Disc. R. Soc. Chem. Faraday Symp.* 15, 139, 1980.
42. **Thuaud, N.,** Mesure des Interactions Macromolecule-Ligand par chromatographie Liquide Haute Performancce. Application à la Fixation Simultanée de Médicaments et d'Acides Gras sur la Serum Albumine Humaine, Ph.D. thesis, Universite Paris Val de Marne, 1980.
43. **Turkova, J.,** Application of affinity chromatography to the quantitative evaluation of specific complexes, in *Affinity Chromatography*, Turkova, J., Ed., Elsevier, Amsterdam, 1978, chap. 4.
44. **Lagercrantz, C., Larson, T., and Karlsson, H.,** Binding of some fatty acids and drugs to immobilized bovine serum albumin studied by column affinity chromatography, *Anal. Biochem.*, 99, 352, 1979.
45. **Nichol, L. W., Ogston, A. G., Winsor, D. J., and Sawyer, W. H.,** Evaluation of equilibrium constants by affinity chromatography, *Biochem. J.*, 143, 435, 1974.
46. **Lagercrantz, C., Larson, T., and Denfors, I.,** Stereoselective binding of the enantiomers of warfarin and tryptophan to serum albumin from some different species studied by affinity chromatography on columns of immobilized serum albumin, *Comp. Biochem. Physiol.*, 69C, 375, 1981.
47. **Nakano, N. I., Shimamori, Y., and Yamaguchi, S.,** Mutual displacement interactions in the binding of two drugs to human serum albumin by frontal affinity chromatography, *J. Chromatogr.*, 188, 347, 1980.
48. **Nakano, N. I., Shimamori, Y., and Yamaguchi, S.,** Binding capacities of human serum albumin monomer and dimer by continuous frontal affinity chromatography, *J. Chromatogr.*, 237, 225, 1982.
49. **Dunn, B. M. and Chaiken, I. M.,** Quantitative affinity chromatography. Determination of binding constants by elution with competitive inhibitors, *Proc. Natl. Acad. Sci. U.S.A.*, 71, 2382, 1974.
50. **Sportsman, J. R.and Wilson, G. S.,** Chromatographic properties of silica-immobilized antibodies, *Anal. Chem.*, 52, 2013, 1980.
51. **Sebille, B., Thuaud, N., and Tillement, J. P.,** A retention data method for the determination of drug-protein binding parameters by high-performance liquid chromatography, *J. Chromatogr.*, 204, 285, 1981.
52. **Uekama, K., Hirayama, F., Nasu, S., Matsuo, N., and Irie, T.,** Determination of the stability constants for inclusion complexes of cyclodextrins with various drug molecules by high performance liquid chromatography, *Chem. Pharm. Bull.*, 26, 3477, 1978.
53. **Uekama, K., Hirayama, F., and Irie, T.,** The new method for determination of the stability constants of cyclodextrin-prostaglandin inclusion complexes by liquid chromatography, *Chem. Lett.*, 641, 1978.

LIPOPROTEIN SEPARATIONS BY HIGH PERFORMANCE GEL PERMEATION CHROMATOGRAPHY

Mitsuyo Okazaki and Ichiro Hara

INTRODUCTION

It is well known that all serum lipids circulate in association with specific proteins (apolipoproteins) to form lipid-protein complexes with remarkable hydrophilic properties, despite their high lipid content. Thus, water-insoluble lipids are transported from their sites of synthesis to their sites of ultilization. These macromolecular lipid-protein complexes are called "lipoproteins". Most serum lipoproteins range from approximately 200,000 to 10,000,000 mol wt and contain from 40 to 95% lipid. The lipids are principally classified into four groups: free cholesterol, triglyceride, cholesteryl ester, and phospholipid. Serum lipoprotein levels are closely related to atherosclerosis and its clinical manifestations, and an understanding of the factors affecting lipoprotein levels is of great importance in health and disease.

CLASSIFICATION

Serum lipoproteins are classified according to their hydrated densities into five major lipoprotein classes: chylomicrons, very low density lipoprotein (VLDL), low density lipoprotein (LDL), high density lipoproteins (HDL_2 and HDL_3), and very high density lipoprotein (VHDL). The general classification of serum lipoproteins and their properties is summarized in Table 1. The major classes can be separated by ultracentrifugal flotation or electrophoresis. Since there is an inverse relationship between lipoprotein density and particle size, they can be classified according to their particle size. Table 1 also shows classification according to particle diameter, which is achieved by using high-performance aqueous gel permeation chromatography.

Several techniques are currently used in the clinical and investigative laboratory for separation and quantitation of human serum lipoproteins. These techniques include preparative ultracentrifugation,[1-3] gel filtration,[1,4] selective precipitation with polyanion/divalent cation reagents,[1,5] electrophoresis,[1,6] affinity chromatography,[7] and ultrafiltration.[8]

For effective separation of all major classes of lipoproteins, a combination procedure consisting of several techniques mentioned above are frequently used and this presents some difficulties, that is, these procedures are time consuming and often result in the loss and alteration or decomposition of lipoprotein components. Although agarose gel chromatography has been used to isolate the major classes for the characterization of serum lipoproteins,[4,9,10] attempts to use it analytically have been limited because of the need for slow elution, large sample injection, low resolution, and poor reproducibility.

High-performance liquid chromatography (HPLC) has been widely utilized for the separation and fractionation of many biological substances because of its inherent speed and excellent reproducibility.

HPLC ANALYSIS METHODS

Recently, a simple and rapid method for the separation of serum lipoproteins has been developed using HPLC with aqueous gel permeation columns (TSK-GEL, type SW and PW, Toyo Soda, Japan) by Hara et al.[11-13] Properties of these gel permeation columns and

Table 1
GENERAL CLASSIFICATION OF HUMAN SERUM LIPOPROTEINS

Classification						
electrophoresis	Origin	pre-β	β	α	α	pre-α
hydrated density	chylomicrons	VLDL	LDL	HDL_2	HDL_3	VHDL
particle diameter(Å)[17]	(>750)	(300—750)	(150—300)	(100—150)	(75—100)	(60—75)
Properties						
Mean hydrated density[28]	0.93	0.97	1.035	1.09	1.15	1.155
Mean molecular weight[28]	504×10^6	19.6×10^6	2.3×10^6	0.36×10^6	0.18×10^6	0.15×10^6
Mean particle radius (Å)[28]	600	200	95.9	50.8	39.2	37.5
Solvent density for isolation	<1.006	<1.006	1.006—1.063	1.063—1.125	1.125—1.210	>1.210
Flotation rate						
S_f(1.063 g/mℓ)[1]	400—5000	12—400	0—12	Sediment	Sediment	Sediment
F (1.20 g/mℓ)[1]	—	43—770	20—43	3.5—9.0	0—3.5	Sediment
K_{av} value for TSK-GEL[17]						
G5000PW	0—0.05	0.05—0.21	0.21—0.34	0.34—0.41	0.41—0.46	0.46—0.52
G4000SW	0	0—0.14	0.14—0.39	0.39—0.53	0.53—0.63	0.63—0.71
G4000SW + G3000SW	0	0—0.05	0.05—0.24	0.24—0.38	0.38—0.46	0.46—0.52
G3000SW	0	0	0—0.08	0.08—0.22	0.22—0.33	0.33—0.41
Chemical composition[29] (% by weight)						
Protein	2	8	21	41	55	62
Triglyceride	84	50	11	5	4	5
Cholesterol	7	19	45	21	15	3
Phospholipid	7	18	22	30	23	28

the molecular weight separation range in aqueous systems are summarized (based on data from the supplier Toyo Soda) in Table 2. The values of molecular weight and particle diameter for lipoprotein analysis in Table 2 were determined from calibration curves using standard proteins and lipoproteins with known particle diameters. All major classes of serum lipoproteins can be separated by combining these columns and using the $d < 1.21$ fraction prepared from human serum by ultracentrifugation. Excellent reproducibility for the lipoprotein separation was obtained, while no decomposition of the lipoproteins during gel permeation under high pressure (about 100 kg/cm^2) was detected.[13] The recovery of lipoproteins from various gel permeation columns was very high (94 to 99%).[14] The differences in the separation profiles of lipoproteins are mainly due to the type of column but also are partly due to the properties of eluent. The effect of pH and salt concentration of eluent on the separation is very small,[12] and consequently 0.15 *M* NaCl is used as the eluent for lipoprotein separation.

The combination of gel permeation columns consisting of G5000SW + G3000SW(x2) has been reported to give the best separation of all the major lipoprotein classes in the early studies of 1980 to 1981.[11-13,15,16] Subsequently, a high resolution G4000SW column has become available from the supplier. Evaluation of gel permeation columns (TSK-GEL) for lipoprotein analysis was done by using the standard globular proteins with a known Stokes' radius as well as spherical lipoproteins. The particle diameter of the lipoproteins was determined by electron microscopy using the negative staining method.[17] The calibration curves for various gel permeation columns are presented in Figure 1. A linear relationship could be obtained by plotting the elution volume parameter (K_{av}) against the logarithm of the particle diameter. This indicates that the precise diameter of spherical lipoproteins can be estimated from the elution volume by using these calibration curves. Classification of particle diameter for the major lipoprotein classes, which were previously classified both by electrophoresis and ultracentrifugal flotation, is defined by the high-performance gel permeation chromatography parameters in Table 1. The range of the K_{av} values for each class is also summarized for various TSK-GEL column systems.[17] At the present time, the best column system for the separation of all major classes may be the combined system of G4000SW + G3000SW. This combination achieves the separation of serum lipoproteins into the following major classes: chylomicron + VLDL, LDL, HDL_2, and HDL_3.

A direct quantitation method for cholesterol in each lipoprotein class from a very small amount of whole serum (10 to 20 $\mu\ell$) was developed by Hara et al.,[11,15,16] who combined the two methods of (1) the separation by HPLC with gel permeation columns and (2) the selective detection of cholesterol in the post-column effluent by the use of a commercial enzyme reaction kit. This technique was applied to other lipid components in serum, i.e., triglycerides[18] and choline-containing phospholipids.[14,19,20] Quantitation of cholesterol[15] and choline-containing phospholipids[14] in each lipoprotein class has been established using this technique.

This analytical method was carried out with a high speed chemical derivatization chromatograph (HLC 805, Toyo Soda) equipped with a reactor (a stainless steel tube of 0.4 mm i.d. × 10,000 mm or a Teflon® tube of 0.5 mm i.d. × 20,000 mm), and two detectors as shown in Figure 2. This equipment can be used for detection and quantitation of cholesterol, triglycerides, and choline-containing phospholipids by using appropriate commercial enzyme kits: Determiner TC"555" (Kyowa Medex, Tokyo, Japan) for cholesterol, Determiner TG (Kyowa Medex) for triglycerides, and PL kit K"f" (Nippon Shoji, Osaka, Japan) for choline-containing phospholipids. The elution patterns of serum lipoproteins can be directly monitored from whole serum by measuring the A_{550} (cholesterol and triglycerides) or A_{500} (choline-containing phospholipids) of the mixed eluate and enzyme solution after passing it through a reactor at the appropriate temperature. Optimum experimental conditions for human serum lipoprotein analysis in various analytical systems and their available data are summarized

Table 2
PROPERTIES OF VARIOUS GEL PERMEATION COLUMNS (TSK-GEL) IN AQUEOUS SYSTEMS

Grade	Particle size (μm)	Mol wt (particle diameter, Å) applicable range for analysis			Exclusion limit	Number of theoretical plate (TP/ft)
Type SW		Protein	Dextran	Lipoprotein[a]	Lipoprotein[a]	
G2000SW	10 ± 2	5×10^5—1×10^5	1×10^3—3×10^4			>5000
G3000SW	10 ± 2	1×10^4—5×10^5	2×10^3—7×10^4	2×10^4—7×10^5 (40—160)	1×10^6 (180)	>5000
G4000SW	10 ± 2	2×10^4—7×10^6	4×10^3—5×10^5	2×10^4—8×10^6 (40—400)	1×10^7 (440)	>5000
Type PW		PEG[b]	Dextran	Lipoprotein[a]	Lipoprotein[a]	
G1000PW	10 ± 2	-1×10^3				5000
G2000PW	10 ± 2	-5×10^3				5000
G3000PW	13 ± 2	-5×10^4	-1×10^5			5000
G4000PW	13 ± 2	2×10^3—3×10^5	4×10^3—6×10^5	3×10^4—5×10^6 (50—350)	1×10^7 (400)	3000
G5000PW	17 ± 2	4×10^3—1×10^6	8×10^3—2×10^6	3×10^4—5×10^7 (50—750)	2×10^8 (1000)	3000
G6000PW	17 ± 2	4×10^4—8×10^6	8×10^4—2×10^7	2×10^5—5×10^8 (100—2000)	2×10^9 (3000)	3000

[a] From Reference 17

[b] PEG: polyethylene glycol.

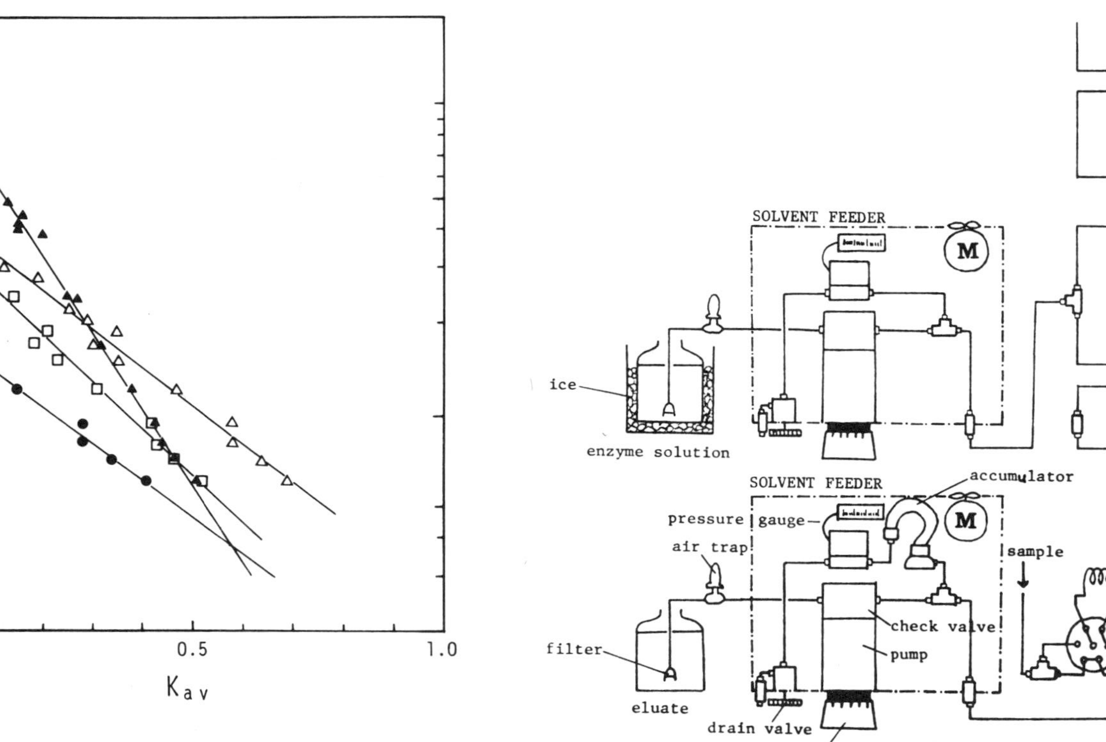

FIGURE 1. Calibration curves of TSK-GEL columns for lipoprotein analysis. Column: G5000PW (▲), G4000SW (△), G4000SW + G3000SW (□), G3000SW (●). Eluent 0.15 1 *M* NaCl. Flow rate: 0.40 to 0.60 mℓ/min. Sample: standard proteins with a known Stokes' radius (Pharmacia Fine Chemicals) and lipoproteins, particle diameters of which were determined by electron microscopy (negative staining). K_{av} is given as follows: $K_{av} = (V_e - V_o)/(V_t - V_o)$, where V_e is the elution volume, V_o, the column void volume; and V_t, the total gel bed volume.

FIGURE 2. Flow diagram of lipoprotein analysis by HPLC developed by Hara et al.

in Table 3. The detection limit for each lipid component under experimental conditions of Table 3 was very high: 0.5 μg for cholesterol,[16] and 1.0 μg for choline-containing phospholipids[20] and triglycerides.[18] Peak broadening in the reaction tube was negligible, provided that the tubing had a diameter of less than 0.5 mm i.d.[14,20]

The elution patterns of human serum (hyperlipidemia) obtained using various analytical systems (No. 1 to No. 9 of Table 3) are shown in Figure 3. The mean elution position corresponding to the major lipoprotein class is shown for the various column systems using numbered arrows. The numbers (1 to 6), correspond to (1) chylomicrons; (2) VLDL; (3) LDL; (4) HDL_2; (5) HDL_3; (6) VHDL. Much data about the size distribution of serum lipoproteins can be obtained from these patterns in various analytical systems. With this knowledge we can select an appropriate combination of columns and detection systems to best analyze any particular sample.

APPLICATIONS

This technique has been successfully applied to the analysis of cholesterol in HDL the two subclasses, HDL_2 and HDL_3,[21,22] and the results obtained for normal male and female groups were consistent with those obtained by the ultracentrifugal method.[21] In addition, a remarkable decrease in serum HDL_3-cholesterol levels in a patient with cirrhosis of the liver has been reported by Okazaki et al.[22] using this technique. Moreover, this technique has been applied to the detection and characterization of abnormal lipoproteins present in serum exhibiting dyslipoproteinemia (LCAT deficiency) by Kodama et al.[23] The heterogeneity of human serum HDL has been examined by a combination of peak frequency analysis and rechromatography of subfractions with this HPLC method. Five subclasses in HDL are clearly present according to the particle size differentiation as shown in Figure 4.[24] The values of HDL-cholesterol obtained by this HPLC method correlated well with those found by the heparin-manganese chloride precipitation method.[15] A degree of high correlation between lipoprotein analysis by this HPLC method to that obtained by the ultracentrifugal flotation method and to that from agarose gel electrophoresis has been reported.[25,26]

More recently, another technique for separation and detection human serum lipoproteins by the use of these gel permeation columns (G3000SW or G4000SW) has been reported by Busbee et al.[27] in which lipoproteins in serum are selectively prestained with formazan dye and the column effluent was monitored at 580 nm so that only the lipoprotein components of serum are detected.

These studies reported by Okazaki et al.[11-26] and Busbee et al.[27] indicate that gel permeation HPLC holds great promise for developing a rapid method of qualitative and quantitative analysis of serum lipoproteins, for both clinical and investigative applications.

Table 3
OPTIMUM EXPERIMENTAL CONDITIONS FOR HUMAN SERUM LIPOPROTEIN ANALYSIS IN VARIOUS ANALYTICAL SYSTEMS AND THEIR AVAILABLE DATA[a]

	Analytical system								
No.	1	2	3	4	5	6	7	8	9
Column[b]	G5PW + G3SWx2	G5PW	G5PW	G4SW	G4SW	G4SW + G3SW	G4SW + G3SW	G3SW ×3	G3SW ×3
Lipid	TC	TG	PL	TC	PL	TC	PL	TC	PL
Loaded volume of whole serum (μℓ)	10—20	10—50	10	5	10	5—10	10—20	10—20	20—30
F_m (mℓ/min)	1.00	0.50	0.50	0.50	0.50	0.60	0.60	0.60	0.60
F_E (mℓ/min)	0.35	0.40	0.20	0.20	0.20	0.20	0.30	0.20	0.30
Reactor									
I.D. (mm)	0.25	0.5	0.5	0.4	0.5	0.4	0.5	0.4	0.5
Length (m)	20	20	20	10	20	10	20	10	20
Reaction temp. (°C)	40	45	39	40	39	40	39	40	39
Time for analysis (min)	54 (28)[c]	48 (26)[c]	48 (26)[c]	53 (30)[c]	55 (30)[c]	71 (36)[c]	73 (36)[c]	88 (40)[c]	90 (40)[c]
Available data	C VLDL LDL HDL_{sub}	C VLDL LDL HDL FG	C VLDL LDL HDL FC	C+VLDL LDL HDL_{sub}	C+VLDL LDL HDL_{sub}	C+VLDL LDL HDL_{sub}	C+VLDL LDL HDL_{sub} VHDL	C+VLDL +LDL HDL_{sub}	C+VLDL +LDL HDL_{sub} VHDL
Ref.	11, 15, 16	18	14	20	14	20	14, 20	20	20

[a] Abbreviations: TC, total cholesterol; TG, triglycerides; PL, choline-containing phospholipids; C, chylomicrons; VLDL, very low density lipoprotein; LDL, low density lipoprotein; HDL, high density lipoprotein; sub, subclasses; FG, free glycerol; FC, free choline; F_m, flow rate of the main path of the column; F_E, flow rate of the enzyme solution (Determiner TC"555" for TC, Determiner TG for TG, and PL kit K"f" for PL).

Table 3 (continued)
OPTIMUM EXPERIMENTAL CONDITIONS FOR HUMAN SERUM LIPOPROTEIN ANALYSIS IN VARIOUS ANALYTICAL SYSTEMS AND THEIR AVAILABLE DATA[a]

[b] G5PW = G5000PW, G4SW = G4000SW, G3SW = G3000SW; each column size, 7.5 mm i.d. × 600 mm; a single protecting column (G3000SW, 7.5 mm i.d. × 75 mm) is connected in the system (No. 4 to 9).

[c] The least interval between samples.

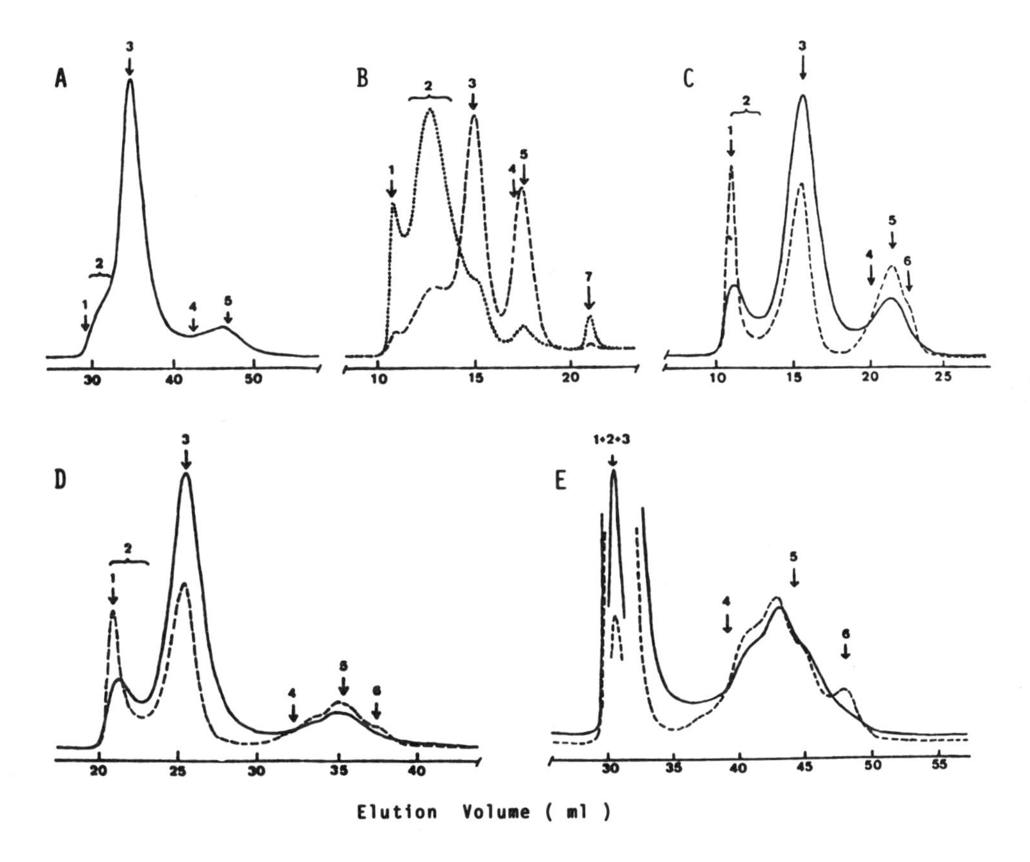

FIGURE 3. Elution patterns of cholesterol (—), triglycerides(....), and choline-containing phospholipids (— —) for human serum in various gel permeation column systems. Column: G5000PW + G3000SW × 2 (A), G5000PW (B), G4000SW (C), G4000SW + G30000SW (D), G3000SW × 3 (E), each column size was 7.5 mm i.d. × 600 mm and a single protecting column (G3000SW, 7.5 mm i.d. × 75 mm) was connected for C, D, and E. Eluent: 0.15 *M* NaCl. Detector: A_{550} for TC and TG, A_{500} for PL (0.1 [ABS] 10 mv for A to D, 0.1[ABS] 5 mv for E). Sample: human whole serum exhibiting hyperlipidemia, (TC = 225, TG = 262, PL = 240 mg/dℓ). Loaded volume: 5 or 10 μℓ for TC, 15 μℓ for TG, 10 or 20 μℓ for PL. Other HPLC conditions as in Table 3. Elution position: 1, chylomicrons; 2, VLDL; 3, LDL; 4 HDL_2; 5, HDL_3; 6, VHDL; 7, total permeation.

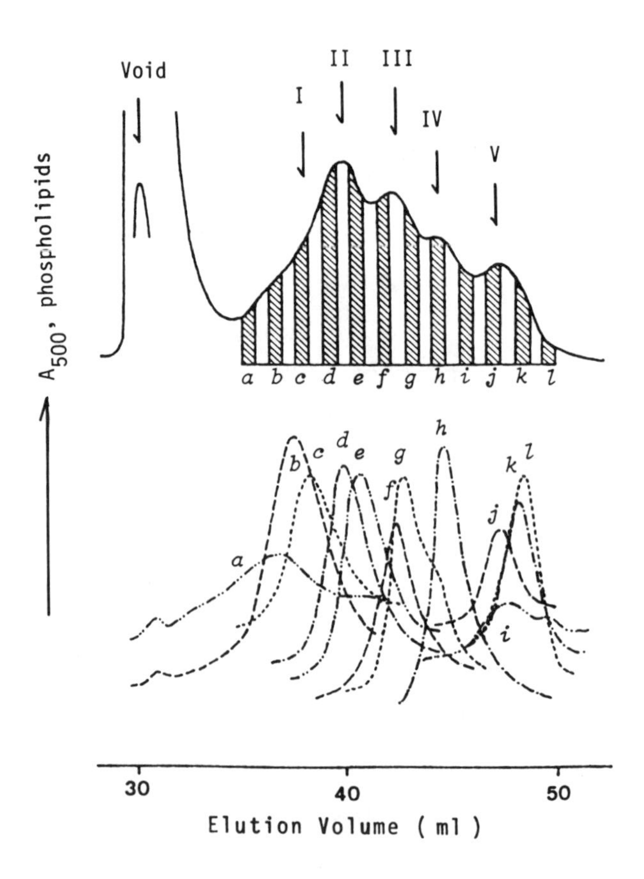

FIGURE 4. Particle distribution monitored by choline-containing phospholipids for HDL subfractions separated by gel permeation columns. Elution pattern of whole serum used for subfractionation (upper) and that of each fraction from whole serum (lower). Analytical system: No. 9(Table 3). Sample: whole serum from healthy male. Other HPLC conditions as in Table 3. The mean elution position of HDL subfractions determined by the peak frequency analysis is shown as arrows labeled I to V (see Reference 24).

REFERENCES

1. **Hatch, F. T. and Lees, R. S.,** Practical methods for plasma lipoprotein analysis, in *Advances in Lipid Research,* Vol. 6, Paoletti, R. and Kritchevsky, D., Eds., Academic Press, New York, 1968, chap. 1.
2. **Havel, R. J., Eder, H. A., and Bragden, J. H.,** The distribution and chemical composition of ultracentrifugally separated lipoproteins, *J. Clin. Invest.,* 34, 1345, 1955.
3. **Chung, B. H., Segrest, J. P., Cone, J. T., Pfau, J., Geer, J. C., and Duncan, L. A.,** High resolution plasma lipoprotein cholesterol profiles by a rapid, high volume semiautomated method, *J. Lipid Res.,* 22, 1003, 1981.
4. **Rudel, L. L., Lee, J., Morris, M., and Felts, J.,** Characterization of plasma lipoproteins separated and purified by agarose-column chromatography, *Biochem. J.,* 139, 89, 1974.
5. **Burstein, M., Scholnick, H. R., and Morfin, R.,** Rapid method for the isolation of lipoproteins from human serum by precipitation with polyanions, *J. Lipid Res.,* 11, 583, 1970.
6. **Cobb, S. A. and Sanders, J. L.,** Enzymatic determination of cholesterol in serum lipoproteins separated by electrophoresis, *Clin. Chem.,* 24, 1116, 1978.

7. **Wicham, A.,** Affinity chromatography of human plasma low- and high-density lipoproteins, *Biochem. J.*, 181, 691, 1979.
8. **Olson, W. P. and Faith, M. W.,** Lipoproteins removed from serum and plasma by membrane filtration, *Prep. Biochem.*, 8, 379, 1978.
9. **Margolis, S.,** Separation and size determination of human serum lipoproteins by agarose gel filtration, *J. Lipid Res.*, 8, 501, 1967.
10. **Sata, T., Estrich, D. L., Wood, P. D. S., and Kinsell, L. W.,** Evaluation of gel chromatography for plasma lipoprotein fractionation, *J. Lipid Res.*, 11, 331, 1970.
11. **Hara, I., Okazaki, M., and Ohno, Y.,** Rapid analysis of cholesterol of high density lipoprotein and low density lipoprotein in human serum by high performance liquid chromatography, *J. Biochem.*, 87, 1863, 1980.
12. **Okazaki, M., Ohno, Y., and Hara, I.,** High-performance aqueous gel permeation chromatography of human serum lipoproteins, *J. Chromatogr.*, 221, 257, 1980.
13. **Ohno, Y., Okazaki, M., and Hara, I.,** Fractionation of human serum lipoproteins by high performance liquid chromatography. I, *J. Biochem.*, 89, 1675, 1981.
14. **Okazaki, M., Hagiwara, N., and Hara, I.,** Quantitation method for choline-containing phospholipids in human serum lipoproteins by high performance liquid chromatography, *J. Biochem.*, 91, 1381, 1982.
15. **Okazaki, M., Ohno, Y., and Hara, I.,** Rapid method for the quantitation of cholesterol in human serum lipoproteins by high performance liquid chromatography, *J. Biochem.*, 89, 879, 1981.
16. **Okazaki, M., Shiraishi, K., Ohno, Y., and Hara, I.,** High-performance aqueous gel permeation chromatography of serum lipoproteins: selective detection of cholesterol by enzymatic reaction, *J. Chromatogr.*, 223, 285, 1981.
17. **Okazaki, M., Kato, H., Nishigai, M., and Hara, I.,** Evaluation of high performance aqueous gel permeation chromatography for separation of human serum lipoproteins, *J. Chromatogr.*, in preparation.
18. **Hara, I., Shiraishi, K., and Okazaki, M.,** High-performance liquid chromatography of human serum lipoproteins: selective detection of triglycerides by enzymatic reaction, *J. Chromatogr.*, 239, 549, 1982.
19. **Hagiwara, N., Okazaki, M., and Hara, I.,** A quantitation reagent specific for choline-containing phospholipids in HPLC for serum lipoproteins, *J. Jpn. Oil Chemists' Soc.*, 31, 262, 1982.
20. **Okazaki, M., Hagiwara, N., and Hara, I.,** High-performance liquid chromatography of human serum lipoproteins: selective detection of choline-containing phospholipids by enzymatic reaction, *J. Chromatogr.*, 231, 13, 1982.
21. **Okazaki, M. and Hara, I.,** Analysis of cholesterol in high density lipoprotein subfractions by high performance liquid chromatography, *J. Biochem.*, 88, 1215, 1980.
22. **Okazaki, M., Hara, I., Tanaka, A., Kodama, T., and Yokoyama, S.,** Decreased serum HDL_3 cholesterol levels in cirrhosis of the liver, *N. Engl. J. Med.*, 304, 1608, 1981.
23. **Kodama, T., Akanuma, Y., Okazaki, M., Aburatani, H., Itakura, H., Takahashi, K., Sakuma, M., Takaku, F., and Hara, I.,** Abnormalities in plasma lipoproteins in familial partial lecithin:cholesterol acyl transferase deficiency, *Biochim. Biophys. Acta*, 752, 407, 1983.
24. **Okazaki, M., Hagiwara, N., and Hara, I.,** Heterogeneity of human serum high density lipoproteins by high performance liquid chromatography, *J. Biochem.*, 92, 517, 1982.
25. **Okazaki, M., Itakura, H., Shiraishi, K., and Hara, I.,** Serum lipoprotein measurement — liquid chromatography and sequential flotation (ultracentrifugation), compared, *Clin. Chem.*, 29, 768, 1983.
26. **Okazaki, M., Takizawa, A., Shiraishi, K., and Hara, I,.** Comparison of serum lipoprotein quantitation high performance liquid chromatography and agarose gel electrophoresis, *J. Jpn. Oil Chemists' Soc.*, 32, 423, 1983.
27. **Busbee, D. L., Payne, D. M., Jasheway, D. W., Carlisle, S., and Lacko, G.,** Separation and detection of lipoproteins in human serum by use of size-exclusion liquid chromatography: a preliminary report, *Clin. Chem.*, 27, 2052, 1981.
28. **Stein, B. W., Scanu, A. M., and Kezdy, F. J.,** Structure of human serum lipoproteins inferred from compositional analysis, *Proc. Natl. Acad. Sci. U.S.A.*, 74, 837, 1977.
29. **Scanu, A. M. and Kruski, A. W.,** in *International Encyclopedia of Pharmacology and Therapeutics*, Vol. 1, Pergamon, Oxford, England, 1975, 21.

MOLECULAR SIEVE HIGH PERFORMANCE LIQUID CHROMATOGRAPHY APPLIED TO THE SEPARATION OF HUMAN PLASMA APOLIPOPROTEINS

Celina Edelstein and Angelo M. Scanu

INTRODUCTION

Because plasma apolipoproteins differ from each other in size (see Table 1), molecular-sieving or size-exclusion chromatography can be successfully utilized for their separations.[1-6] Apolipoproteins also differ in charge and this has permitted their fractionation by ion-exchange chromatography,[7-13] preparative isoelectric focusing,[3] and chromatofocusing[14] using conventional column chromatographic procedures.

Recently high-performance liquid chromatography (HPLC) has been applied to the study of proteins and peptides as well as apolipoproteins.[15-19] The techniques used for apolipoproteins have been reverse-phase,[20-24] ion-exchange,[25] and molecular-sieve-exclusion chromatography[26,27] mostly in the analytical mode with encouraging results.

We wish here to report on the experience gained in this laboratory in separating the apolipoproteins of apo HDL and apo VLDL using molecular-sieve HPLC.

STUDIES ON APO VLDL

Separation of VLDL from Plasma

Blood is collected in a preservative cocktail* in ice and centrifuged at 5000 r/min for 30 min. The plasma supernatant (d <1.006 g/mℓ) is then transferred to a Beckman Ti 60 rotor and centrifuged for 18 hr at 59,000 r/min at 10°C. Following centrifugation, the top fraction (2 to 3 mℓ) is washed twice with saline and recentrifuged under the same ultracentrifugal conditions. The VLDL top fraction is collected and phenylmethyl sulfonyl fluoride (PMSF) at a final concentration of 10^{-3} *M* is added. The VLDL is refrigerated at 4°C no longer than 2 days before delipidation.

Delipidation

VLDL is delipidated with a 2:1 (v/v) chloroform:methanol mixture at room temperature followed by five peroxide-free ethyl ether washes at 4°C. The final apo VLDL is dried under nitrogen and the powder stored at −70°C.

HPLC Procedure[28]

Preparation of sample — Ten milligrams of apo VLDL are dissolved at room temperature in 1 mℓ of 0.01 *M Tris* HCl, 0.01 *M* dithiothreitol, DTT, 6 *M* guanidine hydrochloride (GdmCl) solution, pH 7 containing 10^{-3} *M* PMSF. After a clear solution is obtained (about 2 hr), the sample is filtered through a 0.2 μm polycarbonate filter (Nucleopore, Pleasanton, CA) and 200-μℓ aliquots containing 2 mg apo VLDL are used for each chromatographic injection.

* The preservative cocktail includes EDTA, final concentration 1.5 g/ℓ; sodium azide, 0.1 g/ℓ; chloramphenicol, 50 mg/ℓ; gentamycin sulfate, 0.1 g/ℓ; and aprotinin, 10,000 U/ℓ.

Table 1

Apoprotein	Mol wt
A-I	28,331
A-II (two-chain)	17,400
A-II (one-chain)	8,700
C-I	6,631
C-II	8,837
C-III	8,764
E	34,145
B	200,000—330,000

Chromatographic conditions

Instrument:	Varian LC 5000 (Varian Instruments, Co., Sunnyvale, CA)
Detector:	Varian UV-50 variable wavelength detector
Columns:	Guard Column (7.5 × 75 mm), Bio-Gel® TSK (Bio-Rad, Richmond, CA) Analytical column (7.5 × 300 mm) Bio-Gel® TSK 50 (Bio-Rad) Analytical column (7.5 × 300 mm) Bio-Sil® TSK 400 (Bio-Rad) Analytical column (7.5 × 300 mm) Spherogel TSK 3000 (Altex) All columns are connected in series
Mobile phase:	4 *M* GdmCl solution in 0.01 *M Tris*, 0.01 *M* DTT, pH 7.0.
Flow rate:	0.5 mℓ/min
Equilibration time:	3 hr with mobile phase at a flow rate of 0.5 mℓ/min
Temperature:	23°C
Sample:	Apo VLDL dissolved at room temperature at a concentration of 10 mg/mℓ in a 6 *M* GdmCl solution in 0.01 *M Tris*, 0.01 *M* DTT ph 7.0 containing 10^{-3} *M* PMSF.

A typical HPLC profile of a pooled apo VLDL sample is shown in Figure 1. By SDS-PAGE using known human apoliproproteins as standards, the 34, 48.7-, and 58.3-min components correspond to apo B, apo E, and apo C, respectively. The 44-min component was not studied but its characterization is under way.

Technical Comments

The success of the fractionation shown in Figure 1 is strongly dependent on the completeness of the delipidation step. Repetitive rinsing with peroxide-free diethyl ether insures complete removal of residual lipids and results in a fine white powder which is easily solubilized in the 6 *M* GdmCl solution. The inclusion of DTT for solubilization of apo VLDL and for the mobile phase serves a dual purpose. The first is the achievement of a soluble apo B and the second, an apo E which is no longer disulfide-linked to apo A-II. A loading of 2 mg of apo VLDL is considered to be ideal because the resolution is at an optimum and the total protein recoveries are 93%. Under identical conditions, rechromatography of apo E results in a recovery of 80%.

General Comments

Using the molecular-sieve column system (described above) and GdmCl as a denaturant, we have successfully fractionated the apoproteins of VLDL whose molecular weights range from 250 to 10 K. Of great significance is the excellent yield and purity of apo E which we can obtain in less than 1 hr. Conventional chromatographic techniques for the isolation of apo E are lengthy and highly demanding in manpower.[3-6, 14, 29-32] Thus, this new method should prove valuable not only for the isolation of pure apo VLDL apoproteins but also for their quantitation.

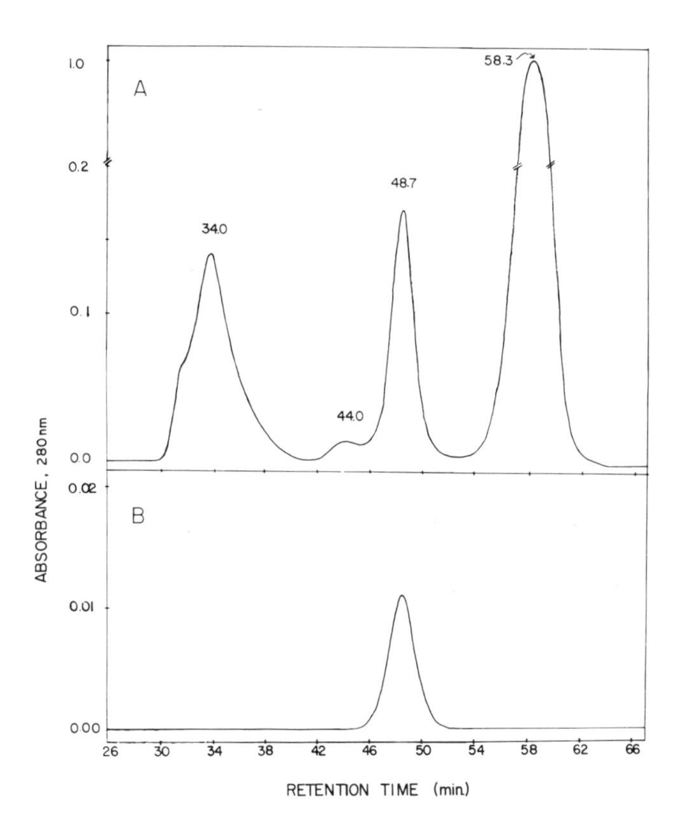

FIGURE 1. (A) HPLC elution profile of apo VLDL (2 mg, injection volume, 200 μℓ). Eluting buffer 4 *M* GdmCl solution in 0.01 *M Tris,* 0.01 *M* DTT, pH 7.0, flowrate, 0.5 mℓ/min. The effluent was continuously monitored at 280 nm. The number on each peak maximum indicates their retention times. Temperature of fractionation, 23°C. (B) Rechromatography of the 48.7-min fraction in Figure 1A. Identical column conditions were used except that 40 μg protein were injected.

STUDIES ON APO HDL

Separation of HDL Species

We routinely use a preliminary screening procedure based on the single-spin methodology to identify the major lipoprotein classes.[33,34] This also allows us to assess subclass heterogeneity if present. It has been shown[35-38] that the HDL subfractions vary in their apoprotein distribution, particularly in terms of apo A-I/apo A-II ratios. Thus HDL fractions can be obtained with either apo A-I as the main protein constituent (HDL_2) or with apo A-I and apo A-II in approximately equimolar ratios, (HDL_3). Figure 2 depicts examples of serum lipoprotein profiles containing both HDL_2 and HDL_3. From this knowledge we proceed in the isolation of HDL_2 and/or HDL_3 by preparative ultracentrifugal techniques described elsewhere.[39] The final HDL or HDL subfractions are stored in d 1.21 g/mℓ solution containing 10^{-3} *M* PMSF at 4°C for no longer than 1 week.

Delipidation

HDL is first dialyzed extensively against 0.15 *M* NaCl, pH 7.0, in the presence of 0.02% sodium azide and 1 m*M* EDTA. Delipidation is carried out at −20°C with 3:2 v/v, ethanol:diethyl ether as previously described.[40] The protein precipitate is then washed over-

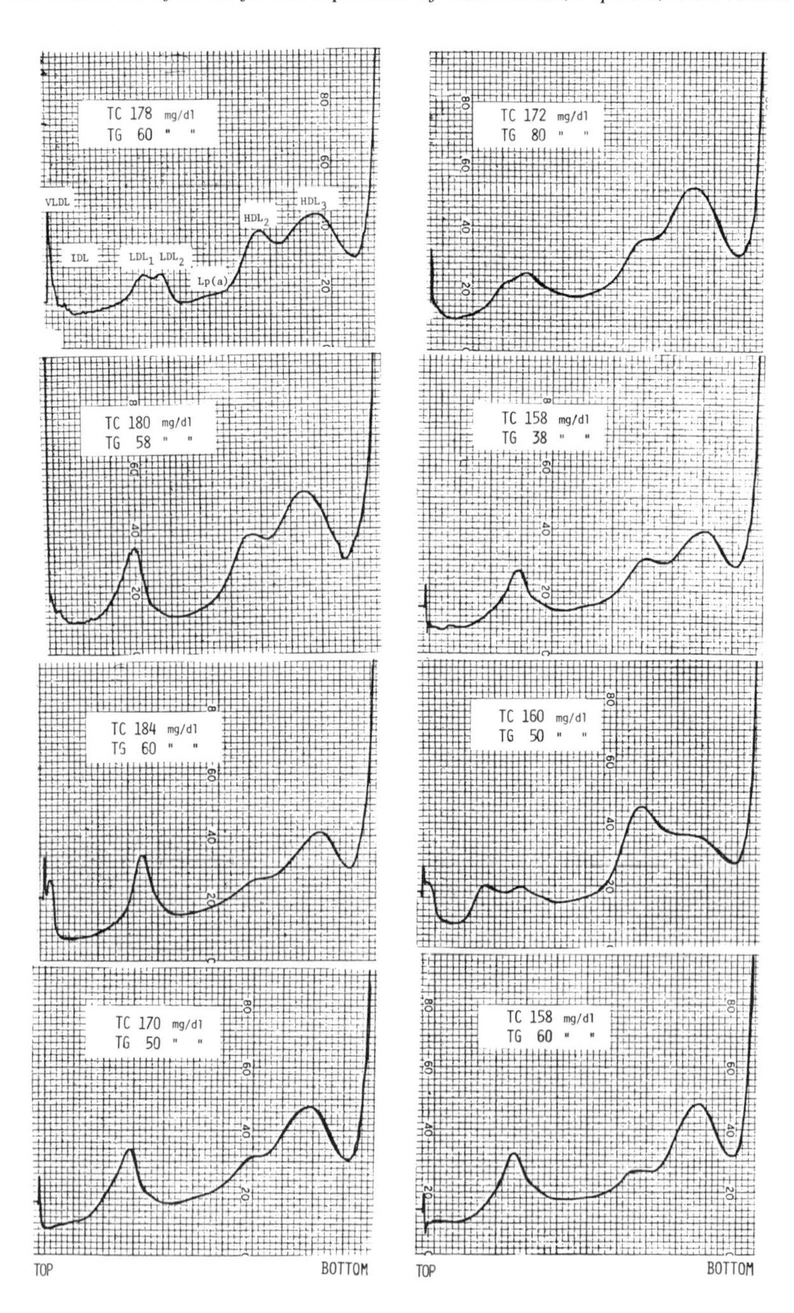

FIGURE 2. Ultracentrifugal single-spin absorbance profiles of 0.2 mℓ of serum obtained from eight normolipidemic subjects. The ordinate is the absorbance at 280 nm and the abscissa the order of fractions from the top to the bottom. The density profile is in the range of d 1.03 to 1.25 g/mℓ. For technical details see References 33 and 34.

night with diethyl ether and the final ether powder is dried and stored under nitrogen or argon gas at −20°C.

HPLC Fractionation

Preparation of sample — Ten milligrams of apo HDL are dissolved at room temperature in 1 mℓ of 6 *M* urea solution previously adjusted to pH 3.15 with formic acid. The solubilized sample is then filtered thru a 0.2-μm polycarbonate filter (Nucleopore, Pleasanton, CA) and 200-μℓ aliquots containing 2 mg apo HDL are used for each injection.

Chromatographic conditions

Instrument:	Varian LC 5000 (Varian Instruments Co., Sunnyvale, CA)
Detector:	Varian UV-50 variable wavelength
Columns:	Guard Column (7.5 × 100 mm), Micro Pak TSK GSWP (Varian), two analytical columns each (7.5 × 300 mm), Spherogel TSK 3000 (Altex, Berkeley, CA)
Mobile phase:	6 *M* urea solution, pH 3.15, adjusted to pH with formic acid
Flow rate:	0.5 mℓ/min
Equilibration time:	4 hr with mobile phase at a flow rate of 0.5 mℓ/min
Temperature:	26°C
Sample:	apo HDL, dissolved in mobile phase buffer at a concentration of 10 mg/mℓ

A typical profile of apo HDL is shown in Figure 3. The bimodal peak (a) eluting between 9 and 11 mℓ is composed of an aggregate of apo A-I and at least two additional unidentified proteins (see gels in Figure 3). The second and third peaks are apo A-I and apo A-II eluting at 11.8 and 14.3 mℓ respectively. A shoulder on the descending portion of the apo A-II peak elutes at 15.5 mℓ. This component can be resolved from apo A-II upon rechromatography on the same column (Figure 3, inset). Currently, this protein is under characterization. The fourth and last peak elutes at 17.5 mℓ and is composed of the whole of the C peptides.

Comments

Under the chromatographic conditions described above a maximal protein load of 3 mg produced a high resolution value (1.7) between the apo A-I and apo A-II peaks. (The resolution, R, is defined as $R = 2\Delta t/(w_2 + w_1)$, where Δt is the retention time between two peaks and w is the peak width in units of time). In general, a value of 1.0 or greater is required for a good separation.[41] A recovery of 90% was obtained for the total protein applied. The column supports (TSK 3000) are manufactured by Toyo Soda, Japan, and are made of spherical silica with a particle diameter of 10 μm and a pore diameter of 240Å. Although the exclusion limit in aqueous media for a globular protein is 5×10^5, this figure decreases in the presence of denaturants.

The adjustment of our urea solutions to pH 3.15 is advantageous in that cyanate formation is avoided. This eliminates possible carbamylation of the protein lysine residues. In this context, similar results can be obtained using 4 *M* GdmCl. As shown in Figure 3, good resolution can be accomplished at a flow rate of 1 mℓ/min; however for a quantitative study where high yields of pure protein fractions are needed, a flow rate of 0.5 mℓ/min is recommended.

GENERAL CONSIDERATIONS

The lifetime of columns, of 6 to 10 months, can be prolonged if certain precautions are followed. In the presence of denaturants such as urea or GdmCl, the pH should not exceed 7, and the columns should be subjected to thorough rinsing with distilled water at the end of the working day. This can be accomplished by constant rinsing at 0.5 mℓ/min overnight. The columns must never be left standing in denaturants. The operating pressure should also be guarded carefully. In our experience, when an increase in pressure is observed (20 to 30% increase) the frit of the guard column is obstructed and needs attention. This can easily be accomplished by disconnecting the top of the guard column which holds the frit, pumping distilled water thru the frit, and observing the pressure. If the pressure is still increased then a quick soak (5 to 10 min) of the frit in 6 *N* nitric acid, followed by flushing with distilled

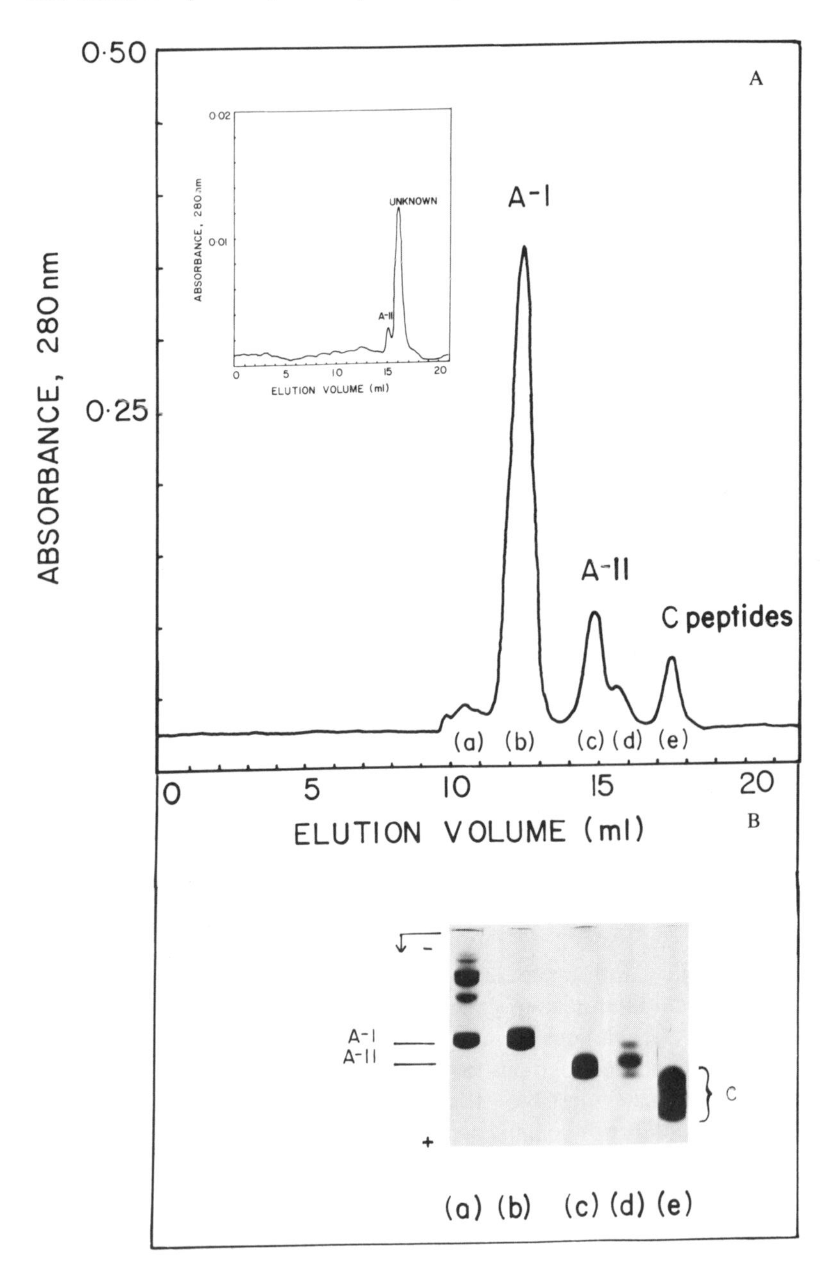

FIGURE 3. (A) elution pattern from the HPLC column of apo HDL in 6 *M* urea pH 3.15. Conditions: injection volume, 100 μℓ at 10 mg/mℓ of apo HDL; flow rate, 1 mℓ/min; chart speed, 1 cm/min; 280-nm abosrbance range, 0.5. Bottom panel: SDS gel electrophoretic patterns of eluted fractions from a through e. Approximately 0.05 mg of protein was applied. The gels were stained with Coomassie® Blue. Inset: elution profile of rechromatographed fraction d (shoulder of apo A-II peak). The chromatographic conditions were essentially the same, except that only 0.02 mg of protein was injected.

water will remedy the problem. Column problems are minimized by using buffers prepared with ultra pure double-distilled water and passed through a 0.4-μm filter. Also ultrafiltration of the sample using either a 0.4-μm or 0.2-μm filter is recommended. In our experience these precautions can significantly increase the lifetime of the column.

CONCLUSIONS

The rapidity, high resolution, and recoveries offered by HPLC in the separation of the apolipoproteins of apo HDL and apo VLDL are attractive reasons for recommending its use and inviting the prediction that this technique will enjoy a primary role in the separation and analysis of lipoprotein apoproteins.

ACKNOWLEDGMENTS

The work by the Authors cited in this review was supported by Program Project USPHS-HL 18577. The authors are grateful to Ms. Barbara Kass for helping in preparing the manuscript.

REFERENCES

1. **Sparks, C. E. and Marsh, J. B.,** Analysis of lipoprotein apoproteins by SDS-gel filtration column chromatography, *J. Lipid Res.,* 22, 514, 1981.
2. **Scanu, A. M., Toth, J., Edelstein, C., Koga, S., and Stiller, E.,** Fractionation of human serum high density lipoprotein in urea solutions. Evidence for polypeptide heterogeneity, *Biochemistry,* 8, 3307, 1969.
3. **Weisgraber, K. H., Rall, S. C., and Mahley, R. W.,** Human E apoprotein heterogeneity. Cysteine-arginine interchanges in the amino acid sequence of the apo E isoforms, *J. Biol. Chem.,* 256, 9077, 1981.
4. **Curry, J. D., McConathy, W. J., Alaupovic, P., Ledford, J. H., and Popovic, J.,** Determination of human apolipoprotein E by electroimmunoassay, *Biochem. Biophys. Acta,* 439, 413, 1976.
5. **Shelburne, F. A. and Quarfordt, S. H.,** A new apoprotein of human plasma very low density lipoproteins, *J. Biol. Chem.,* 249, 1428, 1974.
6. **Utermann, G.,** Isolation and partial characterization of an arginine-rich apolipoprotein from human plasma very low density lipoproteins: apoprotein E, *Hoppe Seyler's Z. Physiol. Chem.,* 356, 1112, 1975.
7. **Lim, C. T., Chung, J., Kayden, H. J., and Scanu, A. M.,** Apoproteins of human serum high density lipoproteins isolation and characterization of the peptides of sephadex fraction V from normal subjects and patients with abetalipoproteinemia, *Biochim. Biophys. Acta,* 420, 332, 1976.
8. **Herbert, P. N., Shulman, R. S., Levy, R. I., and Fredrikson, D. S.,** Fractionation of the C-apoproteins from human plasma very low density lipoproteins. Artifactual polymorphism from carbamylation in urea-containing solutions, *J. Biol. Chem.,* 248, 4941, 1973.
9. **Shore, B. and Shore, V. G.,** Heterogeneity of human plasma VLDL. Separation of species differing in protein components, *Biochemistry,* 12, 502, 1973.
10. **Edelstein, C., Lim, C. T., and Scanu, A. M.,** On the subunit structure of the protein of human serum high density lipoproteins. I. A. study of its major polypeptide component (Sephadex, fraction III), *J. Biol. Chem.,* 247, 5842, 1972.
11. **Scanu, A. M., Lim, C. T., and Edelstein, C.,** On the subunit structure of the protein of human serum high density lipoprotein. II. A study of Sephadex fraction IV, *J. Biol. Chem.,* 247, 5850, 1972.
12. **Shore, B. and Shore, V.,** Isolation and characterization of polypeptides of human serum lipoproteins, *Biochemistry,* 8, 4510, 1969.
13. **Shore, V. and Shore, B.,** Some physical and chemical studies on two polypeptide components of high density lipoproteins of human serum, *Biochemistry,* 7, 3396, 1968.
14. **Weisweiler, P. and Schwandt, P.,** Isolation of apolipoprotein E by chromatography, *Clinica Chemica Acta,* 124, 45, 1982.
15. **Regnier, F. E. and Gooding, K. E.,** High-performance liquid chromatography of proteins, *Biochemistry,* 103, 1, 1980.
16. **Vanecek, G. and Regnier, F. E.,** Variables in the high-performance anion exchange chromatography of proteins, *Anal. Biochem.,* 109, 345, 1980.
17. **Janick, R. A. and Porter, J. W.,** High-performance liquid chromatography of proteins by gel permeation chromatography, *Anal. Biochem.,* 111, 184, 1981.
18. **Hearn, M. T. W., Regnier, F. E., and Wehr, C. T.,** HPLC of peptides and proteins, *Am. Lab.,* Oct., 18.

19. **Regnier, F. E.,** High-performance ion-exchange chromatography of proteins: the current status, *Anal. Biochem.*, 126, 1, 1982.
20. **Hancock, W. S., Capra, J. D., Bradley, W. A., and Sparrow, J. T.,** The use of reversed phase high performance liquid chromatography with radial compression for the analysis of peptide and protein mixtures, *J. Chromatogr.*, 206, 59, 1981.
21. **Hancock, W. S. and Sparrow, J. T.,** Use of mixed-mode high performance liquid chromatography for the separation of peptide and protein mixtures, *J. Chromatogr.*, 206, 71, 1981.
22. **Hancock, W. S., Pownall, H. J., Gotto, A. M., and Sparrow, J. T.,** Separation of apolipoproteins A-I and A-II by ion-paired reversed-phase high performance liquid chromatography, *J. Chromatogr.*, 216, 285, 1981.
23. **Schwandt, P., Richter, W. D., and Weisweiler, P.,** Separation of human C-apolipoproteins by high performance liquid chromatography, *J. Chromatogr.*, 225, 185, 1981.
24. **Hancock, W. S., Bishop, C. A., Gotto, A. M., Harding, D. R., Lamplugh, S. M., and Sparrow, J. T.** Separation of the apoprotein components of human very low density lipoproteins by ion-prepared, reversed phase high performance liquid chromatography, *Lipids*, 16, 250, 1981.
25. **Ott, G. S. and Shore, V. G.,** Anion exchange high performance liquid chromatography of human serum apolipoproteins, *J. Chromatogr.*, 23, 1, 1982.
26. **Wehr, C. T., Cunico, R. L., Ott, G. S., and Shore, V. G.,** Preparative size-exclusion chromatography of human serum apolipoproteins using analytical liquid chromatography, *Anal. Biochem.*, 125, 386, 1982.
27. **Polacek, D., Edelstein, C., and Scanu, A. M.,** Rapid fractionation of human high density apolipoproteins by high performance liquid chromatography, *Lipids*, 16, 27, 1981.
28. **Pfaffinger, D., Edelstein, C., and Scanu, A. M.,** Rapid isolation of apolipoprotein E from human plasma very low density lipoproteins by molecular sieve high performance liquid chromatography, *J. Lipid Res.*, 24, 796, 1983.
29. **Fainaru, M., Havel, R. J., and Imaizumi, K.,** Radioimmunoassay of arginine-rich apolipoproteins of rat serum, *Biochim. Biophys. Acta*, 490, 144, 1977.
30. **Avila, E. M., Holdsworth, G., Sasaki, N., Jackson, R. L., and Harmony, J. A. K.,** Apoprotein E suppresses phytohemagglutinin-activated phospholipid turnover in peripheral blood mononuclear cells, *J. Biol. Chem.*, 257, 5900, 1982.
31. **Weisgraber, K. H. and Mahley, R. W.,** Apoprotein (E--A-II) complex of human plasma lipoproteins. I. Characterization of this mixed disulfide and its identification in a high density lipoprotein subfraction, *J. Biol. Chem.*, 253, 6281, 1978.
32. **Utermann, G., Jacschki, M., and Menzel, J.,** Familial hyperlipoproteinemia type III: deficiency of a specific apolipoprotein (apo E-III) in the very low density lipoproteins, *FEBS Lett.*, 56, 352, 1975.
33. **Foreman, J. R., Karlin, J. B., Edelstein, C., Juhn, D. S., Rubenstein, A., and Scanu, A. M.,** Fractionation of human serum lipoproteins by single-spin gradient ultracentrifugation: quantification of apolipoprotein B and A-I and lipid components, *J. Lipid Res.*, 18, 759, 1977.
34. **Nilsson, J., Mannickarottu, V., Edelstein, C., and Scanu, A. M.,** An improved detection system applied to the study of serum lipoproteins after single-step density gradient ultracentrifugation, *Anal. Biochem.*, 110, 342, 1981.
35. **Kostner, G.M., Patsch, J.R., Sailer, S., Braunsteiner, H. and Holasek, A.,** Polypeptide distribution of the main lipoprotein density classes separated from human plasma by rate zonal ultracentrifugation, *Eur. J. Biochem.*, 45, 611, 1974.
36. **Cheung, M. C. and Albers, J. J.,** The measurement of apolipoprotein A-I and A-II levels in men and women by immunoassay, *J. Clin. Invest.*, 60, 43, 1977.
37. **Cheung, M. C. and Albers, J. J.,** Distribution of cholesterol and apolipoprotein A-I and A-II in human high density lipoprotein subfractions separated by CsCl equilibrium gradient centrifugation: evidence for HDL subpopulations with differing A-I/A-II molar ratios, *J. Lipid Res.*, 20, 200, 1979.
38. **Cheung, M. C. and Albers, J. J.,** Distribution of high density lipoprotein particles with different apoprotein composition: particles with A-I and A-II and particles with A-I but no A-II, *J. Lipid Res.*, 23, 747, 1982.
39. **Scanu, A. M.,** Forms of human serum high density lipoproteins, *J. Lipid Res.*, 7, 295, 1966.
40. **Scanu, A. M. and Edelstein, C.,** Solubility in aqueous solutions of ethanol of small molecular weight peptides of the serum very low density and high density lipoproteins: relevance to the recovery problem during delipidation of serum lipoproteins, *Anal. Biochem.*, 44, 576, 1971.
41. **Snyder, L. R.,** A rapid approach to selecting the best experimental conditions for high speed liquid column chromatography. I. Estimating initial sample resolution and the final resolution required by a given problem, *J. Chromatogr. Sci.*, 10, 200, 1972.

HOMOGENEITY OF CRYSTALLINE INSULIN ESTIMATED BY GPC AND REVERSED PHASE HPLC

B. S. Welinder

INTRODUCTION

Crystalline insulin generally contains about 90% insulin peptide plus a number of contaminants, some of which are insulin-like (desamidoinsulin, proinsulin, etc.) whereas others are noninsulin-like coextraction products from the pancreas (glucagon, pancreatic polypeptide, etc.). Chromatographic purification of crystalline insulin (ion-exchange chromatography, gel filtration chromatography) yields highly purified insulin consisting of more than 99% insulin peptide plus (eventually) small amounts of desamidoinsulin.

Gel filtration chromatography of crystalline insulin under disaggregating conditions separates the compounds with molecular weights >6000 (b-fraction containing pro-insulin, insulin dimer, etc.) from insulin and insulin-like compounds with molecular weight ≈6000 (c-fraction containing desamidoinsulin, arginine insulin, etc.). Since the discovery of proinsulin,[1] gel filtration has usually been performed by low-pressure chromatography (Sephadex®, Bio-Gel® in 1 to 3 *M* acetic acid).

The first separation of the b- and c-components on HP-GPC matrices in aqueous buffers was performed using Bondagel E-125 (Waters), TSK SW2000 (Toyo Soda), or I-125 (Waters) and eluted with urea-containing buffers.[2] Since then, a number of proteins and peptides have been analyzed using primarily TSK- and I-125-matrices and eluted with denaturing as well as more physiological buffers. The present work describes the use of I-125 columns eluted with organic solvents in combination with RP-HPLC for the elucidation of the composition of the b- and c-fractions of crystalline insulin.

MATERIALS AND METHODS

Chemicals

Acetonitrile and isopropanol were obtained from Rathburn (HPLC-Grade S), acetic acid and phosphoric acid from Merck (p.a.-grade), and triethylamine (99%) from Aldrich. Distilled water was drawn from a Millipore Milli-Q-system (conductivity < 0.1 μs) and the eluants were filtered (Millipore 0.45 μm) and vacuum degassed before use. Porcine Na-insulinate (batch G-63) containing about 80% insulin, 10% NaCl, and 10% water was obtained from Nordisk Gentofte, Copenhagen.

HP-GPC

Waters I-125 protein separation columns were eluted using Waters M6000 or M45 pumps, Waters U6K injector, Waters 440 UV-detector, and Waters Data Module.

RP-HPLC

Nucleosil® 10-C_{18} columns (250 mm × 8 mm i.d.) were slurry-packed (in the author's laboratory). Two Waters M6000 pumps were controlled by a Waters 660 solvent controller. Buffer A contained 0.25 *M* triethylammonium phosphate (TEAP) in water, pH 3.00, buffer B a mixture of 50% acetonitrile and 50% buffer A. The starting conditions were 45% A/55% B. After sample application the composition of the eluant was changed linearly over 25 min to 35% A/65% B and was thereafter kept constant for 5 min. The sample was applied using a Waters WISP 710A sample controller and the absorbance of the eluate at 210 nm was measured with a Pye-Unicam UV-detector. Radioimmunological determinations of proinsulin-like immunoreactivity (PLI) was performed as described elsewhere.[3]

RESULTS

Figure 1 shows the HP-GPC analysis of crystalline insulin on 2 × I-125 columns eluted at 0.5 mℓ/min with 20% acetic acid using varying amounts of acetonitrile. The quality of the separation between the b- and c-fractions was found to be dependent on the concentration of acetonitrile. Acetonitrile at a concentration of 20% was chosen for the succeeding experiments with this batch of column packing.

At 20% acetonitrile concentration the b-fraction peak shows a clearly marked shoulder towards the c-fraction. The PLI shows a maximum in this shoulder (see Figure 2). If the flow rate is further reduced to 0.2 mℓ/min, the b-fraction is resolved into two peaks (Figure 3), and the PLI is located in the latter of the two peaks.

The RP-HPLC analysis of crystalline insulin is shown in Figure 4. The major peak corresponds to insulin peptide, the largest minor peak contains monodesamidoinsulin. Thereafter five clearly resolved minor components are eluted and the PLI activity is located in the third one.

The composition of the b- and c-fractions isolated from HP-GPC on I-125 columns was analyzed using RP-HPLC (Figure 5) and compared with the composition of similar fractions obtained by low-pressure chromatography (Sephadex® G-50 (SF), 3 *M* acetic acid) (Figure 6). In both cases similar patterns were obtained: the b-fraction contained mainly two components one of which coeluted with the peak containing PLI activity. The other peak (which probably contained insulin dimer) coeluted with the fourth minor peak in the insulin chromatogram (Figure 4), but this peak was clearly heterogeneous since the chromatogram of the c-fraction (Figures 5 and 6) contained no peak 3 material, whereas peak 4 was still present but in reduced amount.

The recovery of insulin from I-125 columns eluted with acetic acid/acetonitrile was estimated to be 98% by UV measurements of the eluate. Estimations of the amount of b-fraction in insulin preparations are most precisely performed using an electronic integrator. If the amount of b-fraction is low, the most reproducible results are obtained if the integrator performs a GPC analysis with automatic digitization of the chromatogram. A similar separation to the one described here can be performed by substituting acetonitrile with isopropanol at the same concentration,[4] although the column back-pressure becomes considerably higher. If I-125 columns are eluted with phosphoric acid (pH 2.10) or TEAP (pH 2.20) in combination with various concentrations of isopropanol or acetonitrile, a very poor separation is obtained. Addition of salts (e.g., Na_2SO_4 or NaH_2PO_4) has no effect on the separation.

DISCUSSION

Estimations of the content of compounds in insulin preparations with molecular weights higher than that of the insulin monomer must be performed under disaggregating conditions in order to suppress the marked self-aggregation of the insulin molecule. In low-pressure chromatography elution, 1 *M* acetic acid is sufficient, whereas the presently available HP-GPC matrices demand more vigorous eluants: combinations of 7 *M* urea, acids, and nonionic detergents (Bondagel), 7 *M* urea plus acid (I-125, TSK SW 2000) or 7 *M* urea at neutral pH.[16] The present work describes the use of high concentrations of acetic acid in combination with organic solvents commonly used in reversed-phase separations. This eluant is clearly superior to urea-containing buffers since the b-fraction is resolved in the two main components: insulin dimer (molecular weight 12,000) and proinsulin (molecular weight 9000), both well separated from the c-fraction. Intermediate insulin (proinsulin where one of the two connections between the c-peptide and the insulin molecule has been cleaved) probably coelutes with proinsulin in HP-GPC as well as in this RP-HPLC system since only one PLI peak was found in both analyses.

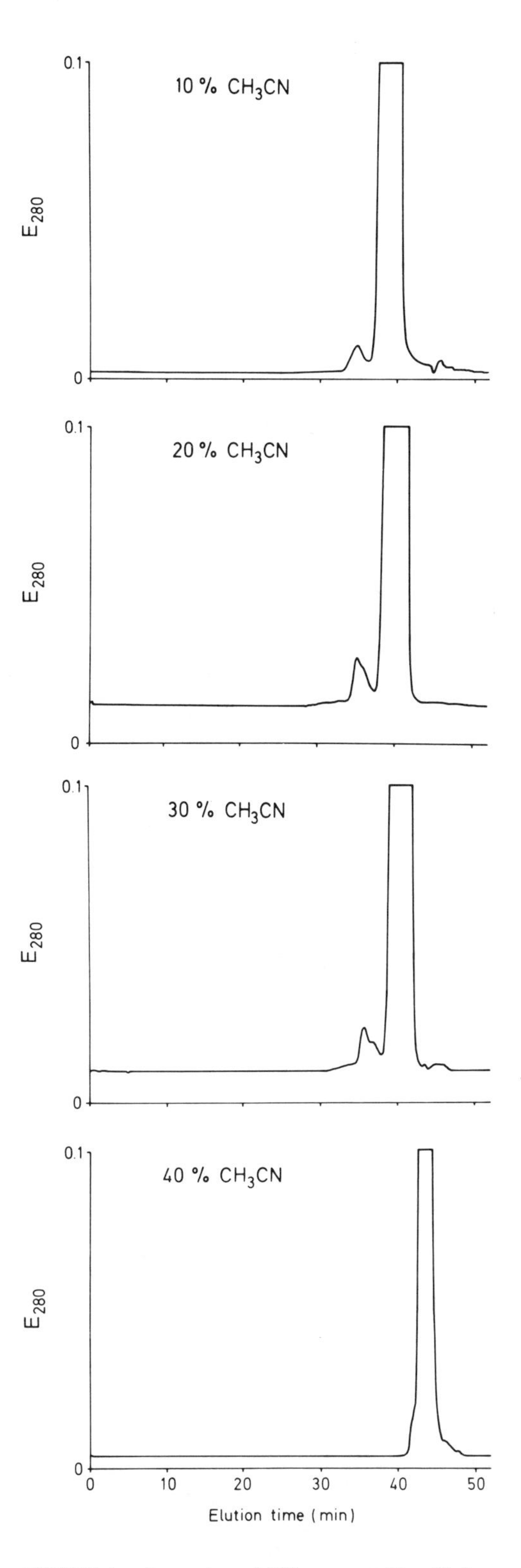

FIGURE 1. Separation of 500 μg crystalline Na-insulinate on 2 × I-125 columns eluted with 20% acetic acid plus varying amounts of acetonitrile. Flow rate: 0.5 mℓ/min.

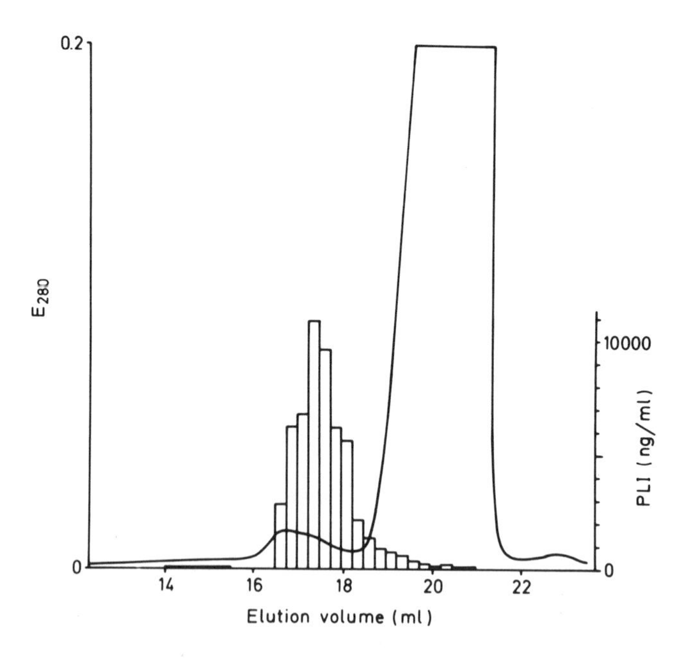

FIGURE 2. Separation of 300 μg crystalline Na-insulinate on 2 × I-125 columns eluted at 0.5 mℓ/min with 20% acetic acid/20% acetonitrile/60% water. The histogram shows the PLI distribution.

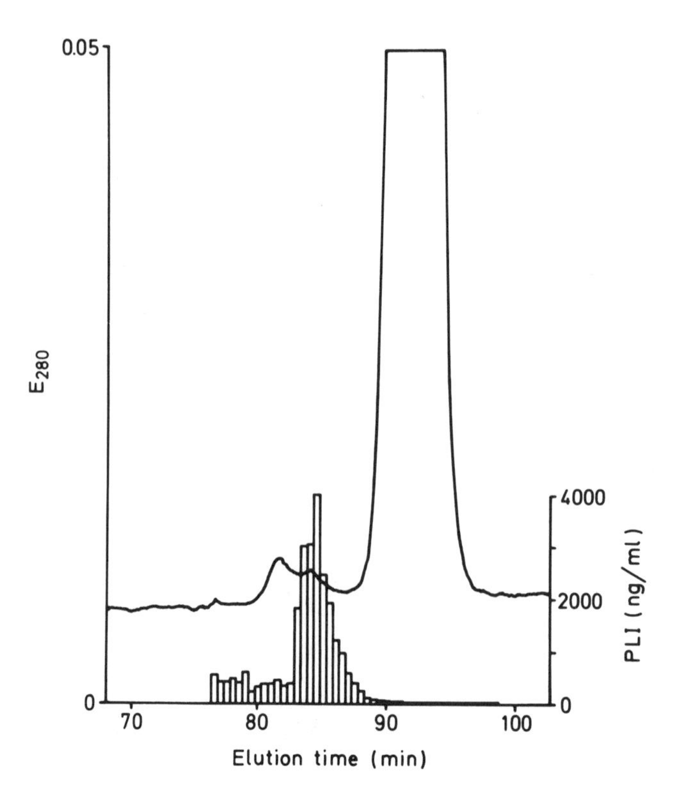

FIGURE 3. Separation of 100 μg crystalline Na-insulinate on 2 × I-125 columns eluted at 0.2 mℓ/min with 20% acetic acid/20% acetonitrile 60% water. The histogram shows the PLI distribution.

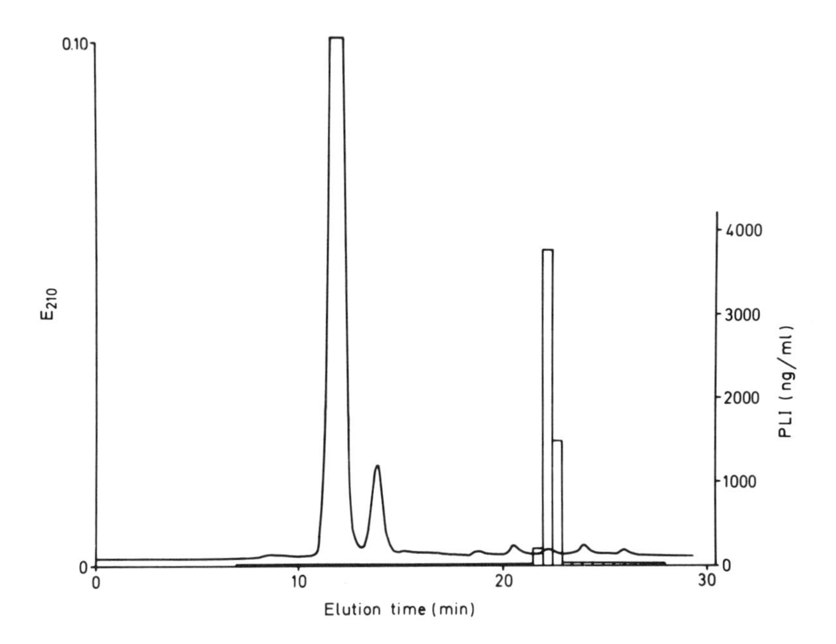

FIGURE 4. RP-HPLC separation of 300 μg crystalline Na-insulinate on a 250 × 8 mm Nucleosil® 10-C_{18} column. For details of elution see text.

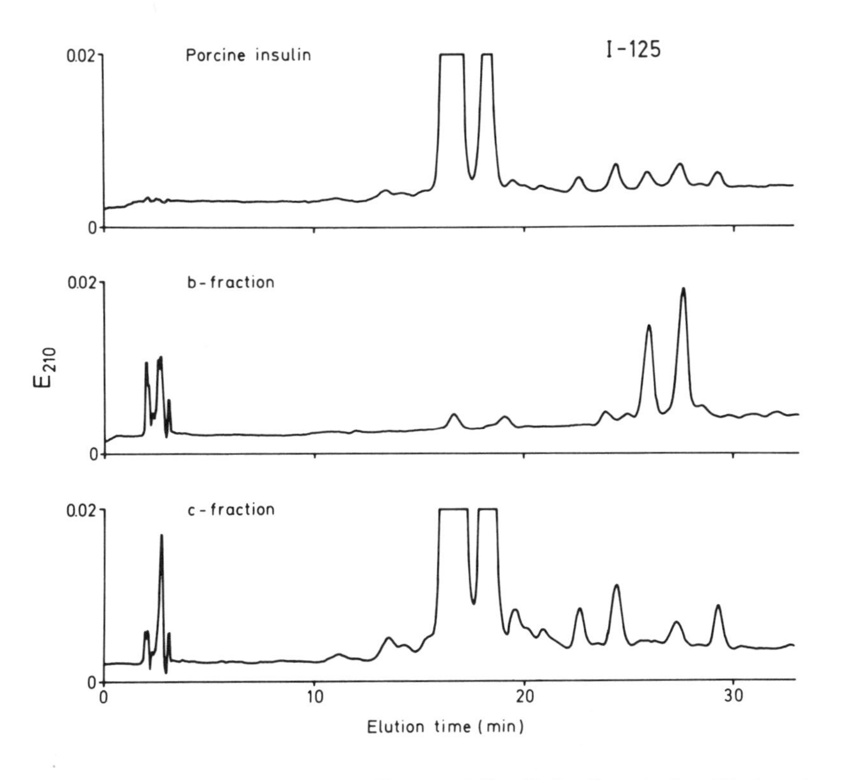

FIGURE 5. RP-HPLC separation of 300 μg crystalline Na-insulinate (top) and the b- and c-fractions isolated under the experimental condition shown in Figure 2.

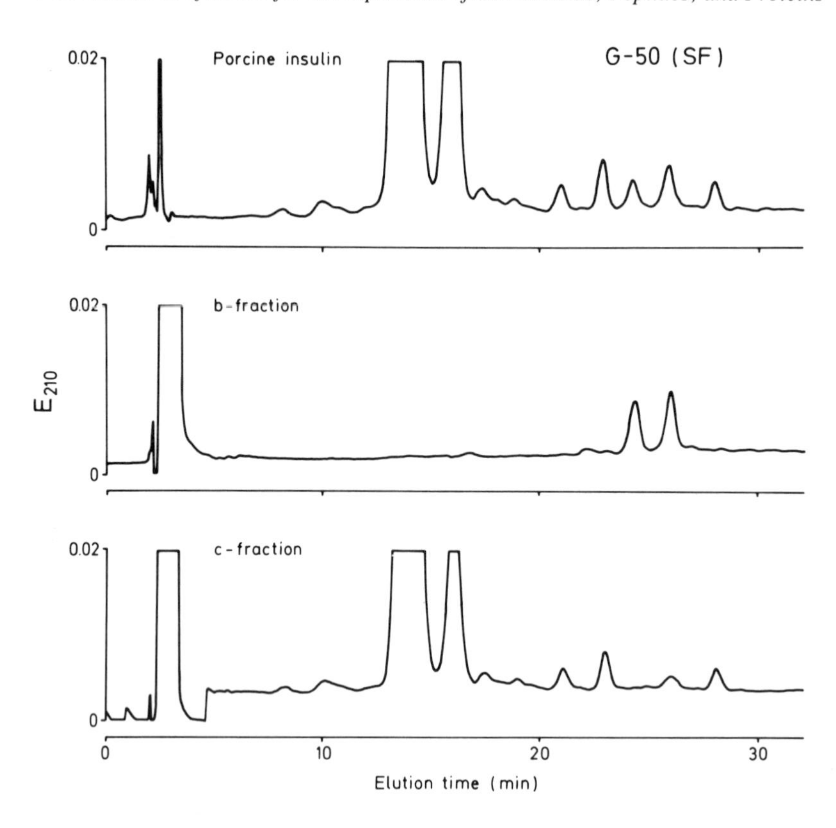

FIGURE 6. RP-HPLC separation of 300 μg crystalline Na-insulinate (top) and b- and c-fractions obtained by separating 50 mg Na-insulinate on a 100 × 2.5 cm G-50(SF) column eluted at 20 mℓ/hr with 3 *M* acetic acid.

The eluant described here has some advantages (good separation potential, lyophilizable) but in other respects it is far from ideal: the pH of the eluant is too close to the dissolving point of silica, it is poisonous, expensive, and not transparent for low wavelength UV detection. These drawbacks are characteristic of problems encountered with presently available matrices for the HP-GPC of peptides and proteins, i.e., reduced choice of pH (2 to 8) and high nonspecific adsorption to the matrix (especially at acid pH and for peptides). If HP-GPC of proteins and peptides as a separation process is to advance to a position comparable to that of low-pressure chromatography these limitations must be overcome. It is also very restrictive that the matrices are available only in expensive prepacked columns, which do not allow semipreparative or preparative separations.

It seems more attractive to focus upon RP-HPLC as a separation method for polypeptides, such as insulin. Reversed-phase systems published so far have demonstrated the possibility of separating insulin from different species (human, bovine, porcine) or insulin from desamidoinsulin or proinsulin.[5-15] A reasonable test substance for the resolving capacity of a RP-HPLC system could be crystalline insulin. In the best of the conventional separation methods normally applied in protein chemistry, e.g., iso-electric focusing in narrow pH gradients, not less than 20 contaminants can be found. Although a similar number of contaminants can be found using RP-HPLC (Figure 5), RP-HPLC is superior to iso-electric focusing with respect to speed, quantitation, and recovery of the separated fractions. Quantitation is especially important with respect to purity estimations of insulin preparation. By careful optimization of the RP-HPLC system, it is possible to estimate the content of a number of contaminants in the range of 0.1 to 0.05%.[16] RP-HPLC seems to be the natural progression from low-pressure gel filtration chromatography/ion-exchange chromatography and radioimmunological estimations. The most serious disadvantages are the characteristics of the eluants, which generally are poisonous, expensive, and nonphysiological. This may be tolerable for purely analytical work, but very limiting for preparative separations (e.g., isolation, characterization, and biological testing of insulin derivatives).

It therefore seems natural, that RP-HPLC analysis (possibly combined with HP-GPC), in the near future will be the method of choice for estimating the extent of contamination of insulin preparations both on the production scale as well as in the pharmacopoeia.

REFERENCES

1. **Steiner, D. F., Hallund, O., Rubenstein, A., Cho, S. and Bayliss, C.,** Isolation and properties of proinsulin, intermediate forms and other minor components from crystalline bovine insulin, *Diabetes,* 17, 725, 1968.
2. **Welinder, B. S.,** Gel permeation chromatography of insulin, *J. Liq. Chromatogr.,* 3, 1399, 1980.
3. **Linde, S., Hansen, B., and Lernmark, Å.,** Stable iodinated polypeptide hormones prepared by polyacrylamide gel electrophoresis, *Anal. Biochem.,* 107, 165, 1980.
4. **Andresen, F. H.,** unpublished method, Nordisk Gentofte, Copenhagen.
5. **Calam, D. H.,** Applications of chromatography in the standardization and controls of biological products, *J. Chromatogr.,* 167, 91, 1978.
6. **Mönch, W. and Dehnen, W.,** High-performance liquid chromatography of polypeptides and proteins on a reversed-phase support, *J. Chromatogr.,* 147, 415, 1978.
7. **Rivier, J. E.,** Use of trialkyl ammonium phosphate (TAAP) buffers in reverse-phase HPLC for high resolution and high recovery of peptides and proteins, *J. Liq. Chromatogr.,* 1, 343, 1978.
8. **Damgård, M. and Markussen, J.,** Analysis of insulins and related compounds, by HPLC, *Horm. Metab. Res.,* 11, 580, 1979.
9. **Dinner, A. and Lorenz, L.,** High performance liquid chromatographic determination of bovine insulin, *Anal. Chem.,* 51, 1872, 1979.
10. **Hearn, M. T. W., Hancock, W. S., Hurrell, J. G. R., Flemming, R. J., and Kemp, B.,** The analysis of insulin related peptides by reverse-phase high-performance liquid chromatography, *J. Liq. Chromatogr.,* 2, 919, 1979.
11. **O'Hare, M. J. and Nice, E. C.,** Hydrophobic high-performance liquid chromatography of hormonal polypeptide and proteins on alkylsilane-bonded silica, *J. Chromatogr.,* 171, 209, 1979.
12. **Terabe, S., Konaka, R., and Inouye, K.,** Separation of some polypeptide hormones by high-performance liquid chromatography, *J. Chromatogr.,* 172, 163, 1979.
13. **Biemond, M. E. F., Sipman, W. A., and Oivié, J.,** Quantitative determination of insulin by gradient elution HPLC, in *Insulin. Chemistry, Structure and Function of Insulin and Related Hormones,* Brandenburg, D. and Wollmer, A., Eds., Walter de Gruyter, Berlin, 1980, 201.
14. **Vigh, G., V.-Puchony, Z., Hlavay, J., and P.-Hites, E.,** Factors influencing the retention of insulins in reversed-phase high-performance liquid chromatographic systems, *J. Chromatogr.,* 236, 51, 1982.
15. **Scepesi, G. and Gazdag, M.,** Improved high-performance liquid chromatographic method for the analysis of insulins and related compounds, *J. Chromatogr.,* 218, 597, 1981.
16. **Welinder, B. S.,** unpublished results.

Examples of Protein Separations by Reversed-Phase HPLC

ENDORPHIN SEPARATIONS BY REVERSED-PHASE HPLC

David D. Gay and Robert A. Lahti

INTRODUCTION

Following their discovery a few years ago, enkephalins and endorphins have been the subject of very active research. The distribution, structures, metabolism, and activities of these endogenous opioid compounds have been studied. HPLC separation of these compounds has been an important tool in this field of research.

To review the structures of the endorphins, the parent compound is β-lipotropin (β-LPH), with 91 amino acid residues. The largest endorphin is β-endorphin (β-EP), which corresponds to β-LPH residues 61-91 and has the following amino acid sequence:

```
 61                                      70
H-Tyr-Gly-Gly-Phe-Met-Thr-Ser-Glu-Lys-Ser-Gln-Thr-Pro-Leu-Val-Thr
             90                                        80
  HO-Glu-Gly-Lys-Lys-Tyr-Ala-Asn-Lys-Ile-Ile-Ala-Asn-Lys-Phe-Leu

                                               (β-endorphin)
```

Other peptides of interest and the corresponding residue numbers are γ-endorphin (γ-EP), 61-77; des-Tyr-γ-endorphin (dTγE), 62-77; α-endorphin (α-EP), 61-76; Met-enkephalin (ME), 61-65, Tyr-Gly-Gly-Phe-Met; Leu-enkephalin (LE), Tyr-Gly-Gly-Phe-Leu.

EXTRACTION AND PREPARATION

In assaying biological samples, the first step is to efficiently extract the compounds from tissue. To assay endorphins or enkephalins in rat or mouse brain tissue, it is probably best to sacrifice the animal by microwave irradiation rather than by decapitation.[1,2] Depending on the peptide and brain area, some degradation may occur after decapitation.

The endorphins may be extracted from homogenized or sonicated tissue with 1 *M* acetic acid,[1-3] 0.4 *N* formic acid,[4] 0.1 *M* HCl,[5-7] acetone acidified with HCl,[5,8-11] or with a combination of acid and butanol.[12,13] The choice of solvent depends partly on the compounds and tissues involved and the subsequent assay system, but acetic acid or acidified acetone have been used most often. It is usually desirable to remove insoluble material by centrifugation, lyophilize the supernatant, and redissolve it in a solution compatible with the next step in the assay.

A biological extract may still require purification or preliminary separation before HPLC analysis. Column chromatography complements reversed-phase (RP) HPLC by separating compounds on the basis of molecular size or ionization rather than polarity. Various Sephadex® or Bio-Gel® columns have been used to remove high-molecular-weight material[3] or salts[14,15] or for preliminary separations.[5,9,16] Carboxymethyl cellulose[9,15] and ion-exchange columns[6,14] may also be used to purify samples.

Fractions from these columns can then be analyzed separately by HPLC. This is particularly useful in identifying peptides in a complex mixture.

SEP-PAK® CARTRIDGES

Sep-Pak® cartridges (Waters Associates, Inc.) are a recent adaptation of HPLC technology,

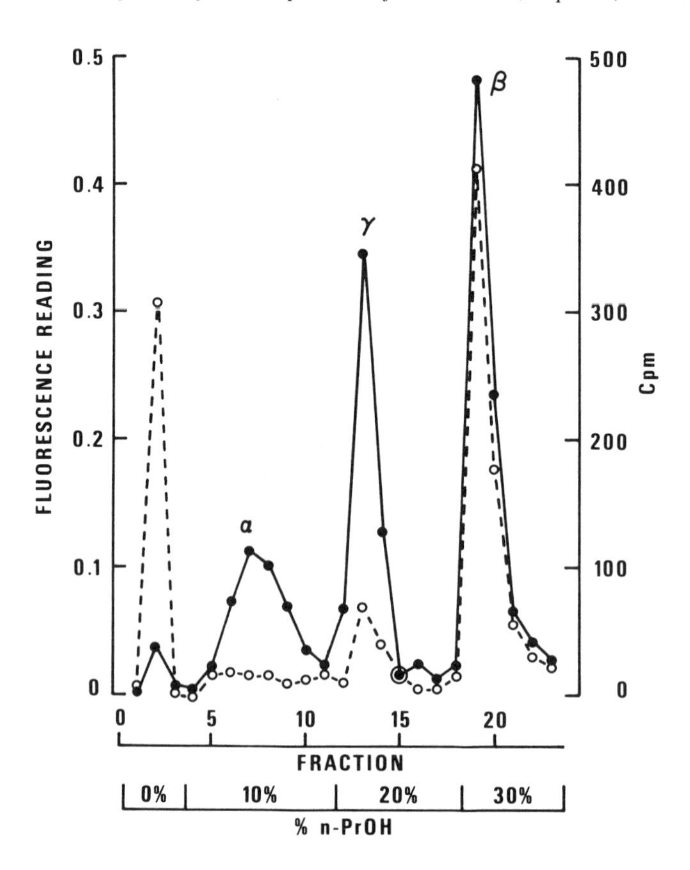

FIGURE 1. Separation of α-, γ-, and β-endorphin on a Sep-Pak® cartridge. The 800-μℓ sample contained 0.42 mg, 0.34 mg, and 19 μCi of α-, γ-, and ^{125}I-β-endorphin, respectively, and was eluted with a step gradient of *n*-propanol in pH 4.0 formate-pyridine buffer. The fluorescence reading (●—●) and cpm (○---○) for an aliquot of each fraction is shown. The first fraction was collected during application of the sample. (From Gay, D. D. and Lahti, R. A., *Int. J. Peptide Protein Res.*, 18, 107, 1981. With permission.)

and we have found them useful for sample preparation or complete separations. The disposable cartridges contain about 1 cc of C_{18} reversed-phase packing and have syringe fittings for solvent application.

An HPLC procedure using a mobile phase gradient[3] was adapated to the Sep-Pak® cartridge. As shown in Figure 1, α-, γ-, and β-endorphin are completely separated using a step gradient of *n*-propanol in a formate-pyridine buffer.[17] Met- and Leu-enkephalin can also be separated by this procedure, and enkephalins can be separated from their metabolic products.[17]

Elution from the cartridges is reproducible with nearly complete recovery. Five replicate samples containing 11 pmol of ^{3}H-ME were eluted from cartridges with a mean recovery of 93.4% and a standard error of the mean of 0.7%.

This technique will only separate a limited number of compounds in a single sample, and not all endorphin separations can be done on a cartridge with this solvent system. However, this simple technique is useful in preparing samples for RIA or HPLC assay, and for desalting or concentrating samples. The cartridges can also be used to purify synthesized peptides, separate metabolic products, or check the purity of standards.

HPLC PROCEDURES

The theories of RP-HPLC separation of peptides and proteins have been discussed in earlier chapters of this handbook. A few comments on the application of these theories to endorphin and enkephalin separations are pertinent however.

Retention is determined primarily by the number of hydrophobic side chains present.[18,19] Polar and neutral residues have little interaction with the stationary phase. However, especially for smaller peptides, lowering the pH and protonating the carboxyl group will increase retention.

Separating the members of the endorphin family of compounds is difficult since they may differ in only one residue in a chain of up to 30 or more amino acids. It is important however, because the deletion or substitution of one amino acid may greatly change the activity of the peptide. Fortunately the change in structure usually also changes the lipophilic properties of the compound and therefore its RP-HPLC retention.

Separating Met- from Leu-enkephalin, or γ-endorphin from α-endorphin and dTγE are examples of such separations. These can be efficiently done by HPLC but would be cumbersome by most other methods. Biological samples, containing a host of peptides, also present a challenge which can be met using HPLC separations.

HPLC separations are summarized in Table 1. While not exhaustive, the list provides a broad selection of pertinent references and an organized introduction to the literature. To be more practical, the references are grouped according to the separations performed. References 1 through 16 each include assays of biological samples.

Methods in the first group have been used to separate compounds ranging in size from ME to β-EP in a single sample. An example of these impressive separations is shown in Figure 2, reprinted from Loeber and Verhoef.[2]

Nice and O'Hare[18] also separated β-EP and β-LPH by HPLC. This is difficult because the larger β-LPH is only a little more lipophilic than the 31-residue β-EP. The effect of chain length apparently begins to be lost in this size range. The β-EP and β-LPH separation can also be done, less efficiently, by column chromatography. The separation must be made before RIA of endogenous β-EP levels because the available β-EP RIAs cross-react with β-LPH.

Other references describe similar methods for separating particular groups of peptides. These procedures emphasize separation of the smaller, rapidly eluting peptides, or the larger, more strongly absorbed peptides.

It is interesting to note the variety of procedures used to separate enkephalins and endorphins. The majority of the references cite Waters μBondapak® C_{18}-columns. However, other brands of microparticulate reversed-phase columns have been used with equal success. Separations done on ion exchange and Fatty Acid Analysis® columns are also listed.

A wide variety of mobile phases and gradients have been used. Most procedures have an acidic mobile phase (HOAc, NH_4OAc, NaH_2PO_4, HCOOH, TFA, or others) with a gradient of organic solvent (CH_3CN, MeOH, or *n*-PrOH). Most gradients are linear or slightly concave. Addition of acid to the organic phase can help maintain a stable UV baseline. Obviously many mobile-phase combinations are possible and can be used effectively to emphasize the particular separation desired. Isocratic conditions are entirely satisfactory for separating a few, or closely related, peptides.

Most of the references do not include internal standards, but α-MSH and ACTH (1—10) are two suggestions.[2] Recovery of peptides has been improved by increasing acid or salt concentration, use of TFA,[13] addition of glycylglycine to the mobile phase,[2] or fraction collection in polystyrene tubes. Use of salt-free buffers may simplify subsequent assay of the fractions, however.

Table 1
SUMMARY OF HPLC PROCEDURES FOR SEPARATING ENKEPHALINS, ENDORPHINS, AND RELATED COMPOUNDS

Column	Mobile phase, gradient	Detection and assay	Ref.
	ME: LE; α-, γ-, β-EP; and Others		
μBondapak®	0.01 *M* NH_4OAc, pH4.15, 30 → 75% MeOH	UV, RIA	1, 2, 8, 20
Lichrosorb® C_{18}	0.05 *M* HOAc/1 *M* pyridine, pH5.5, or 0.5 *M* formic acid/pyridine, pH3 or 4, both with 0 → 20% *n*-PrOH	Fluor, RIA	3, 9
μBondapak®	5% HOAc, *n*-PrOH gradients (e.g., 10 → 40%)	UV, RIA, bioassay	5, 6
μBondapak®	0.01 *M* NH_4OAc, pH4.5, isocratic CH_3CN at 25 to 55%	UV	21
Hypersil ODS	0.1 *M* NaH_2PO_4, pH2.1, 10 → 40% CH_3CN (some at 45°C)	UV, fluor.	18
	ME, LE, Small Peptides		
Dupont SCX	0.1 *M* phosphate, pH 6	Electrochem., bioassay	4, 12
Ultrasphere ODS	0.1 *M* NaH_2PO_4, pH2.1, 0 → 23% CH_3CN	UV	10
Partisil® 5R	30 or 87% aq. CH_3CN, equilibration with $CuSO_4$ and NH_4OH	UV	22
μBondapak®	0.05 *M Tris*, pH7.4, 35 → 100% MeOH, used derivatized peptides	Fluor.	23
Waters Fatty Acid Analysis®	8% HOAc:MeOH (1:1)	UV	24
	β-EP, Fragments		
Altex Ultrasphere ODS	25% HOAc/0.1 *M* NaCl, pH2.5, 5 → 40% CH_3CN bilinearly	UV	11
μBondapak®	5 m*M* TFA, pH2.5, 30 → 50% CH_3CN	UV	13
μBondapak®	10 m*M* HCOOH/MeOH (1:1)	UV, bioassay	14
μBondapak®	0.02 *M* Triethylamine PO_4, pH3, 5 → 70% CH_3CN	UV, LSC	15
	β-LPH		
Partisil®-10 ODS-2	0.05 *M* HOAc/1 *M* pyridine, pH5.5, 10 → 25% *n*-PrOH	Fluor.	16, 19 (See also 15, 18 above)
	γ-EP, dTγE		
μBondapak®	10 m*M* NH_4OAc, pH4.15, 30 → 75% acid MeOH, also 10 m*M* phosphate, pH6.9, 30 → 50% acid MeOH	UV	25
μBondapak®	0.1% HOAc, 0 → 60% MeOH; also 0.1% H_3PO_4, 17% CH_3CN	UV	26
	Dynorphin		
μBondapak®	1 *M* HOAc, pH2.5 or 10 m*M* NH_4COOH, pH4, 5 → 40% CH_3CN	RIA	7

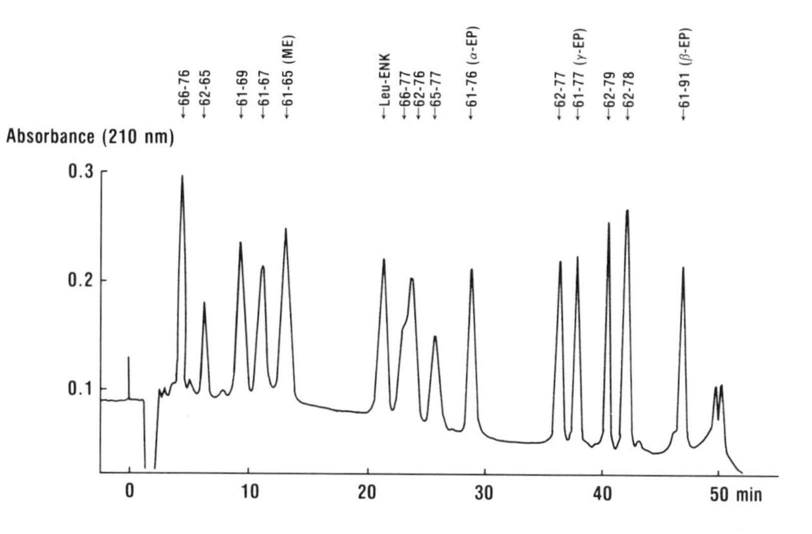

FIGURE 2. Chromatography of a mixture of synthetic human β-lipotropin 61-91 fragments (10 to 20 μg per peptide) on a μBondapak® C_{18}-column. UV absorbance was measured at 210 nm (0.4 absorbance unit full scale). The last two peaks represent organic contaminants of the mobile phase. (From Loeber, J. G. and Verhoef, J., in *Methods in Enzymology*, Vol. 73, Academic Press, New York, 1981, 261. With permission.)

DETECTION AND ASSAY

UV detection is used routinely to monitor column eluates. Fluorescamine assay, of the sample stream or collected fractions, greatly increases sensitivity. These methods are usually sufficient for *in vitro* studies of endorphin and enkephalin metabolism or synthesis.

Assay of biological samples is complicated by the host of peptides present and the small concentration of the compounds of interest. Other studies attempt to find and identify unknown compounds or metabolites in the sample. For such samples HPLC separation and commercially available RIAs are a powerful combination. The RIAs are a sensitive and reasonably specific complement of the HPLC procedure. Without a prior separation step, however, RIAs suffer from cross-reaction with closely related compounds of the same family.

Other sensitive assays of endorphins and enkephalins include bioassays using mouse vas deferens or guinea pig ileum and *in vitro* competitive binding assays. These procedures measure opioid activity.

The several research groups studying enkephalins and endorphins have developed a variety of HPLC separation procedures. Assay techniques specific for this family of compounds have also been developed. In combination with other methods, HPLC separations have made it possible to measure each endorphin compound in almost any sample.

REFERENCES

1. **Verhoef, J., Loeber, J. G., Burbach, J. P. H., Gispen, W.H., Witter, A., and de Wied, D.,** α-Endorphin, γ-endorphin and their des-tyrosine fragments in rat pituitary and brain tissue, *Life Sci.*, 26, 851, 1980.
2. **Loeber, J. G.and Verhoef, J.,** High-pressure liquid chromatography and radioimmunoassay for the specific and quantitative determination of endorphins and related peptides, in *Methods in Enzymology*, Vol. 73, Academic Press, New York, 1981, 261.

3. **Lewis, R. V., Stein, S., and Udenfriend S.,** Separation of opioid peptides utilizing high performance liquid chromatography, *Int. J. Peptide Protein Res.*, 13, 493, 1979.
4. **Bohan, T. P. and Meek, J. L.,** Met-enkephalin: rapid separation from brain extracts using high-pressure liquid chromatography, and quantitation by binding assay, *Neurochem. Res.*, 3, 367, 1978.
5. **Dell, A., Etienne, T., Morris, H. R., Beaumont, A., Burrell, R., and Hughes, J.,** Neuropeptides: high resolution purification procedures and their application to the study of new opioid peptides and opioid peptide precursors, *Molecular Endocrinology*, MacIntyre, I. and Szelke, M., Eds., Elsevier/North Holland, Amsterdam,
6. **Morris, H. R., Etienne, A. T., Dell, A., and Albuquerque, R.,** A rapid and specific method for the high resolution, purification and characterization of neuropeptides, *J. Neurochem.*, 34, 574, 1980.
7. **Seizinger, B. R., Hollt, V., and Herz, A.,** Evidence for the occurrence of the opioid octapeptide dynorphin-(1-8) in the neurointermediate pituitary of rats, *Biochem. Biophys. Res. Commun.*, 102, 197, 1981.
8. **Loeber, J. G., Verhoef, J., Burbach, J. P. H., and Witter, A.,** Combination of high pressure liquid chromatography and radioimmunoassay is a powerful tool for the specific and quantitative determination of endorphins and related peptides, *Biochem. Biophys. Res. Commun.*, 86, 1288, 1979.
9. **Swann, R. W.and Li, C. H.,** Isolation and characterization of β-endorphin-like peptides from bovine brains, *Proc. Natl. Acad. Sci. U.S.A.*, 77, 230, 1980.
10. **Mousa, S., Mullet, D., and Couri, D.,** Sensitive and specific high performance liquid chromatographic method for methionine and leucine enkephalins, *Life Sci.*, 29, 61, 1981.
11. **Evans, C.J., Weber, E., and Barchas, J. D.,** Isolation and characterization of α-*N*-acetyl-β-endorphin(1-26) from the rat posterior/intermediate pituitary lobe, *Biochem. Biophys. Res. Commun.*, 102, 897, 1981.
12. **Meek, J. L. and Bohan, T. P.,** Use of high pressure liquid chromatography (HPLC) to study enkephalins, *Adv. Biochem. Psychopharm.*, 18, 141, 1978.
13. **Dunlap, C. E., III, Gentleman, S., and Lowney, L. I.,** Use of trifluoroacetic acid in the separation of opiates and opioid peptides by reversed-phase high-performance liquid chromatography, *J. Chromatogr.*, 160, 191, 1978.
14. **Gentleman, S., Lowney, L. I., Cox, B. M., and Goldstein, A.,** Rapid purification of β-endorphin by high-performance liquid chromatography, *J. Chromatogr.*, 153, 274, 1978.
15. **Gianoulakis, C., Seidah, N. G., Routhier, R., and Chretien, M.,** Biosynthesis and characterization of adrenocorticotropic hormone, α-melanocyte-stimulating hormone, and an NH_2-terminal fragment of the adrenocorticotropic hormone/β-lipotropin precursor from rat pars intermedia, *J. Biol. Chem.*, 254, 11903, 1979.
16. **Naude, R. J., Chung, D., Li, C. H., and Oelofsen, W.,** β- Lipotropin: primary structure of the hormone from the ostrich pituitary gland, *Int. J. Peptide Protein Res.*, 18, 138, 1981.
17. **Gay, D. D. and Lahti, R. A.,** Rapid separation of enkephalins and endorphins on Sep-Pak® reverse phase cartridges, *Int. J. Peptide Protein Res.*, 18, 107, 1981.
18. **Nice, E. C. and O'Hare, M. J.,** Simultaneous separation of β-lipotropin, adrenocorticotropic hormone, endorphins and enkephalins by high-performance liquid chromatography, *J. Chromatogr.*, 162, 401, 1979.
19. **Yamashiro, D. and Li, C. H.,** Partition and high-performance liquid chromatography of β-lipotropin and synthetic β-endorphin analogues, *J. Chromatogr.*, 215, 255, 1981.
20. **Burbach, J. P. H., de Kloet, E. R., Schotman, P., and de Wied, D.,** Proteolytic conversion of β-endorphin by brain synaptic membranes, *J. Biol. Chem.*, 256, 12463, 1981.
21. **Currie, B. L., Chang, J. K., and Cooley, R.,** High performance liquid chromatography of enkephalin and endorphin peptide analogs, *J. Liq. Chromatogr.*, 3, 513, 1980.
22. **Guyon, A., Roques, B. P., Guyon, F., Foucault, A., Perdrisot, R., Swerts, J. P., and Schwartz, J. C.,** Enkephalin degradation in mouse brain studied by a new H.P.L.C. method: further evidence for the involvement of carboxypeptidase, *Life Sci.*, 25, 1605, 1979.
23. **Wu, K. M., Sloan, J. W., and Martin, W. R.,** Development of a combined high-performance liquid chromatographic-fluorometric quantitative assay for enkepahlins, *J. Chromatogr.*, 202, 500, 1980.
24. **Feldman, J. A., Cohn, M. L., and Blair, D.,** Neuroendocrine peptides — analysis by reversed phase high performance liquid chromatography, *J. Liq. Chromatogr.*, 1, 833, 1978.
25. **Burbach, J. P. H., Schotman, P., Verhoef, J., de Kloet, E. R., and de Wied, D.,** Conversion of des-Tyrosine-γ-endorphin by brain synaptic membrane associated peptidases: identification of generated peptide fragments, *Biochem. Biophys. Res. Commun.*, 97, 995, 1980.
26. **Knight, M., Ito, Y., and Chase, T. N.,** Preparative purification of the peptide des-enkephalin γ-endorphin: comparison of high-performance liquid chromatography and counter-current chromatography, *J. Chromatogr.*, 212, 356, 1981.

NEUROPHYSINS

Jay A. Glasel

The neurophysins are proteins whose monomer molecular weights are approximately 10,000 daltons.[1] They may be obtained from extracts of the mammalian posterior pituitary (neurohypophysis). Neurohistological investigation has localized these proteins in vivo to neurosecretory granules within the neuroendocrine cells containing the peptide hormones oxytocin and vasopressin.[2] Recently, the sequence of a cDNA encoding both arginine vasopressin and neurophysin II in bovine hypothalamus has been determined,[3] confirming earlier suggestions that both hormone and protein are initially formed as a large precurser molecule in the neurohypophesis.[4] The neurophysins are very stable in solution as is consistent with their extraordinarily high disulfide bridge content (seven links per monomer). They contain only one tyrosine residue per monomer. The full sequences of NP-I and NP-II from various species have been reported and compared.[5,6] However, the biological function of the neurophysins are not known. They are known to enter the blood stream during the process of exocytosis. There is no international agreement on the nomenclature of these proteins. For historical reasons many investigators use the terms neurophysins I, II and C (NP-I, NP-II, NP-C) for the three major proteins which are present in material obtained by the usual isolation and purification procedures. However, recently a designation "VLDV" and "MSEL" has been proposed for the former two based on their N-terminal sequences.[7] These proteins become highly aggregated systems in aqueous solution. The biological implications of this are not known but the phenomena has been intensely investigated by physical biochemists.[8]

Open column methods for the preparation of neurophysins have been available for about 15 years. The major difficulty facing early investigators was the fact that unidentified cathepsins attack the proteins during the early steps in their isolation. However, it was found that this could be circumvented by treatment of the starting materials (usually either fresh pituitaries or acetone dried pituitary powder) with 0.1 to 1.0 *M* strong acid.[9] This rather drastic procedure has been almost universally adopted and results in neurophysins which bind oxytocin and vasopressin. However, it has clouded the issue of whether or not these products are the native proteins which are present in the neurosecretory granules. Thus, there have been reports that the neurophysins are lipoproteins or glycolipoproteins.[10,11]

In open column procedures the acid extraction step is followed by salt precipitation, exhaustive dialysis, Sephadex® G-75 gel filtration and final separation by ion-exchange chromatography on DEAE-Sephadex® A-50 using a linear salt gradient. Isoelectric focusing in disc gels shows single bands for NP-I and NP-II from the appropriately pooled fractions with $pI = 4.90 \pm 0.01$ for NP-II and 4.50 ± 0.01 for NP-I.[12]

Recently, in unpublished experiments in our laboratory we have investigated the composition of the neurophysins using an isolation procedure which depends on mild treatment of the starting material with an anticathepsin cocktail consisting of *N*-ethylmaleimide (NEM), phenylmethyl sulfonylfluoride (PMSF) and bacitracin. We found that the treatment is too mild to prevent some catheptic degradation of NP-I. However, no lipid content was found for any of the proteins copurifying with NP-I, or with the NP-II single component, at the detection sensitivity level afforded by Sudan black B or oil red 0 staining of disc gels. Using gas-liquid chromatographic detection following derivitization we found no carbohydrate content in the NP-I preparation. However, in the NP-II preparation which had identical electrophoretic and hormone-binding properties with that found in conventional preparations we found at least two carbohydrate residues per protein molecule. This suggests that at least

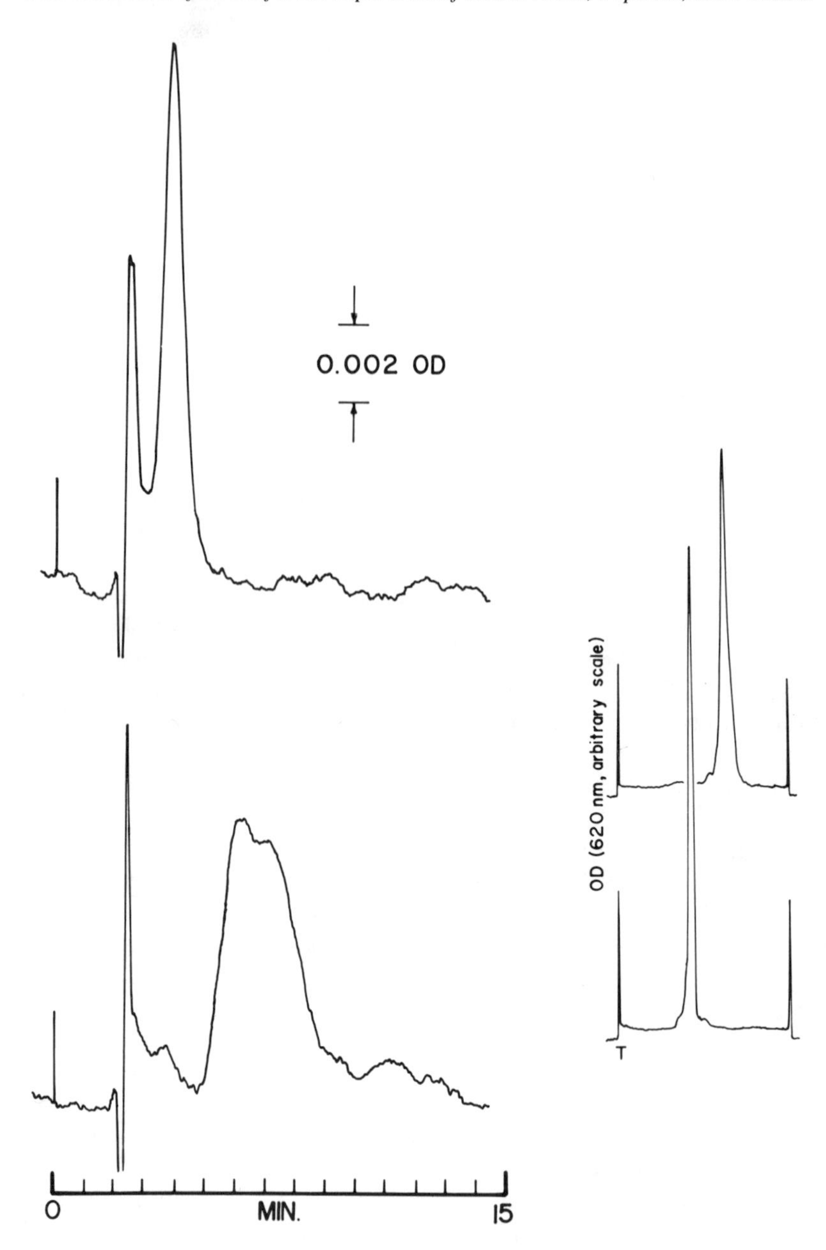

FIGURE 1. HPLC chromatograms, flow rate 1 mℓ/min, detection at 215 nm, eluant composition; 48.4% 0.01 *M* acetate buffer, pH = 5.7, 51.6% methanol. Upper: NP-I, 2 μg injected as phosphate buffer pH = 5 solution. Lower: NP-II, 6 μg injected as phosphate buffer pH = 5 solution. Inset figure: PAGE scans, 7.5% acrylamide gels, *Tris*-glycine buffer, pH = 7.9, stained with amido black. "T" indicates top of gel. Upper: NP-I, 50-μg load. Lower: NP-II, 50-μg load.

one of the neurophysins (or molecules co-electrophoresing with them) is a glycoprotein in the native state.

Investigators in this area have known for some time that the neurophysins, besides being very stable proteins, are somewhat soluble in polar organic solvents such as methanol. Since reverse-phase (RP) HPLC separations basically depend upon the relative interaction strength of solute molecules between eluent and solid phase a possible separation mechanism for the neurophysins can be suggested.

Beginning in 1978 three groups of workers have reported success in the analytical and preparative separation of the neurophysins and their tryptic digests using this method. The comparative details of their procedures are given in Table 1.

The most significant finding has been that neurophysins which are homogeneous under analytical polyacrylamide gel electrophoretic conditions are not necessarily so in an HPLC separation. Thus, it is found that conventionally prepared and purified porcine NP-I is in reality two neurophysins differing by a single carboxy terminus amino acid. The separation of these two peaks under the running conditions given in Table 1 was approximately 12 min.[16]

There appears to be some discrepancy between investigators as to the terminal sequences of NP-I and NP-II. The HPLC separated NP-I$_1$ and NP-I$_2$ give -Phe COOH and -Phe-Leu-COOH termini, respectively, for porcine-derived material.[16] The end sequences of conventionally prepared porcine and bovine "NP-1" are both -Phe-Ser-Glu.[5] In the case of NP-II
91 92 93
sequencing work shows the C-terminal of the 95 residue protein to be -Val in the bovine and ovine case and -Ala for the porcine, equine, and whale cases.[6] The cDNA sequence predicts[3] -Val-Arg-Ala-Asn . . . for the bovine case with the Ala beginning the sequence of a precursor
95 96 97 98
39 residue glycopeptide. This contrasts to the HPLC purified carboxy terminal residues which have been reported. These are Phe and Val for NP-II and the large "NP-III" peak, respectively.[16]

In conclusion, the clear separation of NP-I$_1$ and NP-I$_2$ components, which can only be done imperfectly in analytical gel electrophoresis, shows the power of HPLC separations as applied to the neurophysins. The further application of the technique with subsequent sequencing studies will doubtless clear up the uncertainty concerning the nature of the C-terminus residues.

Table 1
SEPARATION OF NEUROPHYSINS BY RP-HPLC

Column	Packing	Eluent	pH	Flow (mℓ/min)	Detection	Loading (μg)	Material	Ref.
250 × 4.6 mm	RP-18(10 μm)	Methanol/acetate buffer	5.7	1	UV(215 n*M*)	2	Bovine NP-1, NP-II	13
250 × 10 mm	RP-18(10 μm)	Methanol/phosphate buffer	6.0	0.042	UV(210 n*M*)	15	Crude porcine NP preparation	16
260 × 4.6 mm	Zorbax®-CN	TEAP[a]/acetonitrile	3.0	0.8	UV(215)	50—200	Tryptic fragments of oxidized bovine NP-I and NP-II	14

[a] TEAP: triethylamine-phosphoric acid buffer containing sodium azide.

REFERENCES

1. **Breslow, E.,** Chemistry and biology of the neurophysins, *Ann. Rev. Biochem.*, 48, 251, 1979.
2. **Heller, H. and Lederis, K., Eds.,** *Subcellular Organization and Function in Endocrine Tissues,* Part I, Functional Ultrastructure of the Neurohypophysis, Cambridge University Press, Cambridge, U.K., 1971, 233.
3. **Land, H., Schütz, G., Schmale, H., and Richter, D.,** Nucleotide sequence of cloned cDNA encoding bovine arginine vasopressin-neurophysin II precursor, *Nature (London),* 295, 299, 1982.
4. **Sachs, H.,** in *Handbook of Neurochemistry,* Vol. 4, pgs. 373—427, Lajtha, A., Ed., Plenum Press, New York, 1970, chap. 17.
5. **Chauvet, M. T., Codogno, P., Chauvet, J., and Acher, R.,** Comparison between MSEL- and VLDV-neurophysins, *FEBS Lett.*, 98, 37, 1979.
6. **Chauvet, M. T., Codogno, P., Chauvet, J., and Acher, R.,** Phylogeny of neurophysins, *FEBS Lett.*, 88, 91, 1978.
7. **Chauvet, M. T., Chauvet, J., and Acher, R.,** Phylogeny of neurophysins: partial amino acid sequence of a sheep neurophysin, *FEBS Lett.*, 52, 212, 1975.
8. **McKelvy, J. F., Glasel, J. A., and Foreman, M.,** Biochemical aspects of hypothalamic function, in *Handbook of the Hypothalamus,* Vol. 2, Morgane, P. J. and Panksepp, J., Eds.,
9. **Hollenberg, M. D. and Hope, D. B.,** Fractionation of neurophysin by molecular-sieve and ion-exchange chromatography, *Biochem. J.*, 104, 122, 1967.
10. **Mylroie, R. and Koenig, J.,** Soluble acidic lipoproteins of bovine neurosecretory granules. Relation to neurophysins, *Histochem. Cytochem.*, 19, 738, 1971.
11. **Pliska, V. and Meyer-Grass, M.,** Some properties of neurophysins isolated from bovine neurosecretory granules, *Ann. N. Y. Acad. Sci.*, 248, 235, 1975.
12. **Glasel, J. A., McKelvy, J. F., Hruby, V. J., and Spatola, A. F.,** Binding studies of polypeptide hormones to bovine neurophysins, *J. Biol. Chem.*, 251, 2929, 1976.
13. **Glasel, J. A.,** Separation of neurohypophyseal proteins by reversed-phase high-pressure liquid chromatography, *J. Chromatogr.*, 145, 469, 1978.
14. **Chaiken, I. M. and Hough, C.,** Mapping and isolation of large peptide fragments from bovine neurophysins and biosynthetic neurophysin-containing species by high-performance liquid chromatography, *J. Anal. Biochem.*, 107, 11, 1980.
15. **Schwandt, P. and Richter, W. O.,** Purification of porcine neurophysins I and II by high performance liquid chromatography, *Biochem. Biophys. Acta,* 626, 376, 1980.
16. **Richter, W. and Schwandt, P.,** Preparation of porcine neurophysin proteins by high performance liquid chromatography, *J. Neurochem.*, 36, 1279, 1981.

HUMAN GROWTH HORMONE

Choh Hao Li and David Chung

INTRODUCTION

Our initial attempts to purify human growth hormone (HGH), a protein of 21,500 daltons,[1] by commercially available columns with a porosity of 50 to 100 Å gave very poor resolution. Later, by use of reverse-phase column of 10-μm particles, C_8 ligand and 300 Å porosity (a gift of Drs. S. Udenfriend and S. Stein), more promising results were obtained for the purification of various pituitary protein hormones, including HGH. This led us to explore the use of commercial columns with high porosity packings in the study of HGH purification.

EXPERIMENTAL

The reverse-phase column (4.6 × 250 mm, VYDAC, 201TP, C-6000, C_{18}, 300 Å) was purchased from Alltech Associates (Deerfield, IL). The chromatographic system consisted of the Laboratory Control (Riviera Beach, FL), Gradient Master, two Constametric pumps (Constametric I and II G), mixer and Spectromonitor III variable wavelength monitor. The recorder was a Heathkit Model SR-255-B (Benton Harbor, MI). 2-Propanol was obtained from Burdick and Jackson Laboratories (Muskegon, MI) and trifluoroacetic acid (TFA) (redistilled) from Eastman Kodak.

The crude HGH preparation (designated as fraction D or HGH-D) was prepared from fresh human pituitary glands as previously described.[2] The protocol for the preparation of fraction D is presented in Table 1. From 50 fresh glands, 150 mg of fraction D may be obtained.

Amino acid analyses were carried out according to the method of Spackman et al.[3] with an automatic amino acid analyzer (Model 119C, Beckman Instruments). The amino terminal residue was determined by the dansyl-Edman method.[4]

RESULTS AND DISCUSSION

Figures 1A and 1B present the chromatographic patterns of HGH-D obtained by isocratic elution with 43% 2-propanol containing 0.1% TFA when 50 μg (in 50 μℓ 0.1 *M* acetic acid) and 700 μg (in 200 μℓ 0.1 *M* acetic acid) were applied to the column, respectively. It is evident that a main peak at 18.5 min is clearly separated from other components. The contents of each tube over the main peak in Figure 1B were combined and lyophilized. A yield of 140 μg was obtained and shown to be highly purified HGH.

Homogeneity of the purified product was evidenced by the presence of phenylalanine as the sole NH_2-terminal residue[1] and the HPLC pattern (Figure 1C). The amino acid composition is nearly identical to that obtained from the product isolated by the conventional method[2] as shown in Table 2. In addition, the purified product behaved as HGH[2] in polyacrylamide gel electrophoresis (data not shown).

ACKNOWLEDGMENT

We thank Jonathan Roeder for technical assistance. This work was supported in part by the National Institute of Health (AM-18677) and the Hormone Research Foundation.

Table 1
PROTOCOL FOR THE PREPARATION OF HGH FRACTION D

Fraction	Procedure	Weight (g)
	Fresh glands (25 pituitaries)	25
A	Extracted with pH 7 saline; fractionated with $(NH_4)_2SO_4$	
B	1.9 *M* $(NH_4)_2SO_4$ precipitate, dialyzed and lyophilized	1.5
	Extracted with 0.45 *M* $(NH_4)_2SO_4$ in pH 5.1 phosphate buffer	0.5
C	Soluble fraction submitted to chromatography on IRC-50 resin column	
D	Water eluate (active peak), dialyzed and lyophilized	0.15

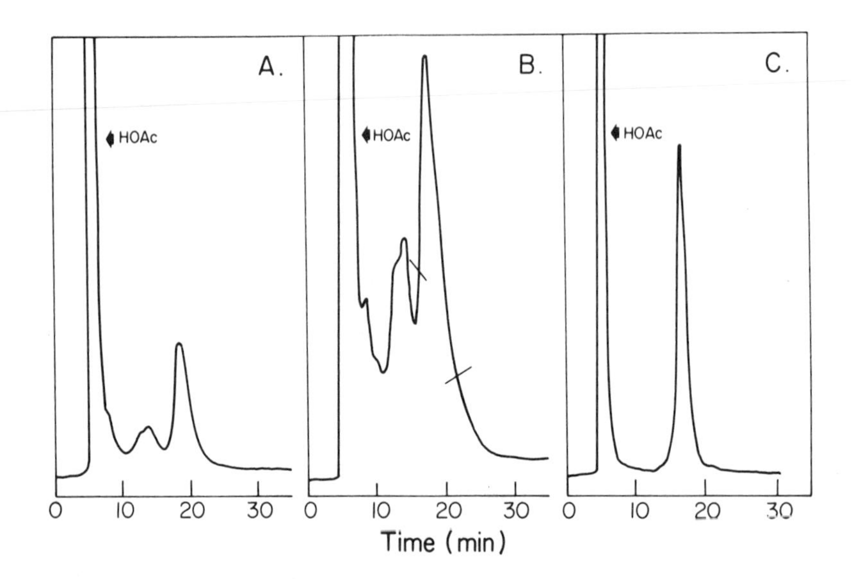

FIGURE 1. (A) HPLC purification of 50 μg (in 50 μℓ 0.1 *M* acetic acid) HGH-D on a reverse-phase VYDAC 201 column (Alltech Associates; C_{18}, 300 Å pore; 4.6 × 250 mm). Isocratic elution with 43% 2-propanol containing 0.1% TFA; 1 mℓ/min; 210 nm 1.0 AUFS. (B) HPLC purification of 700 μg (in 200 μℓ 0.1 *M* acetic acid) HGH-D on the same column under the same conditions as (A). (C) Chromatography of the purified fraction from (B) on the same column under the same conditions as (A).

Table 2
AMINO ACID COMPOSITION OF HGH

	HGH		
Amino acid	**Purified by HPLC**	**Purified as described[2]**	**Primary structure[5]**
Asp	20.9	20.6	20
Thr	9.8[a]	9.8[a]	10
Ser	15.8[a]	15.8[a]	18
Glu	27.3	27.4	27
Pro	9.1	9.0	8
$^{1}/_{2}$ Cys	3.0	3.0	4
Gly	9.3	9.6	8
Ala	7.6	7.8	7
Val	7.0	6.7	7
Met	3.2	3.1	3
Ile	7.0	6.9	8
Leu	25.8	25.8	26
Tyr	7.5	7.6	8
Phe	12.5	12.7	13
His	3.0	3.1	3
Lys	9.4	9.0	9
Arg	11.7	11.9	11
Trp	(1)[b]	(1)[b]	1

[a] Not corrected.
[b] Not determined.

REFERENCES

1. **Li, C. H.,** The chemistry of human pituitary growth hormone: 1967—1973, in *Hormonal Proteins and Peptides,* Vol. 3, Li, C. H., Ed., Academic Press, New York, 1975, 1.
2. **Li, C. H., Liu, W.-K., and Dixon, J. S.,** Human pituitary growth hormone. VI. Modified procedure of isolation and NH_2-terminal amino acid sequence, *Arch. Biochem. Biophys. Suppl.,* 1, 327, 1962.
3. **Spackman, D. H., Stein, W. H., and Moore, S.,** Automatic recording apparatus for use in the chromatography of amino acids, *Anal. Chem.,* 30, 1190, 1958.
4. **Gray, W. R.,** Sequential degradation plus dansylation, in *Methods in Enzymology,* Vol. 11, Hirs, C. H. W., Ed., Academic Press, New York, 1967, 469.
5. **Li, C. H.,** Hormones of the adenohypophysis, *Proc. Am. Philos. Soc.,* 116, 365, 1972.

SOMATOMEDINS*

Marjorie E. Svoboda and Judson J. Van Wyk

The somatomedins are a family of insulin-like polypeptide growth factors which possess four cardinal properties: (1) their concentrations in serum are growth hormone dependent, (2) they possess insulin-like actions in extraskeletal tissues, (3) they promote the incorporation of sulfate into proteoglycans in cartilage, and (4) they are potent mitogens for a wide variety of cell types. The confusing nomenclature of the somatomedins is due to the fact that each of these properties has been used as a basis for their isolation.

Nonsuppressible insulin-like activity (NSILA) is the term that Froesch initially used to designate that portion of the serum insulin-like activity which cannot be neutralized by insulin antibodies. Using a bioassay for insulin which depended on the oxidation of glucose by rat adipose tissue, Rinderknecht and Humbel isolated from human plasma two related peptides which were designated NSILA-I and NSILA-II.[1] After it was recognized that these substances were far more active in growth promotion assays than in assays based on their insulin-like properties, these names were changed to insulin-like growth factor I (IGF-I) and insulin-like growth factor II (IGF-II).[2] Somatomedin-C (Sm-C) was isolated by Van Wyk and associates on the basis of its growth-promoting action in cartilage.[3] Temin first used the term multiplication stimulating activity (MSA) to describe the substances in calf serum which gave rise to its mitogenic activity for chick fibroblasts.[4] Dulak and Temin later used the same term to describe similar or identical mitogenic peptides which were found in conditioned media from cultured rat liver cells.[5] The term MSA is now used exclusively to describe the mitogenic peptides isolated from rat hepatocyte cultures.

Two distinct chemical forms of human somatomedin have been established by sequence analysis. Somatomedin-C (Sm-C) and IGF-I appear to be different names for the same basic peptide (pI 8.0 to 8.7). Sm-C/IGF-I was first sequenced by Rinderknecht and Humbel and found to contain 70 residues in a single chain with three disulfide bridges.[6] In terms of growth hormone dependency and growth-promoting activity, Sm-C/IGF-I is more active than IGF-II, which is a neutral peptide with greater insulin-like biological properties.[7] Those portions of IGF-I/Sm-C and IGF-II corresponding to the A and B chains of insulin exhibit about 50% homology with human proinsulin, but all three peptides are quite different in those portions of the molecule corresponding to the C peptide. Sm-C/IGF-I and IGF II also differ from proinsulin in having eight and six residue extensions, respectively, at the carboxy terminus.[8]

Rats likewise appear to have two forms of somatomedin which correspond to Sm-C/IGF I and IGF-II. A basic growth hormone dependent form corresponding to Sm-C/IGF-I has been isolated from the blood of rats bearing tumors which secrete growth hormone. Partial sequence analysis by Rubin et al. revealed identity of the first 30 residues with Sm-C/IGF-I.[9] Marquardt and Todaro have isolated and sequenced one form of MSA produced by rat hepatocyte cultures and this appears to be the rat counterpart of IGF-II, since these molecules differ in only five residues.[10]

The somatomedins circulate in blood as macromolecular complexes with approximate molecular weights of 140,000 daltons.[11] Somatomedins are not stored in any tissue and it has therefore been necessary to purify them from blood plasma, or from conditioned media.

* All HPLC procedures were carried out with Waters Associates equipment: two Model 6000A solvent delivery systems, a Model 660 solvent programmer and a Model U6K injector. All columns described were from Waters Associates.

Table 1

System[a]	Elution time (min) Sm-C	"Very Basic" Sm	IGF II
μBondapak® C_{18} (3.9 mm × 30 cm)			
Aqueous 0.13% HFBA[b]/CH_3CN, linear gradient 20—60% CH_3CN, 1 mℓ/min for 40 min	27.5	28	27.5
Aqueous 0.1% TFA/CH_3CN, linear gradient 20—60% CH_3CN, 1 mℓ/min for 40 min	24.5	24	27.5
Aqueous 0.045% TFA/CH_3CN, 0—45% CH_3CN, at 1 mℓ/min for 50 min	43.5	42	43.5
μBondapak® phenyl (3.9 mm × 30 cm)			
Aqueous 0.01 *M* KH_2PO_4/CH_3CN linear gradient 20—60% CH_3CN at 1 mℓ/min for 50 min	26.5	26.5, 31, 34	25.5

[a] All systems described were carried out on columns from Waters Associates and employ two Waters Associates solvent delivery systems Model 6000A, Model 660 solvent programmer, and U6K injector.
[b] HFBA, heptafluorobutyric acid.

Since the preparation of only a few milligrams requires the extraction and processing of hundreds of liters of human blood, most investigators have used as starting material large pools of Cohn fraction IV, which contains most of the somatomedins in blood and which is otherwise a waste product of the blood fractionation industry.

Since 1 mg Cohn IV of human plasma contains only about 5 ng of Sm-C and the amount of protein which can be applied to a semipreparative column is 50 to 75 mg, it is imperative that most of the extraneous proteins be removed prior to HPLC. In our laboratory SM-C is extracted from Cohn IV paste by 1 *M* acetic acid and further purified by ion-exchange column chromatography using SP Sephadex®, gel filtration with Sephadex® G50, and flat bed isoelectric focusing.[12] The initial wide range (pH 3 to 10) isofocusing step removes IGF-II (PI 6 to 6.8) from Sm-C (PI 8.0 to 8.7) and also removes a high percentage of contaminating serum proteins which are negatively charged and which may bind Sm-C nonspecifically. A subsequent narrow range (pH 8 to 10) step removes a more basic component of plasma (PI 9 to 9.7) which cross-reacts in our RIA and which we operationally refer to as "Very Basic".

The contaminants remaining after these steps are exceedingly difficult to remove from the pure peptide because of their similar size and charge properties. It is at this stage of purification that reverse-phase high-pressure liquid chromatography (RP-HPLC) has proven to be a particularly useful and powerful tool. Sm-C is an ideal peptide for isolation by RP-HPLC because it is hydrophobic and is stable in organic acids, low pH buffers, and in commonly used organic solvents. Furthermore, it is quantitatively recovered from reverse-phase columns. It is important, however, that IGF-II and "Very Basic" be removed from Sm-C/IGF prior to HPLC since the elution time of all three peptides is similar in all systems that we have tested (Table 1).

An effective first HPLC step is chromatography on an octadecylsilane semipreparative column, (μBondapak® C_{18}, 7.8 mm × 30 cm) with a linear gradient from 20 to 60% acetonitrile in aqueous 0.1% trifluoroacetic acid (TFA) at 3 mℓ/min for 50 min. The efficiency is increased if the gradient is stopped at 31% acetonitrile and the elution continued isocratically until the bulk of the protein has been eluted, after which the gradient is resumed. In our attempts to develop optimum conditions for the C_{18} column, we found that the substitution of acetic acid for TFA or the use of a more polar solvent such as methanol with 0.1% TFA fails to remove Sm-C or contaminating proteins from the column. We also found that a decrease in TFA concentration below 0.1% increases the elution time with no improvement in resolution.

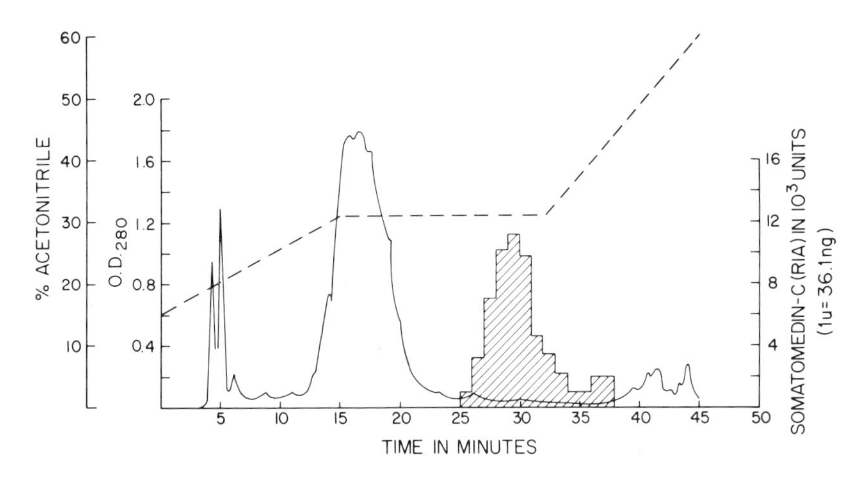

A

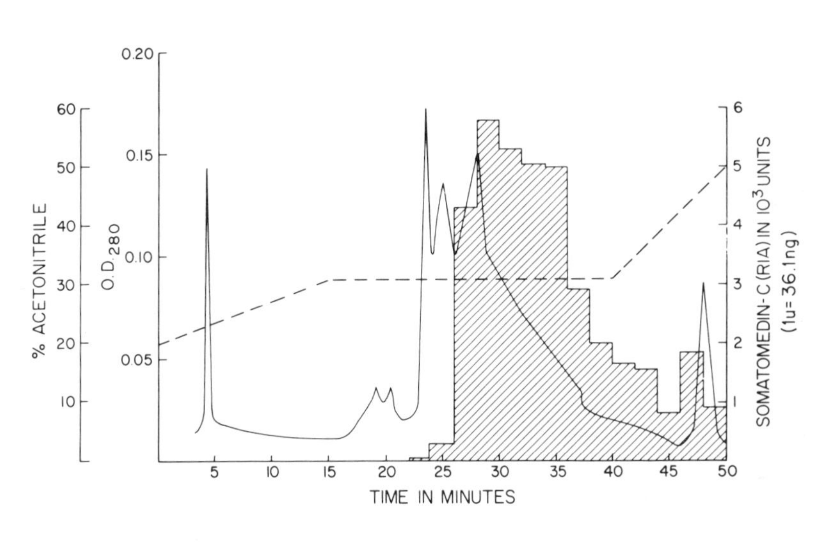

B

FIGURE 1. (A) HPLC of a crude Sm-C preparation (obtained from human Cohn IV extract which had been semipurified by a series of ion-exchange and gel filtration chromatographic procedures) on a semipreparative μBondapak® C_{18} (7.8 mm × 30 cm) column. A linear gradient was run from 15 to 31% acetonitrile in 0.1% TFA in water for 15 min followed by isocratic elution at 31% acetonitrile for 17 min then continuation of linear gradient elution from 31 to 60% acetonitrile for 13 min at a flow rate of 3 mℓ/min; 1-min fractions were collected. (———, O.D.$_{280\ mu}$; - - - - - - -,% acetronitrile; [hatched box], units of Sm-C by RIA.) (B) HPLC of the pooled active fractions eluted from the column shown in A under the same conditions as in A except for the isocratic elution time of 25 min followed by a 10-min linear gradient of 31 to 60% acetonitrile in 0.1% TFA in water for 10 min; 1-min fractions were collected. (———, O.D.$_{280\ mu}$; - - - - - - -,% acetronitrile; [hatched box], units of Sm-C by RIA.)

A single HPLC step is sufficient only if the purity of the starting material is relatively high. A typical chromatograph using this method as a first HPLC step is shown in Figure 1A. Figure 1B shows a repeat passage over the same column under similar conditions. The elution profile shows that the preparation is still heterogeneous but an increase in specific activity has been effected.

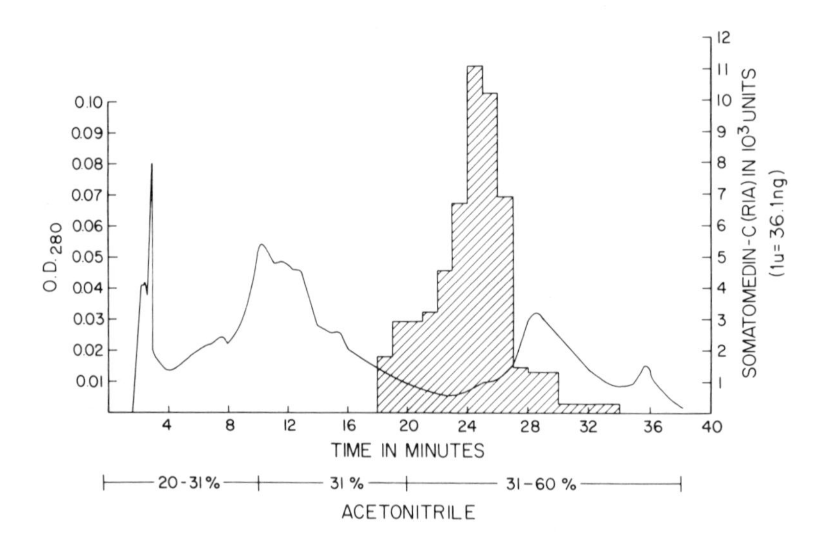

FIGURE 2. HPLC of crude Sm-C (as used in Figure 1) on a semipreparative μBondapak® phenyl (7.8 mm × 30 cm) column using a nonlinear gradient, program #7 on Waters' Model 660 solvent programmer, at 5 mℓ/min from 20 to 31% acetonitrile in 0.01 *M* KH_2PO_4 in water for 10 min followed by a 10-min elution at 31% and continuation of the gradient from 31 to 60% acetonitrile in 18 min, 1-min fractions were collected. (———, O.D.$_{280\ mu}$; ▨, units of Sm-C by RIA.)

As an alternative first step for use with highly impure materials, we employ the μBondapak®-phenyl semipreparative column (7.8 mm × 30 cm) and a gradient system of 20 to 60% acetonitrile in aqueous 0.01 *M* KH_2PO_4. Varying these conditions by increasing the pH of the aqueous phase causes earlier elution of all proteins, whereas the converse occurs with a decrease in pH. In either case the resolution decreases. Substitution of methanol for acetonitrile when using a μBondapak®-phenyl column with 0.01 *M* KH_2PO_4 fails to elute Sm-C and contaminating proteins from the column. As shown in Figure 2, considerable purification of Sm-C is achieved using a nonlinear gradient (program #7 on a Water's solvent programmer) at a flow rate of 5 mℓ/min, interrupted by isocratic elution of the bulk of proteins at 31% acetonitrile. When the active fractions from the above chromatographic step were desalted on Sephadex® G-50 in 1 *M* acetic acid and the lyophilized product was rechromatographed on a semipreparative μBondapak® C_{18}-column using 0.1% TFA/CH_3CN, a sharp protein peak containing immunoreactive Sm-C was eluted followed by a small peak of immunoreactive Sm-C. (Figure 3A.)

Although coincidence of radioimmunoactivity with a sharp protein peak is often offered as evidence of purity, it is essential that this conclusion be verified by some other independent criteria. The high resolution of SDS-PAGE coupled with the exquisite sensitivity of the silver stain has proven invaluable for this purpose. By this technique the apparently homogeneous main peak fractions shown in Figure 3A in fact contained two contaminants whereas the minor peak of activity showed only a single Sm-C band. Rechromatography of the impure pool on an analytical μBondapak® C_{18}-column resulted in the elution pattern shown in Figure 3B. Sodium dodecyl sulfate-polyacrylamide gel electrophoresis (SDS-PAGE) of the eluted fractions now revealed that essentially pure Sm-C was present in fractions which eluted from 45 to 59 min whereas the contaminants were present in earlier and later eluting fractions.

Although we are aware that other laboratories involved in the isolation and characterization of Sm-C and other somatomedins are using HPLC methods, to date only the work of

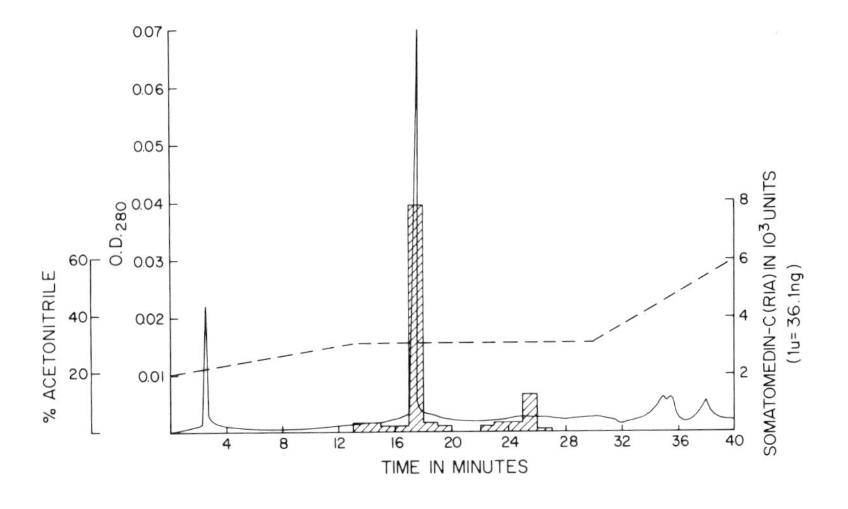

A

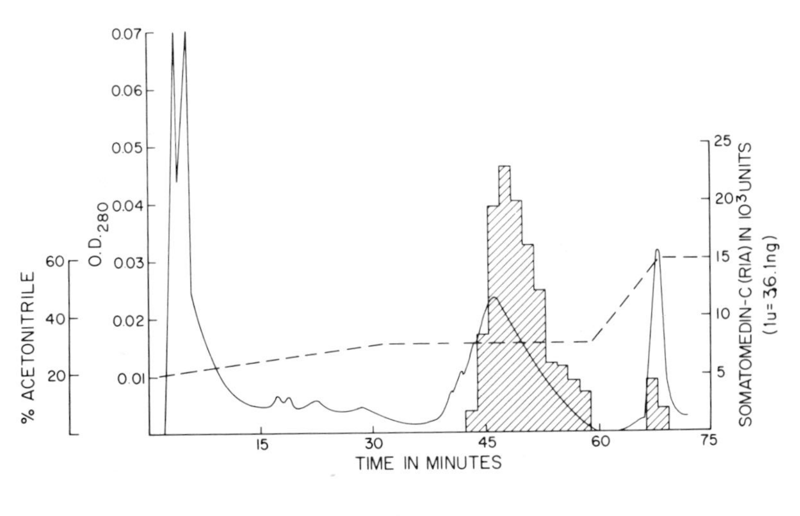

B

FIGURE 3. (A) The pooled active fractions from HPLC shown in Figure 2 after desalting was applied to a μBondapak® C_{18} (7.8 mm × 30 cm) and eluted at 5 mℓ/min using a linear gradient program from 20 to 31% acetonitrile in 0.1% TFA in water for 13 min, isocratic elution at 31% acetonitrile for 15 min, and linear gradient of 31 to 60% acetonitrile for 10 min; 1-min fractions were collected. (———, O.D.$_{280\ mu}$; - - - - - - -,% acetronitrile; ▨, units of Sm-C by RIA.) (B) HPLC of pool of fractions eluted from 17 to 18 min shown in Figure A on an analytical μBondapak® C_{18} (3.9 mm × 30 cm) using a linear gradient of 20 to 31% acetonitrile in 0.1% TFA acid in water for 31 min, an isocratic elution at 31% acetonitrile for 28 min, and a linear gradient of 31 to 60% acetonitrile in 11 min at a flow rate of 1 mℓ/min; 1-min fractions were collected. (———, O.D.$_{280mu}$; - - - - - - -,% acetronitrile; ▨, units of Sm-C by RIA.)

Marquardt et al.[10] has appeared in the literature. This group employed a μBondapak® C_{18}-column (3.9 mm × 30 cm) as a final step in the purification of MSA from buffalo rat liver-conditioned medium. After an initial 8-min elution with 0.05% TFA in water, a linear gradient elution at 40°C was performed from 0 to 43% CH_3CN in aqueous 0.045% TFA.

The product recovered was of sufficiently high purity that the complete sequence could be ascertained.

ACKNOWLEDGMENT

This research supported by NIH Research Grant R01 AM01022.

REFERENCES

1. **Rinderknecht, E. and Humble, R. E.,** Polypeptides with nonsuppressible insulin-like and cell growth-promoting activities in human serum: isolation, chemical characterization and some biological properties of forms I and II, *Proc. Natl. Acad. Sci. U.S.A.,* 73, 2365, 1976.
2. **Rinderknecht, E. and Humble, R. E.,** Amino terminal sequences of two polypeptides from human serum with nonsuppressible insulin-like and cell growth promoting activities: evidence for structural homology with insulin B chain. *Proc. Natl. Acad. Sci. U.S.A.,* 73, 4379, 1976.
3. **Van Wyk, J. J., Underwood, L. E., Baseman, J. B., Hintz, R. L., Clemmons, D. R., and Marshall, R. N.,** Explorations of the insulin-like and growth promoting properties of somatomedin by membrane receptor assays, in *Advances in Metabolic Disorders,* Vol. 8, Luft, R., and Hall, K., Eds., Academic Press, New York, 1975, 127.
4. **Temin, H. M., Pearson, R. W., and Dulak, N. C.,** The role of serum in the control of multiplication of avian and mammalian cells in culture, in *Growth, Nutrition and Metabolism of Cells in Culture,* Vol. I, Rothblat, G. H. and Cristofalo, V. U., Eds., Academic Press, New York, 1972, 137.
5. **Dulak, N. C. and Temin, H. M.,** A partially purified polypeptide fraction from rat liver cell conditioned medium with multiplication stimulating activity for embryo fibroblasts, *J. Cell Physiol.,* 81, 153, 1973.
6. **Rinderknecht, E. and Humble, R. E.,** The amino acid sequence of human insulin-like growth factor I and its structural homology with proinsulin, *J. Biol. Chem.,* 253, 2769, 1978.
7. **Zapf, J., Schoenle, E., and Foresch, E. R.,** Insulin-like growth factors I and II: some biological actions and receptor binding characteristics of two purified constituents of nonsuppressible insulin-like activity of human serum, *Eur. J. Biochem.,* 87, 285, 1978.
8. **Rinderknecht, E. and Humble, R. E.,** Primary structure of human insulin-like growth factor II, *FEBS Lett.,* 89, 283, 1978.
9. **Rubin, J. S., Mariz, I., Jacobs, J. W., Daughaday, W. H., and Bradshaw, R. A.,** Isolation and partial sequence analysis of rat basic somatomedin, *Endocrinology,* 110, 734, 1982.
10. **Marquardt, H., Todaro, G. J., Henderson, L. E., and Oroszlan, S.,** Purification and primary structure of a polypeptide with multiplication-stimulating activity from rat liver cell culures, *J. Biol. Chem.,* 256, 6859, 1981.
11. **Hintz, R. L. and Liu, F.,** Demonstration of specific plasma protein binding sites for somatomedin, *J. Clin. Endocrinol. Metab.,* 45, 988, 1977.
12. **Svoboda, M. E., Van Wyk, J. J., Klapper, D., Fellows, R., Grissom, F., and Schleuter, R. J.,** Purification of somatomedin-C from human plasma: chemical and biological properties, partial sequence analysis and relationship to other somatomedins, *Biochemistry,* 19, 790, 1980.

CALCIOTROPIC HORMONES

P. H. Corran and Joan M. Zanelli

INTRODUCTION

Parathyrin and calcitonin are the two peptide hormones which, together with the steroid metabolites of vitamin D, are thought to be involved in the regulation of calcium homeostasis in mammals.[1] Some of the generally agreed aspects are summarized briefly below: further details may be found in current literature. Both hormones are synthesized as large precursor forms and are processed to their final dimensions before secretion. Parathyrin may be further cleaved within the gland to fragments, some of which retain biological activity.[2]

Parathyrin is produced by the parathyroid glands which are only found in land-dwelling animal species. To date, the sequences of parathyrins from three mammalian species have been determined.[3] All are 84 amino acid residues in length, and show close sequence homology. None of the peptides contains disulfide bonds, all have tryptophan at position 23 and methionine at position 8, and the bovine and human hormones have a second methionine at position 18 (Table 1). All are rich in basic residues. Evidence from compositional analysis suggests that isohormones may exist in the bovine species,[1] but heterogeneity also arises as a result of deamidation or oxidation during extraction and purification procedures. The known biological activities described for parathyrin appear to be associated with the amino-terminal one third of the sequence, and are altered or decreased by oxidation of the methionine residues to the sulfoxide or sulfone forms. To date no known biological activity has been identified for the remaining carboxyl-terminal two thirds of the molecule, yet this apparently functionless region does not show the sequence divergence one might expect for an evolutionarily unconstrained peptide.

After secretion from the parathyroid gland into the blood circulation in response to changes in blood calcium concentrations, the parathyrin is short lived.[4] The kidney and liver are believed to be the primary organs responsible for cleavage of the peptide into an amino-terminal third and carboxyl-terminal two thirds fragment, but the subsequent metabolic cleavage processes involved and the nature, role, and fate of the peptide fragments produced are not fully understood.[1-4]

In contrast to parathyrin, calcitonin appears to be an ancient peptide and is found throughout the animal kingdom.[5] The sequences of seven species of calcitonin are shown in Table 2. All are 32 residues long and contain an almost invariant seven-residue amino-terminal disulfide loop. The different sequences fall naturally into three groups: human/rat, artiodactyl (pig, sheep, cow) and piscine (salmon, eel).[6] In general the interspecies amino acid substitutions are conservative in chemical characteristics, and almost all can be explained by a single DNA base change. Isohormones have been described for salmon calcitonin.[7] Although the biological activity of the calcitonins is generally measured by the pharmacological effect of lowering blood calcium in mammals, the physiological role of calcitonin as a calciotropic peptide in mammalian species is still uncertain.[8] In submammalian species, calcitonin may be more closely associated with neurotransmitter peptides.[5] Paradoxically, despite the apparent lack of involvement of calcitonin with calcium metabolism in species such as salmon and eel, both salmon and eel calcitonins are approximately 40 times more bioactive than mammalian calcitonins when tested in mammalian systems. The proportion of hydrophobic amino acids in the different species of calcitonin differs considerably, the piscine peptides being the most polar. Only mammalian species of calcitonin contain methionine, and a decrease of biological potency is associated with oxidation of the methionine residue. Only porcine calcitonin contains tryptophan.

Table 1
THE SEQUENCES OF HUMAN, BOVINE AND PORCINE PARATHYRIN

```
1               5                   10                  15                  20                  25
Ser.Val.Ser.Glu.Ile.Gln.Leu.Met.His.Asn.Leu.Gly.Lys.His.Leu.Asn.Ser.Met.Glu.Arg.Val.Glu.Trp.Leu.Arg.Lys.Lys.Leu.
Ala.                    Phe.                                Ser.
                                                            Ser.    Leu.

        30                  35                  40                  45                  50                  55
Gln.Asp.Val.His.Asn.Phe.Val.Ala.Leu.Gly.Ala.Pro.Leu.Ala.Pro.Arg.Asp.Ala.Gly.Ser.Gln.Arg.Pro.Arg.Lys.Lys.Glu.
                                            Ser.Ile.    Tyr.        Gly.Ser.
                                            Ser.Ile.Val.His.        Gly.

                60                  65                  70                  75                  80
Asp.Asn.Val.Leu.Val.Glu.Ser.His.Glu.Lys.Ser.Leu.Gly.Glu.Ala.Asp.Lys.Ala.Asp.Val.Asn.Val.Leu.Thr.Lys.Ala.Lys.Ser.Gln Human
                                Gln.                                            Asp.        Ile.            Pro.    Bovine
                                Gln.                                    Ala.    Asp.        Ile.            Pro.    Porcine
```

Note: Only the residues of the bovine and porcine sequences which differ from human are shown.

Table 2
CALCITONIN SEQUENCES

```
                1                5                 10                15                20                25              30
Human    Cys.Gly.Asn.Leu.Ser.Thr.Cys.Met.Leu.Gly.Thr.Tyr.Thr.Gln.Asp.Phe.Asn.Lys.Phe.His.Thr.Phe.Pro.Gln.Thr.Ala.Ile.Gly.Val.Gly
Rat                                                               .Leu                                    .Ser

Ovine       .Ser                    .Val.   .Ser.Ala.   .Trp.Lys.   .Leu.   .Asn.Tyr.   .Arg.Tyr.Ser.Gly.Met.Gly.Phe.   .Pro.Glu.
Bovine      .Ser                    .Val.   .Ser.Ala.   .Trp.Lys.   .Leu.   .Asn.Tyr.   .Arg.   .Ser.Gly.Met.Gly.Phe.   .Pro.Glu.
Porcine     .Ser.                   .Val.   .Ser.Ala.   .Trp.Arg.Asn.Leu.   .Asn.       .Arg.   .Ser.Gly.Met.Gly.Phe.   .Pro.Glu.

Salmon      .Ser.                   .Val.       .Lys.Leu.Ser.   .Glu.Leu.His    .Leu.Gln.   .Tyr.   .Arg.   .Asn.Thr.   .Ser.
Eel         .Ser.                   .Val.       .Lys.Leu.Ser.   .Glu.Leu.His    .Leu.Gln.   .Tyr.   .Arg.   .Asp.Val.   .Ala.
```

20		31	Activity (Eq.I.V./mg)	Sum Hϕ[13]	Sum Hϕ[20]	Sum π vals.[38]	Predicted retention time[21]
21	Human	.Ala.Pro.NH2	c.150	15.25	51.7	16.33	77.1
33	Rat		400	13.91	50.5	15.85	70.2
35	Ovine	.Thr.	c.150	15.29	64.1	13.5	75.0
36	Bovine	.Thr.	c.150	15.83	64.5	14.32	80.9
37	Porcine	.Thr.	c.150	14.0	68.6	15.03	85.3
39	Salmon	.Thr.	5000	8.34	36.5	10.76	42.8
30	Eel	.Thr.	3000	12.18	43.7	12.3	54.4

Note: The sequences of selected calcitonins together with the sum hydrophobic contribution of the amino acid side chains are calculated according to four systems[13,20,21,38], two[13,38] based on octanol-water partition coefficients and two[20,21] on RP-HPLC retentions of peptides. The specific biological activities quoted are approximate. Cys^1 and Cys^7 are joined by a disulfide bridge.

The synthesis of the 32 residue calcitonins and the 34 residue amino-terminal fragment of parathyrin on a commercial scale is technically feasible and synthetic salmon calcitonin has been marketed for clinical purposes for some years. Synthetic human calcitonin and an analog of eel calcitonin are also under clinical trial.

Because salmon calcitonin is chemically more stable and has a higher biological potency in man it is preferred by manufacturers to mammalian calcitonins as a therapeutic product for the long-term treatment of Paget's disease of bone. There is no long-term therapeutic use for the synthetic parathyrin fragment (although it has been used in clinical trials for the treatment of osteoporosis)[9] but it may be needed to replace the natural bovine parathyrin as a clinical diagnostic agent for the differential diagnosis of hypercalcemia.

PARATHYRINS

Practical Considerations

All parathyrins contain tryptophan, so that column effluents may be monitored with reasonable sensitivity at 280 nm. The limit of detectability at 210 nm is about 40-fold greater however and a 10-fold further increase in sensitivity is possible if endogenous tryptophan fluorescence is used.[10] A practical limit for detection of parathyrin at 210 nm in a gradient system is about 250 ng.

Where very limited amounts of parathyrin are available (as with human parathyrin) recovery of the material applied to the HPLC column may be of great importance. Losses may occur either on the column or in sample collection and processing for subsequent analysis. Both the type and brand of chromatographic packing and the mobile-phase composition, particularly the concentration and nature of the salt content, may be of significance. These factors should be borne in mind when adapting a chromatographic system for a particular application.

Bovine and human parathyrins are the only molecular species for which HPLC data have been published, as summarized in Table 3. Experience has shown that the manufacturer of a packing may be as important an item of information for reproducing a separation as the composition of the mobile phase, so appropriate details of the packings used are given. Systems 1,[11] 3,[12] and 4[13] have been reproduced satisfactorily.

Purification and Characterization of Natural Parathyrin and its Cleavage Products

RP-HPLC provides a simple and rapid means of purifying milligram amounts of parathyrin from tissue extracts. In one publication,[11] successive cycles of HPLC in systems 1, 2, and 1 again (Table 3) were used to purify human parathyrin to homogeneity following an initial group fractionation of a crude tissue extract on octadecyl silica cartridges. However other workers have found that it is difficult to process human parathyrin to homogeneity because of its susceptibility to alterations such as oxidation during handling.[14] An electrophoretically homogeneous preparation of bovine parathyrin could be obtained from a single gradient HPLC purification of the 5% TCA precipitate from a parathyroid extract.[10] Both these groups of workers were able to demonstrate the presence of components which were eluted earlier than authentic parathyrin and reacted with C-terminal specific antiparathyrin antisera but not N-terminal. These presumed fragments could be demonstrated in patients with elevated circulatory levels of parathyrin as a result of various disorders[11] and in ampuled preparations of human parathyrin and bovine parathyrin.[10,14] Several do not correspond in retention and immunoreactive characteristics to any of the fragments produced by the accepted pathways of parathyrin degradation.[14] Figure 1 shows a typical chromatogram of the extract from a single bovine parathyroid, fractionated on ODS silica and applied directly to a RP column. RP-HPLC has also been used to purify bovine parathyrin labeled by synthesis in vitro with (^{3}H)Tyr and (^{35}S) Met.[15] The labeled product and bovine parathyrin separately labeled with

Table 3
HPLC SYSTEMS USED FOR THE ANALYSIS OF PARATHYRINS

System	Packing[b]	System[a]	Comments	Ref.
		Human		
1	D	0.1% TFA-CH_3CN	Purified from tissue extract entirely by HPLC	11
2	B	0.05 *M* phosphate, pH 2, or 2.6/0.1 *M* Na_2SO_4-CH_3CN	Synthetic 1-84, many synthetic fragments and analogs	16, 17
3	C	5 m*M* pyridine acetate, pH 3, to 0.5 *M* pH 5	Synthetic 1-34: post-column detection with fluorescamine	18
4	D	Ammonium acetate, pH 4.5-methanol	Synthetic 1-34[c]: isocratic	19
5	D	Triethylammonium phosphate-CH_3CN	Synthetic 1-34[c]: isocratic	19
6	A	0.155 *M* NaCl, pH 2.1-CH_3CN	Reference preparations: also fragments	14
		Bovine		
1	D	0.1% TFA -CH_3CN	^{125}I, (^3H) Tyr and ^{35}S Met labeled: generation of fragments by Kupffer cells	15
6	A	0.155 *M* NaCl, pH 2.1-CH_3CN		

[a] TFA, trifluoroacetic acid; HFBA, heptafluorobutyric acid
[b] Packings: A, Hypersil ODS; B, Nucleosil® C_{18}; C, Partisil® SCX; D, μBondapak® C_{18}.
[c] Retention times very short: some doubt about identification of peptide and interpretation of diagrams.

^{125}I were incubated with Kupffer cells, and the fragments produced were separated by RP-HPLC. The labels were chosen to allow N-terminal and C-terminal fragments to be distinguished by automated sequencing of the purified fragments.

Synthetic Parathyrin and Fragments

Most efforts at synthesis of parathyrins have been concentrated on the biologically active 1-34 fragments of human and bovine parathyrin. However, in a notable synthesis of the complete human parathyrin sequence, extensive and telling use was made of RP-HPLC at almost all stages.[16,17] The authors were able to ascertain the extent of racemization at certain critical coupling stages and were also able to follow the isomerization of the asparagine residues at positions 45 and 56 to α-amido succinyl (α-asc) and β-aspartyl residues. Finally the complete 84 residue peptide was purified by HPLC from its three methionine sulfoxides (all resolved) to yield a product which gave a single peak when examined analytically. These authors also report[16] the HPLC behavior of the following synthetic fragments: 39-84, 44-51, and its (β-asp^{45}) and (α-asp^{45}) analogs, 52-59 and its (β-asp^{56}) and (α-asp^{56}) analogs, 69-84 and its (D-ala^{73}), (D-leu^{78}), and (D-ala^{73}, D-leu^{78}) analogs, and 74-84. Bovine fragments, whose behavior have been reported, include the presumptive C-terminal sequences 34-84, 35-84, 37-84, 38-84, and 41-84 and their (^{125}I-Tyr^{43}) analogs, identified by (^{3}H)Tyr^{43} release on Edman degradation[15] and 53-84.[10]

The human (1-34) synthetic fragment was one of the intermediates examined by HPLC in the total synthesis mentioned above. This peptide has also been used as a chromatographic marker for the elution positions of the expected natural fragments of parathyrin,[11,14] and similarly for the bovine sequence.[10] The human 1-34 peptide was also included in an evaluation of Partisil® SCX, a strong anion-exchange bonded phase.[18] The stepwise gradient system, increasing both the pH and the concentrations of the pyridine-acetate buffer used, was closely based on those used for peptide separation on conventional polystyrene ion

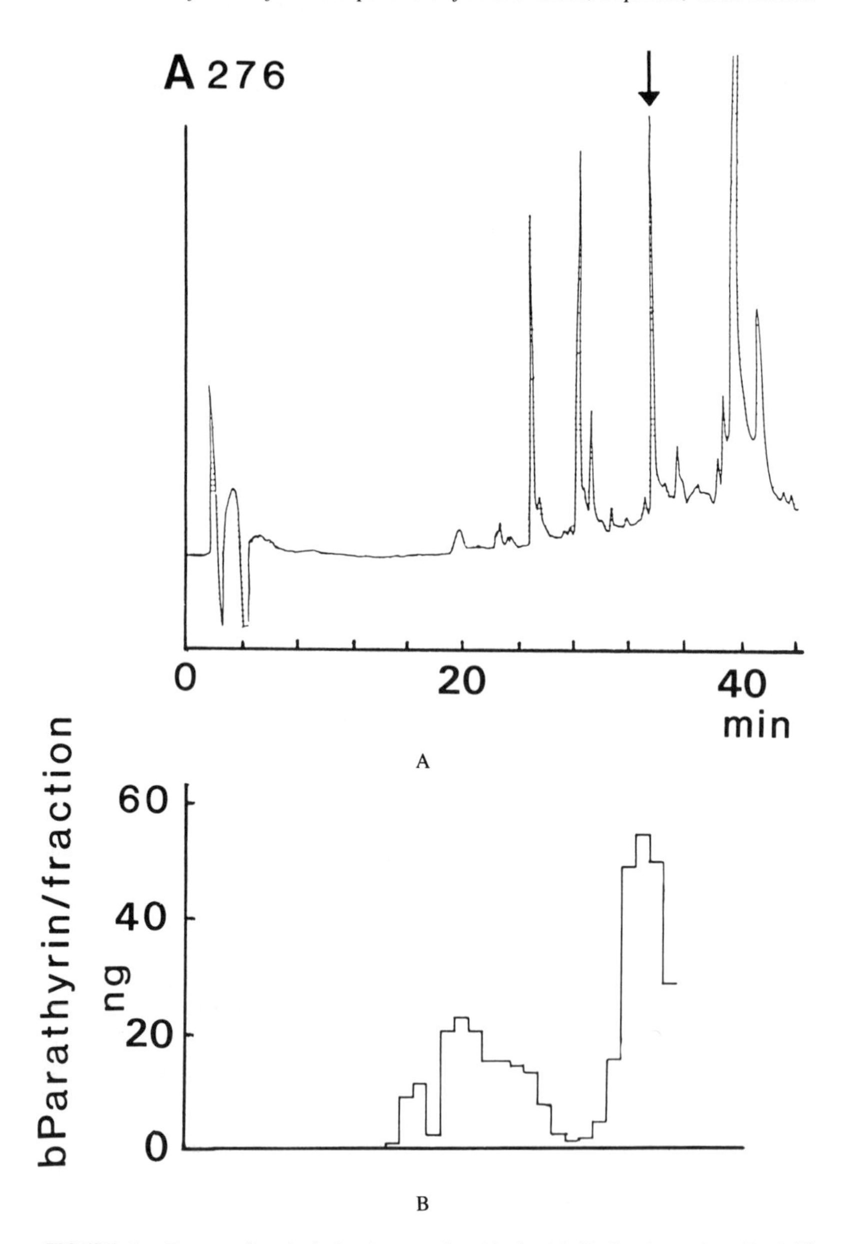

FIGURE 1. Extract of a single bovine parathyroid gland bulk fractionated on Partisil® ODS.[12,14] After dilution with TFA the 20 to 60% CH_3CN fraction was lyophilized, dissolved in 0.1% TFA and analyzed on a column (100 × 4.6 mm) of Nucleosil® 5 C_8 with a linear gradient of 17 to 52% CH_3CN in 0.1% TFA. (A) absorption at 276 nm. (B) parathyrin RIA of fractions with antibody directed at C-terminal region. The arrow shows the normal elution position of bovine parathyrin, but significant amounts of immunoreactive material, presumably carboxyl terminal cleavage products, are associated with earlier fractions.

exchangers. The peptides were detected by post-column reaction with fluorescamine. An additional system[19] has not been discussed because of the low retention time reported and confusion about interpretation of diagrams. Figure 2 shows the result of hydrogen peroxide oxidation of a preparation of human (1-34) parathyrin. All three expected oxidation products are resolved.

Reference and Pharmaceutical Preparations

Reference preparations may be ampuled with mannitol or, where the amount of material

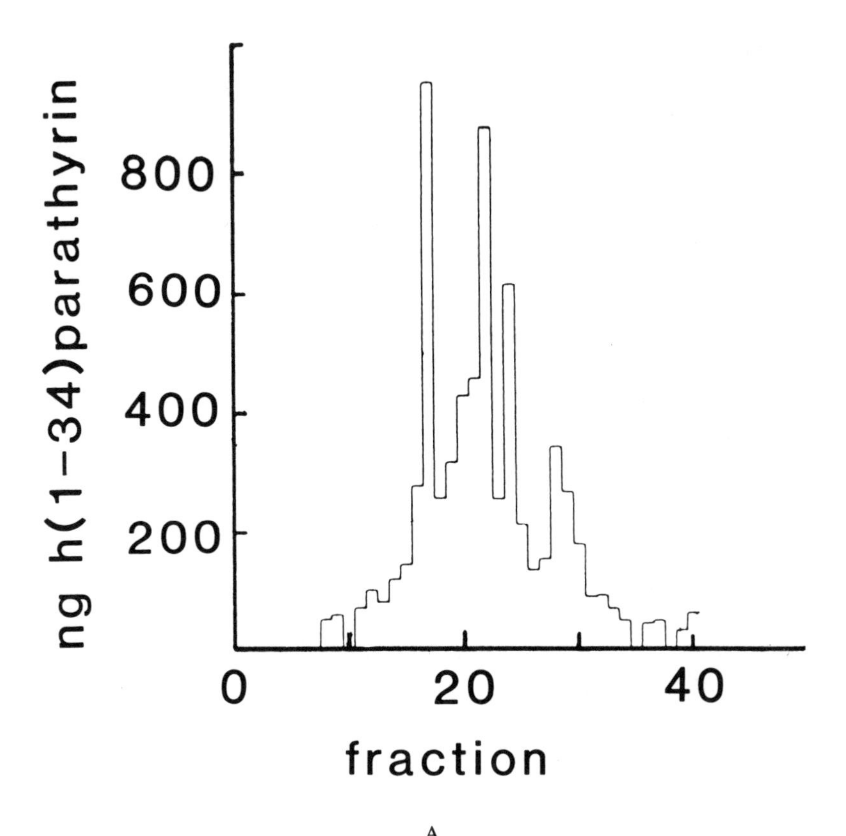

A

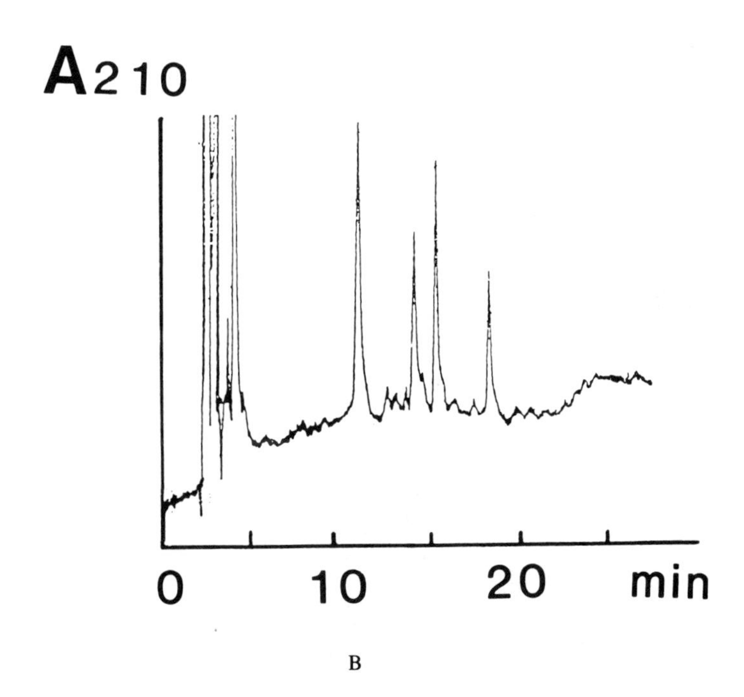

B

FIGURE 2. Chromatogram of a partially H_2O_2-oxidized sample of human (1-34) parathyrin (B) and RIA of fractions (A). All three expected oxidation products are present. The peak eluted at 18 min corresponds to native (1-34) parathyrin. Conditions: linear gradient 29 to 39% CH_3CN in 0.1% TFA (v/v). Flow 1 mℓ/min. Column 135 × 4.6 mm Nucleosil® 5 C_8.

is small, with human albumin. In the latter case the background due to the albumin may be sufficient to swamp peaks due to the parathyrin, and specific detection methods such as radioimmunoassay (RIA) are necessary.[10]

A preparation of natural bovine parathyrin used clinically at one time was formulated with hydrolyzed gelatin, a combination which proved difficult to analyze by conventional RP-HPLC.

Future Prospects

Because of the importance of proteolytic cleavage in the normal and pathological metabolism of parathyrin, HPLC should prove of further value in characterizing patterns of fragmentation (e.g., References[14,15]). Another field in which it will be further used is obviously in the characterization and purification of synthetic products.[16] Finally HPLC should see increased exploitation for preparative or semipreparative purification of natural parathyrins.

CALCITONINS

The structures of natural calcitonins (Table 2) are much more variable than those of the parathyrins, and the content of hydrophobic residues varies significantly between calcitonins of different species. The sum of the hydrophobic contributions of the amino acid side chains calculated by four different methods is shown in Figure 3. Salmon calcitonin is, by common consent, the least hydrophobic sequence, yielding values about half those obtained for bovine or porcine, the most hydrophobic species. The predicted order of elution, based on these calculations, is not observed in practice (Figure 3,[12]). Although human calcitonin is clearly more hydrophobic in amino acid content, it is eluted before salmon calcitonin. Two of the predictive methods[20,21] place human calcitonin correctly before porcine calcitonin, but the other two, both based on octanol/water partition coefficients, give a predicted order which is the complete opposite of that actually observed. It is interesting to speculate whether the anomalously hydrophobic character shown by salmon calcitonin may be related to its biological behavior, i.e., to its increased membrane receptor binding, an interaction in which hydrophobic forces are also thought to place an important part. Table 4 lists all the HPLC systems described for calcitonin as derived from several abstracting databases.

Characterization of Native and Synthetic Preparations

The impurities observed in these preparations might be expected to vary widely between the opposite poles of a traditional fragment condensation synthesis and a Merrifield type stepwise solid-phase synthesis, and between preparations of natural and synthetic origin. In practice this does not seem to impose very different demands on the chromatographic system, as can be seen from Figures 3 and 4, which show preparations of natural porcine calcitonin and a "classical" and a "solid-phase" synthesized sample of human calcitonin.

A number of the references in Table 4 deal only with the examination of synthetic calcitonin preparations. One of these will not be discussed:[19] see footnote to Table 4. Hirt and coworkers[22] report that they were able to use the same volatile solvent system for analysis and purification of synthetic human calcitonin. The methionine sulfoxide and several impurities were separated. This separation has been studied in more detail.[23] The large change in polarity on oxidation of methionine makes RP-HPLC a particularly simple and effective means of estimating the extent of methionine oxidation. A similar, less detailed, report[24] describes the use of an aqueous phosphoric acid-methanol system to purify a tritiated synthetic human calcitonin preparation. A conventional ion-exchange purification was complementary to the RP-HPLC one since each removed impurities not removed by the other. The relative merits of HPLC and TLC for comparing different preparations of salmon calcitonin[25] have been discussed. The HPLC system was pH 7 phosphate-methanol but a lower pH has given

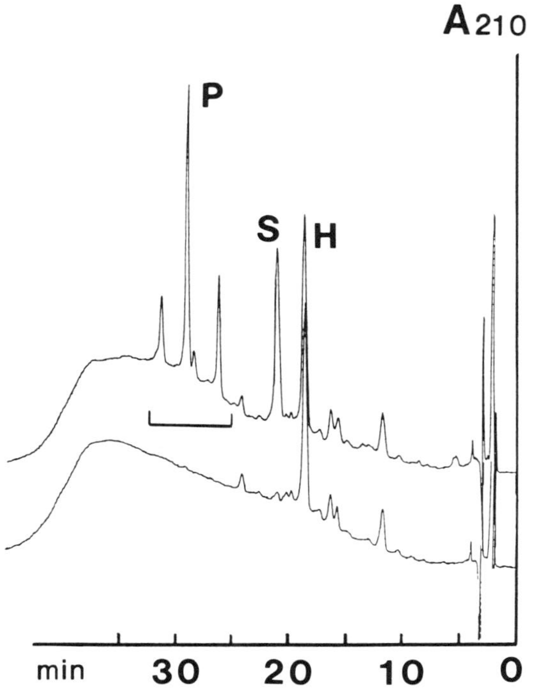

FIGURE 3. Separation of human (H), salmon (S), and porcine (P) calcitonins (upper trace). The four peaks above the bar are components of the porcine calcitonin sample (the British Pharmacopoeia chemical reference preparation). The human sample is also shown run separately (lower trace, 70/234: see Figure 4.). Gradient: linear 18 to 36% acetonitrile-0.1 *M* NaH_2PO_4/H_3PO_4 pH 2. Flow rate: 1 mℓ/min. Detection UV 210 nm. Column: 150 × 4.6 mm Nucleosil® 5 C_8.

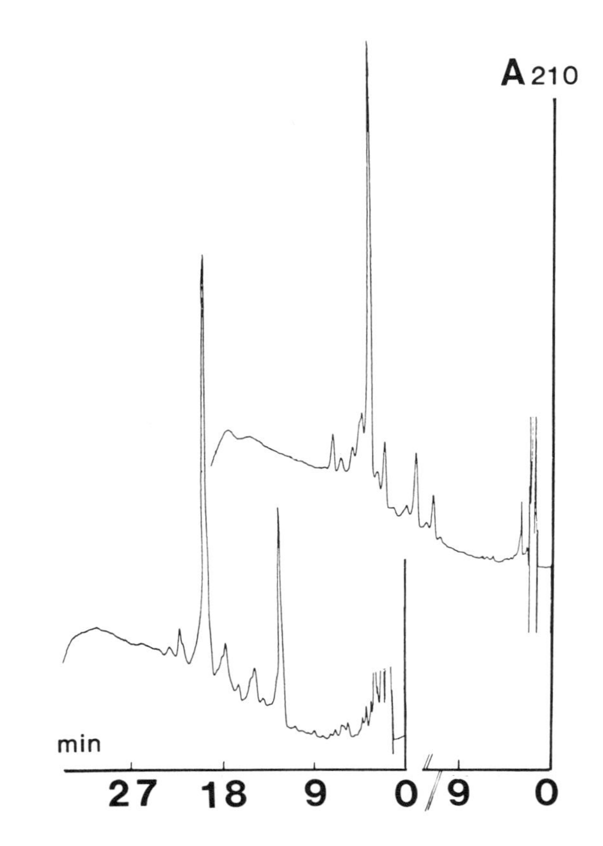

FIGURE 4. Synthetic human calcitonin preparations. Fragment condensation ("classical") (upper) 70/234 and solid phase ("Merrifield") (lower) 77/574 samples. Column Nucleosil® 5 C_8, 135 × 4.6. mm. Other conditions as Figure 3.

Table 4
HPLC SYSTEMS USED FOR THE ANALYSIS OF CALCITONINS

Species[b]	Packing[a]	System	Comments	Ref.
		Systems for the Examination of Purified Preparations		
Human	D	0.1% H_3PO_4/CH_3OH gradient	Preparative: 5 mg on 50 × 0.7 cm column	24
	C	1% TFA-H_2O/CH_3OH gradient	Met SO separated	22
	G	5 m*M* TBAP pH 7.5/MeOH or 5 m*M* Hexane SO_3H-pH 3.5/MeOH	10 m*M* Ammonium acetate, 10 m*M* *Tris* 50 m*M* $NaPO_4$ at pHs 4.5—7.5 inferior Chloramine T iodinated CT not resolved from native: 75% recovery	26
	C, D	TFA (various %)-H_2O/MeOH 0.01, 0.02 *M* KH_2PO_4/MeOH	Also CN, Phenyl phases; Met SO separated	23
	E	0.1 *M* NaH_2PO_4-H_3PO_4 pH 2/CH_3CN	Met SO separates	Figures 3, 4
Salmon	B	Na phosphate pH 6.5/MeOH	Synthetic preparations compared: isocratic	25
	G	Ammonium acetate/CH_3CN,	See text: k′1.3; no pictures: isocratic	19
		0.25 *N* Triethylamine PO_4/25% CH_3CN	See text: k′ = 0: isocratic	
Rat	G	5 m*M* TBAP pH 7.5/MeOH Hexane SO_3H pH 3.5/MeOH	Failed to resolve iodinated and native: see above	26
Porcine	E	0.1 Na_2PO_4-H_3PO_4 pH 2/CH_2CN	Natural: four peaks resolved in BP CRS	Figure 3
		Systems for the Characterization of CT in Natural Extracts		
Human	NA	1% TFA/MeOH	Characterization of circulating form	29
	A	NaCl pH 2.1/CH_3CN	= Reference 35	36
	D	?NaCl pH 2.1/CH_3CN	Characterization of forms synth. by carcinoma	35
	D	0.1% HFBA/CH_3CN	CT from normal human thyroid = synthetic	
		Systems in which CT has been Included as Marker		
For identification of CT-like immunoreactive activity in other spp.			Spp. examined	
Marker CT				
hCT	NA	1% TFA/MeOH	Hagfish, lancelet, sea squirt	34
hCT	F	1% TFA/MeOH	Frog	30
hCT, sCT	F	1% TFA/MeOH	Eel, salamander, rat, ox	31
hCT, sCT	F	1% TFA/MeOH	Pigeon, chicken	32

sCT	NA	1% TFA/MeOH	Lizard	27
sCT	NA	1% TFA/MeOH	Pigeon	28
For other molecules				
hCT	G	0.01 *M* TFA or 0.01 *M* HFBA/CH_3CN	Calibration marker for human ACTH	37

[a] Packings: A, Hypersil ODS; B, Hypersil SAS; C, Lichrosorb® RP18; D, Nucleosil® C_{18}; E, Nucleosil® C_8; F, Partisil® ODS; G, μBondapak® C_{18}; NA, not available.

[b] Abbreviations: sCT, Salmon calcitonin; hCT, human calcitonin; CT, calcitonin.

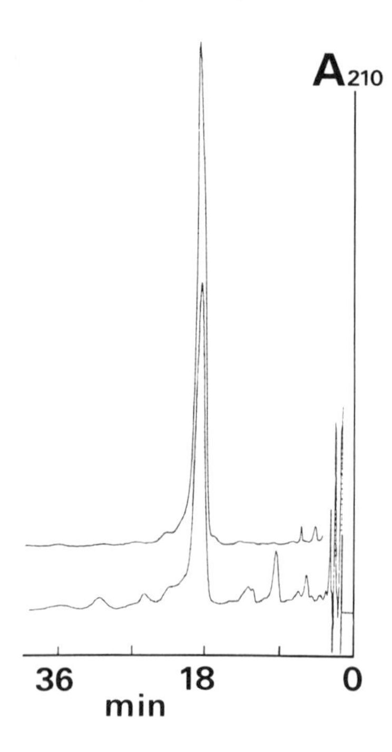

FIGURE 5. Synthetic salmon calcitonin samples. Candidate research standard 81/576 (upper trace) and International Standard for Bioassay 72/158. Column 150 × 4.6 mm Nucleosil® 5 C_8. Conditions: isocratic 28% CH_3CN/72% 0.1 *M* NaH_2PO_4, pH 2. Flow, 1 mℓ/min; detection UV 210 nm.

superior resolution (Figure 5). A study of the optimum conditions for the HPLC of synthetic rat and human calcitonin preparations and their ^{125}I-iodinated derivatives[26] evaluated five systems over the pH range 3.5 to 7.5. The authors concluded that an ion-pairing system using *t*-butylammonium phosphate (TBAP) pH 7.5 or one using sodium hexane sulfonate pH 3.5 was most useful, but that the former allowed detection at a lower wavelength. The radioiodination was carried out with Chloramine-T as oxidizing agent, but the radioactivity coeluted with the UV peak attributed to the native calcitonin with each species. Since Chloramine-T would certainly have oxidized the methionine residue, this implies that the oxidation of methionine was exactly counter-balanced by the incorporation of iodine (which would increase hydrophobicity). It is difficult to see how this coincidence may be turned to advantage.

Purification and Characterization of Calcitonins from Natural Sources

Tissues secreting calcitonin do not appear to store large amounts of the hormone. The chief problem in examining tissue extracts is thus the detection of small amounts of a specific peptide in a complex mixture. This problem has been tackled by two approaches.

The first approach, collection of fractions for RIAs of the effluent, has been effectively exploited by McIntyre and co-workers in their comparative studies of calcitonins and calcitonin-like substances.[27-32] They have been able to partially characterize substances which cross-reacted immunologically with salmon calcitonin or human calcitonin in tissues from a wide variety of vertebrate species, including some of the most primitive. These substances corresponded to known calcitonins in molecular size and in possessing a similar retention on RP-HPLC. In all cases the peptides were extracted from tissue homogenates by adsorption onto octadecylsilanized silica and the fraction eluted with 80% methanol was characterized by Sephadex® chromatography followed by gradient RP-HPLC in 1% TFA-methanol. Since the amounts of peptide involved corresponded to, at most, some tens of nanograms of calcitonin it was important that the mobile phase should be totally volatile to reduce interference with the RIA.

A similar use of HPLC was made to demonstrate that the forms of human calcitonin obtained from a medullary thyroid carcinoma was identical in immunoreactive and chromatographic properties to one of the forms present in normal human thyroid[33] and in the circulation of normal human subjects.[34]

The second common approach for the detection of small amounts of a peptide of interest is to incorporate a radioactive label by biosynthesis in vitro. In a study of calcitonin-like material produced by cells derived from a variety of tumors, label was incorporated by growing cells with ^{35}S- or ^{14}C-labeled amino acids.[35] Certain high-molecular-weight species which cross-reacted immunologically with human calcitonin were characterized by HPLC. The authors were able to demonstrate that some radioactive peptides characteristic of a tryptic digest of human calcitonin were formed.

Other Systems

The ready availability of human calcitonin and salmon calcitonin have made them popular markers both for the calcitonins of different species, as mentioned above, and for peptides of unrelated sequence. Human calcitonin has been used to calibrate chromatographic systems for the purification of rat corticotrophin[37] and as a marker in the fractionation of human parathyrin.[10]

Pharmaceutical and Reference Preparations

Some therapeutic preparations of calcitonin are formulated in saline without preservative. A few are coformulated with hydrolyzed gelatin, a vehicle which contributes a high UV background and makes the analysis of the calcitonin very difficult. A number of ampuled reference preparations of human calcitonin, porcine calcitonin, and salmon calcitonin are lyophilized with mannitol (for instance those illustrated in Figure 4).

Future Prospects

One area in which developments can be expected is in the field of analysis of pharmaceutical preparations. Despite the interest in the clinical use of calcitonins, little has been published about the control and assay of such preparations. Pharmaceutical manufacturers often seem reluctant to publish their analytical methods, and it is possible that the HPLC systems which must have been developed within the industry may surface first in an appropriate Pharmacopoeia. Certainly the simplicity, cheapness, and economical use of peptide of RP-HPLC methods, compared to some traditional "wet chemistry" analytical methods, must provide a powerful incentive for its inclusion in official specifications.

A literature survey of publications over the last 10 years shows more interest in the pharmacological and therapeutic uses of calcitonins than in physiological and endocrinological aspects. Recent studies using recombinant DNA analysis for tissue distribution of precursor and certain associated products have awakened interest in these two fields.[39-41] It seems likely that the most recent developments in size-exclusion and ion-exchange microparticulate packings as well as RP-HPLC will be of great use.

REFERENCES

1. **Birge, S. J., Hahn, T. J., Whyte, M. P., and Avoli, L. V.,** Hormonal regulation of mineral metabolism, in *Int. Rev. Physiol.*, McCann, S. M., Ed., University Park Press, Baltimore, 1981, 201.
2. **Cohn, D. V. and MacGregor, R. R.,** The biosynthesis, intracellular packaging and secretion of parathyroid hormone, *Endocrine Rev.* 2, 1, 1981.
3. **Keutmann, H. T.,** The chemistry of parathyroid hormone, *Clin. Endocrinol. Metab.*, 3, 173, 1974.
4. **Parsons, J. A. and Zanelli, J. M.,** Physiological role of the parathyroid glands, in *Handbuch der Inneren Medizin VI/1A,* Kuhlencordt, F. and Bartelheimer, H., Eds., Springer-Verlag, New York, 1980, 135.
5. **McIntyre, I. and Steven, J. C.,** Calcitonin: a modern view of its physiological role and interrelation with other hormones, in *Calcitonin 80 Chemistry, Physiology and Clinical Aspects*, Pecile, A., Ed., Exerpta Medica, Princeton, 1981.
6. **Guttman, S.,** Chemistry and structure-activity relationship of natural and synthetic calcitonins, in *Calcitonin 80 Chemistry, Physiology and Clinical Aspects,* Pecile, A., Ed., Excerpta Medica, Princeton, 1981, 11.
7. **Potts, J. T., Jr. and Aurbach, G. D.,** Chemistry of the calcitonins, in *Handbook of Physiology and Endocrinology,* Vol. 7, Greep, R. O., Astwood, E. B., and Aurbach, G. D., Eds., American Physiological Society, Washington D.C., 1976, 423.
8. **Austin, L. A. and Heath, H.,** Calcitonin-physiology and pathophysiology, *N.Eng. J. Med.,* 304, 269, 1981.
9. **Reeve, J., Meunier, P. J., Parsons, J. A., Bernat, M., Bijvoet, O. L. M., Courpron, P., Edouard, C., Klenerman, L., Neer, R. M., Renier, J. C., Slovik, D., Vismans, F. J. F. E., and Potts, J. T., Jr.,** Anabolic effects of human parathyroid fragment on trabecular bone in involutional osteoporosis: a multicentre trial, *Br. Med. J.*, 280, 1340, 1980.
10. **Zanelli, J. M., O'Hare, M. J., Nice, E. C., and Corran, P. H.,** Purification and assay of bovine parathyroid hormone by reversed-phase high performance chromatography, *J. Chromatogr.*, 223, 59, 1981.
11. **Bennett, H. P. J., Solomon, S., and Goltzman, D.,** Isolation and analysis of human parathyrin in parathyroid tissue and plasma, *Biochem. J.*, 197, 391, 1981.
12. **Nice, E. C., Capp, M., and O'Hare, M. J.,** Use of hydrophobic interaction methods in the isolation of proteins from endocrine and paraendocrine tissues and cells by high-performance liquid chromatography, *J. Chromatogr.,* 185, 413, 1979.
13. **Molnar, I. and Horvath, C.,** Separation of amino acids and peptides on non-polar stationary phases by high-performance liquid chromatography, *J. Chromatogr.,* 142, 623, 1977.
14. **Zanelli, J. M., Kent, J. C., Rafferty, B., Nissenson, R. A., Nice, E. C., Capp, M., and O'Hare, M. J.,** High-performance liquid chromatography methods for the analysis of human parathyroid hormone in reference standards, parathyroid tissue and biological fluids, *J. Chromatogr.*, 276, 55, 1983.
15. **Bringhurst, F. R., Segre, G. V., Lampman, G. W., and Potts, J. T., Jr.,** Metabolism of parathyroid hormone by Kupffer cells: analysis by reverse-phase high-performance liquid chromatography, *Biochemistry,* 21, 4252, 1982.
16. **Kimura, T., Morikawa, T., Takai, M., and Sakakibara, S.,** Total synthesis of human parathyroid hormone (1-84), *J. Chem. Soc. Chem. Commun.*, p. 340, 1982.
17. **Kimura, T., Takai, M., Masui, Y., Morikawa, T., and Sakakibara, S.,** Strategy for the synthesis of large peptides: an application to the total synthesis of human parathyroid hormone (hPTH (1-84)), *Biopolymers,* 20, 1823, 1981.
18. **Radhakrishnan, A. N., Stein, S., Licht, A., Gruber, K. A., and Udenfriend, S.,** High-efficiency cation-exchange chromatography of polypeptides and polyamines in the nanomole range, *J. Chromatogr.*, 132, 552, 1977.
19. **Rivaille, P., Roulais, D., and Milhaud, G.,** High performance liquid chromatographic analysis of peptide hormones, *Chromatogr. Sci.*, 12, 273, 1979.
20. **Wilson, K. J., Honegger, A., Stotzel, R. P., and Hughes, G. J.,** The behaviour of peptides on reverse-phase supports during high-pressure liquid chromatography, *Biochem. J.*, 199, 31, 1981.
21. **Meek, J. L. and Rossetti, Z. L.,** Factors affecting retention and resolution of peptides in high-performance liquid chromatography, *J. Chromatogr.*, 211, 15, 1981.
22. **Hirt, J., Kranenburg, P., and Beyerman, H. C.,** A convenient synthesis of human calcitonin mainly via the repetitive excess mixed anhydride method, *Recl. Trav. Chim. Pays-Bas,* 98, 143, 1979.
23. **Voskamp, D., Olieman, C., and Beyerman, H. C.,** The use of trifluoroacetic acid in the reverse-phase liquid chromatography of peptides including secretin, *Recl. Trav. Chim. Pays-Bas,* 99, 105, 1980.
24. **Brundish, D. E. and Wade, R.,** Synthesis of (3,5-$^3H_2$-Tyr^{12})- and (2,5-$^3H_2$-His^{20})-human calcitonin, *J. Chem. Soc. Perkin,* 1, 318, 1981.
25. **Corran, P. H. and Calam, D. H.,** The characterisation of pharmaceutical peptides, in *Recent Advances in Chromatography and Electrophoresis,* Frigerio, A. and Renoz, J., Eds., Elsevier, Amsterdam, 1979, 341.

26. **Lambert, P. W. and Roos, B. A.,** Paired-ion reversed-phase high-performance liquid chromatography of human and rat calcitonin, *J. Chromatogr.*, 198, 293, 1980.
27. **Galan, F. G., Rogers, R. M., Girgis, S. I., Arnett, T. R., Ravazzola, M., Orci, L., and MacIntyre, I.,** Immunochemical characterization and distribution of calcitonin in the lizard, *Acta Endocrinol.*, 97, 427, 1981.
28. **Galan, F. G., Rogers, R. M., Girgis, S. I., and MacIntyre, I.,** Immunoreactive calcitonin in the central nervous system of the pigeon, *Brain Res.*, 212, 59, 1981.
29. **Girgis, S. I., Galan, F. G., Arnett, T. R., Rogers, R. M., Bone, Q., Ravazzola, M., and MacIntyre, I.,** Immunoreactive human calcitonin-like molecule in the nervous system of protochordates and a cyclostome, *Myxine*, *J. Endocrinol.*, 87, 375, 1980.
30. **Perez-Cano, R., Galan, F. G., Girgis, S. I., Arnett, T. R., and MacIntyre, I.,** A human calcitonin-like molecule in the ultimobranchial body of the amphibia *(Rana pipiens)*, *Experientia*, 37, 1116, 1981.
31. **Perez-Cano, R., Girgis, S. I., Galan, F. G., and MacIntyre, I.,** Identification of both human and salmon calcitonin-like molecules in bird suggesting the existence of two calcitonin genes, *J. Endocrinol.*, 92, 351, 1982.
32. **Perez-Cano, R., Girgis, S. I., and MacIntyre, I.,** Further evidence for calcitonin gene duplication: the identification of two different calcitonins in a fish, a reptile and two mammals, *Acta Endocrinol.*, 100, 256, 1982.
33. **Tobler, P. H., Jöhl, A., Born, W., Maier, R., and Fischer, J. A.,** Identity of calcitonin extracted from normal human thyroid glands with synthetic human calcitonin-(1-32), *Biochim. Biophys. Acta*, 707, 59, 1982.
34. **Girgis, S. I., McMartin, C., and MacIntyre, I.,** Nature of normal human calcitonin in the circulation, *J. Endocrinol.*, 82, 55P, 1979.
35. **Lumsden, J., Ham, J., and Ellison, M. L.,** Purification and partial characterisation of high-molecular-weight forms of ectopic calcitonin from a human bronchial carcinoma cell-line, *Biochem. J.*, 191, 239, 1980.
36. **Nice, E. C. and O'Hare, M. J.,** Analyical and preparative high-performance liquid chromatography of polypeptide and protein hormone from normal and pathological calcitonin-containing samples, *Anal. Proc.*, 17, 526, 1980.
37. **Bennett, H. P. J., Browne, C. A., and Solomon, S.,** Purification of the two major forms of rat pituitary corticotropin using only reversed-phase liquid chromatography, *Biochemistry*, 20, 4530, 1981.
38. **Pliška, V., Schmidt, M., and Fauchère, J-L.,** Partition coefficients of amino acids and hydrophobic parameters π of their side chains as measured by thin-layer chromatography, *J. Chromatogr.*, 216, 79, 1981.
39. **Amara, S. G., David, D. N., Rosenfeld, D. N., Roos, B. A., and Evans, R. M.,** Characterization of rat calcitonin mRNA, *Proc. Natl. Acad. Sci. U.S.A.*, 77, 444, 1980.
40. **Jacobs, J. W., Goodman, R. H., Chin, W. W., Dee, P. C., Habener, J. F., Bell, N. H., and Potts, J. T., Jr.,** Calcitonin messenger RNA encodes multiple peptides in a single precursor, *Science*, 213, 457, 1981.
41. **Craig, R. K., Hall, L., Edbrooke, M. R., Allison, J., and MacIntyre, I.,** Partial nucleotide sequence of human calcitonin precursor mRNA identifies flanking cryptic peptides, *Nature (London)*, 295, 345, 1982.

CALCIUM-BINDING PROTEINS AND THEIR FRAGMENTS

Allan S. Manalan and Claude B. Klee

INTRODUCTION

Many of the physiological effects of calcium ions are mediated by a group of homologous intracellular calcium-binding proteins (see References 1 and 2 for review). The members of this group, calmodulin, troponin C, parvalbumin, myosin light chain, intestinal calcium-binding protein, and S-100 protein, exhibit striking homologies in amino acid sequence.[1-4] These proteins also share several similar physical and chemical characteristics, including small size, as well as heat and acid stability. In addition, the proteins undergo conformational changes upon binding calcium ions.[1,2,4] The unique properties of the calcium-binding proteins have facilitated their isolation from the bulk of other intracellular proteins, typically by a combination of heat and acid precipitation and ion-exchange chromatography (see Reference 4 for review). However, the close similarities among the various members of this group have made their separation from each other more difficult to achieve. Because the calcium-binding proteins are similar in both size and charge, neither gel filtration nor ion-exchange techniques have been employed in initial efforts to separate these proteins by high-performance liquid chromatography (HPLC). Instead, potential differences in the ability of the proteins to engage in hydrophobic interactions[5] have provided the basis for their separation by reversed-phase (RP) HPLC.[6-8] Advantages of this last method include reproducibility, short elution times, high sensitivity, good recovery, excellent resolution, and preservation of the biological activity of the proteins.

METHODS

The HPLC system consists of two model 6000A solvent delivery systems equipped with a model 660 solvent programmer, two wavelength detectors, and a Data Module for data recording and processing (Waters Associates, Inc., Milford, MA). All aqueous solvents are filtered through 0.45-μm Millipore filters and degassed under vacuum with stirring for 30 min prior to use. The calcium-binding proteins,[6-8] as well as large peptides derived by limited proteolysis,[8] can be separated using a μBondapak®-phenyl reversed-phase column (Waters Associates). The elution characteristics of the protein under study are first assessed by an analytical run. The material is loaded onto the column equilibrated with 0.01 *M* potassium phosphate buffer, pH 6.1, 2 m*M* EGTA, and 5% acetonitrile (v/v) (Buffer A). Elution is accomplished at room temperature with a 20-min linear acetonitrile gradient from 100% Buffer A to a final mixture of 65% acetonitrile and 35% Buffer A. Flow rate is maintained at 1.5 mℓ/min. Mixtures containing several calcium-binding proteins or large peptide fragments of these proteins are then loaded on the column equilibrated with the solvent containing the appropriate concentration of acetonitrile (usually 15 to 25% acetonitrile and 85 to 75% Buffer A as indicated in legends to Figure 1 and 2) and eluted at room temperature at constant flow rate (1.5 mℓ/min) by appropriate stepwise increases in acetonitrile concentration. Small-scale preparative (up to 5 to 10 mg) or analytical runs are carried out with a 3.9 mm × 30 cm column made up of 10-μm beads. Large columns (7.8 mm × 30 cm) are used for large-scale preparations (> 10 mg).

Small peptides derived by more complete proteolytic digestion of calcium-binding proteins can be separated using a μBondapak® C_{18} reversed-phase column (3.9 mm × 30 cm, 10-μm beads, Waters Associates) as described by Fullmer and Wasserman.[9] Complete proteolytic digests are loaded onto the column equilibrated with 0.1% orthophosphoric acid, pH

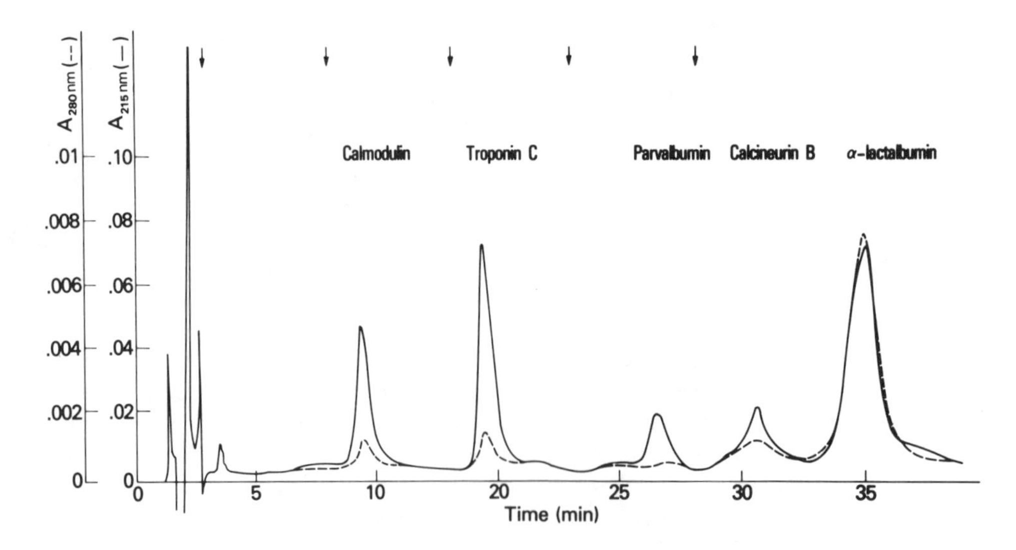

FIGURE 1. Elution profile of calcium-binding proteins. The proteins were separated using an alkylphenyl column as described in the text. The arrows indicate stepwise increases in acetonitrile concentrations: 25, 27, 28, 32, and 34%. (From Klee, C. B., Oldewurtel, M. D., Williams, J. F., and Lee, J. W., *Biochem. Int.*, 2, 485, 1981. With permission.)

2.2 (solvent B). The peptides are eluted at room temperature with a 50-min linear gradient from 100% solvent B to a mixture of 50% solvent B and 50% acetonitrile at a constant flow rate of 2 mℓ/min.

All column eluates are monitored simultaneously for UV absorption at 215 and 280 nm to identify the various components. Fractions containing the individual components are pooled, flash evaporated to remove the solvent, dissolved in 0.05 *M* NH_4HCO_3, and desalted by gel filtration on Sephadex® G 25.

SEPARATION OF CALCIUM-BINDING PROTEINS BY HPLC

Chromatographic separation of the calcium-binding proteins using a μBondapak®-alkylphenyl reversed-phase column is shown in Figure 1. Because of the similar retention times exhibited by these homologous proteins upon gradient elution a complete separation of these proteins from each other cannot be achieved by this method (Table 1). Nevertheless, the small differences in retention times can be exploited by using small stepwise increases in acetonitrile concentration as shown in Figure 1 to accomplish complete separation of the members of this homologous series of proteins. Under these conditions, the calcium-binding proteins are recovered with unaltered biological activity in 80 to 90% yield from samples ranging from 35 μg to 3 mg protein. Recovery depends upon the characteristics of the protein applied. Proteins which bind more tightly are recovered in lower yield, and in certain cases, are not recovered at all under these elution conditions. Each of the calcium-binding proteins exhibits unique spectral properties (summarized in Table 1) which provide means for identification of each protein present in the eluent. Use of dual wavelength monitoring allows identification of the proteins by their characteristic A_{280}/A_{215} ratio. Retention times and acetonitrile concentrations required for elution can also be used to identify the proteins (Table 1).

PURIFICATION OF CALMODULIN PEPTIDES

Limited proteolysis of calmodulin with trypsin generates large calmodulin fragments.[8,10]

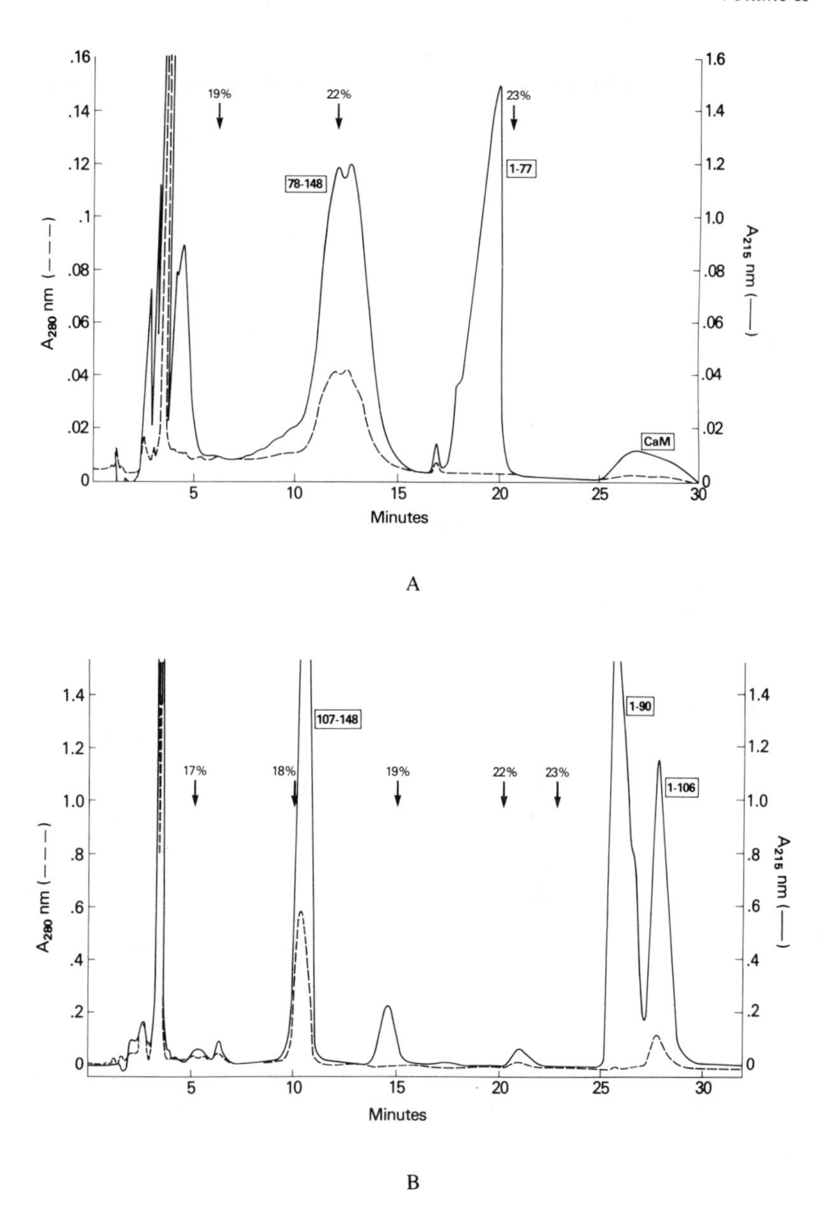

FIGURE 2. HPLC elution profile of tryptic fragments of calmodulin. Calmodulin was subjected to limited proteolysis with trypsin in the presence of calcium (A) or EGTA (B) by a modification of the method of Drabikowski et al.[10] After addition of soybean inhibitor the peptides were resolved by HPLC using an alkylphenyl column with stepwise increases in acetonitrile concentration (arrows) as described in Methods. The peptides were identified by amino acid and end group analysis. (From Oldewurtel, M. D., Krinks, M. H., Lee, J. W., Williams, J. F., and Klee, C. B., in *Protides Biol. Fluids Proc. Colloq.*, 30, 713, 1982. With permission.)

Several of these fragments retain calcium-binding properties and certain of the biological activities of intact calmodulin.[8,11,12] Separation of these fragments by conventional techniques is hampered by the high level of sequence homology shared by the peptides. However, fractionation of the large calmodulin peptides can readily be achieved with the μBondapak® alkylphenyl column, using small stepwise increases in acetonitrile concentration[8] (Figure 2

Table 1
ELUTION CHARACTERISTICS OF CALCIUM-BINDING PROTEINS[a]

Proteins	Retention times[b]	Elution solvent[c]	$\epsilon^{1\%}_{215\ nm}$	$\epsilon^{1\%}_{280\ nm}$
	(min)	(% CH_3CN)		
Calmodulin (bovine brain, bovine testes)*	15.8	25 (7 min)	107	1.8
Calmodulin (*Acanthamoeba castellani*)[d]	13.7	25 (13 min)	150	0.9
Troponin C (Rabbit skeletal muscle)*[e]	16.1	26 (8 min)	126	1.8
Parvalbumin (*Rana esculenta* pI, 4.88)*[e]	16.5	32 (7 min)	118	0.6
Parvalbumin (Carp)*[e]	16.3	29 (7 min)	155	
Parvalbumin (*Xenopus laevis*)[f] I*		27 (9 min)		
II and III*		27 (11, 12 min)		
Calcineurin B*[g]	16.5	28 (8 min)	140	3.5
α lactalbumin* (bovine milk)	16.6	34 (8 min)	200	20.0

[a] The proteins were fractionated on alkylphenyl columns as described in Methods.
[b] Elution was performed with a linear (0 to 65% acetonitrile) gradient. Variations in retention times with different columns were up to 3 min. The proteins marked with an asterisk were tested on the same column with the same solvent.
[c] Concentration of acetonitrile needed to elute the indicated protein when stepwise elution was used. The retention times at the indicated acetonitrile concentration are shown in parenthesis.
[d] Calmodulin was isolated by the two-step procedure described in the text.[34]
[e] Troponin C was a gift of Dr. P. Laevis, carp parvalbumin was from Dr. J. Potter, and *Rana esculenta* parvalbumin was from Dr. J. Haiech.
[f] Partially purified *Xenopus laevis* parvalbumin prepared by the method of Haiech et al.[28] was a gift of Dr. Chien. Three components were detected by HPLC.
[g] Calcineurin B, the small Ca^{2+}-binding subunit of calcineurin,[29] was resolved from calcineurin A by DEAE cellulose chromatography in 6 *M* urea.

Table 2
ELUTION CHARACTERISTICS OF CALMODULIN PEPTIDES[a]

Fragment	$\epsilon^{1\%}_{280\ nm}$	$\epsilon^{1\%}_{215\ nm}$	Retention times (min)[b]
1—106	1.4	120	17.6
1—77	0.2	96	16.5
1—90	0.4	80	16.1
107—148	2.3	100	8.4
78—148	4.9	100	13

[a] The calmodulin fragments were resolved on an alkylphenyl column by gradient elution.
[b] The digests were loaded at 15% acetonitrile in buffer A. Elution was started with a 15 to 28% gradient of acetonitrile in buffer A for 20 min at 1.5 mℓ/min.

and Table 2). Calmodulin fragments 1-77, 78-148, 1-90, and 107-148 have been isolated from tryptic digests by this method in 30 to 50% yield based on the amount of calmodulin

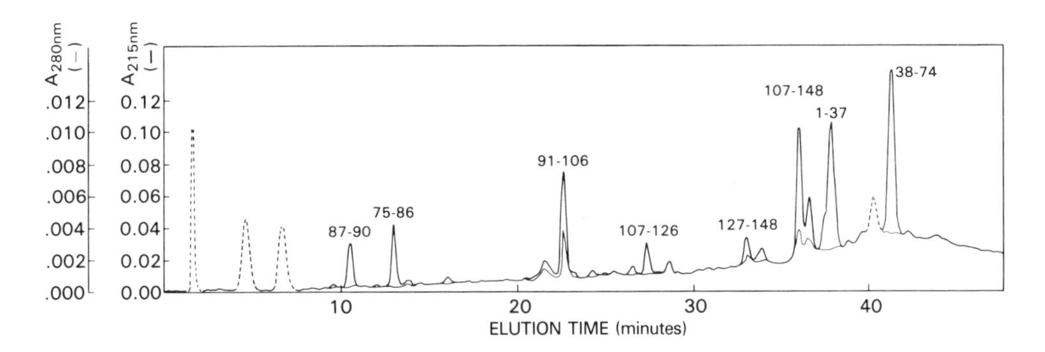

FIGURE 3. HPLC elution profiles of peptides generated by digestion of calmodulin with clostripain. Calmodulin was digested with clostripain in the presence of calcium. Following addition of *N*α-*p*-tosyl-L-lysine chloromethyl ketone, peptides were separated by HPLC using a C_{18} μBondapak® column as described in the text. Peptides were identified by amino acid analysis. Broken lines indicate UV-absorbing peaks contributed by additional reagents present in the digestion mixture.

subjected to digestion. Fragment 1-106, which is susceptible to further tryptic cleavage at Arg 90, is recovered in lower yield. In addition, this fragment is frequently contaminated by intact calmodulin, which elutes at only a slightly higher concentration of acetonitrile. Complete separation of fragment 1-106 from contaminating calmodulin is accomplished by rechromatography on the same column. Low levels of contamination (less than 0.1% w/w) of the peptides by calmodulin remaining adsorbed to the column and slowly released on subsequent runs is generally observed but can be overcome by rechromatography of the peptides on a new column.

Complete digestion of calmodulin with trypsin or the sulfhydryl protease, clostripain, produces a specific series of small peptides which can be easily separated by HPLC.[13-16] A C_{18} μBondapak® column, which exhibits greater affinity than the alkylphenyl column, has been employed in the separation of these small peptides.[9,30,31] Digests of 15 to 600 μg of calmodulin can be applied to a 3.9 mm × 30 cm C_{18} column. Upon elution with a linear acetonitrile gradient, peptides 1-37, 38-74, 75-86, 87-90, 91-106, 107-126, 127-148, and 107-148 are well separated and easily identified by their characteristic, reproducible elution times (Figure 3). This HPLC method, which provides a rapid means for the preparation of peptide maps of calmodulin, has been applied to studies of the reactivity of specific residues in several covalent modification reactions.[31] The retention times exhibited by the peptides are sensitive to the effects of specific chemical modifications. For example, following the performic oxidation of methionyl residues of calmodulin, the retention times of all peptides which contain methionyl residues are decreased. Retention times for peptides devoid of methionyl residues remain unchanged. Guanidination of lysyl residues of calmodulin also produces selective alterations in retention times exhibited by lysine-containing peptides.[31]

APPLICATIONS

HPLC has been successfully employed as the final step in the preparation of homogeneous calcium-binding proteins. Calmodulin and S100b, another calcium-binding protein, have been resolved using these methods.[7] Calcineurin B,[32] the different forms of *Xenopus laevis* parvalbumins,[33] and *Acanthamoeba castellani* calmodulin[34] have also been purified to homogeneity. A two-step procedure, consisting of phenothiazine-Affi-gel 10 or phenyl Sepharose affinity chromatography followed by HPLC, has been developed for the purification of calmodulin. Purified calmodulins are fully active in the cAMP phosphodiesterase activation assay. ^{35}S-calmodulin from chicken embryo fibroblasts cultivated in media containing ^{35}S-methionine,[31] and ^{35}S-calmodulin synthesized in vitro by translation of crude mRNA preparations[33] have also been purified to homogeneity by this two-step method. This method is also applicable to the quantification of low levels of calmodulin (0.2 to 10 μg/mℓ) in partially purified preparations or solutions of chemically modified calmodulin. The calmodulin content of preparations of phosphorylase kinase can be determined using aliquots of boiled enzyme, without prior fractionation of enzyme subunits.[35] A modification of this method has also been developed to quantitate the covalent attachment of various ligands to calmodulin or homologous calcium-binding proteins.[36]

DISCUSSION

In recent years, HPLC systems, based upon partitioning of solute between a hydrophobic stationary phase and a polar mobile phase, have been successfully employed in the resolution of proteins and peptides.[17-24] For these separations, alkyl-bonded phases, notably ethyl-, octyl-, or octadecyl-bonded phases, have received widest attention.[17,20] The alkylphenyl-bonded phase, which has seen only limited use in protein and peptide separations,[24] is well suited to the resolution of the calcium-binding proteins. Use of a mobile phase containing 0.01 *M* phosphate buffer, pH 6.0,[17] has facilitated elution of the calcium-binding proteins at neutral pH with relatively low concentrations of acetonitrile, allowing recovery of these proteins in a biologically active state.

The molecular interactions involved in the retention of the calcium-binding proteins by the alkylphenyl column are incompletely understood. Multivalent attachment of calcium-binding proteins to the solid phase is suggested by the failure of the proteins to elute from the column under isocratic elution conditions. Upon elution with a linear gradient of acetonitrile, the presence of calcium in the eluting buffer induces a decrease in the retention times of calmodulin and troponin C when compared to retention times seen with buffers containing EGTA. Earlier elution of the calcium-protein complex compared to the calcium-free protein is unexpected in view of results utilizing extrinsic fluorescent probes, which suggested an increase in hydrophobicity of calmodulin induced by the binding of calcium.[5] It should be noted that responses of the fluorescent probes are likely to reflect regional changes in the hydrophobicity of the protein surface, while retention on the phenyl column may reflect only more general surface characteristics. The silica support of the stationary phase is not completely capped. Thus, interactions with exposed regions of unmodified silica as well as interactions with phenyl side chains may contribute to solute adsorption. Nevertheless, at present, the cause of the anomalous chromatographic behavior of these proteins in the presence of calcium is not clear. From a practical standpoint, it is obviously desirable to avoid the apparent heterogeneity of these proteins resulting from the different retention times exhibited in the calcium-complexed and calcium-free states. Since addition of calcium to the phosphate buffer induces precipitation, EGTA has been included in the buffer system to insure homogeneity of elution of the proteins in their calcium-free state.

Because each of the calcium-binding proteins possesses unique spectral properties, dual wavelength monitoring provides a sensitive means for detection, quantification, and tentative identification of these proteins in the column eluent.[6] Selection of a monitoring wavelength of 210 nm would provide a high sensitivity in detection of absorbance of the protein backbone, with minimal contribution from aromatic residues. In practice, monitoring is generally performed at 215 nm, a compromise allowing high sensitivity, while avoiding the potentially significant contribution of the absorbance of the eluting buffer. Simultaneous monitoring at 280 nm provides an independent assessment of the absorption of aromatic residues and allows tentative identification of the calcium-binding proteins as they elute from the column.

HPLC has also proven to be a rapid and highly efficient method for separation of peptides derived by limited or complete proteolytic digestion.[25,26] Tryptic digests of calmodulin from various sources have been analyzed by HPLC using an alkylphenyl reversed-phase column.[13-15] Recently, rapid peptide mapping of calmodulin has also been accomplished using a macroreticular anion-exchange resin.[16] In the present analysis, a C_{18} column, which exhibits greater hydrophobic character than the phenyl column, has been selected for separation of small peptides. Addition of phosphoric acid, a hydrophilic ion-pairing solvent, to the mobile phase has been shown to facilitate elution of underivatized peptides from the C_{18} column at significantly lower concentrations of acetonitrile.[17] Based upon this approach, analytical peptide mapping of intestinal calcium-binding protein has been accomplished by Fullmer and Wasserman[9] using a linear gradient from 0.1% phosphoric acid (pH 2.2) to acetonitrile. This method also provides an efficient, reproducible separation of peptides derived from calmodulin.

Although larger peptides generally elute from the C_{18} column later, retention times for the calmodulin peptides do not demonstrate a simple correlation with size. Peptides containing tyrosyl residues appear to elute earlier than would be predicted on the basis of size alone. Based upon correlation of retention times with amino acid composition, Meek[27] has suggested that, for peptides up to 20 residues, retention is primarily due to partition processes that involve all residues. Thus, retention of a given peptide might be predicted by the sum of retention coefficients of the amino acids. Interestingly, with the exception of the small peptide 87-90 and the tyrosine-containing peptide 91-106, which elute early, the order of elution of the calmodulin peptides from the C_{18} column is accurately predicted by this analysis.

Application of HPLC to the purification of the calcium-binding proteins has been limited to terminal stages of the purification scheme. Potential applications of HPLC based on gel filtration and ion-exchange technology for the purification of these proteins has not yet been explored in depth. The excellent resolving power of HPLC provides tremendous potential for accomplishing rapid, reproducible purification of at least some of these proteins in high yields. In view of the recent advances in HPLC technology, it should be possible to design a preparative scheme to purify the calcium-binding proteins from crude extracts entirely by HPLC.

ADDENDUM

In the past year, examples of the application of HPLC to the study of calcium-binding proteins and peptides have dramatically increased. Several skeletal muscle parvalbumins have been resolved from each other using Aquapore RP-300 (C-8) columns.[37] This fractionation has been applied to the quantification of parvalbumin levels in skeletal muscle.[38] The purification of neuronal parvalbumin has been achieved entirely by HPLC using four consecutive reversed-phase chromatographic steps.[39] The purification and yield of more hydrophobic proteins such as calcineurin B and a novel Ca^{2+}-binding protein of brain using alkylphenyl columns are improved by increasing the pH and salt concentration of the mobile phase. The use of CN-μBondapak® columns has allowed the rapid purification and quantification of covalent adducts of calmodulin with hydrophobic drugs such as phenothiazines or fluorescent probes such as rhodamine.[36]

Several new examples of the rapid fractionation of proteolytic digests of calcium-binding proteins for sequence determination have been described. Fractionations by HPLC using C-18 or alkylphenyl columns have been employed in determination of the amino acid sequences of rat skeletal muscle parvalbumin,[40] an α-parvalbumin from frog skeletal muscle,[41] and calcineurin B.[42] Late elution of the amino-terminal peptide blocked with myristic acid as in calcineurin B may be generally useful to detect this type of posttranslational modification in other proteins.[42] Sequential preparative steps using a SynChropak AX 300 ion-exchange column followed by reversed-phase chromatography on a C-18 column has been used to isolate a synthetic 13-residue fragment representing the third Ca^{2+}-binding site of rabbit skeletal muscle troponin C.[43]

The high resolving power of HPLC on C-18 reversed-phase supports has been exploited in the fractionation of peptides derived from calmodulin and its chemically modified derivatives: performic acid oxidized, iodinated, guanidinated, and acetylated.[31] Phosphorylated and unphosphorylated myosin light chains have also been resolved by HPLC on alkylphenyl columns.[32]

REFERENCES

1. **Kretsinger, R. H.,** Structure and evolution of calcium-modulated proteins, *CRC Crit. Rev. Biochem.*, 8, 119, 1980.
2. **Van Eldik, L. J., Zendegui, J. G., Marshak, D. R., and Watterson, D. M.,** Calcium-binding proteins and the molecular basis of calcium action, *Int. Rev. Cytol.*, 77, 1, 1982.
3. **Goodman, M., Pechere, J. F., Haiech, J., and Demaille, J. G.,** Evolutionary diversification of structure and function in the family of intracellular calcium-binding proteins, *J. Mol. Evol.*, 13, 331, 1979.
4. **Klee, C. B. and Vanaman, T. C.,** Calmodulin, *Adv. Protein Chem.*, 35, 213, 1982.
5. **LaPorte, D. C., Wierman, B. M., and Storm, D. R.,** Calcium-induced exposure of a hydrophobic surface on calmodulin, *Biochemistry*, 19, 3814, 1980.
6. **Klee, C. B., Oldewurtel, M. D., Williams, J. F., and Lee, J. W.,** Analysis of Ca^{2+}-binding proteins by high performance liquid chromatography, *Biochem. Int.*, 2, 485, 1981.
7. **Marshak, D. R., Watterson, D. M., and Van Eldik, L. J.,** Calcium-dependent interaction of S100b, troponin C, and calmodulin with an immobilized phenothiazine, *Proc. Natl. Acad. Sci. U.S.A.*, 78, 6793, 1981.
8. **Oldewurtel, M. D., Krinks, M. H., Lee, J. W., Williams, J. F., and Klee, C. B.,** Isolation and characterization of intracellular Ca^{2+}-binding proteins by high performance liquid chromatography, *Protides Biol. Fluids, Proc. Colloq.*, 30, 713, 1982.
9. **Fullmer, C. S. and Wasserman, R. H.,** Analytical peptide mapping by high performance liquid chromatography, *J. Biol. Chem.*, 254, 7208, 1979.
10. **Drabikowski, W., Kuznicki, I., and Grabarek, Z.,** Similarity of Ca^{2+}-induced changes between troponin-C and protein activator of 3′, 5′-cyclic nucleotide phosphodiesterase and their tryptic fragments, *Biochim. Biophys. Acta*, 485, 124, 1977.
11. **Kuznicki, J., Grabarek, Z., Brzeska, H., Drabikowski, W., and Cohen, P.,** Stimulation of enzyme activities by fragments of calmodulin, *FEBS Lett.*, 130, 141, 1981.
12. **Manalan, A. S. and Klee, C. B.,** Interaction of calmodulin with its target proteins, *Chemica Scripta*, 21, 137, 1983.
13. **Autric, F., Ferraz, C., Kilhoffer, M.-C., Cavadore, J. C., and Demaille, J. G.,** Large-scale purification and characterization of calmodulin from ram testis, *Biochim. Biophys. Acta*, 631, 139, 1980.
14. **Kilhoffer, M. C., Gerard, D., and Demaille, J. G.,** Terbium binding to octopus calmodulin provides the complete sequence of ion binding, *FEBS Lett.*, 120, 99, 1980.
15. **Cartaud, A., Ozon, R., Welsh, M. P., Haiech, J., and Demaille, J. G.,** *Xenopus laevis* oocyte calmodulin in the process of meiotic maturation, *J. Biol. Chem.*, 255, 9494, 1980.
16. **Takahashi, N., Isobe, T., Kasai, H., Seta, K., and Okuyama, T.,** An analytical and preparative method for peptide separation by high-performance liquid chromatography on a macroreticular anion-exchange resin, *Anal. Biochem.*, 115, 181, 1981.
17. **Hancock, W. S., Bishop, C. A., Prestidge, R. L., Harding, D. R. K., and Hearn, M. T. W.,** High-pressure liquid chromatography of peptides and proteins. II. The use of phosphoric acid in the analysis of underivatised peptides by reversed phase high-pressure liquid chromatography, *J. Chromatogr.*, 153, 391, 1978.
18. **Larsen, B., Fox, B. L., Burke, M. F., and Hruby, V. J.,** The separation of peptide hormone diastereoisomers by reverse phase high pressure liquid chromatography, *Int. J. Peptide Protein Res.*, 13, 12, 1979.
19. **Nice, E. C., Capp, M., and O'Hare, M. J.,** Use of hydrophobic interaction methods in the isolation of proteins from endocrine and paraendocrine tissues and cells by high-performance liquid chromatography, *J. Chromatogr.*, 185, 413, 1979.

20. **Hearn, M. T. W., Grego, B., and Hancock, W. S.,** High-performance liquid chromatography of amino acids, peptides, and proteins. XX. Investigation of the effect of pH and ion-pair formation on the retention of peptides on chemically-bonded hydrocarbonaceous stationary phases, *J. Chromatogr.*, 185, 429, 1979.
21. **Rubinstein, M.,** Preparative high-performance liquid partition chromatography of proteins, *Anal. Biochem.*, 98, 1, 1979.
22. **Regnier, F. E. and Gooding, K. M.,** High-performance liquid chromatography of proteins, *Anal. Biochem.*, 103, 1, 1980.
23. **Lewis, R. V., Fallon, A., Stein, S., Gibson, K. D., and Udenfriend, S.,** Supports for reverse-phase high-performance liquid chromatography of large proteins, *Anal. Biochem.*, 104, 153, 1980.
24. **Yuan, P-M., Pande, H., Clark, B. R., and Shively, J. E.,** Microsequence analysis of peptides and proteins. I. Preparation of samples by reverse-phase liquid chromatography, *Anal. Biochem.*, 120, 289, 1982.
25. **Hancock, W. S., Bishop, C. A., Prestidge, R. L., and Hearn, M. T. W.,** The use of high pressure liquid chromatography (hplc) for peptide mapping of proteins. IV, *Anal. Biochem.*, 89, 203, 1978.
26. **Takagaki, Y., Gerber, G. E., Nihei, K., and Khorana, H. G.,** Amino acid sequence of the membranous segment of rabbit liver cytochrome b_5: methodology for separation of hydrophobic peptides, *J. Biol. Chem.*, 255, 1536, 1980.
27. **Meek, J. L.,** Prediction of peptide retention times in high-performance liquid chromatography on the basis of amino acid composition, *Proc. Natl. Acad. Sci. U.S.A.*, 77, 1632, 1980.
28. **Haiech, J., Derancourt, J., Pechere, J. F., and Demaille, J. G.,** A new large-scale purification procedure for muscular parvalbumins, *Biochimie*, 61, 583, 1979.
29. **Klee, C. B., Crouch, T. H., and Krinks, M. H.,** Calcineurin: a calcium- and calmodulin-binding protein of the nervous system, *Proc. Natl. Acad. Sci. U.S.A.*, 76, 6270, 1979.
30. **Oldewurtel, M. D. and Klee, C. B.,** unpublished observations.
31. **Manalan, A. S. and Klee, C. B.,** manuscript in preparation.
32. **Klee, C. B.,** unpublished observation.
33. **Chien, Y. H., Klee, C. B., Dawid, I. B.,** unpublished observations.
34. **Klee, C. B. and Korn, E.,** unpublished observations.
35. **Cohen, P. and Klee, C. B.,** unpublished data.
36. **Newton, D. L., Burke, T. R., Jr., Rice, K. C., and Klee, C. B.,** Ca^{2+}-dependent covalent modification of calmodulin with norchlorpromazine isothiocyante, *Biochemistry*, in press.
37. **Berchtold, M. W., Heizmann, C. W., and Wilson, K. J.,** Ca^{2+}-binding proteins: A comparative study of their behavior during high performance liquid chromatography using gradient elution on reverse-phase supports, *Anal. Biochem.*, 129, 120, 1983.
38. **Heizmann, C. W., Berchtold, M. W., and Rowlerson, A. M.,** Correlation of parvalbumin concentration with relaxation speed in mammalian muscles, *Proc. Natl. Acad. Sci. U.S.A.*, 79, 7243, 1982.
39. **Berchtold, M. W., Wilson, K. J., and Heizmann, C. W.,** Isolation of neuronal parvalbumin by high performance liquid chromatography. Characterization and comparison with muscle parvalbumin, *Biochemistry*, 21, 6552, 1982.
40. **Berchtold, M. W., Heizmann, C. W., and Wilson, K. J.,** Primary structure of parvalbumin from rat skeletal muscle, *Eur. J. Biochem.*, 127, 381, 1982.
41. **Jauregui-Adell, J., Pechere, J. F., Briand, G., Richet, C., and Demaille, J. G.,** Amino-acid sequence of an α-parvalbumin, pI = 4.88, from frog skeletal muscle, *Eur. J. Biochem.*, 123, 337, 1982.
42. **Aitken, A., Cohen, P., Santikarn, S., Williams, D. H., Calder, A. G., Smith, A., and Klee, C. B.,** Identification of the NH_2-terminal blocking group of calcineurin B as myristic acid, *FEBS Lett.*, 150, 314, 1982.
43. **Gariepy, J., Sykes, B. D., Reid, R. E., and Hodges, R. S.,** Proton nuclear magnetic resonance investigation of synthetic calcium-binding peptides. *Biochemistry*, 21, 1506, 1982.
44. **Gariepy, J., Sykes, B. D., Reid, R. E., and Hodges, R. S.,** Proton nuclear magnetic resonance investigation of synthetis calcium-binding peptides. *Biochemistry*, 21, 1506, 1982.
45. **Newton, D. and Klee, C. B.,** manuscript in preparation.
46. **Manalan, A. S. and Klee, C. B.,** manuscript in preparation.
47. **Klee, C. B.,** unpublished observation.

WHEY PROTEINS

Rex Humphrey

PROTEIN CONSTITUENTS OF WHEY

Bovine whey or milk serum, the fluid that remains when the casein proteins have been removed from skim milk contains five "major" proteins or protein groups: β-lactoglobulin, α-lactalbumin, immunoglobulins (Ig), bovine serum albumin (BSA), and the proteose-peptones. β-Lactoglobulin and α-lactalbumin are synthesized by the lactating mammary gland and constitute 50% (approximately 3 g/ℓ) and 12% of the whey proteins, respectively.[1] The β-lactoglobulin monomer consists of a single polypeptide chain of 162 amino acid residues with two disulfide bridges and one sulfhydryl group. In solution at room temperature between pH 3 and 7, β-lactoglobulin exists as a dimer with a molecular weight about 36,000. It has five known genetic variants; the two more common variants, A and B, differ from one another by A/B substitutions of aspartic/glycine and valine/alanine. α-Lactalbumin consists of 123 amino acid residues with four disulfide bridges; it has a molecular weight of approximately 14,000. Of the minor proteins, BSA constitutes 5% of the whey proteins and is identical with that in bovine serum. It has a molecular weight of about 68,000. The immunoglobulins (IgG1, IgG2, IgA, and IgM) are high-molecular-weight proteins (greater than 150,000) containing disulfide cross-links; their content in the whey varies greatly during lactation. The proteose-peptone fraction comprises several proteins, most of which are considered to be proteolytic degradation products of caseins.

SEPARATION PROCEDURES

Salt Fractionation

A number of methods have been utilized to separate and purify the whey proteins. Various salt fractionations at different pHs[2-5] yielded β-lactoglobulin of high purity but preparations of α-lactalbumin were generally heterogeneous.[6] Yields from these procedures were high and provided useful material for further purification by chromatography.

Conventional Chromatography

Gel permeation chromatography on Sephadex® gels has been used to capitalize on the significant differences in molecular size between the whey proteins and separate them into four[7] or five[8] fractions. Davies[9] has described the application of chromatography on Sephadex® G-100 to the quantitative determination of whey proteins in the albumin fraction of milk serum.

DEAE-Sephadex® and DEAE-cellulose ion exchangers eluted with sodium chloride gradients have been extensively used for fractionation of the whey proteins.[10] These have been especially useful for separation of the β-lactoglobulin variants using enriched fractions from salt and pH fractionations as starting materials.[11-12] Pearce[13] has recently reported the separation of β-lactoglobulin variants A, B, and C and α-lactalbumin using chromatofocusing. However, the β-lactoglobulin B fraction is contaminated with BSA.[14]

Reversed-Phase (RP) HPLC

Samples of whey obtained from skim milk after acid precipitation of the casein proteins at pH 4.6, were dialyzed against starting buffer prior to elution on μBondapak® C_{18} and Lichrosorb® RP-8 stationary phases packed in 8 mm i.d. × 100 mm radial compression

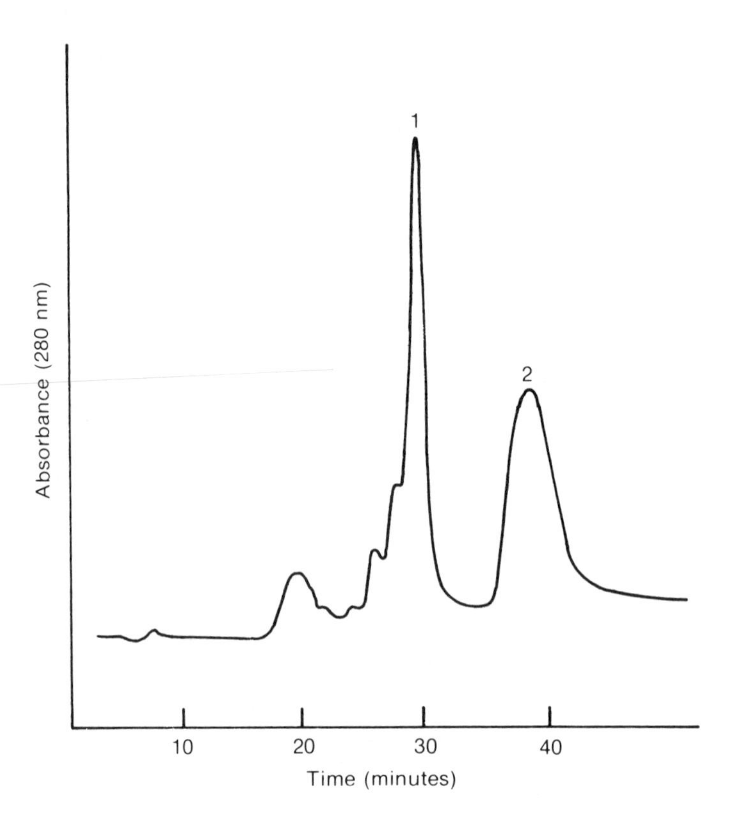

FIGURE 1. Chromatography of dialyzed acid whey on Lichrosorb® RP-8. Elution with linear gradients of 2-PrOH in 0.05 *M* NaH_2PO_4-H_3PO_4, pH 2.1: 0 to 25% in 15 min, 25-40% in 35 min. Flow rate: 0.7 mℓ/min. Peaks: 1 = α-lactalbumin; 2 = β-lactoglobulin.

columns. Four buffer systems were evaluated: 0.05 *M* NaH_2PO_4–H_3PO_4,pH2.1; 0.1*M* NaH_2PO_4–H_3PO_4,pH 2.1; 0.25*M* H_3PO_4-triethylamine, pH 2.1 (0.25 *M* TEAP); and 0.5 *M* acetic acid-triethylamine, pH 4.0. These initial aqueous buffers were modified with gradients of 2-propanol (2-PrOH) in the corresponding buffer. Elution conditions are given in the figure captions.

Separation of β-lactoglobulin and α-lactalbumin was achieved with all buffers on both the μBondapak® C_{18} and the Lichrosorb® RP-8 columns, although resolution was better with the phosphate buffers (see Figures 1 to 5). An increase in phosphate concentration improved the elution behavior of β-lactoglobulin on the RP-8 packing (compare Figures 1 and 2), although little effect was noticeable for the μBondapak® C_{18}. This probably reflects the difference in end-capping of the two packings. Resolution of the whey proteins is similar for both phosphate and TEAP buffers on μBondapak® C_{18} (Figures 3 and 4); however TEAP elutes the proteins at lower concentrations of 2-PrOH. The separation using acetic acid-triethylamine at pH 4.0 (Figure 5) was promising but problems were experienced with protein precipitation, presumably because of α-lactalbumin insolubility at that pH. Only partial separation of the A and B variants of β-lactoglobulin was achieved with any of the packings or buffer systems. The best separation for a mixture of β-lactoglobulin A and B was obtained using 0.05 *M* NaH_2PO_4-H_3PO_4, pH 2.1 on μBondapak® C_{18} (Figure 6); however resolution deteriorated progressively with column use. Protein recoveries on reversed-phase packings in the elution systems evaluated were between 65 and 85%.

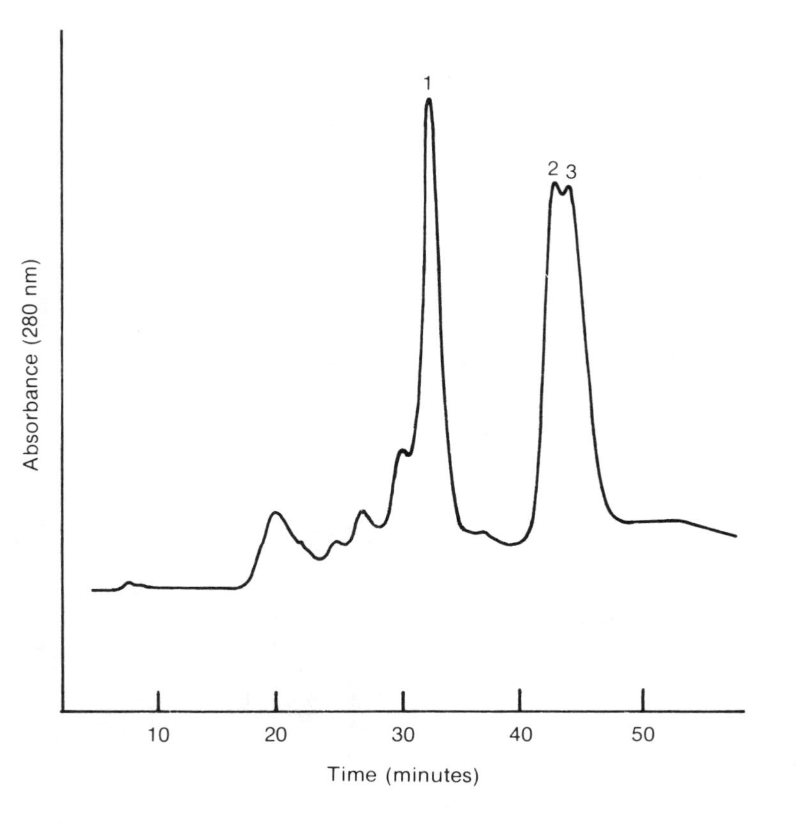

FIGURE 2. Chromatography of dialyzed acid whey on Lichrosorb® RP-8. Elution with linear gradients of 2-PrOH in 0.1 *M* NaH_2PO_4-H_3PO_4, pH 2.1: 0 to 40% in 15 min, 40 to 50% in 45 min. Flow rate: 0.7 mℓ/min. Peaks: 1 = α-lactalbumin; 2 = β-lactoglobulin B; 3 = β-lactoglobulin A.

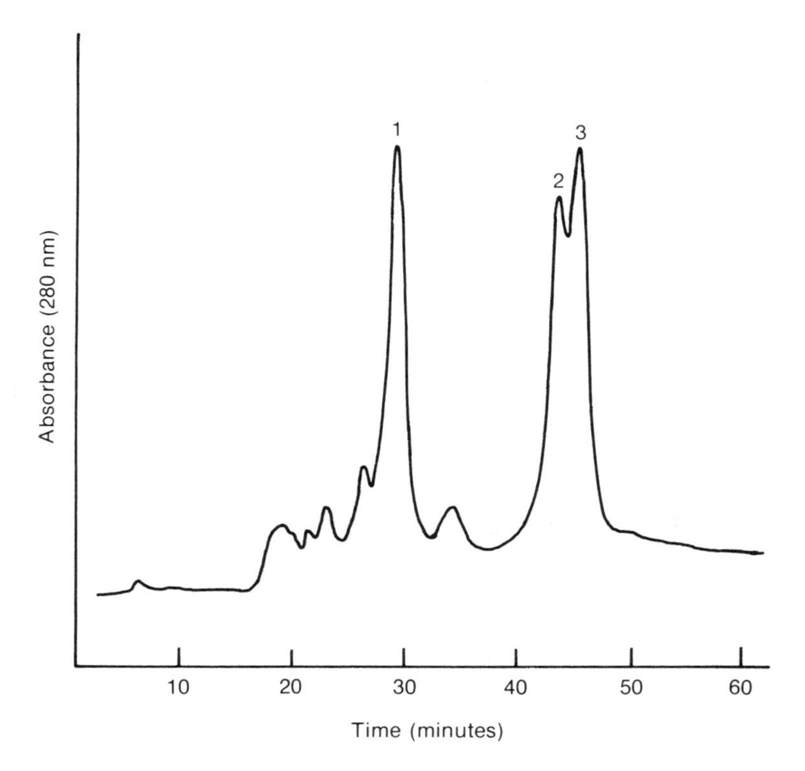

FIGURE 3. Chromatography of dialyzed acid whey on μBondapak® C_{18}. Elution with linear gradients of 2-PrOH in 0.1 *M* NaH_2PO_4-H_3PO_4, pH 2.1: 0 to 38% in 10 min, 38 to 50% in 55 min. Flow rate: 0.7 mℓ/min. Peaks as for Figure 2.

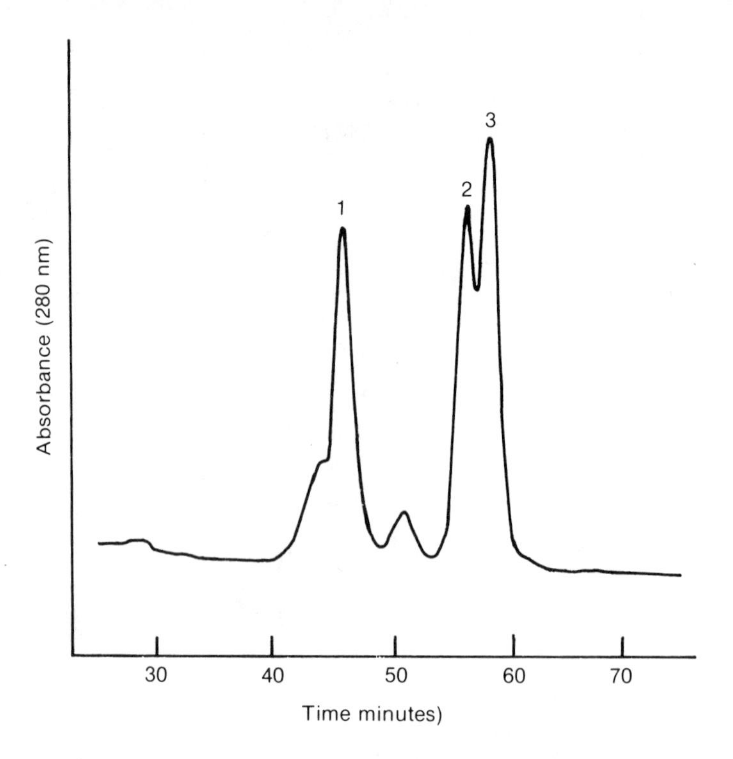

FIGURE 4. Chromatography of dialyzed acid whey on μBondapak® C_{18}. Elution with linear gradients of 2-PrOH in 0.25 *M* TEAP, pH 2.1: 0 to 30% in 20 min, 30 to 40% in 60 min. Flow rate: 0.7 mℓ/min. Peaks as for Figure 2.

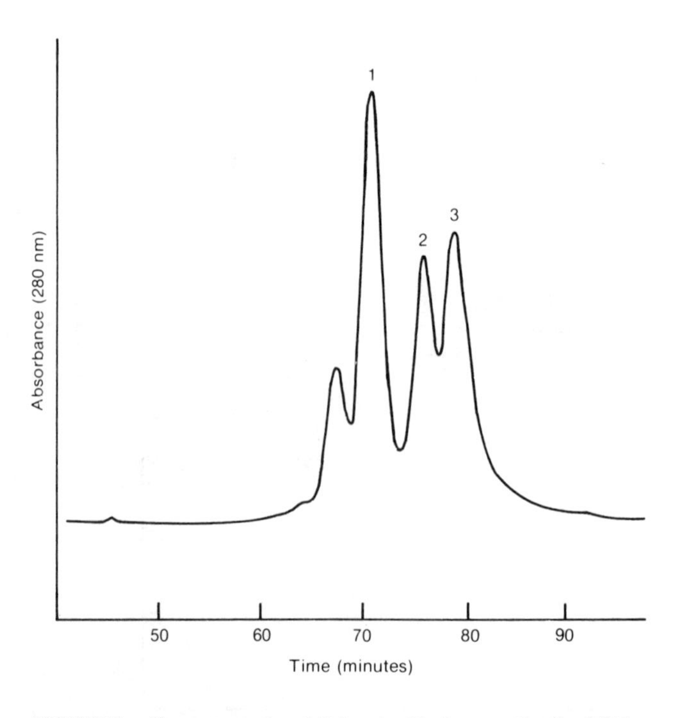

FIGURE 5. Chromatography of dialyzed acid whey on μBondapak® C_{18}. Elution with linear gradients of 2-PrOH in 0.5 *M* acetic acid-triethylamine pH 4.0: 0 to 28% in 20 min, 28 to 40% in 105 min. Flow rate: 0.7 mℓ/min. Peaks as for Figure 2.

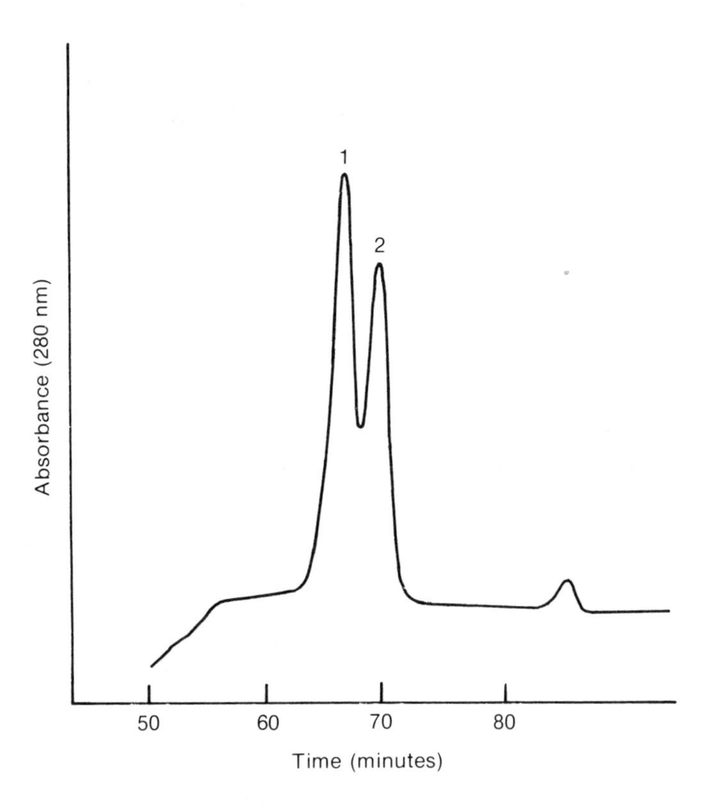

FIGURE 6. Chromatography of a mixture of β-lactoglobulin variants A and B on μBondapak® C_{18}. Elution with linear gradients of 2-PrOH in 0.05 *M* NaH_2PO_4-H_3PO_4, pH 2.1: 0 to 33% in 10 min, 33 to 40% in 45 min. Flow rate: 0.7 mℓ/min. Peaks: 1 = β-lactoglobulin B; 2 = β-lactoglobulin A.

Gel Permeation HPLC

Samples of acid whey (see reversed-phase methods) were dialyzed against buffer and eluted on a 7.5 mm × 600 mm TSK-G3000SW (LKB Ultropac) column preceded by a 7.5 mm × 75 mm TSK-GWSP guard column. Buffers of 0.1 *M* NaH_2PO_4-Na_2HPO_4, pH 7.0, or 0.05 *M* NaH_2PO_4-Na_2HPO_4, 0.1 *M* NaCl, pH 7.0, were used at flow rates of 0.2, 0.5, and 1.0 mℓ/min. Elution profiles of whey at flow rates of 1.0 and 0.2 mℓ/min are shown in Figures 7 and 8. There were no detectable differences in elution behavior for the two buffer systems and as expected resolution improved with decreasing flow rate. Protein recoveries were greater than 96%. Samples (20 μℓ) of nondialyzed wheys (120 μg total protein) adjusted to pH 7.0 were able to be chromatographed with no loss of separation efficiency (see Figure 9) thus permitting separation and analysis of the dialyzable lower molecular weight peptides present in the whey.

Gel permeation HPLC using a TSK-G3000 SW column provides separations that are both faster (1 hr compared to 16 hr) and better resolved than gel permeation chromatography using conventional softer gel materials.

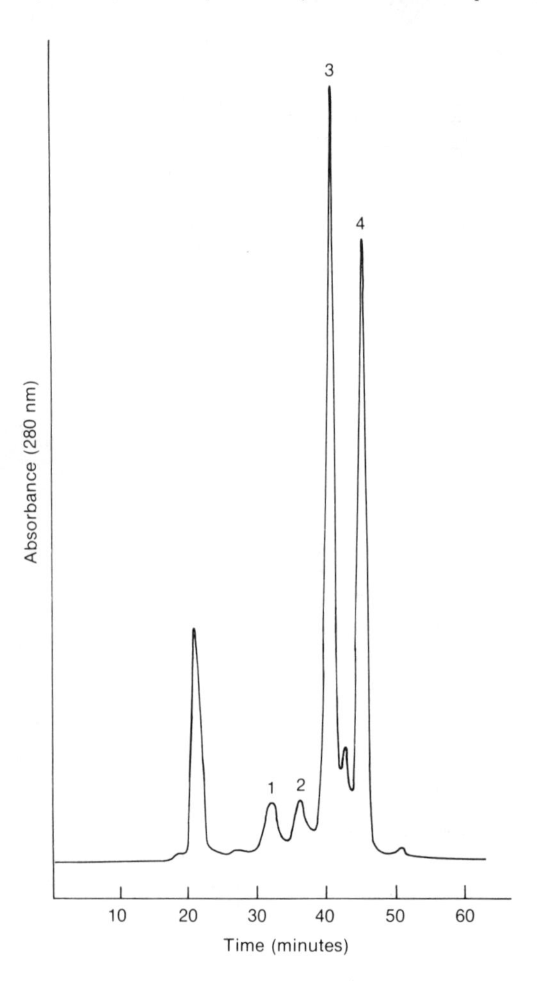

FIGURE 7. Chromatography of dialyzed acid whey on TSK-G-3000SW using 0.1 *M* NaH_2PO_4-Na_2HPO_4, pH 7.0 buffer. Flow rate: 1 mℓ/min. Peaks: 1 = immunoglobulin; 2 = BSA; 3 = β-lactoglobulin A and B; 4 = ∝-lactalbumin.

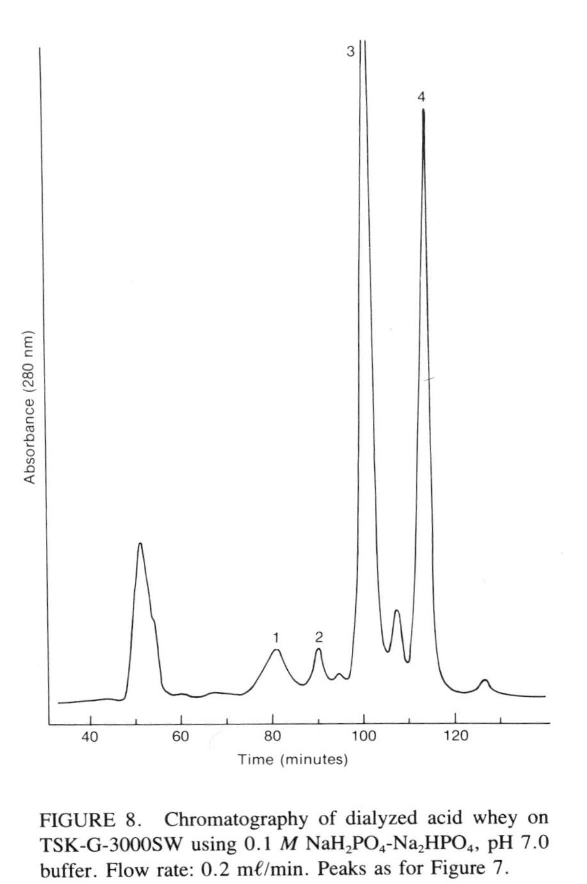

FIGURE 8. Chromatography of dialyzed acid whey on TSK-G-3000SW using 0.1 *M* NaH_2PO_4-Na_2HPO_4, pH 7.0 buffer. Flow rate: 0.2 mℓ/min. Peaks as for Figure 7.

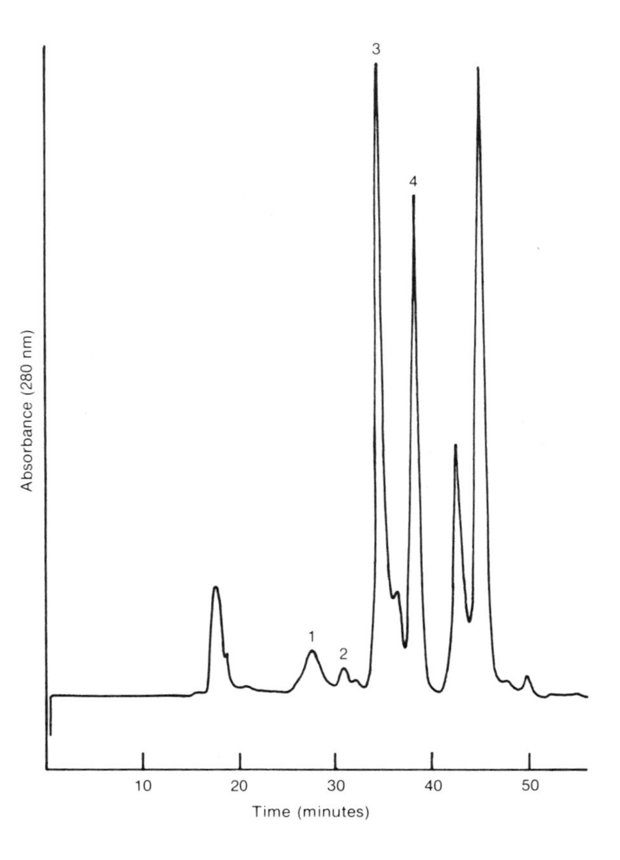

FIGURE 9. Chromatography of nondialyzed acid whey on TSK-G3000SW using 0.1 *M* NaH_2PO_4-Na_2HPO_4, pH 7.0 buffer. Flow rate: 0.5 mℓ/min. Peaks as for Figure 7.

REFERENCES

1. **Evans, M. T. A. and Gordon, J. F.,** Whey proteins, in *Applied Protein Chemistry,* Grant, R. A., Ed., Applied Science Publishers, London, 1980.
2. **Aschaffenburg, R. and Drewry, J.,** Improved method for the preparation of crystalline β-lactoglobulin and α-lactalbumin from cow's milk, *Biochem. J.,* 65, 273, 1957.
3. **Fox, K. K., Holsinger, V. H., Posati, L. P., and Pallansch, M. J.,** Separation of β-lactoglobulin from other milk serum proteins by trichloroacetic acid, *J. Dairy Sci.,* 50, 1363, 1967.
4. **McKenzie, H. A.,** Milk proteins, *Adv. Protein Chem.,* 22, 55, 1967.
5. **Aschaffenburg, R.,** Preparation of α-lactalbumin from cow's or goat's milk: a method improving the yield, *J. Dairy Sci.,* 51, 1295, 1968.
6. **McKenzie, H. A.,** β-lactoglobulins, in *Milk Proteins,* Vol. 2, McKenzie, H. A., Ed., Academic Press, New York, 1971, 2166.
7. **Preaux, G. and Lontie, R.,** Fractionation of milk serum proteins on a Sephadex column, *Arch. Int. Physiol. Biochem.,* 69, 100, 1961.
8. **Armstrong, J., Hopper, K. E., McKenzie, H. A., and Murphy, W. H.,** On the column chromatography of bovine whey proteins, *Biochim. Biophys. Acta,* 214, 419, 1970.
9. **Davies, D. T.,** The quantitative partition of the albumin fraction of milk serum proteins by gel chromatography, *J. Dairy Res.,* 41, 217, 1974.
10. **Yaguchi, M. and Rose, D.,** Chromatographic separation of milk proteins: a review, *J. Dairy Sci.,* 54, 1725, 1971.
11. **Basch, J. J., Kalan, E. B. and Thompson, M. P.,** Preparation of β-lactoglobulin C, *J. Dairy Sci.,* 48, 604, 1965.
12. **Bell, K., McKenzie, H. A., and Murphy, W. H.,** Isolation and properties of β-lactoglobulin Droughtmaster, *Aust. J. Sci.,* 29, 87, 1966.
13. **Pearce, R. J. and Shanley, R. M.,** Analytical and preparative separation of whey proteins by chromatofocusing, *Aust. J. Dairy Technol.,* 36, 110, 1981.
14. **Humphrey, R. S.,** unpublished results.

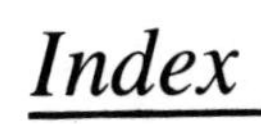

Index

INDEX

A

B

C

D

E

F

G

H

J

K

L

M

N

O

P

Q

R

T

U

V

W

X

Y

Z